Green Energy and Technology

Climate change, environmental impact and the limited natural resources urge scientific research and novel technical solutions. The monograph series Green Energy and Technology serves as a publishing platform for scientific and technological approaches to "green"—i.e. environmentally friendly and sustainable—technologies. While a focus lies on energy and power supply, it also covers "green" solutions in industrial engineering and engineering design. Green Energy and Technology addresses researchers, advanced students, technical consultants as well as decision makers in industries and politics. Hence, the level of presentation spans from instructional to highly technical. **Indexed in Scopus**.

More information about this series at http://www.springer.com/series/8059

Michael O. Daramola • Augustine O. Ayeni
Editors

Valorization of Biomass to Value-Added Commodities

Current Trends, Challenges, and Future Prospects

Springer

Editors
Michael O. Daramola
Department of Chemical Engineering,
Faculty of Engineering, Built Environment
and Information Technology
University of Pretoria
Hatfield, Pretoria, South Africa

Augustine O. Ayeni
Department of Chemical Engineering
Covenant University
Ota, Nigeria

ISSN 1865-3529 ISSN 1865-3537 (electronic)
Green Energy and Technology
ISBN 978-3-030-38031-1 ISBN 978-3-030-38032-8 (eBook)
https://doi.org/10.1007/978-3-030-38032-8

This Springer imprint is published by the registered company Springer Nature Switzerland AG
The registered company address is: Gewerbestrasse 11, 6330 Cham, Switzerland

This book is dedicated to my beautiful and dedicated wife, Omotayo, and our two handsome boys, Timmy and Temmy, for their love and sacrifice during the period of developing this book

– Michael O. Daramola

To my dearest wife, Afolajaye Omoyeni, and our children, Oluwasemilore, Ifeoluwa, Toluwalase, and Boluwatife, for their immense contributions to my academic successes especially during the development of this book

– Augustine O. Ayeni

Preface

The increasing world energy demands have brought along the effect of climate change. Natural resources for energy are getting depleted along with their attendant danger of pollution that threatens vegetation and human and animal health. Therefore, appropriate waste to value-added product technologies offer benefits of avoiding fossil fuels for energy generation, utilization, and development of industrial products. Since energy growth is linked to the comfort of man, reliable and affordable energy is very important to sustainable existence of human on the planet Earth.

In the area of valorization of biomass into value-added products, especially the use of lignocellulosic biomass residues, comprehensive and dependable information on characterization, pretreatment/fractionation, and valorization of biomass is essential. Of utmost importance is the information on the life cycle assessment, predictive molecular modelling and economic assessment of the pretreatment process, and valorization strategies of biomass to value-added products, but this information is very scarce in literature. This book, *Valorization of Biomass to Value-Added Commodities: Current Trend, Challenges and Future Outlook*, provides information on these areas/aspects, thereby bridging the gap in knowledge in these areas in the field, and an overview of alternatives to the traditional sources of fuels and chemical production. It also discusses a few up-to-date technologies for biomass transformation to value-added products, challenges currently encountered during the transformation, and ways to mitigate these challenges. Consequently, this book brings to fore some biomass transformational strategies yet to be explored. In addition, biological and non-biological (physical, physical-chemical, chemical) biomass waste valorization technologies are highlighted, and the direct impacts of these commodities to human well-being are discussed. Comprehensive information on characterization and fractionation of various types of biomass, life cycle assessment, and predictive molecular modelling approaches during valorization of biomass into commodities are provided. Different sources of biomass, such as wood and wood processing waste, agricultural crops and waste materials, food, yard and wood-related materials, animal manure, and human excreta, among others, were used to enumerate the different processing routes from feedstock to value-added products in this book.

Therefore, this book contributes significantly in the area of biomass characterization techniques, pretreatment strategies, and valorization of biomass to value-added products. Various studies documented in this book are thought-provoking and could provide a platform for developing further research efforts toward developing a sustainable holistic approach to treatment and beneficiation of biomass. Therefore, this book will be of interest to stakeholders in the area of waste treatment and beneficiation, as well as to graduate students and researchers in industry and in academia working in the field of waste to energy, waste to wealth, chemical and environmental engineering, and chemistry.

In many ways, editing this book has been a privilege and a unique experience. Thanks firstly to our excellent contributors without whose support this book would not have materialized. It is most fitting that this technological work is published from contributors around the globe and is founded on the spirit of free enquiry coupled with hard work and imagination. It has indeed been a great pleasure to be in touch with all the contributors in the last 8 months to 1 year. Thanks also for their patience and understanding. It is noteworthy that the chapters of this book went through a thorough double-blind peer-review process before their acceptance. Therefore, we would like to express our unparalleled appreciations to the reviewers of these chapters. Their contributions, through the review process, have enhanced the scientific quality of the chapters. We would be utterly remiss if we did not acknowledge our colleagues who have provided us with the inspiration, motivation, and never-ending encouragement during the process of developing this book.

Pretoria, South Africa Michael O. Daramola
Ota, Nigeria Augustine O. Ayeni

Contents

About the Editors

Michael O. Daramola is Professor and Head of Department of Chemical Engineering, University of Pretoria, South Africa. He is a chartered Chemical Engineer (C.Eng.) with Engineering Council of UK, a Registered Engineer with Council for Regulation of Engineering in Nigeria, and a C-rated researcher by the National Research Foundation (NRF) South Africa. He has more than 15 years of teaching and research experience in academia and research institutes nationally and internationally (e.g. in Nigeria, The Netherlands, France; Singapore, South Africa). His research interest is within chemical and biochemical engineering, but with main activities in nanomaterials and bioresource engineering. He has authored/co-authored more than 100 scientific journal articles; produced 2 patents; authored/co-authored 46 referred conference proceedings, produced 10 book chapters and 2 edited books; and made more than 60 oral/poster presentations at national and international conferences as contributions to the above-mentioned research domain. He has trained more than 25 postgraduate students (PhD and MSc) and 5 postdoctoral fellows within the research area. He has served on the editorial board of International Journal of Oil, Gas & Coal Technology (IJOGCT) and Modern Research in Catalysis (MRC). Currently, he serves as an Associate Editor of Chemical Engineering Communications (CEC) and an Associate Editorial Board Member of the Open Chemical Engineering Journal.

Augustine O. Ayeni graduated from Obafemi Awolowo University, Ile – Ife, Osun State, Nigeria with a Second Class Honours division in Chemical Engineering in 1997. He proceeded to the University of Lagos, Lagos, Nigeria where he obtained his MSc in Chemical Engineering in 2002. He obtained his PhD degree (Chemical Engineering) from Covenant University, Ota, Nigeria in 2013. He is currently a Senior Lecturer and the Acting Head, Department of Chemical Engineering, Covenant University, Ota, Nigeria. Among his academic achievements is the award of PhD research grant of Chief S.L. Edu memorial research grant 2008 sponsored by the Chevron Oil Nigeria in association with Nigerian Conservation Foundation,

Lekki, Ajah, Lagos. He was also awarded an Advanced Research and Training fellowship 2009 by the Academy of Sciences for the Developing World (TWAS). This gave him a research stay at the National Environmental Engineering Research Institute, Environmental Biotechnology Division, Nagpur, India from 2010 -2011. He is a Task Team member of the science plan (Sustainable Energy) of International Council for Scientific Union - Region of Africa (ICSU-ROA) based in Pretoria, South Africa. He is a member of the Nigerian Society of Engineers as well as South African Institute of Chemical Engineers. He is a Registered Engineer (COREN: R.26927). He was awarded a Postdoctoral Research Fellowship tenable at the University of Witwatersrand, School of Chemical and Metallurgical Engineering, Johannesburg, South Africa (2015-2017). He is a frequent reviewer for many high-impact scientific journals. He has also reviewed for National Research Foundation (NRF) of South Africa.

Part I
Characterization of Biomass Feedstock

Chapter 1
Application of Lignocellulosic Biomass (LCB)

O. Olatunji, S. Akinlabi, and N. Madushele

1.1 Introduction

The twenty-first century has ushered in a major epoch in the history of humankind due to the level of awareness about climate change and a drive towards sustainable global solution. The research and development, which are tailored towards conversion of lignocellulosic biomass (LCB) to biofuel and other value-added products, have witnessed various levels of successes. As the awareness increases across the globe, the application of renewable resources is taking the centre stage in almost all areas that are relevant to our daily lives. This chapter discusses the LCB feedstock applications and motivations for their exploration. References were made to common lignocellulosic biomass and their structural composition vis-a-vis cellulose, hemicellulose and lignin. Various value-added products, which can be derived from LCB and its residue, are outlined. The value creation pathway for lignocellulosic is discussed.

1.2 Biomass and Climate Change

Biomass-based economy may eventually be the way out for a global economy that is immensely dependent of fossil fuel despite the effect of climate change and dwindling oil resources. Henry Ford has forecasted this scenario in the early nineteenth century when he suggested that the implementation of a bio-based economy is a

O. Olatunji · N. Madushele
Mechanical Engineering Science, University of Johannesburg, Johannesburg, South Africa

S. Akinlabi (✉)
Department of Mechanical Engineering, Faculty of Engineering & Technology, Butterworth Campus, Walter Sisulu University, Eastern Cape, South Africa

M. O. Daramola, A. O. Ayeni (eds.), *Valorization of Biomass to Value-Added Commodities*, Green Energy and Technology,
https://doi.org/10.1007/978-3-030-38032-8_1

rational and essential choice we have to make in order to advance human civilization [1]. The oil boom was the reason why this advice was neglected since the oil was cheaper than other commodities. The environmental concern due to massive fossil fuel consumption has become the main subject of discussion in most global gathering and events, given worrisome messages from most countries as per dwindling oil resources amidst a growing population. Recent events have shown that global warming is rather a reality and not just a mirth that we can wish away [2]. The damning report and the omnium message from UN Intergovernmental Panel on Climate change demand more systematic approach to the exploration of renewable energy. As shown in Fig. 1.1, the average fossil fuel consumption was highest in 2015 over the space of 15 years [3]. By 2030, global warming will reach the decisive pre-industrial threshold of 1.5 °C [4]. By implication, the global net emission of CO_2 will be required to fall by 45% from 2010 levels by 2030 and reach a zero value by approximately 2050 in order to keep global warming at bay of 1.5 °C [5]. There is a technical possibility of achieving this mileage; however, it demands a holistic and wholesale change in product consumption and manufacturing in energy, industrial, building and transportation sectors [4]. The European Union (EU) has already taken a proactive step by approving environmental law against environmentally abusive products and directing more efforts to sourcing green materials [6]. As at 2011, the European Commission (EU) had set a long-term goal to develop a competitive, resource efficient low carbon economy by 2050, which is substantially structured around bioresources [7]. From various outlooks which were analysed by EU, it was proposed that a cost-effective economic scenario demands around 40% decrease in greenhouse gas (GHG) emissions from household consumption for 2030 compared to 1990 level, and about 80% by 2050.

For instance, in the scheme which may eventually help in receding global warming, energy will play a central role since all other economy sector substantially depend on it [8, 9]. The energy sector was acknowledged as vital to achieving cost-effective economy [7]. Bioenergy generation may constitute around 57% of the renewable energy likely to be consumed in 2020, of which 45% will be consumed in the form of heat and electricity, and 12% will be provided in the form of biofuels applicable in transportation and other allied industries [10, 11].

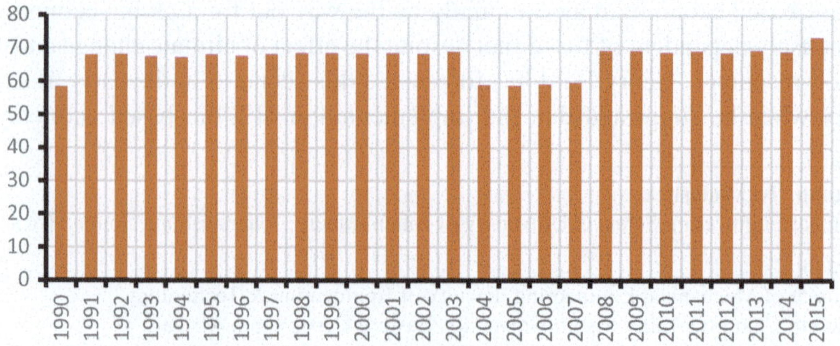

Fig. 1.1 Average global fossil fuel consumption [3]

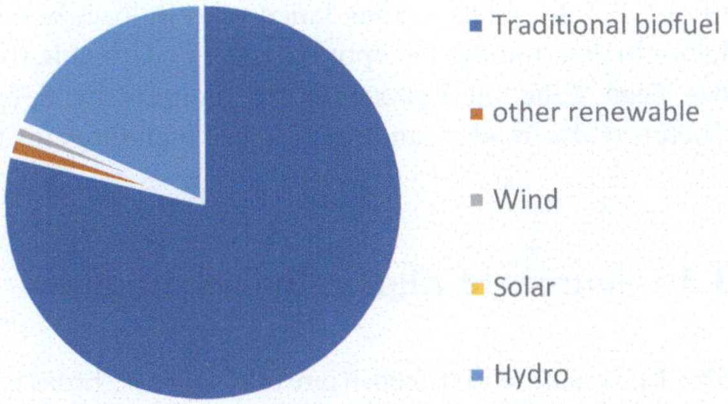

Fig. 1.2 Average global renewable energy consumption (1980–2016) [14]

Interestingly, there is a vast amount of biomass resources, which could possibly satisfy the increasing demand for green products. Biomass is the oldest renewable energy, which has been consumed in the history of humankind. It has the capacity to substantially reduce the over-reliance on fossil fuel [12] and promote the production of sundry industrial-scale value-added product which have found their way to international markets [13]. It is also playing a significant role in the strategic push for sustainable development. As shown in Fig. 1.2, traditional biofuel accounts for 79% of the total renewable energy consumed between 1980 and 2016 [14]. Biomass can be converted to several value-added products. Lignocellulosic biomass and the residue generated from their processing represent the major amount of biomass resources, which cut across different sources and has been put to several uses. Many investigations have proved that LCB holds tremendous potential for sustainable production of chemicals and other highly demanded products [15, 16].

LCB materials are natural, renewable, readily accessible, environmentally friendly, green and low-cost resources with valuable features and noteworthy relevance to the industrial sector [16, 17]. Biomass exploration has the potential to employ thousands of people along the entire value chain. The application of LCB for development of value-added products provides a leeway to utilize huge raw materials, which are readily available in nature. This will ensure the reduction of waste volume and complete biodegradability.

1.2.1 Lignocellulosic Biomass

LCB feedstocks are richly accessible at low cost. The worldwide annual LCB production is approximately 181.5 billion tonnes [18]. Approximately half of worldwide biomass resources are from LCB which is estimated at 3×10^{14} kg [19]. It has been globally acknowledged that LCB holds immense potential for sustainable commodity manufacturing, be it in the developing countries or developed countries [1, 6, 12, 20, 21]. There are three principal components in LCB: cellulose, hemicellulose and lignin. These biopolymers are interlinked to each other in a hetero-matrix and at varying relative composition depending on the type, species and even the origin of the LCB. The relative abundance of cellulose, hemicellulose and lignin is a key factor in determining the optimum energy derivable from LCB. Extensive discussion has been done on lignocellulosic biomass by several researchers [18, 22–27]. Therefore, the reader can refer to these literatures for further reading.

1.3 Sources of Lignocellulosic Biomass

The LCB can be obtained from two sources, broadly classified as herbaceous and woody biomass. Each of these broad classes has several categories based on a standard classification of EN ISO 17225-1:20149 [15]. As shown in Table 1.1, different sources of LCB and their structural composition are detailed. This is not the entire list of the LCB, and there are some others which are not captured in this table. The readers can consult [15] for further reading on the sources of LCB.

Table 1.1 LCB structural characterization [24, 25, 27, 28]

LCB class	Source	Hemicellulose (%)	Cellulose (%)	Lignin (%)
Woody biomass	Oak	21.9	43.2	35.4
	Eucalyptus	18.4	54.1	21.5
	Pine	24.0	45.6	26.8
	Spruce	21.2	50.8	27.5
	Poplar	26.2–28.7	58.8–53.3	15.5–16.3
	Douglas fir	11.0	44.0	27.0
	Ailanthus wood	22.6	46.7	22.2
	Albizia	6.7	59.5	33.8
	Birchwood	40.0	25.7	15.7
	Beechwood	31.8	45.8	21.9
	Furniture sawdust	32.63	37.23	22.16
	Subabul wood	24.0	39.8	24.7
	Oak	21.9	43.2	35.4
	Pine	24.0	45.6	26.8
	Spruce	21.23	50.81	27.5
	Chips from wood	31.8	31.82	19.0
	Wood bark	47	22	31.0
Agricultural residue	Corn cob	28.7–35	40.3–45	15–16.6
	Cornstalk	43.01	22.82	15.59
	Corn stover	30.7	51.2	14.4
	Bagasse	22.6	41.3	18.3
	Cashew nutshell	18.6	41.3	40.1
	Banana waste	14.8	13.2	14.0
	Barley straw	29.7	48.6	27.7
	Tea waste	19.9	30.2	40.0

(continued)

Table 1.1 (continued)

LCB class	Source	Hemicellulose (%)	Cellulose (%)	Lignin (%)
	Rice husk	24.3	31.3	14.3
	Rice straw	22.7	37.0	13.6
	Millet husk	26.9	33.3	14.0
	Sorghum bagasse	24.0	41.0	10.0
	Hazelnut shell	15.7	22.9	51.5
	Hazelnut seed coat	15.7	29.6	53.0
	Groundnut shell	18.7	35.7	30.2
	Coconut shell	25.1	36.3	28.7
	Nuts shell	25–30	25–30	30–40
	Flax straw	34.40	36.70	28.9
	Grape residue	34.40	36.30	12.1
	Tobacco leaf	28.2	42.4	27.0
	Tobacco stalk	25	47	23
	Bast fibre seed flax	18–21	45–53	21–26
	Bast fibre jute	22–23	31–39	15–19
	Bast fibre kenaf	46.3	35	18.8
	Coffee pulp	17.3	60.8	8.8
	Leaf fibre Abaca Leaf fibre Sisal	21–24	43–56	7–9
Energy grasses	Giant reed	29.5–30.0	36.1–37.0	17.6–19.0
	Switchgrass	27.2–27.8	36.5–38.2	17.8–19.1
	Pennisetum	22.53–21.93	41.8–40.9	18.5–16.5
	Silver grass	26.2–25.6	44.1–43.3	17.5–17.1
	Cat grass (orchard)	32	40	4.70
	Reed canary grass	42.6	29.70	7.6
	Medicago Sativa	27–32	33–38	17–19
	Willow copies	49.3	14.1	20
	Orchard grass	40	32	4.7
	Water hyacinth	48.7–49.2	18.2–18.4	3.5–3.55
Energy crop	Bamboo	15–26	26–43	21–31
	Sugarcane	31.3	45.8	22.9
	Jerusalem artichoke (October)	25.99 20.95	4.50 5.48	5.70 5.05
	Jerusalem artichoke (September)	53.86 39.27	5.18 25.96	8.76 9.02
Municipal waste	Urban greening	22.96	6.86	22.73
	Sludge	29.2	50.6	24.7
	Newspaper	25–40	40–55	18–30
	Kraft paper	9.9	57.3	20.8
	General MSW	9–16	33–49	10–14
	Food waste	7.2	55.4	11.4
	Sorted refuse	20	60	20

1.4 Motivations for the Application of Lignocellulosic Biomass (LCB)

The exploration of LCB for the production of various value-added products is motivated by the desire to create value in term of environmental benefits, energy security, socioeconomic advantages and educational benefits. Each of these motivating factors are discussed below.

1.4.1 Environmental Benefits

The ultimate motivation that is leading the drive towards renewable energy is a need to protect the environment from damaging the effect of greenhouse gas (GHG) which has a potential for global warming. In line with this, LCB plays a crucial role in environmental security, climate protection and climate change mitigation. For instance, bioenergy component of biomass has been identified as a major player towards meeting the 2 °C global temperature reduction as set by Paris accord [29]. Several authors have outlined the environmental benefits of LCB [18, 23, 24, 27–31]. For instance, assuming a case of a completed LCB life cycle, the total CO_2 emission due to the combustion of bio-ethanol is zero, because the amount of CO_2 absorbed from the atmosphere is equal to the amount emitted. Also, bio-ethanol do not produce SO_2 which can lead to the formation of acid rain. This will definitely reduce the emission of volatile organic substances. Some value-added products of LCB processing have significant impact on soil and water. Bio-ethanol and biodiesel are biodegradable, and therefore their potential impact on the environment is drastically reduced. The special constituent, which is obtained after ethanol production from sugarcane, can be a source of essential nutrient to the soil. These organic nutrients can reduce the application of chemical-based fertilizer while mitigating water pollution. All these will be of overall benefits to the environment.

1.4.2 Energy Security

From the look of things, the global energy security may substantially depend on the sources other than fossil fuel. Around 75% of world proven oil reserve are located in seven countries [32]. Given the concerns over the sustainable and reliable future supply of fossil fuel, which is hinged on the recurring volatile global oil price, the developing countries have to pay high prices for the importation of fuel and allied products. This means that there are limited resources available for these countries to execute other developmental projects. Meanwhile, biomass could provide about 25% of the global energy demand in a year, and it is the only renewable energy, which can be directly transformed to liquid fuel [33–35]. The energy derived from

LCB (liquid, solid, gaseous) can be used in all sectors of the society for production of electricity, transportation, heating and cooling, and industrial processes. In 2012, about 2.6 billion people depended on traditional biomass to meet their energy demands [36]. Of the total 18% renewable energy that was consumed in 2016, bioenergy contributed 14% [37, 38], and this underlines the increasing significance of biomass in energy security.

1.4.3 Socioeconomic Advantages

Apart from energy security and the environmental impact of LCB, more employment opportunity and economic prosperity abound as the application of LCB progresses across the entire production chain of a value-added product. There would be increase in the labour required to cultivate feedstocks, the processing industries will be built, and small-scale farmers and entrepreneur will increase with the increase in transportation and processing demands of biomass. The residents at proximate locations to the biomass cultivating and processing facilities could also derive income across the supply chain. Also, for waste-derived LCB, waste recycling will be another source of employment creation. The cultivation of bioenergy crops on wasteland will create employment opportunity in rural areas, while processing of biofuel waste products to value-added materials such as soap, fertilizer, catalyst and cattle cake will also be sources of income. For instance, in the year 2016, bioenergy sector provided 130,677 jobs in the United State [39] while biomass energy which are substantially made of LCB produces 4.7 million gross domestic product (GDP) and above 42,000 jobs in Iowa [40]. Also, Urbanchuk [40] reported that biomass energy industry has progressively boosted Iowa's economy for more than 20 years.

1.4.4 Educational Benefits

Sustainable development is impossible without proper education and information dissemination. Jennings and Lund [41] rightly concluded that education is important for the development of renewable energy of which LCB is a vital source. The notable effect of the rapid growth in LCB exploration is an increasing demand for skilled professionals who can design and maintain biomass cultivating and processing facilities. This means that there will be a need for training and retraining of personnel along this line. In addition, in a case where feedstock is to be cultivated and produced on a large scale, a huge workforce is required. At the higher education level, LCB has opened a new frontier for acquiring knowledge on different areas, such as conversion technologies, processing methods, characterization techniques and modelling methods. So, the researcher and experts will have to dig deeper in order to update their knowledge within this scope.

1.5 Value-Added Products of LCB and Their Application

There are several products which have been manufactured and could still be produced from LCB. The followings are some of the products which have been reported in the literature:

1. Nanostructure LCB: The nanostructure LCB mainly made of lignin and cellulose had been applied in various areas. It has been applied to reinforce polymers in order to produce nanocomposites due to its exceptional mechanical strength, aspect ratio and large surface area. These reinforced polymers have been used for orthopaedic applications, building materials, paint additives, catalytic degradation of organic pollutants, monitoring of waterborne pathogens and organic contaminants, cosmetic products and so on [16, 17, 42].
2. Transportation fuel: Rosillo-Calle [43] reviewed the biomass energy potential and usages. There was a detailed reporting of the state of application of liquid fuel derived from LCB as a transportation fuel which cut across aviation and road transportation. Biofuel in liquid form can be used to remove carbon in transportation sector which is still 90% oil dependent [44]. In 2016, 4% of global road transport fuel demand was met by biofuel. Biofuel production is projected to rise to 159 billion litres in the next 5 years [44].
3. Heat and power generation: Most recently, bioenergy derived from LCB has been processed for heating and electric power generation. It has been reported that 6% of global heat consumption in 2015 was provided by biomass [44]. Modern bioenergy has been applied directly in space heating and district heating scheme. Also, approximately 500 TWh of electricity has been generated from biomass in 2016, and this translates to 2% of world electricity generation [44]. This development is largely due to the higher levels of policy support across the nations.
4. Lubricating additive: Lubricants have been very essential to human activity. Lignin as an additive can enhance anti-wear property due to the formation of lubrication film and strong adhesion to the interacting surface [45]. Since lignin is a part of an active structural component of LCB, lubricants or lubrication additives can be extracted from LCB.
5. Biopolymers and fibres: LCB-derived polymers have been used in several commercial applications, such as automotive, packaging, horticulture medical and biomedical equipment, purification and pharmaceuticals [13]. Biopolymer modification based on green chemistry technique leads to a variety of products that can be used at home and industries. For packaging industry, a novel flexible film in semblance of plastic packaging film was engineered from chitin and cellulose from the tree [46]. This may be a major breakthrough for renewable and biodegradable packages given the importance of product packaging across the industries.
6. LCB-based construction materials: Natural fibres can be processed or hybridized in order to produce building construction materials. For instance, bamboo has been used in building construction due to its strength and durability. It has

often served as a green scaffolding in building construction. Also, Giglio [47] proposed Hemp bio-composites as an alternative to masonry. France allows the usage of hemp-reinforced plaster in their construction sector [48]. The composite obtained from hemp fibre has been commercialized in form of hemp block structure. Also, cellulose nanocrystals can be added to concrete in order to obtain stronger matrix through a chemical reaction. Cellulose nanocrystal can be blended with plastic and other synthetic materials to produce industrial fillers. This will reduce the weight of materials and subsequently reduce the energy consumption.

7. Pharmaceutical products, commodity chemical and monomers: Pharmaceutical chemicals include medicinal drugs, veterinary drugs, hygiene materials, diagnostic agents, nutraceuticals, disinfectants, antioxidants and personal care products [49]. It has been reported that 30–90% of the active agents in pharmaceuticals are excreted unchanged after consumption [50]. These unprocessed metabolites and degradation products contaminate the water and soil, thereby posing a threat to human health and wildlife. The extraction of the active agents in these products from LCB will be of benefits in term of rapid bio-digestion in the human body and biodegradability of the metabolites in the environment.
8. Food and feed production: As the world population increases, ensuring the sustainability of healthy food and feed production to meet the demand is one of the greatest challenges in the food industry. Among other applications of LCB in the food industry, proteins from green leaves and algae can be deployed as food and feed ingredient. This will reduce the consumption of fossil fuel-based food additives.
9. Nanocrystal from LCB in healthcare delivery: There are ongoing research on the use of high-performance nanocrystals and optical films as a thickening agent and also in drug delivery in pills and by injection. However, the ongoing concern around the environmental impact and human health is still under investigation [51]. Nanocrystal has also been touted as a potential candidate for vaccine production [52], and it can also be used in scaffolding in order to grow bones for orthopaedic applications [53].
10. Water treatment and antiscalant: The biggest challenge facing water-based industries is the need for the prevention of scale build-up in water-handling equipment. A novel type of cellulose nanoparticles which was invented by Sheikhi et al. [54] has shown great promise in addressing this challenge. As discussed, most of the present anti-scaling agents which are used are high in phosphorous derivatives which could contribute hazardously to environmental pollution and could result in damages of aquatic ecosystems. Nanoengineered hairy cellulose was discovered to work better than phosphonated molecules.
11. Future use: The future application of LCB nanocrystals may include optical chips production, protein analysis, smart materials production, manufacturing of bio-tags for gene identification, bio-sensor and bio-imaging [55], medical imaging and opto-isolators.

1.6 Factors That Affect the Application of LCB

The factors that can affect the application of a certain class of LCB can be broadly divided into mechanical behaviour, thermal behaviour, structural behaviour and chemical behaviour. These factors have been thoroughly discussed in the literature, so readers are referred to these articles for further readings [23, 27, 56].

1.7 Lignocellulosic Biomass Conversion Pathway

The major pathways through which LCB can be transformed into value-added products have been discussed by several authors [23, 24, 27]. A direct combustion method is the most widely used since almost 97% of the LCB which is harnessed across the world are applied as in this form [57]. This method is particularly significant in the developing countries [58]. Combustion can be divided into three stages, which are pyrolysis, drying and reduction. The reduction stage can generate up to 70% of the total heat compared to other combustion stages [24]. These massive heat generations can be channelled into different purposes. Other conversion methods include biochemical conversion and thermochemical conversion. Table 1.2 shows the processes through which value-added chemicals and fuel products can be obtained from LCB.

1.8 Value Creation Pathway

In order to create a value-added product from LCB, there are pathways which the feedstock must be subjected. The sequence has been identified as shown in Fig. 1.3, and it is briefly discussed as follows:

Table 1.2 Value addition process and products

Value addition processes	Value-added products	References
Gasification	Methane	[59, 60]
	Biohydrogen	[61]
	Biogas	[62–64]
	Microbial fuel cell	[65–67]
Biochemical conversion	Methane, biodiesel, ethanol, compost	[68–70]
Direct liquefaction	Methanol, liquid fuel, adhesives, biopolymers, phenolic resin, polyurethane	
Pyrolysis	Bioethanol, methanol, liquid fuel, adhesives, biopolymers, phenolic resin, polyurethane	[71, 72]
	Eelectricity	[73–76]
Direct combustion	Electricity, heat	

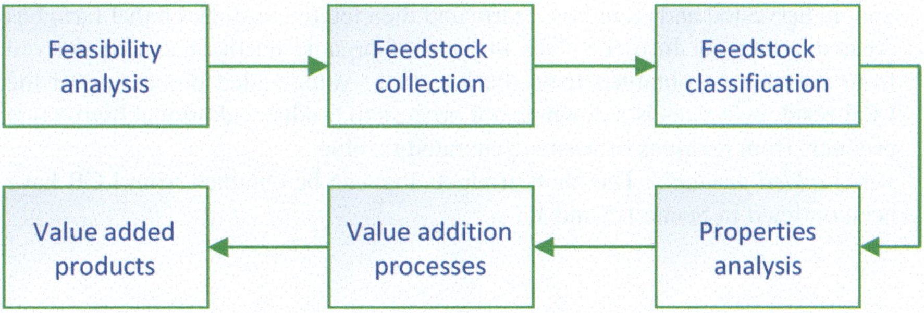

Fig. 1.3 Schematic diagram of the value creation pathway

1. *Feasibility analysis*: First, it should be studied to confirm if there is a sufficient amount of a given LCB which can be applied in producing value-added products. Also, the environmental consequences of such exploration should be established. For instance, the exploration of LCB should not lead to food and water competition; therefore, it is advisable to apply residues and waste in the production. Also, the supply chain for LCB must be analysed in order to allow a strategic planning that is aimed at reducing the cost of production.
2. *Feedstock collection*: Since the biomass materials are grown in different places, the LCB would be collected from different sources and aggregated before application. The studies have shown that charges on the collection, inventory and transportation of LCB feedstock can account for about 50% of the production cost [77]. This must be put into consideration in the overall planning process. Critical path analysis of different sources may help in determining the minimal cost of LCB collection.
3. *Feedstock classification*: The LCB is from various sources and origin; therefore, there is a need for classification to determine the ones that can be processed together for a specific application. This is because the source of a particular LCB may affect its properties and behaviour [78].
4. *Property analysis*: The properties of LCB can be broadly categorized in proximate, compositional, structural and thermal properties. The properties of the biomass determine the type of conversion process which may be applicable to a specific biomass. For instance, LCB with high moisture content is a better candidate for a biochemical conversion process [23, 25, 79]. These properties have been reviewed and reported in the literature [23, 24, 27, 80, 81].
5. *Value addition processes*: Value addition is a process of converting a product, in this case, LCB from its original state to an economically more useful phase. Generally speaking, value addition means to economically improve the usability of a commodity by changing its current phase and altering a set of characteristics to other characteristics that are more acceptable in the market or more efficient in terms of its functionality. The main value addition process through which LCB is converted to useful products has been discussed in Sect. 1.7. Many raw feedstocks have intrinsic value in their unprocessed state. For instance, field corn

grown, harvested and stored on a farm and then fed to livestock on that farm has created value, but further values in form of organic fertilizer can be derived from the manure obtained from the livestock. Value-added processing of the LCB residues/wastes is a downstream process to produce additional high-value products from residues or wastes generated residue.
6. *Value-added products*: The main products that can be obtained from LCB have been outlined in Sects. 1.5 and 1.7.

1.9 Challenges Associated with the Exploration of Lignocellulosic Biomass (LCB)

Notwithstanding the enormous benefits and several motivation for the application of LCB in the development of value-added products, some difficulties need to be surmounted in order to derive the maximum value. These challenges are associated with different stages along the value creation pathway (see Sect. 1.8), from the sourcing of LCB to the final use. Also, the extent of these challenges depends on the kind of feedstock, the location across the globe and other related indicators. Broadly speaking, the challenges can be classified into operational dimension, socioeconomic dimension, policy and regulation, and population explosion. These are briefly discussed below:

1. *Operational challenges*: The expansion of the LCB valuation chain has been affected by the availability of resources and standardization of processing equipment to cope with feedstocks from diverse sources. These equipment are applied during harvesting, transportation, storage and final use. While there are several pre-treatment approaches which have been devised to retain the properties of LCB transported over a long distance [82, 83], there may be increase in the cost associated with the production. Useful data, which are needed for accurate prediction of the quantity, quality and potential of different LCB sources and technologies, are not yet enough. The seasonal variation in the production LCB means that the price of the product base may vary, while the moisture content may lead to an increasing pressure on the transportation infrastructure. Also, there is a further need for the rural communities who are often the major source of the feedstock to understand the technicalities needed to maintain and operate the plants to produce LCB-based products. Despite all these challenges, the advancement in research and innovation is opening a new page towards reduction in operation challenges, which has been highlighted. For instance, the intelligent management of supply chain will reduce the cost associated with supply chain management.
2. *Socioeconomic challenges*: The socioeconomic dimension of the challenges related to value creation from LCB is centred around feedstock acquisition cost, high investment and start-up cost, conflicts related to decision making, land use issues and environmental impact. Since the LCB resources are dispersed across different locations, the investor may want to build the plant close to the most abundant source, leading to the centralization of projects, which may have

engendered development in other locations if the plant were to be cited there. The high operation cost associated with biomass processing technologies is also a source of concern to the investors. Although the decision related to the location, route and processing technologies is noticeable, responsible leadership structure, which ensure proper communication with the stakeholders, will guarantee the proper understanding of socioeconomic benefits of resources utilization. Also, the job creation through the entire value chain should be enough to offset the negative impact of the LCB value chain.

3. *Policy and regulatory challenges*: Policy and regulation in several developing countries have created a bottleneck, which hampered the growth of biomass value chain. Deliberate policy, which promote green pricing in biomass industry, may be an effective tool to enhance market acceptability of biomass value-added products. In some countries, the biomass resources as a renewable alternative have not received sufficient attention, because the government is more focused on fossil fuel. More so, there are no well-structured rules to regulate the utilization of biomass resources, especially in most developing countries. Again, there is no well-established mechanism, which manage the development of biomass resources industry, where the national policy and standards exist, and there is no specialized department that is designated to monitor the implementation.
4. *Growing population and competitive demand for land*: With increasing population, there is more demand for land for both human shelter, recreation centres, religious centres, motor park and agricultural practices. In some part of Africa where the effect of climate change is leading to decreasing availability of green pasture and water, the nomads often transverse the country in search of greener pasture for their livestock [84, 85]. This has led to severe ethnic clashes in the countries. Also, it has been projected that human population will grow to 9 billion by around 2050, and this will result in increasing global food production to about 60% [86–88]. The expected increase in global demand for food means that more croplands need to be cultivated, which means that the agricultural production will need to rise. Thankfully, the residue, which are generated from the agricultural practices, is a good source of biomass feedstock, which is available at lower cost and environmentally sustainable. These provide dual benefits: while the food demand is been met, the product residue is further converted to value-added product for human use. Land use can lead to the interruption of ecosystem and natural habitat of the people and animals. The marginal and degraded lands can be used for the production of LCB, and this will reduce the pressure on the arable land [89, 90].

1.10 Conclusion

It is possible to secure sufficient access to LCB if the resource-constrained world would make it a priority. Among the latest research endeavour, the value-added materials of LCB extract are of interest due to its sustainability, high efficiency, biodegradability and availability. Various value-added products, which can be sourced from LCB, were discussed. In addition, the conversion processes were outlined while the motivation for this latest LCB drive was explained. The future scope of LCB application is expected to further increase, given the advancement of pre-treatment technologies and other processing technique.

References

1. F. Isikgor, C.R. Becer, Lignocellulosic biomass: A sustainable platform for production of bio-based chemicals and polymers. Polym. Chem. **6**, 4497–4559 (2015)
2. P. Phanthong, P. Reubroycharoen, X. Hao, G. Xu, A. Abudula, G. Guan, Nanocellulose: Extraction and application. Carbon Resour. Convers. **1**, 32 (2018)
3. World Bank, Fossil fuel energy consumption (% of total) (2018), https://data.worldbank.org/indicator/EG.USE.COMM.FO.ZS?end=2015&start=1960&view=chart&year_high_desc=true. Accessed 23 Oct 2018
4. C. Brandon Miller, J. Croft, Planet has only until 2030 to stem catastrophic climate change, experts warn (2018), https://edition.cnn.com/2018/10/07/world/climate-change-new-ipcc-report-wxc/index.html. Accessed 23 Oct 2018
5. D. Wuebbles, D. Fahey, K. Hibbard, D. Dokken, B. Stewart, T. Maycock, Climate science special report: fourth national climate assessment, volume I. repository.library.noaa.gov (2017)
6. L.A. Lucia, Lignocellulosic biomass: A potential feedstock to replace petroleum. Bioresources **3**(4), 981–982 (2008)
7. N. Scarlat, J.-F. Dallemand, F. Monforti-Ferrario, V. Nita, The role of biomass and bioenergy in a future bioeconomy: Policies and facts. Environ. Dev. **15**, 3–34 (2015)
8. O. Obafemi et al., Electric power crisis in Nigeria: A strategic call for change of focus to renewable sources. IOP Conf. Ser. Mater. Sci. Eng. **413**(1), 012053 (2018)
9. P.K. Adom, P.A. Kwakwa, Effects of changing trade structure and technical characteristics of the manufacturing sector on energy intensity in Ghana. Renew. Sust. Energ. Rev. **35**, 475–483 (2014)
10. C. Hamelinck, I. De Lovinfosse, M. Koper, C. Beestermoeller, C. Nabe, M. Kimmel, Renewable energy progress and biofuels sustainability. Report for the European Commission, Ecofys, Utrecht, Netherlands, Tech. Rep. ENER/C1/463-2011-Lot2, 2012
11. M. Banja, N. Scarlat, F. Monforti-Ferrario, J.-F. Dallemand, *Renewable Energy Progress in EU 27 (2005–2020)* (Joint Research Centre, Ispra, Italy, 2013)
12. S. Fatma et al., Lignocellulosic biomass: A sustainable bioenergy source for the future. Protein Pept. Lett. **25**(2), 148–163 (2018)
13. S.A. Ashter, *Technology and Applications of Polymers Derived from Biomass* (William Andrew Applied Science Publishers, Oxford, 2017)
14. H. Ritchie, M. Roser, Energy production and changing energy sources (2018), OurWorldInData.org
15. W. Elbersen, T.M. Lammens, E.A. Alakangas, B. Annevelink, P. Harmsen, B. Elbersen, Lignocellulosic biomass quality: Matching characteristics with biomass conversion requirements. In Modeling and Optimization of Biomass Supply Chains (pp. 55–78) (2017). Academic Press
16. L. Brinchi, F. Cotana, E. Fortunati, J. Kenny, Production of nanocrystalline cellulose from lignocellulosic biomass: Technology and applications. Carbohydr. Polym. **94**(1), 154–169 (2013)
17. E. Fortunati, W. Yang, F. Luzi, J. Kenny, L. Torre, D. Puglia, Lignocellulosic nanostructures as reinforcement in extruded and solvent casted polymeric nanocomposites: An overview. Eur. Polym. J. **80**, 295–316 (2016)
18. S. Paul, A. Dutta, Challenges and opportunities of lignocellulosic biomass for anaerobic digestion. Resour. Conserv. Recycl. **130**, 164–174 (2018)
19. J. Lynch, Utilization of lignocellulosic wastes. J. Appl. Microbiol. **63**, s16 (1987)

20. C. Somerville, H. Youngs, C. Taylor, S.C. Davis, S.P. Long, Feedstocks for lignocellulosic biofuels. Science **329**(5993), 790–792 (2010)
21. Y.Y. Tye, K.T. Lee, W.N.W. Abdullah, C.P. Leh, The world availability of non-wood lignocellulosic biomass for the production of cellulosic ethanol and potential pretreatments for the enhancement of enzymatic saccharification. Renew. Sust. Energ. Rev. **60**, 155–172 (2016)
22. J.B. Sluiter, R.O. Ruiz, C.J. Scarlata, A.D. Sluiter, D.W. Templeton, Compositional analysis of lignocellulosic feedstocks. 1. Review and description of methods. J. Agric. Food Chem. **58**(16), 9043–9053 (2010)
23. J. Cai et al., Review of physicochemical properties and analytical characterization of lignocellulosic biomass. Renew. Sust. Energ. Rev. **76**, 309–322 (2017)
24. V. Dhyani, T. Bhaskar, A comprehensive review on the pyrolysis of lignocellulosic biomass. Renew. Energy **129**, 695–716 (2018)
25. Y.D. Singh, P. Mahanta, U. Bora, Comprehensive characterization of lignocellulosic biomass through proximate, ultimate and compositional analysis for bioenergy production. Renew. Energy **103**, 490–500 (2017)
26. O.O. Olatunji, S.A. Akinlabi, M.P. Mashinini, S.O. Fatoba, O.O. Ajayi, Thermo-gravimetric characterization of biomass properties: A review. IOP Conf. Ser. Mater. Sci. Eng. **423**(1), 012175 (2018)
27. O. Obafemi, A. Stephen, O. Ajayi, P. Mashinini, M. Nkosinathi, Experimental investigation of thermal properties of lignocellulosic biomass: A review. IOP Conf. Ser. Mater. Sci. Eng. **413**(1), 012054 (2018)
28. I. Boumanchar et al., Effect of materials mixture on the higher heating value: Case of biomass, biochar and municipal solid waste. Waste Manag. **61**, 78–86 (2017)
29. C.A. Joly, L. Verdade, *Bioenergy & Sustainability: Bridging the Gaps* (SCOPE, Paris, 2015)
30. C.-H. Zhou, X. Xia, C.-X. Lin, D.-S. Tong, J. Beltramini, Catalytic conversion of lignocellulosic biomass to fine chemicals and fuels. Chem. Soc. Rev. **40**(11), 5588–5617 (2011)
31. Q. Kang, L. Appels, T. Tan, R. Dewil, Bioethanol from lignocellulosic biomass: Current findings determine research priorities. Sci. World J. **2014**, 1 (2014)
32. R. Ren21, Global status report, 2011 (2011)
33. C. Briens, J. Piskorz, F. Berruti, Biomass valorization for fuel and chemicals production–A review. Int. J. Chem. React. Eng. **6**(1), 1542–6580 (2008)
34. S. Tewfik, Biomass utilization facilities and biomass processing technologies. Energ. Educ. Sci. Technol. **14**, 1–19 (2004)
35. Y. Yanli, Z. Peidong, Z. Wenlong, T. Yongsheng, Z. Yonghong, W. Lisheng, Quantitative appraisal and potential analysis for primary biomass resources for energy utilization in China. Renew. Sust. Energ. Rev. **14**(9), 3050–3058 (2010)
36. I. D. w. o. r. e. e. W., Global bioenergy statistics (2013). Accessed 26 Dec 2017
37. World Energy Council (WEC), Waste to energy (2016), https://www.worldenergy.org/wpcontent/uploads/2017/03/WEResources_Waste_to_Energy_2016.pdf. Accessed 16 Jan 2018
38. BP, Statistical review of world energy (2016), https://www.bp.com/en/global/corporate/energy-economics/statistical-review-of-world-energy.html. Accessed 26 Mar 2018
39. U. S. D. o. Energy, U.S. energy and employment report (2018). Available online: https://www.energy.gov/downloads/2017-us-energy-and-employment-report. Accessed 25 Nov 2018
40. J.M. Urbanchuk, Contribution of the renewable fuels industry to the economy of Iowa, in *Agricultural and Biofuels Consulting*, (LLP Doylestown, Pennsylvania, 2016)
41. P. Jennings, C. Lund, Renewable energy education for sustainable development. Renew. Energy **22**(1–3), 113–118 (2001)
42. R. Mohammadinejad, S. Karimi, S. Iravani, R.S. Varma, Plant-derived nanostructures: Types and applications. Green Chem. **18**(1), 20–52 (2016)
43. F. Rosillo-Calle, A review of biomass energy–shortcomings and concerns. J. Chem. Technol. Biotechnol. **91**(7), 1933–1945 (2016)
44. I. E. Agency, Bioenergy and biofuels (2018), https://www.iea.org/topics/renewables/bioenergy/. Accessed 27 Nov 2018
45. L. Mu et al., Lignin from hardwood and softwood biomass as a lubricating additive to ethylene glycol. Molecules **23**(3), 537 (2018)

46. C.C. Satam et al., Spray-coated multilayer cellulose nanocrystal—Chitin nanofiber films for barrier applications. ACS Sustain. Chem. Eng. **6**(8), 10637–10644 (2018)
47. F. Giglio, The use of materials from biomass as construction materials. Open J. Civ. Eng. **3**(02), 82 (2013)
48. C. Stevens, *Industrial Applications of Natural Fibres: Structure, Properties and Technical Applications* (John Wiley & Sons, Chichester, 2010)
49. A. Valavanidis, T. Vlachogianni, Pharmaceutical industry and green chemistry: New developments in the application of green principles and sustainability. Pharmakeftiki **24**(3), 44–56 (2012)
50. Pharmaceutical Technology, Green pharma: The growing demand for environmentally friendly drugs (2017), https://www.pharmaceutical-technology.com/comment/comment-green-pharma-the-growing-demand-for-environmentally-friendly-drugs-5937344/. Accessed 27 Nov 2018
51. M. Roman, Toxicity of cellulose nanocrystals: A review. Ind. Biotechnol. **11**(1), 25–33 (2015)
52. L. Davis, Nanocrystal research could lead to new vaccine and computer inks (2007)
53. D.M. Fox et al., Simultaneously tailoring surface energies and thermal stabilities of cellulose nanocrystals using ion exchange: Effects on polymer composite properties for transportation, infrastructure, and renewable energy applications. ACS Appl. Mater. Interfaces **8**(40), 27270–27281 (2016)
54. A. Sheikhi, A. Kakkar, T.G. van de Ven, Nanoengineering colloidal and polymeric celluloses for threshold scale inhibition: Towards universal biomass-based crystal modification. Mater. Horiz. **5**(2), 248–255 (2018)
55. E. Lizundia, T.-D. Nguyen, J.L. Vilas, W.Y. Hamad, M.J. MacLachlan, Chiroptical luminescent nanostructured cellulose films. Mater. Chem. Front. **1**(5), 979–987 (2017)
56. L. Jiménez, F. González, Study of the physical and chemical properties of lignocellulosic residues with a view to the production of fuels. Fuel **70**(8), 947–950 (1991)
57. A.V. Bridgwater, Renewable fuels and chemicals by thermal processing of biomass. Chem. Eng. J. **91**(2–3), 87–102 (2003)
58. L. Zhang, C.C. Xu, P. Champagne, Overview of recent advances in thermo-chemical conversion of biomass. Energy Convers. Manag. **51**(5), 969–982 (2010)
59. G. Kanat, A. Saral, Estimation of biogas production rate in a thermophilic UASB reactor using artificial neural networks. Environ. Model. Assess. **14**(5), 607–614 (2009)
60. B. Ozkaya, A. Demir, M.S. Bilgili, Neural network prediction model for the methane fraction in biogas from field-scale landfill bioreactors. Environ. Model Softw. **22**(6), 815–822 (2007)
61. J. Wang, W. Wan, Application of desirability function based on neural network for optimizing biohydrogen production process. Int. J. Hydrog. Energy **34**(3), 1253–1259 (2009)
62. H. Abu Qdais, K. Bani Hani, N. Shatnawi, Modeling and optimization of biogas production from a waste digester using artificial neural network and genetic algorithm. Resour. Conserv. Recycl. **54**(6), 359–363 (2010)
63. D.P. Strik, A.M. Domnanovich, L. Zani, R. Braun, P. Holubar, Prediction of trace compounds in biogas from anaerobic digestion using the MATLAB Neural Network Toolbox. Environ. Model Softw. **20**(6), 803–810 (2005)
64. L. Alejo, J. Atkinson, V. Guzmán-Fierro, M. Roeckel, Effluent composition prediction of a two-stage anaerobic digestion process: Machine learning and stoichiometry techniques. Environ. Sci. Pollut. Res. **25**(21), 21149–21163 (2018)
65. A. Tardast et al., Use of artificial neural network for the prediction of bioelectricity production in a membrane less microbial fuel cell. Fuel **117**, 697–703 (2014)
66. A. Tardast, M. Rahimnejad, G. Najafpour, K. Pirzade, N. Mokhtarian, Prediction of bioelectricity production by neural network. J. Biotechnol. Pharm. Res. **3**(3), 62–68 (2012)
67. Y. Sewsynker, E.B.G. Kana, A. Lateef, Modelling of biohydrogen generation in microbial electrolysis cells (MECs) using a committee of artificial neural networks (ANNs). Biotechnol. Biotechnol. Equip. **29**(6), 1208–1215 (2015)
68. Y. Sewsynker-Sukai, F. Faloye, E.B.G. Kana, Artificial neural networks: An efficient tool for modelling and optimization of biofuel production (a mini review). Biotechnol. Biotechnol. Equip. **31**(2), 221–235 (2017)

69. V.B. Furlong, R.D. Pereira Filho, A.C. Margarites, P.G. Goularte, J.A.V. Costa, Estimating microalgae Synechococcus nidulans daily biomass concentration using neuro-fuzzy network. Food Sci. Technol. **33**, 142–147 (2013)
70. N. Sarkar, S.K. Ghosh, S. Bannerjee, K. Aikat, Bioethanol production from agricultural wastes: An overview. Renew. Energy **37**(1), 19–27 (2012)
71. E. Betiku, A.E. Taiwo, Modeling and optimization of bioethanol production from breadfruit starch hydrolyzate vis-à-vis response surface methodology and artificial neural network. Renew. Energy **74**, 87–94 (2015)
72. M. Esfahanian, M. Nikzad, G. Najafpour, A.A. Ghoreyshi, Modeling and optimization of ethanol fermentation using Saccharomyces cerevisiae: Response surface methodology and artificial neural network. Chem. Ind. Chem. Eng. Q. **19**(2), 241–252 (2013)
73. H. Uzun, Z. Yildiz, J.L. Goldfarb, S. Ceylan, Improved prediction of higher heating value of biomass using an artificial neural network model based on proximate analysis. Bioresour. Technol. **234**, 122–130 (2017)
74. G.E. Acquah, B.K. Via, O.O. Fasina, S. Adhikari, N. Billor, L.G. Eckhardt, Chemometric modeling of thermogravimetric data for the compositional analysis of forest biomass. PLoS One **12**(3), e0172999 (2017)
75. S. Ghugare, S. Tiwary, V. Elangovan, S. Tambe, Prediction of higher heating value of solid biomass fuels using artificial intelligence formalisms. Bioenergy Res. **7**(2), 681–692 (2014)
76. S.B. Ghugare, S.S. Tambe, Genetic programming based high performing correlations for prediction of higher heating value of coals of different ranks and from diverse geographies. J. Energy Inst. **90**(3), 476–484 (2017)
77. J.K. Kurian, G.R. Nair, A. Hussain, G.V. Raghavan, Feedstocks, logistics and pre-treatment processes for sustainable lignocellulosic biorefineries: A comprehensive review. Renew. Sust. Energ. Rev. **25**, 205–219 (2013)
78. A. Khan, W. De Jong, P. Jansens, H. Spliethoff, Biomass combustion in fluidized bed boilers: Potential problems and remedies. Fuel Process. Technol. **90**(1), 21–50 (2009)
79. M.V. Kok, E. Ozgur, Characterization of lignocellulose biomass and model compounds by thermogravimetry. Energy Sources Part A **39**(2), 134–139 (2017)
80. I. Abed, M. Paraschiv, K. Loubar, F. Zagrouba, M. Tazerout, Thermogravimetric investigation and thermal conversion kinetics of typical North-Africa and middle-east lignocellulosic wastes. Bioresources **7**(1), 1120–1220 (2012)
81. L. Kou, Y. Song, X. Zhang, T. Tan, Comparison of four types of energy grasses as lignocellulosic feedstock for the production of bio-ethanol. Bioresour. Technol. **241**, 424–429 (2017)
82. J.S. Tumuluru, *Biomass Preprocessing and Pretreatments for Production of Biofuels: Mechanical, Chemical and Thermal Methods* (CRC Press, Boca Raton, 2018)
83. K. Karimi, M.J. Taherzadeh, A critical review of analytical methods in pretreatment of lignocelluloses: Composition, imaging, and crystallinity. Bioresour. Technol. **200**, 1008–1018 (2016)
84. C.S. Teitelbaum, T. Mueller, Beyond migration: Causes and consequences of nomadic animal movements. Trends Ecol. Evol. **34**, 569 (2019)
85. G. Sivini, Resistance to modernization in Africa: journey among peasants and nomads. Routledge, New York, USA (2017)
86. A.M. Thomson et al., Sustainable intensification in land systems: Trade-offs, scales, and contexts. Curr. Opin. Environ. Sustain. **38**, 37–43 (2019)
87. D. Tilman, C. Balzer, J. Hill, B.L. Befort, Global food demand and the sustainable intensification of agriculture. Proc. Natl. Acad. Sci. U.S.A. **108**(50), 20260–20264 (2011)
88. N. Alexandratos, J. Bruinsma, World agriculture towards 2030/2050: The 2012 revision (2012)
89. S.A. Edrisi, P. Abhilash, Exploring marginal and degraded lands for biomass and bioenergy production: An Indian scenario. Renew. Sust. Energ. Rev. **54**, 1537–1551 (2016)
90. R.A. Nalepa, A.G.S. Gianotti, D.M. Bauer, Marginal land and the global land rush: A spatial exploration of contested lands and state-directed development in contemporary Ethiopia. Geoforum **82**, 237–251 (2017)

Chapter 2
A Short Overview of Analytical Techniques in Biomass Feedstock Characterization

D. C. Okafor and M. O. Daramola

2.1 Introduction

For a revolutionary transformation in human life, the use of energy is required. Climate change concerns due to over dependence and exploitation of fossil fuel have necessitated the need for alternative energy source. Greenhouse gases (GHGs) from fossil fuel production and utilization are the main causative agent of climate change. Fossil fuel resources include coal, petroleum (oil), and natural gas, while resources of bio-based materials are carbohydrates, lipids, and proteins. The risks and challenges to fossil fuel-based materials are depleting fossil fuel resources, climate changes, water and atmospheric pollution, and environmental concerns [1, 6, 49, 67, 93]. These must be avoided and addressed, and renewable energy from biomasses is the known solution to these present problems that the entire universe is facing.

Biomass refers to any organic material that is from plant- or animal-derived source which are converted into different types of bioenergy using various processing techniques [16, 35, 96]. It includes biomaterials and bio-based materials. Biomaterials are materials from biological (bio/life) biomass sources to make devices to replace a part or a function of the bio, while bio-based material is any material made from living or nonliving organism or biomass or bioresource either

D. C. Okafor
School of Engineering and Engineering Technology, Department of Food Science and Technology, Federal University of Technology, Owerri, Imo State, Nigeria
e-mail: damaris.okafor@futo.edu.ng

M. O. Daramola (✉)
Department of Chemical Engineering, Faculty of Engineering, Built Environment and Information Technology, University of Pretoria, Hatfield, Pretoria, South Africa

M. O. Daramola, A. O. Ayeni (eds.), *Valorization of Biomass to Value-Added Commodities*, Green Energy and Technology,
https://doi.org/10.1007/978-3-030-38032-8_2

animal or plant bio-source. Sources of biomass are municipal solid waste (MSW), aquatic crops, animal residues, agricultural crop residues, forestry crop residues, industrial wastes and residues, and dedicated crops and residues. Biodegradability, sustainability, environment friendliness, increases in fossil fuel prices, environmentally conscious consumers, and carbon neutral are the drivers and motivations of biomass.

Energy sourced from biomass feedstocks is called bioenergy. Many conversion techniques are needed for proper conversion of raw biomass feedstocks into energy and value-added chemicals. Physical, chemical, biological, thermal, thermochemical, biochemical, and physiochemical are series of conversion technologies available for biomass processing. Thermal processes involve combustion under excess air, under partial air known as gasification, and under no air which is pyrolysis. Biochemical processing techniques are of two major categories which are anaerobic digestion with the objective of getting biogas and fermentation for ethanol production. Catalytic chemical conversions involve hydrogenations, dehydroxylation/hydrogenolysis, and oxidations. Hydrogenations involve monosaccharides, furanic compounds, fatty compounds, and phenolic compounds. Dehydroxylation/hydrogenolysis involves sorbitol, xylitol, erythritol, and glycerol etcetera [17, 57, 97, 98].

In oxidative conversions, sugars, carbohydrate, and derivatives are involved. This chapter, therefore, gives an overview on the different biomass conversion technique for ensuring all biomass conversion puzzles that affect biomass characterization as well as the standard analytical tools/equipment used for characterization of diverse biomass attributes. Tools for chemical or elemental variable composition, morphology and particle size distribution, analytical equipment, and standard tools for proper characterization of products of converted biomass feedstocks were thoroughly addressed.

2.2 Biomass Conversion Processes

Conversion processes for valorization of biomass into value-added products are basically grouped into four categories: physical, thermal (thermochemical), biological, and chemical processes.

2.2.1 Thermal Processes

Thermal conversion process is all about the use of heat in the presence or absence of oxygen to ensure biomass transformation into different energy forms. Thermal conversion technologies (TCT) are also known as thermochemical conversion processes and encompass direct combustion, gasification, pyrolysis, and torrefaction. Basically, direct energy production via thermal process is achieved by controlled conditions of temperature and oxygen. By this means, original/raw biomass feedstocks are conveniently transformed into energy carriers (methanol, producer gas, or oils).

2.2.1.1 Direct Combustion Techniques

When biomass is burnt in excess oxygen (air), it is direct combustion. Here, biomasses mostly used are residues like wood waste, wastepaper, cardboard, corn stover, bark, black liquor, bagasse, straw, municipal/global solid waste (MGSW), food industry wastes, etc. In equipment like ovens, spreader-stoker, and fuel cell furnaces, drying (and possibly gasification when done at higher temperature using high-temperature ovens) takes place. Modern and advanced forms of these equipment use rotational or vibration grates or both to ensure proper ash removal with water coolant retrofitted to the system.

For the use of fluidized-bed combustors, biomass particles are mixed in turbulent preheated air and are injected to suspension furnaces where the samples in the preheated air are burnt. The use of fluidized bed combustors requires a preheated sand bed at temperature ≥500 but ≤900 °C that provides the heating medium for the biomass particles. Surely, these systems prevent grate usage as it promotes and permits other techniques for air and sand preheating. For biomass feedstocks that are in liquid form but not bulky, injection systems with water coolant are compulsorily required when any of the suspension furnaces or fluidized bed combustors is employed in thermal processing of biomass feedstocks [2, 63].

2.2.1.2 Co-firing

Co-firing (co-combustion) of biomass with fossil fuels is not a long-term option for renewable energy production. Though it is a cheap process that involves co-firing of biomass with coal or any other fossil fuel, its consideration is on the fact that it reduces 100% usage of fossil fuel (mainly coal) by replacing a small portion of the fossil fuel with renewable biomass fuels. Co-combustion has offered biomass feedstocks an entry point into the energy market. It has stimulated construction of new plants dedicated solely to waste conversion to energy. It has also become a less expensive way of reducing CO_2 emission attributed to utilization of fossil, and biomass fuels contain low sulfur content with nearly zero to zero SOx emission [2].

2.2.1.3 Pyrolysis

In pyrolysis (a thermochemical process), biomass sources are subjected to high temperatures ≥400 °C at low oxygen levels to prevent complete combustion and possibly under controlled pressure to convert them to solid, liquid, and gaseous fuels. Biomass is degraded to carbon molecules (methane (CH_4) and carbon mono oxide

(CO)) and hydrogen (H_2) producing a gaseous mixture known as producer gas. This does not mean that carbon dioxide is not produced during the process but it might have been degraded further under the pyrolytic conditions of the reactor back to CO and H_2O. The proportion of the solid fraction, liquid fraction, and gaseous fraction depends on the conditions of the pyrolysis.

Liquid-phase products are favored by low temperatures which are too low to crack all the long-chain carbon molecules resulting in the production of tars, oils, methanol, acetone, etc., but much lower temperature results in solid-phase formation. As soon as all the volatiles (gas fraction) have been driven off, whatever remains (the residual biomass) is in the form of char (pure carbon). Pyrolysis is an alternative to gasification but has received great deal of attention recently because it is simple, cheap to operate, and has the ability to produce liquid pyrolytic products. The production of liquid fuels from cellulosic feedstocks is only achievable with fast and flash pyrolysis. Liquid fraction is easy to transport and store and possesses chemicals that are economically viable, and they are recovered at high amount [63]. Flash pyrolysis refers to short-time residence in seconds (0.5–3 s) heating of the biomass in the pyrolytic system at high temperature (400–600 °C) in the absence of (air) oxygen [103]. Heating rate of about 300 °C/min in fast pyrolysis is used for the purpose of obtaining high-grade bio-oil. Examples of fast pyrolytic systems with different reactor configurations include circulating fluidized bed reactor, entrained flow reactor, fluidized-bed reactors, rotating reactor, vacuum furnace reactor, vortex reactor, and wire mesh reactor. Extensive studies have been carried out by Bressler's group on pyrolysis, and the group has utilized 15 mL to 5 L microreactor, constructed from stainless steel and equipped with a TC-D8 controller, for pyrolysis of different biomass feedstocks at a temperature range of 350–450 °C under constant agitation over a range of time (0.5–8 h). The desired product fraction of interest and purpose for the pyrolysis determine the optimum pyrolysis temperature and duration of time for the pyrolysis process [8–10, 30, 31, 53, 64]. To maximize the yield of liquid products from biomass pyrolysis, a low temperature with a high heating rate and a short residence time for the gas process were appropriate, while a high temperature with low heating rate and long gas residence time process were found to be ideal to obtaining a maximum yield of fuel gas from biomass pyrolysis [42].

2.2.1.4 Torrefaction

Torrefaction is also a thermal process like pyrolysis [18, 22, 83, 102]. It is a technique that converts biomass materials to bioenergy with the supply of heat still in the absence of oxygen but at temperature lower than those ones used in a pyrolytic process. Torrefaction temperatures range from 200 to 320 °C, and the process removes water with partial decomposition of lignin, cellulose, and hemicellulose, thereby resulting in an energy dense solid fuel as final product. It has better fuel properties than unprocessed (raw) biomass.

2.2.1.5 Carbonization

This is an old method of producing char (coke), a carbon from plant and animal (organic materials) by pyrolytic process. In this case, the pyrolysis process is solely set at optimum conditions favorable to produce charcoal [2]. One disadvantage of carbonization is that most of the volatile components of the biomass are lost during the carbonization process, and because of this, carbonization is at times referred to as dry distillation process. Its major product is accumulation of carbon due to drastic lost/reduction in the level of hydrogen and oxygen gases. At temperatures of 100 and 170 °C, water is mostly evaporated, but between 170 and 270 °C, CO and CO_2 gases are developed. CO and CO_2 are condensable vapors, which are long-chain carbon molecules, and they tend to form pyrolysis oil. Pyrolysis oils are suitable to produce fuel or value-added chemicals after cooling and scrubbing. Temperatures around 270 and 280 °C result in exothermic reactions, which are detected by cogeneration of heat and power [2, 73, 74, 110].

2.2.1.6 Gasification

Gasification employs the use of high temperatures (≥700 °C) heating of biomass in a controlled environment with limited or partial oxygen or steam or both to ensure that all the raw materials are converted to gases (hydrogen, carbon monoxide, and carbon dioxide). Initial stage is the partial combustion of biomass to produce gas and charcoal, and in the final stage, charcoal utilizes the CO_2 and H_2O produced initially to form CO and H_2 via chemical reduction [2, 12, 71, 109, 110, 78, 79]. The resulting mixture of gases is called syngas which is a fuel. More hydrogen can be produced from the syngas via Water Gas Shift (WGS) reaction, and the produced hydrogen, a clean source of energy, can be used in gas turbines to generate electricity or in fuel cell application. Furthermore, the syngas can be converted to liquid fuel over a catalyst via the Fischer-Tropsch process for liquid hydrocarbon [24, 25].

2.2.1.7 Catalytic Liquefaction

Catalytic liquefaction technology (CLT) is a thermal process of liquefaction that requires the use of a catalyst that must be dispersed in a liquid medium (liquid phase) [103]. The use of this technique gives high-quality products of greater energy density. High hydrogen partial pressure may be used instead of a catalyst [2]. Hydrolysis and depolymerization of cellulose, hemicelluloses, and lignin molecules into unit/smaller/simpler building blocks (fragments) are achieved. These fragments need to be broken down further to simpler building blocks of the macromolecules through deoxygenation, hydrogenation, dehydroxylation/hydrogenolysis, dehydration, and oxidation reactions. Unlike torrefaction/carbonization, catalytic liquefaction technologies aim at liquid fuels production to simulate with petroleum products and several high value-added chemicals. The limiting factors to commercialization of biomass liquefaction via CLT are low overall yield of oil it produces which is between 20% and 55% w/w as against up-to-date methods like pyrolysis [2]. Low-quality oil from CLT with heavy coke could be attributed to high operating temperature and pressure in CLT and absence of catalysts and other reactants like CO, propanol, butanol, and glycerin.

2.2.2 *Biochemical Processes*

Any conversion technique that breaks down biomass feedstocks and lignocellulosic materials to bioethanol, biogas (gases and liquid fuel) via the use of microorganisms and/or enzymes is known as biochemical conversion technique (BCT). Anaerobic digestion and fermentation are two main categories of BCT. Microorganisms depolymerize biomass/lignocellulosic materials in the absence of air (anaerobic medium or environment) into fossil fuel known as biogas in anaerobic digestion known also as biomethanation [2, 107]. These replicants (biogas) can be processed further to get cogenerated heat and power (CHP). Bioconversion processes are friendly unlike chemical conversion processes that are hazardous and offer more products for various transformation and usage. Bioreactors are used in bioconversion operated at different modes: continuous, semicontinuous, or batch mode, etc. Fermentation usually occurs in an anaerobic environment, and yeast or bacteria are used as biocatalysts in the process to produce ethanol from first-generation feedstocks (carbohydrate and sugar-rich biomass) that are rich in carbon six (C6) sugars. Many microorganisms are endowed with glucose hydrolytic pathway; this predisposes them to utilize only the C6 sugars. When the lignocellulosic biomasses are hydrolyzed, the hydrolysates are rich in C5 sugars as well as other forms of sugars. Bioconversion technique includes the microorganisms that must utilize non-C6 hydrolysates as carbon source to form ethanol, hydrocarbon fuel, and other intermediates.

2.3 Biomass Composition

Biomass consists of starch, sugar, cellulose, hemicellulose, lignin, proteins, and lipid (extractives). They have different monomers that polymerized (built up) to give each biomass component with different functional groups. Functional groups give different characteristics to molecules and are often involved in bond formation between monomers to give macromolecules/complex molecules.

(a) *Cellulose:* Cellulose is the most abundant, and it is crystalline and amorphous in nature. The crystalline part of the structure makes it unreactive. Cellulose is a long unbranched polymer of only glucose with molecular formula of $(C_6H_{12}O_6)n$. Complete hydrolysis of cellulose gives monosaccharides, for example, glucose, but partial hydrolysis of cellulose results in disaccharide, for example, cellobiose [111].

(b) *Hemicellulose:* Hemicellulose contains different sugar monomers (mixed polymer), including rings of carbon five sugars (C5), for example, xylose, and carbon six (C6) sugars, for example, glucose (pentoses and hexoses). Lignocellulosic hydrolysates are very rich in pentoses and hexoses. In the hydrolysates, the quantity of the pentoses is higher than the hexose sugars [14, 75]. Unlike cellulose, hemicellulose is amorphous in nature, making it more reactive. Its molecular formula is $(C_5H_8O_4)_n$ [11, 15, 41, 66].
(c) *Starch:* Like cellulose, starch is a polymer of glucose but more reactive because of the highly branched and linear structure (amylopectin and amylose) α-glycosidic bonds [36].
(d) *Lignin:* Lignin has a random, complex, and cross-linked network structure (undefined structure), easy to undergo both oxidation and condensation reactions. Building blocks of lignin are p-coumaryl alcohol, coniferyl alcohol, and sinapyl alcohol [50].
(e) *Fats:* Fats, steroids, and phospholipids are lipids which comprise mainly hydrocarbon which makes lipids hydrophobic. Fats building blocks are glycerol and fatty acids. Composition of fatty acids differs in length and number of double bonds which determine their saturation and unsaturation level. Steroids are lipids with four fused rings of carbon skeleton. Cholesterol is a steroid [88, 89, 106]. Phospholipid is fats with one of the fatty acids replaced with a phosphate group and that is why it is referred to as "false" fat [29].
(f) *Proteins:* Proteins are polymers of amino acids monomers (organic molecules with carboxyl and amino groups). Amino acid properties differ due to the differing side chains called the R groups. Amino acids are linked by peptide bonds, and peptide bond is formed as a result of the carboxyl end and the amino end known as C terminus and N terminus, respectively. Polymerization of amino acids gives a polypeptide. Levels of protein structure can be primary, secondary, tertiary, or quaternary depending on the sequence of amino acids, coiling and folding due to hydrogen bonds, total shape of two sets polypeptide, and relationship that exists among several polypeptides of a protein [7, 60].

2.4 Chemical Composition

Chemical composition of biomass is categorized into water and dry substance. Dry substance is subdivided into organic and inorganic parts. Organic biomass components are cellulose, hemicellulose, lignin, extractives, sugar, starch, and proteins, while the inorganic component gives ash content. Components of the organic are further degraded to carbon, hydrogen, oxygen, nitrogen, and sulfur (volatile matter and fixed carbon), while the inorganic content gives ash elements. Further, under complete combustion, the volatile matter and fixed carbon constituents are converted to carbon dioxide, NO_x, SO_x, and water, whereas the ash elements produce ash particulates [84].

2.5 Analytical Techniques for Biomass Characterization

Biomass characterization addresses important feedstock material quality and performance issues such as chemical/elemental composition and variability, thermochemical feedstock properties, particle size characteristics and morphology, particle density and properties, hydrolysis, fermentation, and conversion performance. Standard laboratory analytical procedures established by the National Renewable Energy Laboratory (NREL) for chemical composition analysis are usually employed to determine these properties. Other scientific analytical methods employed include X-ray photoelectron spectroscopy (XPS) or electron spectroscopy for chemical analysis (ESCA) to determine the composition and summative mass closure of biomass feedstock materials. By combining the appropriate laboratory analytical procedures, the biomass sample is fractionated into the constituents that add up to 100% of the starting weight. XPS measurement is based on binding energy of electron in matter via X-ray photoelectron. XPS is also known as ESCA technique, used specifically for the chemical composition investigation of surfaces. It identifies the functional groups present in a composite sample like biomass. XPS is an important and versatile analytical technique to analyze the chemical nature of compounds [55, 61, 69].

2.6 Thermochemical Properties of Biomass

Analysis of thermochemical properties of biomass includes proximate and ultimate analysis, to evaluate the thermal efficiency and energy content of a given biomass. Ultimate analysis gives composition of the biomass in weight percent of carbon, sulfur, oxygen, nitrogen, hydrogen, etc. Analytical capabilities can readily determine volatiles, ash content, fixed-carbon content, elemental composition (e.g., carbon, nitrogen, hydrogen, sulfur, and oxygen), and the higher heating value (HHV) and lower heating value (LHV). Thermogravimetric analyzer/differential scanning calorimetry, coupled with chromatographic techniques, is used to determine the caloric behaviors (such as transformation temperatures, enthalpies, evaporation, and DE volatilization mass changes), to better understand the fuel's properties [17].

2.7 Analytical Techniques for Elemental Analysis

All samples for elemental analysis must be homogenized for the portion taken to be a true representation of the whole biomass. Techniques for elemental analysis include flame atomic absorption spectrometry (FAAS) [13, 26, 82, 91], inductively coupled plasma-optical emission spectrometry (ICP-OES), inductively coupled plasma atomic emission spectrometry (ICP-AES), inductively coupled plasma mass spectrometry (ICP-MS) [58, 72, 82, 92], graphite furnace atomic absorption spectrometry (GFAAS) [27, 38, 82], instrumental neutron activation analysis (INAA) [82, 90, 95], X-ray fluorescence analysis [87], CHNSO elemental analyzer [68, 80, 99], voltammetry [82], and ion chromatography (IC) [13, 26, 82, 91].

Atomic absorption occurs when light of a certain wavelength is absorbed which makes an electron to transit from the primary ground state to an excited state (higher energy level). At this stage, light absorbed is equal to concentration of element. Atomic emission takes place as the electron in the excited state falls back and lose the absorbed energy at a specific wavelength. That lost energy in the form of light emission is known as atomic emission, which is also proportional to the elemental concentration. At the higher energy level, if the light and energy are high that an electron from the shell is ejected, ionization has taken place (free electron and atom with positive charge). Mass spectrometer measures free electron and atom with positive charge (ion) and is known as mass spectrometry. Flame atomic absorption requires atomization of the components containing the element for them to be analyzed. The solution that contains the element as part of the biomass undergoes desolvation, vaporization, atomization, and ionization. Atomization temperatures for FAAS, GFAAS, and ICP-AES are between 1700–3150 °C, 1200–3000 °C, and 6000–8000 °C, respectively [39, 40, 47, 59, 70, 82]. The components used in these techniques include radiation source, atomizer, monochromator, detector, and data processor for successful operation.

Unlike FAAS and GFAAS that use a chemical flame, in ICP, the plasma is an electrical discharge. Gas used in ICP is argon, and plasma is low at atmospheric pressure and high temperature. Through rapid inductive coupling of free electrons with oscillating magnetic field, plasma is generated, and energy is transferred to the argon molecules by means of collision. The flowing gas contains the plasma that transports the sample aerosol to be analyzed to the torch in the ICP-MS for atomization and ionization and extraction into the mass spectrometer. ICP-MS is highly sensitive and fast (less than 5 min) with the capacity for simultaneous detection of elements. Liquid chromatography (LC) is a type of high-performance liquid chromatography or one of the high-performance liquid chromatography (HPLC), can be attached to inductively coupled plasma mass spectrometry (ICP-MS) to give LC-ICP-MS for elemental analysis. Likewise, gas chromatography (GC) can be coupled to inductively coupled plasma mass spectrometry (ICP-MS) giving rise to GC-ICP-MS depending on the properties (nonpolar, high thermal stability, and high volatility for GC-ICP-MS and solubility, nonvolatile, and thermo labile for LC-ICP-MS) of the sample to be analyzed [47, 59, 70]. Some operations involve more than one MS (Fig. 2.1). Like GC-MS, LC-MS, GC-MS-MS or LC-MS-MS (tandem or quadruple) for selected reaction monitoring (SRM). Electron ionization (EI) and chemical ionization (CI) methods are usually used for mass spectrometry techniques for biomass feedstock experiments. However, some samples (analytes) can be analyzed via co-crystallization with a large molar excess of a matrix compound (which in most cases is a weak UV-absorbing organic acids) to generate a solid sample via matrix-assisted laser desorption/ionization (MALDI). In MALDI,

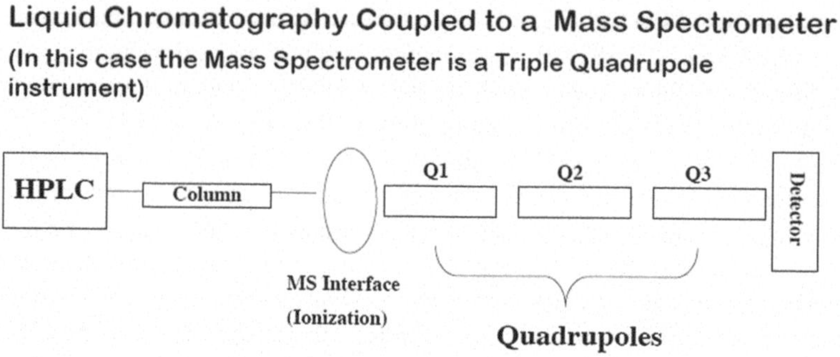

Fig. 2.1 HPLC/MS/MS and triple quadrupole technology. (Source: Applied Biosystems. Diagram courtesy of University of Alberta)

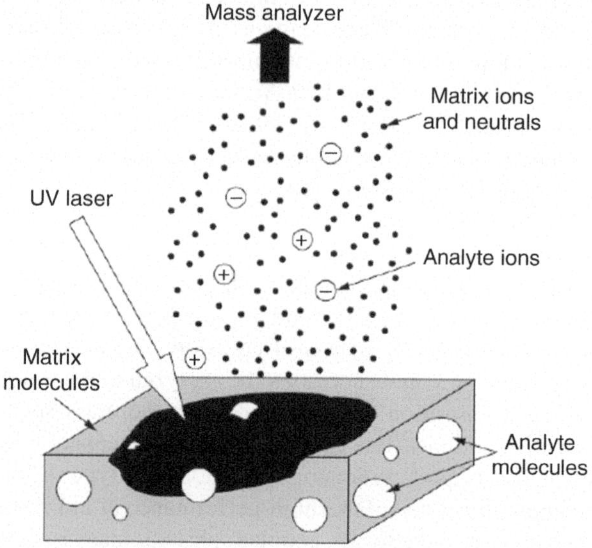

Fig. 2.2 Schematic of the MALDI (matrix-assisted laser desorption/ionization) process. (Source: CRC Press)

irradiation of the solid solution by a pulsed UV laser occurs leading to sublimation of the matrix that in the process carries the samples with it into the gas phase (Fig. 2.2). The matrix is assisted by the laser in the sense that it absorbs laser energy and uses this energy to lift the sample species into the gas phase without destroying them. The matrix serves as a proton donor or acceptor as the case may be in the plasma to ionize the molecules of the samples. Singly protonated $[M + H]^+$ and multiple protonated $[M + nH]n^+$ ion species are generated depending on the type of analyte.

Instrumental neutron activation analysis (INAA) produces its beam of neutrons by a reactor, and the neutron capture is followed by a radioactive decay that liberates gamma rays that is detected by a spectrometer. This way both quantitative and qualitative information about the element is gotten without dissolving or digesting or atomizing the sample. INAA is a very sensitive and accurate analytical technique but has the inability to detect and quantify very light elements.

2.8 Morphology and Particle Size Analysis

Different methods and types of instrumentation are usually employed to determine the size distribution, shape, and general particle characteristics of intermediate feedstock materials. One approach is the use of automated digital imaging, which can measure particle size, particle size distribution, particle shape, density, and other parameters including geometric mean diameter width, geometric mean diameter length, sphericity, and the aspect ratio of feedstock materials [17]. Particle characteristics (shape, size, and morphology of biomass feedstock particles) affect to a large extent its mixing and flow properties as well as heat and mass transfer mechanisms which in turn determine conversion efficiency and energy requirement [17, 23, 85]. Pretreatment of biomass feedstocks plays an important role to ameliorate the problems associated with particle size and shape effects. Selection of the accurate conversion processing technique mostly depends on the particle size of biomass feedstocks and type of product desired at the end of the conversion [17]. Even the design of storage facility and handling methods of biomass feedstocks are affected by biomass particle size and shape, indicating how serious the issue of characterization of particle size and morphology is to renewable energy industries. American National Standards Institute/American Society of Agricultural Engineering's (ANSI/ASAE) standard S424.1 or British Standard/European Standard/International Organization for Standardization (BS EN ISO) standard 17827-1:2016 has direction and guidelines required for sieving analysis [17, 81]. Latest technique for proper analysis of particle characteristics is microscopy digital imaging.

2.9 Microscopy Imaging for Particle Characterization

Advanced microscopic tools detect surface characteristics topography, shape, and size (morphology), element and compounds and their relative amount (composition), as well as crystallographic information (arrangement of atoms in the biomass) of biomass feedstock. Understanding changes in feedstock structure at the tissue and cellular level is increasingly important. To investigate and develop mechanical preprocessing, preconversion, and densification options that impact the chemical and physical attributes of feedstock materials, digital microscopy, confocal laser microscopy, scanning electron microscopy (SEM), and Fourier-transform infrared microscopy are often used to investigate such changes. SEM analyzes all sorts of sample types based on surface interaction whether conducting or nonconducting without the need of electron transparency. It is a three-dimensional view (3D) imaging and semi nondestructive. Demerits of SEM are the need for vacuum compatibility of specimen, low resolution, and requirement for stain coating with metals for some samples. Unlike SEM, atomic force microscopy (AFM) provides high-resolution microscopic images via the use of scanning probe microscope [19, 37, 86]. It does not require optical method and vacuum but can provide image in both air and liquid with high resolution to subnanoscale (Angstrom). AFM is based on almost same principle as SEM with imaging modes at contact and noncontact modes (static and dynamic/tapping modes). AFM performs phase imaging. Transmission electron microscope (TEM) on the other hand utilizes high-power energetic electrons to provide crystallographic, morphological, and compositional characteristics/information on samples (Fig. 2.3a, b). Via TEM, molecular level characteristics can be analyzed, making it possible for structure and texture analysis. TEM images (2D) are of high quality and detailed. TEM has most powerful magnification and a wide range of application. It can be utilized in the industrial field and educational sector. Basically, drawbacks of TEM are laborious sample preparation, black and white image formation, and very large and expensive equipment; special training is required for equipment operation and TEM special housing and maintenance needs.

Imaging particle analysis (digital imaging) determines size and morphological characteristics and dimensions of biomass. Particle characteristics have direct effect on dissolution rates, biomass compliance to treatment, and conversion activities. Gas sorption analysis and pycnometry are techniques used for determining the true (skeletal) density, surface area, pore volume, average pore size, and pore size

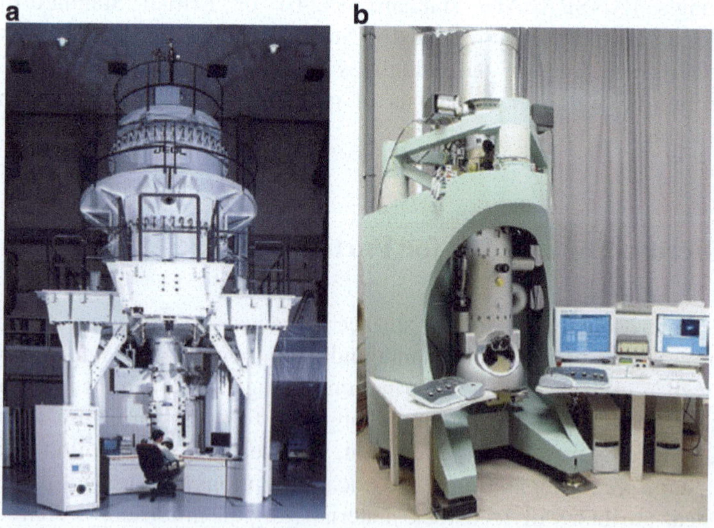

Fig. 2.3 A selection of different commercial TEMs: (**a**) TEM 1.25 MeV HVEM and (**b**) Zeiss HRTEM. (Sources: Kohl and Reimer (2008, 2013). Transmission electron microscopy: physics of image formation. Springer)

distribution of microporous (pore size 2 nm) and mesoporous (pore size 2–50 nm) solids using nonadsorbing helium. It is a classical method (helium void volume method) and there are commercial helium-free methods. Chipped and particulate solid biomass causes handling, logistics, and transfer problems [76, 82, 94, 104].

2.10 Biomass Storage Simulation

Storage simulation reactors are used to monitor the behavior of feedstocks as they are stored in a variety of storage conditions. Raw biomass is naturally susceptible to microbial degradation. Depending on the type of feedstock, its moisture content, and the conditions in which it is stored, storage losses can be extreme. To better understand these relationships and develop engineered storage solutions aimed at minimizing biomass loss and quality changes in storage, series of highly instrumented and automated storage reactors have been developed to monitor stored biomass feedstocks. Microbial action within the reactors is monitored through evolution of CO_2 over time, allowing dry matter losses to be calculated throughout the entire storage period.

Thermogravimetric analysis (TGA) Thermal analytical equipment employs a group of techniques where one or more of properties of a sample are studied while the sample under consideration is under a cntrolled temperature program (constant heating or cooling or both, heating isothermally, heating in response to the sample behavior, and a stepwise or complex program heating). Each thermal method is responsible for examining and measuring a property as a function of temperature. When a substance is under a controlled temperature program and the mass change of the substance is studied and measured as a function of temperature, the technique is called thermogravimetric analysis. Differential or derivative thermogravimetric analysis (DTA) studies rate of mass change as a function of time at a temperature to ascertain the reaction property of the biomass sample in terms of heat absorption or emission (exothermic or endothermic). By means of TGA-DTA biomass, gaseous fraction (volatiles and water) and solid fraction (char/coke) can be quantified. Combustion inhibition studies, chemical composition, weight loss measurement, thermal stability, degradation or decomposition studies, transition studies, and effects of varying factors are studied using TGA-DTA [45, 46, 51, 52].

Differential scanning calorimetry (DSC) DSC measures heat energy that is transferred to a sample that is in a sample pan compared to heat transferred to a reference pan (pan without sample) in a heating medium equipped with a computer which ensures that the heating rate stays absolutely the same throughout the entire operation at the same temperature [33, 65]. Thermal properties are measured in DSC. Physical or chemical changes (changes in properties) may occur in the sample by either heat absorption (endothermic) or heat emission (exothermic). DSC measures these changes based on energy difference to the sample pan (biomass) and reference sample (empty pan). By this means, characteristics of the biomass like degradation or oxidation temperature, glass transition temperature, cross-linking (cure) point, crystallization, and melting points are confirmed. Glass transition temperature is the midpoint temperature that lies between the rubbery or super cooled state [which is time before macroscopic manifestation of changes in mobility occurs (glass transition onset)] and glass shinny molten-like state [which refers when molecular mobility is fully in place (glass transition endpoint temperatures)]. Underlying factors that affect glass transition are heat, molecular mobility, and plasticizer (water). These make materials stretchable, workable, and flexible, a desired characteristic for various applications. Other measurement methods are dynamic mechanical analysis (DMA) and thermomechanical analysis (TMA).

2.11 Advanced Instrumentation Techniques – Hybrid Rapid-Screening Techniques

These are rapid-screening techniques to determine feedstock composition, ash and elemental content, and bench-scale conversion performance of biomass feedstocks. These cost-effective, alternative screening techniques allow for characterization and analysis of feedstock materials in a matter of minutes, instead of days or weeks. Methods include predictive NIR spectroscopy, laser-induced breakdown spectroscopy, and an automated dilute acid pretreatment method for assessing the recalcitrance and reactivity of feedstock materials. Vibrational spectroscopy methods for biomass analysis are nondestructive.

2.11.1 Fourier-Transform Infrared Spectroscopy (FTIR)

FTIR is widely used for qualitative and quantitative analysis based on the measurement of absorption of different frequencies of infrared radiation. Radiation is not a dispersed instrument but fast with a good signal-to-noise ratio. All wavelengths arrive at the detector simultaneously, and by means of a mathematical model (chemometrics) treatment or Fourier transformation (FT), the interferometer readings detected are converted to the typical infrared spectrum giving absorbance vs. spectrum [77].

2.11.2 Fourier-Transform Mid-Infrared Spectroscopy (FT-MIR)

Mid-infrared region is 2.5–15 μm (4000–650 cm^{-1}) known with fundamental vibrations as the characteristic transitions. Sample's ability to absorb light in the mid infrared is the fundamental absorptions measured. Organic functional groups have clear absorption bands in the Mid-IR region which characterizes them. Molecules differ from each other by having different sets of functional groups; in the Mid-IR, this is used to identify them and character their structure. When a mixture of sample (biomass) is subjected to Fourier-transform mid infrared (FT-Mid-IR), absorption bands for individual components in the mixture are delineated from each other and are used to characterize and quantify the components [33]. This technique is very reproducible, highly sensitive, and low cost of operation. FTIR-attenuated total reflectance (ATR) measures total energy reflected from the surface of sample in contact with the IR transmitting crystal. The sample can be solids, pastes, or viscous liquids. Gas chromatography interfaced with FTIR uses sealed glass cell with IR transparent windows. Mid-IR is super special because its spectra can be compared to world libraries of standard spectra.

2.11.3 Fourier-Transform Near-Infrared Spectroscopy (FT-NIR)

Near-infrared region is 0.8–2.5 μm (12,500–4000 cm^{-1}) with overtones and combinations bands as its vibrational modes which are weaker than fundamental vibrations. Fourier-transform near-infrared spectroscopy (FT-NIR) is first calibrated with sample set, then a mathematical model is built to correlate with the data set from primary/standard method result. Using the validation set, the FT-NIR is validated, and an entire new set of samples is predicted, and the result is validated using the primary method. If there is agreement, the method is validated. Once the instrument is calibrated, sample constituents can be rapidly measured without sample weighing, and no hazardous reagents are used. Multivariate modeling (chemometrics) is new analytical methods that is reproducible, can be applied for online or offline biomass analysis, and is used to measure various constituents. Limitations of this technique are laborious calibration and high initial cost.

2.11.4 Ultraviolet–Visible Spectroscopy and Fluorescence

In ultraviolet–visible spectroscopy method, atom and molecules are analyzed in the UV–visible region. Radiation to matter interaction is by absorption, emission, and diffraction. Electromagnetic radiation is transferred to molecule or atom by electronic transition. Chromophores (colored substances, though many compounds absorb strongly in UV but not colored) absorb the characteristic wavelengths. A chromophore is a structural feature that gives rise to light absorption in the UV–visible region. UV–visible spectroscopy measures the emission or absorption in the UV (200–350 nm)–visible (350–700 nm) range. Based on the amount of light absorbed or transmitted from a reference beam as it passes the analyte, the concentration is measured and quantified [54, 101].

2.11.5 High-Performance Liquid Chromatography (HPLC)

Hydrolysis of lignocellulosic biomass results in hydrolysates which are mainly sugars. Appropriate quantification mechanism for sugar separation, detection, and quantification is done using HPLC. Typically, chromatography is based on the portioning of a sample which is the solute between a mobile or moving phase and a static/solid/fixed/stationary phase. HPLC involves a liquid mobile phase and a solid or liquid stationary phase. As the solute (hydrolysates) interacts with the two phases, separation of the mixture occurs, and they are quantified. Appropriate columns are used for the characterization and quantification of the sugars. This is essential as these sugars will be subjected to biofuel and/or ethanol production as desired.

2.11.6 Hydrothermal Pretreatment (HTP) Recalcitrance Screening

Biomass recalcitrance is always as a result of complexity of lignocellulosic matrix. These recalcitrance substances are difficult to quantify but HTP changes the recalcitrant structural characteristics of lignocellulosic matrix which renders available for ease of quantifying biomass destruction. HTP pyrolysis molecular beam mass spectrometry (py-MBMS) is ideal for the quantification and qualification of recalcitrance biomass. Techniques for chemical analysis of hydrothermal pretreated biomass include Fourier-transform infrared spectroscopy (FT-IR), near infrared spectroscopy (NIR), nuclear magnetic resonance (NMR) (Fig. 2.4), mass spectrometry (MS), and monoclonal antibody microarrays, which have binding specificities

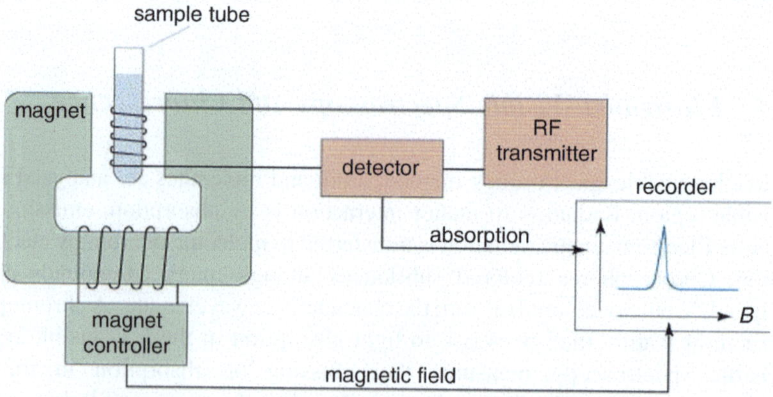

Fig. 2.4 Schematic of the NMR spectrometer. (Source: Diagram courtesy of University of Alberta)

for cell-wall components [5, 21, 33, 56, 62, 87, 105]. However, these techniques produce complex data sets that are challenging to interpret on large sample sets. Chemical fingerprinting of a range of materials that are in the mass spectra of pyrolyzed library is the only way to interpret the complex data set generated from large samples [3, 4, 32–34, 54, 100, 101].

2.12 Analytical Methods in Hybrid Technologies

Hybrid techniques are two or more techniques coupled together as one operated, mostly separated in written form by a hyphen [28, 43, 44, 48]. Some of the hyphenated techniques known as hybrid technologies are LC-ICP-AES, GC-MS-MS-MS, gas chromatography-mass spectrometry (GC-MS), LC-ICP-OES, gas chromatography-infrared spectroscopy (GC-IRS), liquid chromatography-mass spectrometry (LC-MS), liquid chromatography-nuclear magnetic resonance (LC-NMR) spectroscopy, FTIR-ATR, liquid chromatography-infrared spectroscopy, LC-MS-MS (Fig. 2.1), and capillary electrophoresis-mass spectrometry, to mention but a few.

2.13 Other Tests for Characterizing Biomass

Dimensional analysis The dimensional characteristics are evaluated for three major perpendicular dimensions of biomass as length, breadth, and thickness.

Ash content is the inorganic residue remaining after ignition (dry-ashing) or complete oxidation of organic matter (wet-ashing). Dry-ashing requires little operator time and is very safe, many samples can be analyzed at once, and no blanks are required, but there are ample issues associated with it. To resolubilize, the minerals can be hard, it takes longer time to get results, incomplete combustion is bound to happen often, and loss of volatile components of biomass is not prevented. Wet-ashing, on the other hand, takes short time and allows minerals to be in solution, and no little amount of volatiles are lost because temperature of operation is less than or equal to 200 °C. Wet ashing is hazardous, requires operator constant attention, and is able to analyze small number of samples. Microwave ashing uses microwave for the ashing operation in a short time (30 min), but it is very costly and does not have ability for analyzing large number of sample.

Volatile matter is the component of the carbon present in the biomass, which when heated converts to vapor (gas fraction). This is mostly analyzed by using gas chromatography coupled with a thermal conductivity detector (GC-TCD). Comparison of the sample peak retention time with the peak retention time from the standards shows what the content of the vapor or gas fraction is.

Density Density of biomass is bulk and true density which biomass weight per unit volume depending on the size and shape of the biomass. Densimeter is one of the measuring tools.

Angle of repose It is the flow property of the biomass material which gives idea on how it can be piled without falling or descend in a system. Automated heap analyzer can be used to detect angle of repose.

Moisture content Hot air oven drying is good for moisture content determination, though depending on the type of feedstock under consideration. Hand-held moisture analyzers are rapid detection tools. Distillation procedure, chemical method – Karl Fischer titration, physical methods, and freeze-drying (laboratory method) are other suitable methods for moisture content determination depending on the feedstock and suitability purpose.

Elemental analysis is also known as ultimate analysis (see above). The elements are determined by the use of CHNSO analyzer, and oxygen can be determined by the difference.

Lipid determination is done by solvent extraction methods (pretreatment by acid hydrolysis or with base for fat liberation, soxhlet method, total fat by GC, goldfish method, etc.), nonsolvent wet extraction methods (Babcock method and Gerber method), or instrumental methods (infrared, specific gravity, nuclear magnetic resonance (NMR), etc.).

Protein analysis is most commonly done using Kjeldahl and Dumas (N combustion). Dumas is a rapid technique and gives protein results in 4 min.

Fixed carbon content is obtained by mass balance calculations.

Total carbohydrate is not measured but calculated by difference, while reducing sugars are determined by Lane–Eynon method, Munson–Walker method, or Somogyi–Nelson method. The difference in the three methods is on the method of measuring cuprous ions generated. Monosaccharides and oligosaccharides can be measured using HPLC with different stationary phases and different types of detectors. Read more on analytical techniques in Nielsen [82]. When sugars are converted to volatile derivatives (e.g., trimethylsilyl derivatives or alditol peracetates), gas chromatography equipped with flame ionization detector (GC-FID) can be used to analyze the monosaccharides and oligosaccharides. When derivatized, GC-MS can also be utilized for their quantification. Enzymatic methods for carbohydrate analysis are available, but they have low detection limits and are not useful for total carbohydrate determination.

Calorific value Under ideal combustion conditions, heat released by the biomass fuel (biofuel) is the caloric value. Bomb calorimetry determines the sample's (biomass) calorific value.

2.14 Challenges to Biomass Characterization and Analysis

Complexity of biomass, seasonal availability, different origin/sources, conversion process, and type of biofuel are the critical challenges for bio-energy production to be resolved carefully through biomass characterization. The major challenges involve in biomass characterization for bioenergy production are as follows: Nature of biomass feedstocks to be utilized, right conversion/multiphase conversion processes, type of biofuel desired, technological advancement, experiment validation, process optimization, and by-products utilization. Availability of the several equipment, required for characterization of biomass feedstock, is another challenge heating their utilization. Lack of technological know-how does not allow for proper and optimum usage of the advanced techniques that are available in some countries. Product (liquid, gas, and solid) mass balances of some techniques pose a challenge.

2.15 Conclusion and Outlook

Without analytical techniques, quantification and qualification of conversion and characterization of biomass feedstock are impossible. Each desired product has specific features or characteristics expected from it to meet up with the testing standards like American standard testing material (ASTM). For example, biodiesel obtained from biomass must meet up with the expected biodiesel standard properties documented in The Biodiesel Handbook [8, 63, 64]. Typically, some of the standard properties of biodiesel include cetane number which should be comparable to that of the petroleum diesel (indicator of diesel combustion and quality to be used as regular diesel engine with little or without engine modification) and lubricity (low or no sulfur) just to mention a few [63]. Fourier-transform infrared reflectance (FTIR) spectroscopy can predict properties of biomass feedstock, a very heterogeneous feedstock material. Fast and high throughput analytical methods for biomass lignocellulosic qualification and quantification are FTIR, GC-MS, GC-MS-MS, HPLC, DSC, HTP, FT-MIR, FT-NIR, FPLC, NMR, LC-MS-MS, and TGA-DTA. All the mass spectrometry techniques are best for accurate quantification. Topography, morphology, and particle size determination are best done with microscopic imaging techniques.

Although extensive studies have been carried out and documented for characterization and conversion of biomass, which resulted into standard techniques, there is still room for improvement. Some techniques are cumbersome and time demanding, requiring specialized operators and huge capital. Efforts should be made to reduce time of sample preparation and analysis without compromising the quality of results obtained from these techniques. Improvement on ability to analyze large number of samples at a time is essential to reducing time and cost of analysis. Furthermore, most advanced techniques are still at their infancy (e.g., laboratory and pilot plant stage), and their industrial application is yet to be realized and most likely will happen soon in the future. Effective, accurate, and reliable data acquisition should be critically considered as an improvement to these analytical techniques. Optimization of existing operation is an added advantage and an issue to be looked into. Present methods lack the ability to differentiate products made from C5 and C6 sugars [14, 20, 75, 108]. It is a big gap, and this should be looked into as an improvement to the existing techniques. This is essential because in a mixed sugar biomass of the same carbon length, overcorrection of a sugar type and under correction of another type cannot be prevented. So, development of new standard analytical equipment/improvement of existing ones is required to enhance lignocellulosic biomass conversion and quantification.

Acknowledgments DCO acknowledges Professor J.M. Curtis, Professor D. C. Bressler, and Dr. A. Ullah of the University of Alberta for their shared knowledge on advanced analytical techniques, advanced BioCatalysis, renewable biomaterials, and respectively.

References

1. T. Abbasi, S.A. Abbasi, Renewable' hydrogen: Prospects and challenges. Renew. Sust. Energ. Rev. **15**(6), 3034–3040 (2011). https://doi.org/10.1016/j.rser.2011.02.026
2. P. Adams, T. Bridgwater, A. Lea-Langton, A. Ross, A. Watson, *Biomass Conversion Technologies. Book name- Greenhouse Gas Balances of Bioenergy Systems* (Academic Press, Elsevier, Amsterdam, 2018), pp. 107–139
3. F. Agblevor, H.L. Chum, D.K. Johnson, Compositional analysis of NIST biomass standards from the IEA feedstock round Robins, in *Energy from Biomass and Wastes XVI*, ed. by D. L. Klass, (Institute of Gas Technology, Orlando, 1992), pp. 395–421
4. F. Agblevor, R. Evans, K. Johnson, Molecular-beam mass spectrometric analysis of lignocellulosics materials, herbaceous biomass. J. Anal. Appl. Pyrolysis **30**, 125–144 (1994)
5. A. Alonso-Simon, J.B. Kristensen, J. Øbro, et al., High-throughput microarray profiling of cell wall polymers during hydrothermal pre-treatment of wheat straw. Biotechnol. Bioeng. **105**(3), 509–514 (2010)
6. C.R. Álvarez-Chávez, S. Edwards, R. Moure-Eraso, K. Geiser, Sustainability of bio-based plastics: General comparative analysis and recommendations for improvement. J. Clean. Prod. **23**(1), 47–56 (2012). https://doi.org/10.1016/j.jclepro.2011.10.003
7. H. Angellier-Coussy, P. Chalier, E. Gastaldi, V. Guillard, C. Guillaume, N. Gontard, S. Peyron, Protein-based nanocomposites for food packaging, in *Biopolymer Nanocomposites*, vol. 2013, (Wiley, Berlin, 2013), pp. 613–654
8. J. Asomaning, P. Mussone, D.C. Bressler, Thermal deoxygenation and pyrolysis of monounsaturated fatty acids. J. Anal. Appl. Pyrolysis **105**, 1–7 (2013). https://doi.org/10.1016/j.jaap.2013.09.005
9. J. Asomaning, P. Mussone, D.C. Bressler, Thermal deoxygenation and pyrolysis of oleic acid. J. Anal. Appl. Pyrolysis **105**, 1–7 (2014a)
10. J. Asomaning, P. Mussone, D.C. Bressler, Two-stage thermal conversion of inedible lipid feedstocks to renewable chemicals and fuels. Bioresour. Technol. **158**, 55–62 (2014b). https://doi.org/10.1016/j.biortech01.136
11. R.H. Atalla, Celluloses, in *Comprehensive Natural Products Chemistry*, ed. by D. Barton, O. Meth-Cohn, vol. 3, (Elsevier, Amsterdam, 1999), pp. 529–598
12. A.M. Azeez, D. Meier, J. Odermatt, T. Willner, Fast pyrolysis of African and European lignocellulosic biomasses using Py-GC/MS and fluidized bed reactor. Energy Fuel **24**(3), 2078–2085 (2010). View at Publisher • View at Google Scholar • View at Scopus

13. H. Bağ, A.R. Türker, M. Lale, Determination of Cu, Zn, Fe, Ni and Cd by flame atomic absorption spectrophotometry after preconcentration by Escherichia coli immobilized on sepiolite. Talanta **4**(3), 689–696 (2000). https://doi.org/10.1016/S0039-9140(00)00289-7
14. J.B. Binder, R.T. Raines, Fermentable sugars by chemical hydrolysis of biomass. Proc. Natl. Acad. Sci. **107**(10), 4516–4521 (2010)
15. C.D. Blasi, Heat transfer mechanisms and multi-step kinetics in the ablative pyrolysis of cellulose. Chem. Eng. Sci. **51**(10), 2211–2220 (1996). View at Google Scholar • View at Scopus
16. A.V. Bridgwater, Review of fast pyrolysis of biomass and product upgrading. Biomass Bioenergy **38**, 68–94 (2012). https://doi.org/10.1016/j.biombioe.2011.01.048
17. J. Cai, Y. He, X. Yu, S.W. Banks, Y. Yang, X. Zhang, Y. Yu, R. Liu, A.V. Bridgewater, Review of physicochemical properties and analytical characterization of lignocellulosic biomass. Renew. Sust. Energ. Rev. **76**, 309–322 (2017)
18. K.B. Cantrell, T. Ducey, K.S. Ro, P.G. Hunt, Livestock waste-to-bioenergy generation opportunities. Bioresour Technol **99**(17), 7941–7953 (2008). View at Publisher • View
19. B. Cappella, G. Dietler, Force-distance curves by atomic force microscopy. Surf. Sci. Rep. **34**, 1–104 (1999)
20. A. Chareonlimkun, V. Champreda, A. Shotipruk, N. Laosiripojana, Reactions of C5 and C6-sugars, cellulose, and lignocellulose under hot compressed water (HCW) in the presence of heterogeneous acid catalysts. Fuel **89**(10), 2873–2880 (2010). https://doi.org/10.1016/j.fuel.2010.03.015
21. S. Chundawat, V. Balan, B.E. Dale, High-throughput microplate technique for enzymatic hydrolysis of lignocellulosic biomass. Biotechnol. Bioeng. **99**(9), 1281–1294 (2008)
22. J.A. Conesa, R. Font, A. Fullana, et al., Comparison between emissions from the pyrolysis and combustion of different wastes. J. Anal. Appl. Pyrolysis **84**(1), 95–102 (2009). View at Publisher • View at Google Scholar • View at Scopus
23. D.J. Dagel, Y. Liu, L. Zhong, et al., In situ imaging of single carbohydrate-binding modules on cellulose microfibrils. J. Phys. Chem. B **115**, 635–641 (2011)
24. M.O. Daramola, *A catalytic membrane reactor for mitigating water-induced catalyst deactivation in Fischer-Tropsch synthesis*, Chemical Technology, August Issue, (2016), pp. 19–25
25. M.O. Daramola, K. Matamela, O.O. Sadare, Effect of CO_2 CO-feeding on the conversion of syngas derived from waste to liquid fuel over a bi-functional co/H-ZSM-5 catalyst. J. Environ. Chem. Eng. **5**, 54–62 (2017)
26. A.T. Duarte, M.B. Dessuy, M.G.R. Vale, B. Welz, J.B. de Andrade, Sequential determination of Cd and Cr in biomass samples and their ashes using high-resolution continuum source graphite furnace atomic absorption spectrometry and direct solid sample analysis. Talanta **115**, 55–60 (2013). https://doi.org/10.1016/j.talanta.2013.04.036
27. Á.T. Duarte, A.R. Borges, A.V. Zmozinski, M.B. Dessuy, B. Welz, J.B. de Andrade, M.G.R. Vale, Determination of lead in biomass and products of the pyrolysis process by direct solid or liquid sample analysis using HR-CS GF AAS. Talanta **146**, 166–174 (2016). https://doi.org/10.1016/j.talanta.2015.08.041
28. L.A. Ellis, D.J. Roberts, Chromatographic and hyphenated methods for elemental speciation analysis in environmental media. J. Chromatogr. A **774**(1–2), 3–19 (1997). https://doi.org/10.1016/S0021-9673(97)00325-7
29. E.D. Enger, F.C. Ross, *Concepts in Biology*, 10th edn. (McGraw-Hill, New York, 2003)., Chapters 3, 4
30. I. Espinosa-Gonzalez, J. Asomaning, P. Mussone, D.C. Bressler, Two-step thermal conversion of oleaginous microalgae into renewable hydrocarbons. Bioresour. Technol. **158**, 90–94 (2014a)
31. I. Espinosa-Gonzalez, A. Parashar, D.C. Bressler, Hydrothermal treatment of oleaginous yeast for the recovery of free fatty acids for use in advanced biofuel production. J. Biotechnol. **187**, 10–15 (2014b)
32. M. Foston, A.J. Ragauskas, Changes in the structure of the cellulose fiber wall during dilute acid pretreatment in Populus studied by 1H and 2H NMR. Energy Fuel **24**(10), 5677–5685 (2010)

33. M. Foston, A.J. Ragauskas, Biomass characterization: Recent progress in understanding biomass recalcitrance. J. Ind. Biotechnol. **8**(4), 191–208 (2012). https://doi.org/10.1089/ind.2012.0015
34. M. Foston, C. Hubbell, A.J. Ragauskas, Cellulose isolation methodology for NMR analysis of cellulose ultrastructure. Materials **4**, 1985–2002 (2011)
35. I. Gajda, J. Greenman, C. Melhuish, I. Ieropoulos, Self-sustainable electricity production from algae grown in a microbial fuel cell system. Biomass Bioenergy **82**, 87–93 (2015). https://doi.org/10.1016/j.biombioe.2015.05.017
36. A. Gandini, M.N. Belgacem, Chapter 1 – The state of the art, in *Monomers, Polymers and Composites from Renewable Resources*, ed. by M. N. Belgacem, A. Gandini, (Elsevier, Amsterdam, 2008), pp. 1–16
37. R. García, R. Pérez, Dynamic atomic force microscopy methods. Surf. Sci. Rep. **47**, 197–301 (2002)
38. J.L. Gardea-Torresdey, K.J. Tiemann, J.G. Parsons, G. Gamez, M.J. Yacaman, Characterization of trace level au (III) binding to alfalfa biomass (Medicago sativa) by GFAAS. Adv. Environ. Res. **6**, 313–323 (2002). https://doi.org/10.1016/S1093-0191(01)00064-8
39. G. Gauglitz, T. Vo-Dinh, *Handbook of Spectroscopy*, vol 1 (Wiley, Hoboken, 2003), pp. 213–1115
40. G. Gauglitz, T. Vo-Dinh, *Handbook of Spectroscopy*, vol 2 (Wiley, Weinheim, 2014), pp. 300–1900
41. P. Giudicianni, G. Cardone, R. Ragucci, Cellulose, hemicellulose and lignin slow steam pyrolysis: thermal decomposition of biomass components mixtures. J. Anal. Appl. Pyrolysis **100**, 213–222 (2013). Google Scholar
42. H.B. Goyal, D. Seal, R.C. Saxena, Biofuels from thermochemical conversion of renewable resources: A review. Renew. Sust. Energ. Rev. **12**, 504–517 (2008). https://doi.org/10.1016/j.rser.2006.07.014. CrossRefGoogle Scholar
43. G. Guetens, G. De Boeck, M.S. Highley, M. Wood, R.A.A. Maes, A.A.M. Eggermont, A. Hanauske, E.A. De Bruijn, U.R. Tjaden, Hyphenated techniques in anticancer drug monitoring. J. Chromatogr. A **976**(1–2), 239–247 (2002a). https://doi.org/10.1016/S0021-9673(02)01227-X
44. G. Guetens, G. De Boeck, M. Wood, R.A.A. Maes, A.A.M. Eggermont, M.S. Highley, A.T. Van Oosterom, E.A. De Bruijn, U.R. Tjaden, Hyphenated techniques in anticancer drug monitoring. J. Chromatogr. A **976**(1–2), 229–238 (2002b). https://doi.org/10.1016/S0021-9673(02)01228-1
45. P.J. Haines, Introduction, in *Principles of Thermal Analysis and Calorimetry*, (The Royal Society of Chemistry, Cambridge, 2002), pp. 1–9
46. G.R. Heal, Thermogravimetry and derivative thermogravimetry, in *Principles of Thermal Analysis and Calorimetry*, (The Royal Society of Chemistry, Cambridge, 2002), pp. 10–54
47. R.F.M. Herber, M. Stoeppler, *Trace Element Analysis in Biological Specimens*, vol 15, 1st edn. (Elsevier, London, 1994), pp. 130–379
48. R.M. Holt, M.J. Newman, F.S. Pullen, D.S. Richards, A.G. Swanson, High-performance liquid chromatography/NMR spectrometry/mass spectrometry: Further advances in hyphenated technology. J Mass Spectrom **32**(1), 64–70 (1997). https://doi.org/10.1002/(SICI)1096-9888(199701)32:1<64:AID-JMS450
49. A.B.M.S. Hossain, K. Mseddi, *Biomass Derived Biomaterials and Nanobiomaterials* (LAP LAMBERT Academic Publishing GmbH & Co, Saarbrucken, 2012). http://www.umass.edu/molvis/workshop/prot1234.htm. Accessed on 16 May 2019
50. P. Huo, S. Tatsiana, I. Reznikov, D. Hrynshpan, N. Tsygankova, G. Telysheva, A. Arshanitsa, Hydrolysis lignin as a sorbent and basis for solid composite biofuel. Adv. Biosci. Biotechnol. **07**, 501–530 (2016). https://doi.org/10.4236/abb.2016.711046
51. S.S. Idris, N.A. Rahman, K. Ismail, A.B. Alias, Z.A. Rashid, M.J. Aris, Investigation on thermochemical behaviour of low rank Malaysian coal, oil palm biomass and their blends during pyrolysis via thermogravimetric analysis (TGA). Bioresour. Technol. **101**(12), 4584–4592 (2010). https://doi.org/10.1016/j.biortech.2010.01.059

52. S.S. Idris, N.A. Rahman, K. Ismail, Combustion characteristics of Malaysian oil palm biomass, sub-bituminous coal and their respective blends via thermogravimetric analysis (TGA). Bioresour. Technol. **123**, 581–591 (2012). https://doi.org/10.1016/j.biortech.2012.07.065
53. E. Jenab, P. Mussone, G. Nam, D.C. Bressler, Production of renewable hydrocarbons from thermal conversion of abietic acid and tall oil fatty acids. Energy Fuel **28**, 6988–6994 (2014)
54. S.S. Kelley, R. Rowell, M. Davis, et al., Rapid analysis of the chemical composition of agricultural fibers using near infrared spectroscopy and pyrolysis molecular beam mass spectrometry. Biomass Bioenergy **27**, 77–88 (2004)
55. H. Konno, Chapter 8 – X-ray photoelectron spectroscopy, in *Materials Science and Engineering of Carbon*, ed. by M. Inagaki, F. Kang, (Butterworth-Heinemann, Oxford, 2016). https://doi.org/10.1016/B978-0-12-805256-3.00008-8
56. P. Kumar, D.M. Barrett, M.J. Delwiche, P. Stroeve, Methods for pretreatment of lignocellulosic biomass for efficient hydrolysis and biofuel production. Ind. Eng. Chem. Res. **48**(8), 3713–3729 (2009). View at Publisher • View at Google Scholar • View at Scopus
57. M. Kumar, A. Olajire Oyedun, A. Kumar, A review on the current status of various hydrothermal technologies on biomass feedstock. Renew. Sust. Energ. Rev. **81**, 1742–1770 (2018)
58. H. Lachas, R. Richaud, A.A. Herod, D.R. Dugwell, R. Kandiyoti, Determination of trace elements by inductively coupled plasma mass spectrometry of biomass and fuel oil reference materials using milligram sample sizes. Rapid Commun. Mass Spectrom. **14**(5), 335–343 (2000)
59. L.H.J. Lajunen, P. Peramaki, *Spectrochemical Analysis by Atomic Absorption and Emission*, 2nd edn. (Royal Society of Chemistry, Cambridge, 2004), pp. 16–328
60. R.A. Laskowski, Protein structure databases. Mol. Biotechnol. **48**(2), 183–198 (2011). https://doi.org/10.1007/s12033-010-9372-4. PMID 21225378
61. C. Lee, Y. Hyun, J. Choi, Chapter seven – Synthesis and electrochemical performance of transition metal-coated carbon nanofibers on Ni foam as anode materials for lithium secondary batteries, in *Carbon-Based Nanofillers and Their Rubber Nanocomposites*, ed. by S. Yaragalla, R. Mishra, S. Thomas, N. Kalarikkal, H. J. Maria, (Elsevier, Amsterdam, 2019), pp. 226–242. https://doi.org/10.1016/B978-0-12-813248-7.00007-9
62. O. Lerouxel, T. Choo, M. Seveno, et al., Rapid structural phenotyping of plant cell wall mutants by enzymatic oligosaccharide fingerprinting. Plant Physiol. **130**, 1754–1763 (2002)
63. K.D. Maher, D.C. Bressler, Pyrolysis of triglyceride materials for the production of renewable fuels and chemicals. Bioresour. Technol. **98**, 2351–2368 (2007)
64. K.D. Maher, K.M. Kirkwood, M.R. Gray, D.C. Bressler, Pyrolytic decarboxylation and cracking of stearic acid. Ind. Eng. Chem. Res. **47**, 5328–5336 (2008)
65. T.C. Maloney, H. Paulapuro, Hydration and swelling of pulp fibers measured with differential scanning calorimetry. Nord. Pulp. Pap. Res. J. **13**, 31–36 (1981)
66. S. Mansfield, C. Mooney, J. Saddler, Substrate and enzyme characteristics that limit cellulose hydrolysis. Biotechnol. Prog. **15**, 804–816 (1999)
67. D. Marjadi, N. Dharaiya, A.D. Ngo, Bioplastic: a better alternative for sustainable future. Everyman's Sci. **15**, 90 (2010)
68. M.S. Masnadi, R. Habibi, J. Kopyscinski, J.M. Hill, X. Bi, C.J. Lim, N. Ellis, J.R. Grace, Fuel characterization and co-pyrolysis kinetics of biomass and fossil fuels. Fuel Guildford **117**, 1204–1214 (2014). https://doi.org/10.1016/j.fuel.2013.02.006
69. D.M. Mattox, Chapter 2 – Substrate ("Real") surfaces and surface modification, in *Handbook of Physical Vapor Deposition (PVD) Processing*, ed. by D. M. Mattox, (William Andrew Publishing, Boston, 2010). https://doi.org/10.1016/B978-0-8155-2037-5.00002-2
70. G. McMahon, *Analytical Instrumentation* (Wiley, Chichester, 2007).: S. Caroli, *The Determination of Chemical Elements in Food* (Wiley, 2007)
71. X. Miao, Q. Wu, High yield bio-oil production from fast pyrolysis by metabolic controlling of Chlorella protothecoides. J. Biotechnol. **110**(1), 85–93 (2004). View at Publisher • View at Google Scholar • View at Scopus
72. I. Michalak, K. Chojnacka, K. Marycz, Microchim. Acta **172**, 65 (2011). https://doi.org/10.1007/s00604-010-0468-0

73. T.A. Milne, H.L. Chum, F. Agblevor, D.K. Johnson, Standardized analytical methods. Biomass Bioenergy **2**(1–6), 341–366 (1992)
74. T.A. Milne, F. Agblevor, M. Davis, S. Deutch, D. Johnson, A review of chemical composition of fast pyrolysis oils, in *Developments in Thermochemical Biomass Conversion*, ed. by A. V. Bridgwater, (Blackie Academic and Professional, London, 1997), pp. 409–424. View at Google Scholar
75. V. Mittal, 5-Hy droxymethyl furfural based-polymers, in *Renewable Polymers – Synthesis, Processing, and Technology*, (Wiley – Scrivener, New York, 2012), pp. 2–305
76. O. Monticelli, A. Fina, A. Ullah, P. Waghmare, Preparation, characterization, and properties of novel PSMA–POSS systems by reactive blending. Macromolecules **42**(17), 6614–6623 (2009)
77. G. Mouille, S. Robin, M. Lecomte, et al., Classification and identification of Arabidopsis cell wall mutants using Fourier-Transform InfraRed (FT-IR) microspectroscopy. Plant J. **35**, 393–404 (2003)
78. C.A. Mullen, A.A. Boateng, N.M. Goldberg, I.M. Lima, D.A. Laird, K.B. Hicks, Bio-oil and bio-char production from corn cobs and stover by fast pyrolysis. Biomass Bioenergy **34**(1), 67–74 (2010a). View at Publisher • View at Google Scholar • View at Scopus
79. C.A. Mullen, A.A. Boateng, K.B. Hicks, N.M. Goldberg, R.A. Moreau, Analysis and comparison of bio-oil produced by fast pyrolysis from three barley biomass/byproduct streams. Energy Fuels **24**(1), 699–706 (2010b). View at Publisher • View at Google Scholar • View at Scopus.
80. D.K. Naik, K. Monika, S. Prabhakar, R. Parthasarathy, B. Satyavathi, Pyrolysis of sorghum bagasse biomass into biochar and bio-oil products: A thorough physicochemical characterization. J. Therm. Anal. Calorim. **127**, 1277 (2016). https://doi.org/10.1007/s10973-016-6061-y
81. D. Neves, H. Thunmanb, A. Matos, L. Tarelhoa, A. Gómez-Bareac, Characterization and prediction of biomass pyrolysis products. Prog. Energy Combust. Sci. **37**, 611–630 (2011). https://doi.org/10.1016/j.pecs.2011.01.001. CrossRefGoogle Scholar
82. S.S. Nielsen, *Food Analysis*, 5th edn. (Springer, Boston, 2017), pp. 9–632
83. A. Oasmaa, Y. Solantausta, V. Arpiainen, E. Kuoppala, K. Sipilä, Fast pyrolysis bio-oils from wood and agricultural residues. Energy Fuels **24**(2), 1380–1388 (2010). View at Publisher • View at Google Scholar • View at Scopus
84. M. Pande, A. Bhaskarwar, Biomass conversion to energy, in *Biomass Conversion*, ed. by C. Baskar, S. Baskar, R. S. Dhillon, (Springer, Berlin, Heidelberg, 2012), pp. 1–90
85. T. Phuphuakrat, N. Nipattummakul, T. Namioka, S. Kerdsuwan, K. Yoshikawa, Characterization of tar content in the syngas produced in a downdraft type fixed bed gasification system from dried sewage sludge. Fuel **89**(9), 2278–2284 (2010). View at Publisher • View at Google Scholar • View at Scopus
86. L. Reimer, H. Kohl, Transmission electron microscopy: Physics of image formation. Springer Ser. Opt. Sci. **36**, 16–192 (2008)
87. A.R. Robinson, S.D. Mansfield, Rapid analysis of poplar lignin monomer composition by a streamlined thioacidolysis procedure and near-infrared reflectance-based prediction modeling. Plant J. **58**, 706–714 (2009)
88. J.C. Ronda, G. Lligadas, M. Galià, V. Cádiz, Vegetable oils as platform chemicals for polymer synthesis. Eur. J. Lipid Sci. Technol. **113**(1), 46–58 (2011)
89. D. Roy, A. Ullah, B.S. Sumerlin, Rapid block copolymer synthesis by microwave-assisted RAFT polymerization. Macromolecules **42**(20), 7701 (2009)
90. A. Rupali, D. Lagad, R. Alamelu, S. Acharya, K. Aggarwal, Instrumental neutron activation analysis for multi-elemental determination in Indian tea samples. J. Radioanal. Nucl. Chem. **288**, 613–662 (2011). https://doi.org/10.1007/s10967-010-0978-6
91. T.D. Saint'Pierre, T.A. Maranhão, V.L. Frescura, A.J. Curtius, R.Q. Aucélio, Determination of Cd and Pb in fuel ethanol by filter furnace electrothermal atomic absorption spectrometry. Quim Nova **31**(7), 1626–1630 (2008). https://doi.org/10.1590/S0100-40422008000700005

92. M.F. Sawalha, J.L. Gardea-Torresdey, J.G. Parsons, G. Saupe, J.R. Peralta-Videa, Determination of adsorption and speciation of chromium species by saltbush (atriplex canescens) biomass using a combination of XAS and ICP–OES. Microchem. J. **81**(1), 122–132 (2005). https://doi.org/10.1016/j.microc.2005.01.008
93. R.F. Sawyer, Science based policy for addressing energy and environmental problems. Proc. Combust. Inst. **32**, 45–56 (2009). https://doi.org/10.1016/j.proci.2008.07.003
94. L.C. Sawyer, D.T. Grubb, G.F. Meyers, *Polymer Microscopy*, 3rd edn. (Springer, New York, 2008), pp. 27–434
95. A.T. Showler, R.M. Knaus, T.E. Reagan, Foraging territoriality of the imported fire ant, Solenopsis invicta Buren, in sugarcane as determined by neutron activation analysis. Ins. Soc. **36**(3), 235–239 (1989). https://doi.org/10.1007/BF02226306
96. R. Slade, A. Bauen, Micro-algae cultivation for biofuels: Cost, energy balance, environmental impacts and future prospects. Biomass Bioenergy **53**, 29–38 (2013). https://doi.org/10.1016/j.biombioe.2012.12.019
97. J. Sluiter, A. Sluiter, *Summative Mass Closure, Laboratory Analytical Procedure (LAP) Review and Integration: Feedstocks. Technical Report NREL/TP-510-48087* (National Renewable Energy Laboratory, Golden, 2010), pp. 1–13
98. J.B. Sluiter, R.O. Ruiz, C.J. Scarlata, A.D. Sluiter, D.W. Temple-ton, Compositional analysis of lignocellulosic feedstocks. 1. Review and description of methods. J. Agric. Food Chem. **58**(16), 9043–9053 (2010)
99. P. Sugumaran, S. Seshadri, Evaluation of selected biomass for charcoal production. J. Sci. Ind. Res. **68**, 719–723 (2009)
100. R. Sykes, M. Yung, E. Novaes, M. Kirst, G. Peter, M. Davis, High-throughput screening of plant cell-wall compostion using pyrolysis molecular beam mass spectroscopy, in *Biofuels: Methods and Protocols, Methods in Molecular Biology*, ed. by J. R. Mielenz, (Humana Press, New York, 2009), pp. 169–183
101. C. Tessini, R. Romero, M. Escobar, A. Gordon, M. Flores, Development of an analytical method for the main organic compounds derived from thermochemical conversion of biomass. J. Chil. Chem. Soc. **61**(1), 2837–2842. ISSN 0717-9707. (2016). https://doi.org/10.4067/S0717-97072016000100017
102. W.T. Tsai, J.H. Chang, K.J. Hsien, Y.M. Chang, Production of pyrolytic liquids from industrial sewage sludges in an induction-heating reactor. Bioresour. Technol. **100**(1), 406–412 (2009). View at Publisher • View at Google Scholar • View at Scopus
103. M. Verma, S. Godbout, S.K. Brar, O. Solomatnikova, S.P. Lemay, J.P. Larouche, Biofuels production from biomass by thermochemical conversion technologies. Int. J. Chem. Eng. **2012**, 1–18 (2012). https://doi.org/10.1155/2012/542426
104. S. Wang, L. Baxter, F. Fonseca, Biomass fly ash in concrete: SEM, EDX and ESEM analysis. Fuel **83**(3), 372–379 (2008). https://doi.org/10.1016/j.fuel.2007.05.024
105. J.L. Ward, C. Harris, J. Lewis, M.H. Beale, Assessment of ^{1}H NMR spectroscopy and multivariate analysis as a technique for metabolite fingerprinting of Arabidopsis thaliana. Phytochemistry **62**, 949–957 (2003)
106. Y. Xia, R.C. Larock, Vegetable oil-based polymeric materials: Synthesis, properties, and applications. Green Chem. **12**(11), 1893–1909 (2010)
107. S. Zafar, Biomass conversion technologies – biochemical conversion of biomass; biomass pyrolysis process. Power. Clean Energy Future **1**(1–2), 10–19 (2018)
108. J. Zeng, Y. Zheng, X. Yu, L. Yu, D. Gao, S. Chen, Lignocellulosic biomass as a carbohydrate source for lipid production by Mortierella isabellina, in *Microorganisms in Biorefineries. Microbiology Monographs*, ed. by B. Kamm, vol. 26, (Springer, Berlin, Heidelberg, 2013), pp. 74–104. https://doi.org/10.1016/j.biortech.2012.10.079
109. R. Zhang, H.M. El-Mashad, K. Hartman, et al., Characterization of food waste as feedstock for anaerobic digestion. Bioresour. Technol. **98**(4), 929–935 (2007). View at Publisher • View at Google Scholar • View at Scopus

110. H. Zhang, R. Xiao, H. Huang, G. Xiao, Comparison of non-catalytic and catalytic fast pyrolysis of corncob in a fluidized bed reactor. Bioresour. Technol. **100**(3), 1428–1434 (2009). View at Publisher • View at Google Scholar • View at Scopus
111. C. Zhou, Q. Wu, Recent development in applications of cellulose nanocrystals for advanced polymer-based nanocomposites by novel fabrication strategies, in *Nanocrystals – Synthesis, Characterization and Applications*, ed. by S. Neralla, (IntechOpen, Rijeka, 2012), pp. 103–120. https://doi.org/10.5772/48727

Chapter 3
Compositional Analysis of Zimbabwean Sugarcane Bagasse Ash Towards Production of Nano Silicon for Solar Cell Application

F. Farirai, C. Shonhiwa, M. Mupa, and M. O. Daramola

3.1 Introduction

Silicon (Si) is a valuable metalloid whose applications range from being an additive in Aluminium and Ferrous Alloys to semiconductors for electronics [7]. Metallurgical Grade Silicon (MG-Si) is produced by carbo-thermal reduction of quartz at approximately 1900°C, as given in Eq. (3.1):

$$SiO_2(s) \xrightarrow[1900^{\circ}C]{Charcoal} Si(s) + CO_2(g) \quad (3.1)$$

The energy input for the production of MG-Si is approximately 50 kWh/kg [25]. The Siemens process requires 200 kWh/kg for further processing of (MG- Si) to get solar-grade silicon. Globally, production of MG-Si is estimated at 600,000 tons/year, releasing several million tons of CO_2, into the atmosphere [25]. The use of plant biomass as the source of silicon instead of quartz sand may address this environmental concern.

Sugarcane is grown commercially in the Lowveld of Zimbabwe which includes Chisumbanje, Nyanyadzi and Chiredzi [17]. Sugarcane absorbs silicon in the form silicic acid (H_4SiO_4) better than any mineral nutrients from the soil, with the capacity to accumulate 380 kg/ha of Silicon (Si) in a year-old crop [23]. However, the content of silica in sugarcane bagasse ash is highly dependent on the geographical

F. Farirai · M. O. Daramola (✉)
School of Chemical and Metallurgical Engineering, Faculty of Engineering and the Built Environment, University of the Witwatersrand, Wits 2050, Johannesburg, South Africa
e-mail: michael.daramola@wits.ac.za

C. Shonhiwa · M. Mupa
Department of Physics and Engineering, Department of Chemistry, Faculty of Science and Engineering, Bindura University of Science Education, Bindura, Zimbabwe

M. O. Daramola, A. O. Ayeni (eds.), *Valorization of Biomass to Value-Added Commodities*, Green Energy and Technology,
https://doi.org/10.1007/978-3-030-38032-8_3

location and the environment where the sugarcane is grown as well as the farming practices [14]. Ethanol and sugar production create waste by-products such as sugarcane bagasse that can still be tapped for added profitability [26]. For every 260 kg of bagasse, approximately 5–7% of ash is obtained [24]. The use of bagasse in boilers leaves an ash residue called sugarcane bagasse ash (SCBA), another by-product that has been used as a pozzolanic to reinforce building materials and as a fertiliser in sugarcane fields [26]. Silica, a potential constituent of the solid waste (sugarcane bagasse ash), has a wide variety of industrial applications [18]. In this chapter, results of investigation into the composition of Zimbabwean sugarcane bagasse ash are explored and its suitability for the production of nano silicon for solar cell application is presented.

3.2 Methodology

Sugarcane bagasse (SCB), sugarcane bagasse bottom ash (SCBBA) and sugarcane bagasse fly ash (SCBFA) were collected from Green Fuel, a local bio-ethanol manufacturing private company, located in the south eastern part of Zimbabwe. The sugarcane bagasse was rinsed in deionised water to remove dust and sand particles. The as-received ashes of SCBBA and SCBFA and dried SCB were further ground using a pulveriser. The three samples were sieved using a 40-μm sieve in preparation for analysis. The ash content of raw sugarcane bagasse was determined according to ASTM D2974-8 standard (Standard Test Methods for Moisture, Ash, and Organic Matter of Peat and other Organic soils). Thermal behaviour of samples was investigated by using a thermogravimetric analyser (TGA) (Q600 SDT Thermal Analyzer TA Instrument, USA). The samples were heated in air at the rate of 10° C/min till a temperature of 800° C. Chemical analysis of SCB, SCBBA and SCBFA was carried out via X-Ray Fluorescence Spectroscopy (Panalytical PW2404 X-ray Fluorescence Spectrometer). For determination of surface chemistry, Fourier Transform Infrared Spectroscopy, FTIR (Perkin Elmer 100 FTIR Spectrometer), was used. X-Ray Diffractometer (German Diffractometer D2 PHASER Bruker X-Ray with Cu Kα radiation) was used to determine the degree of crystallinity of the samples. Field Emission Scanning Electron Microscope (Carl Zeiss Sigma FESEM Oxford X-ACT EDX detector) equipped with energy-dispersive X-ray was used to check morphology and elemental composition.

3.3 Results and Discussion

3.3.1 Ash Content, Calorific Value and Elemental Composition

In Table 3.1, the results for moisture, ash, calorific value, carbon, nitrogen, hydrogen, sulphur, volatile matter and fixed carbon are shown. It is observed that the lowest moisture content was found in SCBFA with 1.90% compared to SCBBA and SCB with

Table 3.1 Results for physicochemical characterisation of sugarcane bagasse samples

Parameter	Sample		
	SCB	SCBBA	SCBFA
Ash content (%)	6.92 (ASTM) 5.20 (TGA)	–	–
Moisture content (%)	6.40	2.97	1.90
Volatile mater (%)	58.54	5.46	4.89
Fixed carbon (%)	29.86	–	–
Calorific value (MJ/kg)	17.36	1.79	0.31
Elemental analysis			
Carbon (%)	42.42	11.35	4.06
Hydrogen (%)	5.23	–	–
Nitrogen (%)	0.55	0.10	0.09
Sulphur (%)	–	–	–

2.97% and 6.40%, respectively. SCB retains moisture after the extraction of juice for ethanol production while SCBBA and SCBFA are by-products from the boiler; therefore, moisture content is lower. The content of ash in the SCB obtained from ASTM method and TGA is 6.92% and 5.2%, respectively. Both values fall in the range between 5% and 7% as reported by [9]. Norsuraya et al. [18] reported that the percentage ash content of sugarcane bagasse differs with the source of the bagasse. Variation in SCB ash content is due to farming practices during growing of sugarcane and harvesting techniques [1, 8]. SCB showed higher volatile matter of 58.54% compared to SCBBA and SCBFA with 5.46% and 4.89%, respectively, as these two are by-products from burning of bagasse in the boiler. High volatile matter and fixed carbon for SCB positively influence the biomass calorific value (CV) ([5, 13]). The CV for SCB, SCBBA and SCBFA were 17. 36, 1.79 and 0.31 MJ/kg, respectively. The CV obtained in this study is supported by other previous studies. Vieira [28] reported calorific value of 18.25 MJ/kg for sugarcane bagasse from Minais Gerais, Brazil, and Machado et al. [13] reported a CV of 15.55 MJ/kg and 17.15 MJ/kg for bagasse from Parana and Mato Grosso do Sui in Brazil, respectively [13, 28].

The higher the moisture and ash content, the lower the calorific value. If volatiles and fixed carbon are present in large quantities, it results in increased calorific value [21]. SCB exhibits high carbon, hydrogen and nitrogen content of 42.42%, 5.23% and 0.5%, respectively. SCBBA and SCBFA lost carbon, hydrogen and nitrogen during combustion of SCB in the boiler. For all the three samples, sulphur content is negligible. The source of bagasse influences its elemental composition [4].

3.3.2 *Thermogravimetric Analysis of SCB, SCBBA and SCBFA*

The thermogravimetric analysis for sugarcane bagasse (SCB), sugarcane bagasse bottom ash (SCBBA) and sugarcane bagasse fly ash (SCBFA) is shown in Fig. 3.1. Thermogravimetric analysis makes it possible to follow the weight loss of samples

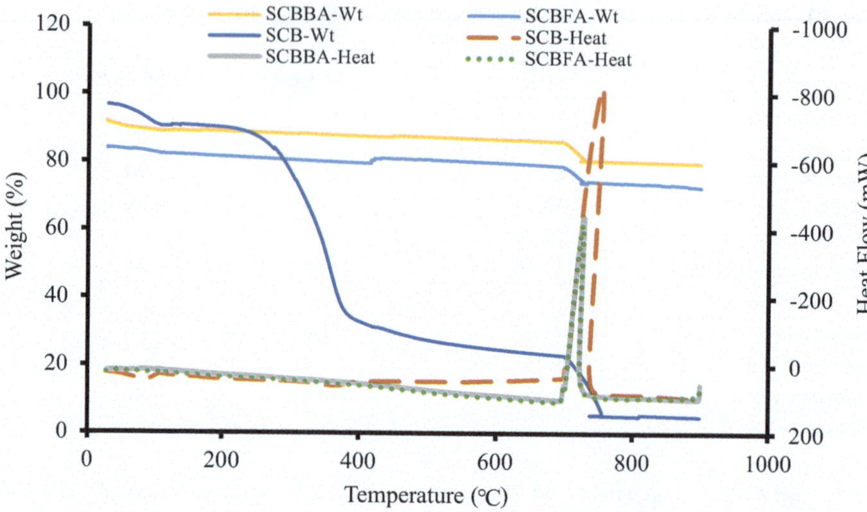

Fig. 3.1 Thermogravimetric curve of SCB, SCBBA and SCBFA samples

as function of temperature Through the TGA curve, it is possible to relate different regions of weight losses as function of temperature with different biomass components.

Sugarcane bagasse exhibits maximum weight loss of up to 80 wt% between 300° C and 700° C and subsequently a sharp exothermic peak in the DTA curve. This could be due to volatilisation of unburnt carbon and other organic matters present.

The weight loss at various temperatures corresponds to decomposition of components. The weight loss at 100° C is a result of water loss while the weight loss around 380° C is as a result of decomposition of volatile matter. The formation of ash between 500° C and 700° C resulted in negative heat flow, showing that energy is consumed for ash formation. Mohomane et al. [15] studied thermal degradation of sugarcane bagasse. The results evidenced the presence of three zones of weight loss in the thermogravimetric analysis, the first one at 100° C attributed to moisture content loss. The second region between 350° C and 500° C was due to organic matter decomposition. The third zone they observed between 550° C and 700° C was due to ash formation. This behaviour was also observed in this current study for sugarcane bagasse [15].

For SCBBA and SCBFA powders, the total weight loss was 9 wt% and 7 wt%, respectively. The weight loss was due to loss in moisture content and unburnt organic matter since incomplete combustion also occurs in the boilers. The lower weight loss of the ashes is evidenced by lower loss on ignition (LOI) as seen in the chemical composition shown in Table 3.2. From the DTA curves of SCBBA and SCBFA, an endothermic peak around 60° C corresponds to moisture content loss. Some researchers, including Torres Agredo et al. [26], Frias et al. [10], Frías et al. [11]

Table 3.2 Chemical composition of SCB, SCBBA and SCBFA

Compound	SCB	SCBBA	SCBFA
SiO_2	60.75	71.49	69.67
Al_2O_3	6.66	6.2	4.69
Fe_2O_3	4.84	3	2.67
CaO	7.54	3.3	3.06
MgO	3.63	1.79	1.37
K_2O	7.04	4.79	8.03
Na_2O	4.34	4.61	4.2
SO_3	0.28	0.11	0.82
TiO_2	1.41	0.80	0.44
MnO	0.17	0.14	0.17
P_2O_5	0.97	0.61	1.02
LOI	1.26	0.31	0.48

and Torres Agredo et al. [26], reported endothermic peaks attributed to decomposition of carbonates between 500° C and 600° C. In study, the same peaks are evidenced for the SCBBA and SCBFA.

3.3.3 XRF Analysis of SCB, SCBBA and SCBFA

The chemical composition of samples determined by XRF is shown in Table 3.2. Silica (SiO_2) is predominant in all the three samples. This is also supported by XRD analysis, which shows the presence of silica in the form of quartz, cristobalite and tridymite as the major components in the samples Table 3.2. The results are also supported by the work reported by Payá and co-workers in 2002. There are oxides of magnesium, calcium, phosphorous, iron and potassium, sodium, titanium, manganese and aluminium present.

The sugarcane bagasse bottom ash (SCBBA) sample exhibited high silica content of 71.49 wt% compared to the sugar cane bagasse fly ash (SCBFA), 69.67 wt% and sugar cane bagasse (SCB), 60.75 wt%. Literature suggests that the silica content in sugarcane bagasse ashes ranges from as low as 31.41 wt% [3] to as high as 96.2 wt% [22]; thus, the values obtained fall within the range. However, the percentage of oxide impurities is on the higher side; therefore, there will be need to conduct acid leaching prior to the extraction of silica to eliminate these oxide impurities [19].

3.3.4 FTIR Analysis for SCB, SCBBA and SCBFA

The FTIR spectra of SCB, SCBBA and SCBFA samples are shown in Fig. 3.2. The strong absorption bands at 1011–1080 cm^{-1} were observed in all the three samples. It depicts Si-O-Si (siloxane) functional group presence which affirms the presence of silica.

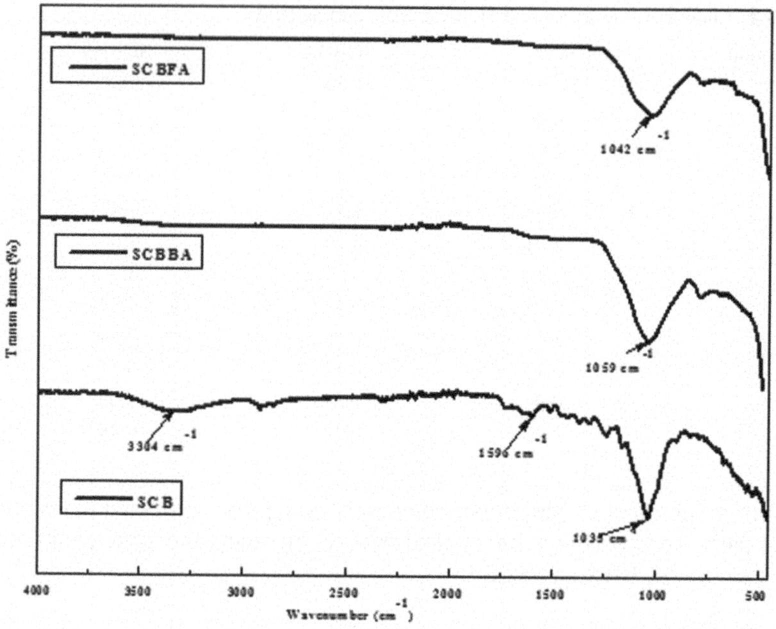

Fig. 3.2 FTIR spectra for SCB, SCBBA and SCBFA

At 3304 cm^{-1}, the vibration is due to both asymmetric and symmetric vibration of (O-H) water bound in sugarcane bagasse due to its hygroscopicity. The vibration at 1596 cm^{-1} due to carboxyl groups shows that the sample is rich in carbon. Arif et al. [2] and Jagadesh et al. [16] reported different stretching groups in sugarcane samples, one strong vibration between 500 cm^{-1} and 1100 cm^{-1}, which indicates symmetric stretching vibration of the silicon bonds. The second stretch at 3400 cm^{-1} affirmed the presence of hydroxyl group, confirming availability of water molecules in the samples [2, 16, 20].

3.3.5 XRD Analysis for SCB, SCBBA and SCBFA

The three samples SCB, SCBBA and SCBFA presented peaks (Fig. 3.3) that are mainly characteristics of the following mineral phase quartz, cristobalite and tridymite with predominance of quartz. It is known that quartz transforms to tridymite and cristobalite with increasing temperature. Since SCBBA and SCBFA are by products from the boiler cristobalite and tridymite form of silica is predominant compared to SCB with silica which is highly amorphous as seen by the broad peak between 10° and 30° [27].

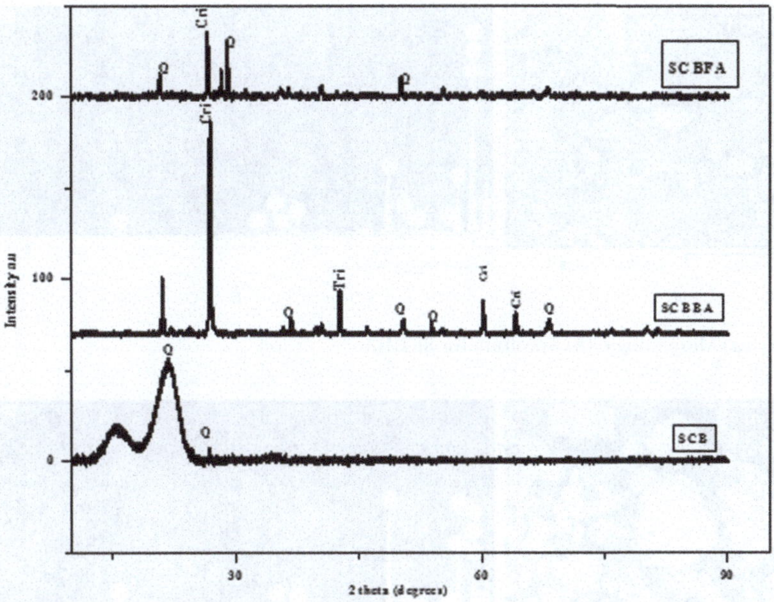

Fig. 3.3 XRD pattern for SCB, SCBBA and SCBFA (Q-Quartz, Cri-Cristobalite and Tri-Tridymite)

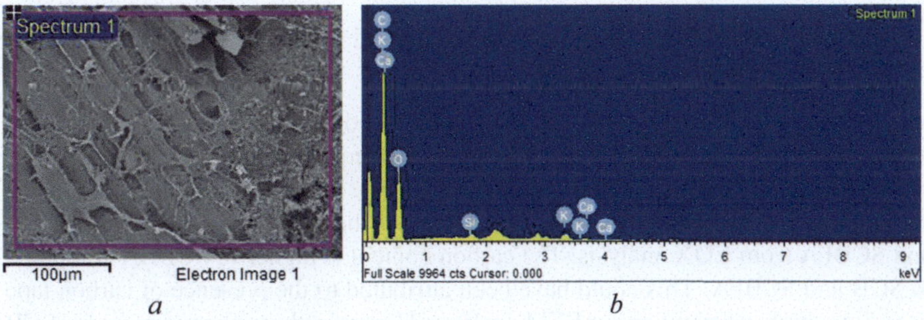

Fig. 3.4 (**a**) Morphology. (**b**) Spectrum for SCB

3.3.6 FE-SEM Analysis for SCB, SCBBA and SCBBA

Microscopic observation shows particle shape and structure variation in all the three samples. The SEM images and EDX spectra of the three samples SCBBA, SCBFA and SCB waste observed by Field Emission Scanning Electron Microscope and Energy Dispersive X-ray analysis are shown in Figs. 3.4, 3.5, and 3.6, respectively. The samples are rich in mainly carbon and silicon and other elements such

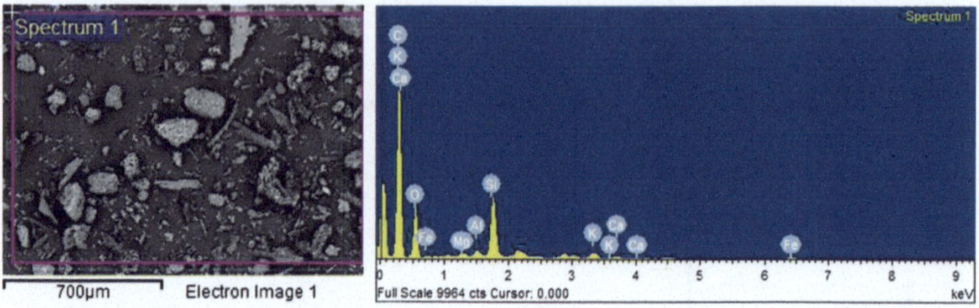

a *b*

Fig. 3.5 (**a**) Morphology. (**b**) Spectrum for SCBBA

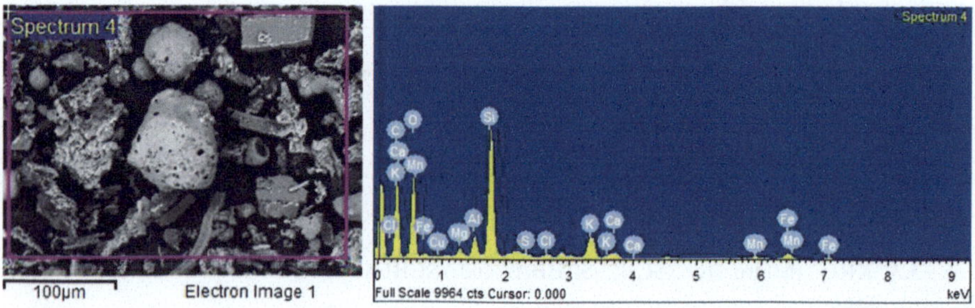

a *b*

Fig. 3.6 (**a**) Morphology. (**b**) Spectrum for SCBFA

as aluminium, calcium, magnesium, oxygen and iron. The spectra are comparable to those in literature confirming that the samples contain silicon and unburnt carbon [16, 20].

Table 3.3 shows the elemental composition of the three samples SCB, SCBBA and SCBFA from EDX analysis. The carbon content is higher in SCBBA compared to SCB and SCBFA. This could have been attributed to the presence of carbon tape used for sample preparation for SEM analysis. However, the carbon content for SCB (42.42%) as shown in Table 3.3 is higher than SCBBA and SCBFA comparable to 43.35% reported by Machado et al. [13]. The silicon content in the three samples SCB, SCBBA and SCBFA is 0.49, 6.04 and 12.32 wt %, respectively. Both ashes exhibited higher silicon content than raw SCB because of the accumulation of silicon during combustion of sugarcane bagasse. Xu et al. [29], Moraes et al. [16] and Embong et al. [6] reported that reactive amorphous silica in sugarcane comes from groundwater. Silicic acid in groundwater is absorbed into sugarcane cells where polymerisation occurs, forming amorphous silica. During usage as fuel in thermal power plants, the burning of sugarcane bagasse produces reactive amorphous silica which accumulates in the ash product [6, 16, 29].

Table 3.3 EDX elemental composition of SCB, SCBBA and SCBFA

Element	Weight composition (%)			Atomic composition (%)		
	SCB	SCBBA	SCBFA	SCB	SCBBA	SCBFA
C	56.96	63.26	40.63	64.41	72.25	53.92
O	41.01	27.63	33.96	34.82	23.69	33.83
Mg	–	0.29	0.67	–	0.16	0.44
Al	–	0.50	1.69	–	0.25	1.00
Si	0.49	6.04	12.32	0.24	2.95	6.99
S	–	–	0.20	–	–	0.10
Cl	–	–	0.64	–	–	0.29
K	1.30	0.98	3.93	0.45	0.34	1.60
Ca	0.24	0.36	1.07	0.08	0.12	0.43
Mn	–	–	0.99	–	–	0.29
Fe	–	0.94	3.88	–	0.23	1.11
Cu	–	–	0.02	–	–	0.00
Total	100	100	100	100	100	100

3.4 Conclusions

The compositional analysis study of sugarcane bagasse and its ashes confirms that silica is the predominant compound in the samples. The scientific and technical study on SCB, SCBBA and SCBFA using XRF, XRD, SEM- EDX, FTIR and TGA enabled the conclusions that follow:

- Silica in amorphous form is present in SCB, while in SCBBA and SCBFA, it is present in the form of cristobalite and tridymite form.
- Incomplete calcination of SCBBA and SCBFA is observed in SEM images and EDX.
- It is evident through XRF and EDX that the samples contain iron, aluminium, calcium, magnesium, phosphorous, potassium oxides.
- The strong vibrations observed by FTIR analysis confirm that silicon bonds are present in the samples.
- The extraction of silicon from sugarcane bagasse ash is worth pursuing, as the ashes are rich in silica.

References

1. S. Anwar, Determination of moisture content of bagasse of jaggery unit using microwave oven. J. Eng. Sci. Technol. **5**, 472–478 (2010)
2. E. Arif, M.W. Clark, N. Lake, Sugar cane bagasse ash from a high-efficiency co-generation boiler as filler in concrete. Construct. Build Mater. **151**, 692–703 (2017)

3. V. Castaldelli, J. Akasaki, J. Melges, M. Tashima, L. Soriano, M. Borrachero, J. Monzó, J. Payá, Use of slag/sugar cane bagasse ash (SCBA) blends in the production of alkali-activated materials. Materials **6**, 3108–3127 (2013)
4. P. Das, A. Ganesh, P. Wangikar, Influence of pretreatment for deashing of sugarcane bagasse on pyrolysis products. J. Biomass. Bioenerg. **27**, 445–457 (2004)
5. A.A. Dos Reis, T.d.P. Protásio, I.C.N. Alves de Melo, P.F. Trugilho, A.d.C. Oliveira Carneiro, Composição da madeira e do carvão vegetal de Eucalyptus urophylla em diferentes locais de plantio. Pesqui. Florest. Bras. **32**, 277–290 (2012)
6. R. Embong, N. Shafiq, A. Kusbiantoro, M.F. Nuruddin, Effectiveness of low-concentration acid and solar drying as pre-treatment features for producing pozzolanic sugarcane bagasse ash. J. Clean. Prod. **112**, 953–962 (2016)
7. E. Epstein, Silicon. Annu. Rev. Plant Biol. **50**, 641–664 (1999)
8. Y. Fang, A. Ellis, M. Uchimiya, T.J. Strathmann, Selective oxidation of colour-inducing constituents in raw sugar cane juice with potassium permanganate. J. Food Chem. **298**, 125036 (2019)
9. V. Ferreira-Leitao, L.M.F. Gottschalk, M.A. Ferrara, A.L. Nepomuceno, H.B.C. Molinari, E.P. Bon, Biomass residues in Brazil: availability and potential uses. Waste Biomass Valori. **1**, 65–76 (2010)
10. M. Frias, E. Villar-Cociña, E. Valencia-Morales, Characterisation of sugar cane straw waste as pozzolanic material for construction: calcining temperature and kinetic parameters. Waste Manag. **27**, 533–538 (2007)
11. M. Frías, E. Villar, H. Savastano, Brazilian sugar cane bagasse ashes from the cogeneration industry as active pozzolans for cement manufacture. Cement Concrete Comp. **33**, 490–496 (2011)
12. R.P. Inácio, A.J. Teixeira Senna, Characterization of the wood production chain in Rosario do Sul-RS, Brazil. (English). Organizações Rurais & Agroindustriais **16**, 291 (2014)
13. G. Machado, F. Santos, D. Faria, T.N. de Queiroz, F. Zinani, J.H. de Queiroz, F. Gomes, Characterization and potential evaluation of residues from the sugarcane industry of Rio Grande do Sul in biorefinery processes. Nat. Resour. J. **9**, 175–187 (2018)
14. G. Martines-Filho, H.L. Burnquist, C.E.d.F. Vian, Bioenergy and the rise of sugarcane-based ethanol in Brazil. Mag. Food Farm Resour. **21**, 91–96 (2006)
15. S.M. Mohomane, T.E. Motaung, N. Revaprasadu, Thermal degradation kinetics of sugarcane bagasse and soft wood cellulose. Materials **10**, 1246 (2017)
16. J.C.B. Moraes, M.M. Tashima, J.L. Akasaki, J.L.P. Melges, J. Monzó, M.V. Borrachero, L. Soriano, J. Payá, Increasing the sustainability of alkali-activated binders: the use of sugar cane straw ash (SCSA). Construct. Build Mater. **124**, 148–154 (2016)
17. T. Nhiwatiwa, T. Dalu, L. Brendonck, Impact of irrigation-based sugarcane cultivation on the Chiredzi and Runde Rivers quality, Zimbabwe. Sci. Total Environ. **587**, 316–325 (2017)
18. S. Norsuraya, H. Fazlena, R. Norhasyimi, Sugarcane bagasse as a renewable source of silica to synthesize santa barbara amorphous-15 (SBA-15). Procedia Eng. **148**, 839–846 (2016)
19. F.C. Pa, A. Chik, M.F. Bar, Palm Ash as an Alternative Source for Silica Production. MATEC Web of Conferences **78**, 01062 (2016)
20. J. Payá, J. Monzo, M. Borrachero, L. Díaz-Pinzón, L.M. Ordonez, Sugar-cane bagasse ash (SCBA): studies on its properties for reusing in concrete production. J. Chem. Technol. Biotechnol. **77**, 321–325 (2002)
21. W. Quirino, Calorific power of wood and ligno-cellulosic materials. Revista da Madeira **15**, 100–106 (2005)
22. A. Sales, S.A. Lima, Use of Brazilian sugarcane bagasse ash in concrete as sand replacement. Waste Manag. **30**, 1114–1122 (2010)
23. N.K. Savant, G.H. Korndörfer, L.E. Datnoff, G.H. Snyder, Silicon nutrition and sugarcane production: a review. J. Plant Nutr. **22**, 1853–1903 (1999)
24. A. Souza, S. Teixeira, G. Santos, F. Costa, E. Longo, Reuse of sugarcane bagasse ash (SCBA) to produce ceramic materials. J. Environ. Manage. **92**, 2774–2780 (2011)

25. M. Tao, Inorganic photovoltaic solar cells: silicon and beyond. Elecrochem. Soc. Interface **17**, 30–35 (2008)
26. J. Torres Agredo, R. Mejía de Gutiérrez, C. Escandón Giraldo, L. González Salcedo, Characterization of sugar cane bagasse ash as supplementary material for Portland cement. Ing. Invest. **34**, 5–10 (2014)
27. K. Umamaheswaran, V.S. Batra, Physico-chemical characterisation of Indian biomass ashes. Fuel **87**, 628–638 (2008)
28. A. Vieira, Caracterização Da Biomassa Proveniente De Resíduos Agrícolas. Dissertação (Mestrado). Universidade Estadual do Oeste do Paraná, 28–47 (2012)
29. Q. Xu, T. Ji, S.J. Gao, Z. Yang, N. Wu, Characteristics and applications of sugar cane bagasse ash waste in cementitious materials. J. Mater. **12**, 1–19 (2018). https://doi.org/10.3390/ma12010039

Chapter 4
Application of Artificial Intelligence in the Prediction of Thermal Properties of Biomass

O. Olatunji, S. Akinlabi, and N. Madushele

4.1 Introduction

The use of lignocellulosic biomass (LCB) as value-added products in energy generation requires detailed characterization of its properties. The higher heating value (HHV) is a key criterion in evaluating the energy content of lignocellulosic biomasses [1–4]. However, the experimental procedure for these analyses demands instruments that are very complex, exorbitant and requires a stable electricity supply. Tremendous progress has been made in artificial intelligence; this is opening a new window of opportunity in the study of biomass properties towards the utilization of bioresources. Artificial intelligence (AI) has been applied in several field of the studies such as supply chain management, risk assessment, medical and health service delivery, manufacturing, energy prediction and so on [5–7].

Artificial neural network (ANN) is demonstrating to be a crucial tool, with a potential to enhance the research development in biomass energy prediction [8]. ANN and other intelligent models have been used for the experimental simulation, with a view to improve researches on biomass processing and related development. Hence, this chapter discusses the stages in ANN model development, training algorithm, transfer functions and error measures. The recent findings regarding the application of AI in LCB properties prediction are discussed. Taking a cue from the status of AI in biomass exploration, an outlook is chatted for further studies. It was noted that some of the disadvantages of ANN such as high processing time are being continuously addressed through different hybridized models [9, 10].

O. Olatunji · N. Madushele
Mechanical Engineering Science, University of Johannesburg, Johannesburg, South Africa

S. Akinlabi (✉)
Department of Mechanical Engineering, Faculty of Engineering & Technology, Walter Sisulu University, Eastern Cape, South Africa

M. O. Daramola, A. O. Ayeni (eds.), *Valorization of Biomass to Value-Added Commodities*, Green Energy and Technology,
https://doi.org/10.1007/978-3-030-38032-8_4

4.2 Renewable Energy and the Application Artificial Intelligence

Artificial intelligence has begun to transform the way many industries create and deliver their products and services including energy provision. A recent study from Roland Berger [11] predicts that AI can improve the efficiency in utility companies by five time within 5 years [11], but more than three-quarters of the companies who responded to the survey said that their company had no immediate plan to explore AI technologies. However, capital commitment to digital technologies by energy companies has swiftly increased in the course of time, with worldwide financial commitment to digital electricity infrastructure and software grown by more than 20% every year since 2014 and climbing to $47 Billion in 2016. This investment, in 2016, was almost 40% greater than global investment in gas-fired power generation which amount to $34 billion [11]. This shows the extent to which AI has been accepted in energy industry. Table 4.1 presents the merits and demerits of the ANN, which possibly may have engendered the rapid acceptance of this technology.

4.3 Kinds of Data for Model Development

Data are the fuel needed for the development of any kinds of model. Two kinds of data are required for a model development. These are [12]:

- Predictor Data: Also known as predictor variables. This data type is used in making prediction. In this case, the proximate values such as fixed carbon, volatile matter, moisture content, M and ash contents are the predictor variables
- Target Data: This is the function or behaviour that is intended to be predicted. In this instance, HHV is the target variable since it determines the enthalpy of combustion of LCB.

A relevant mathematical or statistical technique is applied to establish the correlation between predictor functions and the target functions. Once the relationship from the resulting models is captured, it can be applied to determine the target function of a new case if the predictor data are known [12].

Table 4.1 Merits and demerits of ANN

Merits	Demerits
Can perform nonlinear programming task	It requires training
Self-learning capability	Require high processing time
Easy implementation	
Robust at solving complex problems	
Can be implemented in any application	
Nonlinearity	

4.4 Model Development Techniques

Predictive modelling is an analysis technique through which a future outcome or behaviour is determined based on the past and current data at hand [13]. It can also be defined as a procedure that uses data mining and probability to predict future [14]. Each model comprises several predictor variables, which may possibly affect the future forecast. Data gathering is the first level before model formulation. Once the data are gathered, a model, which could be linear, multiple regression or deep artificial intelligent model, is developed. Predictive models have been variously described and applied by different authors in biomass and non-biomass applications [12, 14–22]. There are many methods, which can be applied to develop a model, but the applicability depends on the shapes and sizes of the data. Most commonly, models can be broadly classified into six, which are [12, 23–25]:

I. Linear models
II. Decision tree also known as CART
III. Kernel Nearest Neighbour (K-NN)
IV. Artificial Neural Networks (ANN)
V. Support vector machine (SVM)
VI. Multiclass SVM (MCSVM)
VII. Cluster model
VIII. Expert systems (ES)

4.5 Model Evaluation Criteria

There are several evaluation methods, which have been developed in the literature. Some of them were highlighted by Konishi and Kitagawa [26]. In addition, there are other model-specific evaluation criteria, for instance, when dealing with classification problems, metrics such as accuracy, error rate, precision, and kappa statistics that can be applied. Therefore, selection of relevant evaluation criteria is very germane to the development of versatile model.

4.5.1 Assessment Criteria for Good Prediction Models

In biomass energy resources and properties modelling, most researchers often focus on prediction accuracy just as if it is the only criterion [12] which can be used to determine the goodness or suitability of a model. However, there are other criteria, which are outlined below and detailed in [12, 27–29]:

I. Simplicity
II. Sensitivity

III. Acceptability
IV. Stability
V. Explicability

4.5.2 *Linear and Nonlinear Equation for Biomass HHV Estimation*

Several linear and nonlinear empirical equations have been formulated for the estimation of the enthalpy of combustion of biomass as shown in Table 4.2. While some were developed based on the ultimate properties, others were derived from the proximate analysis. It must be noted that no single type of model can be used for all prediction problems [12]; therefore, most of the previously developed models for other applications may not be suitable for all biomass data [30–32].

An overview of the studies on prediction models for biomass properties, especially the heating value and elemental composition, is majorly based on proximate analysis. This is understandable since it is cheaper to determine the proximate properties compared to other properties [3, 51, 52]. Up until year 2000, most models developed for LCB properties prediction are either made of linear or multilinear regression, but it has been noted that the dependencies between proximate constituents of biomass and HHVs are sometimes nonlinear. Therefore, the prediction based on linear regression may not be versatile to accurately predict various attributes of biomass, especially when different experimental data are presented to them [2, 53].

4.5.3 *Stages where ANN Can be Applied in Value Creation Cycle*

At different stages in the production of value-added product, which in this case are different kinds of energy products especially fuels, ANN are applied to achieve different outputs using different parameters. Presented in Table 4.3 are the areas of application of ANN towards the production of value-added products from biomass.

4.6 Categories of Models for Biomass Properties Prediction

When developing a model to assist in the design process needed for the exploration of value-added products, it is of necessity to apply the right type of model. This is because a wrongly developed model could waste computing time and could either provide non-commensurate information or an entirely wrong estimation [10, 70–74]. The choice between the existing models depends largely on the level of details required. The three categories of models are [74]:

Table 4.2 Linear and nonlinear empirical equation for biomass HHV prediction

	Equation (HHV × 10^{-1} MJ/kg) based on ultimate analysis[a]	Reference
1	2.30C + 7.61H + 12.47N + 142.59	[33]
2	3.71C + 0.01112H – 1.391O + 3.178N + 1.391S	[34]
3	4.912C – 9.119H + 1.177O	[35]
4	4.925C – 9.26H – 1.176O + 0.193S	
5	4.148C – 1.841H + 1.789O – 21.595	
6	4.259C – 0.698H + 1.817O – 1.805N + 22.77	
7	4.302C – 1.867H + 1.842O – 1.274N + 1.786S – 23.799	
8	4.064C – 2.106H + 1.603O – 1.513N + 1.547S – 0.238A + 0.000034	
9	4.064C – 2.105H + 1.604O – 1.512N + 1.548S – 0.238A	
10	3.959C – 4.471H + 1.543O + 1.543N + 2.555S – 0.181A + 0.217	
11	3.956C – 4.46H + 1.54O + 1.54N + 2.545S – 0.219A	
12	1.343C – 15.021H – 27(O2/(1 – A/100)) + 291.328(1 – A/100)	
13	2.798C – 8.491H + 7.249O – 92(O2/(1 – A/100)) – 1.135S	
14	6.838(0.328C + 1.419H) + 0.928S + 1.546O – 0.000238N – 0.331A	
15	(2.949C + 8.50H)	[1]
16	(8.790C + 3.214H + 0.56O – 248.26)	[36]
17	4.373C – 16.701	[37]
18	3.35C + 14.2.3H – 1.54O	[38]
19	3.699C + 13.178	[39]
20	3.856(C + H) – 16.938	[39]
21	3.259C + 34.56	[40]
22	3.137C + 7.009H + 0.318O – 13.675	[40]
23	3.491C + 11.783H – 1.034O – 0.151N + 1.005S – 0.211A	[41]
24	3.491C + 11.783H – 1.034O – 0.151N + 1.005S	[41]
25	3.01C + 5.25H + 0.64O – 7.63	[42]

(continued)

Table 4.2 (continued)

	Equation ($HHV \times 10^{-1}$ MJ/kg) based on ultimate analysis[a]	Reference
26	3.578C + 11.357H – 0.845O + 0.59N + 1.119S	[43]
27	3.515C + 11.617H – 1.109O + 0.6276N + 1.046S	[44]
28	3421.5C + 4461.3H + 3519.8N + 255.7O – 130.5A	[45]
	Equation ($HHV \times 10^{-1}$ MJ/kg) based on proximate analysis	
29	HHV = 3.133(VM + FC) – 108.141	[46]
30	141.19 + 1.96FC	[47]
31	3.543FC + 1.708VM	[48]
32	3.536FC + 1.559VM – 0.078A	[49]
33	–(99.14 + 2.324A)	[40]
34	2.218VM + 2.601FC – 30.368	[40]
35	1.905VM + 2.521FC	[1]
36	192.880 – 2.135VM/FC – 19.584A/VM + 0.234FC/A	[50]
37	$207.999-\frac{112.277A}{VM}-3.214VM/FC+44.953\left(\frac{A}{VM}\right)^2+0.051\left(\frac{VM}{FC}\right)^2-7.223\left(\frac{A}{VM}\right)^3+0.383\left(\frac{A}{VM}\right)^4+\left(0.076FC/A\right)$	[50]
43	$(34.215C + 44.613H + 35.198N + 2.557O - 1.305A)\times 10^2$	[45]

[a]*FC* Fixed Carbon, *A* Ash, *VM* Volatile matter, *H* Hydrogen, *O* Oxygen, *C* Carbon, *S* Sulphur, *N* Nitrogen

Table 4.3 Areas of ANN application in value creation

Value creation processes	Value-added products	Areas of ANN application	Reference
Gasification	Methane	Methane production, chemical oxygen demand, methane fraction	[54, 55]
	Biohydrogen	PH determination, hydrogen yield substrate degradation	[56]
	Biogas	Biogas yield, H_2S and NH_3 in biogas	[57–59]
	Microbial fuel cell	Current generation, power density, hydrogen yield	[60–62]
Biochemical	Biodiesel	Lipid productivity, biomass concentration	[63–65]
Pyrolysis	Bioethanol	Yield, bioethanol concentration, bioethanol production	[66, 67]
	Electricity	Feedstock characterization	[2, 53, 68, 69]
Direct combustion	Electricity, heat	Energy content, optimization	

I. *Black Box Model*: This represent the class of models whose internal working knowledge is not known. This model simply solves problems without making reference to the working principles [70, 71]. This means there is no prior knowledge of the system they represent. Examples are: neural network, NN and fuzzy systems.
II. *Grey Box Model*: These models are driven by the combinations of mathematical equations and experiential knowledge. In this model, certain elements can be approximated by rules [10, 72, 74]. Examples; Hidden Markov Model (HMM).
III. *White Box Model*: This kind of models are wholly driven by mathematical equations. It contains very detailed simulation without any approximation. They are only applied in situations where the results are expected to closely reflect reality; however, they often consume large chunk of computing power [70, 74]. It should be noted that a pure white box model does not exist in reality. Table 4.4 shows the differences between different categories of model which can be applied in biomass feedstock evaluation.

4.7 Elements of ANN Architecture

Some model applications may require binary input properties along with their summing operations; therefore, these types of operations can be built on relevant elements of ANN [75]. To this end, there are seven main components in ANN which are valid irrespective of the layer (input, output and hidden) in which the neurons are used. These have been discussed in detail in [75–77]. They are:

I. Weighted Factor
II. Summation functions
III. Transfer functions

Table 4.4 Different categories of models [10, 70–74]

Black box models	White box models	Grey box models
Low computation time	Large computation time	More flexibility
Minimum computational power	Large memory requirement	Ability to approximate
Less flexible	Very high flexibility	
It lacks physical meaning	Greater realism	

IV. Scaling and limiting
V. Output functions
VI. Error functions
VII. Learning functions

4.8 Mathematical Modelling of Artificial Neural Network (ANN)

ANN is a complex network, which consists of interconnected basic processing units called neurons. The general component of ANN can be categorized into structure, learning algorithm and activation functions [8]. The multilayer perceptron ANN (MLP-ANN) in Fig. 4.1 is used for the prediction of HHV of biomass; the mathematical expression is highlighted in Fig. 4.2.

ANNs are trained with data sets to learn the pattern. Once trained, new patterns may be presented to them for prediction or classification [78]. ANN behaves like the human brain in two ways: the knowledge is attained by the network through a learning procedure and inter-neuron linking strengths known as synaptic weights are used to store the knowledge [79, 80]. Irrespective of the nature of the problem which needs to be solved by ANN, the connection between the neuron governs the network architecture; therefore, the neuron must be carefully selected [8].

Consider a single artificial neural network whose neurons have input variables $X_1, X_2, X_3, \ldots, X_N$. As shown in Fig. 4.2, the fixed input X_0 is assigned +1 as the bias input. The inputs are then categorized based on their synaptic weights W_1, W_2, W_3, …, W_N which estimate their significance. Recall that the synapse of the biological neuron is the one that interconnects the neural network and determines the strength of the connection and they are modelled as weights. In an ANN, when the weight is negative, it is an indication that there is an inhibitory connection, while positive weight represents excitatory connections.

The entire inputs are modified by weight and then added together. The activation function which controls the value of the output is discussed in Sect. 4.11 and its values range between 0 and 1 or −1 and 1. The mathematical expression for a neuron is given as

$$v_k = \sum_{j=1}^{p} w_{kj} x_j$$

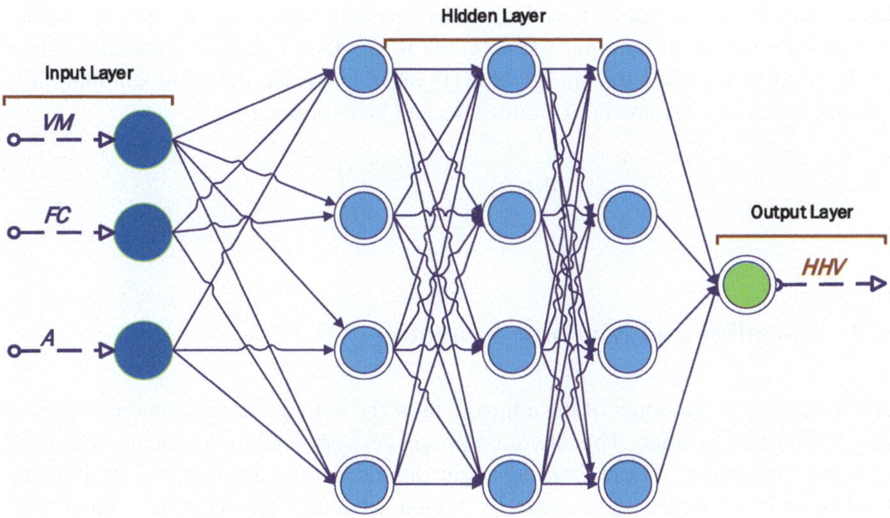

Fig. 4.1 Schematic diagram of multilayer propagation ANN for HHV prediction

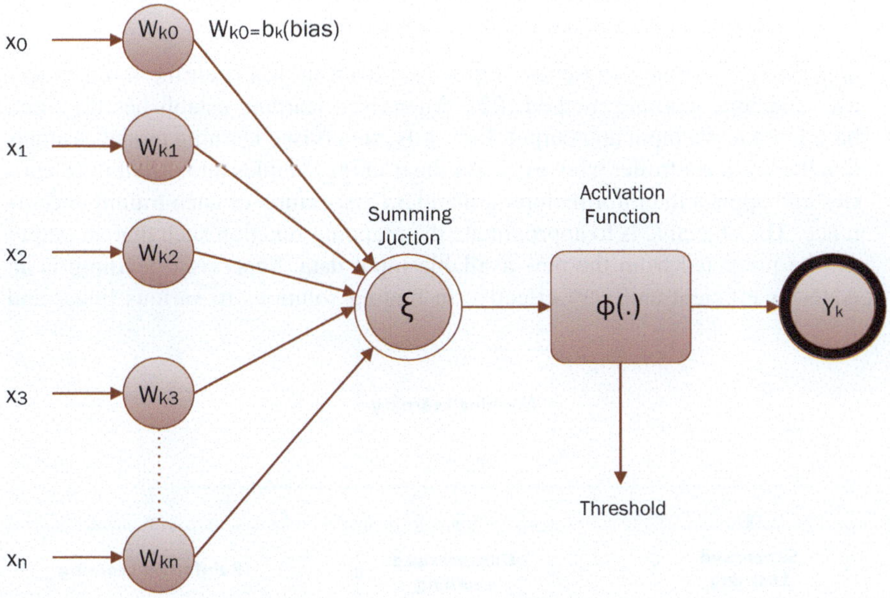

Fig. 4.2 Model of a single artificial neuron

Based on this, the output of Y_k is determined by the activation function v_k. Consider a threshold activation function whose value is less than 0 provided the summation of the input is less than threshold value (v) and 1 provided the input summation is above or the same as threshold value. This can be expressed as

$$\Phi(v) = \begin{cases} 0 & : v < 0 \\ 1 & : v \geq 0 \end{cases}$$

4.9 Paradigms of Machine Learning (ML)

ANN learning is a product of the adjustment of the network criteria that are adapted at ANN-embedded zone. The alteration process is very vital in the categorization of learning algorithms. These learning algorithms can give detailed insight into the pattern and properties of biomass by taking advantage of their different kinds. Learning algorithms are promoted by various learning principles such as error-correcting rule, memory-based rule, Q-learning rule, Boltzmann rule, competitive rule and Hebbian rule [81]. Principally, the machine learning paradigm can be classified into supervised learning, unsupervised learning and reinforced learning. This is shown in Fig. 4.3 [79, 80] and briefly discussed below.

1. *Supervised Learning*: Presently, supervised learning has been the most widely used machine learning method [82]. Supervised learning establishes the trend based on known input and output. Literarily, supervised learning model assumes that there is a controller who organizes the training samples into different categories and applies the information on the class association of each training occurrence. The objective is to approximate the mapping function such that an output can be predicted from the new available input data. Supervised learning is an ANN, is efficient and very effective in finding solutions to various linear and

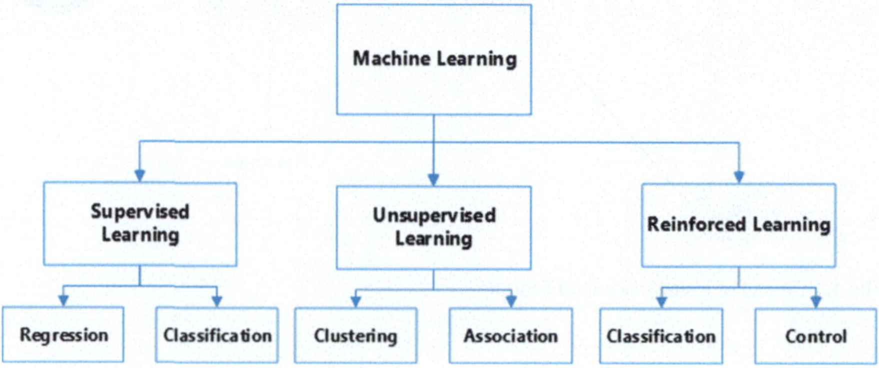

Fig. 4.3 Paradigm of machine learning

nonlinear problems which include biomass classification, plant control, forecasting, heating value prediction [45, 83–85], elemental analysis [86] biomass estimation, modelling of ecological data [82] bioprocess optimization, heat output of biomass boiler, etc. [68, 82, 87–91]. As shown in Fig. 4.3, supervised learning problems can be further divided into regression and classification. Some common cases of supervised machine learning algorithms include linear regression for regression problems [92], random forest for classification [93] and support vector machines for classification problems [24, 94].

2. *Unsupervised Learning*: It denotes the capacity to learn and categorize information even when there is no error signal to estimate the likely solution. As against supervised learning, the neural network is expected to find the pattern on its own in an unsupervised learning. In this learning instance, there is only input data without any corresponding target variable. So, the objective of an unsupervised machine learning is to model the underlying architecture in such a way to get further information about the data. Unsupervised learning can be further divided into clustering and association problems. Literarily, unsupervised learning can be likened to a learning process without teacher or correct answer. Although it is believed that unsupervised learning lacked direction, this lack of direction can be of advantage in some cases since it allows the algorithm to establish patterns that have not been formerly considered [95]. Some common instances of unsupervised learning algorithms are principal component analysis (PCA) [96, 97], K-means for clustering problems [98–100] and apriori algorithm for association rule learning problems. *K*-means clustering is a method whereby data are partitioned into sets based on the information found in data that defines the objects and associations among them, their feature values which can be used in many applications [101, 102]. Further details on *K*-mean clustering can be obtained from [102–105]. Apriori algorithm for association rule learning problems attempts to operate on database records. It applies 'bottom-up approach' to incrementally compare complex data set and it is useful in today's complex artificial intelligence applications which include the biomass. The apriori algorithm may be used in concurrence with other algorithms to efficiently sort and contrast data to show an improved picture of how complex systems reflect patterns and trends. It is computationally costly so much so that the computation time can extend to several days for a huge data set [106].
3. *Reinforced Learning (RL)*: Reinforced learning (RL) belongs to the field of machine learning which investigate how the historical data can be best applied in order to enhance future manipulation of a system [107]. Reinforcement learning automatically allows the ANN to determine the ideal behaviour under a specific condition or learning environment. It should be noted that RL is based on trial-and-error interaction within certain environment. Its unique feature lies in the fact that there is no exhibition of output or input data sets [107, 108]. Another difference is that the assessment of the system is often synchronized with learning; therefore, online performance is necessary. There are two principal approaches to solving reinforcement-learning problems. The first requires the

searching in the space of behaviours in order to perform well under given data conditions. This method has been used in genetic algorithms (GA) and genetic programming (GP) as well as some more novel search methods [109]. The other technique is based on statistical and dynamic programming techniques [109]. Indeed, given the recent dramatic progress in reinforcement learning, a tremendous opportunity lies in deploying its data-driven systems to the prediction of biomass properties. It should be noted that there is still a lot of research that needs to be done for RL to become a mainstream technique in biomass energy prediction. Deep reinforced learning is proving decisive in addressing associated challenges with multiple learning, information storage, and effective data exploration [110–113].

Different ML methods have been explored in biomass-related models with applications ranging from biomass estimation [114, 115], soil moisture retrieval [116], forest carbon mapping [93], key state variable prediction for bioprocess [117], density and viscosity of biofuel [118], prediction of syngas composition during biomass gasification [119], high heating value (HHV) prediction [9, 68, 97, 120, 121], prediction of elemental composition [51] and so on. Most of these researches were based on supervised learning with quite a few applications of unsupervised learning and reinforced learning. The progress in other class of machine learning would infiltrate into the prediction of thermal properties and other vital properties of biomass.

4.10 Sources of Noise in Biomass Data

Biomass data acquired from several sources are prone to noises due to their heterogeneous properties. Noise can influence the model either positively or negatively in terms of error rate, precision, computation time and so on. The problems associated with learning in noisy environments have been the focus of serious attention in ML, so much that most of the inductive learning algorithms were designed with noise-handling mechanisms [122, 123]. Noise handling before hypothesis formulation in data processing ensured that the noise does not affect the hypothesis construction; therefore, data cleansing is performed through a logical solution [124]. Noise should not always be bad; in some instance, deliberate introduction of noise to a data set may be very useful. For instance, noise can be introduced to a training model in order to smear out some data point and remove overfitting [125]. Manual cleansing of life biomass data is not pragmatic or efficient because it very laborious, time-consuming and error-prone. Automation of data purging process is the most practical and cost-effective way to achieve appreciable qualitative data set [126]. In an attempt to enhance the data quality, several approaches have been evolved in data prepossessing [126–128] and to deal with noisy data [124, 127, 129, 130]. The noise source in data analytics can be classified into [131] attribute noise and class noise. The significance and sources of the noise in biomass data have been discussed by

Obafemi et al. [8]. Suffice to say that attribute noise in biomass data is due to erroneous, missing or incomplete values, while class noise may be due to misclassified data or conflicting instances.

4.11 Peculiarities of Activation Functions

Activation function is a decision-making function, which has the capability to deliver the output based on input data. It is very important in ANN learning since it converts the activation level of neurons into an output [89, 132, 133]. Table 4.5 shows standard activation functions that are often used in ANN. Among other things, activation function determines the relevance of the incoming information to the neuron under set conditions. In the prediction of biomass properties, some authors have applied various activation functions, though there are still some others which have not been applied.

4.12 Classification of ANN

Table 4.6 briefly describes various types of ANN with their characteristics. Also, their applications related to biomass heating value prediction are referenced.

4.13 Training Algorithm

ANN has been classified and implemented based on mathematical operation and the set of parameters required to generate the output [8, 62]. Table 4.7 presents different kinds of training algorithms, some of which have been used in biomass heating value prediction and for some other purposes.

4.14 Some Common Terms in ANN

1. Algorithms: These are set of rules given to Neural Network (NN) in order to ensure that it can learn independently. It may be in the form of expressions or programming commands which allow NN to solve a problem intelligently. The algorithms may be developed for data clustering, classification, recommendation or regression.
2. *Classification*: This can be defined as an arrangement of data according to their observed similarities or proximity. The algorithm allows the NN to allot a class to a data point based on model generated from training.

Table 4.5 Typical activation function and their properties [134, 135]

Activation function	Graph	First derivative $f'(x)$	Output $f(x)$	Span
Linear	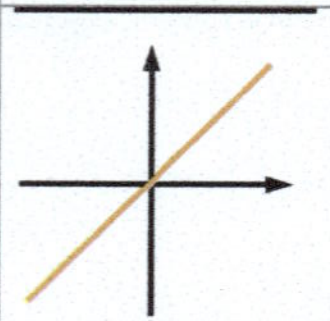	c	$x \times c$	$-\infty < f(x) < \infty$
Sigmoid	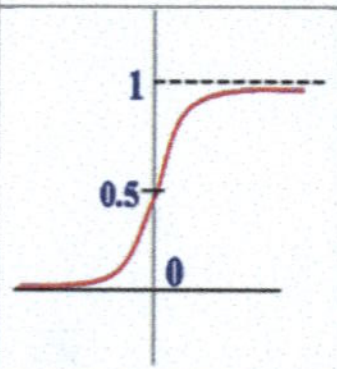	c	$\frac{1}{\left(1+e^{(-2cx)}\right)}$	$0 < f(x) < 1$
Sigmoid stepwise	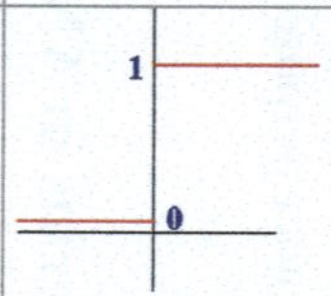	Same as sigmoid	Same as sigmoid but less precise	Same as sigmoid
Sigmoid symmetric activation	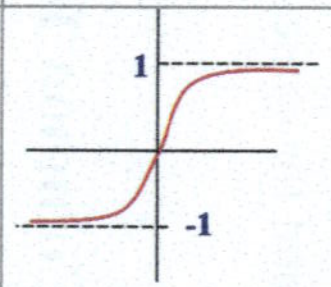	$\frac{c}{1-f(x)^{-2}}$	$2(1 + e^{2cx}) - 1$	$-1 < f(x) < 1$
Gaussian activation function	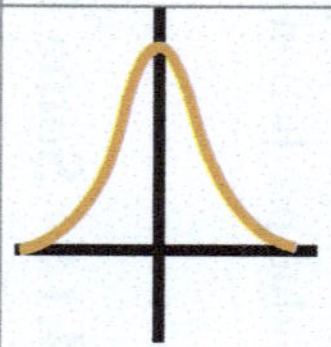	$-2x(f(x))c^2$	$y = e^{(-x^2} \times c^2)$	$0 < f(x) < 1$

Gaussian symmetry	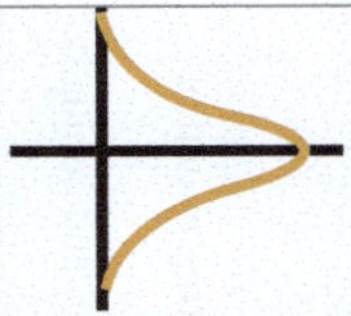	$-2xc^2(f(x) + 1)$	$2e^{(-c^2x^2)} - 1$	$-1 < f(x) < 1$
Sigmoid symmetry stepwise	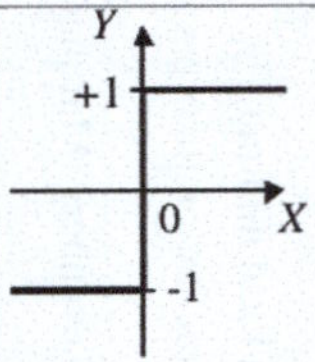	Same as sigmoid stepwise	Piecewise linear approximation of tanh sigmoid function. Faster than symmetry but less precise.	$-1 < f(x) < 1$
Elliot	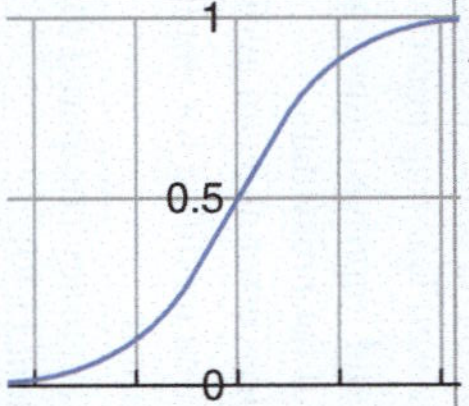	$\frac{c}{2\left[1+abs\left(x^2c^2\right)\right]}$	$y = \frac{cx}{2\left(1+abs\left(x+c\right)\right)} + 0.5$	$0 < f(x) < 1$
Saturating linear	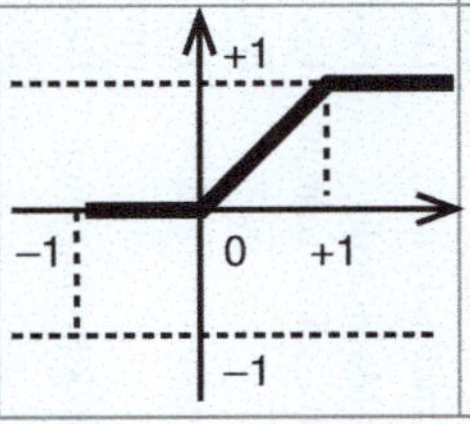	c	cx	$0 < f(x) < 1$

Table 4.6 Classes of ANNs

Networks	Characteristics	Reference example
Kohonen	Multilayered, non-recurrent, unsupervised	
Hopfield	Non-multilayered, recurrent, supervised	
Radial basis function (RBF)	It can be applied to a supervised learning problem	[136]
Generalized regression networks	RBF for function approximation can be varied	[32]
Boltzmann machine	It makes a stochastic decision about whether to be on or off	No known used
Feedforward backpropagation	Supervised, multilayer, non-recurrent	[53, 57, 137–139]

Table 4.7 Training algorithm methods

Training algorithm	Characteristics
Qasi -Newton method [140]	It builds up an approximation to the inverse hessian at every iteration, which is computed using the information on the first derivatives of the loss function.
Gradient descent method [141]	Not path-driven but population-driven. High iteration time. Likely to be trapped in local minima.
Newton's method [142, 143]	Quite expensive in terms of computation. Better training direction with second derivative of loss function
Scaled conjugate gradient [144, 145]	Faster convergence than steepest descent.
Levenberg–Marquardt [142]	Can calculate gradient and Jacobian matrix Efficient for small- and medium-sized data High convergence speed Not able to avoid local minimum
Modified Levenberg–Marquardt method [146]	Good convergence. Reduced amount of oscillation in learning. Decreased learning iteration.
Levenberg–Marquardt trained with cuckoo search [147].	Fast and improved convergence. Useful for noisy data Reduces error and can avoid local minima
Genetic Algorithm for RBF [148]	It can be modified for implementation incorporating mutation operation
Resilient back propagation [149]	Best accuracy compared to Levenberg–Marquardt and conjugate gradient [150].
Hybrid learning (least square and gradient descent) [151]	Combines gradient rule and least squares estimate.

3. Clustering: Clustering is a classification method used to classify groups of similar objects in a multivariate data set gathered from fields because similarity of objects within a cluster plays a significant role. The objective of a good clustering technique is to provide optimal similarity among its objects. This similarity may be determined by the distance among the data set [152]. Many clustering methods use distance functions to determine the resemblance or otherwise between couple of objects [153]. Generally, data clustering can be sectioned into hierarchical and partitional [154].
4. Training Data: These are data which are applied to estimate model parameters or showing instances in data distribution such that the resulting model is able to reasonably generalize the behaviour of the data when supplied with new sets which have not been previously used. Training generally aims to reach convergence of a metric, such as accuracy or a cost function, to a stable value as the algorithm sees more data. If the algorithm converges, that indicates that showing the agent more data does not improve the performance any longer.
5. Testing Data: In some cases, it can also be referred to as validation data. These data can be used to get an independent assessment of the model efficacy. These data should be clearly demarcated and not be used during model training.
6. Hidden Layer: It is the intermediate layer where artificial neurons take on a weight input to produce an output through an activation function output layer. The hidden layer is the engine room of ANN since it contains the transfer functions, weight, bias and training functions.
7. Output Layer: The output layer in an artificial neural network is the last layer of neurons that produces given outputs for the program.
8. Overfitting: This is one of the major problems which may be observed during NN training. Overfitting occurs due to the inability of the training data and assumption to reach generalization level. This means the algorithms perform well such that the error is driven to the lowest level, but it performs poorly when a new data set is presented to it [155]. Overfitting may be aggravated or reduced by the presence of noise [156, 157], like the noise which is present in biomass feedstock of municipal waste origin.
9. Underfitting: It is an attribute of a model which can neither model the training data nor generalize with the new data set presented to it. Regularization methods can be used to rectify overfitting and underfitting through the optimization of the number of hidden layers and the training time.
10. Generalization: The goal of training an ML algorithm is to be able to generalize a hypothesis based on the examples presented to it. Remembering the training data is not hard, but the algorithm should be able to apply learned knowledge to previously unseen data. To do this generalization, data alone are not enough [12]. Without any assumptions outside of the training data, it is impossible to train a learner who performs better than random guessing.
11. Assumption: When designing a system or requirements for a system, different assumptions are made either explicitly or implicitly. A violation of these underlying assumptions could potentially invalidate any analysis and verification

done. To prevent unexpected violations, the assumption made during the model development must be clearly stated.

12. Normalization: In data analytics, normalization is done to improve the coherence of the data entities. The objective of data normalization is to reduce and even get rid of data redundancy since this is the main consideration for application developer. The opposite of data normalization is un-normalization. At this stage, the data are returned to its original form for data analysis and reportage.

4.15 Data Division Methods for ANN Modelling

Most of the ANN-based modelling for biomass usually bases their data division on random selection with no defined criteria. Perhaps, this is so due to its simplicity [158]. Optimal division of data into a training data set, validation set, and an independent test subset is a vital procedure in ANN modelling for complex data analysis. Different authors have proven that data division can have a significant impact on ANN performance [159, 160]. Usually, an ANN model can only interpolate data that fall within the range of data set used for training. So, if the testing data set does not sufficiently fall with the range of training data set, poor prediction model may have been developed [161]. Also, data division is important in the case of an expensive experimental procedure, since much data may not be obtained to minimize the cost. The following data division methods have been identified:

I. Random selection [158, 162]
II. Korhonen self-organizing map (SOM) [159, 162]
III. D optima design [163]
IV. Genetic Algorithms [159]
V. Kennard–Stone algorithm [158, 164–166]
VI. Modified Kennard–Stone Algorithm [167]

Trial and error method is often applied in biomass data division for the training and testing of ANN models. This division method is without any underlining principles which may somewhat be of disadvantages, as the data division has been shown to have a significant influence on the accuracy of a model. An intelligent data division method should be formulated to further enhance the versatility of the ANN model (Table 4.8).

4.16 Sensitivity Analysis (SA)

Sensitivity analysis (SA), which is also known as what-if analysis, is a method which can be deployed to determine the comparative significance of input parameters and how it affects certain outputs under a given set of assumptions. The significance of sensitivity analysis has been well documented ever since the beginning of

Table 4.8 Reported data set division and division methods for ANN

Data division method	Training set %	Test set %	Validation set %	Total data size	Reference
Random method	60	20	20	55	[168]
Random method	60	20	20	Not stated	[31]
Not stated	Not stated	Not stated	Not stated	350	[169]
	50	25	25	444	[9]
Not stated	75	25	The same data set As testing	100	[136]
Not stated	70	15	15	350	[170]
Trial and error		Not stated	Not stated	Not stated	[171]
Trial and error	70	15	15	250	[172]
Not stated	67	33	Used the same data set for testing	131	[68]
Not stated	70	15	15	350	[97]

scientific investigation and addressed in mathematical solutions [173]. SA has been applied in the decision-making process, verification of the robustness of a solution, to identify the critical parameter in a process modelling or design, to identify sensitive or import variables to investigate sub-optimal solution, in comparative analysis of solutions in risk assessment of a strategy, in error verification, in model calibration and simplification [174] and so on.

4.16.1 Methods for Sensitivity Analysis

Several approaches have been adopted to measure the sensitivity, which include the following:

I. Net weight matrix [175]
II. Backward stepwise [176]
III. Partial derivation method
IV. Perturbation method
V. The profile method [28]

Also, the advantages and disadvantages of each sensitivity analysis method are discussed in Table 4.9.

In order to underline the significance of SA in biomass exploration, the optimal design and global what-if analysis of biomass supply chain network for biofuel was conducted under uncertainty conditions using Monte Carlo simulation [178]. Also, a detailed SA was performed on the fixed bed downdraft biomass gasification model. The essence was to analyse the effect of heat transfer mechanism and the rate of reaction on the progression of the reaction front along the bed at the main

Table 4.9 Advantages and disadvantages of SA methods

Sensitivity analysis method	Advantages	Disadvantages
Net weight matrix [175]	It is very easy to implement	Not effective in calculating comparative impact of input variables on the outputs
Stepwise	It allows a stepwise study of the influence of each input variables	It is not very suitable for ANN models
	Better understanding of model behaviour	
Partial derivation method [177]	Computationally efficient	Complex and intensive equation-solving approach which makes the implementation impractical
		Only valid for small change in the value of parameters
		More demanding to implement compared to other techniques
Perturbation method [28]	It is widely used	Subjective scope in noise addition
		Inconsistency and non-uniform definition of cost function
The profile method [28]	This method is stable irrespective of the scale	Not as coherent as partial derivative method in computation

reaction phases [179]. Overall, the partial derivation method is the most applied in ANN, partly due to its ability to compute the effect of the input on the output value.

4.17 Loss Functions in ANN

The knowledge of the past result is vital for accurate learning. The loss function is used to evaluate the consistency between the predicted value and actual value [180]. The robustness of the model is improved as the value of the loss function decreases. The idea is to run the data points forward through the network and then notice the change in output. The parameter is changed based on the derivatives of the network. Moreover, the process is repeated until the lowest possible error is attained in such a way that the overfitting is avoided [180]. Different loss functions will give varying errors for the same prediction; therefore, it can be concluded that loss function has a weighty consequence on the behaviour of a model. It should be noted that there is no one-size-fits-all loss function for algorithms in ANN [181]. Broadly speaking, loss functions can be categorized into two major classes based on the learning task we are solving. These are regression losses and classification losses.

The following loss functions have been identified [180, 181]:

I. Mean squared error (MSE)
II. Mean squared logarithmic error (MSLE)
III. Mean absolute deviation (MAD)

IV. Mean absolute percentage error (MAPE)
V. Mean bias error (MBE)
VI. Kullback–Leibler (KL) divergence
VII. Cross-entropy
VIII. Poisson
IX. Cosine proximity
X. Max-margin
XI. Square hinge loss

4.18 Stages in ANN Modelling

The stages in ANN model development can be illustrated with the following (Fig. 4.4):

I. Preliminary development of models involves data collection, normalization of data, data division into training and testing set. However, data division may also

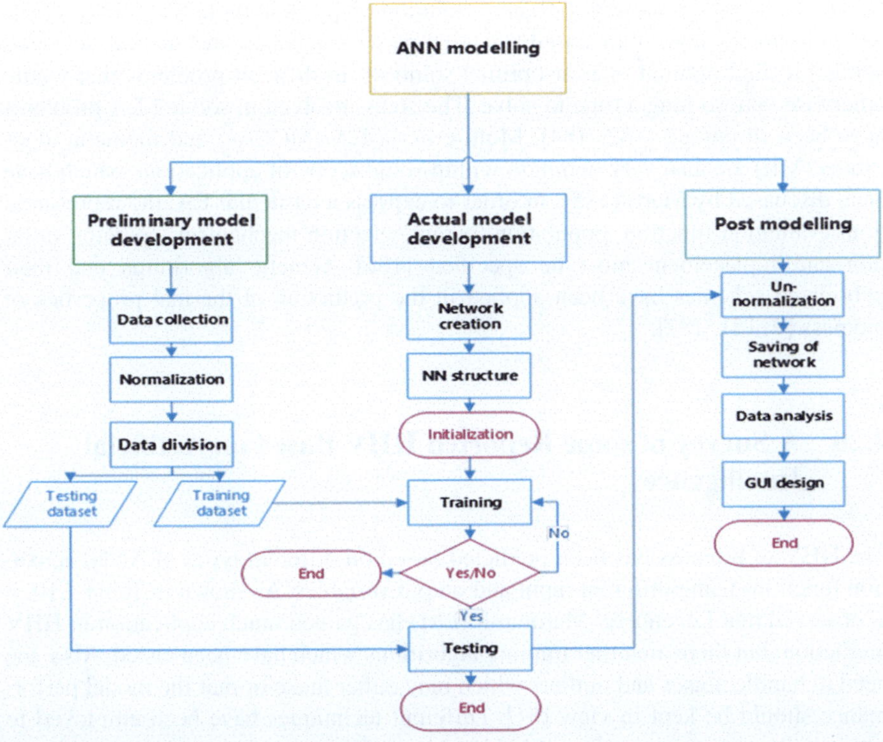

Fig. 4.4 Stages in ANN model development

fall under actual model development in a case where the software is designed to randomly divide the data.

II. Actual Model Development: At this stage, network is being created after initialization, and the structure which is expected to be most suitable for the data type is selected, while the training and testing are carried out and loss functions are calculated to determine the accuracy of the model developed.

III. Post-Modelling: The data are un-normalized by returning it to the initial state before normalization, then the network can be saved, following a successful analysis; a graphical user interface (GUI) may be developed in a case where the model is designed for specific application or prediction.

4.19 Application of Evolutional Algorithm for Biomass Properties Prediction

Historically, the design of evolutionary algorithms was promoted by natural observation of biological evolution. EA belongs to the set of memetic heuristic-based method. Meta-heuristic algorithm is governed by the concept of population [182, 183]. The EA can be subdivided into evolutionary strategies, genetic algorithms (GA), genetic programming (GP) and evolutionary programming (EP) [184]. GA is an evolutionary algorithm based on principles of genetics and natural selection applied to find optimal or near-optimal solutions to difficult problems that would otherwise take so long a time to solve. The steps involved in solving EA problems have been discussed [182, 184]. Multi-criteria EA (MCEAs) and memetic algorithms (ME) are also very common within some scope of applications which have been discussed by Marti [185]. In order to express a particular EA, the representation, evaluation function, population, parent selection mechanism, operator variation and replacement must be specified [186]. Genetic algorithms and their hybridized schemes have been applied in the prediction of thermal properties of biomass [9, 151, 187].

4.20 A Survey of some Reported HHV Based on Artificial Intelligence

The HHV of biomass has been predicted based on different types of ANN, activation functions using different input and output variables. As shown in Table 4.10, it is observed that Levenberg–Marquardt (LM) has gained much application in HHV prediction, but there are other training algorithms which have been tested. Also, the need to handle noises and outliers which may either make or mar the model performance should be kept in view [51]. Different techniques have been employed to estimate the optimum number of hidden layers. However, they are mostly tedious,

Table 4.10 ANN-based HV models [8]

Input variables	Output variables	Type of ANN	R^2	ANN Architecture	Activation functions	Reference
Fatty acid feedstock	HV	LM backpropagation	0.997	5-2-4	Linear and sigmoid symmetry	[168]
%C, %H, %O, %N, %S on the dry basis	HHV	LM backpropagation	0.9231	5-4-1	Linear and sigmoid symmetry	[31]
FC, VM, MC, A	HHV	LM	0.9852	3-7-1	Sigmoid symmetric	[169]
FC, VM, MC, A	HHV	LM	0.9591	1-23-1-1 1-21-1-1 1-25-1-1	Hyperbolic tangent sigmoid and linear	[170]
Heating rate, blending ratio and mass loss (%)	Gross heating value (GHV)	LM backpropagation	0.9959	3-3-19-1	Not stated	[172]
C, H_2O, O, H, N, S, Ash	HHV	Radial basis function combined with LM	0.997		Radial basis function	[136]
FC, VM, Ash	GHV	LM backpropagation	0.9478	3-12-1	Sigmoid	[9]
FC, Ash, MC	HHV	Linear	0.963 and 0.962	3-3-1 3-20-1	Sigmoid and sigmoid symmetry	[97]

with many uncertainties. Standard methods for the selection of hidden layers should be investigated to optimize the prediction process. By and large, detailed information about model development should be encouraged in order to foster clear understanding of the heating value (HV) of LCB.

4.21 Case Studies on the Application of AI for Properties of Biomass

4.21.1 Case Study 1: Improved Prediction of HHV of Biomass Using an ANN Model Based on Proximate Analysis

This study was conducted by Uzun et al. [68]. Several sigmoid architectures were constructed to determine the HHV of biomass feedstock based on their proximate values and to determine the optimal model. Data were collected from the literature for different classes of biomass. In order to execute the model, the proximate analysis data set was divided such that 67% was used for training and 33% for testing. The sigmoid transfer function was applied at the hidden layer while linear transfer function was used at the output layer. This modelling was implemented on MATLAB 2015 software. The network architecture and performance metrics are presented in Table 4.11. The minimum error was reported for network ANN24 with tangent sigmoid transfer function and 3-20-1 architecture.

4.21.2 *Case Study 2: Application of ANFIS-PSO Algorithm as a Novel Method for Prediction of HHV of Biomass*

This study which is purely computational was carried out by Suleymani et al. [187] using 350 experimental data harvested from several literatures. Data were divided at a ratio of 4:1 for training and testing of the model, respectively [49, 50]. In this investigation, particle swarm optimization (PSO) was hybridized with adaptive neuro-fuzzy inference system (ANFIS). The statistical indices, which show the prediction errors, are presented in Table 4.12.

Table 4.11 Network architecture and performance metrics

Model	Hidden layer	Network	RMSE	MAD	MBE	R^2
ANN-24	Tangent sigmoid	3-20-1	0.375	0.328	0.01	0.963
ANN-8	Logistic sigmoid	7-1	0.392	0.344	0.016	0.962
ANN-38	Logistic sigmoid	5-20-1	0.439	0.369	0.064	0.958
ANN-25	Logistic sigmoid	3-25-1	0.413	0.352	0.054	0.953
ANN-6	Logistic sigmoid	5-1	0.419	0.347	0.003	0.953
ANN-16	Logistic sigmoid	3-5-1	0.416	0.353	0.041	0.952
ANN-36	Logistic sigmoid	5-15-1	0.451	0.367	0.075	0.950
ANN-5	Tangent sigmoid	5-1	0.454	0.382	0.015	0.944
ANN-14	Tangent sigmoid	3-3-1	0.471	0.413	0.025	0.939
ANN-27	Tangent sigmoid	3-30-1	0.484	0.411	0.041	0.938
ANN-43	Tangent sigmoid	7-7-1	0.479	0.411	0.046	0.936
ANN-45	Tangent sigmoid	7-10-1	0.413	0.159	0.065	0.749
ANN-23	Tangent sigmoid	3-20-1	0.419	0.201	0.037	0.636
ANN-42	Logistic sigmoid	5-30-1	0.47	0.264	0.069	0.466

Table 4.12 Evaluation criteria of ANFIS-PSO algorithm

	Training	Testing	Overall
RMSE	1.171	1.228	1.179
MAD	5.179	5.767	5.266
R^2	0.911	0.886	0.908

The PSO was utilized for optimization of ANFIS and for augmentation of the precision of the predicting model. The outcome disclosed that the proposed algorithm is applicable in the estimation of HHV with great accuracy. However, the author did not highlight some vital statistical parameters such as MAPE and accuracy ratio (AR) which could determine the coverage of the model.

4.22 Conclusion

Application of intelligent methods for the computation of LCB properties promises to enhance the value addition process. ANN models and other intelligent optimization methods have been reported. Different rules have been proposed to estimate the optimum number of hidden nodes. However, they are mostly tedious, with much uncertainty. Standard methods for the selection of hidden layers should be investigated to fast track the prediction process. Further research should lead to the development of a robust stand-alone and open access software, which can be used to predict the main properties of LCB.

References

1. C.-Y. Yin, Prediction of higher heating values of biomass from proximate and ultimate analyses. Fuel **90**(3), 1128–1132 (2011)
2. S. Ghugare, S. Tiwary, V. Elangovan, S. Tambe, Prediction of higher heating value of solid biomass fuels using artificial intelligence formalisms. Bioenergy Res. **7**(2), 681–692 (2014)
3. O.O. Olatunji, S.A. Akinlabi, M.P. Mashinini, S.O. Fatoba, O.O. Ajayi, Thermo-gravimetric characterization of biomass properties: A review. IOP Conf. Ser. Mater. Sci. Eng. **423**(1), 012175 (2018)
4. O. Obafemi, A. Stephen, O. Ajayi, P. Mashinini, M. Nkosinathi, Experimental investigation of thermal properties of lignocellulosic biomass: A review. IOP Conf. Ser. Mater. Sci. Eng. **413**(1), 012054 (2018)
5. Harsh, D.S.K. Singal, Integration of renewable energy sources using artificial intelligent system. Int. J. Innov. Res. Sci. Eng. Technol. **03**(11), 17291–17305 (2014)
6. C.H. Dagli, *Artificial Neural Networks for Intelligent Manufacturing* (Springer, Dordrecht, 2012)
7. B.-H. Li, B.-C. Hou, W.-T. Yu, X.-B. Lu, C.-W. Yang, Applications of artificial intelligence in intelligent manufacturing: A review. Front. Inf. Technol. Electron. Eng. **18**(1), 86–96 (2017)
8. O. Obafemi, A. Stephen, O. Ajayi, M. Nkosinathi, A survey of artificial neural network-based prediction models for thermal properties of biomass. Proc. Manuf. **33**, 184–191 (2019)
9. E. Akkaya, ANFIS based prediction model for biomass heating value using proximate analysis components. Fuel **180**, 687–693 (2016)
10. F.M. Gray, M. Schmidt, A hybrid approach to thermal building modelling using a combination of Gaussian processes and grey-box models. Energy Build. **165**, 56–63 (2018)
11. International Energy Agency, Digitalization and Energy (2017), http://www.ia.org/digital/. Accessed 12 Sept 2018
12. S. Finlay, *Predictive Analytics, Data Mining and Big Data: Myths, Misconceptions and Methods* (Springer, London, 2014)

13. Technopedia, Artificial Neural Network (2018), https://www.techopedia.com/definition/5967/artificial-neural-network-ann. Accessed 30 Mar 2018
14. D. Rizopoulos, Max Kuhn and Kjell Johnson. Applied predictive modeling. New York, Springer. Biometrics **74**(1), 383–383 (2018)
15. M. Kuhn, K. Johnson, *Applied Predictive Modeling* (Springer, New York, 2013)
16. S.J. Cranmer, B.A. Desmarais, What can we learn from predictive modeling? Polit. Anal. **25**(2), 145–166 (2017)
17. B. Ratner, *Statistical and Machine-Learning Data Mining: Techniques for better Predictive Modeling and Analysis of Big Data* (Chapman and Hall/CRC, Boca Raton, 2017)
18. E. Siegel, *Predictive Analytics: The Power to Predict Who Will Click, Buy, Lie, or Die* (Wiley, Hoboken, 2013)
19. J.I. Ebert, The state of the art in "inductive" predictive modelling: Seven big mistakes (and lots of smaller ones), in *Practical Applications of GIS for Archaeologists: A Predictive Modelling Toolkit*, (Taylor and Francis, London, 2000), pp. 129–134
20. E. Vonesh, V.M. Chinchilli, *Linear and Nonlinear Models for the Analysis of Repeated Measurements* (CRC Press, Boca Raton, 1996)
21. J. Pinheiro, D. Bates, S. DebRoy, D. Sarkar, Linear and nonlinear mixed effects models. R Package Version **3**, 1–117 (2014)
22. P.A. Adedeji, S. Akinlabi, O. Ajayi, N. Madushele, Non-linear autoregressive neural network (NARNET) with SSA filtering for a university energy consumption forecast. Proc. Manuf. **33**, 176–183 (2019)
23. G. Guo, H. Wang, D. Bell, Y. Bi, K. Greer, KNN model-based approach in classification, in *OTM Confederated International Conferences On the Move to Meaningful Internet Systems*, (Springer, Berlin, 2003), pp. 986–996
24. M. Pal, Multiclass approaches for support vector machine based land cover classification. *arXiv preprint arXiv:0802.2411* (2008)
25. C.-W. Hsu, C.-J. Lin, A comparison of methods for multiclass support vector machines. IEEE Trans. Neural Netw. **13**(2), 415–425 (2002)
26. S. Konishi, G. Kitagawa, Various model evaluation criteria. Inf. Crit. Stat. Model. **2008**, 239–254 (2008)
27. A.Z. Hatem Ksibi, Numerical optimization of biogas production through a 3-steps model of anaerobic digestion: Sensitivity of biokinetic constants values. J Bioremediat. Biodegrad. **06**(04), 302 (2015)
28. R. Bai, H. Jia, P. Cao, Factor sensitivity analysis with neural network simulation based on perturbation system. J. Comput. **6**(7), 1402–1407 (2011)
29. R. Mac Nally, R.P. Duncan, J.R. Thomson, J.D. Yen, Model selection using information criteria, but is the "best" model any good? J. Appl. Ecol. **55**(3), 1441–1444 (2018)
30. C. Huang, L. Han, Z. Yang, X. Liu, Ultimate analysis and heating value prediction of straw by near infrared spectroscopy. Waste Manag. **29**(6), 1793–1797 (2009)
31. L.K. Abidoye, F.M. Mahdi, Novel linear and nonlinear equations for the higher heating values of municipal solid wastes and the implications of carbon to energy ratios. J. Energy Technol. Policy **4**(5), 14–27 (2014)
32. E. Tosun, K. Aydin, M. Bilgili, Comparison of linear regression and artificial neural network model of a diesel engine fueled with biodiesel-alcohol mixtures. Alex. Eng. J. **55**(4), 3081–3089 (2016)
33. A. Abbas, A. Ibrahim, M. Muttalib, M. Aris, Fuel characterization and energy prediction of Malaysian poultry processing. Asian J. Sci. Res. **6**(3), 498–507 (2013)
34. L. Meraz, A. Domínguez, I. Kornhauser, F. Rojas, A thermochemical concept-based equation to estimate waste combustion enthalpy from elemental composition☆. Fuel **82**(12), 1499–1507 (2003)
35. P. Thipkhunthod, V. Meeyoo, P. Rangsunvigit, B. Kitiyanan, K. Siemanond, T. Rirksomboon, Predicting the heating value of sewage sludges in Thailand from proximate and ultimate analyses. Fuel **84**(7–8), 849–857 (2005)

36. R. Elneel, S. Anwar, B. Ariwahjoedi, Prediction of heating values of oil palm fronds from ultimate analysis. J. Appl. Sci. **13**(3), 491–496 (2013)
37. D.A. Tillman, *Wood as an Energy Resource* (Elsevier, Saint Louis, 2012)
38. A. Demirbas, D. Gullu, A. Caglar, F. Akdeniz, Estimation of calorific values of fuels from lignocellulosics. Energy Sources **19**(8), 765–770 (1997)
39. A. Demirbaş, Relationships between lignin contents and heating values of biomass. Energy Convers. Manag. **42**(2), 183–188 (2001)
40. C. Sheng, J.L.T. Azevedo, Estimating the higher heating value of biomass fuels from basic analysis data. Biomass Bioenergy **28**(5), 499–507 (2005)
41. S. Channiwala, P. Parikh, A unified correlation for estimating HHV of solid, liquid and gaseous fuels. Fuel **81**(8), 1051–1063 (2002)
42. B.M. Jenkins, J.M. Ebeling, Correlation of physical and chemical properties of terrestrial biomass with conversion, in *Symposium Papers-Energy from Biomass and Wastes*, (Institute of Gas Technology, Chicago, 1985)
43. W.G. Lloyd, D.A. Davenport, Applying thermodynamics to fossil fuels: Heats of combustion from elemental compositions. J. Chem. Educ. **57**(1), 56 (1980)
44. W. Boie, Fuel technology calculations. Energietechnik **3**, 309–316 (1953)
45. K. Phichai, P. Pragrobpondee, T. Khumpart, S. Hirunpraditkoon, Prediction heating values of lignocellulosics from biomass characteristics. Int. J. Chem. Mol. Nucl. Mater. Metall. Eng. **7**(7), 532–535 (2013)
46. L. Jiménez, F. González, Study of the physical and chemical properties of lignocellulosic residues with a view to the production of fuels. Fuel **70**(8), 947–950 (1991)
47. A. Demirbaş, Calculation of higher heating values of biomass fuels. Fuel **76**(5), 431–434 (1997)
48. T. Cordero, F. Marquez, J. Rodriguez-Mirasol, J. Rodriguez, Predicting heating values of lignocellulosics and carbonaceous materials from proximate analysis. Fuel **80**(11), 1567–1571 (2001)
49. J. Parikh, S. Channiwala, G. Ghosal, A correlation for calculating HHV from proximate analysis of solid fuels. Fuel **84**(5), 487–494 (2005)
50. D.R. Nhuchhen, P. Abdul Salam, Estimation of higher heating value of biomass from proximate analysis: A new approach. Fuel **99**, 55–63 (2012)
51. O.O. Olatunji, S. Akinlabi, N. Madushele, P.A. Adedeji, Estimation of the elemental composition of biomass using hybrid adaptive neuro-fuzzy inference system. Bioenergy Res. **12**, 1–11 (2019)
52. O. Olatunji, S. Akinlabi, N. Madushele, P.A. Adedeji, Estimation of Municipal Solid Waste (MSW) combustion enthalpy for energy recovery. EAI Endorsed Trans. Energy Web **6**(23), 159119 (2019)
53. S.B. Ghugare, S.S. Tambe, Genetic programming based high performing correlations for prediction of higher heating value of coals of different ranks and from diverse geographies. J. Energy Inst. **90**(3), 476–484 (2017)
54. G. Kanat, A. Saral, Estimation of biogas production rate in a thermophilic UASB reactor using artificial neural networks. Environ. Model. Assess. **14**(5), 607–614 (2009)
55. B. Ozkaya, A. Demir, M.S. Bilgili, Neural network prediction model for the methane fraction in biogas from field-scale landfill bioreactors. Environ. Model Softw. **22**(6), 815–822 (2007)
56. J. Wang, W. Wan, Application of desirability function based on neural network for optimizing biohydrogen production process. Int. J. Hydrog. Energy **34**(3), 1253–1259 (2009)
57. H. Abu Qdais, K. Bani Hani, N. Shatnawi, Modeling and optimization of biogas production from a waste digester using artificial neural network and genetic algorithm. Resour. Conserv. Recycl. **54**(6), 359–363 (2010)
58. D.P. Strik, A.M. Domnanovich, L. Zani, R. Braun, P. Holubar, Prediction of trace compounds in biogas from anaerobic digestion using the MATLAB Neural Network Toolbox. Environ. Model Softw. **20**(6), 803–810 (2005)

59. L. Alejo, J. Atkinson, V. Guzmán-Fierro, M. Roeckel, Effluent composition prediction of a two-stage anaerobic digestion process: Machine learning and stoichiometry techniques. Environ. Sci. Pollut. Res. **25**, 1–15 (2018)
60. A. Tardast et al., Use of artificial neural network for the prediction of bioelectricity production in a membrane less microbial fuel cell. Fuel **117**, 697–703 (2014)
61. A. Tardast, M. Rahimnejad, G. Najafpour, K. Pirzade, N. Mokhtarian, Prediction of bioelectricity production by neural network. J. Biotechnol. Pharm. Res. **3**(3), 62–68 (2012)
62. Y. Sewsynker, E.B.G. Kana, A. Lateef, Modelling of biohydrogen generation in microbial electrolysis cells (MECs) using a committee of artificial neural networks (ANNs). Biotechnol. Biotechnol. Equip. **29**(6), 1208–1215 (2015)
63. Y. Sewsynker-Sukai, F. Faloye, E.B.G. Kana, Artificial neural networks: An efficient tool for modelling and optimization of biofuel production (a mini review). Biotechnol. Biotechnol. Equip. **31**(2), 221–235 (2017)
64. V.B. Furlong, R.D. Pereira Filho, A.C. Margarites, P.G. Goularte, J.A.V. Costa, Estimating microalgae Synechococcus nidulans daily biomass concentration using neuro-fuzzy network. Food Sci. Technol. **33**, 142–147 (2013)
65. N. Sarkar, S.K. Ghosh, S. Bannerjee, K. Aikat, Bioethanol production from agricultural wastes: An overview. Renew. Energy **37**(1), 19–27 (2012)
66. E. Betiku, A.E. Taiwo, Modeling and optimization of bioethanol production from breadfruit starch hydrolyzate vis-à-vis response surface methodology and artificial neural network. Renew. Energy **74**, 87–94 (2015)
67. M. Esfahanian, M. Nikzad, G. Najafpour, A.A. Ghoreyshi, Modeling and optimization of ethanol fermentation using Saccharomyces cerevisiae: Response surface methodology and artificial neural network. Chem. Ind. Chem. Eng. Q/CICEQ **19**(2), 241–252 (2013)
68. H. Uzun, Z. Yildiz, J.L. Goldfarb, S. Ceylan, Improved prediction of higher heating value of biomass using an artificial neural network model based on proximate analysis. Bioresour. Technol. **234**, 122–130 (2017)
69. G.E. Acquah, B.K. Via, O.O. Fasina, S. Adhikari, N. Billor, L.G. Eckhardt, Chemometric modeling of thermogravimetric data for the compositional analysis of forest biomass. PLoS One **12**(3), e0172999 (2017)
70. R. Guidotti, A. Monreale, S. Ruggieri, F. Turini, F. Giannotti, D. Pedreschi, A survey of methods for explaining black box models. ACM Comput. Surv. (CSUR) **51**(5), 93 (2018)
71. L. Ljung, Black-box models from input-output measurements, in *Instrumentation and Measurement Technology Conference, 2001. IMTC 2001. Proceedings of the 18th IEEE*, vol. 1, (IEEE, Piscataway, 2001), pp. 138–146
72. G. Reynders, J. Diriken, D. Saelens, Quality of grey-box models and identified parameters as function of the accuracy of input and observation signals. Energy Build. **82**, 263–274 (2014)
73. C.D. Prada Moraga, D. Hose, G. Gutierrez, J.L. Pitarch, Developing grey-box dynamic process models. IFAC PapersOnLine **51**(2), 523–528 (2018)
74. M.E. Khan, F. Khan, A comparative study of white box, black box and grey box testing techniques. Int. J. Adv. Comput. Sci. Appl. **3**(6), 12–15 (2012)
75. B.C. Bangal, Automatic Generation Control of Interconnected Power Systems Using Artificial Neural Network Techniques, 2009
76. R. Shariff, Q. Zhang, A. Cudrak, D. Smith, S. Stanley, *Real-Time Artificial Intelligence Control and Optimization of a Full-Scale WTP* (American Water Works Association, Denver, 2006)
77. S.J. Stanley, *Process Modeling and Control of Enhanced Coagulation* (American Water Works Association, Denver, 2000)
78. S.A. Kalogirou, *Artificial Intelligence in Renewable Energy Systems Modelling and Prediction* (Proceedings of the World Renewable Energy Congress VII, Cologne, 2002)
79. S. Haykin, *Neural Networks: A Comprehensive Foundation* (Prentice Hall PTR, New York, 1994)
80. S.S. Haykin, S.S. Haykin, S.S. Haykin, S.S. Haykin, *Neural Networks and Learning Machines* (Pearson, Upper Saddle River, 2009)

81. R. Sathya, A. Abraham, Comparison of supervised and unsupervised learning algorithms for pattern classification. Int. J. Adv. Res. Artif. Intell. **2**(2), 34–38 (2013)
82. C. Crisci, B. Ghattas, G. Perera, A review of supervised machine learning algorithms and their applications to ecological data. Ecol. Model. **240**, 113–122 (2012)
83. G.D. Gillespie, C.D. Everard, C.C. Fagan, K.P. McDonnell, Prediction of quality parameters of biomass pellets from proximate and ultimate analysis. Fuel **111**, 771–777 (2013)
84. A. Özyuğuran, S. Yaman, Prediction of calorific value of biomass from proximate analysis. Energy Procedia **107**, 130–136 (2017)
85. N. Soponpongpipat, D. Sittikul, P. Comsawang, Prediction model of higher heating value of torrefied biomass based on the kinetics of biomass decomposition. J. Energy Inst. **89**(3), 425–435 (2016)
86. S.B. Ghugare, S. Tiwary, S.S. Tambe, Computational intelligence based models for prediction of elemental composition of solid biomass fuels from proximate analysis. Int. J. Syst. Assur. Eng. Manag. **8**(4), 2083–2096 (2017)
87. I. Bayram, I.W. Selesnick, A dual-tree rational-dilation complex wavelet transform. IEEE Trans. Signal Process. **59**(12), 6251–6256 (2011)
88. L. Guan, Z. Fan, R. Tibshirani, Supervised learning via the "Hubnet" procedure. Stat. Sin. **28**, 1225–1243 (2018)
89. Q. Zhang, Y. Yang, Y. Liu, Y.N. Wu, S.-C. Zhu, Unsupervised learning of neural networks to explain neural networks. arXiv preprint arXiv **1805**, 07468 (2018)
90. X. Zhu, Semi-supervised learning literature survey. Comput. Sci. Univ. Wisconsin-Madison **2**(3), 4 (2006)
91. P. Tóth, A. Garami, B. Csordás, Image-based deep neural network prediction of the heat output of a step-grate biomass boiler. Appl. Energy **200**, 155–169 (2017)
92. A.Y. Sendjaja et al., Regression based state space adaptive model of two-phase anaerobic reactor. Chemosphere **140**, 159–166 (2015)
93. J. Mascaro et al., A tale of two "forests": Random forest machine learning aids tropical forest carbon mapping. PLoS One **9**(1), e85993 (2014)
94. Y. Liu, H. Wang, H. Zhang, K. Liber, A comprehensive support vector machine-based classification model for soil quality assessment. Soil Tillage Res. **155**, 19–26 (2016)
95. T. Kohonen, E. Oja, O. Simula, A. Visa, J. Kangas, Engineering applications of the self-organizing map. Proc. IEEE **84**(10), 1358–1384 (1996)
96. A.S. Hess, J.R. Hess, Principal component analysis. Transfusion **58**, 1580 (2018)
97. S. Hosseinpour, M. Aghbashlo, M. Tabatabaei, Biomass higher heating value (HHV) modeling on the basis of proximate analysis using iterative network-based fuzzy partial least squares coupled with principle component analysis (PCA-INFPLS). Fuel **222**, 1–10 (2018)
98. A. Ahmad, L. Dey, A k-mean clustering algorithm for mixed numeric and categorical data. Data Knowl. Eng. **63**(2), 503–527 (2007)
99. V. Prasath, H.A.A. Alfeilat, O. Lasassmeh, A. Hassanat, Distance and similarity measures effect on the performance of K-nearest neighbor classifier-a review. *arXiv preprint arXiv:1708.04321* (2017)
100. K. Teknomo, K-means clustering tutorial. Medicine **100**(4), 3 (2006)
101. P. Vora, B. Oza, A survey on k-mean clustering and particle swarm optimization. Int. J. Sci. Mod. Eng. **1**(3), 1–14 (2013)
102. A.K. Jain, M.N. Murty, P.J. Flynn, Data clustering: A review. ACM Comput. Surv. (CSUR) **31**(3), 264–323 (1999)
103. J.M. Lattin, J.D. Carroll, P.E. Green, *Analyzing Multivariate Data* (Thomson Brooks/Cole, Pacific Grove, 2003)
104. A.K. Jain, Data clustering: 50 years beyond K-means. Pattern Recognit. Lett. **31**(8), 651–666 (2010)
105. S. Azizpour, K. Giesecke, G. Schwenkler, Exploring the sources of default clustering. J. Financ. Econ. **129**(1), 154–183 (2018)

106. Z.K. Baker, V.K. Prasanna, Efficient hardware data mining with the Apriori algorithm on FPGAs, in *Field-Programmable Custom Computing Machines, 2005. FCCM 2005. 13th Annual IEEE Symposium on*, (IEEE, Piscataway, 2005), pp. 3–12
107. R.S. Sutton, A.G. Barto, *Introduction to Reinforcement Learning* (MIT Press, Cambridge, 1998)
108. L.P. Kaelbling, M.L. Littman, A.W. Moore, Reinforcement learning: A survey. J. Artif. Intell. Res. **4**, 237–285 (1996)
109. J. Schmidhuber, A general method for multi-agent learning and incremental selfimprovement in unrestricted environments, in *Evolutionary Computation: Theory and Applications*, (Scientific Publishing Corporation, Singapore, 1996)
110. P. Van Wesel, A.E. Goodloe, *Challenges in the Verification of Reinforcement Learning Algorithms* (NASA Langely Research Center, Hampton, 2017)
111. R.S. Sutton, A.G. Barto, *Reinforcement Learning: An Introduction* (MIT Press, Cambridge, 1998)
112. A.L. Strehl, L. Li, E. Wiewiora, J. Langford, M.L. Littman, PAC model-free reinforcement learning, in *Proceedings of the 23rd International Conference on Machine Learning*, (ACM, New York, 2006), pp. 881–888
113. K. Arulkumaran, M.P. Deisenroth, M. Brundage, A.A. Bharath, A brief survey of deep reinforcement learning. *arXiv preprint arXiv:1708.05866* (2017)
114. D. Lu et al., Aboveground forest biomass estimation with Landsat and LiDAR data and uncertainty analysis of the estimates. Int. J. For. Res. **2012**, 1–16 (2012)
115. N.R. Jachowski, M.S. Quak, D.A. Friess, D. Duangnamon, E.L. Webb, A.D. Ziegler, Mangrove biomass estimation in Southwest Thailand using machine learning. Appl. Geogr. **45**, 311–321 (2013)
116. I. Ali, F. Greifeneder, J. Stamenkovic, M. Neumann, C. Notarnicola, Review of machine learning approaches for biomass and soil moisture retrievals from remote sensing data. Remote Sens. **7**(12), 16398–16421 (2015)
117. Y. Li, J. Yuan, Prediction of key state variables using support vector machines in bioprocesses. Chem. Eng. Technol. **29**(3), 313–319 (2006)
118. D.A. Saldana, L. Starck, P. Mougin, B. Rousseau, N. Ferrando, B. Creton, Prediction of density and viscosity of biofuel compounds using machine learning methods. Energy Fuel **26**(4), 2416–2426 (2012)
119. A.Y. Mutlu, O. Yucel, An artificial intelligence based approach to predicting syngas composition for downdraft biomass gasification. Energy **165**, 895 (2018)
120. B. Najafi, M.A. Fakhr, S. Jamali, Prediction of heating value of vegetable oil-based ethyl esters biodiesel using artificial neural network. Tarım Makinaları Bilimi Dergisi **7**(4), 20140807 (2011)
121. Z. Yildiz, H. Uzun, S. Ceylan, Y. Topcu, Application of artificial neural networks to co-combustion of hazelnut husk-lignite coal blends. Bioresour. Technol. **200**, 42–47 (2016)
122. X. Zhu, X. Wu, Class noise vs. attribute noise: A quantitative study. Artif. Intell. Rev. **22**(3), 177–210 (2004)
123. C. Yuan, C. Wang, Performance of deterministic learning in noisy environments. Neurocomputing **78**(1), 72–82 (2012)
124. D. Gamberger, N. Lavrac, S. Dzeroski, Noise detection and elimination in data preprocessing: Experiments in medical domains. Appl. Artif. Intell. **14**(2), 205–223 (2000)
125. G. An, The effects of adding noise during backpropagation training on a generalization performance. Neural Comput. **8**(3), 643–674 (1996)
126. T.C. Redman, The impact of poor data quality on the typical enterprise. Commun. ACM **41**(2), 79–82 (1998)
127. S. García, J. Luengo, F. Herrera, *Data preprocessing in data mining* (Springer, Berlin, 2015)
128. J.I. Maletic, A. Marcus, Data cleansing: Beyond integrity analysis, in *Iq*, (Citeseer, Boston, 2000), pp. 200–209
129. X. Zhu, X. Wu, Q. Chen, Eliminating class noise in large datasets, in *Proceedings of the 20th International Conference on Machine Learning (ICML-03)*, 2003, pp. 920–927

130. X. Zhu, X. Wu, Q. Chen, Bridging local and global data cleansing: Identifying class noise in large, distributed data datasets. Data Min. Knowl. Disc. **12**(2–3), 275–308 (2006)
131. X. Wu, *Knowledge Acquisition from Databases* (Intellect Books, Norwood, 1995)
132. M. Leshno, V.Y. Lin, A. Pinkus, S. Schocken, Multilayer feedforward networks with a non-polynomial activation function can approximate any function. Neural Netw. **6**(6), 861–867 (1993)
133. B. Karlik, A.V. Olgac, Performance analysis of various activation functions in generalized MLP architectures of neural networks. Int. J. Artif. Intell. Expert Syst. **1**(4), 111–122 (2011)
134. F. Mazhar, A.M. Khan, I.A. Chaudhry, M. Ahsan, On using neural networks in UAV structural design for CFD data fitting and classification. Aerosp. Sci. Technol. **30**(1), 210–225 (2013)
135. P. Sibi, S.A. Jones, P. Siddarth, Analysis of different activation functions using back propagation neural networks. J. Theor. Appl. Inf. Technol. **47**(3), 1264–1268 (2013)
136. S. Gong, J. Sasanipour, M.R. Shayesteh, M. Eslami, A. Baghban, Radial basis function artificial neural network model to estimate higher heating value of solid wastes. Energy Sources A **39**(16), 1778–1784 (2017)
137. E.B. Gueguim Kana, J.K. Oloke, A. Lateef, M.O. Adesiyan, Modeling and optimization of biogas production on saw dust and other co-substrates using artificial neural network and genetic algorithm. Renew. Energy **46**, 276–281 (2012)
138. S. Jacob, R. Banerjee, Modeling and optimization of anaerobic codigestion of potato waste and aquatic weed by response surface methodology and artificial neural network coupled genetic algorithm. Bioresour. Technol. **214**, 386–395 (2016)
139. K. Nath, D. Das, Modeling and optimization of fermentative hydrogen production. Bioresour. Technol. **102**(18), 8569–8581 (2011)
140. R. Setiono, L.C.K. Hui, Use of a quasi-Newton method in a feedforward neural network construction algorithm. IEEE Trans. Neural Netw. **6**(1), 273–277 (1995)
141. G.-B. Huang, Q.-Y. Zhu, C.-K. Siew, Extreme learning machine: a new learning scheme of feedforward neural networks, in *Neural Networks, 2004. Proceedings. 2004 IEEE International Joint Conference on*, vol. 2, (IEEE, Piscataway, 2004), pp. 985–990
142. M.T. Hagan, M.B. Menhaj, Training feedforward networks with the Marquardt algorithm. IEEE Trans. Neural Netw. **5**(6), 989–993 (1994)
143. B. Ghobadian, H. Rahimi, A. Nikbakht, G. Najafi, T. Yusaf, Diesel engine performance and exhaust emission analysis using waste cooking biodiesel fuel with an artificial neural network. Renew. Energy **34**(4), 976–982 (2009)
144. G. Karatas, O.K. Sahingoz, Neural network based intrusion detection systems with different training functions, in *Digital Forensic and Security (ISDFS), 2018 6th International Symposium on*, (IEEE, Piscataway, 2018), pp. 1–6
145. M.F. Møller, A scaled conjugate gradient algorithm for fast supervised learning. Neural Netw. **6**(4), 525–533 (1993)
146. A.A. Suratgar, M.B. Tavakoli, A. Hoseinabadi, Modified Levenberg-Marquardt method for neural networks training. World Acad. Sci. Eng. Technol. **6**(1), 46–48 (2005)
147. N.M. Nawi, A. Khan, M.Z. Rehman, A new Levenberg Marquardt based back propagation algorithm trained with cuckoo search. Procedia Technol. **11**, 18–23 (2013)
148. E.P. Maillard, D. Gueriot, RBF neural network, basis functions and genetic algorithm, in *Neural Networks, 1997., International Conference on*, vol. 4, (IEEE, Piscataway, 1997), pp. 2187–2192
149. F. Günther, S. Fritsch, Neuralnet: Training of neural networks. R. J. **2**(1), 30–38 (2010)
150. Ö. Kişi, E. Uncuoğlu, Comparison of three back-propagation training algorithms for two case studies. Indian J. Eng. Mater. Sci. **12**(5), 434–442 (2005)
151. J.-S. Jang, ANFIS: Adaptive-network-based fuzzy inference system. IEEE Trans. Syst. Man Cybern. **23**(3), 665–685 (1993)
152. A. Saxena et al., A review of clustering techniques and developments. Neurocomputing **267**, 664–681 (2017)

153. L. Rokach, O. Maimon, Clustering Methods, in *Data Mining and Knowledge Discovery Handbook*, (Springer, Boston, 2005), pp. 321–352
154. C. Fraley, A.E. Raftery, How many clusters? Which clustering method? Answers via model-based cluster analysis. Comput. J. **41**(8), 578–588 (1998)
155. P. Domingos, A few useful things to know about machine learning. Commun. ACM **55**(10), 78–87 (2012)
156. Z. You, J. Ye, K. Li, P. Wang, Adversarial Noise Layer: Regularize Neural Network By Adding Noise. arXiv preprint arXiv **1805**, 08000 (2018)
157. N. Srivastava, G. Hinton, A. Krizhevsky, I. Sutskever, R. Salakhutdinov, Dropout: A simple way to prevent neural networks from overfitting. J. Mach. Learn. Res. **15**(1), 1929–1958 (2014)
158. R. Kocjančič, J. Zupan, Modelling of the river flowrate: the influence of the training set selection. Chemom. Intell. Lab. Syst. **54**(1), 21–34 (2000)
159. G.J. Bowden, G.C. Dandy, H.R. Maier, Data transformation for neural network models in water resources applications. J. Hydroinf. **5**(4), 245–258 (2003)
160. H.R. Maier, A. Jain, G.C. Dandy, K.P. Sudheer, Methods used for the development of neural networks for the prediction of water resource variables in river systems: Current status and future directions. Environ. Model. Softw. **25**(8), 891–909 (2010)
161. I. Flood, N. Kartam, Neural networks in civil engineering. I: Principles and understanding. J. Comput. Civ. Eng. **8**(2), 131–148 (1994)
162. K. Rajer-Kanduč, J. Zupan, N. Majcen, Separation of data on the training and test set for modelling: A case study for modelling of five colour properties of a white pigment. Chemom. Intell. Lab. Syst. **65**(2), 221–229 (2003)
163. A. Atkinson, Beyond response surfaces: Recent developments in optimum experimental design. Chemom. Intell. Lab. Syst. **28**(1), 35–47 (1995)
164. R.K.H. Galvao, M.C.U. Araujo, G.E. Jose, M.J.C. Pontes, E.C. Silva, T.C.B. Saldanha, A method for calibration and validation subset partitioning. Talanta **67**(4), 736–740 (2005)
165. R.W. Kennard, L.A. Stone, Computer aided design of experiments. Technometrics **11**(1), 137–148 (1969)
166. A. Saptoro, H.M. Yao, M. Tadé, H. Vuthaluru, Prediction of coal hydrogen content for combustion control in power utility using neural network approach. Chemom. Intell. Lab. Syst. **94**(2), 149–159 (2008)
167. A. Saptoro, M.O. Tadé, H. Vuthaluru, A modified Kennard-Stone algorithm for optimal division of data for developing artificial neural network models. Chem. Prod. Process. Model. **7**(1), 1–16 (2012)
168. S.O. Giwa, S.O. Adekomaya, K.O. Adama, M.O. Mukaila, Prediction of selected biodiesel fuel properties using artificial neural network. Front. Energy **9**(4), 433–445 (2015)
169. I. Estiati, F.B. Freire, J.T. Freire, R. Aguado, M. Olazar, Fitting performance of artificial neural networks and empirical correlations to estimate higher heating values of biomass. Fuel **180**, 377–383 (2016)
170. S. Hosseinpour, M. Aghbashlo, M. Tabatabaei, M. Mehrpooya, Estimation of biomass higher heating value (HHV) based on the proximate analysis by using iterative neural network-adapted partial least squares (INNPLS). Energy **138**, 473–479 (2017)
171. J. Chen et al., Investigation of co-combustion characteristics of sewage sludge and coffee grounds mixtures using thermogravimetric analysis coupled to artificial neural networks modeling. Bioresour. Technol. **225**, 234–245 (2017)
172. U. Ozveren, An artificial intelligence approach to predict gross heating value of lignocellulosic fuels. J. Energy Inst. **90**(3), 397–407 (2017)
173. A.V. Fiacco, *Introduction to Sensitivity and Stability Analysis in Nonlinear Programming* (Elsevier, New York, 1983)
174. D.J. Pannell, Sensitivity analysis: Strategies, methods, concepts, examples. Agric. Econ. **16**, 139–152 (1997)
175. G.D. Garson, Interpreting neural-network connection weights. AI Expert **6**(4), 46–51 (1991)

176. M.H. Shojaeefard, M. Akbari, M. Tahani, F. Farhani, Sensitivity analysis of the artificial neural network outputs in friction stir lap joining of aluminum to brass. Adv. Mater. Sci. Eng. **2013**, 1 (2013)
177. D. Hamby, A review of techniques for parameter sensitivity analysis of environmental models. Environ. Monit. Assess. **32**(2), 135–154 (1994)
178. J. Kim, M.J. Realff, J.H. Lee, Optimal design and global sensitivity analysis of biomass supply chain networks for biofuels under uncertainty. Comput. Chem. Eng. **35**(9), 1738–1751 (2011)
179. J.F. Pérez, P.N. Benjumea, A. Melgar, Sensitivity analysis of a biomass gasification model in fixed bed downdraft reactors: Effect of model and process parameters on reaction front. Biomass Bioenergy **83**, 403–421 (2015)
180. K. Janocha, W.M. Czarnecki, On loss functions for deep neural networks in classification. *arXiv preprint arXiv:1702.05659* (2017)
181. T. Hastie, R. Tibshirani, J. Friedman, Unsupervised learning, in *The Elements of Statistical Learning*, (Springer, New York, 2009), pp. 485–585
182. A.E. Eiben, J.E. Smith, *Introduction to Evolutionary Computing* (Springer, Berlin, 2003)
183. N.M. Al-Salami, Evolutionary algorithm definition. Am. J. Eng. Appl. Sci. **2**(4), 789–795 (2009)
184. P.A. Vikhar, Evolutionary algorithms: A critical review and its future prospects, in *Global Trends in Signal Processing, Information Computing and Communication (ICGTSPICC), 2016 International Conference on*, (IEEE, New York, 2016), pp. 261–265
185. R. Martí, P. Pardalos, M. Resende, *Handbook of Heuristics* (Springer, Cham, 2018)
186. C.A.C. Coello, G.B. Lamont, D.A. Van Veldhuizen, *Evolutionary Algorithms for Solving Multi-Objective Problems* (Springer, Boston, 2007)
187. M. Suleymani, A. Bemani, Application of ANFIS-PSO algorithm as a novel method for estimation of higher heating value of biomass. Energy Sources A **40**(3), 288–293 (2018)

Chapter 5
Thermochemical Characterization of Biomass Residues and Wastes for Bioenergy

T. E. Odetoye, S. F. Ibarhiam, and J. O. Titiloye

5.1 Introduction

Biomass is an organic substance defined mostly in terms of plant and agricultural wastes containing carbon, hydrogen and oxygen components resulting from the process of photosynthesis [20]. Biomass materials are sources of bioenergy. Prior to biomass conversion and transformation into useful bioenergy, the material needs to be characterized in order to maximize its potential yield and applications [18]. Characterization of biomass to determine the physical, chemical and thermochemical properties is an important step to be taken for a successful biofuel production. [5]. It is indeed significant to the successful design and operation of the biomass conversion process.

Biomass can be sourced from forestry, agricultural products, solid waste and landfill site. Biomass derived from forestry is mainly wood or wood by-products such as saw dust or bark, wood logs and wood chip [1]. They are mainly utilized for heat energy, electrical energy generation and bio-fuel purposes. Agricultural products that are sources of bioenergy can be classified into energy crops, food production by-products as well as agricultural wastes and residues [28]. Energy crops are mainly grassy, aquatic crops cultivated and harvested seasonally for bioenergy purposes. This includes biomasses such as miscanthus, algae and sugar cane. [28]. Waste and by-products resulting from food production are also a source of biomass. Examples include corn cob, date palm rachis and stone, parinari fruit shells, straw, cocoa shell and oat husks. Various types of agricultural wastes including

T. E. Odetoye (✉)
Department of Chemical Engineering, University of Ilorin, Ilorin, Nigeria
e-mail: odetoye.te@unilorin.edu.ng

S. F. Ibarhiam · J. O. Titiloye
Chemical and Environmental Engineering, College of Engineering, Swansea University, Swansea, UK

M. O. Daramola, A. O. Ayeni (eds.), *Valorization of Biomass to Value-Added Commodities*, Green Energy and Technology,
https://doi.org/10.1007/978-3-030-38032-8_5

animal waste, sewage sludge and crop cultivation waste are commonly used as biomass [14]. These wastes undergo various types of thermal processing including incineration and pyrolysis to produce bio-energy.

Solid domestic wastes produced on a daily basis are also suitable source of biomass after contaminants have being removed and can be processed for energy production. Biogases from landfill sites are also considered suitable biomass for generating energy in the form of heat or electrical power. The gas consists chiefly of methane and carbon dioxide formed as a result of biological activity on biodegradable wastes through anaerobic digestion.

Biomass residues and wastes are ideal feedstock for conversion using thermochemical process methods. Biomass is classified into four broad categories, namely herbaceous, woody and aquatic plants and manure [16]. Biomass is composed of three main polymeric components of interest [11]; lignin, cellulose and hemicellulose which exist in varying proportions depending on the type of biomass. Most forest and agricultural residues are regarded as lignocellulosic biomass based on their composition [13]. Biomass particle sizes, shape, regularity and surface area have been found to have effect on the thermochemical characteristics of the material [3]. Eventually, the thermochemical properties of a particular biomass sample are part of the factors that influence the characteristics of the resultant bio-oil.

5.2 Biomass Characterization Methods

For a successful characterization of the material in terms of its thermochemical properties, various analyses are carried out on biomass samples [10, 23]. The thermochemical characterization methods available include physical and chemical methods of proximate and ultimate analysis, thermogravimetric method and structural composition analysis.

5.2.1 Particle Size Distribution

Different biomass particle sizes exhibit different characteristics, and have influence on the rate of drying and the quality of product produced. A sieve analysis is usually carried out to distribute the particles into different sizes using a standard set of sieves with different aperture sizes. This helps to determine relationship between design technique, production control requirements and verification specifications.

5.2.2 Proximate Analysis

Proximate analysis is a popular procedure for biomass characterization which gives information concerning the moisture, volatile ash, and the fixed carbon contents of biomass samples. Proximate analysis usually indicates the volatile, fixed carbon and ash content on dry basis. This information is quite useful for the determination of the appropriate and efficient method, processing conditions as well as the suitability of the biomass for the pyrolysis process. In proximate analysis, the moisture contents of biomass samples are very crucial. This specifies the amount of water present in a sample stated as a percentage of the biomass sample weight.

The moisture content of biomass is dependent on the conditions to which biomass is exposed, such as climate and period of cultivation which is determined by using standard methods. The ash content of a sample is defined as the amount of incombustible material remaining after burning a given sample, expressed as a percentage of its initial mass. The understanding of a sample's ash content can help improve char production. Volatile matter which is the gaseous product given off by biomass sample, excluding moisture, at temperatures of up to 900° C is also required as part of the proximate analysis. The fixed carbon of a sample can be calculated after determining the samples moisture, volatile and ash content.

5.2.3 Compositional Analysis of Biomass

The compositional analysis is carried out for identification of the cellulose, hemicellulose, extractives and lignin content of biomass samples. Cellulose constitutes the major part of biomass and makes up majority of its cell wall. Hemicellulose is similar to cellulose. It is a polysaccharide that is found in plant cell walls and consists of 5 and 6 carbon sugar polymers [4]. The third major component in a typical biomass is lignin. Its composition varies from one biomass sample to another. Lignin is a greatly branched, polyphenolic substance with chemical structure more complex when compared to cellulose and hemicellulose.

The chemical composition of a biomass is known to determine the mechanism of degradation, thermal stability as well as nature of the primary reaction products obtainable from it. Hence, it is important to determine the composition of the biomass intended for production of biofuel. Basically, lignins are of interest because the lignin content of a biomass is a factor that indicates the type of catalyst needed during catalytic pyrolysis processes [22].

VanSoest method is popularly used to determine hemicelluloses, cellulose and lignin contents. The method involves the neutral detergent fibre (NDF), acid detergent fibre and acid detergent lignin determinations from which the percentage values of the fibre are calculated. The less digestible plant cell wall comprises hemicelluloses, cellulose and lignin, while the mostly digestible contents include starch and sugars. NDF indicates the hemicellulose, cellulose and lignin, cutin and silica contents of the cell wall. The NDF is obtained after digesting with the neutral detergent solution which is expressed as a percentage dry matter. The determination of hemicelluloses, cellulose and lignin composition of biomass can also be done using thermogravimetric method [6, 25].

5.2.4 Ultimate Analysis

Ultimate analysis of a biomass sample defines the elemental composition of the biomass. It involves determination of the carbon, hydrogen, nitrogen, oxygen, sulphur, chlorine, and trace elements in the biomass sample. Ultimate analysis, measured in wt.% on dry basis, is a partial representation of the elemental content (C, H, N, O and S) within the feedstock. Results obtained from ultimate analysis can be used to evaluate the molecular formula and to predict calorific value of the biomass [24]. The calorific value of the biomass samples indicates the potential of the sample as a fuel source. Bomb calorimeter is used for the determination of the biomass calorific value.

5.2.5 Thermogravimetric Analysis

Biomass thermal properties are best determined by thermogravimetric analysis. Much more, the resultant thermal data acquired can also be used for verifying the proximate analysis method. Thermogravimetric analysis (TGA) has to do with the technique for measuring the weight loss of a sample under controlled atmosphere. The results obtained are used to plot a weight loss against temperature or time graph. TGA helps to understand the kinetic and physical properties of a biomass in line with its characteristic composition. The behaviour of such samples during pyrolysis can be predicted from the TGA analysis since biomass samples are thermally decomposed in similar conditions to the conditions they undergo during the pyrolysis experiments. The method focuses on the decomposition behaviour of the lignin, hemicellulose as well as cellulose contents of the biomass samples. Correlations between heating value of biomass samples and TGA parameters have been reported [26] (Table 5.1).

Table 5.1 Typical ultimate analysis results of jatropha seed coat biomass [18]

Parameters	Nigeria Odetoye et al. [18]	Japan Murata et al. [31]	Indonesia Wever et al. [33]	Indonesia Manurung et al. [30]	Thailand[b] Sricharoenchaikul and Atong [32]	India Kratzeisen and Müller [29]
C	48.34	48.15	48.5	50.3	45.5	50.9
H	5.74	6.48	5.7	6.6	7.2	5.8
N	1.17	1.39	0.67	1.8	4	0.8
O[a]	44.03	–	41.0	38.3	43.3	39.5
S	<0.10	–	0.01	n.d.	–	0.1
Cl	0.62	–	–	–	–	0.1
H/C	0.12	0.13	0.12	0.13	0.16	0.11
HHV(MJ/kg)	20.06	–	–	–	–	16.5

[a]by difference, [b]mixed waste

5.2.6 *Pyrolysis Gas-Chromatography-Mass Spectrometry (Py-GC-MS)*

Biomass thermal decomposition process and products are investigated using Py-GC-MS. Py-GC-MS serves as a fast screening technique for investigating potential bioenergy materials. Heating rates and process temperature can be designed to be representative of potential industrial process parameters with the benefit of investigating the potential pyrolysis products and composition within the biomass.

5.2.7 *Other Methods of Characterization: Nanoscale*

Biomass has been found to be difficult to analyse at nanoscale because the biomass structure is sensitive to water removal. Meanwhile, biomasses are mostly pre-treated for water removal prior to thermochemical conversions. Evaluating the mass fraction and size of crystals has been achieved using infrared (IR), NMR, Raman and X-ray diffraction methods [2, 8, 10]. Raman and IR spectroscopy were recommended as ways to assess the ratio of amorphous to crystalline cellulose [2] (Fig. 5.1 and Table 5.2).

5.3 Evolving Methods of Biomass Characterization, Challenges and Future Outlook

Biomass characterization methods are still evolving as a result of a wide variety of plants that are being searched for as potential feedstock for biofuel production [27]. Quite a variety of genetically altered plants are developed and characterized for

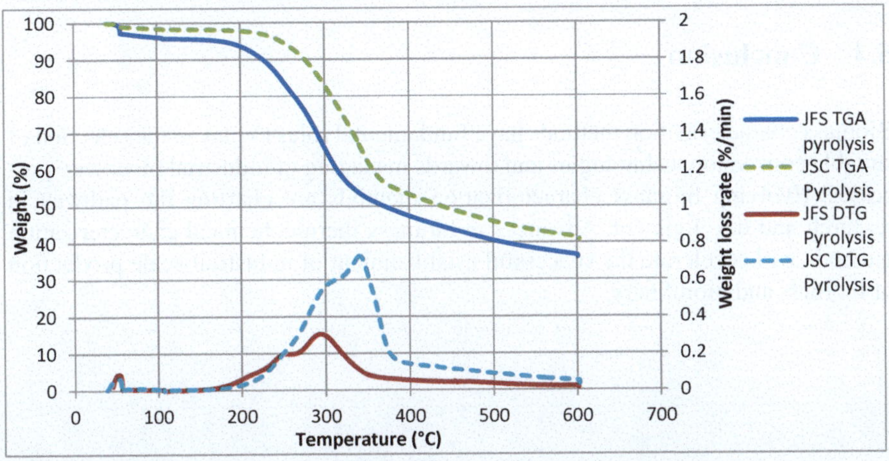

Fig. 5.1 Typical pyrolysis curves for jatropha fruit shell (JFS) and jatropha seed coat [18]

Table 5.2 Typical compositional analysis results of some biomass [17]

Components[a] (%)	Parinari fruit shell	Jatropha seed coat	Jatropha fruit shell	Empty fruit bunch[b]	Wheat straw[c]	Palm shell[d]	Rape straw[c]	Switch grass[c]
Extractives	18.1	13.3	42.3	25.5	27.7	6.7	9.7	26.3
Cellulose	45.4	34.0	32.5	23.7	33.2	27.7	37.6	36.0
Hemicellulose	6.4	40.0	10.5	21.6	24.0	21.6	31.4	31.6
Lignin	30.1	12.7	5.7	29.2	15.1	44.0	21.3	6.1

[a]Wt % on dry basis
[b]Omar et al. [19]
[c]Greenhalf et al. [12]
[d]Garcia et al. [9]

bioenergy purpose [8]. Hence, there is a need for efficient high-throughput methods to analyse biomass for thermochemical conversion [15, 21]. Therefore, researchers are in the use of various analyses to develop efficient, rapid methods to enable fast evaluation of diverse varieties of plants.

One of the main challenges of biomass analysis is the tedious and time-consuming process, particularly when a large set of samples is to be analysed [15]. Recent advances in biomass characterization methods attempt to overcome the challenge by adopting various techniques culminating in high-throughput technology [7]. These include High Throughput Plant Cell Wall Compositional Analysis [7], rapid analysis of varieties of samples using Raman spectroscopy [15]. Automation of elemental analyser, proximate analysers that pick sets of samples from the rack for analysis have been done [27]

Basically, the methods have been found to be applicable when handling a relatively large sample set. Although the high-throughput technology is quite time-saving, reduces the use of harsh reagents, the cost of acquiring such equipment needed for the analyses is still relatively high [7].

5.4 Conclusion

Biomass characterization methods have fundamental roles to play in the selection of suitable conversion technologies and towards managing of industrial bioenergy processes. Evolving biomass characterization methods are charting the pathway in research and development. Advances in biomass thermochemical characterization methods will accelerate the successful establishment of industrial-scale production of biofuels and biorefinery.

References

1. P. Akhator, A. Obanor, A. Ugege, Nigerian wood waste: a potential resource. J. Appl. Sci. Environ. Manage. **21**(2), 246–251 (2017)
2. J.F. Beecher, C.G. Hunt, J.Y. Zhu, Tools for the characterization of biomass at the nanometer scale, in *The Nanoscience and Technology of Renewable Biomaterials*, ed. by L. A. Lucia, O. J. Rojas, (Wiley, West Sussex, 2009), pp. 61–90
3. T.G. Bridgeman, L.I. Darvell, J.M. Jones, P.T. Williams, R. Fahmi, A.V. Bridgwater, T. Barraclough, I. Shield, N. Yates, S.C. Thain, I.S. Donnison, Influence of particle size on the analytical and chemical properties of two energy crops. Fuel **86**, 60–72 (2007)
4. R.C. Brown, Biorenewable resources: engineering new products from agriculture. Renew. Sust. Energ. Rev. **1**(1), 59–72 (2003)
5. J. Cai, Y. He, X. Yu, S.W. Banks, Y. Yang, X. Zhang, Y. Yu, R. Liu, A. Bridgwater, Review of physicochemical properties and analytical characterization of lignocellulosic biomass. Renew. Sust. Energ. Rev. **76**, 309–322 (2017)
6. M. Carrier, A. Loppinet-Serani, D. Denux, J. Lasnier, F. Ham-Pichavant, F. Cansell, C. Aymonier, Thermogravimetric analysis as a new method to determine the lignocellulosic composition of biomass. Biomass. Bioenerg. **35**(1), 298–307 (2011)
7. S.R. Decker, A.E. Harman-Ware, R.M. Happs, E.J. Wolfrum, G.A. Tuskan, D. Kainer, G.B. Oguntimein, M. Rodriguez, D. Weihill, P. Jones, D. Jacobson, High throughput screening technologies in biomass characterization. Front. Energ Res. **6**, 120 (2018). https://doi.org/10.3389/fenrg.2018.00120. Accessed 14 Feb 2019
8. M. Foston, A.J. Ragauskas, Biomass characterization: recent progress in understanding biomass recalcitrance. Ind. Biotechnol. **8**(4), 191–208 (2012). https://doi.org/10.1089/ind.2012.0015. Accessed 15 Nov 2018
9. R. Garcia, C. Pizarro, A.G. Lavin, J.L. Bueno, Characterization of Spanish biomass waste for energy use. Bioresour. Technol. **103**, 249–258 (2012)
10. P. Ghetti, L. Ricca, L. Angelini, Thermal analysis of biomass and corresponding pyrolysis products. Fuel **75**(5), 565–573 (1996)
11. H.B. Goyal, D. Seal, R.C. Saxena, Bio-fuels from thermochemical conversion of renewable resources: a review. Renew. Sust. Energ. Rev. **12**(2), 504–517 (2008)
12. C.E. Greenhalf, D.J. Nowakowski, A.V. Bridgwater, J. Titiloye, N. Yates, A. Richie, I. Shield, Thermochemical characterisation of straws and high yielding perennial grasses. Ind. Crop. Prod. **36**, 449–459 (2012)
13. S.K. Khanal, R.Y. Surampalli, T.C. Zhang, B.P. Lamsal, R.D. Tyagi, C.M. Kao, *Bioenergy and Biofuel from Biowastes and Biomass* (American Society of Civil Engineers Publishers, Reston, 2010), pp. 190–197
14. H. Kominko, K. Gorazda, Z. Wzorek, K. Wojtas, Sustainable management of sewage sludge for the production of organo-mineral fertilizers. Waste Biomass. Valori. **9**, 1817–1826 (2018)
15. J. Lupoi, B. Simmons, R. Henry, Editorial: biomass modification, characterization, and process monitoring analytics to support biofuel and biomaterial production. Front Bioeng. Biotech. **4**, 25 (2016). Available via https://www.ncbi.nlm.nih.gov/pmc/articles/PMC4785235/. Accessed 14 Feb 2019
16. P. McKendry, Energy production from biomass (part 1): overview of biomass. Bioresour. Technol. **83**, 37–46 (2002)
17. T.E. Odetoye, K.R. Onifade, M.S. AbuBakar, J.O. Titiloye, Pyrolysis of *Parinari polyandra* Benth fruit shell for bio-oil production. Biofuel. Res. J. **1**(3), 85–90 (2014)
18. T.E. Odetoye, M.S. AbuBarkar, J.O. Titiloye, Thermochemical characterization of Nigerian *Jatropha curcas* fruit and seed residues for biofuel production. Energ. Ecol. Environ. **3**(6), 330–337 (2018)
19. R. Omar, A. Idris, R. Yunus, K. Khalid, M.I. Isma, Characterization of empty fruit bunch for microwave-assisted pyrolysis. Fuel **90**, 1536–1544 (2011)

20. C.S. Park, R. Sarothi, S.H. Kim, Current Developments in thermochemical conversion of biomass to fuels and chemicals. (2018), Available via https://www.intechopen.com/books/gasification-for-low-grade-feedstock/current-developments-in-thermochemical-conversion-of-biomass-to-fuels-and-chemicals. Accessed 14 Feb 2019
21. S. Pattathil, U. Avci, T. Zhang, C.L. Cardenas, M.G. Hahn, Immunological approaches to biomass characterization and utilization. Front Bioeng. Biotech. **4**, 25 (2015)
22. A. Pattiya, J.O. Titiloye, A.V. Bridgwater, Fast pyrolysis of cassava rhizome in the presence of catalysts. J. Anal. Appl. Pyrol. **81**, 72–79 (2008)
23. J. Pulka, P. Manczarski, J.A. Koziel, A. Białowiec, Torrefaction of sewage sludge: kinetics and fuel properties of biochars. Energies **12**(3), 565 (2019)
24. R. Saidur, E.A. Abdelaziz, A. Demirbas, M.S. Hossain, S. Mekhilef, A review on biomass as a fuel for boilers. Renew. Sust. Energ. Rev. **15**, 2262–2289 (2011)
25. L. Sanchez-silva, D. Lopez-Gonzalez, J. Villasenor, P. Sanchez, J.L. Valverde, Thermogravimetric-mass spectrometric analysis of lignocellulosic and marine biomass pyrolysis. Bioresour. Technol. **109**, 163–172 (2012)
26. Z. Sebestyén, F. Lezsovits, E. Jakab, G. Várhegyi, Correlation between heating values and thermogravimetric data of sewage sludge, herbaceous crops and wood samples. J. Therm. Anal. Calorim. **110**, 1501–1509 (2012)
27. P. Tanger, J. Field, C.E. Jahn, M.W. DeFoort, J.E. Leach, Biomass for thermochemical conversion: targets and challenges. Front Plant. Sci. **4**, 218 (2013). https://www.ncbi.nlm.nih.gov/pmc/articles/PMC3697057/. Accessed 14 Feb 2019
28. J.O. Titiloye, M.S. AbuBarkar, T.E. Odetoye, Thermochemical characterisation of agricultural wastes from West Africa. Ind. Crop. Prod. **47**, 199–203 (2013)
29. M. Kratzeisen, J.Müller, Energy from seed shells of Jatropha curcas, Landtechnik 64 (6) 391–393 (2009)
30. R. Manurung, R.D.A.Z. Wever, J. Wildschut, R.H. Venderbosch, H. Hidayat, J.E.G. van Dam, E.J. Leijenhorst, A.A. Broekhuis , H.J. Heeresa, Valorisation of Jatropha curcas L. plant parts: Nut shell conversion to fast pyrolysis oil, Food Bioproduct Proc. 87,187–196 (2009)
31. K. Murata, P. Somwongsa, S. Larkiattaworn, Y. Liu, M. Inaba, I. Takahara Analyses of liquid products from catalytic pyrolysis of jatropha seed cakes. Energy Fuels. 25, 5429–5437 (2011)
32. V. Sricharoenchaikul, D. Atong, Thermal decomposition study on Jatropha curcas L. waste using TGA and fixed bed reactor. J Anal. Appl. Pyrol. 85, 155–162 (2009)
33. D.A.Z. Wever, H.J. Heeres, A.A. Broekhuis, Characterization of physic nut (Jatropha curcas L.) shells. Biomass Bioenerg. 37,177–187 (2012)

Chapter 6
Evaluation of Methods for the Analysis of Untreated and Processed Lignocellulosic Biomasses

A. O. Ayeni, M. O. Daramola, D. T. Oyekunle, O. O. Sadare, O. Agboola, and A. A. Ayoola

6.1 Introduction

Diverse kinds of fuels and chemicals can be obtained from lignocellulosic materials. Biomass, trees and plants, including marine ones, are reservoirs for solar energy (conversion of sunlight to cellulosic materials via photosynthesis). Lignocelluloses are domestically available energy resources, and they are abundant and renewable. Renewable energy from lignocelluloses is inexpensive compared to the known conventional energy sources from fossil fuels. Biofuels is used to describe renewable sources of energy derived from wood cellulosic waste materials, marine plants, crops and some other related lignocellulosic materials. The biomass is made up of cellulose, hemicellulose, and lignin in appropriate abundance average of 50: 25: 25, respectively, and other extractable components. Most often, cellulose has the largest composition in the lignocellulosic complex make-up. Cellulose is made up of individual D-glucose units linked together by β-1,4 glucosidic bonds. A highly well-arranged form of crystal areas, which are not permissible to water, are made possible by the hydrogen bonding between the cellulosic molecules. The crystalline part of cellulose is estimated to be about 50–90% of the total biomass with the remaining

A. O. Ayeni (✉) · D. T. Oyekunle · O. Agboola · A. A. Ayoola
Covenant University, College of Engineering, Department of Chemical Engineering, Ota, Nigeria
e-mail: augustine.ayeni@covenantuniversity.edu.ng

M. O. Daramola
Department of Chemical Engineering, Faculty of Engineering, Built Environment and Information Technology, University of Pretoria, Hatfield, Pretoria 0028, South Africa

O. O. Sadare
School of Chemical and Metallurgical Engineering, Faculty of Engineering and the Built Environment, University of the Witwatersrand, Johannesburg, South Africa

M. O. Daramola, A. O. Ayeni (eds.), *Valorization of Biomass to Value-Added Commodities*, Green Energy and Technology,
https://doi.org/10.1007/978-3-030-38032-8_6

10% made up of the amorphous cellulose [37, 39]. According to Puls and Schusiel [69], the hemicellulose part of the lignocellulose is made up of the heterogeneous polymer which varies from plant to plant. Lignin, which is the third polymer in plant cell walls, acts as a binder of the lignocellulosic complex. Lignin composition varies according to species and sources. Lignin is composed of the phenyl propanoid polymer [17, 34]. For example, some gymnosperms have dehydrogenation polymers of coniferyl alcohols. Angiosperm lignin has de-hydrogenation polymeric alcohols of the coniferyl and sinapyl [28, 38]. The primary sources of energy in Africa, including South Africa, are petroleum products, thermal and hydroelectric power, and fuel wood. Processes in the transformation of lignocelluloses to desired products include biological (using fungi, bacteria and archeal) and non-biological (physical, physico-chemical and chemical). However, methods can be interrelated. A physical method can be used for the pretreatment while a biological method can be used for the remaining downstream processing. A complete biological processing can be evaluated for the entire transformations. For example, during pyrolysis of lignocelluloses, products after initial treatments include liquid, gaseous and solid materials. These intermediates can then be transformed further biologically or chemically to desired products. Discussions in this paper enumerate methodologies for compositional analysis of lignocelluloses before and for post pretreatments; physical and chemical measurements such as proximate, ultimate and higher heating values determination, and spectroscopy techniques of measurements.

6.2 Compositional Analysis

6.2.1 Gravimetric Determination

During gravimetric compositional analysis, samples are quantitatively determined based on weight of a solid. Often, dried solids are measured using high-precision analytical weighing balance. Gravimetric analysis is generally used to quantify the solid fraction of the raw and treated lignocelluloses (extractives, cellulose, hemicellulose, lignin and ash contents). This form of analysis hardly encounters instrumental error and does not involve the use of costly equipment [8]. Numerous approaches to gravimetric analysis of lignocelluloses exist in the literature. Application of methods is selective depending on biomass nature as wood, wood chips, wood shavings, cereal grains, herbs, seeds, hays, grasses, by-products from lignocelluloses materials utilization or mixed proportions of lignocelluloses. Gravimetric analysis separates lignocelluloses into extractives, cellulose, hemicellulose, lignin and ash fractions. It is commonly applied on raw and pretreated biomasses. The extractive content (also referred to as non-cell-wall materials of lignocelluloses) determination involves solvent extraction using different solvents. Nonstructural components of lignocellulosic biomass that have the potential of interfering with downstream analysis are referred to as the extractives. Non-structural components (such as protein, ash, chlorophyll, waxes and nitrate/nitrites) removal prior to analysis for carbohydrates and lignin improves precision of the downstream compositional analysis [59, 78]. In one report, for extractive content determination, dried biomass sample can be leached with mixture of benzene/ethanol in a ratio 2:1 at a constant temperature for 3 h. Residues are oven dried at 105–110° C to a constant weight. Weight loss from the raw to extracted sample is the extractive content [9, 46]. Other researchers [12–14, 16] used acetone as the solvent for extraction. The National Renewable Energy Laboratory of USA [78] recommended a one-step or two-step extraction process depending on the nature of lignocelluloses. Extraction set-up can be a simple Soxhlet type or the automatic type [78]. A one-step process involves using a single solvent like ethanol on samples that have little or no water-extractable materials. A two-step extraction of biomass samples involves samples containing a significant amount of water-soluble material or biomass that has water-soluble components of interest, for example, corn stover and likely herbaceous feedstocks [78]. In the NREL method, between 5 and 20 g of biomass is required with a minimum of 8 g of extracted sample needed for compositional analysis. Samples are also refluxed for between 6 and 24 h. Sluiter et al, Soest and Wine [78, 87] also established procedures for estimating the non-cell wall components (extractives) by different steps of chemical washing, sample reflux, heating and drying [86]. The dietary fibre method is also available [31, 86]. The Van Soest orderly fibre solubilization method (Soest 1967) is an established method to quantify the three forms of lignocellulotic contents of plant feedstocks. The predictive nutritive value of lignocelluloses can also be established by this method. The method provides the classification of cell wall components into three insoluble residues: neutral detergent fibre (NDF), acid detergent fibre (ADF) and acid detergent lignin (ADL), as shown in Table 6.1.

However, there are some deficiencies in accuracy of the detergent fibre method. The cellulose and hemicellulose contents are overestimated while the lignin content is underestimated [32, 41, 84]. This is due to the fact that the NDF residue contains some proteins [41, 84]. ADF contains some pectins and also contains some hemicelluloses [41, 42].

Table 6.1 Descriptions of the three detergent methods for the gravimetric analysis of lignocelluloses

Method	Description			References
NDF	Reflux of biomass in hot neutral detergent solution for 1 h	Represents incomplete digestion composed of lignin, cellulose, and hemicellulose		Jung and Lamb [42], Soest and Wine [87]
ADF	Reflux of biomass in a hot aqeous detergent of 0.5 mol/L of H_2SO_4 for 1 h	Components of the cell wall obtained from the insoluble residue are the cellulose and lignin. This step comes before the ADL measurement.	The difference between ADF insoluble residues and the NDF results in hemicelluloses part of the complex.	Soest [86]
ADL	Remaining structural carbohydrate of ADF are solubilized with H_2SO_4 (12.2 mol/L) for 3 h at room temperature	Subtracting the ADF weight from the ADL insoluble residues gives the cellulose content.	ADL insoluble residue is the lignin content	Soest [86], Jung and Lamb [42], Jung [41]

Source: Godin et al. [31]

The dietary fibre method description is as established by Uppsala [79, 84]. This method can quantify animal, human food as well as crop feedstock (a source for biofuels and bio-based chemicals). It is a more accurate method, compared to the detergent method, in the determination of the cellulose and hemicelluloses. It provides the structural polysaccharides based on the monosaccharide components such as D-glucose, D-xylose, L-arabinose, D-mannose and D-galactose [77, 83]. The structural components are initially solubilized with 12.2 mol/l sulphuric acid at the operating temperature of 30 °C for 1 h. During the second stage, hydrolysis occurs with 0.419 mol/l sulphuric acid on the polysaccharides by autoclaving at 121 °C for 1 or 2 h [77, 83]. After the second hydrolysis step, the released monosaccharides are obtained and measured through liquid chromatography (HPLC) or gas chromatography. Also, dietary fibre method is used for the measurement of hemicellulosic glucan. The cellulose solubilization step is not considered during the hemicellulose determination. The difference between the total glucan and hemicellulosic glucan compositions [33] is calculated as the amount of cellulose. The monosaccharide components (D-galactose, D-mannose, L-arabinose, D-glucose and D-xylose) of hemicelluloses are obtained in their polymeric forms [33]. The sum of these polymeric forms gives the hemicellulose contents by using the dietary fibre method. Underestimation of monosaccharides concentrations during acidic degradation can be corrected by the treatment in parallel with the samples a standard mixture of the monosaccharide with sulphuric acid [78, 83]. Liquid fraction after pretreatment (the hydrolysate) can be quantified for the carbohydrates in the form of hemicellulose polymers (arabinoxylan) through dilute sulphuric acid analysis (4 wt.% H_2SO_4, 121 ° C for 10 min or any other appropriate heating time in an autoclave) [15, 76] so as to determine xylose, arabinose and glucose and other minor sugars. A conversion factor is generally used on polymerization to hemicellulose (162/180 for glucose to glucan and 132/150 for xylose to xylan or to arabinan) [77]. Gravimetric analysis strategy is that it is time consuming and most often require lengthy chemical sample preparations. In any method chosen for gravimetric analysis, an initial investigation on biomass should be carried out to know the applicability and reproducibility of method.

6.3 Physical and Chemical Measurements

The suitability of lignocellulosic biomass for fuel, chemicals, and energy production also involves adequate physical and chemical analysis. Physical analysis involves proximate analysis and biomass crystallinity with X-ray diffraction technique. The chemical analysis involves the ultimate analysis, caloric value estimation and Fourier Transform Infra-Red spectroscopy. Estimating properties of proximate, ultimate and caloric values (or higher heating value) of biomasses establish their quality for bioprocessing, assessing the fuel characteristics and bio-commodity capacities.

6.3.1 Proximate Analysis

The analysis is a physical characterization technique that estimates the presence of volatile matter, moisture, fixed carbon, ash, caloric value, total solid using the thermogravimetric analyser (TGA) or any other suitable methods [75] (Fig. 6.1). The proximate analysis reveals the combustion behaviour of biomasses. Proximate analysis involves subjecting the biomass to three logical order of drying, removal of volatile substances (devolatilization) and oxidation in oxygen with the use of a TGA [10, 54]. Low moisture content implies that the biomass is suitable for fuel and chemicals production. High moisture improves microbial growth in biomass, thereby reducing the physical quality of the fuel and pyrolysed gases. Volatile matters are short- and/or long-chain aromatic hydrocarbons: the building blocks for fuels and other platform chemicals. Increasing ash content in biomass decreases yield and heating values [10].

6.3.2 Ultimate Analysis

This is a form of chemical analysis that reveals the elemental compositions of biomass. Mostly, components such as oxygen, carbon, hydrogen, nitrogen and trace amount of sulphur are analysed. Elemental compositions of biomass are commonly estimated using the CHNS/O analyser. Low levels of sulphur and nitrogen are indicative that SO_2 and NO_x emissions are negligible during biomass processing. Significant levels of SO_2 lead to formation of sulphates which generate ashes [10].

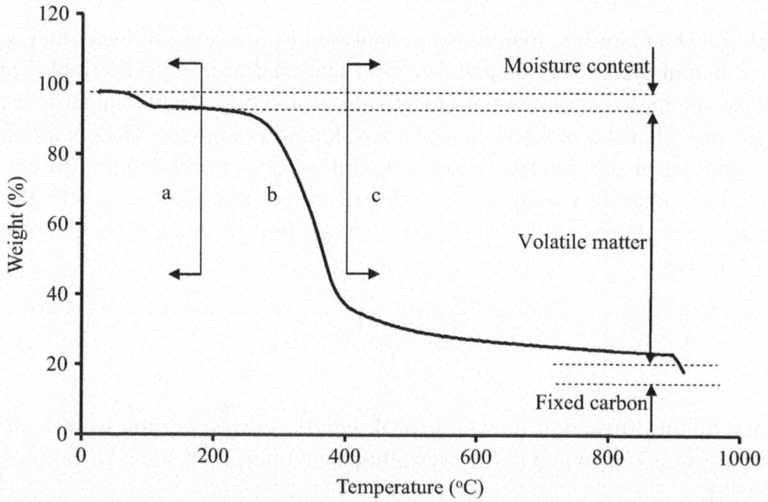

Fig. 6.1 Devolatilization and proximate fractions representations (weight%). Stages a, b and c stand for drying, primary devolatilization and secondary devolatilization, respectively

6.3.3 *Higher Heating Value*

Caloric value or higher heating value (HHV) determines the measurable quantity rather than quality of energy possessed by the lignocellulosic biomasses [65]. HHV is the heat liberated during fuel combustion having the original and generated water in a condensed state [74]. HHV of biomasses can be obtained experimentally through a bomb calorimeter, which determines the difference in enthalpy between reactants and products. Higher heating value of biomass is an essential step during the analysis and designing of bioenergy systems [62].

6.3.4 *Spectroscopy Techniques of Measurement*

Spectroscopy refers to various analytical techniques with the help of radiation in obtaining data on structure and properties of matters. It deals with measuring and interpreting spectra that arise from the interaction of electromagnetic radiation. In the process of the interactions, absorption, emission of electromagnetic radiation by atoms or molecules occur. For the structural and compositional interpretation of multifaceted samples, several instrumental methods have been used. The utilization of spectroscopy can make available a non-invasive, high-throughput, technique demanding little to no sample preparation and qualitative and quantitative information from on- or offline processes [51].

6.3.4.1 X-ray Diffraction (XRD)

Through XRD, crystalline materials are analysed by studying the way they scatter X-rays aimed at them. The intensities of the scattered X-rays and reliable calculations allow the understanding of atomic positions and how the crystal structures are arranged, provided the incident X-ray's wavelength is known. XRD is a common method applied in the determination of cellulose crystallinity index. In common methods, the radiation is employed in the 2ϑ range between 5 and 50°, the time fixed with a step interval of 0.02° [22]. The crystallinity index (Icr) is given empirically [72, 82] as:

$$\%\mathrm{Icr} = \frac{\mathrm{I}_{002} - \mathrm{I}_{\mathrm{am}}}{\mathrm{I}_{002}} \times 100$$

I_{002} = maximum diffraction intensity (peak height) corresponding to 002 plane of cellulose crystals (attributed to the crystalline and amorphous areas of cellulose 1 at 2ϑ = 22.5°).

I_{am} = diffraction intensity for amorphous cellulose (2ϑ = 18° for cellulose 1). This is the height of the minimum between 200 and 110 peaks [40]. Ju et al. [40] studied a modified XRD method by identifying three distinct amorphous peaks in order to calculate CrI. They reported a 2ϑ range of 10–75° to be more suitable for the determination of CrI. Crystalline structural factors such as *d*-spacing and crystalline size were also investigated. They concluded that this method is applicable to other cellulose polymorphs. It can be concluded that attaining information through XRD data is easy; nonetheless, data assessment in some instances is somewhat demanding [43].

6.3.4.2 Visible/Ultraviolet (UV)

The visible/ultraviolet spectroscopy utilizes the fact that atoms are able to emit or absorb visible light. Visible absorption spectroscopy can be combined with ultraviolet (UV) absorption spectroscopy in UV/Vis spectroscopy. The techniques associated with the regions of the electromagnetic spectrum are widely used and very rapid to operate. Colorimetric assay requires a calibration curve to be plotted which should be linear as long as Beer-Lambert law applies. Concentrations of unknown samples are determined (interpolated from the linear plot) from the absorbance values. Anthrone/H_2SO_4 method can be used to quantify for hexoses during lignocelluloses fractionation at wavelength of 620 nm [30]. The orcinol/HCl method is also suitable for pentoses assay [23] with pentoses reading taken at wavelength of 660 nm. A common colorimetry method to quantify reducing sugars in an unknown sample is the dinitrosalicylic (DNS) acid assay [11, 58].

6.3.4.3 Infrared (IR)

Numerous studies have determined and likened the crystallinity data of the pure cellulose via the XRD, FTIR and NMR techniques. It has, however, been reported in literature that FTIR is considered to be the most preferred in the analysis of the cellulose structure [26]. Report has it that FTIR technique requires little or no sample preparation without any specific working experience; nonetheless, it only offers relative data not the absolute data, because the information gathered through the inspection of the amorphous and crystalline portions are found in the Fourier transform infrared spectrum [26].

Fourier transform infrared (FTIR) is a common method for the determination of the main organic groups or the type of bonds present in lignocelluloses structure (hemicellulose, cellulose, and lignin) [35, 55]. Most FTIR analyses are carried out scanning from 400 to around 4000 cm^{-1} using an FTIR spectrophotometer. The crystal information of samples is brought into bear through FTIR spectroscopy. The crystallinity information of lignocellulosic samples can be evaluated using three methods [9, 19, 67]. The total crystallinity index, TCI, corresponds to C–H stretching. The absorbance ratio (1437 cm^{-1} and 899 cm^{-1}) is the lateral order index, LOI, corresponding to a CH_2 bending vibrations. The hydrogen bond intensity, HBI, is the absorbance ratio from 3400 cm^{-1} and 1320 cm^{-1} and due to O–H starching vibrations from alcohols and carboxylic acids present in polysaccharide and lignin [9, 61, 63]. HBI is also related to the crystallinity, the crystal arrangements, and the degree of regularity of molecules [18, 20, 64]. Increases in TCI and LOI values imply that highest degree of crystallinity and the cellulose structure is more ordered. However, decreased values of the two indicate that more amorphous structure is present in the cellulose structure. In other words, LOI increases with the degree of crystallinity. TCI is directly proportional to the degree of crystallinity of the cellulose sample [9, 63].

6.3.4.4 Nuclear Magnetic Resonance (NMR)

NMR observes local magnetic fields around atomic nuclei. Solid-state ^{13}C nuclear magnetic resonance offers a more precise measure of the cellulose crystallinity [43]. The most common types are proton NMR (^{1}H NMR) and carbon-13 NMR (^{13}C NMR) [71]. They can be used to quantify hydroxyl groups in lignin. The most convenient technique used lately is phosphorus-31 NMR (^{31}P NMR) [6]. ^{31}P NMR allows the estimation of hydroxyl, carboxyl groups as well as to differentiate between condensed and non-condensed phenolic groups, aliphatic groups within very short analytical time [25]. Information about the dynamics, structure and chemical environments of atoms can be provided by NMR spectroscopy. ^{31}P NMR is able to reveal the differences between individual lignin molecules within the same lignocellulosic sample, such as some molecules showing high degree of linearity while others revealing high degree of branching [81].

Apart from crystallinity, a thorough structural analysis and interpretation of the lignocelluloses components are achievable via analysis from NMR [68]. Furthermore, nuclear magnetic resonance spectroscopy affords valuable data and facts in light of the ultrastructure of the cellulosic polymer; especially, the interlink between the morphology of cellulose crystals [66]. NMR technique possesses enough capacity to differentiate the chains of the cellulose situated on the surface crystallites of the cellulosic material.

6.3.4.5 Raman Spectroscopy

The spectroscopic method of Raman is a technique due to vibrations centred on the non-conserved kinetic energy of collision due to the internal friction of the monochromatic light, whereby the scattered photons, generated in the course of interaction amid light and matter, are determined [49]. It can be utilized to examine solid, liquid and gaseous samples. When spectroscopic measurement of an analyte occurs, light interactivity with the sample can be scattered, transmitted or absorbed [51]. The spectroscopy measures the light scattered from a molecule when irradiated with a light source, usually a laser [56, 60]. Raman spectroscopy is a non-destructive technique, and it does not necessarily require the preparation of the sample. It delivers high spectral resolution and not subdued by the presence of moisture in samples; it also provides field portability and can generate ample qualitative and quantitative data. When developing Raman applications, the main experimental variable is the proper choice of excitation wavelength [51].

Should there be a alteration in the polarizability of the electron cloud during the interactivity of the molecule with light, a molecule will be considered 'Raman active' [49]. Most of the Raman features of cellulose have been studied and recognized in the spectra of cellulose materials [88]. Nonetheless, many of the vibrational modes are exceedingly coupled by virtue of the cellulose chain comprising of C–C and C–O bonds; it is essential for an enhanced band obligation to occur for some of the bands [53]. Cellulose molecules can be aggregated in an extensive range of secondary and tertiary structures. Hence, Raman spectroscopy is used in providing structural information on lignocellulosic materials for the reason that it can offer in situ determination on the cell wall of plants in the absence of sample preparation. This has led to variations in the intra- and inter-molecular hydrogen bonding and the arrangement of the cellulose chains such as chain polarity [3]. Allomorphs have been studied using Raman spectroscopy. Studies have been done on precise Raman bands related to allomorphs cellulose I, cellulose II and cellulose III [1]. Studies have shown that a mixture of these allomorphs can be quantified by Raman spectroscopy [2]. Hernández-Hernández et al. [36] described the variations that occurred in agave fibres in the course of a green pulping process and evaluated them by spectroscopy techniques. They provided a complete quantitative interpretation of the phenomenon that took place during the delignification of agave fibres. The delignification process took place in two phases: the first phase is related to the superficial delignification of the fibre (0–30 min) and the second phase is related to the elimination interfibrillar lignin of the fibre, which is verified by the vanishing of the band properties of the lignin in Raman spectra and the upsurge in the index of crystallinity.

Despite the advantages of Raman spectroscopy, it is important to consider the ecological effects of the solvents when analysing a lignin sample in solutions with several solvents [45, 48]. In addition, numerous compounds in biomass display noteworthy and damaging spectral contributions from laser-induced fluorescence, which can frequently obscure analyte signal [51]. In order to overcome the interference of the fluorescence, application of near-infrared (NIR) wavelengths is usually used.

6.3.4.6 Near-Infrared Spectroscopy (NIR)

The developments in near-infrared (NIR) and dispersive Raman spectroscopy instrumentation have contributed greatly to lignin structural research [47, 52, 57, 85, 89]. Certain NIR Raman instrument allowed the analysis of a commercial isolated lignin [57]. Furthermore, NIR spectroscopy has been the principal vibrational method in analysing the whole lignin composition. The application of vigorous, systematically assessed multivariate models minimizes the necessity to perform labour-intensive reference techniques for all samples, especially the upsurge of the throughput while reducing experimental and analytical time and expense [50]. Yeh et al. [90] developed a rapid transmittance NIR spectroscopic technique to characterize the lignin composition of solid wood with the aid of simple multiple regression. Furthermore, partial least-square statistical method was employed to measure the lignin components of wood wafers. The results showed strong correlations between the predicted NIR and those obtained from the traditional chemical methods. Spectroscopic evaluations were done on milled wood samples in diffuse reflectance mode of NIR spectrometer and the spectra were employed to produce multivariate calibration models with genetic inverse least squares (GILS). NIR spectroscopic evaluation of these properties provides considerable faster analysis for screening purpose. The results obtained revealed that NIR spectroscopy coupled with multivariate calibration has the capacity to be employed for routine and fast analysis of these quality parameters of wood samples.

6.3.5 *Microscopy Analysis*

Lignocelluloses are known to be 3-D nano-composites and have dynamic mixtures which are multi-functional in orientation. Hence, the analysis of the composition is inadequate for the investigating of the effects of a pretreatment on a lignocellulose. As a way of illustration, it is inadequate to quantify the amount of lignin in biomass; hence, the necessity to determine the location of the lignin in the biomass and how it relates to the other components. Through pretreatments there is also relocation of lignin, delamination of cell walls which are required to enhance the dissolution of lignocellulosic biomasses. [43]. Lignocelluloses with their complex structures (mainly cellulose, hemicellulose, and lignin) are recalcitrance to deconstruction or hydrolysis. Suitable pretreatments improve digestibility of lignocellulosic biomass into useful end products. Microscopy techniques visualize, measure and quantify plant cell wall features as a result of pretreatments or conversion [5]. Generally, microscopy can be used to investigate the morphology of treated and untreated lignocelluloses successfully.

6.3.5.1 Scanning and Transmission Electron Microscopy

Scanning electron microscopy (SEM) is one of the most reliable tools extensively employed for investigating the surfaces of lignocelluloses [4]. SEM is used for surface characterization, morphology and analysis of microstructure. The energy-dispersive spectroscopy (EDS) attached to SEM is also used for the determination of elemental composition with about 1–3% precision [43]. Furthermore, the utilization of SEM reveals that surface erosions of biomasses are critically visualized, disrupted, and the relocalization of cell wall components and giving thorough accessibility and the enhancement of enzymatic hydrolysis [24]. It has been extensively used to study the morphological changes that brought about treatments of lignocelluloses [70, 73]. In a previous work, Ayeni et al. [10], studied the effect of alkaline peroxide oxidation (APO) pretreatment on rice husk biomass. The study showed the relationship between surface deformation and hemicellulose solubilization during biomass processing to valuable products [10]. SO_2 or CO_2 as pretreatment agent on sugarcane bagasse showed disorganization of biomass exposing the fibres [21, 27]. Figure 6.2 reveals the scanning electron microscopy images of raw and APO pretreated wheat straw biomass. With all these benefits of the use of SEM, the materials

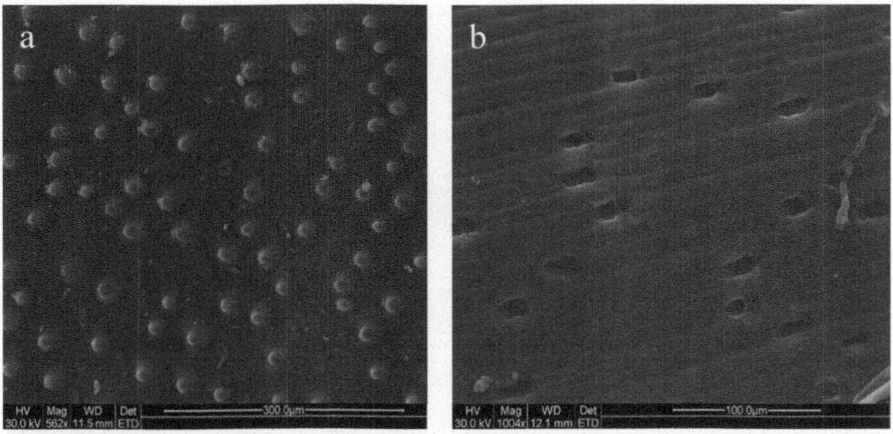

Fig. 6.2 SEM images of (**a**) raw and (**b**) APO-treated wheat straw biomass samples

to be inspected for SEM must not be charged but must possess enough conductivity; in addition, the stream of electrons in the vacuum space of the SEM could destroy the samples. These two disadvantages put a boundary to the application of SEM.

In some other studies, field emission scanning electron microscopy (FESEM) was used because it has higher resolution than conventional SEM. The microfibrillar orientations of biomass are visualized by FESEM [29]. Furthermore, scanning electron microscopy coupled with energy-dispersive x-ray diffraction (SEM-EDX) analysis can be carried out on raw and processed biomass. SEM-EDX analysis can successfully quantify elemental distributions along with morphological changes occurring after pretreatments of lignocelluloses [9]. Through SEM-EDX analysis, the empirical formulas for the raw and processed biomass samples may be established as reported by Ayeni and Daramola [9].

Transmission electron microscopy (TEM) is a technique used to study the ultrastructural behaviours of plant cell walls, especially through thin sections [29]. The ultra-thin biomass internal sections of TEM imaging of treated and untreated biomass distinguish between cell wall layers of primary, secondary and middle lamella. The three layers are bonded together giving a dense structure of the cell walls. Thin sections are prepared by cutting through by an ultra-cut ultramicrotome using a glass knife after staining biomass samples with 1% $KMnO_4$ [10, 29]. TEM can reveal the levels of lignification and delignification in the cell walls after biomass treatments [29, 80]. TEM analysis can also reveal differences in the distribution of lignin remaining after lignocelluloses treatments.

6.3.5.2 Light Microscopy Imaging

Light microscopy using different observation methods may be used on lignocelluloses revealing ultrastructural changes after biomass processing. After cutting thin biomass sections through ultra-cut ultramicrotome and staining (revealing meta-

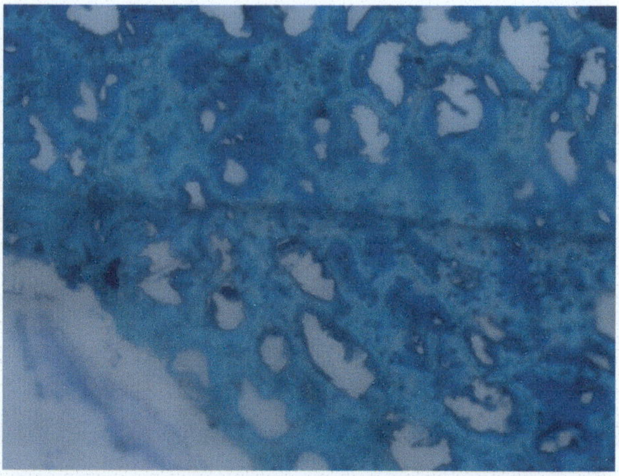

Fig. 6.3 Bright field light microscopy of raw corn cob cell wall showing the deposition of lignin (turquoise colour) and polysaccharides (violet colour). Readers are referred to the online version of this work for proper interpretation of colours

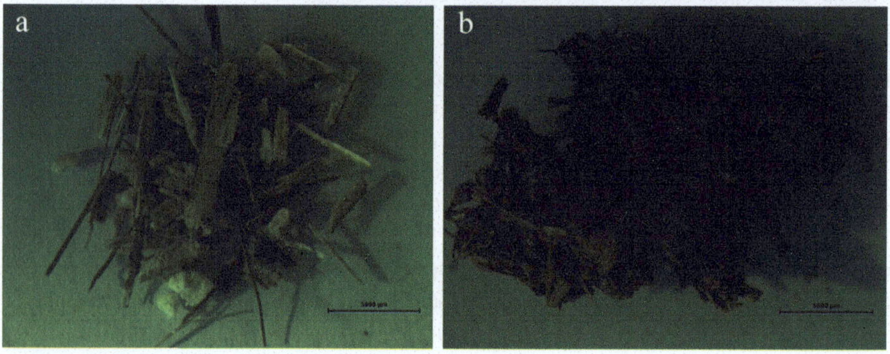

Fig. 6.4 Stereomicroscopy imaging for (**a**) raw and (**b**) APO-treated Siam weed (*Chromolaena odorata*)

chromatic properties of substances present in biomass tissues), light microscopy can be used to reveal lignin deposition and non-deposition, and increase in cellulose and hemicellulose compounds in biomass cell walls [9, 80]. Figure 6.3 shows the light microscopy bright field observation of a toluidine-blue-stained cell wall of raw corn cob biomass. Toluidine-blue stains polysaccharides violet and lignin turquoise (greenish blue).

6.3.5.3 Stereomicroscopy Imaging

This type of imaging shows physical variability after lignocelluloses treatments. Changes in particles sizes, particles shapes and particles colour can be visualized with a stereomicroscope. In a study [44], stereoscope micrographs of native corn stover and pretreated samples showed a general trend with particle variability in sizes, shapes and colour with more of thin, long fibres observed at higher NaOH loadings. Figure 6.4 shows the stereomicroscopy imaging of raw and treated Siam weed.

6.4 Conclusion

Lignocellulosic biomasses have gained much attention by virtue of their alternative use in or transgenic biofuel and biomaterial feedstocks progresses when compared to fossil-based resources. Advanced methods of evaluation for the analysis of untreated lignocellulosic biomasses are important, as these methods of analysis will swiftly and precisely screen large arrays of diverse plants. This chapter highlights and evaluates the methods used for analysing the untreated lignocellulosic biomasses. Methods such as composition analysis, physical and chemical measurements, types of spectroscopy techniques of measurement and types of microscopy analysis are discussed in this chapter. Special care and consideration are prerequisites for the compositional analysis, as this method involves drying of sample, reduction in size, occurrence of non-structural carbohydrates, extractable ash, proteins, starch, and elevated extractable contents. It is therefore, imperative that this method of analysis produce sustainable bioproducts without generating a lot of wastes. Spectroscopy techniques of measurement are used for analysing the structural and compositional interpretation of multifaceted lignocellulosic biomasses. Microscopy analysis is used for revealing some important aspects, such as ignin re-localization and cell wall delamination.

References

1. U.P. Agarwal, 1064 nm FT-Raman spectroscopy for investigations of plant cell walls and other biomass materials. Front. Plant Sci. **5**, 1–12 (2014)
2. U.P. Agarwal, Raman Spectroscopy in the analysis of cellulose nanomaterials, in *Nanocelluloses: Their Preparation, Properties and Applications*, ed. by U. P. Agarwal, R. H. Atalla, A. Isogai, (American Chemical Society, Washington, DC, 2017), pp. 75–90
3. U.P. Agarwal, Analysis of cellulose and lignocellulose materials by Raman Spectroscopy: A Review of the current status. Molecules **24**, 1–16 (2019)
4. H. Amiri, K. Karimi, Improvement of acetone, butanol, and ethanol production from woody biomass by using organosolv pretreatment. Bioprocess Biosyst. Eng. **38**, 1959–1972 (2015)
5. C.S. Anna, W. de Souza, Microscopy as a tool to follow deconstruction of lignocellulosic biomass, in *Current Microscopy Contribution to Advances in Science and Technology*, ed. by A. Mendez-Vilas, (FORMATEX, 2012), pp. 639–645
6. D.S. Argyropoulos, Quantitative phosphorus-31 NMR analysis of lignin, a new tool for the lignin chemist. J. Wood Chem. Technol. **14**, 45–63 (1994)
7. A.A. Awosusi, A. Ayeni, R. Adeleke, M.O. Daramola, Biocompositional and thermodecompositional analysis of South African agro-waste corncob and husk towards production of biocommodities. Asia Pac. J. Chem. Eng. **12**, 960–968 (2017)
8. A.O. Ayeni, O.A. Adeeyo, O.M. Oresegun, T.E. Oladimeji, Compositional analysis of lignocellulosic materials: Evaluation of an economically viable method suitable for woody and non-woody biomass. Am. J. Eng. Res. **4**, 14–19 (2015)
9. A.O. Ayeni, M.O. Daramola, Lignocellulosic biomass waste beneficiation: Evaluation of oxidative and non-oxidative pretreatment methodologies of South African corn cob. J. Environ. Chem. Eng. **5**, 1771–1779 (2017)
10. A.O. Ayeni, M.O. Daramola, A. Awoyomi, et al., Morphological modification of Chromolaena odorata cellulosic biomass using alkaline peroxide oxidation pretreatment methodology and its enzymatic conversion to biobased products. Cogent Eng. **5**, 1509665 (2018)

11. A.O. Ayeni, M.O. Daramola, P.T. Sekoai, et al., Statistical modelling and optimization of alkaline peroxide oxidation pretreatment process on rice husk cellulosic biomass to enhance enzymatic convertibility and fermentation to ethanol. Cellulose **25**, 2487–2504 (2018)
12. A.O. Ayeni, F.K. Hymore, S.N. Mudliar, et al., Hydrogen peroxide and lime based oxidative pretreatment of wood waste to enhance enzymatic hydrolysis for a biorefinery: Process parameters optimization using response surface methodology. Fuel **106**, 187–194 (2013)
13. A.O. Ayeni, R. Ogu, A.A. Awosusi, M.O. Daramola, Alkaline peroxide oxidation pretreatment of corn cob and rice husks for bioconversion into bio-commodities: Part A-Enzymatic convertibility of pretreated rice husks to reducing sugar. Paper presented at the 24th European Biomass Conference and Exhibition, Amsterdam, Netherlands, 6–9 June 2016 (2016)
14. A.O. Ayeni, J.A. Omoleye, F.K. Hymore, et al., Effective alkaline peroxide oxidation pretreatment of shea-tree sawdust for the production of biofuels: Kinetics of delignification and enzymatic conversion to sugar and subsequent production of ethanol by fermentation using Saccharomyces cerevisiae. Braz. J. Chem. Eng. **33**, 33–45 (2016)
15. A.B. Bjerre, A. Plöger, T. Simonsen et al., Quantification of solubilized hemicellulose from pretreated lignocellulose by acid hydrolysis and high-performance liquid chromatography, RISØ National Laboratory, Roskilde, Denmark, pp. 1–39 (1996)
16. C.D. Blasi, G. Signorelli, C. Di Russo, G. Rea, Product distribution from pyrolysis of wood and agricultural residues. Ind. Eng. Chem. Res. **38**, 2216–2224 (1999)
17. N. Carpita, M. McCann, The cell wall, in *Biochemistry and Molecular Biology of Plants*, ed. by B. Buchanan, W. Gruissem, R. Jones, (American Society of Plant Physiologists, Rockville, 2000), pp. 52–108
18. F. Carrilo, X. Colom, J.J. Suñol, et al., Structural FTIR analysis and the thermal characterization of lyocell and viscose-type fibers. Eur. Polym. J. **40**, 2229–2234 (2004)
19. D. Ciolacu, F. Ciolacu, V.I. Popa, Amorphous cellulose-Structure and characterization. Cellul. Chem. Technol. **45**, 13–21 (2011)
20. S.C. Corgié, H.M. Smith, L.P. Walker, Enzymatic transformations of cellulose assessed by quantitative high-throughput fourier transform infrared spectroscopy (QHT-FTIR). Biotechnol. Bioeng. **108**, 1509–1520 (2011)
21. R.C.N.R. Corrales, M.F.M. Teixeira, C.C. Perrone, et al., Structural evaluation of sugar cane bagasse steam pretreated in the presence of CO_2 and SO_2. Biotechnol. Biofuels **5**, 36 (2012)
22. F.B. de Oliveira, J. Bras, M.T. Borges Pimenta, et al., Production of cellulose nanocrystals from sugar cane bagasse fibers and pith. Ind. Crop. Prod. **93**, 48–57 (2016)
23. Z. Dische, Colour reactions of pentoses, in *Methods in Carbohydrate Chemistry*, ed. by R. L. Whistler, M. L. Wolfrom, (Academic Press, New York, 1962), pp. 475–514
24. B.S. Donohoe, T.B. Vinzant, R.T. Elander, V.R. Pallapolu, Y.Y. Lee, R.J. Garlock, V. Balan, B.E. Dale, Y. Kim, N.S. Mosier, M.R. Ladisch, M. Falls, M.T. Holtzapple, R. Sierra-Ramirez, J. Shi, M.A. Ebrik, T. Redmond, B. Yang, C.E. Wyman, B. Hames, S. Thomas, R.E. Warner, Surface and ultrastructural characterization of raw and pretreated switchgrass. Bioresour. Technol. **102**, 11097–11104 (2011)
25. A. Duval, M. Lawoko, A review on lignin-based polymeric, micro- and nano structured materials. React. Funct. Polym. **83**, 78–96 (2014)
26. M. Fan, D. Dai, B. Huang, Fourier transform infrared spectroscopy for natural fibres, in *Fourier Transform – Materials Analysis*, ed. by S. Mohammed Salih, (InTech, 2012). https://doi.org/10.5772/35482
27. V. Ferreira-Leitão, C.C. Perrone, J. Rodrigues, et al., An approach to the utilisation of CO_2 as impregnating agent in steam pretreatment of sugar cane bagasse and leaves for ethanol production. Biotechnol. Biofuels **3**, 1–8 (2010)
28. T. Foyle, L. Jennings, P. Mulcahy, Compositional analysis of lignocellulosic materials: Evaluation of methods used for sugar analysis of waste paper and straw. Bioresour. Technol. **98**, 3026–3036 (2007)
29. J. Fromm, B. Rockel, S. Lautner, Lignin distribution in wood cell walls determined by TEM and backscattered SEM techniques. J. Struct. Biol. **143**, 77–84 (2003)

30. S.C. Fry, *The Growing Cell Wall: Chemical and Metabolic Analysis* (Blackburn Press, New Jersey, 2000), pp. 1–333
31. B. Godin, R. Agneessens, P. Gerin, J. Delcarte, Structural Carbohydrates in a plant biomass: Correlations between the detergent fiber and dietary fiber methods. J. Agric. Food Chem. **62**, 5609–5616 (2014)
32. B. Godin, R. Agneessens, P. Gerin, et al., Composition of structural carbohydrates in biomass: Precision of a liquid chromatography method using a neutral detergent extraction and a charged aerosol detector. Talanta **85**, 2014–2026 (2011)
33. B. Godin, R. Agneessens, S. Gofflot, et al., Revue sur les méthodes de caractérisation des polysaccharides structuraux des biomasses lignocellulosiques. Biotechnol. Agron. Soc. Environ. **15**, 165–182 (2011)
34. B. Godin, S. Lamaudière, R. Agneessens, et al., Chemical composition and biofuel potentials of a wide diversity of plant biomasses. Energy Fuel **27**, 2588–2598 (2013)
35. A.A. Guilherme, P.V.F. Dantas, E.S. Santos, et al., Evaluation of composition, characterization and enzymatic hydrolysis of pretreated sugar cane bagasse. Braz. J. Chem. Eng. **32**, 32–33 (2015)
36. H.M. Hernández-Hernández, J.J. Chanona-Pérez, E. Terrés, A. Vega, P. Ligero, R.R. Farrera-Rebollo, S. Villanueva, Microscopy and spectroscopy tools for the description of delignification. Cell. Chem. Technol. **53**, 87–97 (2019)
37. S.E. Jacobsen, C.E. Wyman, Cellulose and hemicelluose hydrolysis models for application to current and novel pretreatment processes. Appl. Biochem. Biotechnol. **84–86**, 81–96 (2000)
38. H. Janshekar, A. Fiechter, Lignin: Biosynthesis, application and biodegradation. Adv. Biochem. Biotechnol. **27**, 119–178 (1983)
39. J.P. Joseleau, J. Comtat, K. Ruel, Chemical structure of xylans and their interaction in the plant cell walls, in *Progress in Biotechnology – Xylans and Xylanases*, ed. by J. Visser, (Elsevier, 1992), pp. 1–15
40. X. Ju, M. Bowden, E.E. Brown, X. Zhang, An improved X-ray diffraction method for cellulose crystallinity measurement. Carbohydr. Polym. **123**, 476–481 (2015)
41. H.-J. Jung, Analysis of forage fiber and cell walls in ruminant nutrition. J. Nutr. **127**, 810S–813S (1997)
42. H.-J. Jung, J. Lamb, Prediction of cell wall polysaccharide and lignin concentrations of alfalfa stems from detergent fiber analysis. Biomass Bioenergy **27**, 365–373 (2004)
43. K. Karimi, M.J. Taherzadeh, A critical review of analytical methods in pretreatment of lignocelluloses: Composition, imaging, and crystallinity. Bioresour. Technol. **200**, 1008–1018 (2016)
44. E.M. Karp, B.S. Donohoe, M.H. O'Brien, et al., Alkaline pretreatment of corn stover: Bench-scale fractionation and stream characterization. ACS Sustain. Chem. Eng. **2**, 1481–1491 (2014)
45. K.L. Larsen, S. Barsberg, Environmental effects on the lignin model monomer, vanillyl alcohol, studied by Raman spectroscopy. J. Phys. Chem. B **115**, 11470–11480 (2011)
46. S. Li, S. Xu, S. Liu, C. Yang, et al., Fast pyrolysis of biomass in free-fall reactor for hydrogen-rich gas. Fuel Process. Technol. **85**, 1201–1211 (2004)
47. L. Liu, X.P. Ye, A.R. Womac, S. Sokhansanj, Variability of biomass chemical composition and rapid analysis using FT-NIR techniques. Carbohydr. Polym. **81**, 820–829 (2010)
48. Y. Lu, Y.-C. Lu, H.-Q. Hu, F.-J. Xie, X.-Y. Wei, X. Fan, Structural characterization of lignin and its degradation products with spectroscopic methods: Review. J. Spectrosc., 1–15 (2017)
49. J.S. Lupoi, E. Gjersing, M.F. Davis, Evaluating lignocellulosic biomass, its derivatives, and downstream products with Raman spectroscopy: A Review. Front. Bioeng. Biotechnol. **3**, 1–18 (2015)
50. J.S. Lupoi, S. Singh, R. Parthasarathi, B.A. Simmons, R.J. Henry, Recent innovations in analytical methods for the qualitative and quantitative assessment of lignin. Renew. Sust. Energ. Rev. **49**, 871–906 (2015)
51. J.S. Lupoi, S. Singh, B.A. Simmons, R.J. Henry, Assessment of lignocellulosic biomass using analytical spectroscopy: An evolution to high-throughput techniques. Bioenergy Res. **7**, 1–23 (2014)

52. J.S. Lupoi, E.A. Smith, Characterization of woody and herbaceous biomasses lignin composition with 1064 nm dispersive multichannel Raman spectroscopy. Appl. Spectrosc. **66**, 903–910 (2012)
53. M. Makarem, C.M. Lee, K. Kafle, S. Huang, I. Chae, H. Yang, J.D. Kubicki, S.H. Kim, Probing cellulose structures with vibrational spectroscopy. Cellulose **26**, 35–79 (2019)
54. T. Mani, P. Murugan, J. Abedi, et al., Pyrolysis of wheat straw in a thermogravimetric analyser: Effect of particle size and heating rate on devolatilization and estimation of global kinetics. Chem. Eng. Res. Des. **88**, 952–958 (2010)
55. D.A. Matos, I.P. Whitney, M.J. Harrington, et al., Cell walls and the developmental anatomy of the *Brachypodium distachyon* stem internode. PLoS One **8**(11), e80640 (2013). https://doi.org/10.1371/journal
56. R.L. McCreery, *Raman Spectroscopy for Chemical Analysis* (Wiley, New York, 2000), p. 448
57. M.W. Meyer, J.S. Lupoi, E.A. Smith, 1064 nm dispersive multichannel Raman spectroscopy for the analysis of plant lignin. Anal. Chim. Acta **706**, 164–170 (2011)
58. G.L. Miller, Use of dinitrosalicylic acid reagent for determination of reducing sugar. Anal. Chem. **31**, 426–428 (1959)
59. T. Milne, H. Clum, F. Agblevor, D.K. Johnson, Standard analytical methods. Biomass Bioenergy **2**, 341–366 (1992)
60. B.I. Na, S.J. Chang, K.H. Lee, G. Lee, J.W. Lee, Characterization of cell wall structure in dilute acidpretreated biomass by confocal Raman microscopy and enzymatic hydrolysis. Biomass Bioenergy **93**, 33–37 (2016)
61. M.L. Nelson, R.T. O'Connor, Relation of certain infrared bands to cellulose crystallinity and crystal lattice type. Part II. A new infrared ratio for estimation of crystallinity in cellulose I and II. J. Appl. Polym. Sci. **8**, 1325–1341 (1964)
62. D.R. Nhuchhen, P. Abdul Salam, Estimation of higher heating value of biomass from proximate analysis: A new approach. Fuel **99**, 55–63 (2012)
63. R.T. O'Connor, E.F. DuPré, D. Mitcham, Application of infrared absorption spectroscopy to investigations of cotton and modified cottons. Part I: Physical and crystalline modifications and oxidation. Text. Res. J. **28**, 382–392 (1958)
64. S.Y. Oh, D.I. Yoo, Y. Shin, et al., FTIR analysis of cellulose treated with sodium hydroxide and carbon dioxide. Carbohydr. Res. **340**, 417–428 (2005)
65. J. Parikh, S.A. Channiwala, G.K. Ghosal, A correlation for calculating HHV from proximate analysis of solid fuels. Fuel **84**, 487–494 (2005)
66. S. Park, J.O. Baker, M.E. Himmel, P.A. Parilla, D.K. Johnson, Cellulose crystallinity index: Measurement techniques and their impact on interpreting cellulase performance. Biotechnol. Biofuels **3**, 1754–6834 (2010)
67. M. Poletto, H.L. Ornaghi Júnior, A.J. Zattera, Native cellulose: Structure: Characterization and thermal properties. Materials **7**, 6105–6119 (2014)
68. Y. Pu, B. Hallac, A.J. Ragauskas, Plant biomass characterization: Application of solution- and solid-state NMR spectroscopy, in *Aqueous Pretreatment of Plant Biomass for Biological and Chemical Conversion to Fuels and Chemicals*, (Wiley, 2013), pp. 369–390
69. J. Puls, J. Schusiel, Chemistry of hemicelluloses: Relationship between hemicellulose structure and enzymes required for hydrolysis, in *Hemicelluloses and Hemicellulases*, ed. by M. P. Coughlan, G. P. Hazlewood, (Portland Press, 1993), pp. 1–27
70. C.A. Rezende, P. Maziero, E.R. Azevedo, et al., Chemical and morphological characterization of sugarcane bagasse submitted to delignification process for enhanced enzymatic digestibility. Biotechnol. Biofuels **4**, 54 (2011)
71. D.R. Robert, G. Brumow, Quantitative estimation of hydroxyl group in milled wood lignin from spruce and in a dehydrogenation polymer from coniferyl alcohol using ^{13}C NMR spectroscopy. Holzforschung **38**, 85–90 (1984)
72. L. Segal, J.J. Creely, A.E. Martin, et al., An empirical method for estimating the degree of crystallinity of native cellulose using the X-ray diffractometer. Text. Res. J. **29**, 786–794 (1959)
73. M.J. Selig, T.B. Vinzant, M.E. Himmel, et al., The effect of lignin removal by alkaline peroxide pretreatment on the susceptibility of corn stover to purified cellulolytic and xylanolytic Enzymes. Appl. Biochem. Biotechnol. **155**, 397–406 (2009)

74. C. Sheng, J.L.T. Azevedo, Estimating the higher heating value of biomass fuels from basic analysis data. Biomass Bioenergy **28**, 499–507 (2005)
75. Y.D. Singh, P. Mahanta, U. Bora, Comprehensive characterization of lignocellulosic biomass through proximate, ultimate and compositional analysis for bioenergy production. Renew. Energy **103**, 490–500 (2017)
76. A. Sluiter, B. Hames, R. Ruiz et al., Determination of sugars, byproducts, and degradation products in Liquid Fraction Process Samples, National Energy Renewable Laboratory, Golden Colorado, USA, NREL/TP-510-42623 (2008a)
77. A. Sluiter, B. Hames, R. Ruiz et al., Determination of structural carbohydrates and lignin in biomass, National Renewable Energy Laboratory, Golden, CO, USA, pp. 1–16, (2008b)
78. A. Sluiter, R. Ruiz, C. Scarlata et al., Determination of extractives in biomass. National Energy Renewable Laboratory, Golden, CO, USA, NREL/TP-510-42619, pp. 1–9, (2008c)
79. J. Sluiter, R. Ruiz, C. Scarlata, et al., Compositional analysis of lignocellulosic feedstocks. 1. Review and description of methods. J. Agric. Food Chem. **58**, 9043–9053 (2010)
80. S. Tao, S. Khanizadeh, H. Zhang, et al., Anatomy, ultrastructure and lignin distribution of stone cells in two Pyrus species. Plant Sci. **176**, 413–419 (2009)
81. N. Terashima, M. Yoshida, J. Hafrén, et al., Proposed supramolecular structure of lignin in softwood tracheid compound middle lamella regions. Holzforschung **66**, 907–915 (2012)
82. N. Terinte, R. Ibbett, K.C. Schuster, Overview on native cellulose and microcrystalline cellulose I structure studied by X-ray diffraction (waxd): Comparison between measurement techniques. Lenzinger Ber. **89**, 118–131 (2011)
83. O. Theander, P. Aman, E. Westerlund, et al., Total dietary fiber determined as neutral sugar residues uronic acid residues, and Klason lignin (the Uppsala method): Collaborative study. J. AOAC Int. **78**, 1030–1044 (1995)
84. O. Theander, E. Westerlund, Quantitative analysis of cell wall components, in *Forage Cell Wall Structure and Digestibility*, ed. by H. Jung, D. Buxton, R. Hatfield, J. Ralph, (American Society of Agronomy, Madison, WI, 1993), pp. 83–104
85. B. Uner, I. Karaman, H. Tanriverdi, D. Ozdemir, Determination of lignin and extractive content of Turkish Pine (Pinus brutia Ten.) trees using near infrared spectroscopy and multivariate calibration. Wood Sci. Technol. **45**, 121–134 (2011)
86. P.J. Van Soest, Collaborative study of acid-detergent fiber and lignin. J. Assoc. Off. Anal. Chem. **56**, 781–784 (1973)
87. P.J. Van Soest, R.H. Wine, Use of detergents in the analysis of fibrous feeds. IV. Determination of plant cell-wall constituents. J. Assoc. Off. Anal. Chem. **50**, 50–55 (1967)
88. J.H. Wiley, R.H. Atalla, Band assignments in the Raman spectra of celluloses. Carbohydr. Res. **160**, 113–129 (1987)
89. S. Yao, G. Wu, M. Xing, S. Zhou, J. Pu, Determination of lignin content in Acacia spp. using near-infrared reflectance spectroscopy. Bioresources **5**, 556–562 (2010)
90. T.-F. Yeh, C. H-m, J.F. Kadla, Rapid prediction of solid wood lignin content using transmittance near-infrared spectroscopy. J. Agric. Food Chem. **52**, 1435–1439 (2004)

Part II
Pretreatment and Processes for Conversion of Biomass to Value-Added Commodities

Chapter 7
Biological and Non-Biological Methods for Lignocellulosic Biomass Deconstruction

A. O. Ayeni, M. O. Daramola, A. E. Adetayo, P. T. Sekoai, O. C. Nwinyi, and O. Ejekwu

7.1 Introduction

The high dependence on hydrocarbon fuels has resulted in various bottlenecks such as environmental deterioration, escalating energy prices, and health issues [64]. Therefore, scientists are currently searching for non-polluting and cost-effective technologies in order to alleviate these problems. Biofuels are considered as a potential replacement for fossil fuels owing to their numerous and overwhelming socio-economic merits [50]. In recent years, biofuel technologies, such as biohydrogen, bioethanol, biodiesel, and biogas, have been gaining increasing prominence in the literature as a result of their socio-economic benefits [51, 52]. In our present world, a majority of the world's biofuels are generated from various consumable crops such as wheat, sorghum, sugar cane, corn, etc. [10]. However, in order to have

A. O. Ayeni (✉)
Covenant University, College of Engineering, Department of Chemical Engineering, Ota, Nigeria
e-mail: augustine.ayeni@covenantuniversity.edu.ng

O. C. Nwinyi
Covenant University, College of Science & Technology, Department of Biological Sciences, Canaan land, Ota, Nigeria

M. O. Daramola · O. Ejekwu
Department of Chemical Engineering, Faculty of Engineering, Built Environment and Information Technology, University of Pretoria, Hatfield, Pretoria, South Africa

A. E. Adetayo
Department of Material Science, Engineering & Commercialization, Texas State University, San Marcos, TX, USA

P. T. Sekoai
The University of Hong Kong, Hong Kong SAR, China

M. O. Daramola, A. O. Ayeni (eds.), *Valorization of Biomass to Value-Added Commodities*, Green Energy and Technology,
https://doi.org/10.1007/978-3-030-38032-8_7

a viable biofuel-driven economy, this process should make economic and environmental sense. Consequently, researchers are now focusing on second-generation biofuel options because they do not pose a threat to food security and use highly accessible feedstocks like wastes [52].

Lignocellulosic wastes are viewed as suitable feedstocks for biofuels because these materials are highly abundant in nature and inexpensive [53, 54]. Moreover, it has been shown that these materials can produce up to 23 exajoules of energy and meet up to 14.8% of global transport fuel requirements [44]. However, the use of lignocellulosic materials presents some challenges in biofuels. The complex polymeric structure of lignocellulosic biomass makes it difficult for microorganisms to access the fermentable sugars (Fig. 7.1) [43]. This implies that an initial pretreatment process is needed, prior to the fermentation process [2, 27] (Fig. 7.2).

An efficient methodology must meet the requirements so as to effectively break the lignocellulosic structure, have reduced crystallinity, have minimum inhibitory compounds, and have low operational costs. An integrated approach for the lignocellulose

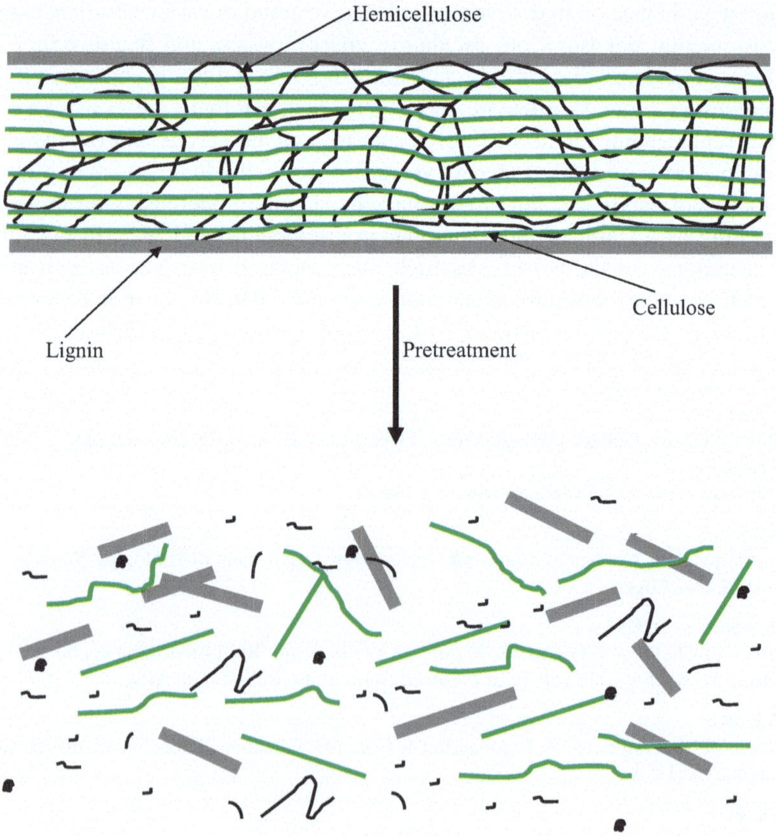

Fig. 7.1 Alteration of structural features of lignocellulose after pretreatment. (Modified from Mosier et al. [43])

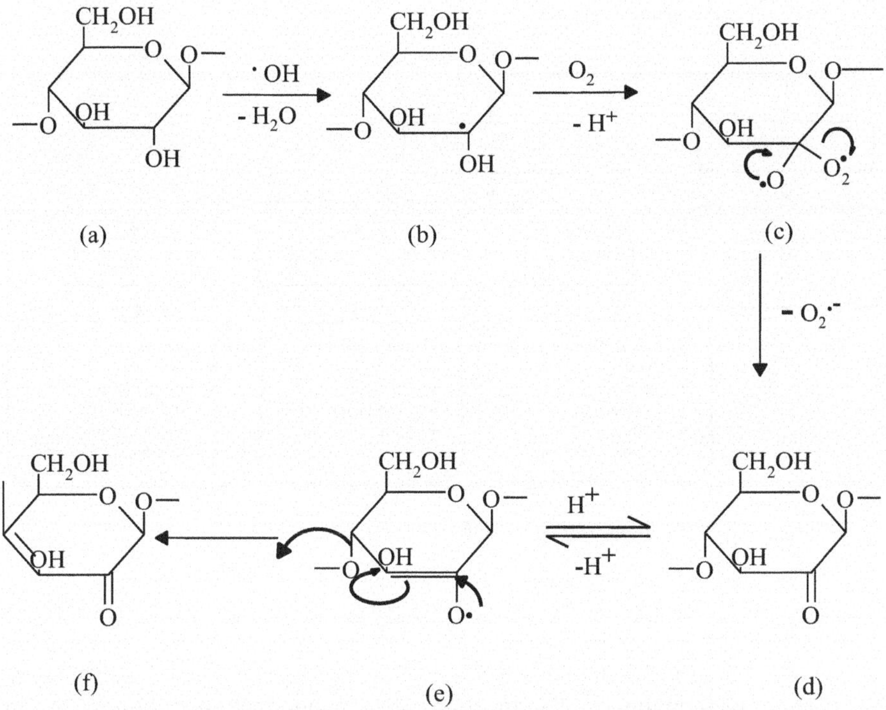

Fig. 7.2 Reaction mechanism during carbohydrate degradation by hydroxyl radical during alkaline oxidative delignification process. (Modified from Gierer [27])

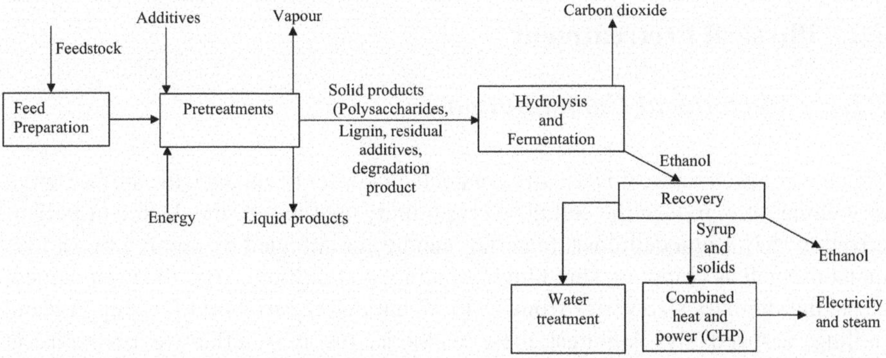

Fig. 7.3 Schematic diagram showing the utilization of lignocellulosic biomass in ethanol production

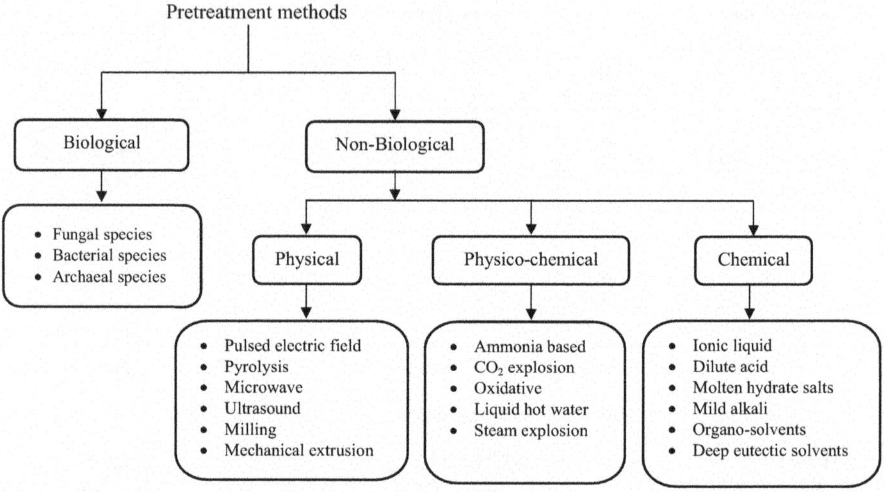

Fig. 7.4 Lignocellulosic biomass pretreatment methodologies

process is shown in Fig. 7.3. Generally, no particular method of pretreatment is absolutely suitable for all lignocelluloses. Each pretreatment is specific depending on choices. These methodologies have their own advantages and disadvantages. Pretreatment methodologies' classification can be divided into two (biological and non-biological), as shown in Fig. 7.4. For example, in the lignocellulosic bioethanol production process, approximately 18–20% of the operational costs is derived from the pretreatment steps [11]. This paper provides an overview of the current methods that are used in the pretreatment of lignocellulosic materials.

7.2 Physical Pretreatment

7.2.1 Mechanical Comminution

This type of pretreatment is usually conducted in order to enhance the surface-area-to-volume ratio, reduce the cellulose crystallinity, and reduce the degree of polymerization [57]. Lignocellulosic materials can be comminuted by combining various methods such as chipping, shredding, grinding, and milling. After the pretreatment steps, the biomass size varies from 10 to 30 nm and 0.2 to 2 mm for chipping and milling, respectively. Amongst these methods, the most effective pretreatment method for lignocellulosic biomass is the vibratory ball milling [41]. However, the effectiveness of mechanical comminution is still dependent on the characteristics of the biomass [17]. A detailed description of the effects of the mechanical comminution method on sizes of lignocellulosic materials and the energies required to achieve these sizes is provided by Cadoche and López [17].

7.2.2 Pyrolysis

Pyrolysis is a fast and flexible method providing high yields of products (in the form of vapours, aerosols, and some char and gas) from lignocelluloses [16]. The pyrolytic process operates in anoxic conditions. The process uses high temperatures to break down biomass residues into gases and vapours, which are used in secondary processes to produce a wide range of products. Pyrolysis uses various pretreatment conditions and this depends on the type of product that is desired (char, liquid, and gaseous material). Consequently, various operational parameters, such as reaction temperature, pressure, catalysts, hot vapour residence time, solids' residence time, etc., influence the overall process performance [15].

A temperature of more than 300 °C has been shown to be effective in the pretreatment of lignocellulosic biomass due to the rapid conversion of the cellulosic material into products that are gaseous and into char as the residue [36, 56]. On the contrary, lower temperature reduces the conversion efficiency and produces less volatile products. A high cellulosic conversion of up to 85% was achieved using mild acid hydrolysis (1 N H_2SO_4, 97 °C, 2.5 h). The pyrolytic processes are also optimized by the addition of oxygen [56]. Furthermore, it has been shown that the energy inputs can be reduced by using catalysts like zinc chloride, zeolites, and sodium carbonate, amongst others [55]. For an efficient pyrolytic process, these procedures must be adhered to: (i) the feed must have a moisture content of less than 10 wt%, (ii) the materials must be finely ground (<3 mm), (iii) the reaction temperature must be carefully controlled, (iv) there should be a rapid removal of product char, (v) there should be rapid cooling of the pyrolytic vapours to minimize thermal cracking, and (vi) the ash content should be below 2.5 wt% [12].

7.2.2.1 High-Energy Radiation

High-energy radiation methods, such as gamma rays, ultrasound, electron beam, microwave, and ultraviolet rays, have also been demonstrated to be useful in improving the digestibility of cellulosic materials in biomass residues [13]. As a result of high energy, there is a decrease in the degree of polymerization and cellulose crystallinity, increase in surface area, and hydrolysis of hemicelluloses.

7.3 Physico-Chemical Pretreatment

7.3.1 Steam Explosion (Autohydrolysis)

During steam explosion, the lignocellulosic materials undergo heating at 160–260 °C. It is a pretreatment method that is widely used and which is operated using a closed system with high pressure ($7 \times 10^5 - 5 \times 10^6$ Pascals) for short pretreatment period (0.5–20 min). The disruption of the polymeric features occurs at high temperature and pressure releasing the fermentable sugars. A study that evaluated the pretreatment of wood chips showed that 90% of enzymatic hydrolysis could be achieved using steam explosion, whereas the untreated wood chips produced only 15% of hydrolysis [28]. This method of pretreatment relies heavily on process parameters such as the residence time, pressure, temperature, moisture, and composition of the lignocellulosic material [23].

In order to achieve very high output for hemicelluloses' solubilization, high temperature versus shorter residence period (190 °C, 10 min) or low temperature versus prolonged residence period (270 °C, 1 min) [23] is recommended. Looking into the literature, it has been reported that low temperature versus prolonged residence period is often time favourable for lignocelluloses' feedstocks [62]. Furthermore, the use of chemicals such as sulphuric acid (H_2SO_4) and carbon dioxide (CO_2) in steam explosion offers many advantages such as enhanced hydrolytic efficiency, reduction in the formation of inhibitory compounds, and improved removal of hemicellulose [42]. A study was conducted in order to assess the effect of steam explosion on sugarcane bagasse by Morjanoff and Gray [42]. The authors obtained a high sugar yield of 65.1 g sugars/100 g biomass at an operational temperature of 220 °C, residence time of 30 seconds, water-to-solids ratio of 2, and H_2SO_4 of 1%, respectively.

Steam explosion is advantageous in the pretreatment of lignocellulosic materials because it requires minimum energy inputs. On the other hand, conventional mechanical approaches use up to 70% of energy during the pretreatment of lignocellulosic biomass [31].

However, it has some limitations just like any other method. It partially removes the lignin–carbohydrate matrix and other compounds that have an inhibitory effect on microbial growth during the upstream process [21, 38].

7.3.2 Ammonia Fibre Explosion (AFEX)

Ammonia fibre explosion (AFEX) is a physico-chemical pretreatment that combines liquid ammonia, high temperatures, and pressure [32]. AFEX is used in the pretreatment of various plant materials such as barley straw, corn stover, grasses, wheat straw, etc. [40]. However, AFEX is not effective in the solubilization of hemicellulose when compared to acid pretreatment and steam explosion [40, 60]. Holtzapple et al. [33] observed a more than 90% hydrolytic efficiency in cellulosic materials (Bermuda grass and bagasse) when using AFEX.

Nonetheless, this method was not effective in feedstocks that had a high lignin content, such as newspaper (18–30% lignin) and aspen chips (25% lignin) [39]. There are also environmental concerns about the utilization of ammonia. Hence, the ammonia must be recycled at the end of each process to prevent environmental pollution [32].

7.3.3 *CO2 Explosion*

Carbon dioxide (CO_2) explosion is also applied in the pretreatment of lignocellulosic biomass. It was previously reported that CO_2 can form carbonic acid and increase the rate of hydrolysis. A study which evaluated this method using alfalfa biomass achieved around 75% of hydrolytic efficiency during 24 h of the enzymatic hydrolysis [22]. Nonetheless, compared to steam or ammonia explosion treatments the yield was higher. However, during the enzymatic hydrolysis, the yield was higher for the CO_2 explosion [65].

7.4 Chemical Pretreatment

7.4.1 *Ozonolysis*

Ozone is also used to hydrolyse a wide variety of lignocellulosic materials such as sugarcane bagasse, wheat straw, corn, grass, sawdust, etc. [46, 59]. Vidal and Molinier [59] reported that this process does not effectively remove lignin and cellulose but is useful towards the removal of hemicellulose. Nonetheless, the pretreatment agent, ozone, is applied in an appreciable amount, thus making the process expensive.

7.4.2 *Acid Hydrolysis*

Acids such as H_2SO_4 and HCl (hydrochloric acid) are commonly used in the pretreatments if there are lignocellulosic wastes.

Although they are effective towards the pretreatments of lignocellulosic materials, they are expensive and toxic as well [25]. Several authors have reported promising results when using acids like H_2SO_4 in the pretreatments of plant-based materials [15, 28, 32, 44]. However, from an environmental and economic perspective, the use of acids is not suitable in bioprocesses.

7.4.3 *Alkaline Hydrolysis*

Bases are also used in the alteration of the polymeric structure of lignocellulosic materials [14, 26, 39]. It has been shown that bases disrupt the intermolecular ester bonds in biomass materials, as shown in Fig. 7.5 [26, 29]. It was observed that the digestibility of hardwood increased up to 55% after being subjected to sodium

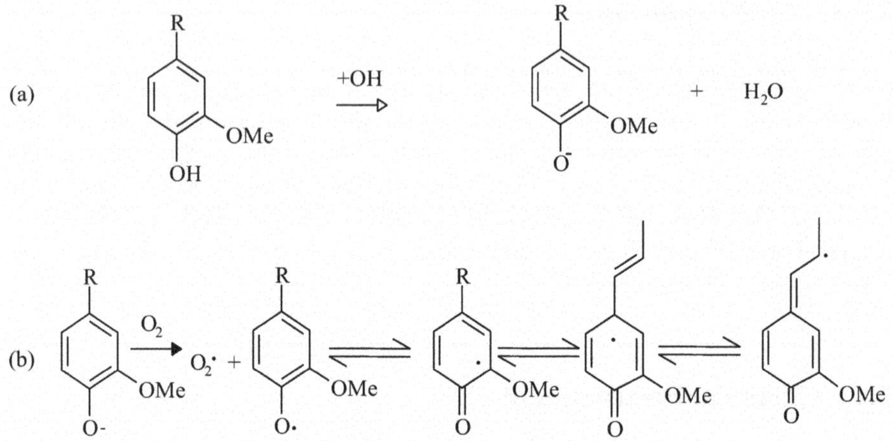

Fig. 7.5 Radical chain reactions during oxygen delignification in alkaline condition

hydroxide (NaOH) treatment, thus demonstrating the effectiveness of this method [41].

The first step in the alkaline hydrolysis mechanism, the lignin reaction is initiated where the free phenolic groups are ionized in order to liberate a phenolate ion (Fig. 7.5a). In the second step, the radical oxygen reacts with the released phenolic lignin, whereby a reactive intermediate (hydroperoxide) is formed (Fig. 7.5b).

7.4.4 Oxidative Delignification

In alkaline peroxide oxidative (APO), delignification takes place in the presence of hydrogen peroxide (H_2O_2) [4, 7, 8]. During this pretreatment process, delignification is enhanced by adjusting the pH of the oxygenated solution or mixture (pH made between 11 and 12 using sodium hydroxide (NaOH)/calcium hydroxide ($Ca(OH)_2$)/or any other related alkaline solutions) [3, 5, 6, 47, 48]. Pretreatments are carried out at fairly low temperatures and pressures for specified durations like varying from few minutes to a day [58], which can often be classified as short-term pretreatments [8]. Oxidative delignification solubilizes lignin, thus, disrupting the polymeric matrix, leading to the accessibility of cellulose. Ayeni and Daramola [6] reported 78% lignin solubilization of corncob pretreated in 1% (v/v) H_2O_2/NaOH solution with a high degree of hemicellulose solubilization (79%). Other authors recorded good results when evaluating this method on different types of lignocellulosic materials [9, 19].

7.4.5 Organosolv Process

In the Organosolv process, a mixture of organic solvent (e.g. acetone, ethylene glycol, methanol, etc.) and inorganic catalyst (HCl or H_2SO_4) is used to disrupt the polymeric structure of lignocellulosic biomass [20]. Organosolv breaks the three components into one part (cellulose-laden part) and finally precipitates the hemicellulose fluid [45]. This process is also advantageous because the concentration of hazardous chemicals is reduced.

7.5 Biological Pretreatment

Microorganisms are now gaining increasing prominence in the pretreatment of lignocellulosic materials, owing to their economic and environmental benefits. Microorganisms of the fungal origin (*Fomitopsis palustris*, *Trichoderma reesei*, etc.) with some cellulosic degrading bacteria (*Thermoanaerobacterium* sp. and *Caldicellulosiruptor, Clostridium cellulolyticum* sp.) are widely used in the hydrolysis of lignocellulosic feedstocks [1, 24, 34, 35, 37, 49, 61, 63]. Moreover, fungal species, such as brown-, white-, and soft-rot fungi, have previously been shown to have a remarkable effect on the delignification of cellulosic wastes [26, 49]. A comparison of the various pretreatment methodologies is given in Table 7.1.

7.6 Conclusion

All the methodologies for lignocelluloses' pretreatment have proven to be efficient, based on the biomass species, geographical location, and target products. Any method chosen with the attendant challenges can, with time, be overcome by advancement in technology. In order to select a particular process, efforts should be geared towards considering the feedstock properties, economy of process, and yield of product. It should be noted that the biomass processing must be treated using a process-integrated approach where the processes from the upstream and downstream steps are incorporated in one single line.

Table 7.1 Comparing known pretreatment methodologies with changes in their composition [66]

	Composition				
Pretreatment	Cellulose	Hemicellulose	Lignin	Advantages	Disadvantages
Ball milling	Decrystallization occurs	Not removed	Not removed	Decrystallization occurs	Consumes energy
Microwave irradiation	Depolymerization occurs	Solubilization is appreciable	Solubilization is appreciable	Operation is easy, with the efficiency of heating being very high	High economic cost
Dilute acid	Depolymerization occurs	Solubilization ranges between 80% and 100%	Not easily digested	Operated at moderate condition with pentose sugar yields	Acid not easily recovered with significant product inhibition
Sodium hydroxide	Biomass shows appreciable swelling	Solubilization is appreciable	Solubilization is appreciable most times to be greater than 50%	The ester bonds in the lignin complex are removed adequately	Reagent could be expensive with little alkali recovery
Lime	Depolymerization is not quite appreciable	Solubilization is appreciable	Incomplete solubilization, not appreciable	Not expensive	The poor solubility of lime makes the method ineffective
Alkaline peroxide	Biomass swelling is appreciable	Solubilization is close to 50%	Solubilization is close to 50%	Lignin removal is efficient	Soluble sugar is efficiently decomposed in slurry
Ozonolysis	Depolymerization hardly occurs	Solubilization is not appreciable	Up to 70% solubilization	Effective delignification, mild conditions	High dosages of ozone needed
Wet oxidation	A little extent of depolymerization can be noticed	Solubilization is high with over 90%	Solubilization is within 40–50%	Lignin and hemicelluloses are effectively removed	Not effective for biomass with high lignin content
Organosolv	Significant swelling	Solubilization is high	Solubilization is high	Lignin removal is effective with high xylose output	Expensive to recover solvent

Steam explosion	Depolymerization is not too appreciable	Solubilization could be in the range of 80–100%	Enough redistribution occurs with lesser solubility	There is high-energy efficiency	By-product inhibition occurs with degradation of xylan
AFEX	Decrystallization	Solubilization could be close to 60%	Solubilization could range between 10% and 20%	Xylan is relatively retained without the formation of inhibitors	This is not effective for high lignin feedstock, while recovery of ammonia is hard to achieve
CO_2 explosion	The swelling of biomass is appreciable	Solubilization is highly appreciable	Solubilization is high	Temperature is low without by-product inhibition	The process is expensive
Hydro-thermolysis	There is appreciable swelling	Solubilization is almost total	Solubilization is appreciable	It is not a corrosive process because no chemical is involved	This is not effective for high lignin biomass
Biological	20–30% depolymerization	Solubilization could be as high as 89%	Lignin removal could be close to 40%	Little energy is required with effective lignin removal	Process is mild, so the hydrolysis rate is slow with loss of cellulose

References

1. D.E. Akin, A. Sethuraman, W.H. Morrison, S.A. Martin, K.E. Eriksson, Microbial delignification with white rot fungi improves forage digestibility. Appl. Environ. Microbiol. **59**, 4274–4282 (1993)
2. A.O. Ayeni, Short-Term Lime Pretreatment and Enzymatic Conversion of Sawdust into Ethanol. PhD Thesis, Covenant University, Ota, Nigeria (2013)
3. A.O. Ayeni, M.O. Daramola, A. Awoyomi, et al., Morphological modification of Chromolaena odorata cellulosic biomass using alkaline peroxide oxidation pretreatment methodology and its enzymatic conversion to biobased products. Cogent Eng. **5**, 1509665 (2018b)
4. A.O. Ayeni, M.O. Daramola, P.T. Sekoai, et al., Statistical modelling and optimization of alkaline peroxide oxidation pretreatment process on rice husk cellulosic biomass to enhance enzymatic convertibility and fermentation to ethanol. Cellulose **25**, 2487–2250 (2018a)
5. A.O. Ayeni, O. Adeeyo, A.A. Awosusi et al., Alkaline peroxide oxidation pretreatment of corn cob and rice husks for bioconversion into bio-commodities: Part B- Enzymatic convertibility of pretreated corn cob to reducing sugar. 25th European Biomass Conference and Exhibition, Stockholm, Sweden, 12–15 June 2017 (2017)
6. A.O. Ayeni, M.O. Daramola, Lignocellulosic biomass beneficiation: Evaluation of oxidative and non-oxidative pretreatment methodologies for south African corn cob. J. Environ. Chem. Eng. **5**, 1771–1779 (2017)
7. A.O. Ayeni, J.A. Omoleye, S. Mudliar, et al., Utilization of lignocellulosic waste for ethanol production: Enzymatic digestibility and fermentation of Pretreated shea tree sawdust. Korean J. Chem. Eng. **31**, 1180–1186 (2014)
8. A.O. Ayeni, S. Banerjee, J.A. Omoleye, et al., Optimization of pretreatment conditions using full factorial design and enzymatic convertibility of shea tree sawdust. Biomass Bioenergy **48**, 130–138 (2013)
9. A.M. Azzam, Pretreatment of cane bagasse with alkaline hydrogen peroxide for enzymatic hydrolysis of cellulose and ethanol fermentation. J. Environ. Sci. Health B **24**(4), 421–433 (1989)
10. V. Balan, Current challenges in commercially producing biofuels from lignocellulosic biomass. ISRN Biotechnol. **2014**, 1–31 (2014)
11. S. Banerjee, S. Mudliar, R. Sen, et al., Commercializing lignocellulosic bioethanol: Technology bottlenecks and possible remedies. Biofuels Bioprod. Biorefin. **4**, 77–93 (2010)
12. S.W. Bank, A.V. Bridgwater, Catalytic fast pyrolysis for improved liquid quality, in *Handbook of Biofuels Production: Processes and Technologies*, ed. by R. Luque, C. S. K. Lin, K. Wilson, J. Clark, (Elsevier, Duxford, 2016), p. 391
13. P. Bajpai, *Pretreatment of Lignocellulosic Biomass for Biofuel Production. Springer Briefs in Green Chemistry for Sustainability* (Springer Science and Business Media, New York, 2016)
14. A.B. Bjerre, A.B. Olesen, T. Fernqvist, Pretreatment of wheat straw using combined wet oxidation and alkaline hydrolysis resulting in convertible cellulose and hemicellulose. Biotechnol. Bioeng. **49**, 568–577 (1996)
15. L. Brennan, P. Owende, Biofuels from Microalgae—A review of technologies for production, processing, and extractions of biofuels and co-products. Renew. Sust. Energ. Rev. **14**, 557–577 (2010)
16. T. Bridgwater, Challenges and opportunities in fast pyrolysis of biomass: Part 1. Johnson Matthey Technol. Rev. **62**, 118–130 (2018)
17. L. Cadoche, G.D. Lopez, Assessment of size reduction as a preliminary step in the production of ethanol from lignocellulosic wastes. Biol. Wastes **30**, 153–157 (1989)
18. D.R. Cahela, Y.Y. Lee, R.P. Chambers, Modeling of percolation process in hemicellulose hydrolysis. Biotechnol. Bioeng. **25**, 3–17 (1983)
19. V.S. Chang, M. Nagwani, C. Kim, et al., Oxidative lime pretreatment of high lignin biomass. Appl. Biochem. Biotechnol. **94**, 1–28 (2001)

20. H.L. Chum, D.K. Johnsoon, S. Black, Organosolv pretreatment for enzymatic hydrolysis of poplars: 1. Enzyme hydrolysis of cellulosic residues. Biotechnol. Bioeng. **31**, 643–649 (1988)
21. T.A. Clark, K.L. Mackie, Steam explosion of the soft-wood *Pinus radiata* with Sulphur dioxide addition. I. Process optimization. J. Wood Chem. Technol. **7** (1987)
22. B.E. Dale, M.J. Moreira, A freeze-explosion technique for increasing cellulose hydrolysis. Biotechnol. Bioeng. Symp. **12**, 31–43 (1982)
23. S.J.B. Duff, W.D. Murray, Bioconversion of forest products industry waste cellulosics to fuel ethanol: A review. Bioresour. Technol. **55**, 1–33 (1996)
24. K.E. Eriksson, L. Vallander, Properties of pulps from thermomechanical pulping of chips pretreated with fungi. Sven. Papperstidn. **6**, 33–38 (1982)
25. A. Esteghlalian, A.G. Hashimoto, J.J. Fenske, et al., Modeling and optimization of the dilute-sulfuric-acid pretreatment of corn stover, poplar and switchgrass. Bioresour. Technol. **59**, 129–136 (1997)
26. L.T. Fan, M.M. Gharpuray, Y.-H. Lee, *Cellulose Hydrolysis* (Biotechnol Monogr Springer-Verlag, Berlin, 1987), p. 57
27. J. Gierer, Formation and involvement of superoxide (O_2/HO_2) and hydroxyl (HO) radicals in TCF bleaching processes: A review. Holzforschung **51**, 34 (1997)
28. W.R. Grous, A.O. Converse, H.E. Grethlein, Effect of steam explosion pretreatment on pore size and enzymatic hydrolysis of poplar. Enzyme Microbiol. Technol. **8**, 274–280 (1986)
29. M. Hausman, A Mechanistic Study of the Degradation of Lignin Model Compounds with Oxygen species. Ph.D Thesis, University of Maine, Maine (1999)
30. N.D. Hinman, D.J. Schell, C.J. Riley, et al., Preliminary estimate of the cost of ethanol production for SSF technology. Appl. Biochem. Biotechnol. **34/35**, 639–649 (1992)
31. M.T. Holtzapple, A.E. Humphrey, J.D. Taylor, Energy requirements for the size reduction of poplar and aspen wood. Biotechnol. Bioeng. **33**, 207–210 (1989)
32. M.T. Holtzapple, J.-H. Jun, G. Ashok, et al., *Ammonia Fiber Explosion (AFEX) Pretreatment of Lignocellulosic Wastes* (American Institute of Chemical Engineers National Meeting, Chicago, IL, 1990)
33. M.T. Holtzapple, J.-H. Jun, G. Ashok, The ammonia freeze explosion (AFEX) process: A practical lignocellulose pretreatment. Appl. Biochem. Biotechnol. **28/29**, 59–74 (1991)
34. H. Itoh, M. Wada, Y. Honda, et al., Bioorganosolv pretreatments for simultaneous saccharification and fermentation of beech wood by ethanolysis and white rot fungi. J. Biotechnol. **103**, 273–280 (2003)
35. M. Kurakake, N. Ide, T. Komaki, Biological treatment with two biological starins for enzymatic hydrolysis of office papers. Curr. Microbiol. **54**, 424–428 (2007)
36. F.J. Kilzer, A. Broido, Speculations on the nature of cellulose pyrolysis. Pyrodynamics **2**, 151–163 (1965)
37. L.G. Ljungdahl, The cellulase/hemicellulase system of the anaerobic fungus *Orpinomyces* PC-2 and aspects of its applied use. Ann. N. Y. Acad. Sci. **1125**, 308–321 (2008)
38. K.L. Mackie, H.H. Brownell, K.L. West, et al., Effect of sulphur dioxide and sulphuric acid on steam explosion of aspenwood. J. Wood Chem. Technol. **5**, 405–425 (2007)
39. J.D. McMillan, Pretreatment of lignocellulosic biomass, in *Enzymatic Conversion of Biomass for Fuels Production*, ed. by M. E. Himmel, J. O. Baker, R. P. Overend, (American Chemical Society, Washington, DC, 1994), pp. 292–324
40. M. Mes-Hartree, B.E. Dale, W.K. Craig, Comparison of steam and ammonia pretreatment for enzymatic hydrolysis of cellulose. Appl. Microbiol. Biotechnol. **29**, 462–468 (1988)
41. M.A. Millet, A.J. Baker, L.D. Scatter, Physical and chemical Pretreatment for enhancing cellulose saccharification. Biotech. Bioeng. Symp. **6**, 125–153 (1976)
42. P.J. Morjanoff, P.P. Gray, Optimization of steam explosion as method for increasing susceptibility of sugarcane bagasse to enzymatic saccharification. Biotechnol. Bioeng. **29**, 733–741 (1987)
43. N. Mosier, C.E. Wyman, B.E. Dale, Features of promising technologies for pretreatment of lignocellulosic biomass. Bioresour. Technol. **96**, 673–686 (2005)

44. S. Niphadkar, P. Bagade, S. Ahmed, Bioethanol production: Insight into past, present and future perspectives. Biofuels **9**(2), 229–238 (2018)
45. C. Nitsos, U. Rova, P. Christakopoulos, Organosolv fractionation of softwood biomass for biofuel and biorefinery applications. Energies **11**(50) (2018)
46. W.C. Neely, Factors affecting the pretreatment of biomass with gaseous ozone. Biotechnol. Bioeng. **20**, 59–65 (1984)
47. B.C. Saha, M.A. Cotta, Enzymatic sacccharification and fermentation of alkaline peroxide pretreated rice hulls to ethanol. Enzyme Microb. Technol. (4), 528–532 (2007)
48. B.C. Saha, M.A. Cotta, Ethanol production from alkaline peroxide pretreated enzymatically saccharified wheat straw. Biotechnol. Prog. **22**, 449–453 (2006)
49. J. Schurz, in *Bioconversion of Cellulosic Substances into Energy Chemicals and Microbial Protein Symposium Proc*, ed. by T. K. Ghose, (IIT, New Delhi, 1978), p. 37
50. P.T. Sekoai, K.O. Yoro, M.O. Daramola, Batch fermentative biohydrogen production process using immobilized anaerobic sludge from organic solid waste. Environments **3**(38), 1–10 (2016)
51. P.T. Sekoai, C.N.M. Oumaa, S.P. du Preez, P. Modisha, N. Engelbrecht, D.G. Bessarabov, A. Ghimire, Application of nanoparticles in biofuels: An overview. Fuel **237**, 380–397 (2019)
52. P.T. Sekoai, K.O. Yoro, M.O. Bodunrin, A.O. Ayeni, M.O. Daramola, Integrated system approach to dark fermentative biohydrogen production for enhanced yield, energy efficiency and substrate recovery. Rev. Environ. Sci. Biotechnol. **17**(3), 501–529 (2018)
53. P.T. Sekoai, E.B. Gueguim Kana, Semi-pilot scale production of hydrogen from Organic Fraction of Solid Municipal Waste and electricity generation from process effluents. Biomass Bioenergy **60**, 156–163 (2014)
54. P.T. Sekoai, E.B. Gueguim Kana, A two-stage modelling and optimization of biohydrogen production from a mixture of agro-municipal waste. Int. J. Hydrog. Energy **38**(21), 8657–8663 (2013)
55. F. Shafizadeh, Y.-Z. Lai, Thermal degradation of 2-deoxy-D-arabino-hexonic acid and 3-deoxy-D-ribo-hexono-1,4-lactone. Carbohydr. Res. **42**, 39–53 (1975)
56. F. Shafizadeh, A.G.W. Bradbury, Thermal degradation of cellulose in air and nitrogen at low temperatures. J. Appl. Poly. Sci. **23**, 1431–1442 (1979)
57. Y. Sun, J. Cheng, Hydrolysis of lignocellulosic materials for ethanol production: A review. Bioresour. Technol. **83**, 1–11 (2002)
58. M.J. Taherzadeh, K. Karimi, Pretreatment o lignocellulosic astes to improve ethanol and biogas production: A Review. Int. J. Mol. Sci. **9**, 1621–1651 (2008)
59. P.F. Vidal, J. Molinier, Ozonolysis of lignin - improvement of *in vitro* digestibility of poplar sawdust. Biomass **16**, 1–17 (1988)
60. E.Y. Vlasenko, H. Ding, J.M. Labavitch, Enzymatic hydrolysis of pretreated rice straw. Bioresour. Technol. **59**, 109–119 (1997)
61. J.K. Weng, X. Li, N.D. Bonawitz, et al., Emerging strategies of lignin engineering and degradation for cellulosic biofuel production. Curr. Opin. Biotechnol. **19**, 166–172 (2008)
62. J.D. Wright, Ethanol from biomass by enzymatic hydrolysis. Chem. Eng. Prog. **84**(8), 62–74 (1988)
63. J.J. Yoon, C.J. Cha, Y.S. Kim, et al., The brown-rot basidiomycete Fomitopsis palustris has the endo-glucanases capable of degrading microcrystalline cellulose. J. Microbiol. Biotechnol. **17**, 800–805 (2007)
64. K.O. Yoro, P.T. Sekoai, A.J. Isafiade, M.O. Daramola, A review on heat and mass integration techniques for energy and material minimization during CO2 capture. Int. J. Energy Environ. Eng.. In press. (2019). https://doi.org/10.1007/s40095-019-0304-1
65. Y.Z. Zheng, H.M. Lin, G.T. Tsao, Pretreatment for cellulose hydrolysis by carbon dioxide explosion. Biotechnol. Prog. **14**, 890–896 (1998)
66. L. Zhu, Fundamental study of structural eatures affecting enzymatic hydrolysis of lignocellulosic biomass, PhD Dissertation (2005), Texas A&M University, College Station, Texas, USA.

Chapter 8
Lignocellulosic Pretreatment Methods for Bioethanol Production

E. F. Aransiola, T. D. Shittu, T. F. Oyewusi, A. O. Adetoyese, O. S. Fagbeyiro, and U. P. Eyibio

8.1 Introduction

Pretreatment has been described as an essential process step required for transforming lignocellulosic biomass to various value-added materials [2, 22]. The biomass composition is a major determinant in evaluating the efficiency of the pretreatment required for enzymatic hydrolysis [52, 65]. The treatment steps become essential due to the nature of the structure of lignocellulosic materials. Generally, on average, their plant cell wall comprises cellulose, lignin, and hemicellulose, with an approximate value of 30–50%, 10–30%, and 15–35%, respectively [22].

They also contain minute quantity of pectin, protein, extractives, and ash [9]. The quantity of the components varies from one species of plant to another, depending on their age and growth stage [9]. Biomass conversion in their raw form to bioethanol without pretreatment will require the use of more enzymes and will limit the production of fermentable sugars on cellulose [52]. This is because their conversion to other products is always tedious as a result of the intractable nature of their structure [22]. Thus, an appropriate measure of pretreatment is necessary for successful transformation to biofuel. The treatment step decomposes the polymeric components and allows the production of monomer sugars; thus, the enzymatic conversion

E. F. Aransiola (✉) · T. D. Shittu · A. O. Adetoyese
Department of Chemical Engineering, Obafemi Awolowo University, Ife, Osun-State, Nigeria

T. F. Oyewusi · O. S. Fagbeyiro
Department of Agricultural and Environmental Engineering, Obafemi Awolowo University, Ile – Ife, Osun State, Nigeria

U. P. Eyibio
Department of Chemical and Petrochemical Engineering, Akwa Ibom State University, Uyo, Akwa Ibom State, Nigeria

M. O. Daramola, A. O. Ayeni (eds.), *Valorization of Biomass to Value-Added Commodities*, Green Energy and Technology,
https://doi.org/10.1007/978-3-030-38032-8_8

of the cellulosic component is enhanced, and it improves the edibility of the substrate for both enzymatic and microbial bioconversion.

Mood et al. [52] considered some major characteristics in choosing the most efficient pretreatment method(s) for a particular substrate. The characteristics highlighted were the (1) content of the lignin, (2) polymerization degree, (3) biomass surface area, (4) cellulose crystallinity index of the cellulose, and (5) acetyl content. Similarly, for an effective pretreatment, Galbe and Zacchi [25] and Alvira et al. [6] reported that the following criteria must be met: (1) there should be maximum recovery of all the carbohydrate present; (2) the enzymatic hydrolysis should lead to high digestibility of the cellulose; (3) fermentation should be possible without detoxification; (4) there should be high concentrations of liberated sugars; (5) the cost of investment and operational cost should be minimized; and (6) energy required should be reduced and, if possible, reused.

Different methods of pretreatment had been employed to promote the conversion of lignocellulosic substrate to value-added products. Majorly, all these pretreatment types are grouped into chemical, physical, biological, and physicochemical methods [6].

8.2 Physical and Physicochemical Pretreatment

This section discusses the pretreatment methods available and the underlining principles of each group will be discussed under different subsections.

8.2.1 Steam Explosion

Among the mostly utilized physicochemical techniques in the pretreatment of lignocellulosic substrate is steam pretreatment. This technique was earlier called steam explosion due to the notion that a violent effect on the substrate was a prerequisite for saccharificatiom [4]. The lapse of the structure of lignocellulosic constituents by the warming up effects can arise through steam explosion acting as a thermo-mechanochemical pretreatment. Organic acids' production in the course of the procedure and shearing forces results in moisture expansion. Steam-explosion process includes two distinct stages: vapor cracking and explosive decompression. The stages result in the change of the material constituents: hydrolysis of the constituents of hemicellulose (mono and disaccharides produced), change in the chemical structure of lignin, and change in the crystalline index of the cellulose. The results of these changes affect the yield obtained from the enzymatic hydrolysis and permit the lignocellulosic structure opening [31].

Principally, the substrate is operated at saturated steam of elevated pressure between 7 and 48 bar and at increased temperatures between 160 and 260 °C for some time. This enhances hydrolysis and the formation of hemicellulose. The steam causes the walls of substrates to expand, giving yield to incomplete hydrolysis with increase in the way the enzymes access the cellulose. Immediately, the pressure is lowered to 1 atmosphere [60]. This breakage into fiber and the noticeable autohydrolysis importantly increase the substrate edibility and biological conversion in conjunction with its reactivity as to other catalytic reactions. The consecutive quick relief of the pressure lowers the temperature, thus, terminating the process.

8.2.1.1 Steam Explosion Drawbacks

This process produces some poisonous spin-off that could hinder the consecutive saccharrification and fermentation steps. The inhibitors produced alongside include derivatives of furan, for example, compounds of phenol, 5-hydroxymethyl-2-furaldehyde, furaldehyde, and derived from depolymerization of lignin. Although these inhibitors can be removed by washing with water, the simultaneous demerits with washing are the reduction in hydrolysis yield, and the removal of soluble sugar derived from hydrolysis of hemicelluloses. Again, this process damages a part of the fraction of xylan, partial rupture of the lignin [42].

8.2.2 Ammonia Fiber Explosion (AFEX)

The AFEX approach is much related to steam explosion. A physicochemical pretreatment method, where lignocellulosic substrate is vulnerable to aqueous NH_3 at elevated pressure and temperature for some time, when, the pressure is without warning decreased, thus, vaporizing the ammonia which is recovered and recycled, is known as ammonia fiber explosion [26, 42]. The most significant determining parameters on the process include ammonia and water loading, temperature of mixture, and time of pretreatment [52]. Usually, a small quantity of ammonia is required for each kg of the dry substrate; the method is done at medium-to-high temperature between 60 and 140 °C, a high pressure between 1.72 and 2.06 Mpa, and the heating is usually done for 5–45 min [22, 52].

Ammonia treatment causes a physical disruption in the biomass and area of the surface is made larger by opening up the fiber structure, resulting in high enzymatic digestibility. A major merit of this method is the reusability of the ammonia for subsequent operation without loss of potency, and the production of inhibitors is eradicated, eliminating the need for detoxification before saccharification and fermentation [22]. Ammonia treatment was applied successfully to different substrates, for example, corn stover, aspen wood, switch grass, and rice and wheat straw [52].

The process of AFEX results in nearly done solids recovery (no part of the liquid was dissolved into fractions), increased of digestibility (decrystallization of cellulose), joining of lignin to carbohydrate linkages, and an improved surface area and wettability.

8.2.2.1 Operating Costs of AFEX

The economics of the process is highly influenced by ammonia loading and residence time. When the pretreatment is over, recovery of the ammonia will take place and it will be recycled. The cost of recovery serves as a limitation for large scale application.

8.2.2.2 Drawback of AFEX

AFEX has not been a much efficient innovation for lignocellulosic substrate with closely high content of lignin, for example, nutshells and woods.

8.2.3 Carbon Dioxide Explosion

This process harnesses CO_2 as a material at pressure and temperature beyond its critical points [26]. To enhance the growth of enhanced lignocelluloses pretreatment methods, the concept of utilizing CO_2 explosion at a pressure and temperature beyond its critical points was developed. This is preferably cheaper than ammonia explosion and supposed to have a lesser temperature than steam explosion. A supercritical fluid occurs when in its gaseous form is compressed to a liquid-like density at a temperature beyond the critical point. A hypothesis showed that since carbon dioxide produces an acid of carbon when dissolution in water takes place, the rate of hydrolysis could be increased by the acid. CO_2 molecules are roughly related in mass to H_2O and NH_3, and should pass through little pores reachable to H_2O and NH_3 molecules. In sugarcane bagasse pretreatment, recycled paper mix, AFEX, was found to be more expensive than CO_2 explosion [88].

8.2.4 Liquid Hot Water (LHW) Pretreatment

The technique is also referred to as unsaturated water which is also related to steam pretreatment approach. It is clear from the suggestion of the name that it utilizes water at an elevated temperature between 170 and 230 ° C and pressure up to 50 bar. Apart from steam, LHW is also called uncatalyzed solvolysis, hydrothermolysis, aquasolv steam, or liquid fractionation [55]. This method renders assistance to avoid the process where fermentation inhibitors are formed at elevated temperatures [80]. The reactor configurations of liquid hot water are of three kinds: cross-current, co-current, and flow-through. Cross-current pretreatment is configured to carry lignocelluloses and water in counter directions along the reactor used for pretreatment.

Co-current pretreatment allows a suspension of substrate and water to be warm up to the expected temperature, and thus remain at the pretreatment terms for a controlled duration of residence before cooling takes place. Also for the flow-through reactor, hot water covers a fixed bed of lignocelluloses, hydrolyzes and dissolves lignocelluloses constituents, and removes them from the reactor [55]. It is crucial to note that all three kinds of LHW are carried out at elevated temperatures between 180 and 230° C and increase pressures with contact times from 15 to 60 min, and the concentration of solids is not up to 20 wt% [28].One of the advantages of LHW is that as a result of the input of high water, it generates reduced concentrations of inhibitory derivatives. It is an easy and eco-friendly approach.

8.2.5 Mechanical Pretreatment (Grinding and Milling)

The mechanical pretreatment technique increases the pore size of the cellulose and area of the surface access, and causes the polymerization degree and degree of crystals formed to reduce. The required power of mechanical pretreatment is quite much and dependent on the kind of substrate and on the size of the final particle; but these pretreatments become economically impossible after a certain particle size [28]. Examples include cracking, granulating, and/or pulverizing methods.

Cracking can decrease the size of substrate to 10–30 mm only while granulating and pulverizing can decrease the particle size up to 0.2 mm. The shortcomings of heat and mass transfer can be reduced by cracking while the size of particle and the degree of cellulose crystals that come from the shear forces formed during pulverizing can be efficiently reduced by granulating. The kind and time taken for pulverizing and type of substrate control the net decrease in the rate of cellulose crystal, final degree of polymerization, and rise in area of the specific surface. Various methods of pulverizing such as colloid pulverizing, two-roll pulverizing, vibratory pulverizing, and hammer pulverizing are utilized to make the edibility of the lignocellulosic substrate better. In minimizing the rate of cellulose crystal of aspen chips and spruce and in enhancing their edibility, vibratory ball pulverizing is discovered to be more durable than ordinary ball pulverizing [42].

8.2.5.1 Mechanical Extrusion

The process of forming a stationary object with a cross-sectional profile pushed along a die of the required cross-section is called an extrusion. The object will enlarge when it leaves the die [87]. An advance process of mechanical comminution whereby the access of enzyme is broadened to hammer much exposed carbohydrates is extrusion. It involves mixing rapidly, reduces residence time, high shear, modest barrel temperature, absence of furfural, rinsing not allowed, and conditioning.

In this method, materials are operated at a temperature above 300° C, and it is after this, both mixing and shearing which cause modification of the cellulose physically and chemically.

8.2.6 Ultrasound

Sonication is comparatively a recent approach utilized during the pretreatment of lignocellulosic substrate. Each of the chemical and physical effects that change the structure of lignocellulosic substrate is generated by ultrasound waves. Ultrasound treatment causes generation of small spherical cavity which suddenly bursts the cellulose and hemicellulose parts by multiplying the easy access to cellulose-diminishing enzymes for proper decomposition into less complex reducing sugars.

8.2.7 Microwave Pretreatment

Irradiation via microwave is a method greatly adopted as a result of its elevated warming-up capacity and is easily operated. The thermal action generated by microwaves in aquatic surroundings is being used by microwaved treatment. Microwave radiation causes the generation of internal heat in the substrate. This is as a result of the oscillation of the polar bonds in the substrate and the aquatic surrounding. Therefore, a high-temperature spot is formed inside the nonhomogeneous substance. This special heating feature gives rise to an explosion effect amidst the fragments and makes better the distortion of unwilling structures of lignocellulose. Microwave irradiation has residence time of 5–20 min. It can alter the ultra-structure of cellulose by diminishing lignin and hemicelluloses and by raising the susceptibility of enzyme for lignocellulosic materials [51].

8.2.8 Pulsed-Electric Field (PEF)

In this process, a short burst of elevated voltage is applied to a specimen inserted amid cathode and anode. PEF pretreatment can adversely affect the shape of plant tissues. An external electric field is used at high intensity and a critical electric potential is stimulated along the cell membrane; this has led to quick electrical degradation, also with changes in the local structure of the cell wall, the cell membrane, and also the plant tissue. The electric field produces a dramatic expansion in mass permeability and, sometimes, mechanicalbreakdown of the plant tissue. The amount of cellulose in the substrate is revealed by this pretreatment by forming pores in the cell membrane, therefore, permitting the carriers entrance that is able to degrade the cellulose into sugar components.

8.2.9 Electron Beam (EB) Irradiation

EB ionizing radiation is a process that makes use of beta radiation, mostly of high energy, to treat an object for a wide application. It is obtained from a linear accelerator. This pretreatment uses accelerated electron beam to expose lignocellulosic biomass to radiation to deform the cell wall polymer structure, that is, lignin, hemicellulose, and cellulose, by such processes as generation of free radicals, inducing chain scission or cross-link formation, decrystallization, and/or decreasing the polymerization degree.

8.2.10 Wet Oxidation

Wet oxidation is among the easy methods, where the air/oxygen along with H_2O or H_2O_2 is operated with the substrates at elevated temperatures beyond 120 ° C for 30 min [71]. Formerly, this technique was utilized for soil remediation and wastewater treatment [20]. The technique is best for substrate wastes that are rich in lignin. The pretreated slurry is filtered to isolate the solid rich in cellulose from the filtrate rich in hemicellulose, and the solid composition is rinsed with deionized water shortly after which enzymatic hydrolysis takes place.

The efficiency of wet oxidation is a factor of these things: pressure, oxygen, reaction time, and temperature. For this technique, the temperature is increased beyond 170 ° C, and water acts like an acid and alters the rate of hydrolytic reactions. The hemicelluloses are decomposed into lower pentose monomers and the lignin undergoes oxidation, while the cellulose is unaffected by wet oxidation pretreatment. However, findings on inclusion of chemical agents such as sodium carbonate and alkaline peroxide in wet oxidation decrease the temperature of the reaction, make better hemicellulose deterioration, and reduce the formation of inhibitory components such as furfurals and furfuraldehydes [12].

8.2.11 Alkaline Wet Oxidation

This method combines both alkaline and oxidation to pretreat the biomass before the enzymatic hydrolysis [22]. The operation employs temperature within the range 170–220° C and the commonly used oxidizing agents are air or oxygen at high pressure and hydrogen peroxide, while the alkaline component is normally sodium carbonate [22, 68]. The alkaline is used to aid the solubility of hemicellulose and to minimize the production of enzyme-inhibiting products. The main important parameters that affect the efficiency of wet oxidation treatment are the temperature of operation, reaction time, and the pressure of the oxidizing agent used [68] (Table 8.1).

Table 8.1 The merits and demerits of selected pretreatment processes of lignocellulosic substrate [26, 42]

Pretreatment Process	Advantages	Limitations and Disadvantages
Mechanical pretreatment	Minimizes cellulose crystallinity and increases surface area	Power utilization usually more than ingrained substrate energy; needs to be combined with other treatments
Steam explosion	Increase of allowable surface area; higher substrate digestibility; depolymerization of lignin; solubilization of hemicellulose	Demolition of a part of the xylan fraction; partial rupture of the lignin–carbohydrate matrix; formation of compounds inhibitory to microorganisms
AFEX	Low formation of inhibitors; increase of accessible surface area	Not suitable for substrates with high content of lignin; expensive plant and ammonia
CO_2 explosion	No toxicity; easy recovery; expansion of accessible surface area; efficient hydrolysis of hemicellulose	High cost of plant; does not modify lignin or hemicelluloses
LHW	Enhanced substrate edibility; low formation of inhibitors; inexpensive plant	High energetic requirements; high water input

8.3 Chemical and Biological Pretreatment

This section presents the different chemical and biological pretreatments used for structural modification of lignocellulosic biomass.

8.3.1 Acid Pretreatments

Acid treatment involves the addition of acidic catalyst like H_2SO_4, H_3PO_4, HNO_3, and HCl to the biomass. The acids help in solubilizing the lignin and hemicellulose content of the biomass and in hydrolyzing the polysaccharides to monosaccharides, thus making it easy for enzymes to hydrolyze the cellulose [22, 52].The most commonly used acid is the H_2SO_4 [52]. The major distinguishing factor among the acids is the concentration of the acid used, either in diluted or in concentrated form, but, they essentially share the same methodology and chemistry [22].

Usually, acid with high concentration is mostly operated at a lower temperature and this helps in saving energy cost while dilute acid is operated at a higher temperature [52]. The use of concentrated acids poses threats like toxicity, corrosion of equipment, and the co-production of such products like furfural, phenolic acids, and aldehydes not desirable for enzymatic hydrolysis and fermentation process [22, 52]. However, dilute acids, mostly H_2SO_4, at concentrations of (0.05–5%) and temperature of 160–220° C can reduce the production of these products and cause a rise in the yield of sugar [22]. Also, the addition of H_2SO_4 removes the need to add hemicellulose hydrolytic enzymes, and this reduces the process cost [22]. The application of dilute acid to biomass has witnessed a lot of success. Hsu et al. [30] treated rice straw with dilute acid, Lu et al. [47] worked on rapeseed, and Y. Sun and Cheng [67] pretreated Bermuda grass with dilute acid, and they all reported a high yield of fermentable sugar at different experimental conditions.

8.3.2 *Alkaline Pretreatments*

Alkaline pretreatment for lignocellulosic substrates is usually operated at a reduced temperature, and it removes any lignin and acetyl groups that may inhibit successful hydrolysis [52]. Mainly, alkaline targets the hemicellulose acetyl groups and the linkages of lignin–carbohydrate ester [22]. Treatment with alkaline digests lignin matrix and ensures easy access of the degradation enzymes to the hemicellulose and cellulose [65]. The treatment causes the cross-links between hemicelluloses and other components to be disrupted, thus enhancing the porosity of the biomass [52]. Also, this can lead to the delignification of the substrates resulting from the disruption of the ester bond that links lignin and xylan [52]. Alkaline treatment improves cellulose digestibility and lignin solubility and offers a higher solubility of cellulosic components than acid treatment [6]. Additionally, sugar degradation of alkaline-treated substrates is less than acid pretreatment. The alkaline that has been successfully employed are NaOH, NH_3, $Ca(OH)_2$, KOH, and NH_4OH [52, 68]. The main drawbacks with this method are the lengthy residence time required and the need for aftermath separation. Alkaline treatment is more effective when used on agricultural materials like residues than on wood residues [6]. Several researchers have confirmed the efficacy of this method [74, 75, 79].

8.3.3 *Organosolv Treatment*

The method involves the mixing of organic acid with lignocellulosic biomass. The organic acids may be used singly in the raw form or used together with inorganic acids or bases [67]. C_2H_5OH, CH_3OH, C_3H_6O, and $C_2H_6O_2$ are among the acids that have resulted in effective biomass pretreatment [52, 86]. Organic acids are capable of degrading substrates having high lignin content and increasing the surface area, thus promoting hydrolysis [52, 68]. The merits of this method rest on the fact that the organic solvent and water mixtures help remove the requirement to burn off the liquor and permit the isolation of the lignin [65]. However, factors like high pressure requirement, high volatility, and flammability of organic acids limit the use of this method [52]. Also, the price of organic acids limits it use for commercial or industrial application [6]. Alvira et al. [6] reported, for economic reasons, lower weight alcohols such as C_2H_5OH and CH_3OH should be used.

8.3.4 Ozone Treatment

The method uses ozone gas to eliminate the lignin component contained in the substrate and to improve enzymatic hydrolysis of the cellulosic content [10, 68]. Ozone gas is a highly efficient oxidizing agent for the removal of lignin [6]. The treatment is mostly carried out at about 25° C and moderate pressure, and it offers the advantage of inhibitory products and toxic by-product formation that may otherwise prevent the hydrolysis and fermentation stages are eliminated [6, 10]. The application of ozone has been previously employed to treat wheat and rye straw with a highly successful production of fermentable sugars after enzymatic hydrolysis [10]. It has also been used on cotton straw, green hay, peanut, and poplar sawdust [42, 67]. However, the problem involved with ozone use is the large quantity required for effective operation [6, 10].

8.3.5 Biological Pretreatment

Biological treatment is an energy-extensive, cost-effective, and eco-friendly process that uses microorganisms for the pretreatment of lignocellulosic substrates and does not involve the addition of chemicals [22, 34, 65, 75]. The microorganisms that possess the potential to degrade the lignin and hemicellulose contents of the biomass are brown rot, whit rot, and soft rot fungi.

The performance of the biological treatment for lignin removal is affected by moisture content, the size of particles, residence time, and the temperature of operation [52]. Some of the fungi that have been successfully employed for this purpose are white rot *basidiomycetes, Phanerochaete chrysosporium,* and certain *actinomycetes*. These fungi possess the ability to produce lignin peroxidases and laccases, which are enzymes that disrupt the biomass cell wall and remove a large proportion of lignin. The organisms convert the biomass in to compounds that are convertible by enzymes for bioethanol production [52]. *Phanerochaete chrysosporium* has been reported to be highly effective among the white rot fungi because of the high potential to degrade lignin and rapid growth rate [52].

However, some drawbacks are inherent in the use of this method. A considerable amount of time is required for the pretreatment to be carried out and some carbohydrates present in the biomass are usually utilized by the microbes for growth, which reduces the yield of sugar obtainable after hydrolysis [22]. Additionally, the method involves large space for operation and the microorganism needs to be monitored during growth [52]. The method has been successfully employed in the production of bioethanol from corn stover, rice husk, wheat straw, and rice straw [52].

8.4 Pretreatment of Selected Biomass

8.4.1 *Physical and Physicochemical Pretreatment of Selected Lignocellulosic Biomass*

8.4.1.1 Switch Grass

Switch grass (Panicum virgatum) originated from North America and it is known as a perennial grass [34]. According to Cherubini and Jungmeier [21], it is a viable bioenergy crop that can grow at different geographical locations. Keshwani and Cheng [34] explained that switch grass is known to be a typical plant for production of energy according to the Department of Energy, USA. The cultivation is economical, in that it requires low nutrient and the demand of water and soil fertility is reduced, and intensive agricultural practices are not involved [21]. The yield of switch grass from the southeastern of USA is approximately between 7 and 15 dry tons per acre [34]. Also, [34, 67] suggested that the appropriate pretreatment must be investigated on the structure of switch grass (45% cellulose, 31.4% hemicellulose, and 12.0% lignin) for enhancement of the hydrolysis (through enzyme) of the cellulose into fermentable sugar.

Various pretreatments like AFEX, dilute H_2SO_4, lime, microwave-based radio frequency-assisted alkali pretreatments have successfully been used on switch grass [34]. Moreover, Keshwani and Cheng [35] described AFEX as the only physicochemical pretreatment that has been applied to enhance hydrolysis for the treatment of switchgrass. As reported by Dale et al. [24], 5–eight-fold improvement in sugar reduction yields was obtained over the raw (untreated) samples after a 48-hour pretreatment of switchgrass with enzymatic hydrolysis of AFEX. Alizadeh et al. [5] carried out a more comprehensive study on switchgrass by optimizing AFEX pretreatment. The impacts of moisture content, reactor temperature, and ammonia loading were examined on the efficacy of the hydrolysis (enzymatic). A 100 ° C optimum pretreatment condition of reactor temperature was reported with NH_3 loading of 1 g g^{-1} of feedstock at 5 min residence time. With all this, a six-fold enhancement in hydrolysis efficacy was obtained.

Moreover, the changes that result from reduction in size of switchgrass attained via ball milling were studied by Bridgeman et al. [16]. It was noted that particle sizes less than 90 μm, cellulose substance was 13.4% lower than that of larger-sized particles. Also, hemicelluloses and lignin losses were significantly less precise at 43% and 4.74%, in the order given. The result is an indication that wide reduction in size is unsuitable, as it gives rise to substantial carbohydrate losses which eventually result in a smaller amount of sugars and reduction in ethanol yield. In a similar study, the energy needed for reduction in the size of switchgrass was examined by employing a hammer mill (Mani et al. [48]. As the size in particles decreased at 8% MC, energy requirements increased linearly and at 12% moisture content, the energy needed for reduction in size was higher. Based on tensile stress, Yu et al. [84]

stated that reduction in size of the switchgrass was not so much influenced by moisture content than tensile stress.

Galletti and Antonetti [26] subjected switchgrass to a high pressure with CO_2–H_2O process and temperatures ranging from 150°C to 250°C for 20–60 sec. The total yields as described by the portion of the hypothetical maximum were discovered for hemicelluloses sugars, glucose, and two degeneration products: furfural and 5-hydroxymethylfurfural. The yield on glucose was 81%, hemicelluloses sugars yield was 13%, and furfural yield was 11.8% at 160 ° C.

A study was carried out on the effect of varying moisture content of 15%, 25%, 35%, and 45% on sugar recovery in prairie cord grass and switch grass at 50, 100, and 150 rpm (screw speed), compaction ratio (2:1 and 3:1), and barrel temperature (50, 100, and 150 ° C), (Karunanithy and Muthukumarappan [32]. After enzymatic hydrolysis pretreatment of the biomass, a highest yield of 65.8% sugar was regained from prairie under 25% at 50 rpm and 50 ° C, while a maximum of 45.2% sugar was recovered from switchgrass under 15% at 50 rpm and 150 ° C. Also, acetic acid and glycerol were produced at low concentrations in the range of 0.02–0.18 g/L. From this study, acetic acid and glycerol were well known as the co-products produced during the biomass lignocellulosic pretreatment. Nevertheless, the by-product formation was significantly lower when compared with pretreatment methods of acid hydrolysis and compressed hot water.

Again, Karunanithy et al. [33] investigated the effect of 50, 75, 100, 150, and 200 ° C barrel temperatures, different 1:1–1:4 of concentrations of cellulose with β-glucosidase and 100, 150, and 200 rpm of screw speeds on selected species of warm season grasses (big bluestem, prairie cord grass, and switch grass). The result showed that when the ratio of 1:4 (cellulase and β-glucosidase) was upheld, maximum reducing sugars were obtained. The reducing sugars from the prairie cord grass treated at 150 rpm and 100 ° C produced 49.2%, while switchgrass pretreated at 200 rpm and 75 ° C yielded 28.2% and big bluestem pretreated at 200 rpm and 150°C produced 66.2%. Though, to some extent, the sugar yields are high, a variety of lignocellulosic biomass with varied lignin, cellulose, and hemicellulose contents cannot suffice pretreatment with mechanical extrusion alone. Therefore, better methods of pretreatment are needed to increase the sugar produced. Also, properties of the feedstocks significantly influenced the amount of sugar produced.

Hu and Wen (2008) studied the pretreatment of switch grass with microwave-based alkali which yielded almost 70–90% of sugar. Also, Keshwani and Cheng [35] studied the pretreatment of coastal Bermuda grass and switch grass by employing microwave-based alkali with different alkalis. Sodium hydroxide was found to be the most suitable alkali for the pretreatment. Meanwhile, coastal Bermuda grass produced 59% xylose and 87% glucose, while switchgrass produced 63% xylose and 82% glucose under optimum condition [35]. Though not important, the disparity noticed on the reducing sugars was linked to the variation in the content of lignin (22% in switchgrass vs 19% in Bermuda grass) in the lignocellulosic biomass as reported by the authors.

8.4.1.2 Wheat Straw

Wheat enjoys major production in Asia, Europe, and North America accounting for 43%, 32%, and 15% of the total world production, respectively [36]. Inevitably, these regions produced the highest amount of wheat straw, with a total global production of 529 Tg, yearly. According to Tomás-Pejó et al. [69], in Europe, an amount of 140 million tons of wheat straw was generated in the year 2006. Wheat straw has a composition including cellulose, hemicellulose, and lignin at 5–25%, 35–45%, and 25–40%, respectively [19, 37, 67]). The high composition of cellulose makes it a viable lignocellulosic feedstock for bioethanol production [69].

Bioethanol production from wheat straw has been encouraged to ease out the competition with the wheat grain itself for human food consumption [36]. According to Kim and Dale [36], with efficient cultivation practices, roughly 354 Tg of wheat straw can be produced and processed globally for ethanol production, with the potential yield of 104 GL. Also, the use of wasted wheat grain and wheat straw can both contribute about 115.2 GL of bioethanol globally. However, the structure and components of wheat straw pose a serious obstruction to the enzymatic accessibility of the hemicellulose, cellulose, and polysaccharides present. Hence, it is required that it undergoes an appropriate pretreatment in such a way to improve the enzyme degradability of the structure and components [37, 69]. Various approaches have been employed to pretreat wheat straw to enhance bioethanol production. Viola et al. [72] performed a research on steam explosion pretreatment of wheat straw with optimization on the bedrock of carbohydrate recovery and 25% improvement on the digestibility was achieved.

Kumar et al. [42] subjected wheat straw to ammonia fiber explosion by maintaining 1–2 kg dosage of NH_3 kg^{-1} of dried feedstock at 30 min residence time 90 °C. A very small quantity of the material (in solid form) was solubilized after pretreatment and no lignin or hemicellulose was removed. The hemicellulose was broken down to de-acetylated and oligomeric sugars. However, there was alteration to the material structure which subsequently resulted in higher digestibility and improved water-holding capacity.

A comparative study on the effect of ball and wet disk milling pretreatments on sugarcane bagasse and straw was performed (da Silva et al. [23]. Pretreatment through ball milling method performed better than wet disk milling method with respect to hydrolysis yields of xylose and glucose. The values of xylose and glucose yielded with ball milling pretreatment are 72.1% and 78.7%, and 56.8% and 77.6%, respectively, for bagasse and straw.

An orthogonal design was developed by Jian Xu et al. [78] to improve the pretreatment of wheat straw using microwave, and an increase in bioethanol yield from 2.68% to 14.8% was obtained.

In another related study on wheat straw, investigation of R. Sun and Tomkinson [66] showed that there was an increase in delignification of 7.6–8.4% when wheat straw was pretreated with sonicated alkaline for 15–35 min. Apart from the interval, the power of sonication was directly determined by the frequency of sonication, a very significant parameter stirring the pretreatment of lignocellulosic biomass.

8.4.1.3 Cassava Peels

Cassava cultivation enjoys a worldwide distribution and it represents the third highest carbohydrate source available for human consumption [39]. In order to get access to the consumable part of cassava (cortex), the thin outer cover with brown color and inner covering with parenchymatous leather and phelloderm needs to be peeled off because most of these materials are not really employed for economic purposes [2, 39]. This processing of cassava generates a lot of by-products that may lead to environmental pollution if they are not utilized or converted to value-added products [70]. There have been various efforts aimed at the preservation of the environment by the total utilization of raw materials from agricultural industries [39]. The main use of cassava peels has been in the production of animal feeds. However, a challenge with the feeds produced is the low protein content, usually less than 6%, and the presence of anti-nutrients (tannic acid, oxalate, and hydrocyanic acid) has discouraged its use for feed production [2]. A viable means of engaging cassava peels is utilizing it for ethanol production, since it is not consumable as food by humans [2, 39].

There has been a successful application of cassava peels for ethanol production, and the major concern has mainly been that the commercial production or usage from this biomass may not be feasible due to the amount that is currently produced [2]. Also, it was suggested that the inexpensive nature of cassava peels would in the long run complement this challenge with increased cassava production. In addition, Adekunle et al. [2] recommended that in a way to increase the amount of sugar produced after enzymatic hydrolysis, cassava peels conversion to bioethanol requires that it first goes through appropriate pretreatments due to its composition in order to increase the amount of sugar yield after enzymatic hydrolysis. The lignin, cellulose, and hemicellulose contents are greater than 50% of the dry weight.

Ethanol production from cassava peels by means of mechanical form of milling pretreatment combined with another form of pretreatment was investigated by Adesanya et al. [3]. Liquid suspension of the milled cassava peel was inoculated with freshly harvested cells followed by hydrolysis. The maximum amount of simple sugar formed in the inoculated milled suspension was 0.88 mg/ml throughout the experiment, while the concentration of the reducing sugar of the inoculated one remained at 0.45 mg/ml. The cell-free cassava peels hydrolysate inoculated with Saccharomyces cerevisiae gives rise to highest yield of bioethanol 3 days after incubation. However, the concentration of bioethanol yielded was somewhat low (1.05%). Regardless of the low yield of bioethanol produced, the results are, however, similar to the facts obtained from other studies in which 2.3% bioethanol yield was reported from cassava peel hydrolysates prepared enzymatically.

8.4.1.4 Rice Straw

Rice straw is an essential bioresource material obtainable from rice production [83]. In accordance with Binod et al. [15], the choice of rice straw for bioethanol production is due to its abundance and composition. Rice straw production is distributed worldwide across Africa, Asia, and Europe at 731 million tons of annual production from which 205 billion liters of bioethanol can be obtained [76]. The compositions of rice straw primarily include lignin hemicellulose, cellulose, and ashes in proportions of 5–24%, 19–27%, 32–47%, and 18.8%, respectively. As suggested by Binod et al. [15], fermentable sugar can be achieved through cellulose and hemicellulose hydrolysis. However, these contents are within the lignin matrix, and it then becomes difficult to access it for enzymatic conversion to fermentable sugar [76].

Rice straw of 1462.69 Tg is available for bioethanol production and that for a fully utilized rice straw, a total weight of 731 Tg could result in the production of 205 GL of bioethanol as reported by Kim and Dale [36]. Also, the study reported that a total of 221 GL of ethanol could be produced from rice straw and wasted rice grain with Asia accounting for 90.5% of the overall production.

Morone et al. [54] scrutinized the impact of some operation factors on lignin removal, hemicellulose solubility, and cellulose recovery with pretreatment of rice straw using an advanced oxidation process of alkaline wet air oxidation (AWAO). This resulted in 32–66% lignin removal, 67–87% hemicellulose solubilization, and 68–90% cellulose recovery with generation of limited inhibitors. The pretreatment method triggered cleavage of carbohydrate lignin linkages, hemicellulose deacetylation, and oxidative delignification, thus, enhancing cellulose accessibility by 42–89% enzymatic cellulose conversion. The findings indicated that chemical input was minimized and potent inhibitors were absent in the liquor. Altogether, this is an indication of reduction in the cost of pretreatment, fresh water requirements and minimization of waste generation.

Hideno et al. [29] compared the impact of conventional ball and wet disk milling pretreatment methods on rice straw. For the optimal conditions obtained, 10 recurrent milling processes were carried out for wet disk milling, whereas 60 min of operation was observed for dry ball milling. The conventional ball milling method yielded maximum glucose of 89.4% and xylose of 54.3% compared to 78.5% and 41.5% obtained for wet disk milling. Nevertheless, wet disk milling did not produce inhibitors, but required lower energy and high effectiveness of enzymatic hydrolysis.

Abedinifar et al. [1] in his work effectively converted rice straw into bioethanol by the separation of enzymatic hydrolysis and fermentation through S. cerevisiae, *Mucor indicus, and Rhizopus oryzae*. The pretreatment was mechanically done by milling, then separated to a size smaller than 20–48 meshes (295–833 μm), followed by another pretreatment with just steam or dilute-acid hydrolysis. The rice straw pretreated with milling and dilute-acid hydrolysis produced 0.72 g per gram of sugar yield in 48-hr enzymatic hydrolysis more than rice straw treated with milling and steam that produced 0.60 g g^{-1} and untreated produced 0.46 g g^{-1}.

8.4.1.5 Corn Stover

Corn stover is an indigenous biomass with ability to yield some amount of biofuel like ethanol and many bioenergy and bioproducts [46]. Wet milling pretreatment of this feedstock was considered to be better than dry milling pretreatment [44].

The optimum conditions for milling were considered to be particle size of 0.5 mm, 20 number of steel balls of 10 mm each, solid/liquid ratio of 1:10, and ground for 30 min with ball speed of 350 rpm/min. It was noticed that when milling was combined with alkaline pretreatment method, better results were produced. Also, when compared with wet milling process, enzymatic hydrolysis efficiency of the corn stover was increased by 110% using alkaline milling pretreatment.

Zhang et al. [85] performed sugar recovery from corn stover using a twin-screw extruder, where 48.79% of glucose, 24.98% of xylose, and 40.07% of sugar recovery were obtained at 80 rpm (screw speed), 0.028 g/g (enzyme dose) of dry feedstock, and moisture content of 27.5%. The values obtained were better (2.2, 6.6, and 2.6 times) more than the values obtained for the untreated feedstock samples.

Also, Karunanithy and Muthukumarappan [32] investigated the impact of screw speed and extruder temperature on the pretreatment of corn stover while varying enzymes and their proportions. Corn stover was pretreated at different screw speeds of 25, 50, 75, 100, and 125 rpm along with different temperatures of 25, 50, 75, 100, and 125 ° C. The results showed that the maximum concentrations obtained for glucose, xylose, and combined sugar were 75%, 49%, and 61% respectively, at screw speed of 75 rpm and temperature of 125° C with a ratio 1:4 of cellulase and β-glucosidase. The values obtained were 2.1, 1.7, and 2.0 times higher than the values obtained for the controls. This is an indication that the overall yields of reducing sugars were synergistically affected by optimization of the requirements for pretreatment method than concentrations of the enzyme.

Galletti and Antonetti [26] reported that glucose yield of corn stover when subjected to carbon dioxide explosion-CO_2-H_2O process at 160 °C was 85% and the hemicelluloses sugar yield was 10% while the furfurals yield was 11.2%.

8.4.1.6 Soybean Hull

Pretreatment through thermomechanical process on soybean hulls was investigated by Yoo [82]. A 95% cellulose was formed to glucose at a barrel temperature of 80 ° C, MC of 40%, and 350 rpm (screw speed) under optimum processing conditions.

8.4.2 *Chemical and Biological Pretreatment of Selected Lignocellulosic Biomass*

8.4.2.1 Rice Straw

The study of Yoswathana et al. [83] on production of bioethanol from rice straw employed fermentation with Saccharomyces cerevisiae. The pretreatment was done using sulfuric acid of concentrations between 1% and 9%, for 15 min and at 121 ° C with proportion of rice straw to acid maintained at 1:10 (w/v). The study concluded that 21.45% sugar w/w was measured once treated with simultaneous reduction in the sugar concentration at a higher rate of acid concentration. The reduction was ascribed to the breakdown of glucose with xylose (monomeric sugars) present in furfural and hydroxymethylfurfural.

Ethanol was produced from rice straw using dilute acid (0.5%) for pretreatment [1]. The rice straw was put into the dilute acid for 20 h before being introduced into a high-pressure reactor for 10 min at 1.5 MPa. The dilute acid pretreatment gave 0.72 g/g yield of sugar, while the untreated gave 0.46 g/g yield. The sugar produced reduced with increase in concentration of the acid. On the ethanol yield, the conversion was done by *Rhizopus oryzae,* Saccharomyces cerevisiae, and *Mucor indicus* with yield of 0.33 to 0.41, 0.37 to 0.45, and then 0.36 to 0.43 g g^{-1}, respectively.

Binod et al. [15] pretreated rice straw with 2% sodium hydroxide for 1 h at 85°C. The treatment brought about 36% reduction in the lignin content and facilitated enzymatic hydrolysis through increase in the external surface area. The treatment was done with sodium hydroxide of concentrations of 1–5%. The mixture was maintained in water bath for 1 h at the temperature of 85°C. The result showed that 0.55% of sugar was measured using this method. The study also reported a high yield of sugar when acid and alkaline treatment was followed by the application of technical enzymes. For instance, the sugar concentration increased from 0.55% to 24.60% when 5% NaOH and enzyme (0.8 w/v) was used. On the ethanol yield, approximately 55–65% of the sugar was converted to ethanol after 3 days of fermentation with Saccharomyces cerevisiae given about 0.42 g/g of ethanol yield.

Belal [14] reported on bioethanol production from rice straw residues using both alkaline and acid pretreatment. Sodium hydroxide (5%) was used as the chosen alkaline and a proportion of 1:10 w/v residues to alkaline was used. The treatment was carried out at 85°C for 1 h. Also, pretreatment with acid was done by maintaining the same biomass-to-solvent ratio in the dilute acid solution (1%). Then, it was heated in an autoclave at 121°C for 15 min. The yield of ethanol was about 10–11 gram per liter after 7 days of Saccharomyces cerevisiae fermentation.

Moradi et al. [53] studied the production of ethanol, butanol, and acetone from rice straw using both concentrated phosphoric acid and sodium hydroxide for pretreatment and *Clostridium acetobutylicum* for fermentation. Alkaline treatment was done with sodium hydroxide (12% w/v) with 5% w/w of solid loading and the sample mixture was maintained for 3 h. Phosphoric acid concentration (85%) was used for the treatment for 30 min and at 50 °C. Quenching of the mixture was done with pre-cold acetone. The mixture of the acetone, phosphoric acid, and the rice straw was centrifuged thrice to ease the separation. The yield on glucose was 163.5 g and 192.3 g per kg of rice straw and the conversion after 72 h of fermentation was ethanol (1.2 g), butanol (45.2 g), and acetone (17.7 g) for alkaline treatment, and ethanol (0.6 g), butanol (44.2 g), and acetone (18.2 g) for phosphoric acid treatment.

A new method of pretreatment that does not involve a solid–liquid separation for bioethanol produced from rice straw was presented by Park et al. [58]. Lime was employed for the pretreatment and it involves release of calcium through carbonation. Desired weight of the feedstock (10 g) was added to a 1 g/100 ml of limewater solution with the resulting blend heated in a high-pressure steam vessel for 1 h at 121 °C. After the treatment period, neutralization was provided with CO_2 by bubbling it through the mixture for 30 min. Both Saccharomyces cerevisiae and Pichia stipites were employed, and 74% theoretical yield of the glucose and xylan was converted to give 19.1 g/l of ethanol.

The investigation of Ko et al. [38] on ethanol produced from rice straw feedstock soaking with aqueous ammonia pretreatment reported an optimal conditions of 21% level of ammonia at 69 ° C and 10 h.

Amiri et al. [7] gave an account of ethanol, butanol, and acetone produced from rice straw with the use of organosolv pretreatment along with clostridium acetobutylicum for fermentation. The pretreatment involved 75% (v/v) of ethanol (aqueous) mixed with 1% (w/w) catalyst (H_2SO_4) at 150°C for 60 min with a feedstock to solvent proportion of 1:8. On the yield of sugar, 31 g/l was reported on enzymatic hydrolysis. The sugar and ethanol yield increased as temperature of mixture was increased to 180°C for 30 min with acetone, butanol, and ethanol given 21.1 g, 80.3 g, and 22.5 g, respectively.

8.4.2.2 Wheat Straw

Nigam [56] reported ethanol produced from wheat straw by Pichia stipites with dilute acid chosen for pretreatment and the yield of ethanol was 0.41 ± 0.01. The comparison effect of organic acid (fumaric and maleic) and H_2SO_4 pretreatment on wheat straw for bioethanol production and the possibility of dilute organic acid producing less sugar degradation products when compared with dilute H_2SO_4 were studied by Kootstra et al. [40]. The acid treatment employed soaking the desired straw in acid for 20–24 h before the mixture was transferred to four reactors with thermocouples attached. The temperature was increased to 130, 150, and 170 ° C and maintained for 30 min prior to hydrolysis (using enzymes). With organic acid treatment, the desired amount of organic acid/wheat straw was prepared and soaked at a temperature higher than room temperature. The report concluded that at 150 ° C, and a dry wheat straw of 20–30% (w/w), the organic acid can serve as a direct substitute for wheat straw pretreatment. The yield of sugar (glucose) of fumaric acid, maleic acid, and sulfuric acid was 86%, 96%, and 98%, respectively. The amount of sugar convertible to ethanol was not stated.

Kuhar et al. [41] investigated the potential of *Phanerochaete chrysosporium*, *Pycnoporus cinnabarinus*, fungal isolate RCK-1, and fungal isolate RCK-3 on the treatment of wheat straw for bioethanol production. A 1:4 substrate-to-moisture level was used and the inoculation was done in an enamel tray using 7.5 (±0.05) mg fungal dry mass per gram of straw. The temperature of the *P. cinnabarinus* and fungal isolate RCK-1 was maintained at 30°C and *P. chrysosporium* and fungal isolate RCK-3 was maintained at 37°C for 20 days with provision of manual agitation after every 3 days. The fungi-treated wheat straw and the untreated sample both underwent sulfuric acid treatment again between concentrations of 0.5 and 4.5% v/v at 121 ° C. After 20 days of treatment, there was a degradation of 18.89%, 19.12%, 22.64%, and 20.17% for the *P. chrysosporium*, P. cinnabarinus, fungal isolate RCK-1, and fungal isolate RCK-3, respectively. The fermentation was done with

Pichia stipites and determination of ethanol concentration was done with gas chromatography. After 36 h, an optimum yield of 0.48 g/g ethanol was reported.

Saha et al. [63] used dilute sulfuric acid to pretreat wheat straw, and the blend was heated in an autoclave at 121 ° C for 1 h. Fermentation was provided using yeast extract and it was reported that, at the acid concentration of 0.75% v/v and wheat straw of 78.3 g, 19.1 g/l of ethanol was obtained. The study concluded that increasing the level of H_2SO_4 from 0.75 to 4% (v/v) reduced some amount of fermentable sugar obtained.

Saha and Cotta [62] studied bioethanol production with wheat straw using pretreatment of alkaline peroxide. The biomass–water mixture used (8.6%, w/v) was added to the solution of hydrogen peroxide with varying concentrations of between 0% and 4.3% v/v. The resultant pH of 11.5 was maintained with the aid of sodium hydroxide and was then taken to an incubator shaker operated at 25 or 35 °C for 3–24 h. Before enzymatic saccharification, hydrogen chloride was mixed with the mixture to lower the pH to 5.0. The study reported the yield of ethanol to be 15.1 g/l with an instance of concurrent saccharification and fermentation with Escherichia coli and 18.9 g/l for the alkaline-pretreated sample hydrolyzate by recombinant E. coli.

8.4.2.3 Jatropha

Jatropha *(*Jatropha curcas*)* belongs to Euphorbiaceous family. It is an equatorial plant that can grow as tall as 5–7 m [8]. The tree is widely distributed and found in Africa, India, Brazil, Argentina, and Paraguay. J. curcas matures in about 5 years with life cycle between 30 and 50 years [8]. Oil extracted from jatropha seed has been engaged majorly for production of biodiesel [8, 49]. However, there have been concerted efforts as to the possibility of deriving more than one biofuel from a particular biomass, and this has led to the investigation of the feasibility of deriving bioethanol from other parts of J. curcas [27]. According to García et al. [27], the fruit of J. curcas is made up of 31.6% of shells and approximately 25 tons of these shells can be produced per hectare. Hence, bioethanol production from J. curcas shell and hull is beginning to gain attention.

From the findings of Marasabessy et al. [49] on the pretreatment of Jatropha fruit hull, the pretreatment was done at a high temperature with sulfuric acid. The sulfuric acid concentrations employed were 0.1%, 0.5%, or 0.9% (w/v) and initially, the soaking time was varied between 20 and 24 h at room temperature. Later, the mixture was transferred into a reactor with temperature raised to 140, 160, and 180 ° C at 30, 45, and 60 min. The optimum conditions were reported to be 0.9% w/v dilute H_2SO_4 at 180 ° C for 45 min; the total sugar yield was 77% after enzymatic hydrolysis. S. cerevisiae was employed for the production of ethanol, and fermentation was examined for 24 h, which produced 8.4 g l^{-1} of bioethanol corresponding to a maximum theoretical yield of 71%.

The saccharification and fermentation of J. curcas were simultaneously studied by utilizing the by-products from biodiesel production (Visser et al. [73]. Dilute sulfuric acid (0.5%) and sodium hydroxide (1%) pretreatment of the solid was done in an autoclave for 1 h at 121 ° C. *S. cerevisiae* was employed for fermentation. In conclusion, the maximum conversion of sugar to ethanol was 41.03% and 40.43% for the sodium hydroxide and sulfuric acid pretreatment, respectively.

García et al. [27] considered the potential of using dilute sulfuric acid pretreatment and enzymatic hydrolysis to produce ethanol from fruit shells of J. curcas. The impact of temperature, acid level, and pretreatment time on the enzymatic hydrolysis with values varied from 110 to 150° C, 0.5 to 2.5%, and 15 to 45 min, respectively, was predicted using Box-behnken designed response surface methodology. The research developed a model which put the maximum conversion of cellulose to ethanol at 82% with experimental conditions of 1.5% of H_2SO_4 for 136° C at 30 min.

8.4.2.4 Cassava Peels

Kongkiattikajorn and Sornvoraweat [39] employed separate strains of yeast to examine the efficiency of dilute-acid, dilute-base, and distilled water pretreatment on bioethanol production from cassava peels. A pretreatment of 1.5% (w/v) cassava peels was done under 1.03 bar pressure for 30 min at 135°C in 0.1 H_2SO_4 or 0.025% NaOH or distilled water. The suspended solution after pretreatment for the fermentation process was neutralized to pH 5.5. The acid treatment was done with 0.1 M of H_2SO_4 added to 1.5% (w/v) of cassava peels, and the blend was heated under 1.03 bar pressure for 30 min at a temperature of 135 ° C. After 24-hr incubation, the reducing sugar produced the highest yield of 0.72 g/g dry cassava peels based on diluted acid treatment. The ethanol yielded 0.418 g/g of dry cassava peels, and this was achieved when saccharification and fermentation (*Saccharomyces diastaticus* 2047) were simultaneously employed. The alkaline treatment was done with sodium hydroxide (0.025%). Addition of 1.5% (w/v) cassava peels was made with alkaline and heated under 1.03 bar at 135 ° C within 30 min. The bioethanol produced was 0.177 g/g of dry cassava peels, and this was achieved when saccharification and fermentation (*Saccharomyces. diastaticus* 2047) were simultaneously employed.

Research on bioethanol production from the nonfood part of cassava (stem, leaf, root, and cassava peels) was conducted by Nuwamanya et al. [57]. The acid treatment was done by adding 1 M hydrochloric acid to 200 g of cassava parts in 200 ml of solution. Upon enzymatic hydrolysis, the sample was analyzed for the reducing sugar present by withdrawing the samples after 1 h for 8 h and withdrawal was changed to 24 h for 5 days. The yield of reducing sugar based on 5 g of substrate was approximately 3 g after 5 days. Approximately 78.5% of the reducing sugar produced from cassava peels was converted to ethanol. The alkaline treatment was done by adding 1 M sodium hydroxide to 200 g of cassava parts in 200 ml of solution. Upon enzymatic hydrolysis, the sample was analyzed for the reducing sugar present by withdrawing the samples after 1 h for 8 h and withdrawal was changed to 24 h for 5 days. The yield of reducing sugar based on 5 g of substrate was approximately 3 g after 5 days.

8.4.2.5 Switch Grass

Keshwani and Cheng [34] pretreated switch grass with sodium hydroxide, calcium hydroxide, and sodium carbonate with the aid of microwave radiation (with 250 W of power level). A loading ratio of 1:10 (solid-to-liquid) was used and treatment time of 10 min at 2% (w/v) concentration. The yields of reducing sugar based on the treated samples were 207, 372, and 446 mg/g for sodium carbonate, calcium hydroxide, and sodium hydroxide, respectively. These values were more than the yield obtained for the untreated sample, 151 mg/g.

Ethanol production from switch grass using dilute H_2SO_4 with concentration of 1.2% (w/w) was reported by Martín and Grossmann [50]. The biomass was heated for 180° C and at a pressure of 12 bar. After enzymatic hydrolysis, fermentation of the liberated sugars was done with *Z. mobilis* and glucose of 92% was converted to bioethanol as reported.

The potential of varying acid concentration on bioethanol yield of switchgrass was investigated by Y. Yang et al. [81]. The experimental settings were at levels of 0.5%, 1.0%, and 1.5% (w/v) H_2SO_4 with solid/liquid proportion of 1:10, heating at 121 ° C and 15 psi was provided and operated in an autoclave. The treatment time used was 30, 45, and 60 min. An optimum condition that gave the highest ethanol yield based on glucose (0.082 g) after fermentation was 1.5% concentration of H_2SO_4 at 60 min residence time.

In a way to improve the enzymatic digestibility, Jiele Xu and Cheng [79] pretreated switchgrass with sodium hydroxide. The experimental conditions are NaOH at 0.5%, 1.0%, and 2.0%, (w/v) levels; temperature of 121 ° C, 50 ° C, and 21 ° C was maintained for the residence time of 0.25–1, 1–48, and 1–96 h, respectively. The optimum experimental conditions were 1.0%, 50 ° C and 12 h and the yield of total fermentable was 453.4 mg per the treated sample, while the untreated sample was 3.78 times less than this value. The study also reported a correlation between the pretreatment and the amount of the lignin removed from the biomass. At the level of 2% (w/v) with longest period of residence, the amount of lignin removed was 85.8%, 77.8%, and 62.9% for 121 ° C, 50 ° C and 21 ° C, respectively.

8.4.2.6 Corn Cob

Corn cob is unique among the most possible lignocellulose feedstocks. According Xie et al. [77], corncob is chosen among other potential raw materials for the production of bioethanol by POET Company (the world's major producer of bioethanol) and some other global energy giants. The biocompatibility of ionic-liquid (1-methyl-3-methyl–imidazolium dimethyl phosphite) of corn cob for saccharification improvement was evaluated by Li et al. [43]. The solvent (1-methyl-3-methyl-imidazolium dimethyl phosphite) was used in the pretreatment, as it is an environmentally friendly solvent due to its biocompatibility with both cellulase activity and lignocellulosic solubility. Pretreatment was done by dissolving the corn cob (0.3 g) in 10 g/L ionic solution using a 50-ml round-bottom bottle with magnetic stirring under nitrogen atmosphere. The resultant solution was incubated in ionic liquid at 130 ° C for 20 min. Precipitation for pretreatment using deionized water was done on the corn cob powder. The sample was centrifuged to remove the supernatant containing ionic liquid recovery and the precipitate was cleaned by adding deionized water. Then, the final washing was done with buffer solution and the regenerated corncob was dried for 10 h at 60°C in an oven and kept for further use. The dimethyl phosphate (DMP) saccharification was employed for effective performance of the corncob in bioconversion to sugars with amount greater than 70% saccharification gained.

Xie et al. [77] examined the outcome of various pretreatment means of corn cob for production of ethanol and enzyme recovery. The pretreatment methods are dilute acid, acid–base coupling (dilute H_2SO_4/aqueous NH_3), and sodium hydroxide, then soaking in aqueous ammonia. Four (4) specimens were differently set up with addition of oven-dried corn cob in four different Pyrex glasses, which were labeled A to D. The samples were prepared in such a way that 2% (wt) H_2SO_4 and 2% (wt) NaOH were put in glasses labelled A and B and 15% (wt) aqueous ammonia in samples C and D to make a proportion of 6 ml of liquid to 1 g of solid. The mixtures were allowed to pass through heat in flasks A and B for 1 h at 121 ° C, with incubation without agitation in water bath for 6 h at 80 ° C and for 12 h at 60 ° C in flasks C and D, respectively. The solutions were filtered and then washed in tap water until a pH of 7 was reached. The filtrate was oven dried at 105 ° C and weighed thereafter. After the glucose analysis, pretreatment with acid–base method was considerably better when compared with solitary pretreatment of acid or alkaline.

8.4.2.7 Corn Stover

Lloyd and Wyman [46] determined the pretreatment condition of dilute acid to maximize glucose yields by using sulfuric acid in a reactor. A sample of corn stover was soaked beforehand for at least 4 h at room temperature in 5% solids (w/w) of dilute H_2SO_4 acid solution. The slurry that was presoaked was then moved to the reactor fitted to the impeller drive motor that was hung with a string-like hoist saddled on a crane with about 100 rpm speed. Then, to the bottom of the reactor, head flange was set a sand bath at 320 ° C inside a vessel. With this pretreatment, about 15% of the full potential sugar in the substance could be discharged in the form of glucose.

Liu et al. [45] studied the pretreatment of corn stover by choosing five salts (inorganic) that are commonly used; these are: two monovalent salts – NaCl (0.03 M) and KCl (0.03 M), two divalent salts – $CaCl_2$ (0.15 M) and $MgCl_2$ (0.15 M), and one trivalent salt – $FeCl_3$ (0.10 M). All the tests were carried out in the same manner (solid/liquid = 1:10, for 20 min at 160 ° C). $FeCl_3$ has a strong influence on the hemicellulose removal (91% of the hemicellulose was removed at 0.1 M $FeCl_3$ for 20 min at 140 ° C). Also, as observed, the pretreatment with $FeCl_3$ at minimum temperature can degrade as low as 9% cellulose.

8.4.2.8 Olive Tree

Olive tree is an affordable tree cultivated mostly in the Mediterranean countries [61]. Pruning is an important operation in olive tree which generates about 3000 kg/ha lignocellulosic biomass annually. Depending on production, culture, local uses, and other conditions, pruning of olive tree consists of thin branches and leaves in various proportion. With respect to the higher proportion of possibly fermentable carbohydrates, it can be taken as an appropriate raw material for bioethanol [11].

The metamorphosis of olive tree residues into fermentable sugars was examined by Cristóbal Cara et al. [18] using pretreatment method of dilute acid. Pretreatment at five levels of temperature ranging from 170 to 210 °C was performed on pruned olive tree for 10 min. There was addition of water to a 200 g quantity of the dry substrate in the ratio 5:1 (v/w) liquid-to-solid. After the pretreatment, cellulase enzyme was used for the hydrolysis. The maximum hydrolysis yield observed was 76.5% at 1.4% acid concentration and 210 ° C. The total sugar produced was given as 36.3 g sugar per 100 g feedstock at 1% H_2SO_4 acid concentration and 180 °C, indicating 75% of all sugars in the raw materials. These results were considerably improved in comparison with water pretreatment method.

Steam explosion pretreatment was studied on olive tree and further delignified by an alkaline peroxide pretreatment ([17]. The steam explosion pretreatment was examined by heating the 200 g of the biomass for 5 min at 190, 210, 230, and 240 °C. The mixture was filtered (after allowing it to cool down to 40 °C) to recover the solid from the liquid. A hot alkaline peroxide pretreatment (1% of hydrogen peroxide solution at 4% (w/v) solids level) was used to delignify the residue from the filtrate. The pretreatment at 80 °C continued for 45 min with addition of 4 M NaOH, then pH was adjusted to 11.5. The suspension was separated and cleaned until a neutral pH was achieved and then dried for sugar analysis. The maximum overall yield obtained by considering both glucose from the steam and sugars available in the liquid from the pretreatment was delignified and hydrolyzed, and 52.6% of the solid was achieved at the minimum steam explosion temperature.

8.4.2.9 Rice Husk

Rice is among the most widely produced crops worldwide, thereby releasing several tons of rice hulls [13]. The leading producer of rice is China, followed by India and Indonesia. According to Potumarthi et al. [59], rice hulls amount to about 28.6%, 28.6%, 24.4%, respectively, for cellulose, hemicellulose, and lignin with 18.4% extractive matter.

Biological pretreatment was studied on rice husk by Potumarthi et al. [59] in order to reduce sugar production using lignin peroxidase sourced from white rot fungus (*Phanerochaete chrysosporium)* and for microbial delignification. The simultaneous pretreatment was done by putting RH (2 g) inside a 250-ml Erlenmeyer flask which had in it mineral salt (100 ml) composed of urea, 0.3; glucose,10; $CaCl_2$ $2H_2O$, 0.4; $(NH_4)_2SO_4$, 1.4; Tween 80, 0.2; $MgSO_4$, 0.9; peptone,1; $MnSO_4$, 0.15;

$ZnSO_4$, 0.03 g; $FeSO_4$, 0.03; $CoCl_2$, 0.002 thiamine hydrochloride 1; and $CuSO_4$, 0.03. The medium was inoculated into the sterilized flasks. The mixtures were incubated for 26 days at 30 ° C and 150 rpm in an incubator for further analysis. The rice husk pretreated with fungal produced 895.9 mg/ml /2 g of rice hull with sugar reduction on day 18 of fungal pretreatment.

Also, the result of H_2SO_4 pretreatment on husks was investigated by Saha et al. [64]. The rice husk was milled and 3.0 g of it was dissolved in 15% (w/v) dilute sulfuric acid. Then, it was pretreated at 121 ° C in an autoclave at 140, 160, and 180 ° C predetermined temperatures. The treated husk was well balanced to pH of 5.0 with 10 M NaOH before it was hydrolyzed enzymatically. The monomeric sugar from rice husk gave a maximum yield of 287 (±3) mg per g with 60% yield base on overall carbohydrate. A 0.43 g bioethanol per gram of sugar was produced through fermentation of the rice husk hydrolysate by recombinant E. coli strain FBR 5.

8.5 Conclusion

The production of ethanol from lignocellulosic biomass has continued to gain more acceptance due to the advantages it offers over petroleum-derived ethanol. The choice of the pretreatment method ultimately defines the product yield and processing cost; therefore, a careful selection of the treatment method is not a mere routine matter. Among the physicochemical methods of biomass treatment, ammonia fiber explosion gives variety of advantages such as ammonia reusability, eradication of inhibitors, the elimination of detoxification stage, complete solid recovery, enhanced surface area, and wet-ability. However, this treatment method is limited by high cost and it is unsuitable for high lignin content biomasses; these deficiencies are eliminated when CO_2 explosion and wet oxidation treatment methods are employed. The major chemical treatments for bioethanol production from lingo-cellulosic biomass are acid and alkaline, and both methods have been extensively researched. The use of dilute acids for treatments is preferred over the concentrated form due to the ability of the former to minimize the production of unwanted side products and eradicate the problems of equipment corrosion and toxins production. The advantages of alkaline treatment over acid treatment are low sugar degradation, higher cellulose digestibility, and lignin solubility. The optimum process conditions are often reached when biomass treatment methods are combined.

References

1. S. Abedinifar, K. Karimi, M. Khanahmadi, et al., Ethanol production by Mucor indicus and Rhizopus oryzae from rice straw by separate hydrolysis and fermentation. Biomass Bioenergy **33**(5), 828–833 (2009)
2. A. Adekunle, V. Orsat, V. Raghavan, Lignocellulosic bioethanol: A review and design conceptualization study of production from cassava peels. Renew. Sust. Energy Rev. **64**, 518–530 (2016)

3. O. Adesanya, K. Oluyemi, S. Josiah, et al., Ethanol production by Saccharomyces cerevisiae from cassava peel hydrolysate. Int. J. Microbiol. **5**(1), 25–35 (2008)
4. V.B. Agbor, N. Cicek, R. Sparling, et al., Biomass pretreatment: Fundamentals toward application. Biotechnol. Adv. **29**(6), 675–685 (2011)
5. H. Alizadeh, F. Teymouri, T.I. Gilbert, et al., Pretreatment of switchgrass by ammonia fiber explosion (AFEX). Appl. Biochem. Biotechnol. **124**(1–3), 1133–1141 (2005)
6. P. Alvira, E. Tomás-Pejó, M. Ballesteros, et al., Pretreatment technologies for an efficient bioethanol production process based on enzymatic hydrolysis: A review. Bioresour. Technol. **101**(13), 4851–4861 (2010)
7. H. Amiri, K. Karimi, H. Zilouei, Organosolv pretreatment of rice straw for efficient acetone, butanol, and ethanol production. Bioresour. Technol. **152**, 450–456 (2014)
8. A. Atabani, A. Silitonga, H. Ong, et al., Non-edible vegetable oils: A critical evaluation of oil extraction, fatty acid compositions, biodiesel production, characteristics, engine performance and emissions production. Renew. Sust. Energy Rev. **18**, 211–245 (2013)
9. F.M. Bagwan, K. Patil, S. Aanekar, et al., Treatment of lignocellulosic biomass to alter lignin content. Int. J. Eng. **10**(1), 2017 (2017)
10. M. Balat, Production of bioethanol from lignocellulosic materials via the biochemical pathway: A review. Energy Convers. Manag. **52**(2), 858–875 (2011)
11. I. Ballesteros, M. Ballesteros, C. Cara, et al., Effect of water extraction on sugars recovery from steam exploded olive tree pruning. Bioresour. Technol. **102**(11), 6611–6616 (2011)
12. S. Banerjee, R. Sen, S. Mudliar, et al., Alkaline peroxide assisted wet air oxidation pretreatment approach to enhance enzymatic convertibility of rice husk. Biotechnol. Prog. **27**(3), 691–697 (2011)
13. A. Bazargan, M. Bazargan, G. McKay, Optimization of rice husk pretreatment for energy production. Renew. Energy **77**, 512–520 (2015)
14. E.B. Belal, Bioethanol production from rice straw residues. Braz. J. Microbiol. **44**(1), 225–234 (2013)
15. P. Binod, R. Sindhu, R.R. Singhania, et al., Bioethanol production from rice straw: An overview. Bioresour. Technol. **101**(13), 4767–4774 (2010)
16. T. Bridgeman, L. Darvell, J. Jones, et al., Influence of particle size on the analytical and chemical properties of two energy crops. Fuel **86**(1–2), 60–72 (2007)
17. C. Cara, E. Ruiz, I. Ballesteros, et al., Enhanced enzymatic hydrolysis of olive tree wood by steam explosion and alkaline peroxide delignification. Process Biochem. **41**(2), 423–429 (2006)
18. C. Cara, E. Ruiz, J.M. Oliva, et al., Conversion of olive tree biomass into fermentable sugars by dilute acid pretreatment and enzymatic saccharification. Bioresour. Technol. **99**(6), 1869–1876 (2008)
19. A.K. Chandel, O.V. Singh, Weedy lignocellulosic feedstock and microbial metabolic engineering: Advancing the generation of 'Biofuel'. Appl. Microbiol. Biotechnol. **89**(5), 1289–1303 (2011)
20. V. Chaturvedi, P. Verma, An overview of key pretreatment processes employed for bioconversion of lignocellulosic biomass into biofuels and value added products. Biotech **3**(5), 415–431 (2013)
21. F. Cherubini, G. Jungmeier, LCA of a biorefinery concept producing bioethanol, bioenergy, and chemicals from switchgrass. Int. J. Life Cycle Assess. **15**(1), 53–66 (2010)
22. L. da Costa Sousa, S.P. Chundawat, V. Balan, et al., 'Cradle-to-grave'assessment of existing lignocellulose pretreatment technologies. Curr. Opin. Biotechnol. **20**(3), 339–347 (2009)
23. A.S.A. da Silva, H. Inoue, T. Endo, et al., Milling pretreatment of sugarcane bagasse and straw for enzymatic hydrolysis and ethanol fermentation. Bioresour. Technol. **101**(19), 7402–7409 (2010)
24. B.E. Dale, C. Leong, T. Pham, et al., Hydrolysis of lignocellulosics at low enzyme levels: Application of the AFEX process. Bioresour. Technol. **56**(1), 111–116 (1996)
25. M. Galbe, G. Zacchi, *Pretreatment of Lignocellulosic Materials for Efficient Bioethanol Production Biofuels* (Springer, 2007), pp. 41–65

26. A.M.R. Galletti, C. Antonetti, *Biomass Pretreatment: Separation of Cellulose, Hemicellulose, and Lignin-Existing Technologies and Perspectives* (From Biomass to Chemicals and Fuels, Biorefinery, 2012)
27. A. García, C. Cara, M. Moya, et al., Dilute sulphuric acid pretreatment and enzymatic hydrolysis of Jatropha curcas fruit shells for ethanol production. Ind. Crop. Product **53**, 148–153 (2014)
28. A. Hendriks, G. Zeeman, Pretreatments to enhance the digestibility of lignocellulosic biomass. Bioresour. Technol. **100**(1), 10–18 (2009)
29. A. Hideno, H. Inoue, K. Tsukahara, et al., Wet disk milling pretreatment without sulfuric acid for enzymatic hydrolysis of rice straw. Bioresour. Technol. **100**(10), 2706–2711 (2009)
30. T.C. Hsu, G.L. Guo, W.H. Chen, et al., Effect of dilute acid pretreatment of rice straw on structural properties and enzymatic hydrolysis. Bioresour. Technol. **101**(13), 4907–4913 (2010)
31. N. Jacquet, G. Maniet, C. Vanderghem, et al., Application of steam explosion as pretreatment on lignocellulosic material: A review. Ind. Eng. Chem. Res. **54**(10), 2593–2598 (2015)
32. C. Karunanithy, K. Muthukumarappan, Influence of extruder temperature and screw speed on pretreatment of corn Stover while varying enzymes and their ratios. Appl. Biochem. Biotechnol. **162**(1), 264–279 (2010)
33. C. Karunanithy, K. Muthukumarappan, W.R. Gibbons, Effect of extruder screw speed, temperature, and enzyme levels on sugar recovery from different biomasses. ISRN Biotechnol. **2013**, 942810 (2012)
34. D.R. Keshwani, J.J. Cheng, Switchgrass for bioethanol and other value-added applications: A review. Bioresour. Technol. **100**(4), 1515–1523 (2009)
35. D.R. Keshwani, J.J. Cheng, Microwave-based alkali pretreatment of switchgrass and coastal bermudagrass for bioethanol production. Biotechnol. Prog. **26**(3), 644–652 (2010)
36. S. Kim, B.E. Dale, Global potential bioethanol production from wasted crops and crop residues. Biomass Bioenergy **26**(4), 361–375 (2004)
37. H.B. Klinke, L. Olsson, A.B. Thomsen, et al., Potential inhibitors from wet oxidation of wheat straw and their effect on ethanol production of Saccharomyces cerevisiae: Wet oxidation and fermentation by yeast. Biotechnol. Bioeng. **81**(6), 738–747 (2003)
38. J.K. Ko, J.S. Bak, M.W. Jung, et al., Ethanol production from rice straw using optimized aqueous-ammonia soaking pretreatment and simultaneous saccharification and fermentation processes. Bioresour. Technol. **10**(19), 4374–4380 (2009)
39. J. Kongkiattikajorn, B. Sornvoraweat, Comparative study of bioethanol production from cassava peels by monoculture and co-culture of yeast. Kasetsart J. (Nat. Sci.) **274**, 268–274 (2011)
40. A.M.J. Kootstra, H.H. Beeftink, E.L. Scott, et al., Comparison of dilute mineral and organic acid pretreatment for enzymatic hydrolysis of wheat straw. Biochem. Eng. J. **46**(2), 126–131 (2009)
41. S. Kuhar, L.M. Nair, R.C. Kuhad, et al., Pretreatment of lignocellulosic material with fungi capable of higher lignin degradation and lower carbohydrate degradation improves substrate acid hydrolysis and the eventual conversion to ethanol. Can. J. Microbiol. **54**(4), 305–313 (2008)
42. P. Kumar, D.M. Barrett, M.J. Delwiche, et al., Methods for pretreatment of lignocellulosic biomass for efficient hydrolysis and biofuel production. Ind. Eng. Chem. Res. **48**(8), 3713–3729 (2009)
43. Q. Li, X. Jiang, Y. He, et al., Evaluation of the biocompatibile ionic liquid 1-methyl-3-methylimidazolium dimethylphosphite pretreatment of corn cob for improved saccharification. Appl. Microbiol. Biotechnol. **87**(1), 117–126 (2010)
44. Z. Lin, H. Huang, H. Zhang, et al., Ball milling pretreatment of corn stover for enhancing the efficiency of enzymatic hydrolysis. Appl. Biochem. Biotechnol. **162**(7), 1872–1880 (2010)
45. L. Liu, J. Sun, C. Cai, et al., Corn stover pretreatment by inorganic salts and its effects on hemicellulose and cellulose degradation. Bioresour. Technol. **100**(23), 5865–5871 (2009)
46. T.A. Lloyd, C.E. Wyman, Combined sugar yields for dilute sulfuric acid pretreatment of corn stover followed by enzymatic hydrolysis of the remaining solids. Bioresour. Technol. **96**(18), 1967–1977 (2005)

47. X. Lu, Y. Zhang, I. Angelidaki, Optimization of H2SO4-catalyzed hydrothermal pretreatment of rapeseed straw for bioconversion to ethanol: Focusing on pretreatment at high solids content. Bioresour. Technol. **100**(12), 3048–3053 (2009)
48. S. Mani, L.G. Tabil, S. Sokhansanj, Grinding performance and physical properties of wheat and barley straws, corn stover and switchgrass. Biomass Bioenergy **27**(4), 339–352 (2004)
49. A. Marasabessy, A.M.J. Kootstra, J.P. Sanders, et al., Dilute H 2 SO 4-catalyzed hydrothermal pretreatment to enhance enzymatic digestibility of Jatropha curcas fruit hull for ethanol fermentation. Int. J. Energy Environ. Eng. **3**(1), 15 (2012)
50. M. Martín, I.E. Grossmann, Energy optimization of bioethanol production via gasification of switchgrass. AICHE J. **57**(12), 3408–3428 (2011)
51. D.P. Maurya, S. Vats, S. Rai, et al., Optimization of enzymatic saccharification of microwave pretreated sugarcane tops through response surface methodology for biofuel. Indian J. Exp. Biol. **51**(11), 992–996 (2013)
52. S.H. Mood, A.H. Golfeshan, M. Tabatabaei, et al., Lignocellulosic biomass to bioethanol, a comprehensive review with a focus on pretreatment. Renew. Sust. Energy Rev. **27**, 77–93 (2013)
53. F. Moradi, H. Amiri, S. Soleimanian-Zad, et al., Improvement of acetone, butanol and ethanol production from rice straw by acid and alkaline pretreatments. Fuel **112**, 8–13 (2013)
54. A. Morone, G. Sharma, A. Sharma, et al., Evaluation, applicability and optimization of advanced oxidation process for pretreatment of rice straw and its effect on cellulose digestibility. Renew. Energy **120**, 88–97 (2018)
55. N. Mosier, C. Wyman, B. Dale, et al., Features of promising technologies for pretreatment of lignocellulosic biomass. Bioresour. Technol. **96**(6), 673–686 (2005)
56. J. Nigam, Ethanol production from wheat straw hemicellulose hydrolysate by Pichia stipitis. J. Biotechnol. **87**(1), 17–27 (2001)
57. E. Nuwamanya, L. Chiwona-Karltun, R.S. Kawuki, et al., Bio-ethanol production from non-food parts of cassava (Manihot esculenta Crantz). Ambio **41**(3), 262–270 (2012)
58. J.Y. Park, R. Shiroma, M.I. Al-Haq, et al., A novel lime pretreatment for subsequent bioethanol production from rice straw–calcium capturing by carbonation (CaCCO) process. Bioresour. Technol. **101**(17), 6805–6811 (2010)
59. R. Potumarthi, R.R. Baadhe, P. Nayak, et al., Simultaneous pretreatment and saccharification of rice husk by Phanerochete chrysosporium for improved production of reducing sugars. Bioresour. Technol. **128**, 113–117 (2013)
60. H. Rabemanolontsoa, S. Saka, Various pretreatments of lignocellulosics. Bioresour. Technol. **199**, 83–91 (2016)
61. E. Ruiz, C. Cara, M. Ballesteros, et al., Ethanol production from pretreated olive tree wood and sunflower stalks by an SSF process. Appl. Biochem. Biotechnol. **130**(1–3), 631–643 (2006)
62. B.C. Saha, M.A. Cotta, Ethanol production from alkaline peroxide pretreated enzymatically saccharified wheat straw. Biotechnol. Prog. **22**(2), 449–453 (2006)
63. B.C. Saha, L.B. Iten, M.A. Cotta, et al., Dilute acid pretreatment, enzymatic saccharification and fermentation of wheat straw to ethanol. Process Biochem. **40**(12), 3693–3700 (2005a)
64. B.C. Saha, L.B. Iten, M.A. Cotta, et al., Dilute acid pretreatment, enzymatic saccharification, and fermentation of rice hulls to ethanol. Biotechnol. Prog. **21**(3), 816–822 (2005b)
65. N. Sarkar, S.K. Ghosh, S. Bannerjee, et al., Bioethanol production from agricultural wastes: An overview. Renew. Energy **37**(1), 19–27 (2012)
66. R. Sun, J. Tomkinson, Comparative study of lignins isolated by alkali and ultrasound-assisted alkali extractions from wheat straw. Ultrason. Sonochem. **9**(2), 85–93 (2002)
67. Y. Sun, J. Cheng, Hydrolysis of lignocellulosic materials for ethanol production: A review. Bioresour. Technol. **83**(1), 1–11 (2002)
68. M.J. Taherzadeh, K. Karimi, Pretreatment of lignocellulosic wastes to improve ethanol and biogas production: A review. Int. J. Mol. Sci. **9**(9), 1621–1651 (2008)
69. E. Tomás-Pejó, J.M. Oliva, M. Ballesteros, et al., Comparison of SHF and SSF processes from steam-exploded wheat straw for ethanol production by xylose-fermenting and robust glucose-fermenting Saccharomyces cerevisiae strains. Biotechnol. Bioeng. **100**(6), 1122–1131 (2008)

70. A. Ubalua, Cassava wastes: Treatment options and value addition alternatives. Afr. J. Biotechnol. **6**(18), 2065–2073 (2007)
71. E. Varga, A.S. Schmidt, K. Réczey, et al., Pretreatment of corn stover using wet oxidation to enhance enzymatic digestibility. Appl. Biochem. Biotechnol. **104**(1), 37–50 (2003)
72. E. Viola, F. Zimbardi, M. Cardinale, et al., Processing cereal straws by steam explosion in a pilot plant to enhance digestibility in ruminants. Bioresour. Technol. **99**(4), 681–689 (2008)
73. E.M. Visser, D. Oliveira Filho, M.R. Tótola, et al., Simultaneous saccharification and fermentation (SSF) of Jatropha curcas shells: Utilization of co-products from the biodiesel production process. Bioprocess Biosyst. Eng. **35**(5), 801–807 (2012)
74. C. Wan, Y. Zhou, Y. Li, Liquid hot water and alkaline pretreatment of soybean straw for improving cellulose digestibility. Bioresour. Technol. **102**(10), 6254–6259 (2011)
75. Z. Wang, D.R. Keshwani, A.P. Redding, et al., Sodium hydroxide pretreatment and enzymatic hydrolysis of coastal Bermuda grass. Bioresour. Technol. **101**(10), 3583–3585 (2010)
76. S.G. Wi, I.S. Choi, K.H. Kim, et al., Bioethanol production from rice straw by popping pretreatment. Biotechnol. Biofuels **6**(1), 166 (2013)
77. N. Xie, N. Jiang, M. Zhang, et al., Effect of different pretreatment methods of corncob on bioethanol production and enzyme recovery. Cellul. Chem. Technol. **48**(3), 313–319 (2014)
78. J. Xu, H. Chen, Z. Kádár, et al., Optimization of microwave pretreatment on wheat straw for ethanol production. Biomass Bioenergy **35**(9), 3859–3864 (2011)
79. J. Xu, J.J. Cheng, Pretreatment of switchgrass for sugar production with the combination of sodium hydroxide and lime. Bioresour. Technol. **102**(4), 3861–3868 (2011)
80. B. Yang, C.E. Wyman, Effect of xylan and lignin removal by batch and flowthrough pretreatment on the enzymatic digestibility of corn stover cellulose. Biotechnol. Bioeng. **86**(1), 88–98 (2004)
81. Y. Yang, R. Sharma-Shivappa, J.C. Burns, et al., Dilute acid pretreatment of oven-dried switchgrass germplasms for bioethanol production. Energy Fuel **23**(7), 3759–3766 (2009)
82. J. Yoo, *Technical and Economical Assessment of Thermo-Mechanical Extrusion Pretreatment for Cellulosic Ethanol Production* (Kansas State University, 2011)
83. N. Yoswathana, P. Phuriphipat, P. Treyawutthiwat, et al., Bioethanol production from rice straw. Energy Res. J. **1**(1), 26 (2010)
84. M. Yu, A. Womac, C. Igathinathane, et al., Switchgrass ultimate stresses at typical biomass conditions available for processing. Biomass Bioenergy **30**(3), 214–219 (2006)
85. S. Zhang, Y. Xu, M.A. Hanna, Pretreatment of corn stover with twin-screw extrusion followed by enzymatic saccharification. Appl. Biochem. Biotechnol. **166**(2), 458–469 (2012)
86. X. Zhao, K. Cheng, D. Liu, Organosolv pretreatment of lignocellulosic biomass for enzymatic hydrolysis. Appl. Microbiol. Biotechnol. **82**(5), 815 (2009)
87. J. Zheng, L. Rehmann, Extrusion pretreatment of lignocellulosic biomass: A review. Int. J. Mol. Sci. **15**(10), 18967–18984 (2014)
88. Y. Zheng, H.M. Lin, G.T. Tsao, Pretreatment for cellulose hydrolysis by carbon dioxide explosion. Biotechnol. Prog. **14**(6), 890–896 (1998)
89. Z. Hu, Z. Wen, Enhancing enzymatic digestibility of switchgrass by microwave-assisted alkali pretreatment. Biochem. Eng. J. **38**(3), 369–378 (2008)

Chapter 9
Extraction of Multiple Value-Added Compounds from Agricultural Biomass Waste: A Review

A. F. A. Chimphango, L. R. Mugwagwa, and M. Swart

9.1 Introduction

There has been a recent drive to utilize agricultural residues including wheat straw, corncobs, potato peels and mango peels as sources of high-value products including biopolymers (cellulose, hemicellulose and lignin) and bioactive compounds (polyphenols, anthocyanins, enzymes, etc.) to replace fossil fuel-based sources that are non-renewable [18, 43, 99]. In addition, the demand for the bio-based products has increased because of the increase in the realization of the healthy benefits these products can offer.

The agricultural residues are in abundance, constituting over 50% of the world's absolute dry matter [115]. For example, Europe alone produced 315 million tons of agricultural residues in 2016 [112]. In Africa, agricultural residues are a major contributor to biomass availability [20]. For example, maize, potatoes and sugarcane residues from five selected African countries (Malawi, Mozambique, South Africa, Zambia and Zimbabwe) account for over 216.5 million tons per annum [47]. Fruit processing residues can contribute as high 50% of the total food waste produced globally [137].

Current disposal methods for the agricultural residues include landfilling, incineration, animal feeding and composting, with no or very little economic value [58]. Landfilling is an environmental liability and a health hazard because it occupies land, which could be utilized for other productive activities such as food production. In addition, landfilling of the residues releases greenhouse gases and agricultural chemicals leachates, and forms a breeding ground for pests and diseases [18, 99]. Furthermore, the high costs of transporting the residues that are often bulky (densities of up to 380 kg/m^3) and have high water content (up to 83.8%) is financial

A. F. A. Chimphango (✉) · L. R. Mugwagwa · M. Swart
Process Engineering Department, Stellenbosch University, Stellenbosch, South Africa
e-mail: achimpha@sun.ac.za

M. O. Daramola, A. O. Ayeni (eds.), *Valorization of Biomass to Value-Added Commodities*, Green Energy and Technology,
https://doi.org/10.1007/978-3-030-38032-8_9

burden. Similarly, incineration as a common practice in many parts of the world, releases substantial amounts of carbon emissions into the atmosphere [34] and may require more energy (combustion temperatures of 400–900 °C) than what is recovered [9, 20, 50, 75]. Therefore, the conversion of agricultural residues into value-added products constitutes an alternative disposal strategy with a double gain that can be realized through the replacement of petroleum-based chemicals and polymers, and from the additional value chains that emerge with the potential to create jobs, increase the financial returns and diversify the product spectrum.

The structure and the chemical composition of the residues are the key determining factors for choosing technologies that enable selective isolation of high-value bioproducts (biopolymers and bioactive compounds) from the residues. About 98% of the residues constitute structural biopolymers [95], predominantly cellulose, hemicellulose, pectin and lignin and non-structural compounds such as polyphenols and anthocyanins [104]. The biopolymers and natural antioxidants have applications in the food and pharmaceutical industries [97]. The market for both the polymers and bioactive compounds is growing. For instance, the projected growth of the biopolymers was 1.5% per annum in 2011 and is expected to double by 2020 [41]. On the other hand, the market share for bioactive compounds is projected to increase from U$757 million in 2015 to U$1.33 billion by 2024 [3, 90], mainly due to increased awareness by consumers of the environmental and the health benefits that can be realized from the bio-based products.

The composition of the plant cell wall compounds varies depending on the source. However, the overall objective of the fractionation is to maximize the value that can be obtained from a particular biomass. The common biomass fractionation objectives include increasing the product spectrum, yield and purity of specific or combination of products as well as increasing the resource-use efficiency. Therefore, the fractionation methods are tailor-made to address the specific objectives through identification of optimal extraction conditions and process configurations for the products and their combinations.

In many of the fractionation processes, the removal of the non-structural components is done as a pre-treatment process for increasing accessibility of chemicals and biological agents to the cell wall compounds, which enhances fractionation of the products in the downstream processes. However, optimizing the pretreatment processes can provide additional value chains while at the same time improving the product quality and efficiency in the downstream processes.

9.2 Multi-Objective Fractionation of Agricultural Residues

Traditional biomass fractionation methods, such as those applied in the pulp and paper making, target a single product from the biomass, which in this case is the cellulose. Consequently, cell wall components such as the hemicelluloses, lignin and the non-structural components are discarded in the waste streams. In a multi-objective fractionation process, the ultimate goal is to unlock the value from the biomass, by recovering as many products as possible in their functional form, thus, minimizing by-products going to waste streams. Therefore, the fractionation methods and the extraction conditions are selected and combined in a strategic manner to selectively obtain as many compounds as possible (based on the residue composition) in high yields, purity and functionality. However, the major challenge is to reduce the degradation of the targeted compounds, while at the same time maximizing the yield, purity and functional properties of the co-products in the downstream processes. Thus, a thorough understanding of the reactivity and optimal extraction conditions of the individual components in the residues is necessary.

The bioproducts such as polyphenols, pectin, and nanocellulose that have often been obtained in a single step from agricultural residues such as banana, mango, orange and passion fruit peels, lemon pomace, wheat straw and corn cob [5, 11, 26, 44]. During fractionation, process parameters including temperature, pH, solid loading, solvent type and concentrations are varied to obtain the optimum extraction conditions with respect to single product yield and functional properties (Table 9.1).

Table 9.1 Single-step extraction methods for pectin from fruit waste

Fruit waste	Extraction conditions	Uronic acid content (%)	DE (%)	Pectin yield (%)	References
Pistachio green hull	Citric acid solution of pH 0.5, temperature of 90 °C, extraction time of 30 min, liquid/solid ratio of 50 (v/w)	65	53	22.1	Chaharbaghi et al. [26]
Citrus peel	0.5 M citric acid treatment at pH 7.0 at 65 °C For 2 h	27.2	8.4	7.4	Kurita et al. [69]
Pomegranate peels	Extraction solvent at pH 2.5, temperature of 88 °C and time of 120 min	80.95	54	11	Pereira et al. [94]
Grapefruit	(ultrasound-assisted extraction + microwave-assisted extraction) solid/liquid ratio 1:30 sonication time – 30 min; microwave heating – 10 min	74.25	82.61	31.88	Bagherian et al. [14]
Grape pomace	Ultrasound-assisted extraction temperature – 75 °C, extraction time 60 minutes, citric acid solution at pH 2	96	55.2	32	Minjares-Fuentes et al. [85]
Passion fruit peel	Hydrochloric acid solution of pH 2.0, solid/liquid ratio of 1:30 (w/v), temperature 98.7 °C and extraction time of 60 min	9.6	88.2	14.8	Kulkarni and Vijayanand [67]
Ambarella peels	0.03 M hydrochloric acid solution at pH of 1.5 at 85 °C for 1 h	55.7	50	19.4	Koubala et al. [65, 66]
Lime	0.03 M hydrochloric acid solution at pH of 1.5 at 85 °C for 1 h	77.6	75	26.9	Koubala et al. [65, 66]

DE Pectin degree of esterification

Experimental design tools including Box Behnken and Central Composite Design [36, 40, 94] have been used to optimize the extraction conditions. In the subsequent section, various objectives pertinent to fractionation of agricultural residues for pectin, polyphenols, cellulose and nanocellulose are discussed to unveil the challenges and opportunities when implemented in a biorefinery set-up.

9.2.1 Fractionation of Fruit Residues for Pectin and Polyphenol Recovery

9.2.1.1 Co-Presence of Pectin and Polyphenols

Pectin is a polymer that is made of galacturonic acid that is bonded together with 1–4 glucosidic bonds in the backbone chain [98] with varying methyl esters. The backbone may also contain rhamnopyranosyl group and side chains in the form arabinan and galactan. The pectin can be highly esterified (>50% degree of esterification) through the carboxyl groups, which are cross linked by calcium [64]. The pectin is present in many fruit residues such as mango, banana and passion fruit peels [46, 64]. In addition, the fruit residues are a rich source of antioxidants [46, 79], attributed to the co-presence of polyphenolic compounds.

The co-presence of pectin and polyphenols is both an opportunity and a challenge in fractionation of agricultural residues. Such coexistence affects extractability and the quality of pectin [79] as well as of the polyphenols, for some applications. For example, the polyphenols that are adsorbed to pectin during the fractionation affect the bioavailability of the polyphenols and the fermentation kinetics of both pectin and polyphenols [101]. The polyphenols mainly affect the colour and the purity of the extracted pectin [18], but their co-presence with pectin can diversify the functional properties of pectin as a source of antioxidants. Therefore, the choice of whether to remove or not to remove the polyphenols before pectin extraction depends on the intended end use of both pectin and polyphenols. Consequently, the co-extraction of pectin and polyphenols from residues such as mango peels requires a sequential fractionation process whereby the polyphenols are pre-extracted prior to pectin extraction. Furthermore, the sequential process can simultaneously be optimized for polyphenol and pectin production, resulting in two value-added product stream from the same fruit residues.

9.2.1.2 Multi-Objective Fractionation of Fruits Residues to Diversify Pectin Functional Properties

Pectin recovery from the fruit residues is done either through a single-step extraction or a sequential extraction process in which the residue is pretreated with alcohol-based solvents to pre-extract the polyphenols [94]. The removal of the polyphenols constitutes a pretreatment process that is normally performed with 75% ethanol at 70 °C, a process that is repeated 4 times [65, 88]. The pretreatment objective is to recover high-quality pectin (the high quality is defined depending on the desired functionality) and in high yields. In addition, the pretreatment is done to obtain the polyphenols as additional products. The sequential process for obtaining pectin and polyphenols is an example of a multi-objective fractionation process whereby the extraction conditions are influenced by the desirable attributes of both products (pectin and polyphenols).

The functional property of pectin that is critical for applications such as in food, biomedical and pharmaceutical applications is the gelling ability. Thus, pectin is used as a thickener, emulsifier and coating [18, 26]. Such functionalities are influenced by the uronic acid content [76] as well as the molecular weight. Consequently, during the fractionation, the rheological properties (viscosity), the capacity of pectin to be acetylated (as reflected by the degree of esterification) and the molecular weight are some of the attributes that should be optimized [76] to enhance pectin application.

Pectin is considered to be of high quality when the galacturonic acid is >65% [65]. In addition, pectin for the formation of gels and films should have a molecular weight between 54,000 and 122,000 Da, an intrinsic viscosity of 22.74 dL/g [116, 136] and pectin degree of esterification >50%, which reduces the need for additives such as sugars and calcium during the formation of gels [122]. Notably, the degree of acetylation is influenced by the uronic acid content, which varies depending on the biomass source, extraction methods (chemical, biological or physical extraction) and their combinations (Table 9.1).

The choice of pectin extraction route depends largely on the objective of the fractionation process and the anticipated functional properties. The pH, temperature and duration of the extraction process affect the extractability, quality and functional properties of pectin differently [30]. Notably, pectin extraction is achieved by use of both organic-based and mineral-based acids. For example, pectin was extracted from mango peels using citric acid, ammonium oxalate/oxalic acid, hydrochloric acid, nitric acid and sulphuric acid (pH 0.5 to pH 4.6) as shown in Table 9.1 [22, 26, 65, 66, 100]. Pectin with highest uronic acid content >65% and degree of methoxylation >50% was obtained from mango peels using the ammonium oxalate/oxalic acid compared to the one obtained using hydrochloric acid and water as an extraction solvent [65].

The variation in pectin yields with different extraction methods was evident when the different acids were used for the extraction of pectin from the same biomass [18]. The pectin extracted from mango peels at pH 2.5, a temperature of 80 °C for 120 min using hydrochloric acid, sulphuric acid and nitric acid gave pectin yields of 13%, 21% and 15%, respectively. However, in another study, pectin was extracted from banana peels [76] in relatively higher yields (approximately 12%) using water (pH 6) than using acidic conditions (pH 1.5), which gave yields of less than 5%. Under acidic conditions, pectin experienced a reduction in molecular weight and viscosity [56, 76]. For example, the molecular weight of the acid extracted pectin from banana peels was in the range 17–29 kDa compared to 21–40 kDa for water-extracted pectin [76]. In addition, the intrinsic viscosity of pectin extracted from banana peels using hydrochloric acid, pH 1.5, reduced from

12 mPa.s to 8 mPa.s, but exhibited a higher degree of methylation (up to 60%) than the pectin extracted using water (<40%) [76]. The effect on the molecular weight and the viscosity are both attributed to the depolymerization of the polygalacturonic acid backbone chain [18].

The pectin yields and functional properties also vary with temperature [67]. Temperatures ranging from 75 °C to 150 °C have previously been used for pectin extraction [130]. However, relatively higher yields are obtained when the extraction is done at <80 °C compared to 150 °C [130]. Higher temperatures enhance pectin hydrolysis, which depolymerize the main galacturonic acid backbone chain. A pectin yield of 2.13% was obtained at 150 °C when compared to 6.75% obtained at 75 °C [130]. Similar trends were reported by Rehman et al. [100]. In the case of mango peels, a higher pectin content of 16% was obtained for extractions performed at 80 °C, which was higher than the yield of 13% obtained at 90 °C, with pH 2.5 and for an extraction period of 60 min [100].

Furthermore, the yield and quality of pectin, extracted at a given temperature and pH, vary with the increase of the duration of the extraction process. Pectin extractions can take 30–180 min to complete depending on the technology being used [76, 100]. Extended extraction durations result in an increase in the pectin yield, but at the expense of its functional properties. For example, a pectin yield from banana peels increased from 8% to 12% when the extraction time was increased from 30 to 60 min [76]. However, an increase to 90 min did not increase the yield significantly. In another study, the maximum yield (21%) of pectin from mango peels was obtained after 120 min extractions at pH 2.5 and a temperature of 80 °C, beyond which, no increase in the yield was observed [100]. Instead, in both cases, the extended exposure of the biomass to the extraction solvents reduced the pectin molecular weight, which subsequently, affected the viscosity [76]. There was a reduction in the molecular weight of pectin from banana peels from 30 kDa to less than 20 kDa when the extraction time was increased from 30 to 120 min [76].

The literature reviewed shows that the pH, temperature and extraction time can be manipulated during the fractionation to favour specific pectin attributes (yield, gelling ability, molecular weight, degree of acetylation and viscosity) that are desired for specific application. In addition, the optimization of the process conditions to obtain a combination of desired attributes for pectin can depend on how easy it is to control or maintain the process parameters with significant effects on the pectin attributes. do Nascimento Oliveira et al. [36] obtained two optimum conditions for extraction of pectin from the Uba mango variety, which were lower temperatures (85.4 °C vs 97 °C), lower acid concentrations (pH 2.4 vs 1.6) and shorter extraction times (35 min vs 85 min). The pectin yield and properties were similar at the two optimum conditions, thus, the choice of the fractionation condition will depend on whether it is the pH or the temperature that is easy to control and maintain. In do Nascimento Oliveira et al. [36], pectin yields >30%, degree of esterification ≥70.0% and viscosity ≥20.0 mPa·s were obtained at the two optimal conditions.

The fractionation conditions aiming at increasing the pectin yield should also consider the impact on the downstream modification processes and the desired functional properties for end-use applications as well as resource use. The optimal extraction conditions should have minimal negative impacts on the environmental. To avert the negative effects in terms of resource use and environmental effects associated with the conventional pectin extraction methods, new advanced technologies have been developed for pectin extraction over the years. The new methods have included microwave-assisted extraction, ultrasound-assisted extraction and subcritical water extraction [49, 119, 123, 126]. These new technologies have been effective in increasing the pectin yields, reducing the extraction time and improving the purity [126, 128]. For example, pectin extracted using subcritical water gave higher yields (18.34%) than pectin extracted under acid-based conditions, 4.88% [130]. The pectin extracted using the subcritical water extraction had a degree of esterification >70% [130]. On the other hand, ultrasound-assisted extraction increased pectin yield by 16.34%, shortened the extraction time by 37.78% and lowered the extraction temperature by 13 °C, when compared to the conventional extraction processes [127]. However, the high extraction efficiencies were not accompanied by energy savings; instead, energy consumption increased, which can have negative impacts on the environment depending on the source. For example, the ultrasound-assisted pretreatments require power input >10 W/cm^2 [24, 127], which is not any better than the energy required in conventional methods.

In general, the resource consumption associated with conventional as well as advanced technologies is rarely evaluated. This is a critical research gap that needs to be explored. Ultimately, the extraction conditions that favour high yields, high purity and high functionality should also minimize negative impacts to the environment by minimizing the energy consumption, biological and chemical agents use.

9.2.1.3 Multi-Objective Fractionation to Enhance Polyphenol Yield and Functional Properties

Fruit residues contain a variety of polyphenols. The common extraction methods involve use of organic solvents including ethanol, methanol and acetone (Table 9.2). Among the solvents, the methanol gives the highest amount of polyphenols [37]. However, due to methanol's toxicity, ethanol is preferred for polyphenol extraction [86]. The ethanol can solubilize polyphenols >100 mg/g with antioxidant activity >90%. However, the major drawback is large quantities of the organic solvents that have to be used due low solid loading (solid-to-liquid ratio of 1:20) and extended extraction periods required (24 h) [31, 90].

The focus on the extraction of polyphenols has been on increasing the yields and antioxidant activity (Table 9.2). However, there are contradicting findings regarding the relationship between polyphenol content and antioxidant activity. Oliveira et al. [42] and Paixao et al. [91] reported that the antioxidant activity of polyphenol extracts has a linear relationship with the polyphenol content of the polyphenol extract. Yet, Berardini et al. [22] indicated that there is no linear relationship between the polyphenol content and antioxidant activity but that the antioxidant activity is dependent on the type and combination of polyphenols in the extract. Therefore, application of selective extraction techniques can target isolation of specific types of polyphenol as

Table 9.2 Single-step methods for polyphenol extraction

Fruit waste	Extraction method	Polyphenol content	Antioxidant activity	Reference
Blue berry pomace	Ultrasound-assisted extraction with 50% ethanol as solvent. Maximum operating power of 35 kHz and 64 W, at 40 °C for 40 minutes	22.3 3 mg/g DM	41.79 mg Trolox/g DPPH (Diphenyl-1-picrylhydrazyl) radical scavenging activity	Bamba et al. [17]
Orange peels	Ultrasound-assisted extraction, sonication power of 150 W, solid/liquid ratio, with 80% ethanol as extraction solvent	275.8 mg of GAE/100 g DM	54% DPPH scavenging activity	Khan et al. [60]
Kiwi fruit seeds	Extraction with 59.45% acetone for 79.65 minutes 38.35 °C and a solid/liquid ratio of 1:11.52 (w/v)	53.73 mg GAE/g DW	63.25% DPPH scavenging capacities	Deng et al. [35]
Camellia oleifera fruit hull	Microwave-assisted extraction- extraction temperature 76 °C, time of 35 minutes, liquid/solid ratio of 15.33:1	15.05%	Not determined	Zhang et al. [133]
Pomegranate peel	Extraction with 60% ethanol, solid/liquid ratio – 1:30, temperature of 50 °C, time of 45 minutes	510 mg Gallic acid	24.54% DPPH radical scavenging activity	Ankita Sood [8]
Mango peels	Homogenization of peels with 80% acetone for 15 minutes	100 mg/g	1.83 μg GAE DPPH scavenging activity	Ajila et al. [5]
Mango peels	Extraction with water, solid/liquid ratio 1:5, temperature 60 °C for 30 minutes	11.66 mg/g dry peels	70.31% DPPH radical scavenging activity	Rojas et al. [103]

suggested by Klavins et al. [63]. For example, mango peels contain both anthocyanins and polyphenols. The former requires acidic extraction conditions (>pH 4) [56], whereas the latter require alcoholic conditions. The differences in the extraction conditions provide an opportunity to obtain two products with antioxidant activity.

9.2.2 Co-Extraction of Pectin and Polyphenol in Fractionation of Agricultural Residues

The recovery of more than one product from agricultural waste using a single-step extraction and sequential extractions is done to realize the full value potential of fruit waste such as mango peels and orange peels that contain both pectin and polyphenols [88, 90]. Pectin and polyphenols are co-extracted by hydrolysing peels with sulphuric acid followed by recovery of the pectin precipitation using ethanol and of polyphenols from the remaining liquid using resins [19, 22, 111]. However, the acid degrades some of the polyphenols (129.4 mg/g vs 71 mg/g). Therefore, to minimize the degradation, the polyphenols can be extracted prior to pectin extraction.

A combination of chemical and physical methods has been exploited for effective extraction of pectin from fruit waste (Table 9.3). The yield of pectin extracted

Table 9.3 Combination of extraction methods for sequential extraction

Fruit waste	Extraction method	Combination of mechanisms	Reference
Grape seed	Two-stage extraction with 50% ethanol at 65 °C at a solid/liquid ratio of 7.5:1 for 1.5 h for each stage	Two chemical extraction methods	Shi et al. [113]
Mango peels	Hydrothermal extraction (pressure of 16.2 psi and temperature of 121 °C) of pectin and recovery of polyphenols and cellulose from residue	Chemical treatments	Banerjee et al. [18]
Carrot, tomato, cucumber, apple pomace	Treatment with hot water for 10 min, followed by extraction of polyphenols with 0.5 M hydrochloric acid for 30 min at 85 °C. extraction of pectin with 1 M sodium hydroxide for 30 min at 85 °C. cellulose extraction after bleaching the residues in 1–2% sodium chlorite for 60 min at 95 °C	Combination of four chemical extraction methods	Szymanska-Chargot et al. [120]
Passion fruit rind	Pressurized solvent-based polyphenol extraction-pressure of 10.34–11.72 MPa at a temperature of 80 °C for 10 min, extraction solvent 60% ethanol, solid/liquid ratio 1:2. Pectin was extracted from residue with 1% citric acid	Chemical, physical and thermal extraction methods	de Souza et al. [32]
Passion fruit rind	Ultrasound-assisted extraction of polyphenols, extraction with 60% ethanol solid/liquid ratio 1:20, ultrasound frequency of 40 KHz and power of 135 W, extraction time of 10 min. Pectin was extracted from residue with 1% citric acid	Chemical methods	de Souza et al. [32]
Apple pomace	Extraction with 70% acetone for 1 h, adsorption of polyphenols onto resins. Pectin left in the liquid extract was then precipitated with alcohols	Chemical method + physical method	Schieber et al. [111]
Mango peels	Hydrolysis of mango peels with 98% sulphuric acid at 90 °C for 2.5 h. *Process 1*: Adsorption of polyphenols from the mango peel hydrolysate onto resins first and then precipitation of pectin using 98% ethanol *Process 2*: Precipitation of pectin from the hydrolysed mango peels with 98% ethanol first then adsorption of polyphenols in the remaining liquid.	Chemical method + physical method	Berardini et al. [22]
Mango peels	Microwave-assisted extraction at 150 °C for 30 min	Chemical extraction	Rojas et al. [103]

from mango peels of 26% was obtained at 80 °C using sonication-assisted acid hydrolysis (hydrochloric acid, pH 2.5) in 20 min [19]. A similar pectin yield from the mango peels required 150 min in a conventional process (acid only) carried out at the same temperature [19]. Thus, 7.5 times of time saving, which can translate into energy costs savings.

However, some combinations of the extraction techniques would negatively affect the pectin. For example, Banerjee et al. [19] combined a hydrothermal extraction and chemical extraction method, which reduced the pectin yield when compared to the use of each of the techniques as a separate method [19]. Therefore, it is important to understand the synergetic effects of the extraction methods on the properties of pectin. However, sequential extraction process could provide such synergetic effect compared to simultaneous application of the extraction methods.

The presence of components like fat hinders the extraction of pectin. Therefore, Jeong et al. [53] proposed a combination of a fat-removing and enzymatic hydrolysis step prior to pectin extraction from rape seed cake. The combination of these pretreatment steps was found to be more effective in producing a pectin fraction richer in galacturonic acid (pectin yield of 6.23% and galacturonic acid content of 64.23%) when compared to pectin extraction without the pretreatment step (approximately 2% pectin yield and 45% galacturonic acid) [53]. They also attempted to include a chemical pretreatment before enzymatic hydrolysis and pectin extraction. However, pectin yields reduced significantly (carbohydrate degradation as high as 75.91%) due to breaking down of the Galacturonic acid chains in the acid medium [53]. A combination of ultrasonic-assisted extraction and microwave-assisted extraction improved the pectin yields (increased yield by almost 100%) and properties when compared to each individual extraction process, although this process took a longer time when compared to individual extraction [14].

Szymanska-Chargot et al. [120] demonstrated the potential to sequentially extraction polyphenols, pectin and cellulose from pomace, although they have different optimal conditions. The sequential extraction increased in severity from using water at 25 °C for 10 min to using sodium chlorite at 90 °C for 60 min. This is mainly due to the individual products susceptibility to degradation when exposed to conditions in the previous stage. Essentially, cellulose is the least affected by chemicals when compared to polyphenols and pectin; hence, it was extracted last [120]. The multiple-stage processes have the advantage of leaving a residue richer in extractable components when compared to using a single-stage extraction process from the biomass. For example, [18] reported that the residue after pectin extraction had 37–39% cellulose when compared to the original mango peels that had 19–23% cellulose. Therefore, the multi-step extraction processes are a step towards multiple products extraction and increasing the value potential of extracted products.

9.2.3 *Multi-Objective Fractionation of Lignocellulosic Agricultural Residues*

Wheat straw, rice straw, sugarcane bagasse, corncob, cotton, wood, jute, flax and hemp are among the lignocellulosic agricultural residues rich in natural fibres mainly of cellulose and hemicellulose [54], but they do also contain varying amounts of lignin, pectin, waxes and other water-soluble compounds [57]. The biomass contains about 35–45% cellulose, 18–30% hemicellulose and 7–25% lignin. The latter hinders effective extraction and purity of the cellulose [141]. The strong interaction of cellulose with these cell wall components in the biomass minimizes accessibility of cellulose to enzymatic and chemical breakdown into nanocellulose [71]. Therefore, there is a need for a biomass pretreatment step, for example, delignification (Fig. 9.2) before nanocellulose can be produced from cellulose with properties that are suitable for downstream modification into products such as nanocelluloses. Therefore, the focus of this section is to review and discuss the challenges and opportunities for the multi-objective fractionation of the lignocellulosic residues to obtain celluloses and at the same time diversify the functionality of the cellulose by transforming it into nanocellulose.

9.2.3.1 Multi-Objective Fractionation to Enhance Cellulose Yield and Diversify Functional Properties

Cellulose is a natural linear polysaccharide consisting of a glucose backbone (Fig. 9.1). The abundancy, biodegradability, low densities, non-abrasiveness, combustibility, non-toxicity and high toughness are the attributes that have made them important natural fibres in many applications including food packing, biomedical treatment and bio reinforcements [57].

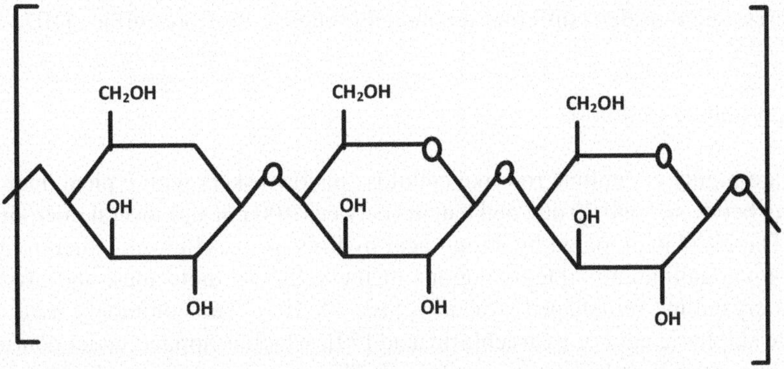

Fig. 9.1 The chemical structure of cellulose. (Drawn using Microsoft Visio)

Cellulose can be deconstructed into smaller constituents, namely nanocellulose providing abundant applications such as biosorbents, biofilms and medical delivery systems [29, 93]. Depending on the cellulosic source, fibre dimensions, functionality and method of preparation or processing conditions, nanocellulosic fibres can be classified according to three main categories: (1) nanocrystalline cellulose (NCC), (2) nanofibrillated cellulose (NFC) and (3) bacterial nanocellulose (BNC) [12, 52]. Biomass suitability for nanocellulose production depends on the cellulose, lignin and hemicellulose composition.

Nanofibrillated Cellulose

Nanofibrillated cellulose (NFC) are the smallest structural units of agricultural plant fibres [59]. They are characterized by entangled bundles of flexible nanofibres, with typical diameters between 1 and 100 nm [74]. These bundles can be several micrometres long, containing alternating amorphous and crystalline regions with vast amount of hydrophilic, reactive surface hydroxyl groups [10]. The NFC is extracted from cellulose by mechanical methods such as grinding, cryocrushing, homogenization and microfluidization [70]. These methods are energy intensive because of the multiple cycles required to defibrillate the cellulose to nanosize [138]. For this reason, chemicals and/or enzymatic pretreatments can be applied before mechanical fibrillation in order to reduce the fibre size, increase the contact surface, break the intracellular Van der Waals forces and consequently reduce the energy consumption necessary for mechanical processing [107, 138].

Chemical pretreatments include alkali bleaching [107], carboxymethylation or 2,2,6,6-tetramethyl-1-piperidinyloxy (TEMPO) mediated oxidation [70]. However, biological methods make use of either whole-cell microorganisms or enzymes and their cocktails. Enzymatic methods are specific due to their substrate specificity and therefore can be applied to customize the pretreatment processes. However, the chemical methods are non-selective methods, which alter the surface chemistry of the fibres, and introduce different functional groups to the fibre surface [107].

Nanocrystalline Cellulose

NCC are highly crystalline rod-like cellulose nanowhiskers with typical diameters ranging between 2 and 20 nm, and lengths between 100 nm and several micrometres [74]. The NCC is prepared by strong acid hydrolysis of cellulose, which removes the non-crystalline, amorphous regions of the cellulose molecules, and produces highly crystalline rod-shaped structures [10, 70, 107]. The commonly used acids include sulphuric acid or hydrochloric acid [59], which is applied under controlled temperature, time and agitation conditions. The acids target removal of the amorphous regions of the cellulose, thus leaving the crystalline constituencies. The degree of crystallinity of the isolated NCC as well as the dimensions and morphology are critical attributes for their application. The degree of crystallinity depends

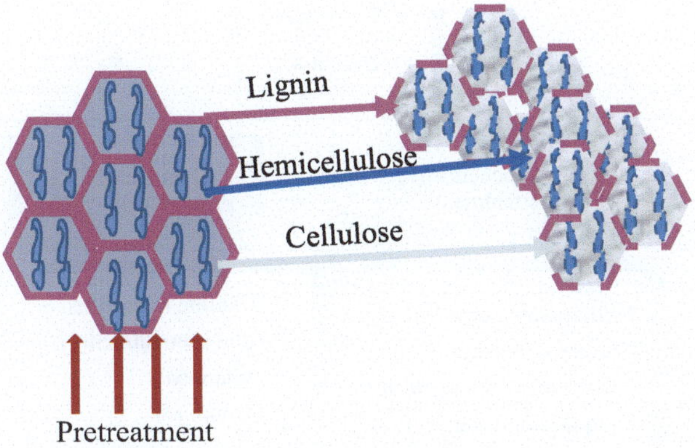

Fig. 9.2 Effect of pretreatment on the cell wall structure

on the cellulosic source and the extraction conditions used in the isolation method. For many applications, the NCC has to be pure, thus, free of impurities from the cell wall components such as lignin, hemicellulose and pectin. Typically, the NCC has a degree of crystallinity between 54 and 88% [138].

9.2.3.2 Lignocellulose Pretreatment to Increase Nanocellulose Yield and Extraction Efficiency

Lignocellulosic materials are recalcitrant due to the presence of lignin (Fig. 9.2). Therefore, the pretreatment is required to breakdown the cellulose–lignin linkages, which exposes the cellulose surface to fractionation agents including acids, alkali and enzymes [21].

Various pretreatment processes have been developed (Fig. 9.3), which include physical, physicochemical, chemical and biological pretreatment methods. The pretreatment methods can be performed as a separate unit operation or simultaneously with main extraction methods. The ultimate goal of the pretreatment is to enhance the fractionation of the desired components, in this case the cellulose, which should further be processed into nanocellulose. Therefore, the degradation of the cellulose can compromise the yields and at the same time release compounds with inhibitory effects on the downstream processes [118].

Physiochemical Pretreatment to Enhance Nanocellulose Extraction from Agricultural Residues

The physical pretreatment methods for enhancing valorization of lignocellulosic materials have included milling and grinding, high-pressure homogenization, ultrasonic, catalysed, and un-catalysed steam explosion methods. Notably, the pretreatment of the biomass is done for a variety of reasons including deconstructing the

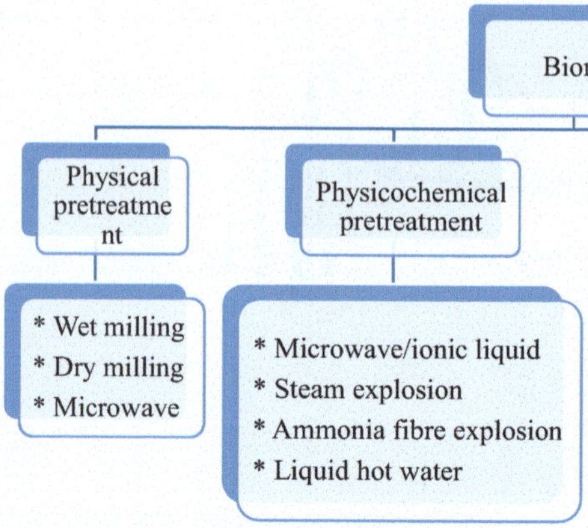

Fig. 9.3 Biomass pretreatment methods

lignocellulosic materials to improve the enzyme digestibility in the production of bioproducts such as biofuels and nanocellulose. In addition, pretreatment is for increasing the recovery of specific products such as lignin and hemicelluloses, and organic acids, and for reducing formation of degraded products with inhibitory effects during fermentation [39, 96]. Furthermore, the pretreatment can be used as a one-stop process to obtain various bioproducts and co-products. However, it is important to note that not all pretreatment methods can fulfil all the objectives at the same time.

Milling, grinding, high-pressure homogenization are prevalent physical pretreatment methods for the production of nanocellulose [1]. Other methods including ultrasonic and steam explosion pretreatment are becoming popular because of the advantages they offer. The ultrasonic method employs sound waves at frequencies between 16 kHz and 100 kHz to deconstruct biomass. A sound transducer transfers the ultrasound through a fluid, which results in pressure differences within the fluid. The ultrasonication can be performed in two ways, direct and indirect. The direct sonication entails transferring of the energy into the cellulose fibres, whereas in the indirect method, the sonication energy is transferred through a vessel containing the biomass. The direct sonication of biomass for cellulose extraction is preferred when compared to indirect sonication because it promotes cavitation, a process whereby bubbles collapse near the biomass surface, forcing liquid to disrupt the biomass structure into smaller pieces [106]. In order to enhance the sonication process, organic solvents, aqueous and ionic liquids are applied [24], which can result in dissolution of different plant cell wall components such as lignin and hemicelluloses and cellulose. Furthermore, the sonication is performed to increase the yields of specific products. González-fernández et al. [48] showed that the methane yield from sonication-assisted pretreated feedstock for was 1.2 times more than the yield from thermal pretreated feedstock. However, the sonication was performed at 85 °C compared to the 80 °C for the thermal treatment. Thus, the relative small gains in the methane yield does not justify the use of the sonication pretreatment method

[48]. Hence, other gains such as increasing the purity and diversification of the product spectrum would justify the extra energy required in ultrasonication.

Advanced pretreatment of lignocellulosic materials for bioethanol production has involved the use of steam explosion process, a process in which saturated steam at a temperature ranging from 160 to 260 °C and pressure ranging from 0.69 to 4.83 MPa are used [68]. Deconstruction of the biomass is achieved during release of the pressure, which results in the dissolution of hemicellulose and lignin in the liquid form, leaving the cellulosic rich solid residues [28, 33]. Various biomasses have been subjected to steam explosion for conversion to nanocellulose and these include pineapple leaf fibres, meadow hay, banana stem and jute stem [2, 28]. The advantage of the method is the short time it takes to deconstruct the biomass. However, at such high temperatures and pressure, water becomes a weak acid, which catalyses the release of organic acids from the biomass, leading to an autocatalytic process. The phenomenon results in the depolymerization and degradation of the hemicellulose fraction. Pretreatment of wheat straw at 200 °C for 4 min produced furfural with a concentration of 0.3 g/L and xylose with a concentration of 1 g/L [7]. Increasing the temperature by 10 °C and the treatment time by 1 min, (210 °C, 5 min) led to the release of xylose and furfural with concentrations of 24.7 g/L and 1.4 g/L, respectively [77], and further increase in the pretreatment time (20 min) at 204 °C degraded the xylan into monomers and oligomers. Some studies include Jönsson and Martín [55], Pielhop et al. [96] and Carrascoa et al. [25]. Carrascoa et al. [25] have shown that steam explosion pretreatment can be catalysed using acids (sulphur dioxide, sulphuric acid and carbon dioxide) and to some extent alkaline agents production [62], in particular, when the hemicellulose fraction is required in a polymeric form.

Steam explosion pretreatment is used to improve enzymatic hydrolysis of the cellulose. Factors affecting steam explosion include residence time, temperature, particle size and moisture content [118]. The pretreatment requires elevated temperatures. For example, alkaline catalysed steam explosion pretreatment meadow hay at150°C was not effective in improving enzymatic hydrolysis [124]. However, increasing the temperature to 200 °C improved the enzymatic hydrolysis significantly [124]. However, depending on the severity of the pretreatment conditions, partial degradation of the cellulose can occur. Therefore, the steam explosion conditions are controlled to minimize loss of cellulose. In addition, the different catalyst can be used to manipulate the pretreatment conditions in order to minimize cellulose degradation. Furthermore, the process parameters and thus residence time, temperature, particle size and moisture content can be manipulated [23, 89, 118]). Energy usage in the steam process is low when compared to other pretreatment methods such as grinding and milling [51].

Ammonia fibre explosion (AFEX) is a pretreatment method similar to steam explosion only that instead of using water, ammonia is used for the fractionation of biomass. The conditions of AFEX pretreatment are solid-to-liquid loading ratio (1–2 kg ammonia to 1 kg biomass), temperature (70–200 °C), pH (<pH 12) and residence time (<30 min). Biomass is exposed to hot ammonia liquid, which is under pressure (100–400 psi) for a defined period followed by sudden release of the pressure [16, 134]. The sudden change in pressure causes the structure of the biomass to break down. The process removes lignin and solubilizes hemicellulose in the ammonia liquid, leaving the cellulose in the solid state [134]. The pretreatment has been done on various biomass including corn stover and switch grass [121]. Pretreatment of biomass with AFEX has been demonstrated to boost enzymatic hydrolysis significantly. Alizadeh et al. [6] have concluded that AFEX pretreatment improved enzymatic hydrolysis of switch grass by 2.5 times [6]. Alizadeh et al. [6] demonstrated this by pretreating switch grass at a temperature of 100 °C with ammonia to switch grass matter loading of 1:1 and a residence time of 5 min prior to enzymatic hydrolysis [6].

The pretreatment increases the surface area, pore size and a number of contact sites in cellulose fibres and at the same time reduces the cellulose crystallinity. In addition, such pretreatment methods have shown to facilitate dissolution of non-polysaccharide compounds such as ferulic acid and phenolic compounds [102], which suggests that if optimized and combined with proper recovery methods for such products, the pretreatment step can diversify the product spectrum for the feedstock in a biorefinery. However, most of these pretreatment methods are known to be energy intensive, for example, about 20,000–30,000 kWh per ton to mechanically convert cellulose to nanocellulose [82]. In addition, the specific energy demand is high for feedstock with high moisture content [82, 140] and for the products that require smaller particle sizes and narrow particle size distribution, both of these characteristics are typical of agricultural residues and their products. For example, the grinding of switch grass, which is similar to wheat straw, to a particle size of 3.2 mm, requires the highest specific energy of Mani et al. [140]). In addition, the study by Ghorbani et al. [45] showed that a hammer mill generating pretreated feedstock with a screen size of 4.76 mm required less energy than the one with the screen size of 1.68 mm [45].

Chemical Pretreatment to Enhance Nanocellulose Extraction from Agricultural Residues

The pretreatment and production methods for nanocellulose involve two main routes, the acid and the alkali routes. Depending on the feedstock and the objectives of the pretreatments, the type and strength of the chemicals can be varied.

The acid hydrolysis involves the use of dilute and concentrated acids for biomass fractionation [23]. Research has been done on the potential of acidic chemicals as biomass fractionation medium for a wide variety of biomass which includes hardwoods, grasses and agricultural residues [87]. The acids used include phosphoric acid, peracetic acid, nitric acid and sulphuric acid; however, the most commonly used acid for nanocellulose production is sulphuric acid [15, 73, 110]. Acids have been discovered to disrupt linkages between hemicellulose, lignin and hemicellulose. They also break down the hydrogen bonds in cellulose and reduce the crystallinity of cellulose. Acid pretreatment increases the surface area for effective digestibility of cellulose by enzymes to nanocellulose minimizing on the required enzyme loading when compared to untreated biomass. The major disadvantages of acid hydrolysis are that acids are highly corrosive and therefore there will be a need to invest in costly acid-resistant equipment and can be an environmental threat if the acidic residues are improperly disposed. [125].

Hot water pretreatment function similar to the acid method depending on the temperature and pressure of biomass involves fractionation of biomass using water at very high temperatures (180–230 °C) well above its boiling temperature [87]. In order to maintain water in liquid state, high pressures of up to 5 MPa are maintained during the pretreatment of biomass [139]. Liquid hot water pretreatment has been used as a pretreatment method on a various type of biomass (corncob, cornhusk and wheat straw) to isolate cellulose [142]. The degeneration of biomass during hot water treatment depends on hydronium ions that are produced during the water auto-ionization process. Additionally, as the hemicellulose solubilizes and is degraded, it releases acetic acid, which is the major contributor to biomass breakdown [135].

Mosier et al. [87] carried out hot water fractionation of biomass at temperatures between 200 and 230 °C for 15 min. Their work demonstrated that liquid hot water treatment at the stated conditions was capable of dissolving between 40% and 60% of biomass. The high temperature resulted in the loss of only about 4–22% of the cellulose with the removal of all the hemicellulose and about 35–60% of lignin from the cellulose [87]. Michelin and Teixeira [84] investigated the effect of this pretreatment method on corncobs, cornhusk, wheat straw and Luffa sponge. Their results showed that the pretreatment worked by solubilizing hemicellulose and leaving the cellulose in solid form. The lignin underwent reactions in which it is simultaneously made soluble and insoluble in the aqueous solution [84]. These reactions precipitate some of the lignin on the surface of the cellulose recovered, producing a barrier to further processing of the cellulose by enzymes or by chemicals [135]. A comparison of untreated biomass to the liquid hot water-treated biomass was done, and it was observed that the treated biomass had higher crystallinity and thermal stability when compared to untreated biomass. This is because when the biomass is treated by liquid hot water, the remaining residue is made mostly of cellulose which is more resistant to thermal degradation when compared to biomass that is composed of a lot of volatile materials [84]. The major drawback of this pretreatment method is that it is energy intensive and uses high volumes of water. However, the advantage of this pretreatment method is that it does not require the addition of chemicals, making it more environmentally friendly.

The alkaline pretreatment involves the use of base chemicals like sodium hydroxide, lime and ammonium salts in fractionation of biomass. This process is mainly affected by the lignin content of the biomass, as the mechanism of operation is by delignification of biomass. Delignification improves the reactivity of the remaining carbohydrates by increasing their surface areas. Therefore, cellulose is made accessible to enzymatic hydrolysis. Michalska and Ledakowicz [83] highlighted these phenomena when they carried out enzymatic hydrolysis of alkali-treated sorghum biomass and untreated biomass. The biomass was treated with 5% (w/w) sodium hydroxide for 30 min at a temperature of 121 °C. They observed that enzymatic hydrolysis to methane increased by 70% from when using untreated sorghum samples and to alkali-treated samples [83]. A combination of alkaline treatment and other biomass treatment methods is considered very efficient. Wang et al. [129] highlighted this fact when they realized an overall enzymatic conversion efficiency of 90.4% after first treating Bermuda grass with alkaline solutions before enzymatic extraction [129].

The alkali hydrolysis is advantageous because it can be carried out at relatively low temperatures when compared to other pretreatment methods such as the steam explosion. In addition, it selectively removes lignin without excessive degradation of the other cell wall components, thereby reducing the formation of inhibitory compounds in downstream processes. However, the alkali process has the limitation of long residence time and discharges corrosive waste waters from the pretreatment, which require to be neutralized prior to release into the environment [15].

Organic Solvent Pretreatment to Enhance Nanocellulose Extraction from Agricultural Residues

Organosolv alcohol pretreatment is a process that employs alcohols to cleavage ester linkages of lignin to cellulose and the bonds between lignin and hemicellulose. The pretreatment is capable of breaking ester bonds within the lignin structure and glycosidic bonds in the hemicellulose, thereby degrading the two polymers. This process produces cellulose fractions of high purity that are suitable for nanocellulose synthesis by enzymatic hydrolysis. Pan et al. [92] highlighted that a solid fraction containing a cellulose fraction as high as 88% was obtained after pretreatment of poplar biomass with ethanol (180 °C, 60 min, 1.25% H_2SO_4 and 60% ethanol) [92]. Organic solvents can be used in combination with inorganic acids such as hydrochloric acid and sulphuric acid which will act as catalysts [68].

Organic alcohol solvents can be divided into high boiling point solvents and low boiling point solvents. The low boiling point solvents (ethanol and methanol) are commonly used for cellulose isolation, as they require less energy for their recovery after pretreatment when compared to high boiling point solvents [81]. Ethanol-based pretreatments of biomass for cellulose recovery are preferred and regarded as safer when compared to methanol-based pretreatments [114]. The high boiling point organic solvents mostly used in pretreatment are polyhydroxy alcohols (ethylene glycol and glycerol) [117]. When compared to low boiling point organic solvent-based pretreatment processes, a high boiling point solvent-based pretreatment process can be performed under atmospheric pressure, but it remains costly, as its recovery after pretreatment is energy intensive [81]. Ethanol-based pretreatments of biomass for cellulose recovery are preferred and regarded as safer when compared to methanol-based pretreatments. The high boiling point organic solvents mostly used in pretreatment are polyhydroxy alcohols (ethylene glycol and glycerol).

Organic acids have also been utilized as solvents for fractionation of biomass. The organic solvents mainly used are formic acid and acetic acid. Generally, the use of acids is restricted by their corrosive tendencies. Their mechanism of operation is by donating a hydrogen ions that facilitate the removal of lignin and hydrolysis of hemicellulose [114] The advantage of this pretreatment is the ability of these acidic solutions to solubilize hemicellulose and lignin, leaving more pure cellulose for nanocellulose production [13].

Biological Pretreatment to Enhance Nanocellulose Extraction from Agricultural Residues

Biological pretreatment of biomass utilizes living organisms to attack lignin and hemicellulose selectively in biomass, leaving cellulose intact. Organisms typically used include fungi that can produce enzymes which can hydrolyse lignin and hemicellulose [21]. The most commonly used organisms are the white and soft rot-fungi, actinomycetes and bacteria. The mechanism by which fungi degrades biomass is either by producing enzymes that attack outer cell wall layer or the polysaccharides inside the cell wall [109]. The most widely studied white-rot organism is *P. chrysosporium* and this microorganism produces enzymes called lignases which break down lignin in the absence of oxygen [108]. Biological methods are attractive pretreatment methods, as they do not require a lot of energy and are environmentally friendlier when compared to physical and thermochemical pretreatment methods. The major limitation of biological pretreatment methods is that enzymes are very expensive and are very unstable (can be denatured if not properly handled) [68].

9.2.4 Integrated Methods for Multi-Objective Fractionation of Agricultural Residues

The recovery of more than one product from agricultural waste using a single-step extraction and sequential extraction has been done to realize the full value potential of fruit waste such as mango peels and orange peels that contain both pectin and polyphenols [90]. Pectin and polyphenols have been co-extracted by hydrolysing peels with sulphuric acid, precipitation of pectin from the hydrolysate with ethanol and recovery of polyphenols from the remaining liquid using resins [19, 22, 111]. However, the major drawback of the sequential extraction using the above method was that polyphenols were degraded during the hydrolysis stage (129.4 mg/g vs 71 mg/g). In order to minimize the degradation of the polyphenols, combination of extraction process in which polyphenols are extracted prior to pectin would be the advantageous. Sequential extraction has been proven to produce flavonoids and pectin with functional properties in the process, reducing the amount of organic liquids used in the extraction [32].

A combination of chemical and physical methods has been exploited for effective extraction of pectin from fruit waste. Banerjee et al. [19] reported that sonication of mango peels with hydrochloric acid (pH 2.5) resulted in the same pectin yield (26%) after only 20 min compared to the conventional extraction after 150 min.

The combination of the extraction process would save on time (20 min vs 150 min) and therefore energy costs since both processes were carried out at 80 °C [19]. However, some combinations of extraction techniques negatively affect pectin. A combination of a hydrothermal extraction and chemical extraction method done by Banerjee et al. [19] showed that the pectin yield was negatively affected by the combination of extraction techniques when compared to the use of each technique separately [19]. Therefore, the selection of a combination of extraction methods should be determined by the effects the extraction processes have on the properties of pectin.

The presence of components like fat hinders the extraction of pectin. Jeong et al. [53] proposed a combination of a fat-removing and enzymatic hydrolysis step prior to pectin extraction from rape seed cake. The combination of these pretreatment steps was found to be more effective in producing a pectin fraction richer in galacturonic acid (pectin yield of 6.23% and galacturonic acid content of 64.23%) when compared to pectin extraction without the pretreatment step (approximately 2% pectin yield and 45% galacturonic acid) [53]. They also attempted to include a chemical pretreatment before enzymatic hydrolysis and pectin extraction. However, pectin yields reduced significantly (carbohydrate degradation as high as 75.91%) due to breaking down of the galacturonic acid chains in the acid medium [53].

A combination of ultrasonic-assisted extraction and microwave-assisted extraction improved the pectin yields (increased yield by almost 100%) and properties when compared to each individual extraction process, although this process took a longer time when compared to individual extraction [14].

Szymanska-Chargot et al. [120] demonstrated the potential to sequentially extract polyphenols, pectin and cellulose from pomace, although they have different optimal conditions. The sequential extraction increased in severity from using water at 25 °C for 10 min to using sodium chlorite at 90 °C for 60 min. This is mainly due to the individual products susceptibility to degradation when exposed to conditions in the previous stage. Essentially, cellulose is the least affected by chemicals when compared to polyphenols and pectin; hence, it was extracted last [120]. The multiple-stage processes have the advantage of leaving a residue richer in extractable components when compared to using a single-stage extraction process from the biomass. For example, [18] reported that the residue after pectin extraction had 37–39% cellulose when compared to the original mango peels that had 19–23% cellulose. Therefore, multi-step extraction processes are a step towards multiple products extraction and increasing the value potential of extracted products.

9.2.5 Optimization of Multi-Objective of Fractionation of Agricultural Residues

Optimization of the extraction conditions for polyphenols, pectin and nanocellulose extraction has been carried out for both the single-step processes and sequential extraction processes [78, 88, 126, 131]. Mathematical techniques such as the Box-Behnken, central composite design, have facilitated obtaining the optimum conditions taking into consideration the significant extraction conditions and their interactions [61, 94]. Multi-objective optimization has since replaced the single factor optimization techniques that have commonly been used by researchers. Multi-objective optimization takes into consideration more than one extraction factor and the properties of the individual products extracted. The most commonly optimized extraction conditions for pectin, polyphenols and nanocellulose are temperature, time and solvent concentration, taking into consideration one or more properties of the product [26, 105]. Sai-Ut et al. [105] optimized the extraction conditions of ethanol concentration (40–80%) temperature (40–80%) and time (60–180%) based on the antioxidant activity of the mango peel polyphenol and tyrosinase inhibitory activity. On the other hand, Mugwagwa and Chimphango [88] optimized a two-step extraction process for the recovery of anthocyanins, a type of polyphenol and pectin. The difference in the optimization of one product and two products from the same biomass includes the number of product properties considered. In the two-step extraction process by Mugwagwa and Chimphango [88], they considered the anthocyanin properties, antioxidant activity and anthocyanin concentration and pectin uronic acid content, pectin yield, neutral sugars content and antioxidant activity during optimization. The optimum conditions for the sequential extraction were observed to enhance the functional properties of the products extracted, for example, a pectin extracted had 23% more galacturonic acid when compared to single-step conventional extraction processes [88]. Therefore, multiple functional properties of products can be improved at the same time increasing the number of products from sequential extraction process when multi-objective techniques are applied.

9.2.6 End Applications of Polyphenols, Pectin and Nanocellulose

The polyphenols, pectin and nanocellulose have different applications attributed to their different physicochemical properties. Due to polyphenols' antioxidant activity, they have been recommended as an additive to food packaging to increase food shelf life [38]. Pectin has film-forming properties and lipid barrier properties and therefore has been applied as a coating or packaging material for fatty foods [132]. Due to nanocellulose stiffness (up to 86 GPa), it has found use as a reinforcement material in polymers such as hemicellulose, low-density polyethylene [4]. Taking into consideration the aforementioned properties of the products that can be sequentially extracted from a single biomass, these products can be integrated to form an active food packaging. Pectin has been reinforced with nanocellulose to increase its mechanical properties [132]. The active food packaging mainly for fatty foods would be based on pectin reinforced with the nanocellulose and have the polyphenols incorporated to increase the packaged food shelf life. Therefore, multi-objective extraction of valuable products from agricultural waste would not only effectively extract the structural and non-structural components in a single biomass but also provide all the raw materials in one goal for the development of bio-based materials such as food packaging.

9.3 Conclusion

Agricultural residues present valuable bioresources for obtaining a variety of bioproducts that can respond to the current environmental and healthy living style demands. The review has shown how multiple fractionation objectives, such increasing yield, purity, number of products and consumption of resources can be achieved in the co-production of pectin, polyphenols and nanocelluloses. A potential viable strategy is to configure strategically the extraction methods in a sequential fractionation process that emulates a biorefinery in order to obtain multiple products with varying optimal extraction conditions. In addition, extraction methods can be combined and configured to diversify the functional properties of the fractionation products to suit various end uses. The chapter has provided a framework for devising strategies that can be used to unlock maximum value from agricultural residues, thus finding multiple optimal conditions (depending on the process parameter that is easy to be controlled). In addition, the chapter has unveiled a critical research gap in the evaluation of the fractionation methods in particular concerning resource use and the environmental impact for the alternative fractionation methods. In particular, the lack of assessing the economic and environmental impacts arising from the use of energy, chemical and biological reagents. Therefore, systems approaches, including use of the life cycle thinking, are required to understand the real value the agricultural residues can offer in a multi-product fractionation process.

References

1. H.P.S. Abdul Khalil, Y. Davoudpour, M.N. Islam, et al., Production and modification of nanofibrillated cellulose using various mechanical processes: a review. Carbohydr. Polym. **99**, 649–665 (2014). https://doi.org/10.1016/j.carbpol.2013.08.069
2. E. Abraham, B. Deepa, L.A. Pothan, et al., Extraction of nanocellulose fibrils from lignocellulosic fibres: a novel approach. Carbohydr. Polym. **86**, 1468–1475 (2011). https://doi.org/10.1016/j.carbpol.2011.06.034
3. O.C. Adebooye, A.M. Alashi, R.E. Aluko, A brief review on emerging trends in global polyphenol research. J. Food Biochem. **42**, e12519 (2018). https://doi.org/10.1111/jfbc.12519
4. Y. Aitomäki, K. Oksman, Reinforcing efficiency of nanocellulose in polymers. React. Funct. Polym. **85**, 151–156 (2014). https://doi.org/10.1016/j.reactfunctpolym.2014.08.010
5. C.M. Ajila, S.G. Bhat, U.J.S. Prasada Rao, Valuable components of raw and ripe peels from two Indian mango varieties. Food Chem. **102**, 1006–1011 (2007). https://doi.org/10.1016/j.foodchem.2006.06.036
6. H. Alizadeh, F. Teymouri, T. Gilbert, B. Dale, Pretreatment of switchgrass by Ammonia Fiber Explosion (AFEX). Appl. Biochem. Biotechnol. **121–124**, 1133–1141 (2005)
7. P. Alvira, M.J. Negro, I. Ballesteros, et al., Steam explosion for wheat straw pretreatment for sugars production. Bioethanol **2**, 66–75 (2016). https://doi.org/10.1515/bioeth-2016-0003
8. M.G. Ankita Sood, Extraction process optimization for bioactive compounds in pomegranate peel. Food Biosci. **12**, 100–106 (2015)
9. R.A.D. Arancon, C.S.K. Lin, K.M. Chan, et al., Advances on waste valorization: New Horizons for a more sustainable society. Energy Sci. Eng. **1**, 53–71 (2013). https://doi.org/10.1002/ese3.9

10. S. Arola, *Biochemical Modification and Functionalization of Nanocellulose Surface* (Aalto University School of Science, VTT Technical Research Centre, Finland 2015)
11. K.M. Azada, Isolation and characterization of pectin extracted from lemon pomace during ripening. J. Food Nutr. Sci. **2**, 30 (2014). https://doi.org/10.11648/j.jfns.20140202.12
12. H.M.C. Azeredo, M.F. Rosa, L.H.C. Mattoso, Nanocellulose in bio-based food packaging applications. Ind. Crop. Prod. **97**, 664–671 (2017). https://doi.org/10.1016/j.indcrop.2016.03.013
13. M. Badiei, N. Asim, J.M. Jahim, K. Sopian, Comparison of chemical pretreatment methods for cellulosic biomass. APCBEE Procedia **9**, 170–174 (2014). https://doi.org/10.1016/j.apcbee.2014.01.030
14. H. Bagherian, F. Zokaee Ashtiani, A. Fouladitajar, M. Mohtashamy, Comparisons between conventional, microwave- and ultrasound-assisted methods for extraction of pectin from grapefruit. Chem. Eng. Process. Process Intensif. **50**, 1237–1243 (2011). https://doi.org/10.1016/j.cep.2011.08.002
15. M. Balat, G. Ayar, Biomass energy in the world, use of biomass and potential trends. Energy Sources **27**, 931–940 (2005). https://doi.org/10.1080/00908310490449045
16. B. Bals, C. Rogers, M. Jin, et al., Evaluation of ammonia fibre expansion (AFEX) pretreatment for enzymatic hydrolysis of switchgrass harvested in different seasons and locations. Biotechnol. Biofuels **3**, 1–11 (2010). https://doi.org/10.1186/1754-6834-3-1
17. B.S.B. Bamba, J. Shi, C.C. Tranchant, et al., Influence of extraction conditions on ultrasound-assisted recovery of bioactive phenolics from blueberry pomace and their antioxidant activity. Molecules **23**, 1–17 (2018). https://doi.org/10.3390/molecules23071685
18. J. Banerjee, R. Singh, R. Vijayaraghavan, et al., A hydrocolloid based biorefinery approach to the valorisation of mango peel waste. Food Hydrocoll. **77**, 142–151 (2017). https://doi.org/10.1016/J.FOODHYD.2017.09.029
19. J. Banerjee, R. Vijayaraghavan, A. Arora, et al., Lemon juice based extraction of pectin from mango peels: waste to wealth by sustainable approaches. ACS Sustain. Chem. Eng. **4**, 5915–5920 (2016). https://doi.org/10.1021/acssuschemeng.6b01342
20. B. Batidzirai, H.M. Junginger, A.P.C. Faaij, Current and future technical, economic and environmental feasibility of maize and wheat residues supply for biomass energy application: illustrated for South Africa. Biomass Bioenergy **92**, 106–129 (2016). https://doi.org/10.1016/j.biombioe.2016.06.010
21. S. Behera, R. Arora, N. Nandhagopal, S. Kumar, Importance of chemical pretreatment for bioconversion of lignocellulosic biomass. Renew. Sust. Energ. Rev. **36**, 91–106 (2014). https://doi.org/10.1016/j.rser.2014.04.047
22. N. Berardini, M. Knödler, A. Schieber, R. Carle, Utilization of mango peels as a source of pectin and polyphenolics. Innov. Food Sci. Emerg. Technol. **6**, 442–452 (2005). https://doi.org/10.1016/j.ifset.2005.06.004
23. G. Brodeur, E. Yau, K. Badal, et al., Chemical and physicochemical pretreatment of lignocellulosic biomass: a review. Enzyme Res. **2011**, e787532 (2011). https://doi.org/10.4061/2011/787532
24. M.J. Bussemaker, D. Zhang, Effect of ultrasound on lignocellulosic biomass as a pretreatment for biorefinery and biofuel applications. Ind. Eng. Chem. Res. **52**, 3563–3580 (2013). https://doi.org/10.1021/ie3022785
25. C. Carrascoa, H. Baudela, J. Sendeliusa, et al., SO2-catalyzed steam pretreatment and fermentation of enzymatically hydrolyzed sugarcane bagasse. Enzym. Microb. Technol. **46**, 64–73 (2010). https://doi.org/10.1007/s12010-013-0097-2
26. E. Chaharbaghi, F. Khodaiyan, S.S. Hosseini, Optimization of pectin extraction from pistachio green hull as a new source. Carbohydr. Polym. **173**, 107–113 (2017). https://doi.org/10.1016/j.carbpol.2017.05.047
27. M. Chaichi, M. Hashemi, F. Badii, A. Mohammadi, Preparation and characterization of a novel bionanocomposite edible film based on pectin and crystalline nanocellulose. Carbohydr. Polym. **157**, 167–175 (2017). https://doi.org/10.1016/j.carbpol.2016.09.062

28. B.M. Cherian, A.L. Leo, S.F. de Souza, et al., Isolation of nanocellulose from pineapple leaf fibres by steam explosion. Carbohydr. Polym. **81**, 720–725 (2010). https://doi.org/10.1016/j.carbpol.2010.03.046
29. C.J. Chirayil, L. Mathew, S. Thomas, Review of recent research in nano cellulose preparation from different lignocellulosic fibers. Rev. Adv. Mater. Sci. **37** (2014)
30. S. Chockchaisawasdee, C.E. Stathopoulos, *Utilisation of Bioactive Compounds from Agricultural and Food Waste*, 1st edn. (CRC Press, Boca Raton, 2017)
31. N. Ćurko, M. Tomašević, M.C. Bubalo, et al., Extraction of proanthocyanidins and anthocyanins from grape skin by using ionic liquids. Food Technol. Biotechnol. (2017). https://doi.org/10.17113/ftb.55.03.17.5200
32. C.G. de Souza, T.H. Rodrigues, L.M. e Silva, et al., Sequential extraction of flavonoids and pectin from yellow passion fruit rind using pressurized solvent or ultrasound. J. Sci. Food Agric. **98**, 1362–1368 (2018). https://doi.org/10.1002/jsfa.8601
33. B. Deepa, E. Abraham, B. Mathew, et al., Bioresource Technology Structure, morphology and thermal characteristics of banana nano fibers obtained by steam explosion. Bioresour. Technol. **102**, 1988–1997 (2011). https://doi.org/10.1016/j.biortech.2010.09.030
34. G.F. Deng, C. Shen, X.R. Xu, et al., Potential of fruit wastes as natural resources of bioactive compounds. Int. J. Mol. Sci. **13**, 8308–8323 (2012). https://doi.org/10.3390/ijms13078308
35. J. Deng, Q. Liu, C. Zhang, et al., Extraction optimization of polyphenols from waste kiwi fruit seeds (Actinidia chinensis Planch.) and evaluation of its antioxidant and anti-inflammatory properties. Molecules **21** (2016). https://doi.org/10.3390/molecules21070832
36. A. do Nascimento Oliveira, D. de Almeida Paula, E. Basílio de Oliveira, et al., Optimization of pectin extraction from Ubá mango peel through surface response methodology. Int. J. Biol. Macromol. **113**, 395–402 (2018). https://doi.org/10.1016/J.IJBIOMAC.2018.02.154
37. E. Dorta, M.G. Lobo, M. Gonzalez, Reutilization of mango byproducts: study of the effect of extraction solvent and temperature on their antioxidant properties. J. Food Sci. **77**, 80–89 (2012). https://doi.org/10.1111/j.1750-3841.2011.02477.x
38. L. Dou, B. Li, K. Zhang, et al., Physical properties and antioxidant activity of gelatin-sodium alginate edible films with tea polyphenols. Int. J. Biol. Macromol. **118**, 1377–1383 (2018). https://doi.org/10.1016/j.ijbiomac.2018.06.121
39. A. Duque, P. Manzanares, I. Ballesteros, M. Ballesteros, Steam explosion as Lignocellulosic biomass pretreatment, in *Biomass Fractionation Technologies for a Lignocellulosic Feedstock Based Biorefinery*, (Elsevier, 2016), Amsterdam, Netherlands pp. 349–368
40. E. Espada-Bellido, M. Ferreiro-González, C. Carrera, et al., Optimization of the ultrasound-assisted extraction of anthocyanins and total phenolic compounds in mulberry (Morus nigra) pulp. Food Chem. **219**, 23–32 (2017). https://doi.org/10.1016/J.FOODCHEM.2016.09.122
41. Eubia, *An Overview on the Biopolymers Market Potential in Europe* (European Biopolymers Summit, London, 2016). http://www.eubia.org/cms/wp-content/uploads/2016/12/EUBIA-ppt-EU-Biopolymer-summit.pdf
42. A. Fernandes de Oliveira, L. Sousa Pinheiro, C. Souto Pereira, et al., Total phenolic content and antioxidant activity of some Malvaceae Family species. Antioxidants **1**, 33–43 (2012). https://doi.org/10.3390/antiox1010033
43. C.M. Galanakis, Food and Bioproducts Processing Short communication Emerging technologies for the production of nutraceuticals from agricultural by-products: a viewpoint of opportunities and challenges. Food Bioprod. Process. **91**, 575–579 (2013). https://doi.org/10.1016/j.fbp.2013.01.004
44. C.H. Geerkens, A.E. Matejka, R.M. Schweiggert, et al., Optimization of polyphenol recovery from mango peel extracts by assessing food-grade adsorbent and ion exchange resins and adsorption parameters using a D-optimal design. Eur. Food Res. Technol. **241**, 627–636 (2015). https://doi.org/10.1007/s00217-015-2489-8
45. Z. Ghorbani, A.A. Masoumi, A. Hemmat, Specific energy consumption for reducing the size of alfalfa chops using a hammer mill. Biosyst. Eng. **105**, 34–40 (2010). https://doi.org/10.1016/j.biosystemseng.2009.09.006

46. E. Girma, T. Worku, Extraction and characterisation of Pectin from selected fruit peel waste. Int J Sci **6**, 447–454 (2016). https://doi.org/10.15406/mojfpt.2016.02.00029
47. K.R. Gomes, A.F.A. Chimphango, J.F. Görgens, Modifying solubility of polymeric xylan extracted from Eucalyptus grandis and sugarcane bagasse by suitable side chain removing enzymes. Carbohydr. Polym. **131**, 177–185 (2015). https://doi.org/10.1016/j.carbpol.2015.05.029
48. C. González-Fernández, B. Sialve, N. Bernet, J.P. Steyer, Comparison of ultrasound and thermal pretreatment of Scenedesmus biomass on methane production. Bioresour. Technol. **110**, 610–616 (2012). https://doi.org/10.1016/j.biortech.2012.01.043
49. A.N. Grassino, M. Brnčić, D. Vikić-Topić, et al., Ultrasound assisted extraction and characterization of pectin from tomato waste. Food Chem. **198**, 93–100 (2016). https://doi.org/10.1016/j.foodchem.2015.11.095
50. R. Gumisiriza, J.F. Hawumba, M. Okure, O. Hensel, Biomass waste-to-energy valorisation technologies: a review case for banana processing in Uganda. Biotechnol. Biofuels **10**, 11 (2017). https://doi.org/10.1186/s13068-016-0689-5
51. P. Harmsen, W. Huijgen, L. López, R. Bakker, Literature review of physical and chemical pretreatment processes for lignocellulosic biomass. Food Biobased Res, 1–49 (2010)
52. M. Ioelovich, Characterization of various kinds of Nanocellulose, in *Handbook of Nanocellulose and Cellulose Nanocomposites*, (Wiley-VCH Verlag GmbH & Co. KGaA, Weinheim, 2017), pp. 51–100
53. H.S. Jeong, H.Y. Kim, S.H. Ahn, et al., Effects of combination processes on the extraction of pectins from rapeseed cake (Brassica napus L.). Food Chem. **139**, 9–15 (2013). https://doi.org/10.1016/j.foodchem.2012.12.056
54. N. Johar, I. Ahmad, A. Dufresne, Extraction, preparation and characterization of cellulose fibres and nanocrystals from rice husk. Ind. Crop. Prod. **37**, 93–99 (2012). https://doi.org/10.1016/j.indcrop.2011.12.016
55. L.J. Jönsson, C. Martín, Pretreatment of lignocellulose: formation of inhibitory by-products and strategies for minimizing their effects. Bioresour. Technol. **199**, 103–112 (2016). https://doi.org/10.1016/j.biortech.2015.10.009
56. V.K. Joshi, M. Preema Devi, Optimization of extraction treatment and concentration of extract on yield and quality of anthocyanins from plum var. "Santa Rosa". Indian J. Nat. Prod. Resour. **5**, 171–175 (2014)
57. S. Kalia, A. Dufresne, B.M. Cherian, et al., Cellulose-based bio- and nanocomposites: a review. Int. J. Polym. Sci. **2011**, 2011 (2011). https://doi.org/10.1155/2011/837875
58. J.Z. Kermani, A. Shpigelman, H.T.T. Pham, et al., Functional properties of citric acid extracted mango peel pectin as related to its chemical structure. Food Hydrocoll. **44**, 424–434 (2015). https://doi.org/10.1016/j.foodhyd.2014.10.018
59. H.P.S.A. Khalil, A.H. Bhat, A.F.I. Yusra, Green composites from sustainable cellulose nanofibrils: a review. Carbohydr. Polym. **87**, 963–979 (2012). https://doi.org/10.1016/j.carbpol.2011.08.078
60. M.K. Khan, M. Abert-Vian, A.S. Fabiano-Tixier, et al., Ultrasound-assisted extraction of polyphenols (flavanone glycosides) from orange (Citrus sinensis L.) peel. Food Chem. **119**, 851–858 (2010). https://doi.org/10.1016/j.foodchem.2009.08.046
61. K.M. Khazaei, S.M. Jafari, M. Ghorbani, et al., Optimization of anthocyanin extraction from saffron petals with response surface methodology. Food Anal. Methods **9**, 1993–2001 (2016). https://doi.org/10.1007/s12161-015-0375-4
62. J.S. Kim, Y.Y. Lee, T.H. Kim, A review on alkaline pretreatment technology for bioconversion of lignocellulosic biomass. Bioresour. Technol. **199**, 42–48 (2016). https://doi.org/10.1016/j.biortech.2015.08.085
63. L. Klavins, J. Kviesis, I. Nakurte, M. Klavins, Berry press residues as a valuable source of polyphenolics: extraction optimisation and analysis. LWT-Food Sci. Technol. **93**, 583–591 (2018). https://doi.org/10.1016/j.lwt.2018.04.021
64. E. Kliemann, K.N. De Simas, E.R. Amante, et al., Optimisation of pectin acid extraction from passion fruit peel (Passiflora edulis flavicarpa) using response surface methodology. Int. J. Food Sci. Technol. **44**, 476–483 (2009). https://doi.org/10.1111/j.1365-2621.2008.01753.x

65. B.B. Koubala, G. Kansci, L.I. Mbome, et al., Effect of extraction conditions on some physicochemical characteristics of pectins from “Améliorée” and “Mango” mango peels. Food Hydrocoll. **22**, 1345–1351 (2008a). https://doi.org/10.1016/j.foodhyd.2007.07.005
66. B.B. Koubala, L.I. Mbome, G. Kansci, et al., Physicochemical properties of pectins from ambarella peels (Spondias cytherea) obtained using different extraction conditions. Food Chem. **106**, 1202–1207 (2008b). https://doi.org/10.1016/j.foodchem.2007.07.065
67. S.G. Kulkarni, P. Vijayanand, Effect of extraction conditions on the quality characteristics of pectin from passion fruit peel (Passiflora edulis f. flavicarpa L.). LWT – Food Sci. Technol. **43**, 1026–1031 (2010). https://doi.org/10.1016/j.lwt.2009.11.006
68. P. Kumar, D.M. Barrett, M.J. Delwiche, P. Stroeve, Methods for pretreatment of lignocellulosic biomass for efficient hydrolysis and biofuel production. Ind. Eng. Chem. Res. **48**, 3713–3729 (2009). https://doi.org/10.1021/ie801542g
69. O. Kurita, T. Fujiwara, E. Yamazaki, Characterization of the pectin extracted from citrus peel in the presence of citric acid. Carbohydr. Polym. **74**, 725–730 (2008). https://doi.org/10.1016/j.carbpol.2008.04.033
70. N. Lavoine, I. Desloges, A. Dufresne, J. Bras, Microfibrillated cellulose – its barrier properties and applications in cellulosic materials: a review. Carbohydr. Polym. **90**, 735–764 (2017). https://doi.org/10.1016/j.carbpol.2012.05.026
71. H.V. Lee, S.B.A. Hamid, S.K. Zain, Conversion of lignocellulosic biomass to Nanocellulose: structure and chemical process. Sci. World J. **2014**, 1–20 (2014). https://doi.org/10.1155/2014/631013
72. M. Li, S. Cao, X. Meng, et al., The effect of liquid hot water pretreatment on the chemical – structural alteration and the reduced recalcitrance in poplar. Biotechnol. Biofuels **10**, 1–13 (2017). https://doi.org/10.1186/s13068-017-0926-6
73. C. Liu, B. Li, H. Du, et al., Properties of nanocellulose isolated from corncob residue using sulfuric acid, formic acid, oxidative and mechanical methods. Carbohydr. Polym. **151**, 716–724 (2016). https://doi.org/10.1016/j.carbpol.2016.06.025
74. H. Liu, B. Geng, Y. Chen, H. Wang, Review on the aerogel-type oil sorbents derived from Nanocellulose. ACS Sustain. Chem. Eng. **5**, 49–66 (2017). https://doi.org/10.1021/acssuschemeng.6b02301
75. P.K. Mandal, Dioxin: a review of its environmental effects and its aryl hydrocarbon receptor biology. J. Comp. Physiol. B Biochem. Syst. Environ. Physiol. **175**, 221–230 (2005). https://doi.org/10.1007/s00360-005-0483-3
76. N. Maneerat, N. Tangsuphoom, A. Nitithamyong, Effect of extraction condition on properties of pectin from banana peels and its function as fat replacer in salad cream. J. Food Sci. Technol. **54**, 386–397 (2017). https://doi.org/10.1007/s13197-016-2475-6
77. A.P. Manfredi, I. Ballesteros, F. Sáez, et al., Integral process assessment of sugarcane agricultural crop residues conversion to ethanol. Bioresour. Technol. **260**, 241–247 (2018). https://doi.org/10.1016/j.biortech.2018.03.114
78. J.P. Maran, B. Priya, Ultrasound-assisted extraction of pectin from sisal waste. Carbohydr. Polym. **115**, 732–738 (2015). https://doi.org/10.1016/j.carbpol.2014.07.058
79. C.D. May, Industrial pectins: sources, production and applications. Carbohydr. Polym. **12**, 79–99 (1990). https://doi.org/10.1016/0144-8617(90)90105-2
80. J.F. Mendes, J.T. Martins, A. Manrich, et al., Development and physical-chemical properties of pectin film reinforced with spent coffee grounds by continuous casting. Carbohydr. Polym. **210**, 92–99 (2019). https://doi.org/10.1016/j.carbpol.2019.01.058
81. L. Mesa, E. González, C. Cara, et al., The effect of organosolv pretreatment variables on enzymatic hydrolysis of sugarcane bagasse. Chem. Eng. J. **168**, 1157–1162 (2011). https://doi.org/10.1016/j.cej.2011.02.003
82. Z. Miao, T.E. Grift, A.C. Hansen, K.C. Ting, Energy requirement for comminution of biomass in relation to particle physical properties. Ind. Crop. Prod. **33**, 504–513 (2011). https://doi.org/10.1016/j.indcrop.2010.12.016
83. K. Michalska, S. Ledakowicz, Alkali pre-treatment of Sorghum Moench for biogas production. Chem. Pap. **67**, 1130–1137 (2013). https://doi.org/10.2478/s11696-012-0298-0

84. M. Michelin, J.A. Teixeira, Bioresource Technology Liquid hot water pretreatment of multi feedstocks and enzymatic hydrolysis of solids obtained thereof. Bioresour. Technol. **216**, 862–869 (2016). https://doi.org/10.1016/j.biortech.2016.06.018
85. R. Minjares-Fuentes, A. Femenia, M.C. Garau, et al., Ultrasound-assisted extraction of pectins from grape pomace using citric acid: a response surface methodology approach. Carbohydr. Polym. **106**, 179–189 (2014). https://doi.org/10.1016/j.carbpol.2014.02.013
86. A. Mokrani, K. Madani, Effect of solvent, time and temperature on the extraction of phenolic compounds and antioxidant capacity of peach (Prunus persica L.) fruit. Sep. Purif. Technol. **162**, 68–76 (2016). https://doi.org/10.1016/j.seppur.2016.01.043
87. N. Mosier, Features of promising technologies for pretreatment of lignocellulosic biomass. Bioresour. Technol. **96**, 673–686 (2005). https://doi.org/10.1016/j.biortech.2004.06.025
88. L.R. Mugwagwa, A.F.A. Chimphango, Box-Behnken design based multi-objective optimisation of sequential extraction of pectin and anthocyanins from mango peels. Carbohydr. Polym. **219**, 29–38 (2019). https://doi.org/10.1016/j.carbpol.2019.05.015
89. X. Ouyang, Effect of simultaneous steam explosion and alkaline depolymerization on corncob lignin and cellulose structure. Chem. Biochem. Eng. Q. **32**, 177–189 (2018). https://doi.org/10.15255/CABEQ.2017.1251
90. B. Ozturka, C. Parkinsona, M. Gonzalez-Miquel, Extraction of polyphenolic antioxidants from orange peel waste using deep eutectic solvents. Sep. Purif. Technol. **206**, 1–13 (2018)
91. N. Paixao, R. Perestrelo, J. Marques, J. Camara, Relationship between antioxidant capacity and total phenolic content of red, rosé and white wines. Food Chem. **105**, 204–214 (2007). https://doi.org/10.1016/j.foodchem.2007.04.017
92. X. Pan, N. Gilkes, J. Kadla, et al., Bioconversion of hybrid poplar to ethanol and co-products using an Organosolv fractionation process: optimization of process yields. Biotechnol. Bioeng. **94**, 851–861 (2006). https://doi.org/10.1002/bit
93. S. Paunonen, Strength and barrier enhancements of cellophane and cellulose derivative films: a review. Bioresources **8**, 3098–3121 (2013)
94. P.H.F. Pereira, T.Í.S. Oliveira, M.F. Rosa, et al., Pectin extraction from pomegranate peels with citric acid. Int. J. Biol. Macromol. **88**, 373–379 (2016). https://doi.org/10.1016/J.IJBIOMAC.2016.03.074
95. P.H.F. Pereira, K.W. Waldron, D.R. Wilson, et al., Wheat straw hemicelluloses added with cellulose nanocrystals and citric acid. Effect on film physical properties. Carbohydr. Polym. **164**, 317–324 (2017). https://doi.org/10.1016/j.carbpol.2017.02.019
96. T. Pielhop, J. Amgarten, P.R. Von Rohr, M.H. Studer, Steam explosion pretreatment of softwood: the effect of the explosive decompression on enzymatic digestibility. Biotechnol. Biofuels **9**, 1–13 (2016). https://doi.org/10.1186/s13068-016-0567-1
97. S. Quideau, D. Deffieux, C. Douat-Casassus, L. Pouységu, Plant polyphenols: chemical properties, biological activities, and synthesis. Angew Chemie – Int. Ed. **50**, 586–621 (2011). https://doi.org/10.1002/anie.201000044
98. S.A. Raj, S. Rubila, R. Jayabalan, T. Ranganathan, A review on pectin: chemistry due to general properties of pectin and its pharmaceutical uses. Sci. Rep. **1**, 1–5 (2012). https://doi.org/10.4172/scientificreports.5
99. R. Ravindran, A.K. Jaiswal, Exploitation of food industry waste for high-value products. Trends Biotechnol. **34**, 58–69 (2016). https://doi.org/10.1016/j.tibtech.2015.10.008
100. Z.U. Rehman, A.M. Salariya, F. Habib, W.H. Shah, Utilization of mango peels as a source of pectin. J. Chem. Soc. Pakistan **26**, 73–76 (2004)
101. C.M.G.C. Renard, A.A. Watrelot, C. Le Bourvellec, Interactions between polyphenols and polysaccharides: mechanisms and consequences in food processing and digestion. Trends Food Sci. Technol. **60**, 43–51 (2017). https://doi.org/10.1016/j.tifs.2016.10.022
102. G. Rocchetti, S. Castiglioni, G. Maldarizzi, et al., UHPLC-ESI-QTOF-MS phenolic profiling and antioxidant capacity of bee pollen from different botanical origin. Int. J. Food Sci. Technol. **54**, 335–346 (2019). https://doi.org/10.1111/ijfs.13941

103. R. Rojas, J.C. Contreras-Esquivel, M.T. Orozco-Esquivel, et al., Mango Peel as source of antioxidants and pectin: microwave assisted extraction. Waste Biomass Valor. **6**, 1095–1102 (2015). https://doi.org/10.1007/s12649-015-9401-4
104. N.A. Sagar, S. Pareek, S. Sharma, et al., Fruit and vegetable waste: bioactive compounds, their extraction, and possible utilization. Compr. Rev. Food Sci. Food Saf. **17**, 512–531 (2018). https://doi.org/10.1111/1541-4337.12330
105. S. Sai-Ut, S. Benjakul, S. Kraithong, S. Rawdkuen, Optimization of antioxidants and tyrosinase inhibitory activity in mango peels using response surface methodology. LWT – Food Sci. Technol. **64**, 742 (2015). https://doi.org/10.1016/j.lwt.2015.06.003
106. M. Saif, U. Rehman, I. Kim, et al., Use of ultrasound in the production of bioethanol from lignocellulosic biomass. Energy Educ. Sci. Technol. Part A Energy Sci. Res. **30**, 1391–1410 (2013)
107. M. Salajkova, *Wood Nanocellulose Materials and Effects from Surface Modification of Nanoparticles* (KTH Royal Institute of Technology: School of Chemical Science and Engineering, Stockholm, 2013)
108. C. Sánchez, Lignocellulosic residues: biodegradation and bioconversion by fungi. Biotechnol. Adv. **27**, 185–194 (2009). https://doi.org/10.1016/j.biotechadv.2008.11.001
109. M. Saritha, A. Arora, Lata, Biological Pretreatment of lignocellulosic substrates for enhanced delignification and enzymatic digestibility. Indian J. Microbiol. **52**, 122–130 (2012). https://doi.org/10.1007/s12088-011-0199-x
110. N. Sathitsuksanoh, Z. Zhu, S. Wi, Y.H. Percival Zhang, Cellulose solvent-based biomass pretreatment breaks highly ordered hydrogen bonds in cellulose fibers of switchgrass. Biotechnol. Bioeng. **108**, 521–529 (2011). https://doi.org/10.1002/bit.22964
111. A. Schieber, P. Hilt, P. Streker, et al., A new process for the combined recovery of pectin and phenolic compounds from apple pomace. Innov. Food Sci. Emerg. Technol. **4**, 99–107 (2003)
112. S.Y. Searle, C.J. Malins, Waste and residue availability for advanced biofuel production in EU Member States. Biomass Bioenergy **89**, 2–10 (2016). https://doi.org/10.1016/j.biombioe.2016.01.008
113. J. Shi, J. Yu, M. Bryan, et al., Optimization of the extraction of polyphenols from grape seed meal by aqueous ethanol solution. Food Agric. Environ. **1**, 42–47 (2003)
114. D.K. Sidiras, I.S. Salapa, Organosolv pretreatment as a major step of Lignocellulosic biomass refining. In: Biorefiinery 1 chemicals and materials from thermochemical biomass conversion and related processes (2015)
115. V. Smil, Nitrogen in crop production: an account of global flows. Global Biogeochem. Cycles **13**, 647–662 (1999). https://doi.org/10.1029/1999GB900015
116. P. Srivastava, R. Malviya, Sources of pectin, extraction and its applications in pharmaceutical industry – an overview. Indian J. Nat. Prod. Resour. **2**, 10–18 (2011)
117. F. SUN, H. CHEN, Enhanced enzymatic hydrolysis of wheat straw by aqueous glycerol pretreatment. Bioresour. Technol. **99**, 6156–6161 (2008). https://doi.org/10.1016/j.biortech.2007.12.027
118. Y. Sun, J. Cheng, Hydrolysis of lignocellulosic materials for ethanol production: a review q. Bioresour. Technol. **83**, 1–11 (2002)
119. A.A. Sundarraj, R. Thottiam Vasudevan, G. Sriramulu, Optimized extraction and characterization of pectin from jackfruit (Artocarpus integer) wastes using response surface methodology. Int. J. Biol. Macromol. **106**, 698–703 (2018). https://doi.org/10.1016/j.ijbiomac.2017.08.065
120. M. Szymanska-Chargot, M. Chylinska, K. Gdula, et al., Isolation and characterization of cellulose from different fruit and vegetable pomaces. Polymers (Basel) **9** (2017). https://doi.org/10.3390/polym9100495
121. F. Teymouri, L. Laureano-perez, H. Alizadeh, B.E. Dale, Optimisation of the ammonia fiber explosion treatment parameters for enzymatic hydrolysis of corn stover. Bioresour. Technol. **96**, 2014–2018 (2005)
122. B.R. Thakur, R.K. Singh, A.K. Handa, Chemistry and uses of pectin – a review. Crit. Rev. Food Sci. Nutr. **37**, 47–73 (1997). https://doi.org/10.1080/10408399709527767

123. K. Thirugnanasambandham, V. Sivakumar, J. Prakash Maran, Process optimization and analysis of microwave assisted extraction of pectin from dragon fruit peel. Carbohydr. Polym. **112**, 622–626 (2014). https://doi.org/10.1016/j.carbpol.2014.06.044
124. M. Tutt, T. Kikas, H. Kahr, et al., Using steam explosion pretreatment method for bioethanol production from floodplain meadow hay. Agron. Res. **12**, 417–424 (2014)
125. A. Verardi, I. De Bari, E. Ricca, V. Calabrò, Hydrolysis of lignocellulosic biomass: current status of processes and technologies and future perspectives. Bioethanol **290** (2012). https://doi.org/10.5772/23987
126. S. Wang, F. Chen, J. Wu, et al., Optimization of pectin extraction assisted by microwave from apple pomace using response surface methodology. J. Food Eng. **78**, 693–700 (2007). https://doi.org/10.1016/j.jfoodeng.2005.11.008
127. W. Wang, X. Ma, Y. Xu, et al., Ultrasound-assisted heating extraction of pectin from grapefruit peel: optimization and comparison with the conventional method. Food Chem. **178**, 106–114 (2015). https://doi.org/10.1016/j.foodchem.2015.01.080
128. Y.C. Wang, Y.C. Chuang, H.W. Hsu, The flavonoid, carotenoid and pectin content in peels of citrus cultivated in Taiwan. Food Chem. **106**, 277–284 (2008). https://doi.org/10.1016/j.foodchem.2007.05.086
129. Z. Wang, D.R. Keshwani, A.P. Redding, J.J. Cheng, Sodium hydroxide pretreatment and enzymatic hydrolysis of coastal Bermuda grass. Bioresour. Technol. **101**, 3583–3585 (2010). https://doi.org/10.1016/j.biortech.2009.12.097
130. H. Xia, A.S. Matharu, Unavoidable food supply chain waste: acid-free pectin extraction from mango peel via subcritical water. Faraday Discuss. **202**, 31–42 (2017). https://doi.org/10.1039/C7FD00035A
131. T. Yoshida, S. Tsubaki, Y. Teramoto, J. Azuma, Optimization of microwave-assisted extraction of carbohydrates from industrial waste of corn starch production using response surface methodology. Bioresour. Technol. **101**, 7820–7826 (2010). https://doi.org/10.1016/j.biortech.2010.05.011
132. W.X. Yu, C.Y. Hu, Z.W. Wang, Release of potassium sorbate from pectin- carboxymethyl cellulose films into food simulant. J. Food Process. Preserv. **41**, 1–8 (2017). https://doi.org/10.1111/jfpp.12860
133. L. Zhang, Y. Wang, D. Wu, et al., Microwave-assisted extraction of polyphenols from camellia oleifera fruit hull. Molecules **16**, 4428–4437 (2011). https://doi.org/10.3390/molecules16064428
134. Y. Zheng, Z. Pan, R. Zhang, Overview of biomass pretreatment for cellulosic ethanol production. Int. J. Agric. Biol. Eng. **2**, 51–68 (2009). https://doi.org/10.3965/j.issn.1934-6344.2009.03.051-068
135. X. Zhuang, W. Wang, Q. Yu, et al., Bioresource Technology Liquid hot water pretreatment of lignocellulosic biomass for bioethanol production accompanying with high valuable products. Bioresour. Technol. **199**, 68–75 (2016). https://doi.org/10.1016/j.biortech.2015.08.051
136. G. Zsivánovits, M. Marudova, S. Ring, Influence of mechanical properties of pectin films on charge density and charge density distribution in pectin macromolecule. Colloid Polym. Sci. **284**, 301–308 (2005). https://doi.org/10.1007/s00396-005-1378-2
137. FAO Global food losses and food waste- Extent, causes and prevention. Edward Elgar Publishing, Rome (2011)
138. H.P.S.A. Khalil, Y. Davoudpour, M.N. Islam, A. Mustapha, K. Sudesh, R. Dungani, & M. Jawaid, Production and modification of nanofibrillated cellulose using various mechanical processes : A review. Carbohydrate Polymers, 99, 649–665 (2017). https://doi.org/10.1016/j.carbpol.2013.08.069
139. AK. Kumar, S. Sharma, Recent updates on different methods of pretreatment of lignocellulosic feedstocks: a review. Bioresour Bioprocess 4: (2017). https://doi.org/10.1186/s40643-017-0137-9
140. S. Mani, LG. Tabil, S. Sokhansanj, Grinding performance and physical properties of wheat and barley straws, corn stover and switchgrass. Biomass and Bioenergy 27:339–352 (2004). https://doi.org/10.1016/j.biombioe.2004.03.007

141. A. Maria, R. Galletti, C. Antonetti, Biomass Pre-treatment: seperation of cellulose, hemicellulose and lignin. Existing technologies and perspectives. Eurobioref 69 (2011). https://doi.org/10.1515/gps-2012-0094
142. M. Michelin, JA. Teixeira, Bioresource Technology Liquid hot water pretreatment of multi feedstocks and enzymatic hydrolysis of solids obtained thereof. Bioresour Technol 216:862–869 (2016). https://doi.org/10.1016/j.biortech.2016.06.018

Chapter 10
Conversion of Lignocellulosic Biomass to Fuels and Value-Added Chemicals Using Emerging Technologies and State-of-the-Art Density Functional Theory Simulations Approach

P. N. Amaniampong, N. Y. Asiedu, E. Fletcher, D. Dodoo-Arhin, O. J. Olatunji, and Q. T. Trinh

10.1 Introduction

Renewable biochemicals and fuels derived from lignocellulosic biomass represent a sustainable approach toward the dwindling amount of fossil fuels. The use of biomass as a sustainable feedstock, however, does not ensure a successful transition toward a sustainable society without adapting processes designed with green chemistry and cost-competitive manufacturing processes. To this end, the basic concept of integrated biorefineries encompassing green chemistry principles for production, enhanced energy and material efficiency, reduced waste generation and

P. N. Amaniampong (✉)
INCREASE (FR CNRS 3707), ENSIP, Poitiers, France

Institut de Chimie des Milieux et Matériaux de Poitiers (IC2MP), Université de Poitiers, CNRS, ENSIP, Poitiers, France
e-mail: prince.nana.amaniampong@univ-poitiers.fr

N. Y. Asiedu
Department of Chemical Engineering, College of Engineering, Kwame Nkrumah University of Science and Technology, Kumasi, Ghana

E. Fletcher
Ottawa Institute of Systems Biology, University of Ottawa, Ottawa, ON, Canada

D. Dodoo-Arhin
Department of Materials Science and Engineering, University of Ghana, Legon-Accra, Ghana

O. J. Olatunji
Faculty of Traditional Thai Medicine, Prince of Songkla University, Hat Yai, Thailand

Q. T. Trinh
Cambridge Centre for Advanced Research and Education in Singapore (CARES), Campus for Research Excellence and Technological Enterprise (CREATE), Singapore, Singapore

M. O. Daramola, A. O. Ayeni (eds.), *Valorization of Biomass to Value-Added Commodities*, Green Energy and Technology,
https://doi.org/10.1007/978-3-030-38032-8_10

toxicity, efficient atom economy and the increased reusability of the products at the end of their lives has drawn particular interests and discussions among academics and industrialists. Greener conversion processes incorporating concept that promotes efficient atom economy such as use of heterogeneous catalysis, water-based reactions, environmentally friendly oxidants substituting for less efficient processes using toxic solvents and materials with high environmental risks are anticipated to be a key driving force of sustainability. Similarly, alternative and unconventional activation methods such as microwave (MW), nonthermal atmospheric plasma (NTAP), ball-milling (BM), and ultrasound technologies should replace energy-intensive heating to facilitate chemical reactions and their productions. Identifying routes of production for both energy and value-added chemicals is imperative, and their idealistic pathways have been discussed and documented in the literature [1–3]. For example, the oxidative polymerization of extract-free lignocellulosic materials can produce polyfunctional monomeric compounds that can be used as an alternative or replacement for fossil fuel–derived building blocks, including amino acids, aromatics, biofunctional molecules (e.g., succinic acid and lactic acid) and fatty acids. Biomass feedstocks often containing valuable extractable chemicals, of interest to pharmaceutical industry, and agricultural commodity chemicals can be isolated using greener processes such as supercritical carbon dioxide extraction. Ironically, to date, relatively little bioenergy and chemicals originate from lignocellulosic biomass as compared to feedstock such as starch and sugarcane, primarily due to high cost of production encompassing biomass pretreatment steps to biomass components (i.e., lignin, cellulose, and hemicellulose) and to disrupt the natural recalcitrant structure of these rigid polymers; complex structure of celluloses renders the molecules resistant to biological degradation. The primary objective of this section, therefore, is to review current trends and technologies such as physical, chemical, and biological processes for biomass conversion to fuels and pharmaceutical platform chemicals.

10.2 Biomass Conversion Technologies to Fuels and Chemicals

10.2.1 Biochemical Conversion of Biomass to Value-Added Chemicals

The debate on the conversion of food sources such as corn and cassava to biofuel production leading to potential spikes in food prices continues unabated [4]. Now, attention is rapidly shifting toward the use of nonfood materials such as agricultural and forestry residues as alternative feedstocks for the production of second-generation biofuels and value-added chemicals. Value-added chemicals obtained from plant biomass conversion include biopharmaceuticals, fragrances, and food flavors [5]. Agricultural and forestry residues are mainly composed of lignocellulose [6], which, when degraded, yields an enormous amount of carbons in the form of

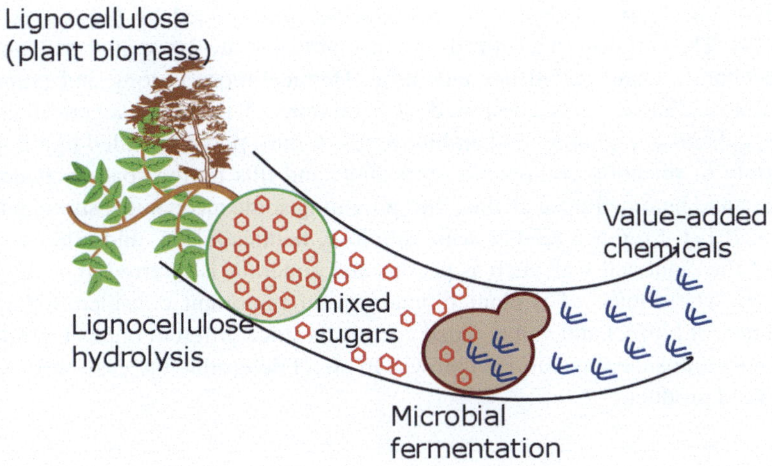

Fig. 10.1 Plant biomass is first hydrolyzed to obtain sugars that are then fermented to produce different value-added chemicals

sugars that can be fermented to produce biofuels and other biochemicals (Fig. 10.1). Other sections in this chapter provide insights into the physicochemical and other related conversion technologies of plant biomass (lignocellulose) to useful chemicals. In this section, the focus is on progress made in the field, so far, using biochemical approaches in the form of enzyme and microbial biotechnology to turn plant biomass to valuable chemicals. While biochemical conversion provides a potentially cheaper route for chemical production from plant biomass, there are several challenges associated with this process that are also highlighted in this chapter.

10.2.2 Biochemical Composition of Plant Biomass

Biomass feedstock used for biofuel production mainly comprises plant and algae materials, including rice straw, corn stover, sugar cane bagasse, and wood shavings, among others. The definition of biomass feedstocks can also be extended to other materials of plant origin such as paper waste and food fractions in municipal waste [7]. Plant biomass serves as an important feedstock for biofuel production because it is cheaply available and abundant. In addition, plants act as carbon sinks by taking up carbon dioxide from the atmosphere. The atmospheric carbon dioxide taken up by plants, in the presence of light and water, is converted to sugars which plants use for growth or are stored in the form of starch. The bulk of the sugar is channeled to the plant cell wall, where they are polymerized into polysaccharides (sugar polymers) including cellulose and hemicellulose to maintain the structural integrity and turgor of the plant [8]. Cellulose has been mentioned as the most abundant biopolymer on earth made up of only glucose units joined by β-1,4 glycosidic bonds [9]. The cellulose microfibrils are insoluble and are associated with another polysaccharide chain called hemicellulose. Hemicellulose is long and branched, but unlike cellulose, it is composed of a mixture of different sugars including xylose, galactose, mannose, and arabinose [9]. A third polymer called lignin, made up mainly of phenolic compounds, crosslinks and fills up the space between the cellulose and hemicellulose chains, thus providing some mechanical strength to the plant wall [9]. Lignin, together with cellulose, hemicellulose, and other components of the plant cell wall such as pectins and minerals, is referred to as lignocellulos [9]. The choice of a plant biomass is an important consideration in the production of biofuel and value-added chemicals since different plants have different lignocellulosic compositions that will in effect determine the ease and yield of the desired product.

10.2.3 Biomass Deconstruction

The biomass is first pretreated with hot water or steam in the presence of dilute acids or bases. This harsh pretreatment step breaks down the lignin, which crosslinks cellulose and hemicellulose. The challenge with harsh pretreatment of biomass is that some of the sugars obtained from hemicellulose, such as xylose, are degraded to form furfural and hydroxyl methyl furfural, which are toxic to bacteria or yeasts used or fermentation. Furthermore, lignin is also broken down to phenolic compounds including ferulic acid, coniferyl aldehyde, vanillin, p-coumaric acid, among others, which are also severely toxic (even in minute concentrations) to microorganisms used for fermentation [10]. With the rapid development of genomic and molecular biology tools, there has been a recent interest in unraveling the biological basis underlying the toxicity of these compounds to microorganisms used for fermentation. These efforts are gradually resulting in the development of robust yeast and bacteria strains suitable for fermentation in the presence of toxic compounds in plant hydrolysates [10].

Following the pretreatment step, the biomass is hydrolyzed enzymatically to release single sugar units (glucose, xylose, arabinose, and mannose) for fermentation. Given the complex nature of lignocellulose, pretreated biomass is hydrolyzed using a cocktail of different enzymes including cellulases, hemicellulose, and lytic polysaccharide monooxygenases [11, 12]. Cellulases degrade cellulose to its component sugars (glucose only). In a classic view, three classes of cellulases (exoglucanases, endoglucanases. and β-glucosidase) act synergistically to hydrolyze cellulose [9]. Exoglucanses target the reducing and nonreducing ends of the cellulosic chain and break the chain by releasing two sugar units called cellobiose. The endoglucanases randomly cleave internal bonds in the cellulosic chain to release new reducing and nonreducing ends available for the exoglucanase to act on. The third component, β-glucosidase, converts the cellobiose produced by the action of the exo- and endoglucanases into single glucose units [12].

Hemicellulose is also broken down to yield several sugar units, mainly xylose and other sugars including xylose, arabinose, galactose, and mannose [13]. This hydrolysis is facilitated by the enzyme class referred to as hemicellulases [14].

Cellulases and hemicellulases are of microbial origin and have been obtained from bacteria and fungi, including *Trichoderma*, *Cellulomonas*, *Bacillus*, and *Aspergillus* found in the soil, compost, or gut of animals such as termites and caterpillars [15]. Cellulases from the fungus *Trichoderma reesei* have been extensively characterized and have been described as the "gold standard" against which all cellulosic cocktail enzymes are compared [12, 16]. Indeed, *T. reesei* has been developed for the industrial production of cellulases and hemicellulases [16]. There is still a lot of interest in discovering new enzymes with improved properties such as thermotolerance and enhanced activity at broader pH ranges. With the development of next-generation DNA sequencing and "omics" tools, the traditional way of bioprospecting new cellulases by culturing environmental microorganisms has now evolved into using metagenomic approaches to find novel enzymes to degrade plant biomass [17]. These novel cellulases and hemicellulases can be cloned and expressed in industrial strains of yeast and bacteria for large-scale biomass degradation. Protein engineering has also been a useful method in recent times to develop or enhance the properties of already-known enzymes. For example, using protein engineering, it has been possible to generate a variant of a *Humicola insolens* cellulase that is stable at high temperatures (75–80 °C) [18].

These strategies are geared toward cheaply producing enzyme cocktails with improved activities and properties for rapid hydrolysis of lignocellulose to produce fermentable sugars, thus, cutting down the cost of production.

More recently, a new class of enzymes called lytic polysaccharide monooxygenases (LPMOs) has been identified as contributing to the cleavage of internal β–glycosidic bonds [11]. Having been recently discovered [11], very little is known about its activity on cellulosic biomass, although it has already been established that the enzyme requires oxygen and an electron source for its activity [19].

10.2.4 Biomass Fermentation into Fuels and Value-Added Chemicals

The final step in the biochemical conversion of biomass to fuels is fermentation of sugars present in the biomass hydrolysate. In this process, microbial hosts such as *Zymomonas mobilis* (bacterium) or *Saccharomyces cerevisiae* (yeast) added to the hydrolysate utilize the sugars as a carbon source, break it down via glycolysis to yield pyruvate, which is then further converted to bioethanol or other biochemicals. Metabolic engineering tools have made it possible to rewire biochemical pathways in microorganisms to produce novel high-value chemicals from sugars. A success story has been the production of artemisinin (antimalarial drug) precursors in engineered yeasts, which is now being commercialized by Amyris [20].

Downstream processing is then used to purify the final product. In an ideal world, all the glucose consumed should be converted to the desired product. However, the theoretical yield (gram of product per gram of biomass) of bioethanol, for example, from of glucose is only 0.51 g ethanol per gram of glucose [21]. This is because a fraction of the sugars is used for yeast growth and the formation of by-products including glycerol, organic acids, and carbon dioxide [22]. Current efforts, using strain engineering, are being used to block the formation of by-products in an effort to direct most of the carbon flux from sugars to bioethanol and other desired products [22].

Although glucose is the main sugar found in plant hydrolysates, other sugars such as xylose, arabinose, and galactose represent significant fraction of sugars in the hydrolysate [23]. However, *S. cerevisiae* that is used to carry out most fermentation is unable to efficiently utilize other sugars apart from glucose, thus further limiting the yield of products from biomass. The use of xylose-fermenting *S. cerevisiae* is still far from an industrial application. This is because although numerous efforts have gone into engineering the pathway for xylose and galactose utilization in *S. cerevisiae*, there is still the need to fine-tune these pathways especially to eliminate xylitol by-products from forming. Others have investigated the use of nonconventional yeasts like *Kluyveromyces marxianus* and *Scheffersomyces (Pichia) stipitis*, which naturally metabolize pentose sugars for ethanol fermentation from biomass. This is still challenging because most nonconventional microorganisms require strict growth requirements, and genetic tools needed to further improve these strains are underdeveloped [24].

Economic viability is a key consideration in the conversion of biomass to biofuels. In order to achieve this, there is the need to refine and fine-tune every aspect of the pipeline right from the choice of biomass to enhanced enzymatic degradation and fermentation. Development of microbial strains with improved tolerance to toxic compounds and conditions in hydrolysates [25] as well as the final product itself (ethanol) will play a key role in enhancing the biochemical conversion of biomass to fuels. Furthermore, improved enzymes used for biomass degradation and sugar utilization, together with reduction of by-product formation in microbial hosts, will result in an efficient conversion of biomass to fuels, thus making the process economically profitable.

10.3 Thermochemical Conversion of Biomass to Fuels and Value-Added Chemicals

Thermochemical conversion is a major method for the production of second-generation liquid biofuels, which involves the use of heat to change biomass from solid to other useful forms. This process possesses numerous advantaged over biological processes, including flexibility of feedstock, conversion of both lignin and carbohydrate into products, ease of transportation, and faster reaction rates. This section will give an overview of the current status of direct thermochemical

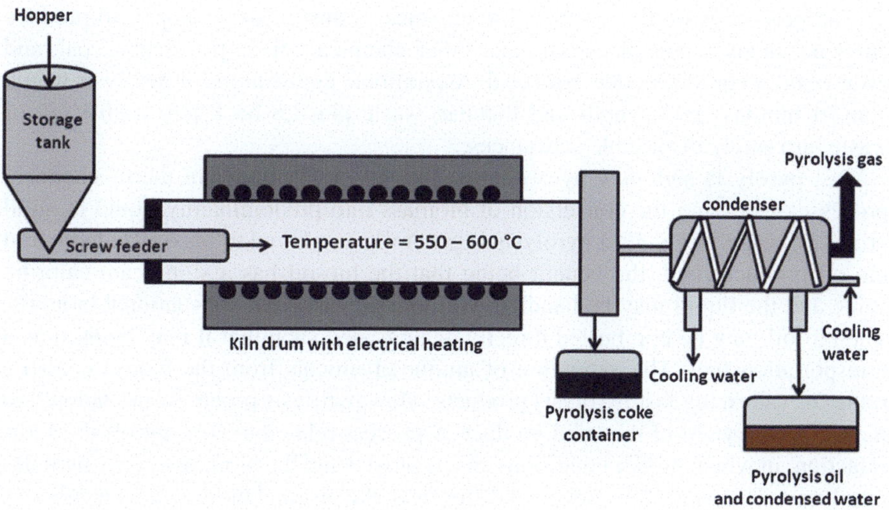

Fig. 10.2 Pyrolysis plant for bio-oil production [26]

processes for the production of liquid biofuels. Particularly, aspects of operating conditions, choice of catalysts, different raw materials, and reaction mechanism will be discussed. A classical reaction pathway for the production of liquid biofuels from lignocellulosic biomass is shown in Fig. 10.2.

Thermochemical conversion technologies focus on the relationship between the change of molecular structures and the reaction parameters (heat and pressure), mimicking the reaction taking place during the formation of petroleum oil. The general analogy is that thermochemical conversion can realize the rearrangement of molecular structures of lignocellulosic biomass to form high-grade biofuels comparable to petroleum fuels. Different thermochemical conversion processes include combustion, gasification, liquefaction, hydrogenation, and pyrolysis. The choice of which actions are used greatly depends upon the type and quantity of the biomass feedstock, the desired type of fuels, the environmental standards, and the economic conditions.

10.3.1 Pyrolysis of Lignocellulosic Biomass for Liquid Fuels

Pyrolysis is the thermal decomposition of materials at elevated temperatures in an inert atmosphere [27]. It involves the change of chemical composition and is irreversible. Pyrolysis is most commonly used for the treatment of organic materials. It is one of the processes involved in charring wood around 200–300 °C (390–570 °F). In general, pyrolysis of organic substances produces volatile products and leaves a solid residue enriched in carbon, char. Extreme pyrolysis, which leaves mostly carbon as the residue, is called carbonization.

The process is used heavily in the chemical industry, for example, to produce ethylene, many forms of carbon, and other chemicals from petroleum, coal, and even wood, to produce coke from coal. Aspirational applications of pyrolysis would convert biomass into syngas and biochar, waste plastics back into usable oil, or waste into safely disposable substances.

Fast pyrolysis and slow pyrolysis (including torrefaction) are thermochemical processes involving the conversion of biomass into predominantly liquid or solid products, respectively. Fast pyrolysis is generally employed to maximize the liquid bio-oil product yield, the benefit being that the bio-oil has a significant calorific value and the liquid may be handled with greater ease than conventional biomass. The bio-oil may be combusted directly or upgraded such that it may be used as a transportation fuel. The extraction of unique chemicals from the bio-oil is also a route for valorizing fast pyrolysis products. Slow pyrolysis processes are tailored to maximize the yield of the solid product. A process related to slow pyrolysis is torrefaction, in which milder conditions are employed and the biomass is only partially pyrolyzed. The solids thus obtained differ from the original biomass in a number of ways, such as increased energy density, hydrophobicity, grindability, and reduced biodegradability. These characteristics allow for improved handling, transport, and utilization of the char and torrefied biomass within existing coal-based processes when compared to biomass. Char may also be used as a soil amendment in the form of biochar or processed further to form activated carbon.

10.3.1.1 Slow Pyrolysis

Slow pyrolysis is a predominantly popular thermochemical conversion process that has been around adopted for several years mainly for the production of charcoal. In slow pyrolysis, biomass is typically heated at about 500 °C at slow heating rates (up to 10–20 °C/min) with a vapor residence time varying between 50 and 30 min. The components in the vapor phase continue to react with each other, as solid char and liquid are formed. Charcoal, which is formed as the main product in slow pyrolysis process, can be used in a wide range of areas, including domestic cooking, fireworks, activated carbon, absorbents, soil conditioners, and as raw materials for chemicals production. Mok et al. reported that a higher yield of charcoal can be obtained from biomass feedstocks with higher lignin contents and lower hemicelluloses contents. In contrast to fast pyrolysis, slow pyrolysis does not necessarily require fine feedstock particles size (smaller than 1 mm).

10.3.1.2 Fast Pyrolysis

Fast pyrolysis is a thermochemical conversion process that converts solid lignocellulosic biomass to a liquid product known as pyrolytic oil or bio-oil. Employing carefully controlled conditions (typically reaction temperature of around 500 °C, vapor phase temperature of 400–450 °C, and short vapor residence times of typically <2 s), it is often possible to generate bio-oil, from whole biomass, in yields of up to

75% of weight on a dry feed basis. However, the presence of high oxygen content is the primary difference between bio-oils and hydrocarbon fuels. The oxygen content in bio-oils is typically within the range of 25–40%. The presence of high oxygen content leads to a lower energy density than fossil fuels by 50% and immiscibility with hydrocarbon fuels. During fast pyrolysis, the choice of pyrolysis reactor is also crucial. Currently, there are three main well-known reactor types utilized for fast pyrolysis conversions of lignocellulosic biomass feedstock: the fluid bed reactor, rotating con reactor, and the ablative fast pyrolysis reactor.

10.3.1.3 Flash Pyrolysis

The rapid movement of biomass substrates through a heated tube under gravity or in a gas flow is often referred to as very fast or flash pyrolysis. Here, higher temperatures and shorter residence times than fast pyrolysis are required. Nonetheless, the main product distributions are similar to those of fast pyrolysis.

10.3.1.4 Hydrogenation

Bio-oil obtained via fast pyrolysis conversions is not produced under conditions of thermodynamic equilibrium, but instead involves reactions at short residence time at a high pyrolysis temperature followed by a rapid cooling or quenching. This results in a rapidly unstable bio-oil chemical composition, which tends to change toward thermodynamic equilibrium. This is the main reason why upgrading processes for bio-oil production are required to improve stability and fuel properties. The removal of oxygen in the form of water in bio-oil has been shown to be efficient via hydrodeoxygenation (HDO) reactions that take place at moderate temperature (300–600 °C). Several successful examples of HDO of bio-oils and bio-derived substrates have been reported in literature. For example, phenolics can be converted to cycloalkanes and alkenes using supported noble and non-noble metal catalysts (Pt, Rh, Ru, and Pd on γ-Al_2O_3 and CeO_2-ZrO_2). Aqueous-phase HDO of carboxylic acids over noble metal catalysts was reported at 300 °C under a pressure of hydrogen (6.4 MPa). Propanoic acid was converted by generation to propanol or dehydration-hydrogenation to propane, or the alcohols reacted with carboxylic acids to form esters (Fig. 10.3).

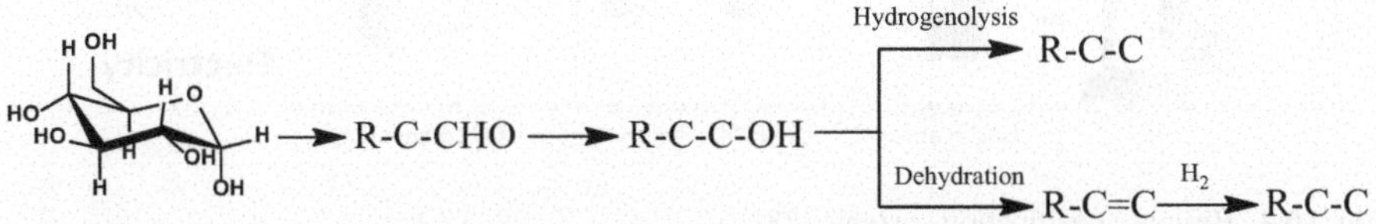

Fig. 10.3 Hydrodeoxygenation (HDO) process pathway from bio-derived substrates

10.3.1.5 Gasification

This refers to the partial oxidation of biomass at conditions of increased pressure at high temperatures (>800 °C). About 70–80% of the biomass is converted to syngas and the remainder, bio-char. Some gasification conditions can yield some amount of bio-oil. Higher yields of syngas can be obtained by performing the gasification in supercritical water conditions or by employing the use of catalysts to increase the efficiency of the process [28].

Major processes in biomass conversion by gasification are shown in Fig. 10.4. Products from gasification can be synthesized to produce fuels such as hydrogen, gasoline, and diesel for energy.

Gasification of biomass produces large amount of pollutant gases, such as CO_2, NO_x, and SO_x. Therefore, there is the need to upgrade the system in order to minimize the emission of toxic gases into the environment (Fig. 10.4b).

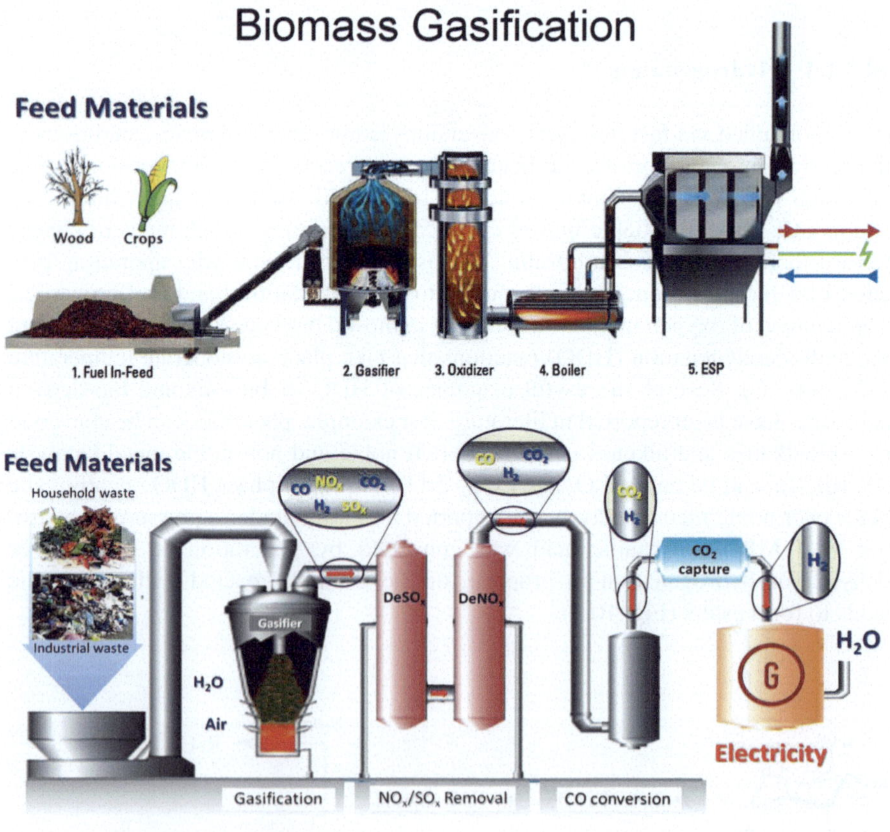

Fig. 10.4 Biomass gasification systems [28]

10.3.2 Hydrothermal Liquefaction (HTL)

Hydrothermal liquefaction (HTL), also referred to as direct liquefaction, hydrothermal upgrading/pyrolysis, depolymerization, and solvolysis, is the production of bio-oil from biomass at low-to-moderate temperatures (250–330 °C) and elevated pressures in the presence of a catalyst, usually sodium carbonate and a solvent such as water as well as an organic solvent (super critical fluid). During the hydrothermal liquefaction process, the biomass macromolecules are first hydrolyzed and/or degraded into smaller molecules. Many of the produced molecules are unstable and reactive and recombine into larger ones. During this process, a substantial part of the oxygen in the biomass is removed by dehydration or decarboxylation [29]. The primary product of HTL is bio-oil or bio-crude, and the by-products include bio-char, and water containing soluble organic compounds. The chemical composition and properties of the products mostly depend on the biomass substrate composition. However, this process is quite expensive [30]. A brief summary of thermochemical conversion of biomass is illustrated in Fig. 10.5.

10.4 Physiochemical Conversion

Physiochemical conversion processes include transesterification (biodiesel production) and involve the physical and chemical synthesis of products from feedstocks. It is primarily associated with the transformation of fresh or used vegetable oils, animal fats, greases, tallow, and other suitable feedstocks into liquid fuels or biodiesel.

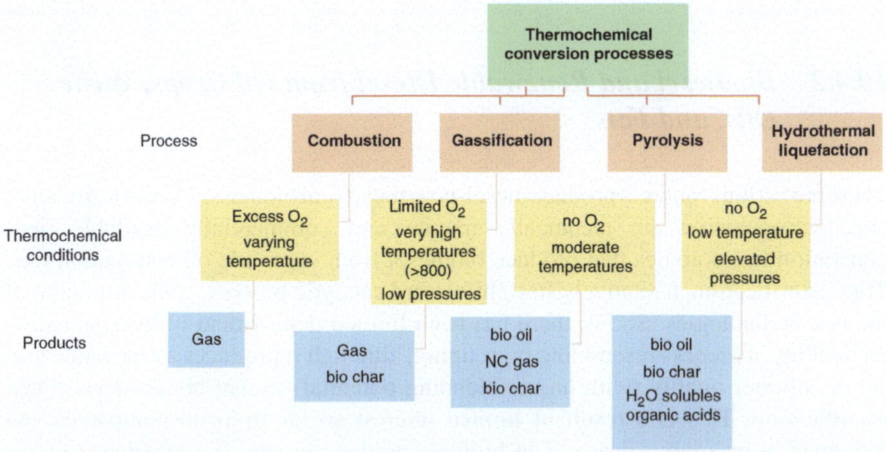

Fig. 10.5 Summary of thermochemical conversion of biomass (NC – non condensable)

Physiochemical conversion processes such as transesterification can remove difficult-to-manage wastes from the solid waste stream, including waste fats, oils, and greases as well as butcher waste and animal carcasses. Products derived by this process include liquid fuels and biodiesel, which can displace fossil fuels. Since feedstocks tend to be wet, use in biodiesel production facilities can provide additional benefits similar to biochemical conversion (reducing water quality impacts from traditional waste management and reducing greenhouse gas emissions from landfills). Renewable materials like polylactic acid derived from cornstarch are increasingly becoming commercially viable and could form an important part of future biorefineries. New concepts of biorefineries are looking at extracting a much broader range of materials and chemicals from the rich variety of biomass building blocks.

10.4.1 Liquid-Phase Catalytic Processing of Biomass-Derived Compounds

Sugars and other carbohydrates extracted from various biomass components can be biochemically or catalytically converted into hydrocarbons (e.g., hydroxymethylfurfural (HMF) or its derivatives). These can serve as substitutes for the petroleum-based building blocks used for the production of fuels, plastics, and fine chemicals (NSF 2008). These routes are potentially interesting, as they could produce high energy density liquid fuels, with potentially high yields via a limited number of chemical reactions, from a potentially wide range of biomass feedstocks. There have been much recent development activities, largely at the applied R&D stage, on these routes both in universities and companies, and in particular start-up companies.

10.4.2 Biodiesel and Renewable Diesel from Oil Crops, Waste Oils, and Fats

There are various routes to produce diesel-type fuels from biomass. Transesterification and hydrogenation are technically mature and commercially available first-generation technologies that produce biodiesel from vegetable oil and animal fats. Transesterification, a relatively straightforward catalytic process, is the dominant of the two technologies. So far, there has been limited deployment of hydrogenation technology, a process resembling oil refining, although it produces a renewable diesel of superior quality (with higher blending potential) to that obtained via transesterification. This is a result of limited interest so far from oil companies and refineries in becoming involved in biofuels production, and the reluctance of the sector due to potential technical risks associated with the degradation of hydrogenation catalysts. However, continued interest in vegetable oils and animal fats as feedstocks could lead to greater deployment of hydrogenation.

10.5 Heterogeneous Catalyst in Biomass Conversion and Sate-of-the-Art Density Functional Theory Simulations Approach

Second-generation lignocellulosic biomass, consisting of lignin, cellulose, and hemicellulose, is the biggest renewable source of carbon [28, 31]. To convert biomass to fuels and chemicals, C–O bonds need to be selectively cleaved, without breaking C–C bonds. Thus, utilizing the processing knowledge base of petroleum may not be possible and novel catalysts, solvents, and processes need to be developed. Among possible pathways to process biomass, the liquid-phase processing of biomass, using inorganic homogeneous and heterogeneous catalysts to converts solid biomass into smaller molecules containing less oxygen, is one of the most promising method and is receiving great attention [28, 31–33]. For example, the transformation of cellulose, the main component of lignocellulosic biomass, into chemicals and fuels in liquid-phase catalytic processing involves multiple reactions like hydrolysis, hydrogenation, oxidation, reduction, isomerization, dehydration etc., [33, 34]. Numerous experimental studies have been performed and are currently undergoing to identify efficient catalysts and processes for these reactions. A schematic of these reactions and experimentally identified catalysts is shown in Fig. 10.6 [31]. Under high-temperature conditions in l*iquid-phase catalytic processes*, cellulose is usually hydrolyzed to glucose, which consequently can be transformed into many high-value-added chemicals. Heterogeneous catalysts, in particular transition metals, have proven to be active catalysts for the selective conversion of glucose [32–35]. Additionally, they are easy to be separated and recycled, have good stability, are noncorrosive and highly selective, making their advantages to overcome these drawbacks of enzymatic and homogeneous catalytic processes [34–36].

However, in the present scenario, liquid-phase catalytic processing of biomass is facing major challenges in achieving high conversion and selectivity because of its highly active nature and the lack of fundamental understanding of catalytic reaction mechanisms [31, 36, 37]. First-principles calculations based on density functional theory (DFT), on the other hand, provide powerful tools to study and design heterogeneous catalysts, especially for complicated, multi-step reactions occurred in liquid-phase catalytic processing of biomass. The typical scale to study catalytic reactions is the molecular scale – a scale that is difficult to access experimentally. First-principles-based molecular modeling is best placed to investigate molecular-level effects and can greatly help to guide and validate the chemical intuition on the molecular scale [31, 38–40]. Using computational simulations, rate and/or selectivity determining steps could be identified to help guide the design of better catalyst. Some recent works on the design of high efficient oxygen reduction reaction catalyst [41], the promotional effect of B to Ni, Cu and Co in C–H activation in tar elimination process [42], methane non-oxidative coupling [43] and Fischer-Tropsch synthesis [38, 44], or the insights into the active sites of CuO catalyst in methane activation [45], glycerol selective oxidation [46] and amine oxidative coupling

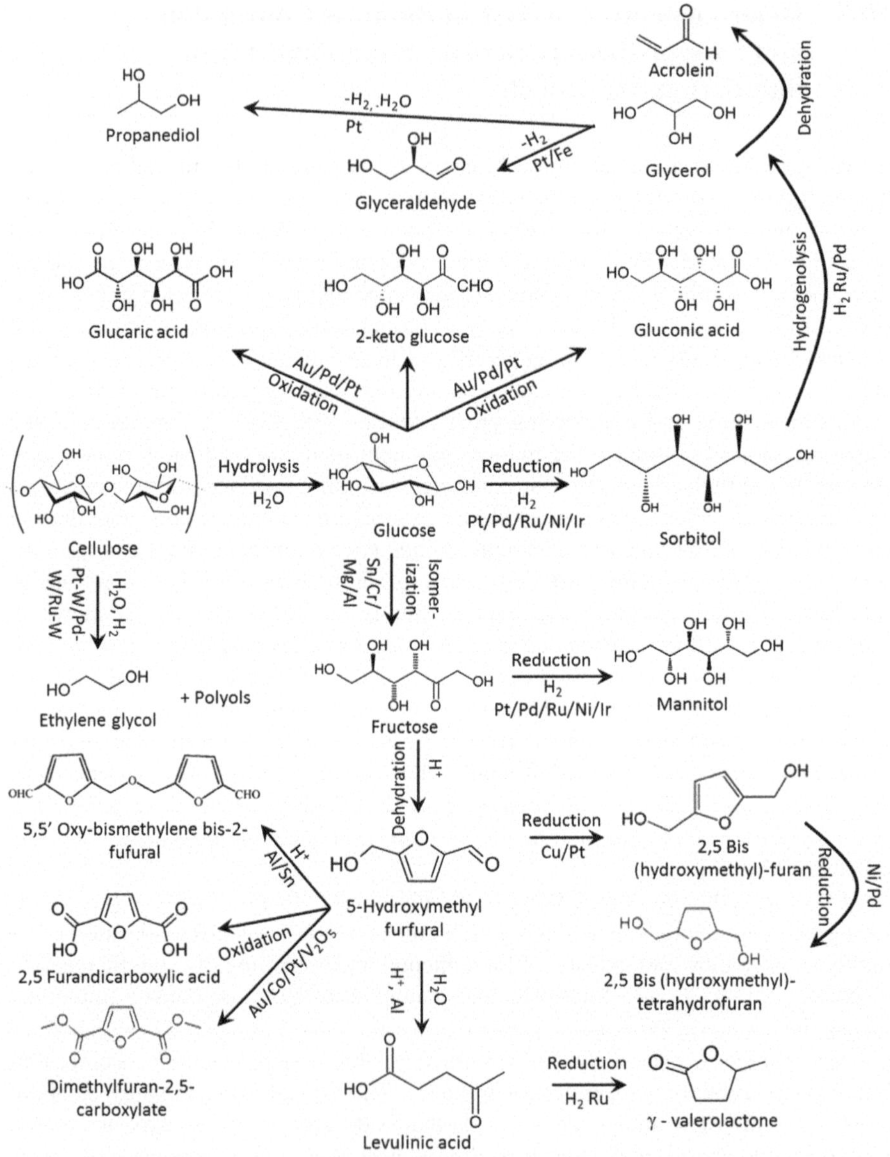

Fig. 10.6 Schematic of some of the key reactions involved in converting carbohydrate-based feedstock into chemicals and fuels. Experimentally tested catalysts for the reactions are also shown. (Reprinted with permission from Elsevier [31])

reactions [47] could illustrate how first-principles-based modeling is able to shed light into the mechanism and guide the design of heterogeneous catalysts with improved activity, selectivity, and stability.

Recently, theoretical investigations in biomass conversion, especially cellulosic biomass, to unravel the catalytic reaction mechanisms at molecular level, were performed on only few selected reactions. The adsorption of cellulosic biomass–derived aldoses on heterogeneous catalysts is the first and governing step in their conversion to platform chemicals, and complete understanding of the adsorption of biomass substrates on transition metals (particularly glucose as a model surrogate) is crucial to further investigate their surface chemistry on the metals, which is very challenging to be accessed by surface science experiments due to the extremely low vapor pressure of solid glucose at room temperature [48, 49]. Trinh et al. used DFT calculations to study the adsorption and activation of glucose on different transition metals and revealed that glucose and its surrogates (glycolaldehyde and glyceraldehyde) coordinated with Pd (111) and Pt (111) surfaces in the η_1(O) configuration via the carbonyl group (cf. Fig. 10.7), where the aldose adsorbed on metal surfaces only via the oxygen lone pair electrons [36, 48, 49]. This η_1(O) adsorption configuration was a result of the transition metal catalyzed ring-opening process, wherein adsorbed glucose in the cyclic form underwent deprotonation and ring-opening, and the resultant open chain configuration closely resembled the η_1(O) structure. Using entropy calculations, the authors also demonstrated that the transformation from η_1(O) to another configuration η_2 (C,O) was not thermodynamically favorable even at higher temperatures [36]. Finally, based on the most stable η_1(O) adsorbed configuration, the catalytic activity of Pd and Pt surfaces toward the decomposition, oxidation, and hydrogenation reactions was evaluated. The dissociation of α-hydroxyl group of glucose is challenging and can only occur at high temperatures. However, the activation of formyl C–H bond is easy on both Pd and Pt, suggesting that Pd and Pt are potential catalysts for the oxidation of glucose toward gluconic acid. The hydrogenation of both carbon and oxygen atoms in the glucose's carbonyl group on Pt (111) is also feasible, while it is more difficult on Pd (111). This work therefore provided some insights into how glucose adsorbs and reacts on those systems, which is very helpful in designing novel catalyst with better activity and selectivity for biomass conversion. Indeed, the difference in the adsorbed conformations of the reactants is believed to affect the activation barriers and selectivities [36, 49]. By tuning the preferred adsorption configuration of glucose on the catalyst surface, its surface chemistry could therefore be changed and controlled accordingly. Successful example based on this strategy is the designing of Zn-Pt (111) catalyst, which has higher deoxygenation chemistry due to the stabilization of reactant adsorbed in an η_2 (C,O) configuration and hindering the formation of acyl intermediates [50, 51].

Besides transition metals, metal oxides are well-known oxidation catalysts for hydrocarbons, and now, in the wake of heightened interests in biomass conversion to fuels and chemicals, are explored for the oxidation of cellulosic biomass [37, 52, 53]. However, the role of lattice oxygen in the oxidation of hydrocarbons is not yet completely understood and in the case of biomass oxidation, investigations have just begun. In the study by Amaniampong et al. [37], using CuO (which is a popular catalyst) as an example, experimental catalyst characterization methods were combined with state-of-the-art quantum mechanical computations to elucidate the mechanistic details of the catalytic reaction. DFT calculations revealed that the key

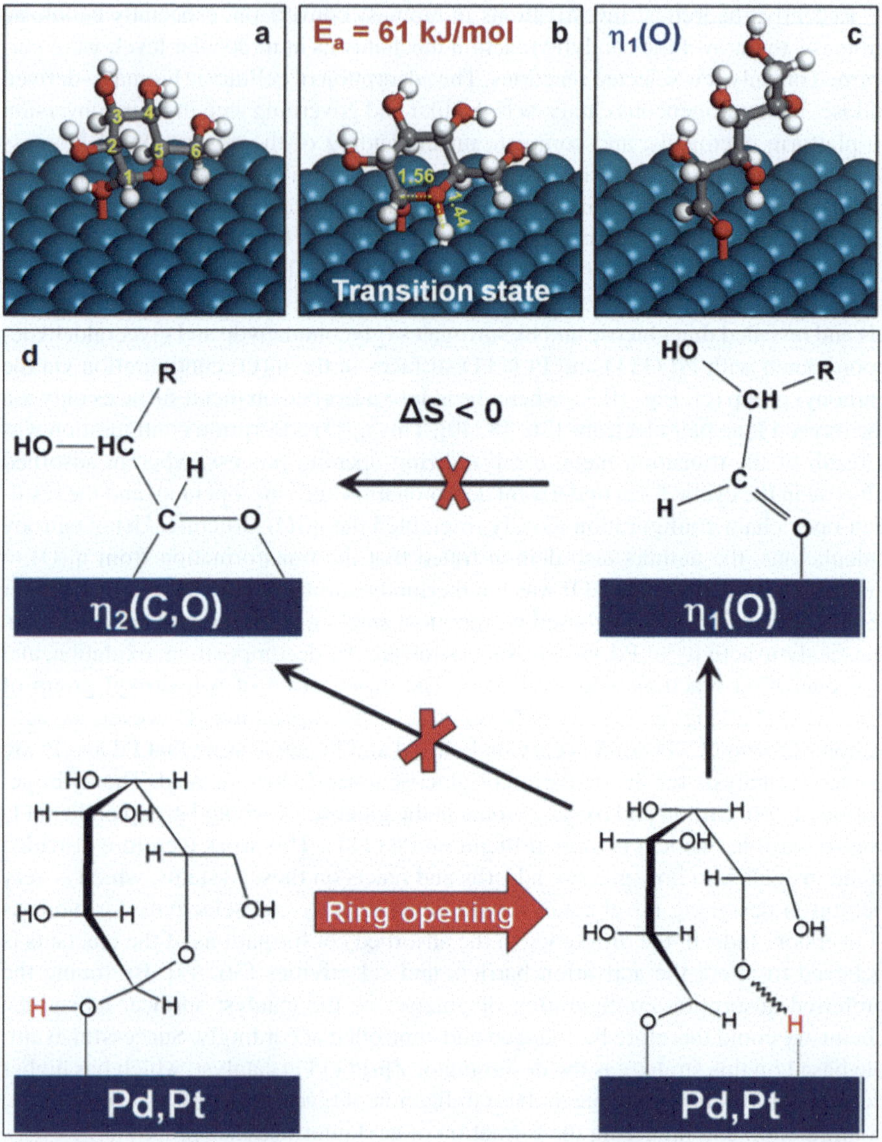

Fig. 10.7 (**a**) Adsorption configurations of the cyclic-form glucose; (**b**) transition state of glucose ring opening, (**c**) adsorption of linear-chain glucose in the η_1(O) configuration, and (**d**) schematic illustration for glucose ring opening an adsorption on transition metals surface [36]

role of the surface lattice oxygen in the metal oxide is to activate the formyl C–H bond. C–H bond cleavage is a crucial step in the oxidation of cellulosic sugars to their respective acids, without breaking C–C bonds. In that study, the authors demonstrated that surface lattice oxygen of the metal oxide selectively catalyzed the dissociation of the C–H bond and also inserted itself into the sugar molecule seamlessly, thus, acting as a catalyst and as an oxygen supplier for the reaction (Fig. 10.8) [37]. Activities of surface lattice oxygen in CuO and of chemisorbed surface oxygen on Cu surface were compared to explain the superior activity of the lattice oxygen and identify a descriptor for biomass oxidation. Additionally, they revealed that CuO, which gets reduced to Cu during the reaction, could be completely regenerated under oxygen flow. All of those theoretical investigations were all consistent with experimental results, which showed almost 100% conversion and 85+% yields were obtained when the catalyst was prepared in the form of 2-dimensional nanostructures [37]. This study is directly related to the conversion of one of the most abundant nonpetroleum feedstock, that is, biomass, available to mankind to manufacture these building blocks. The work, for the first time, explains how lattice oxygen in the metal oxide can be used to selectively activate the aldehyde C–H bond in biomass substrates and thus opens up new avenues to design and develop catalysts for plethora of biomass reactions that require selective activation/dissociation of the C–H bond without cleaving the C–C bond and hence will instigate multiple theoretical and experimental works, which can add a completely new dimension to C–H bond activation chemistry in biomass reactions.

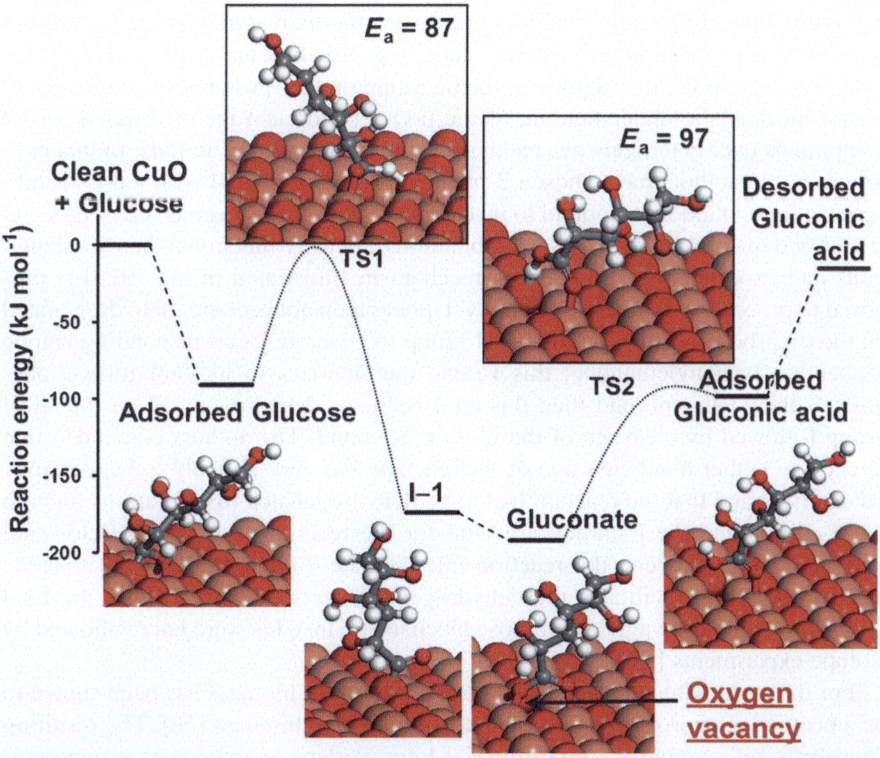

Fig. 10.8 Energy profile for the oxidation of glucose to gluconic acid on CuO(111) surface with reaction intermediates and transition states [37]

Compared to cellulosic biomass, conversion of lignin to chemicals and fuels is more difficult. Lignin is the second most abundant biopolymer on earth and constitutes around 30 wt% of the lignocellulosic biomass and due to its lower O/C ratio compared to carbohydrates, lignin comprises more than 50% of the energy of the total biomass [54, 55]. However, its utilization is still limited due to its recalcitrance to chemical processing, and currently it is simply burned for heat and power in the paper and pulping industry. Furthermore, lignin in lignocellulosic biomass is the only renewable source for aromatic compounds, and effective valorization of lignin remains a significant challenge in biomass [54]. Therefore, there is a strong need for making full use of lignin by converting it into liquid fuels or value-added chemicals. Catalytic depolymerization is one efficient way of utilizing lignin, and breaking the etheric β-O-4 linkage in lignin is the first step to efficiently utilize lignin, since it comprises >45% of total linkages in softwood lignin and >60% in hardwood lignin. In order to understand the mechanism of β-O-4 bond cleavage by catalysts, both homogeneous and heterogeneous catalysts have been employed. However, despite intensive experimental efforts, there was a common agreement that they were unable to provide direct thermodynamic or kinetic evidence to support the reaction mechanism they proposed. First-principles methods, that is, mainly density functional theory (DFT), excel at describing the adsorption behavior of possible surface intermediates during a catalytic reaction and at identifying transition states of all possible elementary steps that link up all the surface intermediates. One of the first reports in the literature on first-principles calculations to gain a mechanistic understanding of the β-O-4 bond cleavage of dilignol model compounds over heterogeneous catalysts is the study of Lu et al. [54]. In that contribution, the authors have chosen 2-phenoxy-1-phenylethanol with a β-O-4 linkage as a lignin model compound to investigate the cleavage mechanism of its C–O ether bond over Pd catalysts by a combination of density functional theory calculations and experiments. The reaction mechanism (illustrated in Fig. 10.9) is proposed to proceed as follows: 2-phenoxy-1-phenylethanol becomes dehydrogenated on the α-carbon and then on the –OH group to generate its corresponding ketone 2-phenoxy-1-phenylethanone; this ketone tautomerizes to its enol form 2-phenoxy-1-phenylethenol, and then this enol becomes dehydrogenated on the –OH group followed by cleavage of the C–O ether bond. The authors concluded that direct C–O ether bond cleavage of the reactant was very unlikely to happen over Pd catalysts and that the reactant had to be dehydrogenated first. H atoms on both the α-carbon and the β-carbon are vital for the reaction to proceed. Keto–enol tautomerization between the reaction intermediate 2-phenoxy-1-phenylethanone and its enol form facilitates the dehydrogenation process and leads to the final C–O ether bond cleavage. All of those mechanistic insights were later validated by isotope experiments [54].

For the sustainable production of fuel, pyrolysis of biomass has been shown to be a promising approach in the context of processing biomass [56]. The resulting "pyrolysis oil" contains a mixture of a large variety of oxygenate compounds [56–58]. For the intended purpose, the high oxygen content of this product

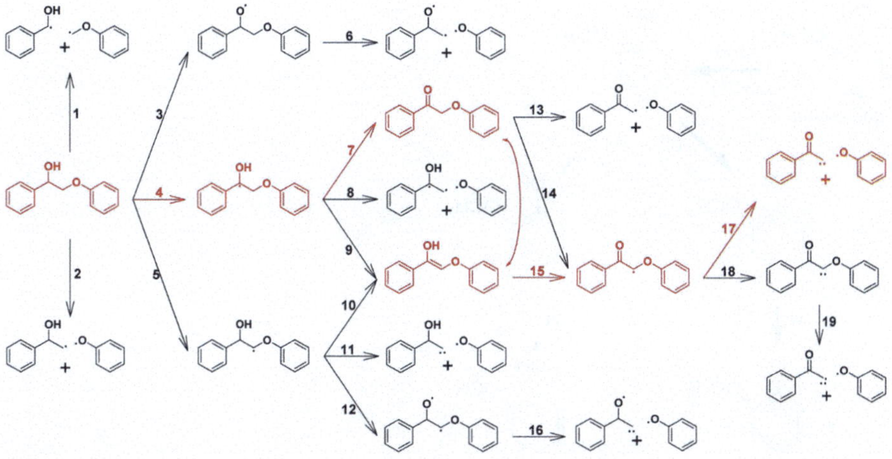

Fig. 10.9 Reaction network for the β-O-4 bond cleavage of 2-phenoxy-1-phenylethanol over Pd(111). The predicted dominant reaction pathway is colored in red [54]

mixture represents a problem, as it is associated with undesired properties, for example, chemical instability, and low heating values. Therefore, removal of oxygen functionalities is essential for increasing the quality of the pyrolysis oil and for compatibility with the current petrochemical infrastructure. One strategy for reaching these goals is hydrodeoxygenation (HDO) in which formally, the O containing groups are replaced by hydrogen atoms [56, 57]. Aromatic oxygenates, which mainly derive from the pyrolysis of lignin, have been shown to present a particular challenge because of the stability of the aromatic structures. Although a rough idea about the mechanism for the HDO of aromatics on various metallic catalysts can be gained from a large number of experiments, including isotope labeling, and the analysis of intermediate species as well as products observed under process conditions, the atomistic processes occurring at the catalyst surface remain unclear because computational studies dealing with aromatic oxygenates are scarcer. Recently, in the work by Chiu et al. [56] the authors reported DFT results for the HDO of guaiacol as a model compound for "lignin pyrolysis oil" over Ru surface to explore the yet-unknown potential energy surface for guaiacol HDO, which is illustrated in Fig. 10.10. According to their computational model, the C_{alkyl}−O bond is activated upon dehydrogenation at the C_{alkyl} center and can be cleaved with barriers as low as 35 kJ/mol. In contrast to the C_{alkyl}−O bond, the C_{aryl}−O bonds have quite high cleavage barriers, above 100 kJ/mol, despite the aromatic C center being unsaturated. Thus, the cleavage of the aromatic C_{aryl}−O bonds is the crucial step for the complete catalytic HDO of guaiacol. The authors also calculated the barriers for cleaving the aromatic C_{aryl}−O bonds to depend notably on the local environment of the C_{aryl}−O bond. The dependence of the barriers on the local geometry indicates that the surface structure of the catalyst can play an important role for the HDO activity, a circumstance that may be exploited

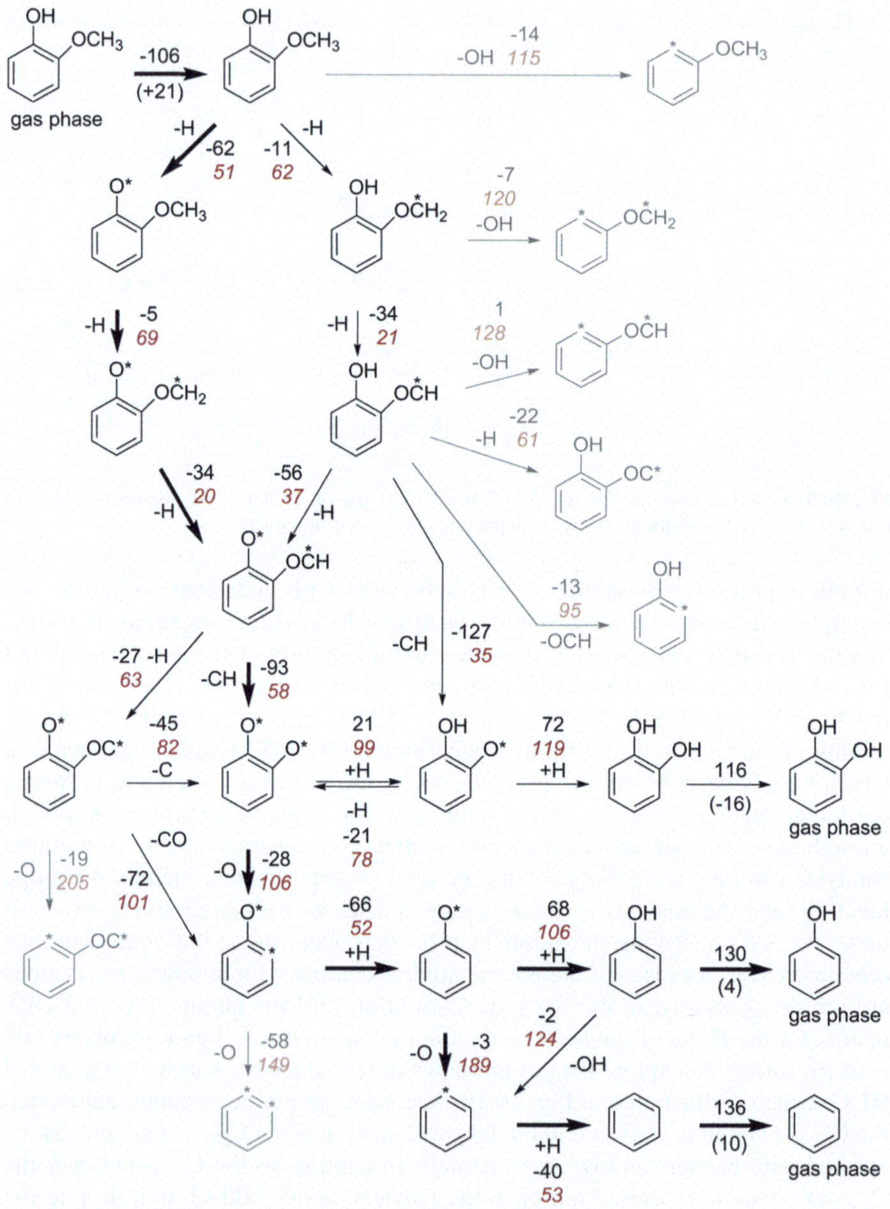

Fig. 10.10 Schematic representation of the reaction pathways during guaiacol decomposition on Ru(0001) surface. Bold black arrows denote the main reaction pathway and the kinetically accessible alternative pathways are marked by thin black arrows [56]

when investigating a new catalyst. Indeed, these insightful understanding on atomistic scale of the HDO reaction on heterogeneous catalysts has been successfully applied for designing core-shell bimetallic alloys NiAu catalyst, which exhibited unprecedented low-temperature activity in lignin hydrogenolysis [58], the use of bimetallics to control the selectivity for the upgrading of lignin-derived oxygenates on PtZn catalyst [59–61] or for Ni-doped Boron catalyst that was able to conduct furfural hydrogenation at lower operating temperatures for selectively producing furan and C_4/C_5 products and could also be considered as a viable alternative for replacing the expensive Ru-based catalysts for the HDO of pyrolysis-derived bio-oil [57].

Those selected examples have illustrated how molecular modeling could serve as a powerful tool to search and design novel catalysts, especially for those complicated multi-step reactions occurring in biomass conversion. Catalyst design and kinetic modeling often start from molecular-scale hypotheses about the reaction mechanism, the structure of the active catalyst, and the nature of the rate and selectivity determining steps. Computational catalysis has become a crucial tool to analyze molecular-scale concepts and elucidate their electronic origin. In combination with characterization and experimental kinetic validation, insights gained from computational catalysis can be translated all the way to the industrial scale. This combination between experiment and theory is becoming the new paradigm in catalyst design and kinetic modeling, both in academia and in industry.

10.6 Future Prospects and Challenges of Processing Biomass to Value-Added Products

With the continued pace of world economic growth, sustainable socioeconomic development will depend upon a secure supply of raw material inputs for agriculture, industry, energy, and related sectors. Currently, there are so many efforts to reduce global dependence on fossil fuels and the reasons include rising of energy demands, cost of petroleum products, fear of fossil fuel exhaustion, and the need to lower greenhouse emission. However, the long-term economic and environmental concerns have resulted in a considerable amount of work on renewable sources of liquid fuels to replace fossil fuels that led to an increasing search by scientists in evaluating the potential use of the inedible part of food crops to produce bioenergy and biomaterial, most especially for biofuel production [62]. Biofuels are fuels manufactured from renewable resources, namely, energy crops, crop residues, and forest and waste biomass. Thus, biofuels include fuels that have been used for millennia, like fuel wood and charcoal, as well as newer fuel like ethanol, methanol, biodiesel, and biogas [63]. Interestingly, researches have shown that biofuel from biomass resources can be produced. However, in order to achieve enhanced production, it was considered very crucial to examine the future prospects and challenges in processing biomass to value-added products.

10.6.1 Economic Importance of Biomass

The use of bio-based renewable resources holds great potential value for industries in many sectors, including energy, organic chemicals, polymers, fabrics, and healthcare products.

In general, a bio-based economy offers many benefits and opportunities such as new areas of economic growth and development for the many regions that have plentiful biomass resources, creation of new innovative business sectors and entrepreneurial skills, improved energy security via reducing dependence on nonrenewable resources, enhanced economic and environmental linkages between the agricultural sector, and a more prosperous and sustainable industrial sector. These also further help in the reduction of greenhouse gas emissions, improved health by reducing exposure to harmful substances through substitution of natural bio-based materials for chemical and synthetic materials as well as job creation and rural development.

In spite of all the benefits elucidated, there are many challenges that remain to be addressed in order to avoid negative impacts that could derail a smoother transition to a bio-based economy. These issues include how to manage competition of land used as raw material production source for industry with other land users, especially in relation to food and animal feed. Others include bioethical issues, where genetically modified crops are used or proposed; potential loss of biodiversity through large-scale and/or contract farming; equitable treatment of farmers in their interaction with bio-based companies; expanded research and development efforts, including potential integration of fossil fuel and bio-based approaches; improving transportation and delivery systems, for example, for raw materials, delivery to/from processing facilities, and final product distribution and use.

10.6.1.1 Biomass

Biomass is the most important of the renewable energy forms in terms of its current and projected consumption on a world scale. It ranks 4th on the importance as an energy source with oil, coal, and gas contributing more energy to the world [64]. According to the International Energy Agency [65], biomass currently provides approximately 14% of the total worldwide need for energy, and it represents an important contributor to the world economy [66, 67]. The major biomass resources come from the residues generated by agricultural, forest, and industrial sources. There are four major classes of biomass that include food crops, nonfood/energy crops, forest residues, and industrial process residues.

10.6.1.2 Food Crops

Most of the arable land in the world is used for food crops. However, food scientists and world leaders agree that the whole world will face a great food crisis in the near future. The main food crops produced are rice, sugar cane, vegetables, wheat, pulses, coconuts, maize, millet, and groundnuts. In most cases, only small percentages of the food crop residues, mainly the lignocellulosic parts, are used for heat energy in rural areas. Crop residues can be collected, mostly by bailing, either at the time the primary crop is harvested or later [68]. But not all field residues are recoverable.

10.6.1.3 Nonfood/Energy Crops

Where crop residues are needed to maintain sustainable production, a more viable option may be crops that could be grown specifically for use as energy crops, including herbaceous energy crops such as switch grass and short-rotation woody crops such as hybrid poplar. These crops are perennials, so they require few field passes and little soil disturbance, which result in their having low erosion rates. Paine et al. [69] recommended growing these crops on marginal lands, such as land that is subject to significant erosion, poorly drained soils, or areas used for wastewater reclamation. This approach would avoid competition with food crops and increase the amount of arable land. Energy crops show higher productivity levels per square acre than their conventional counterparts. In addition, by comparison, such crops are more homogeneous in terms of their physical and chemical characteristics than residual resources that are often described as the biomass resource of the future [70]. Among the most promising of the identified options so far are perennial crops. They are expected to contribute to the goal of increased bioenergy development by exhibiting low input requirements, over 10–15 years of productive life, and high yields [71]. On the other hand, Ericsson et al. [72] suggested a controversial result by stating that energy crops make a fairly small contribution to the biomass supply and that production is dominated totally by annual crops for the production of transportation fuels.

10.6.1.4 Forest Residues

Forest residues are the second largest source of lignocellulosic biomass next to the agro-residues. Much of this residue is uncontrollable; yet, a large portion of it may be reduced. The irregular round shape and varying dimensions of trees must be changed to the required shape and size by removal of extraneous material. Parts of the trees, or entire trees, are rejected for a specific use because of size, shape, quality, or species. Such residue is often increased by poor operating practice or by manufacturing issues and failures. For some wood species, faulty manufacturing is the direct cause of residue in the later stages of processing [73].

Forest residues typically refer to those parts of trees that are unsuitable for saw logs, that is, treetops, branches, small-diameter wood, stumps, dead wood, and even misshapen whole trees, as well as undergrowth and low-value species. Wood-processing residues (e.g., sawmill rejects and sawdust) and recycled wood (e.g., wood derived from the demolition of buildings, pallets, and packing crates) are important sources of LCB [74].

10.6.1.5 Industrial Process Residues

Advances in industrial biotechnology offer potential opportunities for economic utilization of agro-industrial residues. The residues produced in industrial crop processing are abundant sources of cellulosic biomass. Such biomass is 100% collectible and reusable. It is a relatively inexpensive raw material and already has been used for the production of several industrially important chemicals and bioethanol.

10.6.2 Prospects of Processing Biomass Conversion to Value-Added Products

Production of value-added products from biomass offers a number of prospects for development but also pose some challenges, especially in developing economies. The International Energy Agency (IEA) projected that biofuels would be competitive with petroleum at petroleum prices of between US$60 and US$100 a barrel, and that point was reached as at 2006 [63]. This is why biofuel production is generating a lot of interest worldwide. Another reason is that enhanced biofuel production offers a number of prospects that include renewability, enhanced national energy security, and economic growth as a direct consequence of higher energy efficiency and lower cost of production. It also offers environmental sustainability through greenhouse abatement, and again, poverty reduction, especially among the rural farming population through increased profitable employment [63, 75]. The efficiency and costs of biofuel production are largely a function of the type of feedstock, the conversion technology used, as well as the agro-ecological and socioeconomic conditions of production of biomass and the use of biofuel [76]. The major challenge to expanding or enhancing biofuel production continues to be whether crop production for biofuels will compete with and drive out food production, thereby increasing food insecurity [77]. The competitiveness of biofuels depends strongly on the relative prices of petroleum and of agricultural feedstocks for biofuel. When the demand for biofuel increases agricultural prices, the competitiveness of biofuels will begin to diminish.

The future of biofuel as an important source of carbon-neutral renewable energy will lie in reducing the direct competition with the food sector, and instead use feedstock with lower agricultural production costs compared to food and feed crops [63]. Incidentally, it is only cellulosic biofuel production that does not compete strongly with food and feed production, as much of the feedstocks can be supplied by the forestry sector from non-arable land or from byproducts of the agricultural sector. Generally, energy crop production may not lead to increased food insecurity if food crop residues only are utilized for biofuel, energy crops are cultivated on marginal or degraded lands only, farmers rotate food and energy crops on the same lands, crop productivity is increased through research; and if enhanced biofuel production can raise incomes of small farmers and rural laborers in developing economies. Policymakers should ensure that the possible pitfalls of biofuel production are avoided. These pitfalls include negative energy balance, insignificant reduction in greenhouse gases compared with petroleum, socioeconomic inequalities by concentrating benefits on the rich, deforestation, loss of biodiversity, excessive use of fertilizer and chemicals that can degrade the land and water that poor people depend on [63].

10.6.3 Challenges of Processing Biomass Conversion to Value-Added Products

These challenges are presented here under three subheadings, namely, pretreatment processes, cellulolysis, and fermentation processes, and thermochemical systems.

10.6.3.1 Pretreatment Processes

Lignocellulosic feedstock consists of lignin, cellulose, and hemicellulose. Lignin, which contains no sugar, encloses the cellulose and hemicellulose molecules. Like starch, cellulose consists of long chains of glucose molecules, but unlike starch, cellulose has structural features, which coupled with its encapsulation by lignin, make it more difficult to be hydrolyzed. Hemicellulose contains long chains of glucose molecules (6-carbon sugars) plus 5-carbon sugars in varying proportions depending on its source. The research needs for enhanced CE production therefore include developing processes or methods capable of liberating, reducing, and fermenting lignocellulosic feedstocks into ethanol at faster rates, higher yields and overall efficiency both technically and economically. Lignocellulosic materials require more drastic hydrolysis steps to achieve high ethanol conversion yields because of the presence of various amounts of 5-carbon sugars like xylose and arabinose. The costliest aspect of producing bioethanol from lignocellulosic materials is pretreatment to make them accessible to the enzymes or chemicals that will cut the sugars from the polymers before fermentation to ethanol [78]. The following pretreatment processes are highlighted: acid-catalyzed systems, alkaline systems, ozonolysis, ammonia-fiber/freeze explosion, CO_2 explosion, uncatalyzed steam explosion, liquid hot water, and microwave oven heating systems.

10.6.3.2 Cellulolysis and Fermentation Processes

A common disadvantage of enzymatic hydrolysis is the end-product inhibition of the enzymes used to hydrolyze cellulose and hemicellulose. This problem can, however, be reduced by the adoption of SSF or use of immobilized enzymes with a hollow-fiber membrane reactor, in which case the enzymes are confined inside the reactor allowing the separation of substrate and hydrolysis products as well as enabling the reutilization of the enzymes [79]. The adoption of SSF and use of immobilized enzymes, however, create the need for innovative design of bioreactors for integrated processes.

10.6.3.3 Thermochemical Systems Processes

Ethanol yields of up to 50% have been obtained using thermochemical processes [80]. Unfortunately, finding a cost-effective all-thermochemical process has been difficult. Thermochemical conversions of lignocellulosic biomass still present a technological barrier, not only on laboratory-scale research, but also in scaled-up processes. This calls for serious research effort toward achieving this end.

10.7 Conclusion and Recommendation

In conclusion, there is urgent need to develop new technologies capable of increasing efficiency and productivity in crop production and biofuel processing.

1. In developing economies, public-private partnership can work to increase farmers' awareness of opportunities being presented by biofuel production and the potential benefits. What will actually make enhanced CE production a win-win affair for poverty reduction and energy production on the one side, the environment and the economy on the other side, include sound technological innovation, appropriate government policies and support, and sound institutional innovations.
2. Pretreatment of cellulosic and lignocellulosic biomass in a cost-effective manner is a major research and development challenge for enhanced cellulosic ethanol production.
3. There is need for research to address the issues of low sugar yields and high energy consumption of acid pretreatment processes.
4. If less costly enzymes can be developed through research, enzymatic processes possess several advantages that include high efficiency, controllable production of by-products, mild process conditions, less expensive reaction vessels, and relatively low process energy.
5. Research is seriously needed to create super microorganisms that will enhance the adoption of direct microbial conversion (DMC) for CE production.
6. Increased research effort is advocated to develop genetically engineered fungi that can produce large volumes of cellulase, xylanase, and hemicellulose enzymes that can be used to convert agricultural residues into fermentable sugars [81].

References

1. M. Benoit, A. Rodrigues, Q. Zhang, E. Fourré, K. De Oliveira Vigier, J.M. Tatibouët, F. Jérôme, Angew. Chem. Int. Ed. **50**, 8964–8967 (2011)
2. Q. Zhang, F. Jerome, ChemSusChem **6**, 2042–2044 (2013)
3. P.N. Amaniampong, A. Karam, Q.T. Trinh, K. Xu, H. Hirao, F. Jérôme, G. Chatel, Sci. Rep. **7**, 40650 (2017)
4. D.J. Tenenbaum, Environ. Health Perspect. **116**, A254 (2008)
5. E. Fletcher, A. Krivoruchko, J. Nielsen, Biotechnol. Bioeng. **113**, 1164–1170 (2016)

6. M. Irfan, M. Nadeem, Q. Syed, Braz. J. Microbiol. **45**, 457–465 (2014)
7. R. Nair, P.R. Lennartsson, M.J. Taherzadeh, *Current Developments in Biotechnology and Bioengineering* (Elsevier, Amsterdam, 2017), pp. 157–190
8. K.H. Caffall, D. Mohnen, Carbohydr. Res. **344**, 1879–1900 (2009)
9. S.J. Horn, G. Vaaje-Kolstad, B. Westereng, V. Eijsink, Biotechnol. Biofuels **5**, 45 (2012)
10. P.T. Adeboye, M. Bettiga, L. Olsson, AMB Express **4**, 46 (2014)
11. K.S. Johansen, Biochem. Soc. Trans. **44**, 143–149 (2016)
12. E.M. Obeng, S.N.N. Adam, C. Budiman, C.M. Ongkudon, R. Maas, J. Jose, Bioresour. Bioprocess. **4**, 16 (2017)
13. E.M. Rubin, Nature **454**, 841 (2008)
14. A.S. Bhattacharya, A. Bhattacharya, B.I. Pletschke, Biotechnol. Lett. **37**, 1117–1129 (2015)
15. M.E. Himmel, Q. Xu, Y. Luo, S.-Y. Ding, R. Lamed, E.A. Bayer, Biofuels **1**, 323–341 (2010)
16. M. Paloheimo, T. Haarmann, S. Mäkinen, J. Vehmaanperä, *Gene Expression Systems in Fungi: Advancements and Applications* (Springer, Cham, 2016), pp. 23–57
17. C. Yang, Y. Xia, H. Qu, A.-D. Li, R. Liu, Y. Wang, T. Zhang, Biotechnol. Biofuels **9**, 138 (2016)
18. I. Wu, F.H. Arnold, Biotechnol. Bioeng. **110**, 1874–1883 (2013)
19. R.J. Quinlan, M.D. Sweeney, L.L. Leggio, H. Otten, J.-C.N. Poulsen, K.S. Johansen, K.B. Krogh, C.I. Jørgensen, M. Tovborg, A. Anthonsen, Proc. Natl. Acad. Sci. **108**, 15079–15084 (2011)
20. S.H. Kung, S. Lund, A. Murarka, D. McPhee, C.J. Paddon, Front. Plant Sci. **9**, 87 (2018)
21. M.S. Krishnan, N.W. Ho, G.T. Tsao. Appl Biochem Biotechnol. 77–79:373–388 (1999)
22. C. Navarrete, J. Nielsen, V. Siewers, AMB Express **4**, 86 (2014)
23. G.J. Martin, A. Knepper, B. Zhou, N.B. Pamment, J. Ind. Microbiol. Biotechnol. **33**, 834 (2006)
24. A.-K. Löbs, C. Schwartz, I. Wheeldon, Synth Syst Biotechnol. Aug 31;2(3):198–207 (2017)
25. E. Fletcher, A. Feizi, M.M. Bisschops, B.M. Hallström, S. Khoomrung, V. Siewers, J. Nielsen, Metab. Eng. **39**, 19–28 (2017)
26. M. Adjin-Tetteh, N. Asiedu, D. Dodoo-Arhin, A. Karam, P.N. Amaniampong, Ind. Crop. Prod. **119**, 304–312 (2018)
27. A.D. McNaught, A.D. McNaught, *Compendium of Chemical Terminology* (Blackwell Science Oxford, Oxford, 1997)
28. C.-H. Zhou, X. Xia, C.-X. Lin, D.-S. Tong, J. Beltramini, Chem. Soc. Rev. **40**, 5588–5617 (2011)
29. S. Xiu, A. Shahbazi, Renew. Sust. Energ. Rev. **16**, 4406–4414 (2012)
30. S. Xiu, A. Shahbazi, V. Shirley, D. Cheng, J. Anal. Appl. Pyrolysis **88**, 73–79 (2010)
31. S.H. Mushrif, V. Vasudevan, C.B. Krishnamurthy, B. Venkatesh, Chem. Eng. Sci. **121**, 217–235 (2015)
32. Y.-C. Lin, G.W. Huber, Energy Environ. Sci. **2**, 68–80 (2009)
33. M.J. Climent, A. Corma, S. Iborra, Green Chem. **13**, 520–540 (2011)
34. P.L. Dhepe, A. Fukuoka, ChemSusChem **1**, 969–975 (2008)
35. M. Besson, P. Gallezot, C. Pinel, Chem. Rev. **114**, 1827–1870 (2014)
36. Q.T. Trinh, B.K. Chethana, S.H. Mushrif, J. Phys. Chem. C **119**, 17137–17145 (2015)
37. P.N. Amaniampong, Q.T. Trinh, B. Wang, A. Borgna, Y. Yang, S.H. Mushrif, Angew. Chem. Int. Ed. **54**, 8928–8933 (2015)
38. A. Nandula, Q.T. Trinh, M. Saeys, A.N. Alexandrova, Angew. Chem. Int. Ed. **54**, 5312–5316 (2015)
39. C. Yang, Q.T. Trinh, X. Wang, Y. Tang, K. Wang, S. Huang, X. Chen, S.H. Mushrif, M. Wang, Chem. Commun. **51**, 3375–3378 (2015)
40. J. Mondal, Q.T. Trinh, A. Jana, W.K.H. Ng, P. Borah, H. Hirao, Y. Zhao, ACS Appl. Mater. Interfaces **8**, 15307–15319 (2016)
41. Q.T. Trinh, J. Yang, J.Y. Lee, M. Saeys, J. Catal. **291**, 26–35 (2012)
42. Q.T. Trinh, A.V. Nguyen, D.C. Huynh, T.H. Pham, S.H. Mushrif, Cat. Sci. Technol. **6**, 5871–5883 (2016)

43. Q.T. Trinh, A. Banerjee, Y. Yang, S.H. Mushrif, J. Phys. Chem. C **121**, 1099–1112 (2017)
44. Q.T. Trinh, K.F. Tan, A. Borgna, M. Saeys, J. Phys. Chem. C **117**, 1684–1691 (2013)
45. J.J. Varghese, Q.T. Trinh, S.H. Mushrif, Cat. Sci. Technol. **6**, 3984–3996 (2016)
46. P.N. Amaniampong, Q.T. Trinh, J.J. Varghese, R. Behling, S. Valange, S.H. Mushrif, F. Jérôme, Green Chem. **20**, 2730–2741 (2018)
47. R. Singuru, Q.T. Trinh, B. Banerjee, B. Govinda Rao, L. Bai, A. Bhaumik, B.M. Reddy, H. Hirao, J. Mondal, ACS Omega **1**, 1121–1138 (2016)
48. J.R. McManus, W. Yu, M. Salciccioli, D.G. Vlachos, J.G. Chen, J.M. Vohs, Surf. Sci. **606**, L91–L94 (2012)
49. J.R. McManus, M. Salciccioli, W. Yu, D.G. Vlachos, J.G. Chen, J.M. Vohs, J. Phys. Chem. C **116**, 18891–18898 (2012)
50. J.R. McManus, E. Martono, J.M. Vohs, Catal. Today **237**, 157–165 (2014)
51. J.R. McManus, E. Martono, J.M. Vohs, ACS Catal. **3**, 1739–1750 (2013)
52. P.N. Amaniampong, Q.T. Trinh, K. Li, S.H. Mushrif, Y. Hao, Y. Yang, Catal. Today **306**, 172–182 (2018)
53. M.T. Le, V.H. Do, D.D. Truong, E. Bruneel, I. Van Driessche, A. Riisager, R. Fehrmann, Q.T. Trinh, Ind. Eng. Chem. Res. **55**, 4846–4855 (2016)
54. J. Lu, M. Wang, X. Zhang, A. Heyden, F. Wang, ACS Catal. **6**, 5589–5598 (2016)
55. P. Srifa, M.V. Galkin, J.S.M. Samec, K. Hermansson, P. Broqvist, J. Phys. Chem. C **120**, 23469–23479 (2016)
56. C.-C. Chiu, A. Genest, A. Borgna, N. Rösch, ACS Catal. **4**, 4178–4188 (2014)
57. A. Banerjee, S.H. Mushrif, J. Phys. Chem. C (2018). https://doi.org/10.1021/acs.jpcc.8b01301
58. J. Zhang, H. Asakura, J. van Rijn, J. Yang, P. Duchesne, B. Zhang, X. Chen, P. Zhang, M. Saeys, N. Yan, Green Chem. **16**, 2432–2437 (2014)
59. D. Shi, L. Arroyo-Ramírez, J.M. Vohs, J. Catal. **340**, 219–226 (2016)
60. D. Shi, J.M. Vohs, ACS Catal. **5**, 2177–2183 (2015)
61. D. Shi, J.M. Vohs, Catal. Today **302**, 272–276 (2018)
62. J. Tao, C. Dai, Q. Dai, M. Zhao, J. Microbiol. **26**, 48–54 (2006)
63. J. Von Braun, R.K. Pachauri, *The Promises and Challenges of Biofuels for the Poor in Developing Countries: IFPRI 2005–2006 Annual Report Essay* (International Food Policy Research Institute, Washington, DC, 2006)
64. D.O. Hall., Biomass energy. Energy policy, **19**(8), 711–737 (1991)
65. International Energy Agency. World energy outlook. IEA/OECD, (1995)
66. M. Parikka, Biomass Bioenergy **27**, 613–620 (2004)
67. G. Antonopoulou, H.N. Gavala, I.V. Skiadas, K. Angelopoulos, G. Lyberatos, Bioresour. Technol. **99**, 110–119 (2008)
68. A. Al-Muyeed, A. Shadullah, Electrification through biogas. FORUM, Monthly publication of THE DAILY STAR. **3**, 1 (2010)
69. L.K. Paine, T.L. Peterson, D. Undersander, K.C. Rineer, G.A. Bartelt, S.A. Temple, D.W. Sample, R.M. Klemme, Biomass Bioenergy **10**, 231–242 (1996)
70. C. Panoutsou, Energy Policy **35**, 6046–6059 (2007)
71. R. Venendaal, U. Jørgensen, C. Foster, Biomass Bioenergy **13**, 147–185 (1997)
72. K. Ericsson, H. Rosenqvist, L.J. Nilsson, Biomass Bioenergy **33**, 1577–1586 (2009)
73. L.H. Reineke, *Research Note FPL No. 090* (USDA, Forest Service, Forest Products Laboratory, Madison, 1965), 14 pages, 90
74. A. Hossain, O. Badr, Renew. Sust. Energ. Rev. **11**, 1617–1649 (2007)
75. M.A. Oke, M.S.M. Annuar, K. Simarani, Bioenergy Res. **9**, 1189–1203 (2016)
76. J.R. Ziolkowska, Biotechnol. Rep. **4**, 94–98 (2014)
77. N. Dilekli, F. Duchin, J. Ind. Ecol. **20**, 120–131 (2016)
78. K. Sanderson, A field in ferment. Nature 444.7120: 673 (2006)
79. C.A. Cardona, Ó.J. Sánchez, Bioresour. Technol. **98**, 2415–2457 (2007)
80. P. Badger, Trends New Crops New Uses **1**, 17–21 (2002)
81. S. Ezeoha, C. Anyanwu, J. Nwakaire, Nigerian. J. Technol. **36**, 267–275 (2017)

Chapter 11
Production and Processing of the Enzymes from Lignocellulosic Biomass

C. S. Osorio-González, M. Chaali, K. Hegde, S. K. Brar, A. Kermanshahipour, and A. Avalos-Ramírez

11.1 Introduction

Lignocellulosic waste accounts approximately 50% of all biomass produced in the earth, and cellulose and hemicellulose represent approximately 35% and 25% of the agricultural total and 40% and 20% of the total mass of forest residues, respectively [66]. In general, lignocellulosic biomass is composed of three different fractions: cellulose (~30% to 50%), hemicellulose (~20% to 35%), and lignin (~15% to 25%) [64]. The main advantage of lignocellulosic biomass is that it can be hydrolyzed and thus achieve different compounds that serve as feedstock for obtaining different industrial products. In this sense, the enzymatic methods have multiplied interest because no inhibitory compounds or toxic effluents are formed during the hydrolysis process [34].

C. S. Osorio-González · M. Chaali · K. Hegde
INRS-ETE, Université du Québec, Quebec City, QC, Canada

S. K. Brar (✉)
INRS-ETE, Université du Québec, Quebec City, QC, Canada

Department of Civil Engineering, Lassonde School of Engineering, York University, North York, Toronto, ON, Canada
e-mail: satinder.brar@ete.inrs.ca; Satinder.Brar@lassonde.yorku.ca

A. Kermanshahipour
Biorefining and Remediation Laboratory, Department of Process Engineering and Applied Science, Dalhousie University, Halifax, NS, Canada

A. Avalos-Ramírez
INRS-ETE, Université du Québec, Quebec City, QC, Canada

Centre National en Électrochimie et en Technologies Environnementales, Shawinigan, QC, Canada

M. O. Daramola, A. O. Ayeni (eds.), *Valorization of Biomass to Value-Added Commodities*, Green Energy and Technology,
https://doi.org/10.1007/978-3-030-38032-8_11

Currently, the production of enzymes from lignocellulosic biomass, such as corn cobs, sugarcane bagasse and leaves, waste papers, forestry residues, and primary wastewater solid, is one of the main challenges research is facing due to several factors such as high incubation time, difficult control of growing conditions, difficult scaling, and low yields, among others [1]. Hydrolytic enzymes, such as cellulases and xylanases, convert lignocelluloses into sugars that can be fermented by various microbes to biofuels and other value-added products. The relatively high cost of these enzymes remains an important barrier for commercial application in any bio-industry, although in recent years a significant reduction in the cost of these enzymes has been achieved [67]. The areas of interest of the research have been to improve the efficiency of known enzymes, identify new and more active enzymes, find enzyme mixtures optimized for the pretreated lignocellulose, and reduce the cost of enzyme production. Furthermore, for an industrial process to be economically viable, the enzymatic decomposition of lignocellulose into fermentable sugars must occur as quickly as possible, preferably in hours [20]. On the other hand, the lignocellulose conversion requires a pretreatment step to degrade or loosen the recalcitrant and heterogeneous lignin fraction. This multifaceted challenge is being addressed by a growing body of ligninolytic enzymes isolated from various sources. Among these, it is known that ligninolytic enzymes play an important role in the degradation or modification of lignin [58, 76].

11.2 Lignocellulosic Biomass Features

Lignocellulosic biomass is the most abundant renewable organic material on earth. In general, the biomass is composed of a narrow association of three main components: (1) Cellulose is the major structural component of the plant cell wall (40–50%) that contributes to its rigidity and its mechanical strength. (2) Hemicellulose (20–30%) is the second most abundant heterogeneous polymer, which mainly consists of glucuronoxylan, glucomannan, and other polysaccharides. (3) Lignin is the smallest fraction (15–20%) and the most complex one, and it forms a protective seal around the two other components (cellulose and hemicelluloses) [2, 80, 86]. However, these three components individually have different characteristics, which are described in Table 11.1.

However, the amount of these components and their composition could vary through lignocellulosic materials, depending on its source and whether it is derived from [32].

Table 11.1 Main characteristics of the lignocellulosic biomass

Component	Features
Cellulose	Linear polymer linked by β-1,4-glycosidic bonds Composed mainly of glucose units Made up of two regions, a crystalline (2/3 of the total cellulose) and amorphous cellulose Insoluble in water
Hemicellulose	Linear or branched heterogeneous polymer Composed of five different sugar units (xylose, arabinose, mannose, glucose, and galactose) Homopolymers or heteropolymers of short chains linked by β-(1,4)-glycosidic or β-(1,3)-glycosidic linkages The classification is determined by the remains of sugars (xylans, mannans, or glucans) Hemicellulose features are determined by biomass source Structure completely amorphous
Lignin	It is an insoluble amorphous compound It is formed by cumaric alcohol, sinapyl alcohol, and coniferyl alcohol units Gives a very high resistance to the cell wall, provides support, isolation, and protection Difficult to degrade Lignin features are different between species, age, and stage of growth

11.3 Enzyme Production

11.3.1 Cellulases

This group of hydrolytic enzymes is the most prominent, because they promote the catalysis of the β-1,4 bonds present in cellulose, to obtain glucose monomers. These enzymes are produced by different microorganisms, mainly fungi, and bacteria. However, its production has been reported using protozoa, nematodes, and some mollusks [7]. This type of enzyme is made up of multienzyme groups contained three general groups: (i) endoglucanases or endo-β1-4-d-glucan 4-glucanohydrolase, (ii) exoglucanase or exo-β-1-4-d-glucan 4-cellobiohydrolase, and (iii) β-glucosidases. However, depending on the microorganism, some groups can also be found exo-β-1-4-d-glucan 4-glucohydrolases and exo-β-1-4-cellobiosidases [64]. Currently, the industrial production of this enzyme complex is through the use of mesophilic microorganisms and a wide variety of fungi in a temperature range of 30–35 °C [69, 71]. The model microorganisms used for the production of this type of enzyme are the fungi *Trichoderma reesei* and *Aspergillus niger* [72].

However, different investigations carried out recently have shown the use of other potentially promising microorganisms for the obtaining of this enzymatic complex. Examples of these microorganisms are from genera *Clostridium*, *Pseudoalteromonas*, *Thermoanaerobacter*, and *Thermotoga*. Table 11.2 summarizes some microorganisms commonly used for the production of cellulases. Some of the advantages that this type of microorganism offers to the production process

Table 11.2 Microorganisms used for the production of cellulases from lignocellulosic biomass

Microorganism	Substrate	Enzyme fraction	Activity	Reference
Ceratocystis paradoxa	Sugarcane bagasse	β-Glucosidase	1.068 IU/mL	[5]
Aspergillus niger	Apple pomace	FPase, β-glucosidase, endoglucanase, and xylanase	133.68, 6.0, 172.31, and 1412.58 IU/mL, respectively	[13]
Trichoderma reesei	Hemp fiber	Laccase, endoglucanase, exoglucanase, and β-glucosidase	50.4, 26, 5.5, and 15.2 IU/mL, respectively	[38]
Trichoderma reesei RUT C-30	Wheat bran and cellulose	β-Glucosidase	959.53 IU/mL	[29]
Aspergillus niger SCBM1	Biomass sorghum	Exoglucanase, β-glucosidase, and endoglucanase	30.64, 54.90, and 41.47 IU/mL, respectively	[14]
Penicillium janthinellum EMS-UV-8	Wheat bran/Avicel	Cellulase	3.1 IU/mL	[68]
Trichoderma reesei and *Aspergillus niger*	Corn stover	Cellulase, cellobiohydrolase, endoglucanase, and β-glucosidase	12.17, 53.10, 716.11, 1.93 IU/mL, respectively	[83]
Trichoderma koningii	Sugarcane bagasse	Cellulase	8.2 IU/mL substrate	[63]
Bacillus velezensis ASN1	Waste office paper	Cellulase	2.42 IU/mL	[52]

of hydrolytic enzymes is the temperature range in which they grow (5–20 °C). Since, from the process point of view, this reduces the consumption of energy and decreases the total cost of production [81]. In this sense, a wide variety of cellulases can be obtained through different microorganisms, which makes this enzyme complex an attractive method to increase cellulose catalysis.

11.3.2 Hemicellulases

To make the hydrolysis process more efficient by using cellulases, it is necessary to use "complementary" enzymes generally known as hemicellulases. This enzymatic cocktail is synthesized by different microorganisms, such as bacteria, algae, and fungi [33]. Nevertheless, its production on an industrial scale is done through the use of different filamentous fungi such as *Aspergillus niger*, *Ceratocystis paradoxa*, *Aspergillus flavus*, and *Thermoascus aurantiacus*, as well as the use of different types of lignocellulosic biomass (agro-waste mainly) as a renewable substrate for its production. Table 11.3 shows some of the most important microorganisms for the production of these enzymes.

Table 11.3 Microorganisms used for the production of xylanase from lignocellulosic biomass

Microorganism	Substrate	Activity	Reference
Aspergillus niger SCBM1	Biomass sorghum	30 IU/mL	[14]
Ceratocystis paradoxa	Wheat bran	12.728 IU/mL	[5]
Cochliobolus sativus	Wheat straw	146.9 IU/mL	[3]
Aspergillus flavus FPDN1	Pearl millet (bajara) bran	153.0 IU/mL	[87]
Thermoascus aurantiacus	Sugarcane straw	167.9 IU/mL	[47]
Streptomyces sp. P12	Wheat bran	27.77 IU/mL	[11]
Streptomyces sp. ER1	Beechwood xylan	10.220 U/mL	[61]
Bacillus pumilus SV-85S	Wheat bran	2995.20 IU/mL	[51]
Bacillus aerophilus KGJ2	Wheat bran	4.5 IU/mL	[19]

11.3.3 Lignolytic Enzymes

Production of this type of enzymes is through the use microorganisms that include soft rot fungi, filamentous fungi, bacteria, and brown rot fungi using lignocellulosic biomass as a substrate. The production of this enzyme complex is related by the interaction of various factors such as temperature, pH, and culture media composition among others [30]. The most important enzymatic fractions that are present in this complex are lignin peroxidase, laccase, and manganese peroxidase. However other fractions that are involved in the degradation of lignin are vanillate hydroxylase, veratryl alcohol oxidase, aryl alcohol dehydrogenase, or dioxygenase catalase, which act as mediating compounds to increase the catalytic activity of the three main fractions [84]. Table 11.4 shows some of the microorganisms commonly used to obtain lignolytic enzymes using different lignocellulosic materials as a substrate.

So far, enzyme and hyper-enzyme production has been increased by the direct evolution of the microorganisms used for this purpose or by the use of promoters such as *cbhl* obtained from *T. reesei* for cellulase production, inducers such as 1,2-β-xylobiose or 1,3-β-xylobiose of *T. harzianum* for xylanase production, and overexpression of genes such as *pslcc* of *Pycnoporus sanguineus* for increased laccase production. In the same way, another challenge in the specific production of this type of enzymes is the endogenous cellulolytic activity caused specifically when different inductors or promoters are used to increase production. However, a bottleneck in this process is still in production costs, which in turn does not allow scaling to industrial levels for marketing purposes. In this sense, some alternative and low-cost routes have been found to solve some of these obstacles. Nevertheless, to solve these problems, it is necessary to investigate in detail and optimize processes such as reactivation, immobilization, recovery, desorption, and reuse of enzymes, as well as reactors design with better heat and mass transfer during

Table 11.4 Microorganisms used for the production of ligninolytic enzymes from lignocellulosic biomass

Microorganism	Substrate	Enzyme	Activity	Reference
Lentinula edodes	Coffee husks	Laccase	22.2 U/mL	[65]
Trametes pubescens CBS 696.94	Basal media and coffee husk	Laccase	6.07 U/mL	[15]
Pseudomonas sp. S2	Potato peel	Laccase	108.97 U/mL	[9]
Lentinus edodes	A mix of malted barley waste and oak wood chips	Manganese peroxidase	1500 U/mL	[26]
Trametes versicolor IBL-04	Corn cobs	Lignin peroxidase	592 U/mL	[4]
Phanerochaete chrysosporium	Coffee pulp	Peroxidase	3.1 IU/ml	[56]
Schizophyllum commune	Corn stover	Lignin peroxidase, manganese peroxidase, and laccase	1007.39, 614.23, and 97.47 U/mL, respectively	[78]
Ganoderma lucidum	Pineapple leaf	Lignin peroxidase, manganese peroxidase, and laccase	2885.59, 889.71, and 472.31 U/mL, respectively	[24]

production and purification process. Likewise, the impact of this type of improvement on the loss of enzymatic and stability activity has to be considered and how this impacts the overall process cost.

11.4 Challenges in Enzyme Production from Biomass

Hydrolytic enzyme production using lignocellulosic biomass as carbon source has been studied in submerged and solid-state fermentation during the last 20 years. Although many of the factors involved in these two processes have been analyzed in detail, there are still so much to improve in efficiency terms of production and purification cost. Until today, a considerable number of comparative studies have been carried out between both methods. However, from the economic point of view, the solid-state process has shown two greater advantages with respect to submerged fermentation: (i) lower amount of energy during the production process and (ii) high enzyme stability. Nevertheless, one of the major problems with solid-state fermentation process is heat transfer and the lack of substrate homogeneity. Additionally, there are other types of parameters that have been shown to have a great influence in both types of production and purification process, which are described in more detail below.

11.4.1 Environmental Factors

A major challenge in enzyme production from lignocellulosic biomass is the control and optimal conditions of environmental factors that greatly affect microbial growth and therefore production process. Some of these factors are pH, temperature, oxygen level, and water activity [81]. Nevertheless, these parameters are easy to control when submerged culture is used for enzyme production. The above is due to the homogeneity of the nutrient solution and the cell suspension in the liquid medium.

On the other hand, when the production is realized through fermentation in a solid substrate, the control turns out to be more complex. For example, low moisture content has a great effect on microbial growth, causing a longer adaptation phase. This produces a decrease in the specific growth rate and consequently decreases the amount of microbial biomass produced and the products that are obtained from it [88]. A difference in the content of this factor occurs between bacteria and fungi, where the former needs a larger amount of water (>0.9) in order to grow. If this value is compared with that of some fungi species, the amount they use is lower (≤0.6) [6]. In this sense, it has been reported that optimum moisture content for the production of enzymes is between 40% and 70%. However, this value is determined by two factors: (i) the type of microorganism that is used for the production of enzymes and (ii) the type and specific characteristics of the lignocellulosic biomass that is used as a substrate [23]. Another parameter with great importance in the use of lignocellulosic biomass as a substrate for obtaining enzymes is temperature. The importance of this parameter is reflected in the heat transfer within the substrate, which increases for three reasons: (i) high microbial activity, (ii) low moisture content, and (iii) the thermal characteristics of the lignocellulosic biomass [59]. However, one of the critical factors in this process is aeration or oxygen uptake. This factor affects and contributes to the regulation of the two previous factors (moisture and temperature). The above is usually realized through four "stages" or "processes" that are controlled by this factor: (i) the humidity level is regulated, (ii) the temperature of the substrate is controlled, (iii) anaerobic environment is maintained, and (iv) the carbon dioxide of the reactor is desorbed [89]. On the other hand, a low or no control of this parameter can generate problems such as hydric or thermal stress that caused drastic changes in the metabolic processes of microorganisms and their products. Finally, pH is a factor that also influences the metabolic reactions of microorganisms used for enzyme production. The above is because of secretion and assimilation of some organic acids such as citric, lactic, or acetic acid by the microorganisms and/or buffering effect that lignocellulosic material exerts during the microbial growth and enzyme production. In this sense, good control of the pH values during the process is recommended since this factor depends on minimizing or preventing bacterial contamination [40].

11.4.2 Metal Ions as Inducers and Inhibitors

It has been reported the use of different metal ions that act as potentiators or inhibitors of the production of hydrolytic enzymes from lignocellulosic biomass. In general, the most evaluated metal ions as inducing potential or inhibitors of the enzymatic activity of the different enzymatic fractions (cellulases and hemicellulases) are Co^{2+}, Mg^{2+}, Fe^{2+}, Fe^{3+}, Mn^{2+}, Ni^{2+}, Cu^{2+}, Hg^{2+}, Zn^{2+}, Ca^{2+}, Na^{+}, and K^{+} [90]. This is mainly due to the association of metal ions with the protein and the formation of different complexes that act as donors or acceptors electrons that are linked to enzymatic complexes. Generally, these ions have an effect by interacting with the carboxyl and amino groups of the amino acids that make up the structure of the enzymes [57].

Vasconcellos, Tardioli, Giordano, and Farinas [73] evaluated the effect of Ca^{2+}, Co^{2+}, Cu^{2+}, Mg^{2+}, Ni^{2+}, Zn^{2+}, and Mn^{2+}, as well as the effect of the chelating compound ethylenediaminetetraacetic acid (EDTA) and mercaptoethanol as inductors in a range of 2 and 10 mM to increase the activity and stability of hemicellulolytic enzymes produced by *Aspergillus niger*. The best result obtained was the addition of Mg^{2+} with an increase of up to 57% in the enzymatic activity of endoglucanase. On the other hand, the addition during saccharification of sugarcane bagasse with 10 mM of the same ion increased the glucose release by 34%. However, they found an increase in the enzymatic activity of endoglucanase when Cu^{2+} and Fe^{2+} were used at 2 mM concentrations; however, when they used concentrations of 10 mM, the activity was reduced by 60%. Furthermore, the addition of EDTA reduced the enzymatic activity by 12%. Finally, the addition of mercaptoethanol in both concentrations increased the enzymatic activity by 30%, since it promoted changes in the different oxidation states of the thiol groups in the protein.

11.4.3 Enzyme Regulations

Currently, three different metabolic systems have been studied, including regulation (expression at the basal level), secretion (mass secretion of enzymes), and repression (glucose or catabolic repression) of enzymes in filamentous fungi. In different studies of basal expression, it has been observed that activating or repressing proteins do not influence or affect the process of transcription of enzymes [70]. Nevertheless, without an activating protein that codes for the gene that produces the enzyme, transcription is performed partially. Due to the above reason, a small amount of enzyme is secreted at the basal level, which hydrolyzes a small amount of cellulose present in the mycelia that helps fungi to recognize and identify the presence of cellulose [28]. On the other hand, the process of enzyme segregation is influenced principally by the inducers that enter the cell, which activates the transcription of the gene that produces the enzyme. However, this process is regulated by proteins and activating elements that are different and specific for each one of the different fungi species [36]. Finally, another phenomenon observed in enzyme production is the catabolic repression of carbon, where different genes are repressed at the transcriptional level by a carbon source (commonly glucose). This processor mechanism is of great importance during the metabolic process in fungi that produce enzymes for the degradation of biomass, since through this process the assimilation of the carbon source is controlled (Mechanisms, 2001).

Long et al. [43] demonstrated that the creI/creA transcription factors of type Cys2His2 have a mediating effect on the catabolic repression during the production of cellulase enzymes using *Trichoderma orientalis* EU7-22. The above is through the integration of the gene *creA* in the union of the region that encodes the genes (promoter region) with the consensus motive 5′-SYGGRG-3′. After gene introduction, the enzymatic activity of different fractions increased with respect to the activity obtained by the native strain in 1.45-fold (filter paper units), 1.15-fold (endoglucanase), 1.71-fold (cellobiohydrolase), 2.51-fold (β-glucosidase), and 2.72-fold (xylanase) using an inducer medium. On the other hand, the increase was higher when a glucose repressor media was used, with an increase of 6.41-, 7.50-, 10.27-, 11.79-, and 9.25-fold for each of the enzymatic fractions mentioned above. Finally, they found that the expression of the main genes involved in cellulase production such as *cbh1*, *cbh2*, *eg1*, *eg2*, *bgl1*, *xyn1*, and *xyn2* increased in mutant strain in both culture media (inducer and repressor).

11.4.4 Genetic Engineering

Nowadays, industrial production of hydrolytic enzymes is carried out using filamentous fungi, since compared to other microorganisms, they show high levels of production and a more complete enzymatic cocktail. Nevertheless, due to the growing demand for these enzymes, companies have directed their processes to in situ production and genetic improvement of different strains in order to increase yields and decrease the costs of the production process [42]. To be successful in the previous purpose, the genetic improvement and modification of strains are mainly directed to secretion, degradation, folding, and glycosylation process, as well as to the regulation of the genes involved in the production [85].

Guo et al. [22] in conjunction with Dongguan APAC Biotechnology Co., Ltd., developed a patent on the genetic modification and structural analysis of a cellulase enzyme, specifically in the addition of disulfide bonds in primary structure, in order to increase its thermostable capacity, characteristic that increases its economic and industrial value. Cellulase in this invention was obtained from a gene isolated from *T. reesei*. To improve enzyme expression, the gene was optimized by deletion of 91 amino acids at one end of the protein. Cellulase obtained was called wild type (the gene was not mutated) with 984 base pairs and a coding system for 327 amino acids. However, after the structural analysis, the spatial distance between the N and C terminal sites of the enzyme was greater (113 Å) than the distance needed to form a disulfide bond. Because of this, the invention includes the addition of three amino acids (glycine, cysteine, and proline). The cysteine is added at the N and C sites of the protein in order to decrease the terminal distance between the two, and the proline or glycine is used as agents that promote the angular deflection of the main chain at the 5′ end, which gives step to form a disulfide bond. Adding process of three amino acids was carried out by directed mutagenesis, the modified DNA material was transformed by electroporation in *Pichia pastoris* X33, and induction and production were carried out in buffered complex media for expression of the secreted protein (BMMY). Results obtained were a 60% increase in the relative residual activity after treatment at 75 °C/2 min, which shows the improvement in the thermotolerant capacity with respect to the control enzyme (wild).

Nowadays, it has been demonstrated that both fermentation processes have a wide number of advantages and benefits for hydrolytic enzyme production from lignocellulosic biomass. However, due to the different variants in terms of fermentation methods, parameters, microorganisms, and substrates used, research work still is done, with the main target to improve production, purification, and quality of obtained enzymes.

11.5 Performance of Hydrolytic Enzymes on Lignocellulosic Biomass

11.5.1 Cellulase

To carry out the hydrolysis of cellulose to glucose monomers, each of the components of this enzyme acts in a unique and specific place on the structure of cellulose. In Fig. 11.1, the general scheme of the catalysis sites of the specific enzymatic

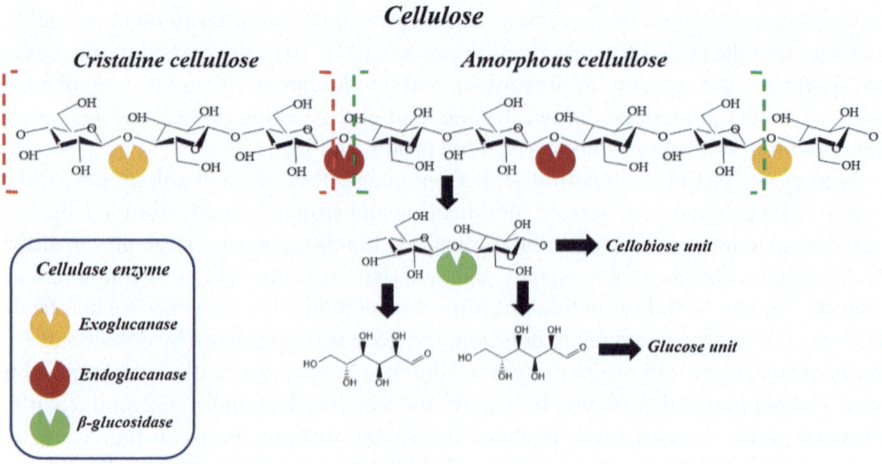

Fig. 11.1 Catalytic sites of different enzymatic fractions in cellulose to obtain glucose

fractions for cellulose hydrolysis is shown. In this way, the endoglucanase fraction, also called carboxymethylcelluloses, catalyzes the internal bonds of the cellulose chains (β-1,4), giving to new reducing and nonreducing residues. Subsequently, exoglucanase or cellobiohydrolase acts on the β-1,4 links of the new nonreducing residues of disaccharides, resulting in units of cellodextrins and cellobioses, which are catalyzed by β-glycosidase, to finally obtain glucose monomer units [64].

11.5.2 Hemicellulase

The most important enzymatic fraction to carry out the degradation of hemicellulose is 1,4-β-D-xylan xylohydrolase, commonly called endoxylanases. This fraction attacks the structure of xylan, thus reducing the degree of polymerization of the hemicellulose [33]. Once this is done, the β-xylosidase catalyzes the hydrolysis of xylobiose and small fractions of xylooligosaccharides; this process gives way to obtaining of xylose, which allows easier hydrolysis of the xylans. However, other enzymatic fractions are also necessary, such as acetyl xylan esterase; this fraction is responsible for hydrolyzing xylans that are partially bound to acetyl groups within the main chain or in some branch. As with the use of cellulases, partial hydrolysis of the hemicellulose generates chains of smaller sugars, which have new substituents, such as arabinoxylans, which when hydrolyzed by another enzymatic fraction called α-1-arabinofuranosidase results in an increase of points potentially accessible for xylanase fraction and the xylan structure [41]. On the other hand, there are lateral residues of the chains that form ester-type bonds with the residues of lignin and uronic acid, in order to obtain 4-O-methyl gluronic acid, product of the degradation of glucuronoxylan with glucuronidases [18, 31, 46, 48, 75]. Figure 11.2 shows a diagram of sites on which each of the enzymatic fractions catalyzes. Together these enzymes act together for the hydrolysis of hemicellulose in its different monomers, mainly sugars of five carbons (pentoses).

11.5.3 Lignolytic Enzymes

This type of enzymes has great importance due to their ability to oxidize non-phenolic and phenolic compounds, present in the lignocellulosic biomass (lignin). Due to the complex nature of lignocellulosic biomass components, it is necessary to use different oxidative enzymes to carry out its partial or complete degradation [53]. In this sense, the three main enzymes to carry out this process are lignin peroxidase, laccase, and manganese peroxidase [21]. In general, lignin peroxidase depolymerizes lignin by the use of free radicals that divide the lateral ends of the lignin structure and thus carry out the oxidation of phenolic and non-phenolic compounds [4]. On the other hand, the oxidative-reductive capacity of the manganese peroxidase enzyme is lower when compared to that of lignin peroxidase, since this is not the

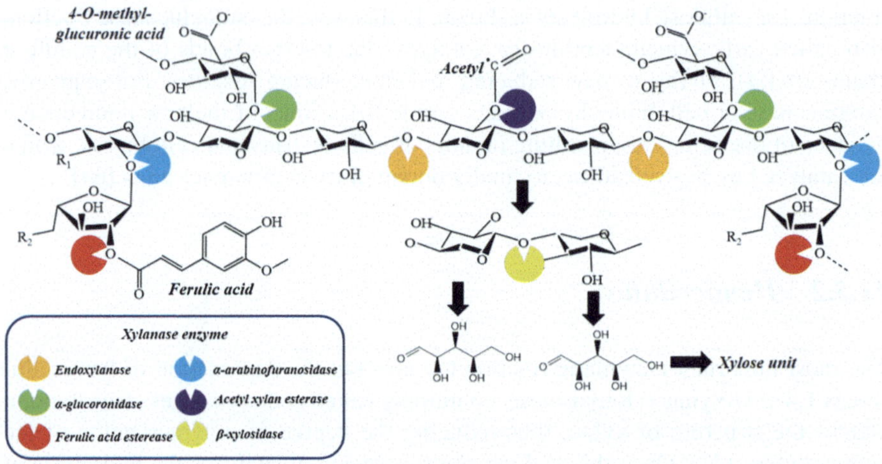

Fig. 11.2 Catalytic sites of different enzymatic fractions in hemicellulose to obtain xylose

non-phenolic compound present in lignin. However, this enzyme fraction shows a strong preference for manganese groups as a reducing substrate and works as a mediator in the degradation by redox synthesis, which allows it to degrade organic compounds through the hydrogen ion [82]. Finally, laccase fraction catalyzes the oxidation of a reducing electron, which leads to a reduction of electrons from molecular oxygen to water. This fraction oxidizes phenolic compounds in the presence of various mediators, which allows them to degrade various compounds that have a greater reductive oxide capacity, even higher to the same enzyme. The most important difference between this and the two previous ones is that laccase does not need hydrogen peroxide to carry out the oxidation of compounds [10].

11.6 Enzymes as Tools for Biomass Treatment and Value-Added Products

Enzymatic hydrolysis is an effective and economical method to achieve fermentable sugars under mild and eco-friendly reaction conditions from the lignocellulosic biomass [77]. The use of hydrolytic enzymes has been of great importance in the transition and formation of new strategies in the development of new by-products and contributing to the improvement of processes from lignocellulosic biomass in three different ways: (i) increasing yields, (ii) making them friendly with the environment, and (iii) decreasing the total cost of the production process [23]. Currently, the production and use of hydrolytic enzymes such as cellulase, xylanase, and ligninases for the pretreatment of lignocellulosic biomass has increased for several decades [45]. Although one of the main uses of these enzymes is to obtain hydrolysates from different lignocellulosic biomass, its use has been diversified and

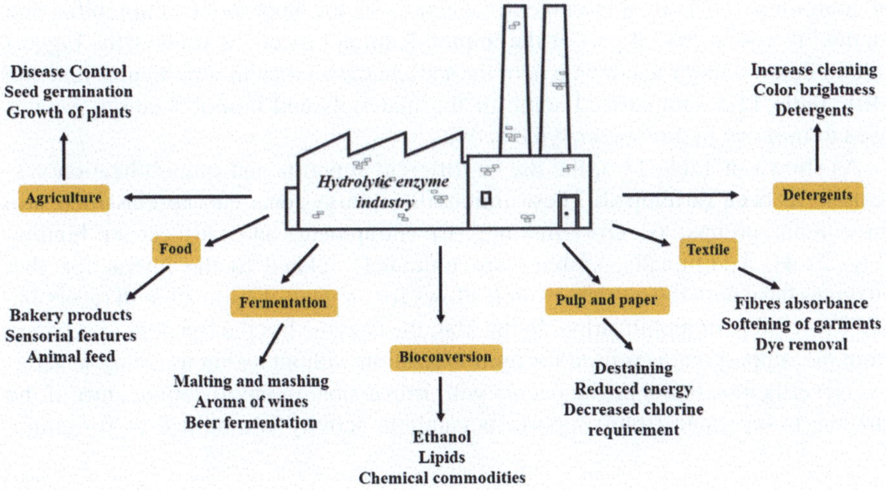

Fig. 11.3 Applications of hydrolytic enzymes in different industries

extended to other types of industries and processes [17, 37]. Figure 11.3 shows the presence of these enzymes in different industries and some of the processes in which they are directly involved in each industry.

Lignocellulosic biomass is a potential source of several bio-based products. Currently, the products made from bioresources represent only a minor fraction of the chemical industry production. However, the interest in the bio-based products has increased because of the rapidly rising barrel costs and an increasing concern about the depletion of the fossil resources in the near future [25].

11.7 Enzymatic Immobilization as a Tool for Improved Performance

A result of the increase in the use of hydrolytic enzymes in different processes, research has increased in technologies that help to improve the yields of the processes in which this type of enzyme is used. One of these alternatives is immobilization systems, which allow to preserve the stability of the enzymes during the process, as well as to increase its reusability in the same [91]. In this sense, it could be defined as an immobilized enzyme as "enzyme that has been physically or chemically attached to physical support without losing its catalytic capacity."

Enzymatic immobilization is a usual prerequisite as a solution to get reusable biocatalysts and thus decrease the price of this relatively expensive compound. However, a proper immobilization technique may permit far more than to get a reusable enzyme. It may be used to improve enzyme performance by improving some enzyme limitations, such as enzyme purity, stability, activity, specificity, selectivity, or inhibitions [16]. In this context, selection of the appropriate immobilization method is a very crucial part of the immobilization process as it plays the biggest role in determining the enzyme activity and characteristics in a particular reaction [46]. Table 11.5 summarized some of the materials and immobilization systems used to improve hydrolytic enzyme activity.

As shown in Table 11.5, the use of different supports and immobilization systems have been developed. These immobilization systems can be classified into three main groups: (i) cross-linking, (ii) entrapment, and (iii) carrier binding (Fig. 11.4). Additionally, systems are intimately related to the interaction that enzymes have with the carriers, which allows for an inter-classification in reversible and irreversible immobilization. In the first, the enzyme has the freedom to separate from the support that was used for immobilization without losing its catalytic activity. Nevertheless, the opposite occurs with irreversible immobilization, since if the enzyme is separated from support, its catalytic activity is modified or forfeiture.

Table 11.5 Immobilization methods used in hydrolytic enzymes for the distinct process

Enzyme	Support	Immobilization system	Product	Reference
Cellulase	Carrageenan gel coated with hyperbranched polyamidoamine and glutaraldehyde	Carrier binding (covalent)	Isolation of cellulose nanofibers (CNF) from bagasse pulp	[79]
	Alginate beads	Entrapment (encapsulation)	Enzymatic digestion of sugarcane bagasse for bioethanol production	[62]
	Amino-functionalized magnetic nanoparticles (MNPs) and glutaraldehyde	Cross-linking	Antioxidant extraction from fruit wastes	[50]
Xylanase	Cross-linked enzyme aggregates	Cross-linking (carrier-free immobilization)	Nutraceutical compounds from sugar cane bagasse	[27]
	Magnetic nanoparticles (MNPs) and glutaraldehyde	Cross-linking	Piperine extraction, sugarcane cell protoplasts, and clarification of papaya juice	[49]
	Glyoxyl-agarose beads attached to dextran polymers and polyethylenimine	Entrapment (encapsulation) and carrier binding (covalent)	Nutraceutical compounds from beechwood, wheat straw, and corncob	[60]
Ligninase	Ca-alginate beads and glutaraldehyde	Cross-linking	Degradation of bisphenol A (BPA)	[39]
	Carbon nanotubes	Carrier binding (adsorption)	Wastewater treatment (dye remotion)	[54]
	Chitosan beads grafted by the genipin	Cross-linking	Degradation of synthetic dyes	[44]

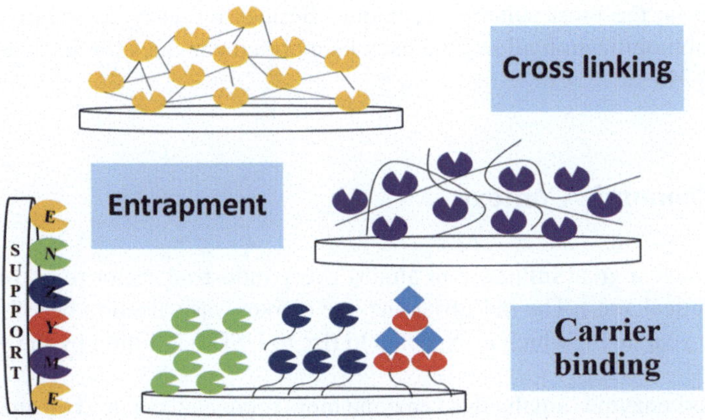

Fig. 11.4 Main systems used for hydrolytic enzymes immobilization

It should be noted that each of the immobilization systems has its own features, advantages, and disadvantages. In this sense, the main task in research is focused on parameters optimization to use and reuse the enzymes without losing their catalytic activity [17].

An example of immobilization is the research carried out by Kirupa-Sankar et al. [35], which evaluated a co-immobilization system using cellulase, laccase, and β-glucosidase to carry out the delignification and saccharification of four different lignocellulosic biomass (*Typha angustifolia*, *Arundo donax*, *Saccharum arundinaceum*, and *Ipomoea carnea*) to obtain hydrolysates that were used to obtain bioethanol. Immobilization was realized on sodium alginate beads (2.5%), with a ratio of 3:1:1:1 (Na-alginate/enzymes). With the use of this immobilization system, they obtained an entrapment efficiency of 84% and enzymatic activities for cellulase of 22.95, laccase of 41.78, and β-glucosidase of 6.79 μmol substrate/min/mg enzyme. However, these values were lower than those obtained in the enzymes in free form (27, 45, and 8 μmol substrate/min/mg enzyme, respectively). One of the factors that can explain a decrease of enzymatic activity as well as substrate hydrolysis is the resistance that Na-alginate spheres present to the substrate and the capacity of obtained product (sugars) to move inside and outside of immobilization matrix. On the other hand, the enzymatic activity decreases by 80% after four cycles and finishes with an activity of 60% after six cycles. In this sense, the enzyme that showed the highest residual enzymatic activity after six cycles was β-glucosidase. Likewise, the immobilization system promoted an improvement in enzymes tolerance at lower process temperatures (30 °C), which from the economic and process point of view is an advantage. Finally, through this pretreatment, the highest sugar content was obtained with *Saccharum arundinaceum* (205 mg/g), and the best bioethanol yield was obtained with *Ipomoea carnea* (63.43%). In literature, only a few papers are available on the cellulose immobilization. This is due to the fact that cellulose is not soluble and some immobilization techniques, such as enzymes entrapment, impede the interaction enzyme-substrate. Immobilization of cellulase via covalent bonds appears to be the most suitable technique. Besides the enzyme stabilization, the covalent immobilization allows the use of supported enzymes for several cycles of reactions [74].

11.8 Economic Context

Enzymes have a great influence in almost every industrial sector (e.g., food, feed, pharmaceutical, etc.). The industrial enzyme market was valued to be USD 4.2 billion in the year 2014, which is expected to rise to USD 6.2 billion by the year 2020 [55, 80].

Cellulase enzymes are the third enzyme more produced and used on an industrial scale in the world. The above is mainly due to its use in industries such as cotton, paper, or food (for human and animals). Nevertheless, this type of enzymes can become the number one worldwide if bioethanol production from lignocellulosic biomass becomes an economically feasible process. Until 2015, the global enzyme market was estimated at $8000 million with a potential annual increase of 7% [12]. Owing to the continuous pressure from the environmental agencies to minimize the pollution, the utilizing of the different wastes such as lignocellulosic residual biomass has been involved for enzyme production [8, 55].

Today there are different companies around the world that lead the production of these enzymes. However, DuPont and Novozymes are the most important when it comes to production, mainly because through research and innovation in the production and technology they use, they have made the price of this type of enzyme more affordable. Both companies have in their portfolio different enzymatic cocktails for industrial processes, whether general or specific [69]. However, Novozymes is the company with the largest presence around the world with 47% of the total production of enzymes and 902 patents development. Figure 11.5 shows some of the brands that these two companies have available to the public. In 2010, new enzymes were produced by two leading companies, Novozymes and Genencor, supported by the USA Department of Energy (DOE). Genencor has launched four new blends: Accelerase®1500, Accelerase®XP, Accelerase®XC, and Accelerase®BG. Accelerase®1500 is a cellulase complex (exoglucanase, endoglucanase, hemicellulase, and β-glucosidase) produced from a genetically modified strain of *T. reesei* [92].

11.9 Conclusions and Future Perspectives

Hydrolytic enzyme production such as cellulases, xylanases, and ligninases using different types of lignocellulosic biomass as a substrate has become an important industry that continues in development. Likewise, this industry has strengthened research to obtain new strains or genetically modified with the purpose to promote

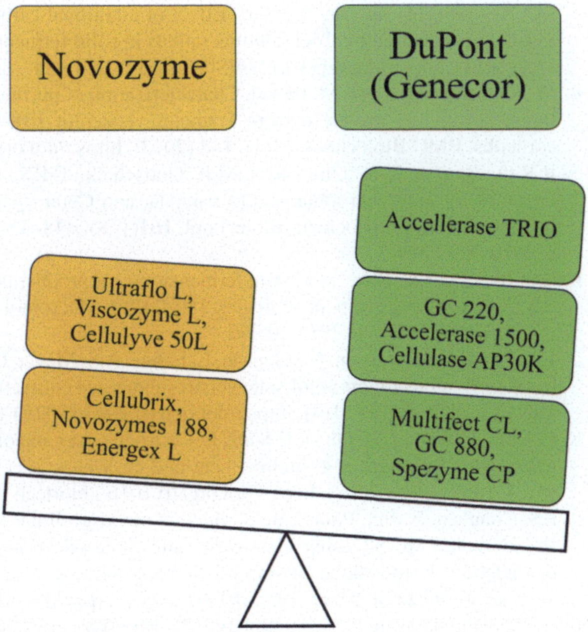

Fig. 11.5 Commercial cellulases cocktail produced by *Trichoderma longibrachiatum* and *Trichoderma reesei*

an increase in production, enzymatic activity, and tolerance toward the different factors that affect them. The use of hydrolytic enzymes for the treatment of lignocellulosic biomass or for another type of process in which they are used is a promising and environmentally friendly alternative. However, within the production, there are different areas of opportunity that have to be taken into account to improve yields, the design of strains focused on obtaining more "complete," thermostable, and pure enzymes. Although best yields in enzyme production are with different strains, it is necessary to make a better characterization and identification of these native enzymes obtained from the new species for better use at the industrial level.

Acknowledgments Financial support from the Natural Sciences and Engineering Research Council of Canada (No. 284111, Discovery; No. 476649-14, Strategic Research Grant) and Institute National de la Recherche Scientifique-Eau Terre Environnement have been acknowledged.

References

1. A. Althuri, A.D. Chintagunta, K.C. Sherpa, R. Banerjee, Simultaneous saccharification and fermentation of lignocellulosic biomass, in *Biorefining of Biomass to Biofuels*, ed. by S. Kumar, R. K. Sani, (Springer, Cham, 2018), pp. 265–285. https://doi.org/10.1007/978-3-319-67678-4_12
2. Z. Anwar, M. Gulfraz, M. Irshad, Agro-industrial lignocellulosic biomass a key to unlock the future bio-energy: A brief review. J. Radiat. Res. Appl. Sci. **7**(2), 163–173 (2014). https://doi.org/10.1016/j.jrras.2014.02.003

3. M.I.E. Arabi, M. Jawhar, Y. Bakri, Effect of additional carbon source and moisture level on xylanase production by Cochliobolus sativus in solid fermentation. Microbiology **80**(2), 150–153 (2011). https://doi.org/10.1134/S0026261711010024
4. M. Asgher, H.M. Iqbal, M. Irshad, Characterization of purified and xerogel immobilized novel lignin peroxidase produced from Trametes versicolor IBL-04 using solid state medium of corncobs. BMC Biotechnol. **12**(1), 1–8 (2012). https://doi.org/10.1186/1472-6750-12-46
5. R.R.O. Barros, R.A. Oliveira, L.M.F. Gottschalk, E.P.S. Bon, Production of cellulolytic enzymes by fungi Acrophialophora nainiana and Ceratocystis paradoxa using different carbon sources. Appl. Biochem. Biotechnol. **161**(1–8), 448–454 (2010). https://doi.org/10.1007/s12010-009-8894-3
6. S.S. Behera, R.C. Ray, Solid state fermentation for production of microbial cellulases: Recent advances and improvement strategies. Int. J. Biol. Macromol. **86**, 656–669 (2016). https://doi.org/10.1016/j.ijbiomac.2015.10.090
7. J.A. Bentil, A. Thygesen, M. Mensah, L. Lange, A.S. Meyer, Cellulase production by white-rot basidiomycetous fungi: Solid-state versus submerged cultivation. Appl. Microbiol. Biotechnol. **102**(14), 5827–5839 (2018). https://doi.org/10.1007/s00253-018-9072-8
8. C. Cara, E. Ruiz, J.M. Oliva, F. Sáez, E. Castro, Conversion of olive tree biomass into fermentable sugars by dilute acid pretreatment and enzymatic saccharification. Bioresour. Technol. **99**(6), 1869–1876 (2008). https://doi.org/10.1016/j.biortech.2007.03.037
9. P.S. Chauhan, B. Jha, Pilot scale production of extracellular thermo-alkali stable laccase from Pseudomonas sp. S2 using agro waste and its application in organophosphorus pesticides degradation: Extracellular thermo-alkali stable laccase from Pseudomonas sp. S2. J. Chem. Technol. Biotechnol. **93**(4), 1022–1030 (2018). https://doi.org/10.1002/jctb.5454
10. A. Chenthamarakshan, N. Parambayil, N. Miziriya, et al., Optimization of laccase production from Marasmiellus palmivorus LA1 by Taguchi method of Design of experiments. BMC Biotechnol. **17**(1), 1–10 (2017). https://doi.org/10.1186/s12896-017-0333-x
11. G. Coman, G. Bahrim, Optimization of xylanase production by Streptomyces sp. P12-137 using response surface methodology and central composite design. Ann. Microbiol. **61**(4), 773–779 (2011). https://doi.org/10.1007/s13213-010-0195-0
12. L.P. de Souza Vandenberghe, J.C. de Carvalho, N. Libardi, C. Rodrigues, C.R. Soccol, Microbial enzyme factories, in *Agro-Industrial Wastes as Feedstock for Enzyme Production*, ed. by G. S. Dhillon, S. Kaur, (Academic Press, London, 2016), pp. 1–22. https://doi.org/10.1016/B978-0-12-802392-1.00001-0
13. G.S. Dhillon, S. Kaur, S.K. Brar, M. Verma, Potential of apple pomace as a solid substrate for fungal cellulase and hemicellulase bioproduction through solid-state fermentation. Ind. Crop. Prod. **38**, 6–13 (2012). https://doi.org/10.1016/j.indcrop.2011.12.036
14. L.M. Dias, B.V. dos Santos, C.J.B. Albuquerque, B.E.L. Baeta, D. Pasquini, M.A. Baffi, Biomass sorghum as a novel substrate in solid-state fermentation for the production of hemicellulases and cellulases by Aspergillus niger and A. fumigatus. J. Appl. Microbiol. **124**(3), 708–718 (2018). https://doi.org/10.1111/jam.13672
15. I.J. Gaitan, S.C. Medina, J.C. González, A. Rodríguez, Á.J. Espejo, J.F. Osma, et al., Evaluation of toxicity and degradation of a chlorophenol mixture by the laccase produced by Trametes pubescens. Bioresour. Technol. **102**(3), 3632–3635 (2011). https://doi.org/10.1016/j.biortech.2010.11.040
16. C. Garcia-Galan, Á. Berenguer-Murcia, R. Fernandez-Lafuente, R.C. Rodrigues, Potential of different enzyme immobilization strategies to improve enzyme performance. Adv. Synth. Catal. **353**(16), 2885–2904 (2011). https://doi.org/10.1002/adsc.201100534
17. I. Garcia-Galindo, R. Gómez-García, S. Palácios-Ponce, J. Ventura, D. Boone, et al., New features and properties of microbial cellulases required for bioconversion of agro-industrial wastes, in *Enzymes in Food Biotechnology*, ed. by M. Kuddus, (Academic Press, London, 2019), pp. 535–550. https://doi.org/10.1016/B978-0-12-813280-7.00031-1
18. F.M. Gírio, C. Fonseca, F. Carvalheiro, L.C. Duarte, S. Marques, R. Bogel-Łukasik, Hemicelluloses for fuel ethanol: A review. Bioresour. Technol. **101**(13), 4775–4800 (2010). https://doi.org/10.1016/j.biortech.2010.01.088

19. D. Gowdhaman, V.S. Manaswini, V. Jayanthi, M. Dhanasri, G. Jeyalakshmi, et al., Xylanase production from Bacillus aerophilus KGJ2 and its application in xylooligosaccharides preparation. Int. J. Biol. Macromol. **64**, 90–98 (2014). https://doi.org/10.1016/j.ijbiomac.2013.11.012
20. G. Guerriero, J.F. Hausman, J. Strauss, H. Ertan, K.S. Siddiqui, Lignocellulosic biomass: Biosynthesis, degradation, and industrial utilization. Eng. Life Sci. **16**(1), 1–16 (2016). https://doi.org/10.1002/elsc.201400196
21. L.R.C. Guimarães, A.L. Woiciechowski, S.G. Karp, J.D. Coral, A. Zandoná-Filho, C.R. Soccol, Laccases, in *Production, Isolation and Purification of Industrial Products*, ed. by A. Pandey, S. Negi, C. R. Soccol, (Elsevier, Amsterdam, 2017), pp. 199–216. https://doi.org/10.1016/B978-0-444-63662-1.00009-9
22. R.T. Guo, Y.S. Cheng, J.W. Huang, T.Z. Wu, H.L. Lai, C.H. Lin, T.Y. Ko, Cellulase Having Improved Thermostability. Patent No.: US 9,944,915 B2, 2018
23. V.K. Gupta, C.P. Kubicek, J.G. Berrin, D.W. Wilson, M. Couturier, et al., Fungal enzymes for bio-products from sustainable and waste biomass. Trends Biochem. Sci. **41**(7), 633–645 (2016). https://doi.org/10.1016/j.tibs.2016.04.006
24. S. Hariharan, P. Nambisan, Optimization of lignin peroxidase, manganese peroxidase, and lac production from Ganoderma lucidum under solid state fermentation of pineapple leaf. Bioresources **8**(1), 250–271 (2012). https://doi.org/10.15376/biores.8.1.250-271
25. R. Hatti-Kaul, U. Törnvall, L. Gustafsson, P. Börjesson, Industrial biotechnology for the production of bio-based chemicals – A cradle-to-grave perspective. Trends Biotechnol. **25**(3), 119–124 (2007). https://doi.org/10.1016/j.tibtech.2007.01.001
26. N. Hatvani, I. Mécs, Production of laccase and manganese peroxidase by Lentinus edodes on malt-containing by-product of the brewing process. Process Biochem. **37**(5), 491–496 (2001). https://doi.org/10.1016/S0032-9592(01)00236-9
27. J.S. Hero, C.M. Romero, J.H. Pisa, N.I. Perotti, C. Olivaro, M.A. Martinez, Designing cross-linked xylanase aggregates for bioconversion of agroindustrial waste biomass towards potential production of nutraceuticals. Int. J. Biol. Macromol. **111**, 229–236 (2018). https://doi.org/10.1016/j.ijbiomac.2017.12.166
28. L.B. Huberman, J. Liu, L. Qin, N.L. Glass, Regulation of the lignocellulolytic response in filamentous fungi. Fungal Biol. Rev. **30**(3), 101–111 (2016). https://doi.org/10.1016/j.fbr.2016.06.001
29. A.S.O. Idris, A. Pandey, S.S. Rao, R.K. Sukumaran, Cellulase production through solid-state tray fermentation, and its use for bioethanol from sorghum stover. Bioresour. Technol. **242**, 265–271 (2017). https://doi.org/10.1016/j.biortech.2017.03.092
30. P.T.L. Ike, A.C. Moreira, F.G. de Almeida, D. Ferreira, W.G. Birolli, A.L.M. Porto, D.H.F. Souza, Functional characterization of a yellow laccase from Leucoagaricus gongylophorus. Springerplus **4**(1), 1–11 (2015). https://doi.org/10.1186/s40064-015-1464-y
31. H. Inoue, C. Kitao, S. Yano, S. Sawayama, Production of B-xylosidase from Trichoderma asperellum KIF125 and its applications in efficient hydrolysis of pretreated rice straw with fungal cellulase. World J. Microbiol. Biotechnol. **32**(186), 1–11 (2016)
32. H.M.N. Iqbal, G. Kyazze, T. Keshavarz, Advances in the valorization of lignocellulosic materials by biotechnology: An overview. Bioresources **8**(2), 3157–3176 (2013). https://doi.org/10.15376/biores.8.2.3157-3176
33. V. Juturu, J.C. Wu, Microbial xylanases: Engineering, production and industrial applications. Biotechnol. Adv. **30**(6), 1219–1227 (2012). https://doi.org/10.1016/j.biotechadv.2011.11.006
34. M. Kapoor, S. Semwal, R. Gaur, R. Kumar, R.P. Gupta, S.K. Puri, The pretreatment technologies for deconstruction of lignocellulosic biomass, in *Waste to Wealth*, ed. by R. R. Singhania, R. A. Agarwal, R. P. Kumar, R. K. Sukumaran, (Springer, Singapore, 2018), pp. 395–421. https://doi.org/10.1007/978-981-10-7431-8_17
35. M. Kirupa-Sankar, R. Ravikumar, M. Naresh-Kumar, U. Sivakumar, Development of co-immobilized tri-enzyme biocatalytic system for one-pot pretreatment of four different perennial lignocellulosic biomass and evaluation of their bioethanol production potential. Bioresour. Technol. **269**, 227–236 (2018). https://doi.org/10.1016/j.biortech.2018.08.091

36. A. Kobakhidze, M. Asatiani, E. Kachlishvili, V. Elisashvili, Induction and catabolite repression of cellulase and xylanase synthesis in the selected white-rot basidiomycetes. Ann. Agrar. Sci. **14**(3), 169–176 (2016). https://doi.org/10.1016/j.aasci.2016.07.001
37. A. Kunamneni, Cellulase in biomedical research, in *Microbial Cellulase System Properties and Applications*, ed. by V. K. Gupta, (Elsevier, Amsterdam, 2016), pp. 267–275
38. M.A. Laadila, K. Hegde, T. Rouissi, S.K. Brar, R. Galvez, et al., Green synthesis of novel biocomposites from treated cellulosic fibers and recycled bio-plastic polylactic acid. J. Clean. Prod. **164**, 575–586 (2017). https://doi.org/10.1016/j.jclepro.2017.06.235
39. F. Lassouane, H. Aït-Amar, S. Amrani, S. Rodriguez-Couto, A promising laccase immobilization approach for Bisphenol A removal from aqueous solutions. Bioresour. Technol. **271**, 360–367 (2019). https://doi.org/10.1016/j.biortech.2018.09.129
40. C. Li, Z. Yang, R. He Can Zhang, D. Zhang, S. Chen, L. Ma, Effect of pH on cellulase production and morphology of Trichoderma reesei and the application in cellulosic material hydrolysis. J. Biotechnol. **168**(4), 470–477 (2013). https://doi.org/10.1016/j.jbiotec.2013.10.003
41. H. Li, X. Chen, L. Xiong, M. Luo, X. Chen, et al., Stepwise enzymatic hydrolysis of alkaline oxidation treated sugarcane bagasse for the co-production of functional xylo-oligosaccharides and fermentable sugars. Bioresour. Technol. **275**, 345–351 (2019). https://doi.org/10.1016/j.biortech.2018.12.063
42. G. Liu, Y. Qu, Engineering of filamentous fungi for efficient conversion of lignocellulose: Tools, recent advances and prospects. Biotechnol. Adv. **37**(4), 519–529 (2019). https://doi.org/10.1016/j.biotechadv.2018.12.004
43. C. Long, Y. Cheng, J. Cui, J. Liu, L. Gan, B. Zeng, M. Long, Enhancing cellulase and hemicellulase production in Trichoderma orientalis EU7-22 via knockout of the creA. Mol. Biotechnol. **60**(1), 55–61 (2018). https://doi.org/10.1007/s12033-017-0046-3
44. H.F. Ma, G. Meng, B.K. Cui, J. Si, Y.C. Dai, Chitosan crosslinked with genipin as supporting matrix for biodegradation of synthetic dyes: Laccase immobilization and characterization. Chem. Eng. Res. Des. **132**, 664–676 (2018). https://doi.org/10.1016/j.cherd.2018.02.008
45. Manisha, S.K. Yadav, Technological advances and applications of hydrolytic enzymes for valorization of lignocellulosic biomass. Bioresour. Technol. **245**, 1727–1739 (2017). https://doi.org/10.1016/j.biortech.2017.05.066
46. N.L. Mohamad, S.M. Mustapa Kamal, M.N. Mokhtar, Xylitol biological production: A review of recent studies. Food Rev. Intl. **31**(1), 74–89 (2015). https://doi.org/10.1080/87559129.2014.961077
47. J.R. Monte, W. Carvalho, A.M.F. Milagres, Use of a mixture of thermophilic enzymes produced by the fungus Thermoascus aurantiacus to enhance the enzymatic hydrolysis of the sugarcane bagasse cellulose. Am. J. Agric. Biol. Sci. **5**(4), 468–476 (2010). https://doi.org/10.3844/ajabssp.2010.468.476
48. F.L. Motta, C.C.P. Andrade, M.H.A. Santana, A review of xylanase production by the fermentation of xylan: Classification, characterization and applications, in *Sustainable Degradation of Lignocellulosic Biomass-Techniques, Applications and Commercialization*, ed. by A. Chandel, (InTech, Rijeka, 2013), pp. 251–275. https://doi.org/10.5772/53544
49. A.B. Muley, A.S. Thorat, R.S. Singhal, K. Harinath Babu, A tri-enzyme co-immobilized magnetic complex: Process details, kinetics, thermodynamics and applications. Int. J. Biol. Macromol. **118**, 1781–1795 (2018). https://doi.org/10.1016/j.ijbiomac.2018.07.022
50. S.S. Nadar, V.K. Rathod, A co-immobilization of pectinase and cellulase onto magnetic nanoparticles for antioxidant extraction from waste fruit peels. Biocatal. Agric. Biotechnol. **17**, 470–479 (2019). https://doi.org/10.1016/j.bcab.2018.12.015
51. S. Nagar, V.K. Gupta, D. Kumar, L. Kumar, R.C. Kuhad, Production and optimization of cellulase-free, alkali-stable xylanase by Bacillus pumilus SV-85S in submerged fermentation. J. Ind. Microbiol. Biotechnol. **37**(1), 71–83 (2010). https://doi.org/10.1007/s10295-009-0650-8
52. A.S. Nair, H. Al-Battashi, A. Al-Akzawi, N. Annamalai, A. Gujarathi, et al., Waste office paper: A potential feedstock for cellulase production by a novel strain Bacillus velezensis ASN1. Waste Manag. **79**, 491–500 (2018). https://doi.org/10.1016/j.wasman.2018.08.014

53. K.N. Niladevi, Ligninolytic enzymes, in *Biotechnology for Agro-Industrial Residues Utilisation*, ed. by P. Singh nee' Nigam, A. Pandey, (Springer, United Kingdom, 2009), pp. 397–414. https://doi.org/10.1007/978-1-4020-9942-7_22
54. S.F. Oliveira, J.M.R. da Luz, M.C.M. Kasuya, L.O. Ladeira, A. Correa-Junior, Enzymatic extract containing lignin peroxidase immobilized on carbon nanotubes: Potential biocatalyst in dye decolourization. Saudi J. Biol. Sci. **25**(4), 651–659 (2018). https://doi.org/10.1016/j.sjbs.2016.02.018
55. P.S. Panesar, R. Kaur, G. Singla, R.S. Sangwan, Bio-processing of agro-industrial wastes for production of food-grade enzymes: Progress and prospects. Appl. Food Biotechnol. **3**(4), 1–20 (2016)
56. K. Parani, M. Eyini, Production of ligninolytic enzymes during solid state fermentation of coffee pulp by selected fungi. Sci. Res. Rep. **2**(3), 202–209 (2012)
57. J.C. Pereira, E.C. Giese, M.M.S. Moretti, A.C.S. Gomes, O.M. Perrone, et al., Effect of metal ions, chemical agents and organic compounds on lignocellulolytic enzymes activities, in *Enzyme Inhibitors and Activators*, ed. by M. Senturk, (InTech, Rijeka, 2017), pp. 140–164. https://doi.org/10.5772/65934
58. A. Piscitelli, C. Del Vecchio, V. Faraco, P. Giardina, G. Macellaro, et al., Fungal laccases: Versatile tools for lignocellulose transformation. C. R. Biol. **334**(11), 789–794 (2011). https://doi.org/10.1016/j.crvi.2011.06.007
59. R.C. Ray, S.S. Behera, Solid state fermentation for production of microbial cellulases, in *Biotechnology of Microbial Enzymes*, ed. by G. Brahmachari, (Academic Press, London, 2017), pp. 43–79. https://doi.org/10.1016/B978-0-12-803725-6.00003-0
60. M. Romero-Fernández, S. Moreno-Perez, S. Martins de Oliveira, R.I. Santamaría, J.M. Guisan, J. Rocha-Martin, Preparation of a robust immobilized biocatalyst of β-1,4-endoxylanase by surface coating with polymers for production of xylooligosaccharides from different xylan sources. New Biotechnol. **44**, 50–58 (2018). https://doi.org/10.1016/j.nbt.2018.04.007
61. E. Rosmine, N.C. Sainjan, R. Silvester, A. Alikkunju, S.A. Varghese, Statistical optimisation of xylanase production by estuarine Streptomyces sp. and its application in clarification of fruit juice. J. Genet. Eng. Biotechnol. **15**(2), 393–401 (2017). https://doi.org/10.1016/j.jgeb.2017.06.001
62. K. Saha, P. Verma, J. Sikder, S. Chakraborty, S. Curcio, Synthesis of chitosan-cellulase nanohybrid and immobilization on alginate beads for hydrolysis of ionic liquid pretreated sugarcane bagasse. Renew. Energy **133**, 66–76 (2019). https://doi.org/10.1016/j.renene.2018.10.014
63. G.S.B. Salomão, J.C. Agnezi, L.B. Paulino, L.B. Hencker, T.S. de Lira, P.W. Tardioli, L.M. Pinotti, Production of cellulases by solid state fermentation using natural and pretreated sugarcane bagasse with different fungi. Biocatal. Agric. Biotechnol. **17**, 1–6 (2019). https://doi.org/10.1016/j.bcab.2018.10.019
64. S.K. Sandhu, A. Mathur, R. Gupta, S.K. Puri, M. Adsul, Cellulosic biomass-hydrolyzing enzymes, in *Waste to Wealth*, ed. by R. R. Singhania, R. A. Agarwal, R. P. Kumar, R. K. Sukumaran, (Springer, Singapore, 2018), pp. 441–456. https://doi.org/10.1007/978-981-10-7431-8_19
65. J.J. Silva, T.T. Santana, A.N. Coronato, P.A. Hirose, et al., Produção de lacase de fungos basidiomicetos por fermentação submersa com cascas de café. Arq. Ciênc. Vet. Zool. **15**(2), 191–196 (2012)
66. M.H.L. Silveira, F.G. de Siqueira, M. Rau, L. Silva, L.R.S. Moreira, E.X. Ferreira-Filho, J. Andreaus, Hydrolysis of sugarcane bagasse with enzyme preparations from Acrophialophora nainiana grown on different carbon sources. Biocatal. Biotransformation **32**(1), 53–63 (2014). https://doi.org/10.3109/10242422.2013.872634
67. R. Sindhu, P. Binod, A. Pandey, Biological pretreatment of lignocellulosic biomass – An overview. Bioresour. Technol. **199**, 76–82 (2016). https://doi.org/10.1016/j.biortech.2015.08.030
68. R.R. Singhania, J.K. Rani-Saini, R. Saini, M. Adsul, A. Mathur, R. Gupta, D.K. Tuli, Bioethanol production from wheat straw via enzymatic route employing Penicillium janthinellum cellulases. Bioresour. Technol. **169**, 490–495 (2014). https://doi.org/10.1016/j.biortech.2014.07.011

69. R.R. Singhania, M. Adsul, A. Pandey, A.K. Patel, Cellulases, in *Current Developments in Biotechnology and Bioengineering*, ed. by A. Pandey, S. Negi, C. R. Soccol, (Elsevier, Amsterdam, 2017), pp. 73–101. https://doi.org/10.1016/B978-0-444-63662-1.00004-X
70. M. Suto, F. Tomita, Induction and catabolite repression mechanisms of cellulase in fungi. J. Biosci. Bioeng. **92**(4), 305–311 (2001)
71. L. Thomas, A. Joseph, R.R. Singhania, A.K. Patel, A. Pandey, Industrial enzymes, in *Current Developments in Biotechnology and Bioengineering*, ed. by A. Pandey, S. Negi, C. R. Soccol, (Elsevier, Amsterdam, 2017), pp. 127–148. https://doi.org/10.1016/B978-0-444-63662-1.00006-3
72. N. Vaishnav, A. Singh, M. Adsul, P. Dixit, S.K. Sandhu, et al., Penicillium: The next emerging champion for cellulase production. Bioresour. Technol. Rep. **2**, 131–140 (2018). https://doi.org/10.1016/j.biteb.2018.04.003
73. V.M. Vasconcellos, P.W. Tardioli, R.L.C. Giordano, C.S. Farinas, Addition of metal ions to a (hemi) cellulolytic enzymatic cocktail produced in-house improves its activity, thermostability, and efficiency in the saccharification of pretreated sugarcane bagasse. New Biotechnol. **33**(3), 331–337 (2016). https://doi.org/10.1016/j.nbt.2015.12.002
74. A. Verardi, I. De, E. Ricca, V. Calabr, Hydrolysis of lignocellulosic biomass: Current status of processes and technologies and future perspectives, in *Bioethanol*, ed. by M. A. Pinheiro Lima, (InTech, Rijeka, 2012), pp. 95–122. https://doi.org/10.5772/23987
75. A. Walia, S. Guleria, P. Mehta, A. Chauhan, J. Parkash, Microbial xylanases and their industrial application in pulp and paper biobleaching: A review. 3 Biotech **7**(1), 1–12 (2017). https://doi.org/10.1007/s13205-016-0584-6
76. C. Wan, Y. Li, Fungal pretreatment of lignocellulosic biomass. Biotechnol. Adv. **30**(6), 1447–1457 (2012). https://doi.org/10.1016/j.biotechadv.2012.03.003
77. C.E. Wyman, B.E. Dale, R.T. Elander, M. Holtzapple, M.R. Ladisch, Y.Y. Lee, Coordinated development of leading biomass pretreatment technologies. Bioresour. Technol. **96**(18), 1959–1966 (2005). https://doi.org/10.1016/j.biortech.2005.01.010
78. Q. Yasmeen, M. Asgher, M.A. Sheikh, H. Nawaz, Optimization of ligninolytic enzymes production through response surface methodology. Bioresources **8**(1), 944–968 (2013). https://doi.org/10.15376/biores.8.1.944-968
79. M.A. Yassin, A.A.M. Gad, A.F. Ghanem, M.H. Abdel Rehim, Green synthesis of cellulose nanofibers using immobilized cellulase. Carbohydr. Polym. **205**, 255–260 (2019). https://doi.org/10.1016/j.carbpol.2018.10.040
80. Y. Yin, Databases for bioenergy-related enzymes, in *Bioenergy Research: Advances and Applications*, ed. by V. K. Gupta, (Elsevier, Amsterdam, 2013), pp. 95–107
81. L.W. Yoon, T.N. Ang, G.C. Ngoh, A.S.M. Chua, Fungal solid-state fermentation and various methods of enhancement in cellulase production. Biomass Bioenergy **67**, 319–338 (2014). https://doi.org/10.1016/j.biombioe.2014.05.013
82. H. Zabed, S. Sultana, J.N. Sahu, X. Qi, An overview on the application of ligninolytic microorganisms and enzymes for pretreatment of lignocellulosic biomass, in *Recent Advancements in Biofuels and Bioenergy Utilization*, ed. by P. K. Sarangi, S. Nanda, P. Mohanty, (Springer, Singapore, 2018), pp. 53–72. https://doi.org/10.1007/978-981-13-1307-3_3
83. C. Zhao, L. Deng, H. Fang, Mixed culture of recombinant Trichoderma reesei and Aspergillus niger for cellulase production to increase the cellulose degrading capability. Biomass Bioenergy **112**, 93–98 (2018). https://doi.org/10.1016/j.biombioe.2018.03.001
84. X.W. Zhou, W.R. Cong, K.Q. Su, Y.M. Zhang, Ligninolytic enzymes from Ganoderma spp: Current status and potential applications. Crit. Rev. Microbiol. **39**(4), 416–426 (2013). https://doi.org/10.3109/1040841X.2012.722606
85. R. Zhuo, H. Yu, X. Qin, H. Ni, Z. Jiang, F. Ma, X. Zhang, Heterologous expression and characterization of a xylanase and xylosidase from white rot fungi and their application in synergistic hydrolysis of lignocellulose. Chemosphere **212**, 24–33 (2018). https://doi.org/10.1016/j.chemosphere.2018.08.062

86. Y. Yin, Databases for Bioenergy-Related Enzymes, in *Bioenergy Research: Advances and Applications*, ed by V.K. Gupta, M.G. Tuohy, C.P. Kubicek, J. Saddler, F. Xu, (Elsevier, United Kingdom, 2014), pp. 95–103. https://doi.org/10.1016/B978-0-444-59561-4.00006-1
87. B. Nikhil, A. Dharmesh, T. Priti, Production of xylanase by Aspergillus flavus FPDN1 on pearl millet bran: optimization of culture conditions and application in bioethanol production, Int. J. Res. Chem. Environ **2**(3), 204–210 (2012)
88. Q. He, H. Chen, Pilot-scale gas double-dynamic solid-state fermentation for the production of industrial enzymes. Food Bioprocess Technol **6**, 2916–2924 (2013). https://doi.org/10.1007/s11947-012-0956-9
89. M. Marín, A. Artola, A. Sánchez, Optimization of down-stream for cellulases produced under solid-state fermentation of coffee husk. Waste Biomass Valor **10**, 2761–2772 (2019). https://doi.org/10.1007/s12649-018-0327-5
90. M. Mandels, E.T. Reese, Inhibition of cellulases. Annual Review of Phytopathology **3**, 85–102 (1965). https://doi.org/10.1146/annurev.py.03.090165.000505
91. K.S. Salem, T.U. Rashid, I. Asaduzzaman, M.N. Khan, S. Sharmeen, M.M. Rahman, P. Haque, Recent updates on immobilization of microbial cellulase, in *New and Future Developments in Microbial Biotechnology and Bioengineering*, ed. By V.K. Gupta, (Elsevier, Amsterdam, 2016), pp. 107–139. https://doi.org/10.1016/B978-0-444-63507-5.00011-3
92. The Novozymes report (2011). https://s21.q4cdn.com/655485906/files/doc_financials/annual_english/The-Novozymes-Report-2011.pdf

Chapter 12
Sustainable Production of Polyhydroxyalkanoates (PHAs) Using Biomass-Based Growth Substrates

D. Kumar and B. Singh

12.1 Introduction

With an annual production of well over 200 million tonnes, a major share of the petroleum-based polymers (mainly plastic) find their way into the environment as industrial, commercial, and household wastes. Their primary disadvantage is their inherent resistivity toward microbial degradation, and as a result, they keep on piling in the environment leading to several undesirable impacts. The harmful effects of plastic and its precursor on the environment are well recognized, and as a result the past few decades have seen dedicated research and development efforts toward the identification and development of their alternatives. Taking note of the urgency, several nations have banned the usage of plastic carry bags in various capacities. Bioplastics are biologically sourced materials having properties comparable to that of petroleum-derived plastics and are reported to have a lower ecological footprint. The global production of bioplastics in 2013–2014 stood at approximately 1200 kilotonnes per annum [20]. Bioplastics can be derived from biomass of various origins including lignocellulosic residues, municipal solid wastes, virgin/used cooking oil, glycerol, starch, and several other renewably sourced forms of carbon. However, the high production cost is the single most significant impediment in the commercialization of bioplastics, which costs approximately 20–80% higher than their petroleum-based counterparts [47]. Lignocellulosic biomass remains to be the single most significant biomass reserve for the production of a range of valuable commodities. However, their valorization through bioprocessing is restricted mainly due to their recalcitrance toward microbial degradation. Accordingly, a biomass pretreatment step is usually required to make lignocellulosic residues amenable to bioprocessing.

D. Kumar · B. Singh (✉)
Department of Environmental Sciences, Central University of Jharkhand, Ranchi, India
e-mail: bhaskar.singh@cuj.ac.in

M. O. Daramola, A. O. Ayeni (eds.), *Valorization of Biomass to Value-Added Commodities*, Green Energy and Technology,
https://doi.org/10.1007/978-3-030-38032-8_12

Polyhydroxyalkanoate (PHA) belongs to a family of intracellular biopolymer that is naturally produced by bacteria (and archaea) primarily as a response to stressful growth conditions. Under nutrient-starved conditions and in an excess supply of carbon, bacteria divert their metabolic machinery toward the production of PHA. A feast-famine regime is the most common strategy for triggering PHA accumulation in the target microorganism. PHA can serve the purpose of a carbon and energy source for bacteria under such stressful conditions and also helps microorganism tolerate the transient environmental assaults including osmotic shock, high temperature, and ultraviolet irradiation [34]. The PHA production and its bioaccumulation (in the form of intracellular lipid granules) trait has been reported in over 70 genera of bacteria and archaea, and these organisms encompass a myriad of PHA anabolic machinery capable of processing various carbon sources. Thus, the PHA production trait is ubiquitous, and research efforts have succeeded in developing culture collections, which is available for further consultations. Since its discovery by Lemogine in 1926 [2], there has been an unprecedented rise in the number of publications on PHA. In addition to PHA, several other biobased polymers are recognized such as polylactic acid (PLA), polybutylene succinate (PBS), polypropylene carbonate (PPC), bio-polyethylene (bPE), bio-polyethylene terephthalate (bPET), bio-propylene (bPP), and bio-poly(trimethylene terephthalate) (bPTT) [12]. However, PHA remains to be the only biopolymer which is entirely of biological origin. Over 150 monomeric forms of PHA have been reported, and thus PHA also offers a diverse range of material properties. Among the most attractive attributes of PHA is its high susceptibility toward microbial degradation under natural settings, biocompatibility, attractive physical and chemical properties, chemical diversity, and its renewability. In addition to these, research efforts have also highlighted its utility in the pharmaceutical industry and more recently as a potential drug with anticancer, anti-HIV, antibiotics, etc. [15, 57]. A typical PHA polymer is made up of 600–35,000 monomer units (R-hydroxy fatty acids). The side chain alkyl group (R) is usually a saturated hydrocarbon but can also appear as an unsaturated/branched/substituted moiety. The properties of PHA can be tailored for more targeted applications by chemical/physical altercation of the naturally available polymer or through genetic engineering-based approaches to produce specialized functional groups. Polyhydroxybutyrate (PHB) is the most commonly found intracellular PHA in the bacterium. Hydroxybutyrate, hydroxyvalerate, hydroxyhexanoate, hydroxyoctanoate, hydroxydecanoate, and hydroxydodecanoate are PHA monomers of frequent occurrence in nature, and these can take the form of either homopolymers or copolymers of various combinations.

The generalized molecular structure of PHA is shown in the form of Fig. 12.1.

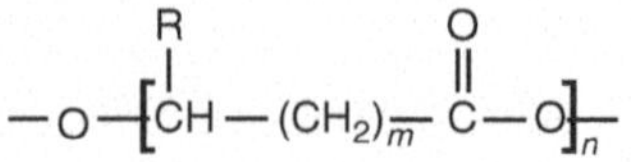

Fig. 12.1 The generalized molecular structure of PHA (m = 1–3, m = 1 is of commonest occurrence, n can range from 100 to several thousands)

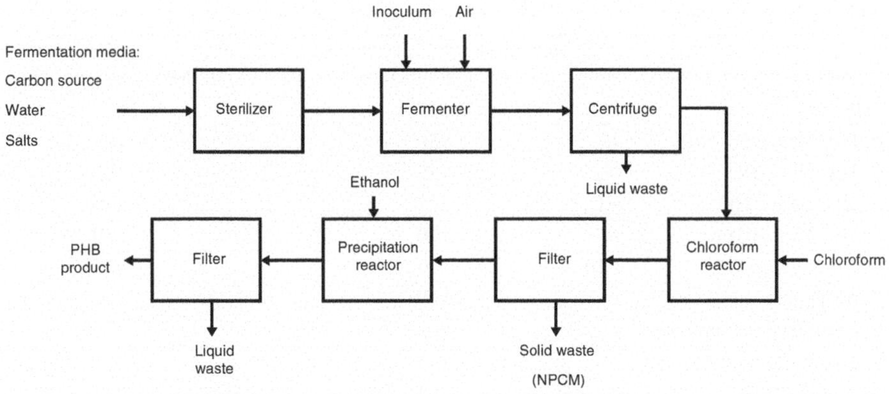

Fig. 12.2 Generalized process flow diagram of the industrial production of PHB using pure culture and chloroform-based extraction of intracellular PHB. (Reproduced from Leong et al. [44])

Regarding biodegradability, complete decomposition of the PHA to CO_2 and H_2O has been reported under composting conditions. The single most substantial deterrent in the successful scale-up of PHA production is unfavorable economics as currently, the production cost of PHA is way higher than their conventional counterparts like polystyrene [47]. The current research in the field is primarily devoted to the efforts aimed at lowering the production cost of PHA. In this regard, the exploration of cheaply available carbon sources has attracted the most attention, and the use of advanced genetic engineering techniques and transgenic plants also hold potential avenues [17]. The industrial scale setup of PHA production involves several sequential operations including the sterilization operation to exclude undesirable species, fermentation of the carbon source using the chosen strain/mixed culture in an aerobic fermentation reactor, concentration of microbial biomass using appropriate technique such as centrifugation, extraction of the intracellular PHA granules, and its further processing/purification [13]. The solid waste recovered after filteration constitutes non-PHA cell mass (NPCM) (Fig. 12.2).

12.2 Biosynthesis and Genetic Manipulation for the Production of PHA

The biosynthetic pathway for PHA production is highly complex as it is intricate to the some of the critical metabolic pathways of the bacterium. These include the Krebs cycle, de novo lipid synthesis, glycolysis, β-oxidation, serine pathway, and the Calvin cycle. The most critical intermediate between these and the PHA pathway is the acetyl-CoA. When the supply of an essential growth nutrient (e.g., nitrogen or phosphorus) is restricted, the acetyl-CoA is channeled toward PHA production, which is otherwise utilized in other metabolic pathways. Under supportive growth

Table 12.1 Important properties of PHA homo-/copolymers and conventional plastics [8]

PHA/conventional plastics	Glass transition temperature (°C)	Melting point temperature (°C)	Tensile strength (MPa)	Elongation to break (%)
PHB	4	177	43	5
P (HB; 10% HV)	–	150	25	20
P (HB; 20% HV)	–	135	20	100
P (HB; 10% HHx)	−1	127	21	400
P (HB; 17% HHx)	−2	120	20	850
Polypropylene	−10	170	34	400
Polystyrene	–	110	50	–
Polyethylene terephthalate	–	262	56	7300
High-density polyethylene	–	135	29	–

HB hydroxybutyrate, *HV* hydroxyvalerate, *HHx* hydroxyhexanoate, – *not Available/Reported*

conditions, the excess supply of acetyl-CoA inhibits a key enzyme called 3-ketothiolase (PhaA), and as a result, the intermediate is channeled toward supporting cell growth and energy production. Assessment of wild strains and recombinant heterologous expression of genes have provided valuable insights into the production of short- to medium-chain-length PHA. The detailed metabolic mechanism for the synthesis and accumulation of PHA has been provided by Madison and Huisman [46], Guoqiang et al. [26], and Francavilla et al. [21]. The important properties of different PHAs (homo-/copolymers) and conventional plastics can be seen in Table 12.1.

Studies on the genetic engineering of a bacterium for the production of PHA were first reported in 1988, and by then the PHB synthesis genes were successfully cloned and expressed from *Ralstonia eutropha* to *E. coli.* It was followed by studies aimed at the expression and stabilization of such genes in *E. coli*. The recombinant *E. coli* was reported to accumulate very high levels of PHB (80% of dry cell weight) within 39 h using glucose as a substrate. In addition to the genetic manipulation of the bacterium, attempts have also been made to introduce genes encoding two key enzymes that convert the acetyl-CoA to PHB from *R. eutropha* to *Arabidopsis thaliana* (a model flowering plant) which accumulated PHB of similar size and appearance as in bacterium. Such studies hold tremendous potential in mass-scale production of cheap PHA, but as of now, the PHA accumulation levels are very low. In an interesting study, Bohmert et al. [7] reported that transgenic *Arabidopsis thaliana* encompassing three crucial genes could accumulate PHB in leaf chloroplast by up to 40% dry weight (4% of fresh weight). Figure 12.3 shows the transmission electron micrograph (TEM) of a normal wild-type mesophyll cell chloroplast of *Arabidopsis thaliana* (a) and that of transgenic *Arabidopsis thaliana* (b). The findings of the study are highly encouraging as unprecedented levels of PHB in plant biomass were recorded. Likewise, PHB production and accumulation have also been reported in transgenic potato and tobacco. However, the accumulation of high levels of PHB in transgenic *Arabidopsis* has been associated with a negative influence on plant growth and development. Numerous recent studies on cloning of PHA synthesis genes from novel stains to different hosts are also available [18, 31].

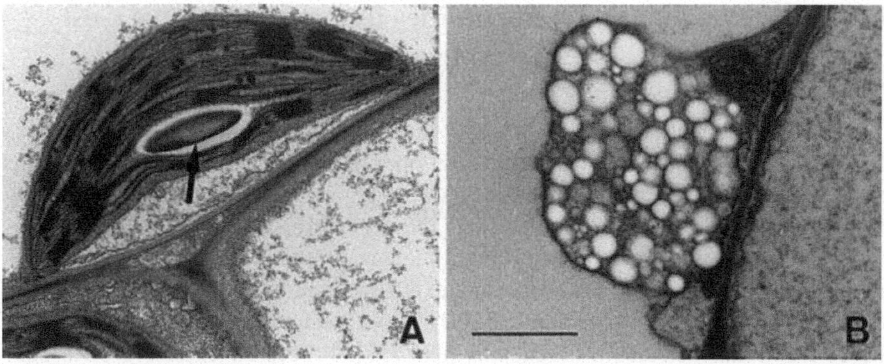

Fig. 12.3 Transmission electron micrograph of a mesophyll cell chloroplast of wild-type *Arabidopsis thaliana* (**a**) and transgenic *Arabidopsis thaliana* with agglomerated PHB granules (**b**). Bar = 1 μm. (Reproduced with permission from Bohmert et al. [7])

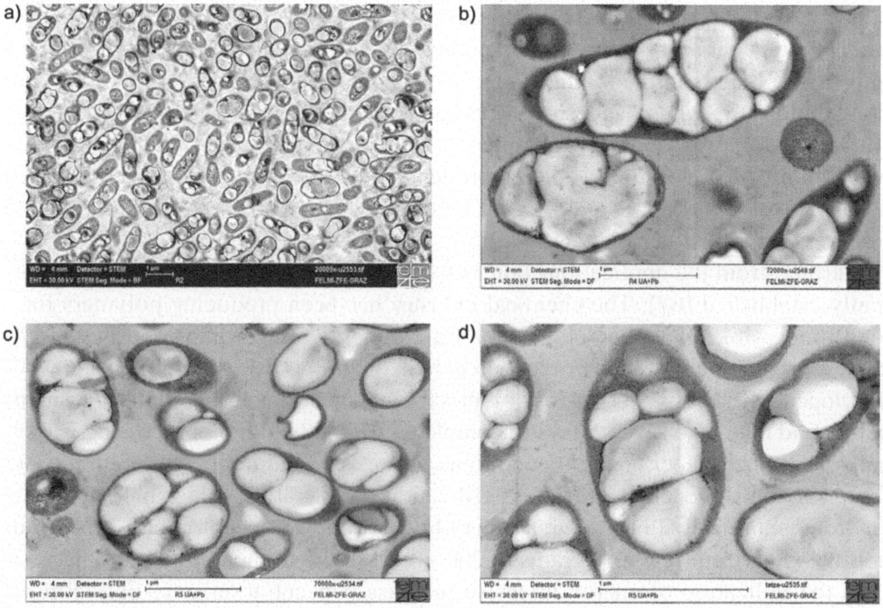

Fig. 12.4 Scanning electron micrograph of *Cupriavidus necator* DSM 545 biomass and intracellular PHA granules: (**a**) magnification 1/20,000, PHA content of 48%; (**b**) magnification 1/72,000, PHA content of 65%; (**c**) magnification 1/70,000, PHA content of 69%; and (**d**) magnification 1/1,50,000, PHA content of 69%. (Reproduced from Koller et al. [42])

Recombinant DNA technology holds promise in broadening substrate utilization, creating novel metabolic pathways, producing PHA with specific copolymers and chain length, and enhancing the PHA production capability.

The scanning electron micrograph of *Cupriavidus necator* DSM 545 biomass and intracellular PHA granules at different resolutions can be seen in Fig. 12.4.

12.3 Challenges and Opportunities in the Competitive Cost Scale-Up of PHA Production

One of the most challenging issues in the successful scale-up of PHA production is the involvement of a number of sequential operations which are often very inefficient [43]. The commercialization aspects of PHA are invariably dependent on the economics of petroleum resources, and the recent discovery of vast shale gas reserves is likely to check any substantial increase in the price of petroleum in the international market [36]. Unless dramatic cost reductions through advanced biotechnological engineering are attained, PHA production is unlikely to compete with the conventional petroleum-based polymers [11]. Processed carbon growth substrates (e.g., glucose) are undesirable from both economic and environmental dimensions [39]. Production of PHA with consistent composition/structure and associated properties are important from an application point of view [53], and the downstream processing operations (post PHA accumulation) lack an efficient, economical, and sustainable technology [30, 73]. Many life cycle impact assessment and techno-economic studies have identified the downstream processing operations to be the hot-spots in the biomass supply chain [20, 37]. Chemical digestion (with or without surfactants) and solvent extraction are the routine strategies for the extraction of PHA [43]. PHAs obtained through chemical digestion procedure have been reported to have slightly altered properties, while the solvent-based operations usually involve hazardous chlorinated solvents whose recovery and reuse adds to the overall production cost [27].

Further, from the applications point of view, not many high-end usages are currently established [67]. The chemical industry has been producing polymers for a long time, while biotechnology is not yet competitive regarding robustness, efficiency, production cost, and production in bulk quantities using simplified and well-developed operations [54]. Most bioprocessing operations are of discontinuous nature and may take several days to complete [70]. Although bioprocessing is typically performed under ambient conditions of pressure and temperature, high water footprint and the energy-intensive sterilization and aeration somewhat negate the environmental friendliness of the system [42]. The low concentration of the products (ranging from few milligrams to 200 g L^{-1}) in the fermentation broth necessitates the utilization of energy-intensive biomass concentration step. Furthermore, the efficiency of the substrate to product conversion is rather low, and the maximum reported efficiency is close to 33% [54], while the conventional chemical industry can reach up to 90–100% [16].

Despite these known limitations, the biopolymers such as PHA have received considerable research attention as several studies have earmarked the potential opportunities which could dramatically improve the environment friendliness and commercialization aspects of these materials [70].

Many of these issues can be overcome if an ideal strain which is suitable for the industrial-scale production of PHA is identified. An industrially suitable strain should ideally exhibit the following characteristics: fast growth rate; high PHA accumulation capacity; no production of toxins; should also be able to produce other high-value compounds; broad substrate utilization potential; well-characterized genetic makeup; easy amenability to genetic manipulation; wide pH, temperature, and salinity tolerance; and self-flocculation tendency among others [70]. Considering the colossal diversity of microorganisms in nature and the availability of advanced biotechnological tools, it is highly likely that many of these demands would be realized [3]. To aid in sustainable downstream extraction and purification, the chosen strain should have a delicate cell wall, large cell size, self-flocculation tendency, or readily inducible cell aggregation and inducible cell lysis [48]. With further advancements in research, it is only a matter of time that many of these demands are met. In terms of the ease of downstream operations, increasing the cell size or the size of PHA granules would be attractive [55]. Furthermore, inducing extracellular production of PHA by using advanced biotechnological tools (e.g., targeting development of a weakened cell wall) could revolutionize the microbial PHA production field [70]. Although research in this regard still has to go a long way, valuable insights have been provided by some of the recent research investigations [58, 64]. Genetic engineering, synthetic biology, and engineering of metabolic pathways hold tremendous potential in realizing these goals. The requirement of sterile conditions can be avoided by using native microbial populations instead of using a pure culture; however, the latter offers the advantage of greater reproducibility [33]. Salt-loving (halophilic) microbes seem to offer several important advantages including the hampered growth of invasive strains; high salt, temperature, and pH window; and the ability to produce PHA in a continuous fermentation process [65]. For driving the commercial market of PHA, it is important that several high-value applications are identified. Currently, the only high-value utility of PHA lies in the field of biomedical implants, and their major application is limited as biodegradable packaging material. However, recent reports have suggested that PHA can also find applications as a high-value product in the field of superior quality textile and for the manufacturing of high-quality fishing nets/lines [60, 66]. Genetic manipulations targeting the production of designer copolymers would further diversify the applications of PHA [51].

12.4 Techno-economic Studies on Microbial PHA Production

The high cost (20–80%) of production remains to be a significant bottleneck in the acceptability of PHA as alternative polyester. Depending on the origin, the carbon substrate may account for 20–60% of the total production cost. Besides the energy requirement for sterilization and fermentation, the yield of PHA and the downstream processing and extraction operations also have a significant effect on the commercialization aspects of PHA. Detailed techno-economic evaluations on pure culture-based PHA production are available in literature such as those by Choi and Lee [13], Akiyama et al. [1], Choi et al. [14], and Van Wegen et al. [68]. Studies such as those by Kim and Dale [37] and Gerngross [23] have identified the pure culture-based

PHA production using processed agricultural resources to be more harmful and resource intensive than polystyrene produced from fossil sources. It was attributed to the requirement of aseptic conditions and process optimization and the high requirement of energy and resources. On the other hand, native microbial population-based PHAs produced on waste resources are likely to outcompete the pure culture-based operations and the conventional routes of polymer production.

Culture collections are available with a description of the strains, their PHA accumulation levels, and suitable growth substrate. The reported growth substrates (carbon source) include carbohydrates (glucose, fructose, xylose, arabinose, maltose, lactose, etc.), lipids (volatile fatty acids, oil and fat), gases (CH_4, CO_2, H_2, etc.), alcohols (methyl alcohol, ethyl alcohol, octyl alcohol, glycerol, etc.), organic acids (acetic acid, butyric acid, oleic acid, lauric acid, propionic acid, valeric acid, etc.), and alkanes (dodecane, hexane, octane, heptane, etc.). The collections can be used to identify a mixed culture of a bacterium capable of utilizing specific growth substrates or a combination thereof. It has been reported that the utilization of a mixed culture has economic and environmental desirability. Fernández-dacosta et al. [20] in their techno-economic study on the microbial community-based production of PHB have emphasized the importance of using native strains present in industrial effluents for economic and environmental gains. The native microbial community could accumulate up to 77% of their dry cell weight as PHB, and the desirability of the process was well pronounced as the requirement of the growth substrate was met from industrial effluents (food industry and paper mills), and the process did not require maintenance of non-aseptic conditions. A mixed culture-based PHA production has recently emerged as an alternative wastewater treatment strategy and has been compared with other conventional approaches such as biogas production [27]. Compared to the pure culture production routes, techno-economic studies on mixed culture PHA production are scarce. Gurieff and Lant [27] compared a pure and mixed culture-based PHA production operation using industrial effluents as carbon substrate and reported the latter to be more cost-effective and ecofriendly, and the process was more efficient than subjecting the industrial effluent to anaerobic digestion for biogas production. Although both pure and mixed culture operations exhibited comparable environmental footprint, they consistently performed better than high-density polyethylene production. The authors suggested that the major economic and environmental burden for the mixed culture operation arise from the energy-intensive downstream processing operations. These findings have also been substantiated by some of the recent studies including those by Wang et al. [71] and Cabrera et al. [9]. Techno-economic and life cycle assessment-based studies are likely to play an important role in efforts aimed at optimization of the operations involved as there is immense scope for improvement. Several studies have reported the environmental impacts of microbial PHA production to be substantially dependent on the choice of the growth substrate and the overall nonrenewable energetic balance of the process. It holds true for processed agricultural substrates such as corn-derived glucose or other high-value materials not only in terms of greenhouse gas (GHG) balance but also regarding acidification, eutrophication, and photochemical smog generation potential. Most of these impacts arise out of the processes involved in the cultivation of corn [37]. Solvent extraction and chemical digestion remain to be the common PHA extraction strategy, and the extraction of PHA from mixed culture is reportedly a more complex and energy-intensive operation [27]. Fernández-dacosta et al. [20] compared three different PHB extraction operations involving (I) a surfactant (sodium dodecyl sulfate) and an alkali (sodium hydroxide)-based digestion, (II) a surfactant (sodium dodecyl sulfate), and (III) a hypochlorite-based digestion and a solvent extraction system using dichloromethane. The process I was identified to be the most favorable operation with an overall production cost of 1.4 € kg^{-1} PHB, CO_2-equivalent GWP (global warming potential) of 2.4 kg kg^{-1} PHB, and an aggregate fossil energy requirement of 106 MJ kg^{-1} PHB produced. The process III (solvent extraction) was reportedly the least attractive approach with an overall production cost, CO_2-equivalent GWP, and nonrenewable energy demand of 1.95 € kg^{-1} PHB, 4.30 kg kg^{-1} PHB, and 156 MJ kg^{-1} PHB, respectively.

Refined sugar-based substrates account for 30–50% of the overall production cost of PHA, and their cheaper and reliable alternatives are likely to affect the cost competitiveness of the system positively. A techno-economic study on industrial-scale production (100,000 tonnes per annum) of PHA suggested that the share of growth substrate in the final cost could be brought down to 22% if methane is utilized in the process. The final cost was estimated to be 4.1–6.8 $ kg^{-1} PHA produced, and the cost could potentially be reduced to 3.2–4.5 $ kg^{-1} PHA if suitable thermophilic methanotrophs are employed in the process [45]. Use of agroindustrial residues and industrial effluents are likely to lower the overall production cost substantially. To this end, several recent studies have focused on diverting waste resources toward PHA production using either pure or mixed culture microorganisms, and some of these have presented encouraging results. Accordingly, there is immense scope for future research and development efforts targeted at improving the commercialization aspects of PHA.

In an exciting study on techno-economic analysis of PHA accumulation by *Haloferax mediterranei* using stillage waste (from rice-based ethanol production unit), Bhattacharyya et al. [6] obtained highly promising results. *Haloferax mediterranei* could tolerate the high levels of salinity present in the growth medium, and it also checked the growth of other non-tolerant species which eliminated the need of maintaining aseptic fermentation conditions. The tendency of the *Haloferax mediterranei* cells to lyse in non-saline conditions simplified the recovery of PHA. The fermentation process was incorporated to a wastewater treatment operation (activated sludge process) for the sake of minimizing capital input which led to a reduction in 5-day biological oxygen demand (BOD) and chemical oxygen demand (COD) by approximately 82%. The *Haloferax mediterranei* cells could accumulate up to 63% of PHA in the form of a copolymer poly-3-hydroxybutyrate-co-hydroxyvalerate (P3HB-3HV). The production of PHA was simulated for a capacity of 1890 tons with an estimated production cost of 2.05 $ kg^{-1} PHA. The desalination operation for the recovery of PHA was identified as the most demanding operation and contributed a significant proportion to the overall production cost. Table 12.2 is a compilation of important studies on PHA accumulating microbe grown on different carbon substrates.

Table 12.2 Important studies on PHA accumulation by microbes grown on different growth substrates

Microbe	Carbon substrate	PHA accumulation (%; dry weight basis)	PHA type	Culture number	Reference
Plasticicumulans acidivorans	Paper mill wastewater	77	P3HB	–	[32]
Bacillus sp. strain COL1/A6	Hydrolyzed wafer residue, hydrolyzed citrus pulp, cane molasses	47.5–62.41	P3HB	–	[59]
Pseudomonas aeruginosa 42A2	Oleic acid	54.6	P3HB	NCIMB 40045	[19]
	Soybean oil-based waste fatty acid	66.1			
	Waste frying oil	29.4			
Azohydromonas lata	Fructose/glucose	76.5–79.4	P3HB	DSM 1123, IAM 12665, LMG 332, ATCC 29714	[24]
	Sucrose	50.0–88.0			[25, 69]
Burkholderia cepacia	Fructose, glucose, sucrose	50.4–59.0	P3HB	NCIB 9085, DSM 50181, ATCC 17759	[52]
	Glycerol	31.3			[74]
	Xylose	58.4			[52]
Cupriavidus necator H16	CO_2	88.9	P3HB	ATCC 17699, DSM 428, KCTC 22496, NCIB 10442	[63]
	Acetate, lactic acid, propionic acid, butyrate	3.9–40	3HB, 3HV		[10]
	Palm oil, olive oil, corn oil, oleic acid	79.0–82.0	P3HB		[22]
	Fructose, glucose	67.0–70.5	P3HB		[24]
Cupriavidus necator	Saccharified waste, potato starch	46.0	P3HB	CECT 4623, KCTC 2649, NCIMB 11599	[28]
	Glucose	76.0			[38]
Pseudomonas putida GO16	Pyrolyzed polyethylene terephthalate (terephthalic acid)	27.0	Medium-chain-length PHA	NCIMB 41538	[35]
Bacillus sonorensis	Glycerol from Jatropha biodiesel	71.8	P3HB	–	[61]
Halomonas hydrothermalis	Glycerol from Jatropha biodiesel	75.0	P3HB		
Haloferax mediterranei	Hydrolyzed whey	72.8	P3HB3HV	ATCC 33500, CCM 3361, DSM 1411	[41]
	Crude and refined glycerol from biodiesel production	75–76			[29]
	Vinasse	50.0–73.0			[5]

(continued)

Table 12.2 (continued)

Microbe	Carbon substrate	PHA accumulation (%; dry weight basis)	PHA type	Culture number	Reference
Methylobacterium extorquens	Methanol	35.0–62.3	P3HB	ATCC 8457, DSM 1340, NCIB 2879, NCTC 2879	[49, 50]
Methylocystis sp. GB25	Methane	51.0	P3HB	DSM 7674	[72]
Novosphingobium nitrogenifigens Y88	Glucose	81.0	P3HB	DSM 19370, ICMP 16470	[62]
Hydrogenophaga pseudoflava	Valerate and hydrolyzed whey	40.0	P3HB3HV	ATCC 33668, DSM 1034	[40]
	Sucrose, lactose	20.2–62.5			[56]
Cupriavidus necator	Molasses	31.0–44.0	P3HB	DSM 545	[4]
	Glucose, propionic acid	80.0	P3HB3HV		[17]

– *not Available/Reported*

12.5 Conclusion

There has been an unprecedented increase in research interests on bioplastics in recent years. The PHA monomers lend themselves toward the production of a wide variety of polymers with tunable properties. Introduction and overexpression of PHA production pathways and the optimization of the PHA synthesis in the recombinant host are particularly attractive opportunities for improving the commercialization potential of PHA. Lowering the energy demand particularly for the sterilization, aeration, and downstream processing operations is likely to improve the sustainability and cost competitiveness of the process substantially. Deconstruction of lignocellulosic residues is yet another challenge in the valorization of lignocellulosic wastes toward the production of PHA. Despite the existing hurdles, further advancements in research in the fields of synthetic biology, biosynthetic pathways, genetic engineering, and molecular biology hold tremendous potential in successful scale-up and in improving the environmental dimensions of PHA production.

Acknowledgments Dipesh Kumar gratefully acknowledges the University Grants Commission (UGC), New Delhi, India for providing senior research fellowship.

References

1. M. Akiyama, T. Tsuge, Y. Doi, Environmental life cycle comparison of polyhydroxyalkanoates produced from renewable carbon resources by bacterial fermentation. Polym. Degrad. Stab. **80**, 183–194 (2003)

2. A.J. Anderson, E.A. Dawes, Occurrence, metabolism, metabolic role, and industrial uses of bacterial polyhydroxyalkanoates. Microbiol. Rev. **54**, 450–472 (1990)
3. A. Anjum, M. Zuber, K.M. Zia, A. Noreen, M.N. Anjum, S. Tabasum, Microbial production of polyhydroxyalkanoates (PHAs) and its copolymers: A review of recent advancements. Int. J. Biol. Macromol. **89**, 161–174 (2016)
4. M. Beaulieu, Y. Beaulieu, J. Melinard, S. Pandian, J. Goulet, Influence of ammonium salts and cane molasses on growth of Alcaligenes eutrophus and production of polyhydroxybutyrate. Appl. Environ. Microbiol. **61**, 165–169 (1995)
5. A. Bhattacharyya, A. Pramanik, S.K. Maji, S. Haldar, U.K. Mukhopadhyay, J. Mukherjee, Utilization of vinasse for production of poly-3-(hydroxybutyrate-co-hydroxyvalerate) by Haloferax mediterranei. AMB Express **2**, 34 (2012)
6. A. Bhattacharyya, K. Jana, S. Haldar, Integration of poly-3-(hydroxybutyrate-co-hydroxyvalerate) production by Haloferax mediterranei through utilization of stillage from rice-based ethanol manufacture in India and its techno-economic analysis. World J. Microbiol. Biotechnol. **31**, 717–727 (2015). https://doi.org/10.1007/s11274-015-1823-4
7. K. Bohmert, I. Balbo, J. Kopka, V. Mittendorf, C. Nawrath, Y. Poirier, G. Tischendorf, R.N. Trethewey, L. Willmitzer, È. Biologie, F. Biologie-chemie-pharmazie, F. Universita, Transgenic Arabidopsis plants can accumulate polyhydroxybutyrate to up to 4% of their fresh weight. Planta **211**, 841–845 (2000)
8. E. Bugnicourt, P. Cinelli, A. Lazzeri, V. Alvarez, Polyhydroxyalkanoate (PHA): Review of synthesis, characteristics, processing and potential applications in packaging. Mater. Sci. **8**, 791–808 (2014). https://doi.org/10.3144/expresspolymlett.2014.82
9. F. Cabrera, Á. Torres, J.L. Campos, D. Jeison, Effect of operational conditions on the behaviour and associated costs of mixed microbial cultures for PHA production. Polymers (Basel) **11**, 191 (2019)
10. P. Chakraborty, W. Gibbons, K. Muthukumarappan, Conversion of volatile fatty acids into polyhydroxyalkanoate by Ralstonia eutropha. J. Appl. Microbiol. **106**, 1996–2005 (2009)
11. G.-Q. Chen, New challenges and opportunities for industrial biotechnology. Microb. Cell Fact. **11**, 111 (2012)
12. G.-Q. Chen, M.K. Patel, Plastics derived from biological sources: Present and future: A technical and environmental review. Chem. Rev. **112**, 2082–2099 (2011)
13. J. Choi, S.Y. Lee, Factors affecting the economics of polyhydroxyalkanoate production by bacterial fermentation. Appl. Microbiol. Biotechnol. **51**, 13–21 (1999)
14. D. Choi, D.C. Chipman, S.C. Bents, R.C. Brown, A techno-economic analysis of polyhydroxyalkanoate and hydrogen production from syngas fermentation of gasified biomass. Appl. Biochem. Biotechnol. **160**, 1032–1046 (2010)
15. G. de Roo, M.B. Kellerhals, Q. Ren, B. Witholt, B. Kessler, Production of chiral R-3-hydroxyalkanoic acids and R-3-hydroxyalkanoic acid methylesters via hydrolytic degradation of polyhydroxyalkanoate synthesized by pseudomonads. Biotechnol. Bioeng. **77**, 717–722 (2002)
16. G.J. Domski, J.M. Rose, G.W. Coates, A.D. Bolig, M. Brookhart, Living alkene polymerization: New methods for the precision synthesis of polyolefins. Prog. Polym. Sci. **32**, 30–92 (2007)
17. G.C. Du, J. Chen, J. Yu, S. Lun, Feeding strategy of propionic acid for production of poly (3-hydroxybutyrate-co-3-hydroxyvalerate) with Ralstonia eutropha. Biochem. Eng. J. **8**, 103–110 (2001)
18. L. Favaro, M. Basaglia, S. Casella, Improving polyhydroxyalkanoate production from inexpensive carbon sources by genetic approaches: A review. Biofuels Bioprod. Biorefin. **13**, 208–227 (2019)
19. D. Fernández, E. Rodríguez, M. Bassas, M. Viñas, A.M. Solanas, J. Llorens, A.M. Marqués, A. Manresa, Agro-industrial oily wastes as substrates for PHA production by the new strain Pseudomonas aeruginosa NCIB 40045: Effect of culture conditions. Biochem. Eng. J. **26**, 159–167 (2005)

20. C. Fernández-dacosta, J.A. Posada, R. Kleerebezem, M.C. Cuellar, A. Ramirez, Microbial community-based polyhydroxyalkanoates (PHAs) production from wastewater: Techno-economic analysis and ex-ante environmental assessment. Bioresour. Technol. **185**, 368–377 (2015). https://doi.org/10.1016/j.biortech.2015.03.025
21. M. Francavilla, P. Manara, P. Kamaterou, M. Monteleone, A. Zabaniotou, Cascade approach of red macroalgae Gracilaria gracilis sustainable valorization by extraction of phycobiliproteins and pyrolysis of residue. Bioresour. Technol. **184**, 305–313 (2015). https://doi.org/10.1016/j.biortech.2014.10.147
22. T. Fukui, Y. Doi, Efficient production of polyhydroxyalkanoates from plant oils by Alcaligenes eutrophus and its recombinant strain. Appl. Microbiol. Biotechnol. **49**, 333–336 (1998)
23. T.U. Gerngross, Can biotechnology move us toward a sustainable society? Nat. Biotechnol. **17**, 541 (1999)
24. J.G.C. Gomez, M.F.A. Rodrigues, R.C.P. Alli, B.B. Torres, C.L.B. Netto, M.S. Oliveira, L.F. Da Silva, Evaluation of soil gram-negative bacteria yielding polyhydroxyalkanoic acids from carbohydrates and propionic acid. Appl. Microbiol. Biotechnol. **45**, 785–791 (1996)
25. E. Grothe, M. Moo-Young, Y. Chisti, Fermentation optimization for the production of poly (β-hydroxybutyric acid) microbial thermoplastic. Enzym. Microb. Technol. **25**, 132–141 (1999)
26. C. Guoqiang, W.U. Qiong, C. Jinchun, Biosynthesis of polyhydroxyalkanoates. Tsinghua Sci. Technol. **6**, 193–199 (2001)
27. N. Gurieff, P. Lant, Comparative life cycle assessment and financial analysis of mixed culture polyhydroxyalkanoate production. Bioresour. Technol. **98**, 3393–3403 (2007)
28. R. Haas, B. Jin, F.T. Zepf, Production of poly (3-hydroxybutyrate) from waste potato starch. Biosci. Biotechnol. Biochem. **72**, 253–256 (2008)
29. C. Hermann-Krauss, M. Koller, A. Muhr, H. Fasl, F. Stelzer, G. Braunegg, Archaeal production of polyhydroxyalkanoate (PHA) co-and terpolyesters from biodiesel industry-derived by-products. Archaea **2013**, 1–10 (2013)
30. N. Jacquel, C.-W. Lo, Y.-H. Wei, H.-S. Wu, S.S. Wang, Isolation and purification of bacterial poly (3-hydroxyalkanoates). Biochem. Eng. J. **39**, 15–27 (2008)
31. M.R. Jangra, R. Batra, N. Passricha, V.K. Sikka, Cloning, sequencing and in silico analysis of phbC gene from Pseudomonas spp. Indian J. Microbiol. **59**, 58–63 (2019)
32. Y. Jiang, L. Marang, J. Tamis, M.C.M. van Loosdrecht, H. Dijkman, R. Kleerebezem, Waste to resource: Converting paper mill wastewater to bioplastic. Water Res. **46**, 5517–5530 (2012)
33. K. Johnson, Y. Jiang, R. Kleerebezem, G. Muyzer, M.C.M. van Loosdrecht, Enrichment of a mixed bacterial culture with a high polyhydroxyalkanoate storage capacity. Biomacromolecules **10**, 670–676 (2009)
34. D. Kadouri, E. Jurkevitch, Y. Okon, S. Castro-Sowinski, Ecological and agricultural significance of bacterial polyhydroxyalkanoates. Crit. Rev. Microbiol. **31**, 55–67 (2005)
35. S.T. Kenny, J.N. Runic, W. Kaminsky, T. Woods, R.P. Babu, C.M. Keely, W. Blau, K.E. O'Connor, Up-cycling of PET (polyethylene terephthalate) to the biodegradable plastic PHA (polyhydroxyalkanoate). Environ. Sci. Technol. **42**, 7696–7701 (2008)
36. L. Kilian, The impact of the shale oil revolution on US oil and gasoline prices. Rev. Environ. Econ. Policy **10**, 185–205 (2016)
37. S. Kim, B. Dale, Life cycle assessment study of biopolymers (polyhydroxyalkanoates)-derived from no-tilled corn (11 pp). Int. J. Life Cycle Assess. **10**, 200–210 (2005)
38. B.S. Kim, S.C. Lee, S.Y. Lee, H.N. Chang, Y.K. Chang, S.I. Woo, Production of poly (3-hydroxybutyric acid) by fed-batch culture of Alcaligenes eutrophus with glucose concentration control. Biotechnol. Bioeng. **43**, 892–898 (1994)
39. M. Koller, R. Bona, G. Braunegg, C. Hermann, P. Horvat, M. Kroutil, J. Martinz, J. Neto, L. Pereira, P. Varila, Production of polyhydroxyalkanoates from agricultural waste and surplus materials. Biomacromolecules **6**, 561–565 (2005)
40. M. Koller, P. Hesse, R. Bona, C. Kutschera, A. Atlić, G. Braunegg, Potential of various archae- and eubacterial strains as industrial polyhydroxyalkanoate producers from whey. Macromol. Biosci. **7**, 218–226 (2007)

41. M. Koller, A. Atlić, Y. Gonzalez-Garcia, C. Kutschera, G. Braunegg, Polyhydroxyalkanoate (PHA) biosynthesis from whey lactose. Macromol. Symp. **272**, 87–92 (2008)
42. M. Koller, A. Salerno, M. Dias, A. Reiterer, G. Braunegg, Modern biotechnological polymer synthesis: A review. Food Technol Biotechnol **48**, 255–269 (2010)
43. M. Koller, H. Niebelschütz, G. Braunegg, Strategies for recovery and purification of poly [(R)-3-hydroxyalkanoates] (PHA) biopolyesters from surrounding biomass. Eng. Life Sci. **13**, 549–562 (2013)
44. Y.K. Leong, P.L. Show, H.C. Lin, C.K. Chang, H.-S. Loh, J.C.-W. Lan, T.C. Ling, Preliminary integrated economic and environmental analysis of polyhydroxyalkanoates (PHAs) biosynthesis. Bioresour. Bioprocess. **3**, 41 (2016)
45. I. Levett, G. Birkett, N. Davies, A. Bell, A. Langford, B. Laycock, P. Lant, S. Pratt, Techno-economic assessment of poly-3-hydroxybutyrate (PHB) production from methane—The case for thermophilic bioprocessing. J. Environ. Chem. Eng. **4**, 3724–3733 (2016)
46. L.L. Madison, G.W. Huisman, Metabolic engineering of poly (3-hydroxyalkanoates): From DNA to plastic. Microbiol. Mol. Biol. Rev. **63**, 21–53 (1999)
47. Market and Market, Market and market (2018), https://www.marketsandmarkets.com/Market-Reports/pha-market-395.html
48. V. Martínez, P. García, J.L. García, M.A. Prieto, Controlled autolysis facilitates the polyhydroxyalkanoate recovery in Pseudomonas putida KT2440. Microb. Biotechnol. **4**, 533–547 (2011)
49. Z.B. Mokhtari-Hosseini, E. Vasheghani-Farahani, A. Heidarzadeh-Vazifekhoran, S.A. Shojaosadati, R. Karimzadeh, K.K. Darani, Statistical media optimization for growth and PHB production from methanol by a methylotrophic bacterium. Bioresour. Technol. **100**, 2436–2443 (2009a)
50. Z.B. Mokhtari-Hosseini, E. Vasheghani-Farahani, S.A. Shojaosadati, R. Karimzadeh, A. Heidarzadeh-Vazifekhoran, Effect of feed composition on PHB production from methanol by HCDC of Methylobacterium extorquens (DSMZ 1340). J. Chem. Technol. Biotechnol. **84**, 1136–1139 (2009b)
51. E.R. Olivera, M. Arcos, G. Naharro, J.M. Luengo, Unusual PHA biosynthesis, in *Plastics from Bacteria*, (Springer, Berlin/Heidelberg, 2010), pp. 133–186
52. W. Pan, J.A. Perrotta, A.J. Stipanovic, C.T. Nomura, J.P. Nakas, Production of polyhydroxyalkanoates by Burkholderia cepacia ATCC 17759 using a detoxified sugar maple hemicellulosic hydrolysate. J. Ind. Microbiol. Biotechnol. **39**, 459–469 (2012)
53. S.J. Park, T.W. Kim, M.K. Kim, S.Y. Lee, S.-C. Lim, Advanced bacterial polyhydroxyalkanoates: Towards a versatile and sustainable platform for unnatural tailor-made polyesters. Biotechnol. Adv. **30**, 1196–1206 (2012)
54. G. Penloglou, C. Chatzidoukas, C. Kiparissides, Microbial production of polyhydroxybutyrate with tailor-made properties: An integrated modelling approach and experimental validation. Biotechnol. Adv. **30**, 329–337 (2012)
55. D. Pfeiffer, D. Jendrossek, Localization of PHB granule associated proteins during PHB granule formation and identification of two new phasins, PhaP6 and PhaP7, in Ralstonia eutropha H16. J. Bacteriol. **194**, 5909–5921 (2012)
56. S. Povolo, M.G. Romanelli, M. Basaglia, V.I. Ilieva, A. Corti, A. Morelli, E. Chiellini, S. Casella, Polyhydroxyalkanoate biosynthesis by Hydrogenophaga pseudoflava DSM1034 from structurally unrelated carbon sources. New Biotechnol. **30**, 629–634 (2013)
57. R. Rai, T. Keshavarz, J.A. Roether, A.R. Boccaccini, I. Roy, Medium chain length polyhydroxyalkanoates, promising new biomedical materials for the future. Mater. Sci. Eng. R Rep. **72**, 29–47 (2011)
58. J.S. Sabirova, M. Ferrer, H. Lünsdorf, V. Wray, R. Kalscheuer, A. Steinbüchel, K.N. Timmis, P.N. Golyshin, Mutation in a "tesB-like" hydroxyacyl-coenzyme A-specific thioesterase gene causes hyperproduction of extracellular polyhydroxyalkanoates by Alcanivorax borkumensis SK2. J. Bacteriol. **188**, 8452–8459 (2006)
59. M.C. Santimano, N.N. Prabhu, S. Garg, PHA production using low-cost agro-industrial wastes by Bacillus sp. strain COL1/A6. Res. J. Microbiol. **4**, 89–96 (2009)

60. L. Sharma, Y. Ogino, T. Kanaya, T. Iwata, Y. Doi, Fiber formation in medium and ultra-high-molecular-weight polyhydroxybutyrate blends under shear flow. Macromol. Mater. Eng. **289**, 1068–1073 (2004)
61. A. Shrivastav, S.K. Mishra, B. Shethia, I. Pancha, D. Jain, S. Mishra, Isolation of promising bacterial strains from soil and marine environment for polyhydroxyalkanoates (PHAs) production utilizing Jatropha biodiesel byproduct. Int. J. Biol. Macromol. **47**, 283–287 (2010)
62. A.-M. Smit, T.J. Strabala, L. Peng, P. Rawson, G. Lloyd-Jones, T.W. Jordan, Proteomic phenotyping of Novosphingobium nitrogenifigens reveals a robust capacity for simultaneous nitrogen fixation, polyhydroxyalkanoate production and resistance to reactive oxygen species. Appl. Environ. Microbiol. **78**(14), 4802–4815 (2012)
63. B. Sonnleitner, E. Heinzle, G. Braunegg, R.M. Lafferty, Formal kinetics of poly-β-hydroxybutyric acid (PHB) production in Alcaligenes eutrophus H 16 and Mycoplana rubra R 14 with respect to the dissolved oxygen tension in ammonium-limited batch cultures. Eur. J. Appl. Microbiol. Biotechnol. **7**, 1–10 (1979)
64. M. Stritzler, C. Berini, C. Jozefkowicz, G. Soto, N. Ayub, Understanding the intracellular-to-extracellular localization switch of polyhydroxybutyrate polymerase in pseudomonas backgrounds as a microevolutionary process. J. Theor. Biol. **456**, 29–33 (2018)
65. D. Tan, Y.-S. Xue, G. Aibaidula, G.-Q. Chen, Unsterile and continuous production of polyhydroxybutyrate by Halomonas TD01. Bioresour. Technol. **102**, 8130–8136 (2011)
66. T. Tanaka, T. Iwata, Y. Aoyagi, Y. Doi, Processing strong fibers of microbial biodegradable aliphatic polyesters-drawing and annealing conditions of polyhydroxyalkanoates fibers. Text. Res. J. **74**, 1031 (2004)
67. M. Tortajada, L. Ferreira, M.A. Prieto, Second-generation functionalized medium-chain-length polyhydroxyalkanoates: The gateway to high-value bioplastic applications. Int. Microbiol. **16**, 1–15 (2013)
68. R.J. Van Wegen, Y. Ling, A.P.J. Middelberg, Industrial production of polyhydroxyalkanoates using Escherichia coli: An economic analysis. Chem. Eng. Res. Des. **76**, 417–426 (1998)
69. F. Wang, S.Y. Lee, Poly (3-hydroxybutyrate) production with high productivity and high polymer content by a fed-batch culture of Alcaligenes latus under nitrogen limitation. Appl. Environ. Microbiol. **63**, 3703–3706 (1997)
70. Y. Wang, J. Yin, G.-Q. Chen, Polyhydroxyalkanoates, challenges and opportunities. Curr. Opin. Biotechnol. **30**, 59–65 (2014)
71. X. Wang, G. Carvalho, M.A.M. Reis, A. Oehmen, Metabolic modeling of the substrate competition among multiple VFAs for PHA production by mixed microbial cultures. J. Biotechnol. **280**, 62–69 (2018)
72. K.-D. Wendlandt, M. Jechorek, J. Helm, U. Stottmeister, Production of PHB with a high molecular mass from methane. Polym. Degrad. Stab. **59**, 191–194 (1998)
73. J. Yu, L.X.L. Chen, Cost-effective recovery and purification of polyhydroxyalkanoates by selective dissolution of cell mass. Biotechnol. Prog. **22**, 547–553 (2006)
74. C. Zhu, C.T. Nomura, J.A. Perrotta, A.J. Stipanovic, J.P. Nakas, Production and characterization of poly-3-hydroxybutyrate from biodiesel-glycerol by Burkholderia cepacia ATCC 17759. Biotechnol. Prog. **26**, 424–430 (2010)

Chapter 13
Production and Applications of Pyrolytic Oil and Char from Lignocellulosic Residual Biomass

J. Argudo-Santamaria, H. A. R. Ortiz, B. D. Cano, I. Auclair, M. L. S. Silva, J. L. V. Palomino, F. D. Fernández, S. L. A. Garcia, T. T. H. Pham, and A. A. Ramírez

13.1 Introduction

Organic matter waste is defined as any residue that decomposes under the action of microorganisms. The terms "compostable or putrescible materials" are reserved for the residential sector. In the industrial, commercial, and institutional (ICI) sector, the term used is "residual organic matter (ROM)" [27].

The main ROM generated by the ICI sector are wood and food processing residues, marine residues, and other commercial residues such as those from food markets and restaurants. The wood residues generated by the construction, renovation, and demolition (CRD) sector come from two main subsectors: infrastructure (e.g., roads) and construction [29]. However, in the industrial sector, the highest quantities of residues are generated by food processing activities. The types and amounts of

J. Argudo-Santamaria · M. L. S. Silva (✉) · J. L. V. Palomino · F. D. Fernández
Universidad Castilla La Mancha, Ciudad Real, Spain
e-mail: MariaLuz.Sanchez@uclm.es

H. A. R. Ortiz · S. L. A. Garcia
Instituto Potosino de Investigación Científica y Tecnológica A.C., San Luis Potosí, Mexico

B. D. Cano · A. A. Ramírez (✉)
Centre National en Électrochimie et en Technologies Environnementales, Shawinigan, QC, Canada
e-mail: aaramirez@cnete.qc.ca

I. Auclair
University of Québec at Trois-Rivières, Trois-Rivières, QC, Canada

T. T. H. Pham
Centre National en Électrochimie et en Technologies Environnementales, Shawinigan, QC, Canada

University of Québec at Trois-Rivières, Trois-Rivières, QC, Canada

M. O. Daramola, A. O. Ayeni (eds.), *Valorization of Biomass to Value-Added Commodities*, Green Energy and Technology,
https://doi.org/10.1007/978-3-030-38032-8_13

ROM vary significantly from sector to sector. To date, there are few data available that would allow us to determine in a significant way the amounts of ROM generated by each industrial sector [27].

13.1.1 Generation of ROM in North America

In 2016, cities generated 2.01 billion tonnes of solid waste worldwide. Waste generation is expected to increase to 3.40 billion tonnes in 2050, an increase of 70.0% from 2016 levels [80]. North America (i.e., Canada, USA, and Mexico) generated around 260 million tonnes of waste in 2012, with an individual contribution of 7.1% for Canada, 77.7% for the USA, and 15.2% for Mexico.

In 2012, 10.18 million tonnes of ROM were generated in Canada, or 299.8 kg/person. The composition of residential waste was food, yard, and paper wastes in a proportion of 28.0%, 38.0%, and 34.0%, respectively. ICI waste generation accounted for a total of 8.3 million tonnes, 242.9 kg/person, with 34.0% of food, 7.0% of yard and wood, and 59.0% of paper waste [9]. The total generation of ROM in Canada in 2012 was 18.4 million tonnes, of which only 24.0% (including both residential and ICI) was diverted from disposal, as shown in Fig. 13.1. This means that close to 14 million tonnes could be valorized.

Canada does not provide a policy framework for solid waste. Instead, each province and territory has its own regulations, guidelines, and policies specifying solid waste treatment and disposal. Municipalities control residential wastes, while the ICI entities manage their wastes by themselves.

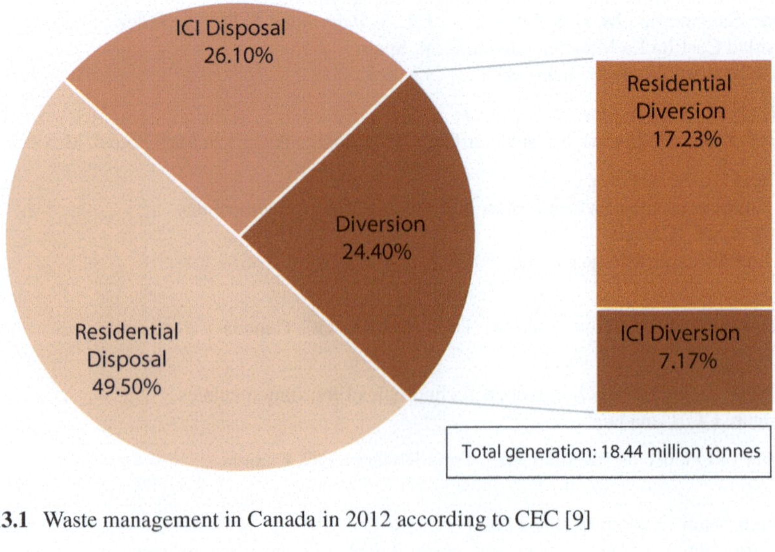

Fig. 13.1 Waste management in Canada in 2012 according to CEC [9]

In the specific case of the province of Quebec, the *Residual Materials Management Policy* was recently approved, intended as the foundation of a green economy. The 2.5 million tonnes of the most commonly recycled residual materials recovered in Quebec in 2006 (metal, paper, cardboard, plastic, and glass) were valued at $550 million and generated over 10,000 direct jobs. This policy is meant to address three main challenges:

- Ending resource waste
- Promoting the achievement of the goals of the Climate Change Action Plan and of the Quebec Energy Strategy
- Making all involved stakeholders responsible for residual material management [30]

In 2015, the generation of ROM in the province of Quebec was estimated at 4.4 million tonnes, excluding the agri-food sector. One million tonnes were valorized by composting processes, corresponding to 24.6% of ROM being diverted from disposal [28], as shown in Fig. 13.2.

Mexico lacks a reliable record of ICI waste generation. In 2012, residential ROM generation was estimated at 27.9 million tonnes, or 238.2 kg/person. The approximate composition was 79.0% of food and yard and 21.0% of paper waste [9]. This indicates a diversion of only 8.9% of generated waste, as shown in Fig. 13.3.

In the USA, the data for 2014 shows that the residential ROM generated represented 222.4 kg/person, for a total of 70.9 million tonnes, the highest total amount in North America. 34.0% of it was food, 32.0% yard, 33.0% paper, and 1.0% wood waste. For ICI, 137.0 million tonnes were generated, 429.8 kg/person. The composition of ICI ROM was 30.0% institutional and commercial food, 25.0% industrial food processing, 6.0% yard, 29.0% paper, and 10.0% wood waste. Figure 13.4 shows the diversion and disposal of ROM in the USA.

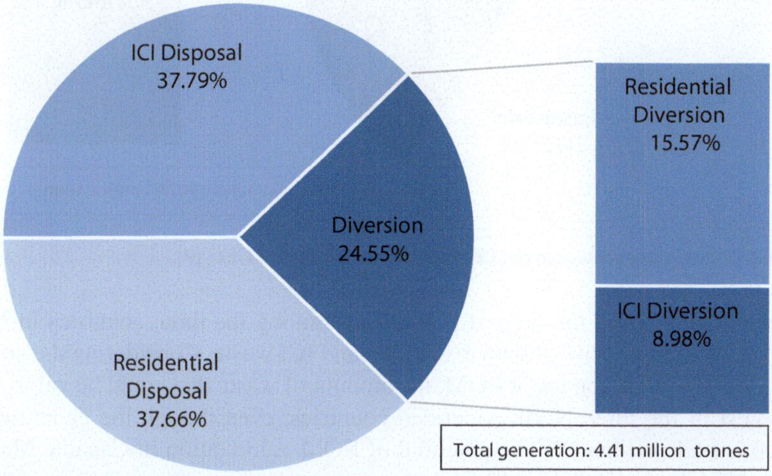

Fig. 13.2 Waste management in Quebec in 2015

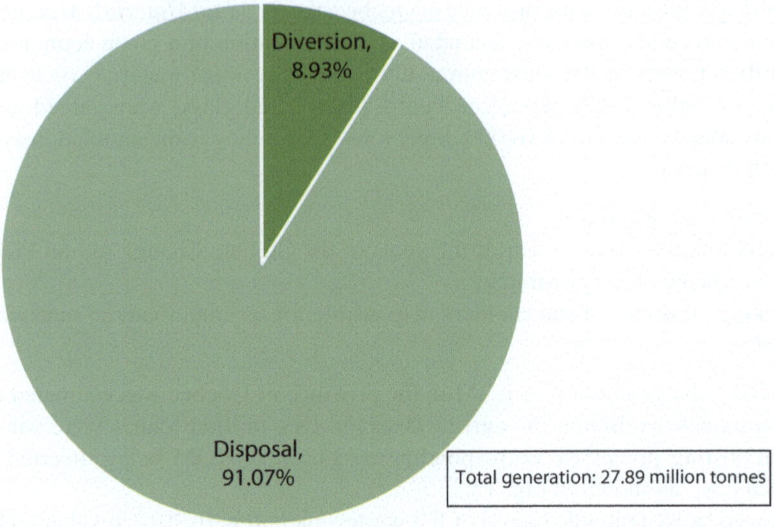

Fig. 13.3 Residential waste management in Mexico in 2012 according to CEC [9]

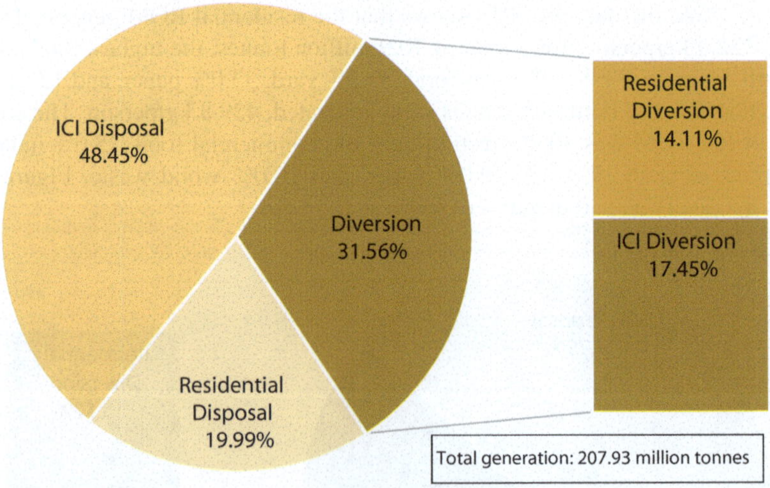

Fig. 13.4 Waste management in the USA in 2014 according to CEC [9]

Diversion achieved the biggest proportions among the three countries in North America, reaching 31.6% of both residential and ICI waste. Considering the generation of 207.9 million tonnes of ROM, the amount of waste that could be valorized is the largest of the three North American countries, even though the generation in Mexico is underestimated. The potential of ROM valorization in Canada, Mexico, and the USA is shown in Fig. 13.5.

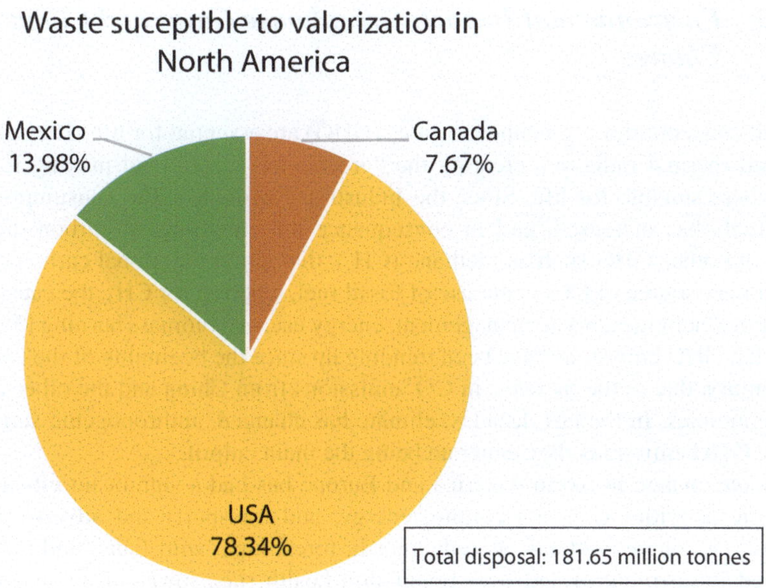

Fig. 13.5 Valorization potential of waste in North America according to CEC [9]

In the case of the European Union, in 2014 the management of solid waste was as follows: 28% of the total solid waste generated in the 27 states was recycled, 27% was incinerated, 16% was composted, and 28% was landfilled.

13.1.2 Socioeconomic Aspect

The byproducts obtained via ROM valorization are varied. They can be used in different industrial processes, contributing to a decrease in the environmental impact of the consumption of fresh raw materials. ROM can be used as feedstock for industrial products such as food supplements, health and beauty products, fertilizers, apparel, and pet food.

In addition, it is important to take into account the socioeconomic benefits that are correlated to these byproducts. For example, renewable natural gas obtained from food waste has a negative life-cycle carbon footprint (−23 g CO_2e/Mj). Captured biogas and digestate produced during anaerobic digestion can be used as energy source (in the form of heat and/or electricity) and biofertilizer, respectively [9].

The uses of compost are dictated by its quality. When the compost has harsh foreign matter or a heightened content of trace elements, it may only be used where human contact is less frequent, whereas high-quality compost can be used in agricultural lands, residential gardens, and horticultural operations.

13.1.3 Environmental Impact: Greenhouse Gases and Climate Change

In small concentrations, greenhouse gases (GHG) are essential for life. They absorb and emit thermal radiation, creating the "greenhouse effect" and making Earth's atmosphere suitable for life. Since the Industrial Revolution, the consumption of fossil fuels has increased, and in consequence the emissions of carbon dioxide (CO_2) and other GHG such as methane (CH_4), disrupting the global carbon cycle. The primary source of CO_2 is the use of fossil fuel, whereas for CH_4, the causes are agricultural activities, waste management, energy use, and biomass burning [35, 71].

Global GHG emissions have been trending up since the beginning of the twenty-first century, due to the increase in CO_2 emissions from China and the other emergent economies. In the last decades, climate has changed, anthropogenic activities and the GHG emissions they generate being the main culprit.

Climate change in North America and Europe has had a significant effect over economic activities (i.e., agriculture, energy, and transport) and adverse social impacts (i.e., on ecosystems, as reduction in forest area, wild fauna and flora, or damaged infrastructure) and over population health (i.e., lower of air and water quality and increase in mortality, chronic and new infectious diseases). According to the climate change evidences, the development of measures and policies allows us to anticipate the risks of climate change and be able to ensure the society adapts to the changes across different sectors [38, 42, 73]. In order to achieve that goal, reducing our dependence on fossil fuels and improving forest conservation as greenhouse gas sinks will be considered [15]. Carbon recycling by means of biomass valorization is one of the most viable options for that purpose. For example, one of the most documented applications of densified roasted biomass is as "renewable" fuel in coal-fired power plants. Notably, the use of 1 tonne of biofuel reduces net CO_2 emissions by almost 3 tonnes and significantly reduces NOx and SOx emissions from mineral coal [26].

13.1.4 Management of Residual Organic Materials

There are several technologies commercially available to valorize ROM, other than thermochemical processes. For example, according to CEC [9]:

- Animal rendering: the cooking and drying processes by which portions of livestock and poultry that are not intended for human consumption are converted into edible and inedible byproducts, thereby providing additional revenue for the meat industry and avoiding costly disposal.
- Anaerobic processing: the natural process that breaks down organic matter in the absence of oxygen to release a gas known as biogas consisting mainly of CO_2 and CH_4, leaving an organic residue called digestate.
- Aerobic processing and treatment: also known as composting, it refers to the transformation of organic materials into humus by aerobic microorganisms.

13.1.4.1 Thermochemical Technologies

Various technologies can be used for the thermochemical conversion of biomass, including gasification [26, 56], pyrolysis [25, 26, 57, 64, 75], roasting [20, 26, 66], carbonization [26], combustion [26, 57], and hydrothermal processes [88].

Gasification is carried out at high temperatures, from 900 to 2000 °C, which favors gas production.

Pyrolysis is the method used to produce more oil than gas and char [26], consisting of the thermal decomposition of biomass under an inert atmosphere (e.g., nitrogen) [58] or in the presence of a minimal amount of oxygen, at an average temperature from 300 to 900 °C [16].

Roasting is considered a mild pyrolysis process, carried out at a lower temperature [26, 60], between 200 and 350 °C [26, 41, 72]. It produces a high ratio of char [66].

Carbonization is a slow thermochemical process that takes place with an air deficit and a pressure equal to or greater than the atmospheric pressure. Unlike pyrolysis, carbonization is used to produce the maximum amount of char at the expense of oil [26].

The combustion of biomass is defined by a complete oxidation of the material in the presence of air. This technique generates energy that can be used for heating, power generation, or both [26].

Hydrothermal processes can be divided into five categories, based on their temperature range: hot water extraction (HWE) at low temperature (<100 °C), pressurized hot water extraction (PHWE), liquid hot water pretreatment (LHW), hydrothermal carbonization (HTC) at medium temperature (100–250 °C), and hydrothermal liquefaction (HTL) at high temperature (>280 °C) [13].

13.1.4.2 Trends and Challenges of Thermochemical Valorization

Each byproduct obtained through thermochemical valorization has specific characteristics that limit its industrial exploitation. These limitations must be overcome in order to develop industrial thermochemical mills. To produce heat, the direct combustion of wood seems to be the best solution from the point of view of economy and ease of use. However, the energy produced must be consumed locally. For electricity generation, gasification has almost reached full development. Pyrolysis, which makes it possible to densify the heat capacity of wood in a liquid (bio-oil) that can be transported, stored, and used for transport, encounters the difficulties of using this liquid in standard equipment designed for petroleum products. Carbonization densifies the calories of wood in a solid biochar, which is transportable and storable and can be used as fuel, alone or in mixture with petroleum fuels (e.g., heating oil) in the form of a slurry (particles in suspension) [26].

The choice among thermochemical technologies depends on the energy needs, the availability and quality of the biomass, its proximity, the costs of transformation and distribution, the environmental context, and the availability of technologies to consume the byproducts generated. For example, the oil produced by pyrolysis has very particular properties that do not allow it to be used in diesel engines, turbines, and standard boilers without upgrading [26].

13.2 Technologies to Produce Pyrolytic Oil and Char

13.2.1 Thermochemical Conversion Definition and Principles

During thermochemical conversion, the chemical structure of the molecules contained in the biomass, such as cellulose, hemicellulose, and lignin, breaks down, causing the reorganization of molecules and the production of gas, oil, and a solid black material rich in carbon, customarily called "char." The char, also called "biochar," differs from mineral "coal" because it is not from fossil origin [16]. In addition, fossil coal contains significant amounts of sulfur and mercury, which makes it highly polluting compared to the char, which is also considered carbon neutral [26].

Three main steps occur during the thermochemical conversion of biomass: (1) loss of moisture in the material; (2) primary production of char, volatiles, and gases; and (3) slow decomposition of char and its chemical rearrangement to form a solid rich in carbon [19].

13.2.2 Main Technologies to Produce Oil

Pyrolysis is the thermochemical process to produce oil. Depending on the resulting products and operating conditions, pyrolysis can be divided into slow, intermediate, and fast pyrolysis [2].

Slow pyrolysis is a batch process carried out at low temperatures, with slow heating rates and long residence times. The initial sample does not need significant pretreatment. The process is more tolerant regarding the initial conditions of the sample. Although liquid fuel can be produced, this technique is mostly used when the desired product is biochar.

Intermediate pyrolysis is carried out at temperatures of between 300 and 500 °C, vapor residence time of a few seconds, and feedstock residence time between 0.5 and 25 min [2].

Nowadays, fast pyrolysis is an advanced technology that is gaining in importance because of the increasing interest in the production of biofuel from biomass. The aim of this process is to prevent further cracking of the pyrolysis products into non-condensable compounds. The main features of fast pyrolysis process to achieve the maximum amount of bio-oil are the following [83]:

- Small particle sizes usually less than 5 mm, to improve the heating rates and quick devolatilization
- Feed moisture content under 10 wt.%

Table 13.1 Operating conditions used for fast, intermediate, and slow pyrolysis and gasification, according to van Swaaij and Palz [83]

Process	Operating conditions	Bio-oil yield (wt.%)	Biochar yield (wt.%)	Gases (wt.%)
Fast pyrolysis	400–600 °C 10^3–10^4 °C/s Short vapor residence time (0.5–15 s)	75	12	13
Intermediate pyrolysis	300–500 °C 25–30 °C/min Moderate vapor residence time (30 s)	50	25	25
Slow pyrolysis	400 °C 5 °C/min Long vapor residence time (hours, days)	30	35	35
Gasification	800–900 °C Moderate hot-vapor time (5 s)	3	1	96

- High heating rates and high heat transfer rates at the biomass-particle reaction interface because the rate of particle heating is usually the limiting step
- Controlling the temperature around 500 °C to maximize the liquid yield
- Short vapor residence time, usually between 0.5 and 15 s, to minimize the secondary reactions
- Rapid cooling of the pyrolysis vapors to give the bio-oil product

Table 13.1 summarizes the operating conditions of each process and typical product weight yields.

However, there are some drawbacks when the fast pyrolysis process is scaled up. The main one is the ability to reproduce the operating conditions. The high heating rates required are currently only possible at laboratory scale. Industrial pyrolysis does not get bio-oil yields as high as those reported for fast pyrolysis. Moreover, the biomass particle size has to very small, leading to a significant amount of energy required to grind and condition the biomass. The second drawback is related to the inability to isolate each final component desired without modifying the rest, because of the complex internal structure of the biomass used.

Several authors have reported studies focused on the flash pyrolytic conversion of lignocellulosic biomass. Zheng et al. [96] carried out catalytic fast pyrolysis of lignocellulosic biomass to promote the aromatic compounds production. Maliutina et al. [55] carried out a comparative study of the flash pyrolysis characteristics of microalgal and lignocellulosic biomasses, with the lignocellulosic biomass providing higher bio-oil yields. Wang et al. [86] studied the microwave-assisted co-pyrolysis of acid pretreated bamboo sawdust and soapstock in order to achieve the highest production of bio-oil. They got a remarkable improvement in the bio-oil yield with the pretreatment. Lestander et al. [48] characterized the bio-oil properties produced through lignocellulosic biomass pyrolysis at different operating conditions. Adjin-Tetteh et al. [1] carried out a fast pyrolysis of cocoa pod husks for direct production of valuable platform chemicals. Charusiri and Vitidsant [12] analyzed the operating conditions to maximize the biofuel production via the pyrolysis of sugarcane leaves.

13.2.3 Main Technologies to Produce Char

A variety of thermochemical processes can be used to convert biomass into char, and pretreatments may be necessary to improve the biomass transformation. Fragmentation and drying are frequently required before the biomass can be used in a thermochemical process. Fragmentation consists in the reduction and homogenization of the size of the biomass as raw material by shredding or grinding. This activity provides a material that is easier to handle and dry. The level of dryness of the biomass is the most important property of the raw material; it helps conserve the biomass during storage. In addition, it has a significant impact on fuel energy content and production costs [26].

13.2.3.1 Available Technologies

The main thermochemical conversion technologies for char production are pyrolysis, roasting, and carbonization. The choice of one technology depends on the energy needs, the availability and the quality of the biomass, the costs of the infrastructures of transformation and distribution, the environmental context, and the market that can consume the products generated [26].

Pyrolysis

There are several enterprises specialized in pyrolysis. Some of the ones active in North America and Europe are presented below.

Pyrovac (Saguenay, QC) has developed a vacuum biomass pyrolysis plant with a capacity of 3.5 t/h of wood bark. The coarsely ground raw material (2–3 cm) enters the reactor after sieving to remove fine particles. A small biomass tank brings the biomass to the upper plateau of the reactor, which is heated with molten salts at temperatures from 500 to 520 °C. At the end of the upper plate, the carbonized raw material falls onto the lower plate, where the char formed fills a vacuum chamber which will supply the pyrolysis reactor under vacuum to produce the pyrolytic oil. Pyrovac's pyrolysis under vacuum system produces bio-oil (30.7%), an aqueous fraction (19.6%), biochar (29.2%), and gas (20.5%) [26].

Ensyn Technologies Inc. (Renfrew, ON) produces bio-oil for use as a food supplement and chemical compounds. The system consumes 33,000 tonnes of sawdust annually (100 t/d) using the Rapid Thermal Processing (RTP)™ technology process, developed at the University of Western Ontario. This process transforms wet fragmented biomass at atmospheric pressure and moderate temperature. Part of the energy generated by the flash pyrolysis is diverted to dry the biomass. Currently, seven commercial production plants use this process in the USA and Canada [26].

Dynamotive Energy Systems Corporation has pilot facilities in Guelph (ON) and a pyrolysis plant in West Lorne (ON), which has the capacity to consume 130 t/d of solid wood flooring factory byproducts and a 2.5 MW turbine which generates electricity [42]. The company is also co-owner of a plant in Guelph with a capacity of 200 t/d of wood residues. Dynamotive is also exploring the possibility of using biochar as amendment for agricultural soils. Dynamotive has also designed a flash pyrolysis process and is developing a Biomass INto GasOil (BINGO) refining process [26].

Biogreen® technology, developed by ETIA, is an innovative thermochemical conversion process. It includes a roasting-pyrolysis system that creates energy and materials of interest. The French company ETIA also produces another pyrolyzer called Spirajoule® [23].

Mobile pyrolysis technologies are being manufactured by a variety of companies such as Agri Term Inc. (a spin-off of the Institute for Chemicals and Fuels from Alternative Resources (ICFAR) at the University of Western Ontario) and Tech Shelter Inc. (formerly BioRefinery Inc.) [26]. Bioénergie La Tuque (QC) uses the mobile mode to develop and implement all the conditions to the development of the bioenergy sector in the La Tuque territory, including the production of biofuel [85].

Pyrolysis technologies produce more oil than char. The solid fraction of the pyrolysis, the biochar which is considered carbon-neutral and theoretically sulfur-free, is usually burned alone or mixed with liquid fuels. In this case, the mixture becomes an interesting biofuel for boilers, coal-fired power plants, and cement plants, to reduce the emissions of SO_2 and CO_2 [26]. The char can be densified to form granules or briquettes or to produce heat and even serve as amendment for agricultural soils.

Roasting

Roasting is a process that is generally characterized by (1) heat treatment of biomass at temperatures ranging from 200 to 320 °C; (2) atmospheric pressure, with little or no oxygen (http://www.airex-energy.com/en/technology); and (3) residence time of less than 30 min. Roasting creates roasted biomass, while biochar is obtained by carbonization [26].

Roasting technologies are grouped into two main types: technologies that use fragmented biomass (e.g., from chippers, shredders, etc.) and those that use granules (densified biomass). The type of roasting reactor is an important feature distinguishing these technologies. The different types of reactor are rotary kiln (RK), fixed bed (FB), cyclonic bed (CB), toroidal fluidized bed (TFB), column reactor (CR), multi-soleus reactor (MSR), reactor heated screw (RHS), conveyor belt (CB), and microwave reactor (MWR) [26].

Around the world, there are various patent holders of roasting technology for fragmented biomass: AB Torkapparater (Sweden) (RK), Airex Industries Inc. (Quebec) (CB), Alterna Biocarbon (British Columbia) (FB), Agri-Tech Producers LLC (USA) (VC), Andritz AG (Austria) (RK), Bio3D Application (France) (RK), Biolake (the Netherlands) (VC), BTG Biomass Technology Group BV (the Netherlands) (VC), CMI NESA (Belgium) (MSR), Energy Research Center of the Netherlands (ECN) (the Netherlands) (CR), Umea University-ETPC (Sweden) (VC), FoxCoal BV (the Netherlands) (VC), NewEarth Renewable Energy Inc. (USA) (BC), Rotaware Ltd. (UK) (MWR), Stramproy Group (the Netherlands) (CB), Thermya (France) (CR), Topell Energy BV (the Netherlands) (TFB), Torr-Coal Groep (the Netherlands) (RK), and Wyssmont Inc. (USA) (MSR) [3, 26]. There are also a few holders of pellet roasting technologies, including Energex Inc. (Quebec) and the University of Georgia (USA) [26].

Carbonization

The carbonization technology is classified according to the heating method: internal, external, and with air recirculation. The internally heated process is the most common system globally. It is a process in which the portion of biomass loaded into the furnace (usually wood) provides the heat necessary to carbonize the rest of the load. The amount of wood consumed depends on the amount of air admitted into the oven through the air intake holes, the wood load, and the enclosure. Ovens are usually made of concrete or brick. Their design is simple and the investment costs are very low. The performance of the Missouri kilns varies from 20% to 30%, depending on the operating conditions and the raw material used. The production cycle, which is a function of the cooling period, varies from 25 to 30 days [26].

The external or indirect heating process (e.g., Van Marion Retort (VMR)) consists of a combustion chamber which is heated by an external energy source (gas or oil). When the pyrolysis begins, the pyro-ligneous vapors are sent to the combustion chamber, where they are burned to generate heat that will be used to heat the second container, placed in the second reactor. At this point, the burner (gas or oil) is stopped. When the carbonization is complete in the first reactor, its container with the char is removed and set to let it cool. Since the VMR process works in an alternating way, the vapors formed in one reactor will serve as a source of heat for the other. The total duration of carbonization of a container varies from 8 to 12 h, and the yield of char, which depends on the raw material used and its moisture content, can reach 30–32%. A 12-reactor unit, operating 24 h a day, has a production capacity of 6000–7000 t/year and requires three shift workers. Several factories use this technology in France, Belgium, and the USA [26].

In the gas recirculation process, the heat transfer is very high. The wood is heated by direct contact with hot inert gases recirculated by fans. Since the gases do not contain oxygen, there is no combustion in the reactor. During cooling, heat can be recovered for use in the system. The Reichert and Lambiotte processes are the main processes that use this principle of carbonization. In the Reichert process reactor, the decomposition vapors flow countercurrent with the raw material, taking out moisture. Incondensable gases pass through the heat exchangers to be heated at the carbonization temperature (450–550 °C). The excess gas is used to pre-dry the raw material. The raw material is introduced into the column with a maximum length of 30 cm and a thickness of 10 cm [26].

A relatively new process called Flash Carbonization™ was developed by the University of Hawaii (UH). This process uses high pressure to produce char from biomass. A commercial-scale demonstration reactor was installed at the UH campus. The fixed carbon yield can reach the thermochemical equilibrium limit in 20–30 min. Not all the carbon found in the raw material is transformed into biochar. There are losses in the form of volatile compounds, CO_2, etc. [26].

Hydrothermal carbonization consists in the transformation of slurry into char [6]. To achieve this, an autoclave is required, where the slurry is contained and heated during 2–24 h at temperatures from 150 to 350 °C. Since the slurry is liquid and due to the temperature, vapor is generated, and the pressure rises, leading to a thermochemical transformation of the raw materials contained in the slurry. With this process, gas emissions are reduced and biochar yields improve [68, 81]. This process gives control of the carbonization process and char chemistry as well as particle morphology and size [8].

13.2.4 Simulation of Pyrolysis and Gasification Process to Produce Oil and Char Using Aspen Plus®

Nowadays, renewable chemical industries have attracted a lot of interest from both the economic and ecological perspectives, as they can reduce the use of fossil sources.

The main barriers for scaling up these technologies have to do with the complexity of the process, the optimization of the operating conditions, and the high investment required. To face this problem, the simulation of processes appears to be a good alternative since it allows performing conceptual designs that can be extrapolated at scale and estimating, through mass and energy balances, thermodynamic models, and chemical equilibriums, the behavior of a process, without having to use an actual pilot plant.

Aspen Plus® software is one of the most widespread commercial simulators of chemical engineering processes. It is a software developed by Aspen Technology Inc. It allows to simulate the whole process and analyze the impact of different operating conditions.

There are few studies focused on the simulation of thermochemical processes using biomass as feedstock. Nilsson et al. [62] simulated a fluidized bed dividing it into three zones according to the type of transformation: pyrolysis of biomass, using homogenous and heterogeneous reactions, each with its corresponding kinetics obtained from literature and experiments. Zhai et al. [95] also proposed a two-stage biomass pyrolysis and gasification scheme for pine sawdust establishing a zero-dimensional thermodynamic equilibrium model. Fernández-López et al. [24] simulated the gasification of animal waste biomass in a dual gasifier where the gasification and the combustion zone were separated. Nur and Syahputra [63] integrated the biomass pyrolysis with organic Rankine cycle for power generation. In this case, the aim is to take advantage of the heat generated during the process to produce power.

On the other hand, Puig-Gamero et al. [67] simulated the pyrolysis and gasification process of pine biomass to produce methanol. In this case, the main steps were gasification process, syngas cleaning, and methanol synthesis (Fig. 13.6). First, the biomass was gasified using steam as the gasifying agent. The gas produced was then fed to a pressure swing adsorption process system (PSA) where it was cleaned and adjusted to achieve a H_2/CO ratio close to 2.4–2.5. Finally, the syngas with the optimal ratio was fed to the methanol synthesis unit. In the gasification process, an equilibrium model based on a Gibbs free energy minimization was used to simulate a dual fluidized bed gasifier where the combustion zone was separated from the gasification zone allowing the use of the combustion heat in the rest of the process. In this case, as the desired product was the purified syngas, a catalyst bed of dolomite was used to reduce the amount of tar produced. The PSA system, composed of four units, was then able to capture CO_2 and CH_4 and simultaneously adjust the stoichiometric ratio of methanol feed. The adsorbent used was the biochar produced in the gasification process. Once the syngas was purified, it was further introduced in the methanol reactor to optimize its operating conditions and achieve the highest methanol yield [67].

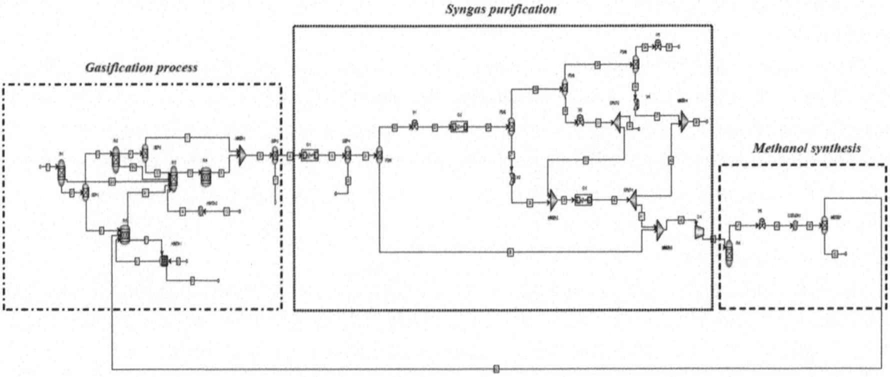

Fig. 13.6 Aspen Plus® flowsheet process

13.3 Applications of the Pyrolytic Oils and Chars

13.3.1 Pyrolytic Oil

The bio-oil contains high amounts of organic compounds such as alkanes, aromatic hydrocarbons, and phenol derivatives and lower amounts of ketones, esters, ethers, sugars, amines, and alcohols with H/C molar ratio higher than 1.5 [36]. It is considered as a potential substitute for petroleum fuel, but its nature is significantly different from petroleum oil because its properties change when stored for long periods. Bio-oil has higher concentrations of oxygenated compounds and is more acidic [2]. The physiochemical properties of the bio-oil depend on several factors such as feed biomass, moisture content in the feed, and the pyrolysis process parameters (vapor residence time, temperature, heating rate, and pressure) [21]. There are advantages to converting lignocellulosic biomass into bio-oil such as easier transportation, higher energy content, and lower impurities (e.g., sulfur) [79].

Many industrially important chemicals can be extracted from bio-oil, such as phenols for the resins industry and hydroxyacetaldehydes and some additives applied in the pharmaceutical industry. Researchers are currently looking for making the best use of the process.

The phenolic compounds present in the bio-oil, such as methylphenols (creosols), methoxyphenols (guaiacol), and methoxypropenylphenol (isoeugenol), have great economic potential in the food, pharmaceutical, cosmetic, and paint industries [21]. Phenols are used as a precursor for the synthesis of bio-plastics, phenol-formaldehyde resins, or epoxy and polyurethane materials [40]. The phenolic compounds have several biological activities such as anti-inflammatory, antimicrobial, antioxidant, antidiabetic, antitumor, and cardioprotective [59].

13.3.2 Char

Char can be used for various purposes, soil amendment, biofuel, adsorbent, and catalytic support being those with the highest potential for exploitation.

13.3.2.1 Soil Amendment

Char has the capacity of improving soil quality and releasing nutrients conserving its carbon structure [5, 11, 16, 34, 39, 53, 77, 78, 84, 88]. Char is a highly densified material containing a large proportion of stable carbon that is very resistant to decomposition and can remain in the soil for several hundred years [47, 54]. For this reason, char can be considered as a method to sequester carbon [4, 43, 46].

It is not possible to predict the performance of biochars, and the whole range of useful characteristics has not been determined, limiting the use of biochar in agriculture and other sectors [76]. From an agronomic point of view, it is important to determine if char (1) modifies the properties of the soil at specific application rates, (2) causes toxic effects, (3) improves nutrient availability, and (4) affects the population and density of the soil flora [22].

13.3.2.2 Biofuel

Regarding bioenergy, biochar was suggested as sustainable source which might be used directly as fuel, serve as substrate or catalyst for gasification processes to product biogases, and replace traditional materials in micro fuel cell (MFC) electrodes and supercapacitor manufacturing (Table 13.2).

As a ready-to-use fuel, biochar is considered a sustainable source of energy in low- and mid-income countries, where the main source of household fuel is still based on limiting fossil coal and wood-derived charcoal. Since lignocellulosic residues (straw, sawdust, fruit shells, phloem, branches, and leaves) from agricultural and forestry activities are abundant, the use of these sources for biochar production would be advantageous for the development of biochar commercialization in these countries. Recent researches suggested that directing production from lignocellulosic waste stream to renewable fuel via pyrolysis might contribute to resolve the laddering demand of energy and pollution issues [37, 52].

In order to produce biochar, lignocellulosic biomass was carbonized through slow pyrolysis under inert atmosphere. After pyrolysis, biochar might be further processed into fuel briquettes that meet commercial product standards [7, 61].

Table 13.2 Application of pyrolysis char in bioenergy fields

Application	Lignocellulosic biomass	Processing method	Traditional source	References
Fuel briquettes	Woody-based high-carbohydrate content biomass	Slow pyrolysis, low-medium temperature	Charcoal, fossil coal	Lohri et al. [52]; Yang et al. [91]; Jafri et al. [37]
Co-combustion	Variable – depending on supply sources	Variable – depend on main product (bio-oil, biogas)	Fossil coal	Yi et al. [93]; Sarkar et al. [74]; Liu and Han [50]
Carbon catalyst for gasification	Rice straw, cotton fiber	Fast pyrolysis, high temperature	Noble metal or transition metal, activate carbon	Lee, Kim and Kwon [44]; Li et al. [49]
MFC electrode's material	Woody-based biomass, microalgal biomass	Fast pyrolysis, high temperature.	Activated carbon or graphite granules	Huggins et al. [33]; Lee et al. [45]
Supercapacitor's material	Woody-based biomass	Variable – including carbonization and activation steps	Chemically synthesized carbon nanomaterials	Gupta et al. [32]; Cheng, Zeng and Jiang [14]

Several properties are required for biochar-based commercial fuel products, including high heating value (HHV), energy density, a fixed carbon content, and ash content [37, 91]. Compared to traditional charcoal, good-quality commercialized biochar should have HHV about 30 MJ/kg with an ash content lower than 5%. Such properties could be improved by using woody-derived biomass [52]. In addition, the competitive price of biochar-derived fuel must be warranted by the low or null cost of the material sources as well as account for the reduced waste management expense. Recent studies suggest that integrating the pyrolysis process into waste management might be a sustainable approach with a reduction of management expenses, an increase in energy efficiency, and lower of biochar production costs [17, 82].

Compared to woody-derived high-carbohydrate content biomass, the biochar generated from pyrolysis of low-carbohydrate-containing biomass (straw, leaves, and municipal waste) has poor yield and thermal properties and is not suitable for commercial fuel application. However, the use of this biochar as feedstock for co-combustion with fossil fuel or other organic wastes was shown to be an interesting solution that might help increase energy efficiency and control polluted air emission [50, 74, 87, 93, 94]. In addition, the biochar can be also used as reserve material for biogas production, via gasification that makes it more convenient to handle and store than raw biomass [89].

Based on its carbon material properties, such as the high electric conductivity and a large pseudo capacitance of char, it can be used as supercapacitor electrodes for energy storage [10].

Another application of biochar in the energy field is its use as a sustainable material for microbial fuel cell electrodes. The biochar-based electrodes were shown to be a promising option to replace the traditional electrodes, which limit the scaling up of microbial fuel cell systems due to their high cost and nonrenewable nature [33, 45].

The porous characteristics of biomass char and the functional groups on its surface make char suitable for use on direct carbon fuel cell (DCFC). It promotes the electrochemical reactions in the low current density region. Graphitized char favors composite conductivity and char can be used as support for electrochemical reactions [10, 69, 90].

13.3.2.3 Adsorbent

Biochar has been tested as adsorbent to remove pollutants like organic dyes, heavy metals, drugs, and gases. The different functional groups present in biochar, as well as its surface area, structure, and mineral content, are factors that influence the adsorption mechanisms [65]. These variables are affected by the feedstock and production conditions (time and temperature), as well as posttreatment such as functionalization or activation.

Biochar has different surface functional groups, mainly consisting of oxygen-containing groups like carboxylate and hydroxyl. These functional groups strongly interact with several charged molecules through processes such as electrostatic interaction, ion exchange, and surface complexation. These interactions can be analyzed by comparing changes in the functional groups of the biochar before and after the adsorption [65].

13.3.2.4 Char-Based Catalyst for Chemical Production

Traditional catalysts and catalyst support can be replaced by char, which is a renewable and low-cost material and presents great physical and chemical properties as a catalyst. It is as a promising alternative to conventional expensive and environmentally unfriendly catalyst sources [49, 70, 92]. There are several applications of biochar-based catalysts in the biofuel field, which relates to biodiesel production and syngas and bio-oil upgrading [44]. The main applications of char as catalyst are the following.

Catalyst for Biogas Production

Tar Reforming Biomass pyrolysis or gasification to produce syngas (H_2 and CO) creates tar (hydrocarbon mixture). Two main methods are used for tar reforming: thermal cracking and catalytic cracking. Tar reforming by thermal cracking takes place at temperatures higher than 1000 °C and is energy intensive. Catalytic cracking allows tar decomposition at lower temperatures (around 700 °C) using a proper catalyst to improve conversion efficiency. The traditional catalysts are dolomite and olivine and metal-based catalysts (i.e., Ni, alkali, or novel metals). However, a promising alternative is the use of char as catalyst or support for an active metallic phase (K, Ca, Ni, Fe, Cu, Ni-Fe) [31, 51, 69]. As metal support, char reduces metal oxides to metallic state, improving catalytic performance related to surface area and pore size but also to mineral content.

Biogas Reforming Hydrogen can be generated by CH_4 conversion in the presence of CO_2 and H_2O. Traditional catalysts used for biogas reforming are based on metals (Ni, Co, Pd, Pt, or bimetallic) supported by alumina or ceria. Recent biochar testing shows improved H_2 production, especially during biomass pyrolysis or gasification. Biochar compounds (K and Ca) act as catalyst; furthermore, some metals can be supported by biochar in order to improve the catalyst performance [90].

Catalyst for Bio-oil Production

Syngas Upgrade Syngas obtained from pyrolysis or gasification of biomass can be upgraded into fuels. Syngas is used to produce liquid hydrocarbons by Fischer-Tropsch synthesis, using an iron-based catalyst supported on alumina or silica.

Char has recently been used to produce carbon-encapsulated iron nanoparticles as catalyst for Fischer-Tropsch synthesis. The conversion of syngas into liquid hydrocarbons showed good efficiency (90% CO conversion and 70% of selectivity to liquid hydrocarbon). Metal-based supported on char presents a great potential as catalyst for syngas upgrading [69, 90].

Esterification/Transesterification Biodiesel (mixture of methyl ester compounds) is commonly produced by esterification and transesterification reactions of vegetable oil or animal fat using acid catalysts (heterogeneous and homogenous). Biochar can be used as a catalyst for biofuel production directly, or it can be functionalized. Functionalization consists in the chemical activation (i.e., KOH) of char in order to increase the specific surface area and sulfonating with concentrated H_2SO_4. The tested catalyst presented high efficiencies and high reusability (seven cycles with no significant loss of esterification activity) [10, 69, 90].

Catalyst for Pollution Control

Selective Catalytic Reduction (SCR) Char can be used as catalyst in low-temperature SCR (selective catalytic reduction) used to control NO_x emissions. Char is a suitable catalyst for SCR because of the abundance of oxygen functional groups on its surface. Furthermore, char can be activated by impregnation using transition metals, which improve the NO_x removal [90].

Photocatalytic Degradation Functionalized char can be used as a photocatalyst for degradation of organic compounds. For example, char functionalized with TiO_2 under ultraviolet irradiation effectively degraded sulfamethoxazole [90].

13.3.2.5 Activated Carbon

Char can be used as raw material for activated carbon. The process has two main parts: carbonization and activation. Carbonization of raw material to obtain char mainly takes place at temperatures between 600 and 1200 °C under inert atmosphere or limited oxygen atmosphere. Char activation can be chemical (using $ZnCl_2$, KOH, H_3PO_4, and K_2CO_3 as agents) or physical (using oxidizing gases such as CO_2, steam, or air). Chemical activation can be a one-step process, where carbonization and chemical activation take place simultaneously, or a two-step process if they happen consecutively. Although physical activation is carried out in two steps, it is cleaner and easier to control than a chemical one [69]. Activated char with a large specific surface area and micropore volume as well as specific functional groups on surface could potentially be used for hydrogen or GHG storage or as catalyst support [18, 90].

References

1. M. Adjin-Tetteh, N. Asiedu, D. Dodoo-Arhin, et al., Thermochemical conversion and characterization of cocoa pod husks a potential agricultural waste from Ghana. Ind. Crop. Prod. **119**, 304–312 (2018)
2. A. Ahmed, M.S. Abu Bakar, A.K. Azad, et al., Intermediate pyrolysis of Acacia cincinnata and Acacia holosericea species for bio-oil and biochar production. Energy Convers. Manag. **176**, 393–408 (2018)
3. Airex Énergie, Technologies (2018), http://www.airex-energy.com/fr/technologie. Accessed 23 Nov 2018
4. N. Anderson, J.G. Jones, D. Page-Dumroese, et al., A comparison of producer gas, biochar, and activated carbon from two distributed scale thermochemical conversion systems used to process forest biomass. Energies **6**(1), 164–183 (2013)
5. H. Asai, B.K. Samson, H.M. Stephan, et al., Biochar amendment techniques for upland rice production in Northern Laos: 1. Soil physical properties, leaf SPAD and grain yield. Field Crops Res. **111**(1–2), 81–84 (2009)
6. D. Basso, E. Weiss-Hortala, F. Patuzzi, et al., In deep analysis on the behavior of grape marc constituents during hydrothermal carbonization. Energies **11**(6), 1379 (2018)
7. A. Bazargan, S.L. Rough, G. McKay, Compaction of palm kernel shell biochars for application as solid fuel. Biomass Bioenergy **70**, 489–497 (2014)
8. E. Çaglar, Y.O. Donar, A. Sinag, et al., Adsorption of anionic and cationic dyes on biochars, produced by hydrothermal carbonization of waste biomass: Effect of surface functionalization and ionic strength. Turk. J. Chem. **42**(1), 86–99 (2018)
9. CEC, Characterization and Management of Organic Waste in North America – Foundational report (2017), http://www3.cec.org/islandora/fr/item/11771-characterization-and-management-organic-waste-in-north-america-foundational-en.pdf. Accessed 20 Sept 2018
10. J.S. Cha, S.H. Park, S.C. Jung, et al., Production and utilization of biochar: A review. J. Ind. Eng. Chem. **40**, 1–15 (2016)
11. K. Chan, L. Van Zwieten, I. Meszaros, et al., Using poultry litter biochars as soil amendments. Aust. J. Soil Res. **46**(5), 437–444 (2008)
12. W. Charusiri, T. Vitidsant, Biofuel production via the pyrolysis of sugarcane (Saccharum officinarum L.) leaves: Characterization of the optimal conditions. Sustain. Chem. Pharm. **10**, 71–78 (2018)
13. W.H. Chen, P.C. Kuo, Torrefaction and co-torrefaction characterization of hemicellulose, cellulose and lignin as well as torrefaction of some basic constituents in biomass. Energy **36**(2), 803–811 (2011)
14. B.H. Cheng, R.J. Zeng, H. Jiang, Recent developments of post-modification of biochar for electrochemical energy storage. Bioresour. Technol. **246**, 224–233 (2017)
15. Conference of the Parties 24th (COP24), Key initiative of the Polish Presidency (2018), http://cop24.gov.pl/presidency/initiatives/. Accessed 15 Dec 2018
16. CQVB-Biotendance, Le biochar: Outil pour la gestion des résidus de biomasse et la fertilité des sols (2011), https://informeaffaires.com/mediatheque/biotendance-le-biochar-outil-pour-la-gestion-des-residus-de-biomasse-et-la-fertilite-des. Accessed 25 May 2018
17. D. Czajczynska, T. Nannou, L. Anguilano, et al., Potentials of pyrolysis processes in the waste management sector. Energy Procedia **123**, 387–394 (2017)
18. B. Delgado, D.L. González, S. Godbout, et al., A study of torrefied cardboard characterization and applications: Composition, oxidation kinetics and methane adsorption. Sci. Total Environ. **593**, 406–417 (2017)
19. A. Demirbas, Effects of temperature and particle size on bio-char yield from pyrolysis of agricultural residues. J. Anal. Appl. Pyrolysis **72**(2), 243–248 (2004)
20. J. Deng, G.J. Wang, J.H. Kuang, et al., Pretreatment of agricultural residues for co-gasification via torrefaction. J. Anal. Appl. Pyrolysis **86**(2), 331–337 (2009)
21. V. Dhyani, T. Bhaskar, A comprehensive review on the pyrolysis of lignocellulosic biomass. Renew. Energy **129**, 695–716 (2018)

22. A. Enders, K. Hanley, T. Whitman, et al., Characterization of biochars to evaluate recalcitrance and agronomic performance. Bioresour. Technol. **114**, 644–653 (2012)
23. ETIA, Procédés de conversion thermochimiques de biomasses et de déchets (2012), http://www.biogreen-energy.com/wp-content/uploads/2013/03/Corporate-Biogreen-2012_05-fr-glc.pdf. Accessed 23 Nov 2018
24. M. Fernandez-Lopez, J. Pedroche, J.L. Valverde, et al., Simulation of the gasification of animal wastes in a dual gasifier using Aspen Plus®. Energy Convers. Manag. **140**, 211–217 (2017)
25. J. Gaskin, C. Steiner, K. Harris, et al., Effect of low-temperature pyrolysis conditions on biochar for agricultural use. Trans. ASABE **51**(6), 2061–2069 (2008)
26. Gouv. du Québec, Profil des produits forestiers: Technologies de bioénergies à base de biomasse forestière (2011), http://agroenergie.ca/pdf/Applications_commerciales/Technologies_bioenergies.pdf. Accessed 20 May 2018
27. Gouv. du Québec, Les matières organiques – Fiches informatives. Recyc-Quebec (2014), https://www.recyc-quebec.gouv.qc.ca/sites/default/files/documents/Fiche-info-matieres-organiques.pdf. Accessed 20 May 2018
28. Gouv. du Québec, Bilan 2015 de La Gestion Des Matières Résiduelles Au Québec (2015), https://www.recyc-quebec.gouv.qc.ca/sites/default/files/documents/bilan-gmr-2015.pdf. Accessed 20 Sept 2018
29. Gouv. du Québec, Résidus de construction, de rénovation et de démolition (CRD) – Fiches informatives (2018a), https://www.recyc-quebec.gouv.qc.ca/sites/default/files/documents/Fiche-info-crd.pdf. Accessed 20 May 2018
30. Gouv. du Québec, Québec residual materials management policy (2018b), http://legisquebec.gouv.qc.ca/en/pdf/cr/Q-2,R.35.1.pdf. Accessed 15 Nov 2018
31. F. Guo, X. Li, Y. Liu, et al., Catalytic cracking of biomass pyrolysis tar over char-supported catalysts. Energy Convers. Manag. **167**, 81–90 (2018)
32. R.K. Gupta, M. Dubey, P. Kharel, et al., Biochar activated by oxygen plasma for supercapacitors. J. Power Sources **274**, 1300–1305 (2015)
33. T. Huggins, H. Wang, J. Kearns, et al., Biochar as a sustainable electrode material for electricity production in microbial fuel cells. Bioresour. Technol. **157**, 114–119 (2014)
34. Husk B, Major J, Commercial scale agricultural biochar field trial in Quebec, Canada over two years: Effects of biochar on soil fertility, biology and crop productivity and quality (2010), https://www.researchgate.net/profile/Barry_Husk/publication/237079745_Commercial_scale_agricultural_biochar_field_trial_in_Quebec_Canada_over_two_years_effects_of_biochar_on_soil_fertility_biology_and_crop_productivity_and_quality/links/552cf7080cf21acb09211085.pdf. Accessed 15 Feb 2018
35. Intergovernmental Panel on Climate Change (IPCC), IPCC guidelines for national greenhouse gas inventories. Prepared by the National Greenhouse Gas Inventories Programme (NGGIP) (2006), https://www.ipcc-nggip.iges.or.jp/public/2006gl. Accessed 28 Sept 2018
36. W.N.R.W. Isahak, M.W.M. Hisham, M.A. Yarmo, et al., A review on bio-oil production from biomass by using pyrolysis method. Renew. Sust. Energ. Rev. **16**(8), 5910–5923 (2012)
37. N. Jafri, W.Y. Wong, V. Doshi, et al., A review on production and characterization of biochars for application in direct carbon fuel cells. Process. Saf. Environ. Prot. **118**, 152–166 (2018)
38. G. Janssens-Maenhout, M. Crippa, D. Guizzardi, et al., *Fossil* CO_2 *and GHG Emissions of all World Countries* (Publications Office of the European Union, Luxembourg, 2017)
39. S. Jeffery, F. Verheijen, M. Van Der Velde, et al., A quantitative review of the effects of biochar application to soils on crop productivity using meta-analysis. Agric. Ecosyst. Environ. **144**(1), 175–187 (2011)
40. J.S. Kim, Production, separation and applications of phenolic-rich bio-oil – A review. Bioresour. Technol. **178**, 90–98 (2015)
41. J. Koppejan, S. Sokhansanj, S. Melin, S. Madrali, Status overview of torrefaction technologies, in IEA Bionergy Task 32 report, Final report (2012), https://www.ieabioenergy.com/publications/status-overview-of-torrefaction-technologies-a-review-of-the-commercialisation-status-of-biomass-torrefaction/. Accessed 25 May 2018

42. R.S. Kovats, R. Valentini, L.M. Bouwer, et al., Europe, in *Climate Change 2014: Impacts, Adaptation, and Vulnerability. Part B: Regional Aspects. Contribution of Working Group II to the Fifth Assessment Report of the Intergovernmental Panel on Climate Change*, ed. by V. R. Barros, C. B. Field, D. J. Dokken, et al., (Cambridge University Press, Cambridge, 2014)
43. Y. Lee, J. Park, C. Ryu, et al., Comparison of biochar properties from biomass residues produced by slow pyrolysis at 500 C. Bioresour. Technol. **148**, 196–201 (2013)
44. J. Lee, K.H. Kim, E.E. Kwon, Biochar as a catalyst. Renew. Sust. Energ. Rev. **77**, 70–79 (2017)
45. J.H. Lee, D.S. Kim, J.H. Yang, et al., Enhanced electron transfer mediator based on biochar from microalgal sludge for application to bioelectrochemical systems. Bioresour. Technol. **264**, 387–390 (2018)
46. J. Lehmann, Bio-energy in the black. Front. Ecol. Environ. **5**(7), 381–387 (2007)
47. J. Lehmann, J. Skjemstad, S. Sohi, et al., Australian climate–carbon cycle feedback reduced by soil black carbon. Nat. Geosci. **1**(12), 832–835 (2008)
48. T.A. Lestander, L. Sandström, H. Wiinikka, et al., Characterization of fast pyrolysis bio-oil properties by near-infrared spectroscopic data. J. Anal. Appl. Pyrolysis **133**, 9–15 (2018)
49. L. Li, H. Wang, X. Jiang, et al., Microwave-enhanced methane combined reforming by CO_2 and H_2O into syngas production on biomass-derived char. Fuel **185**, 692–700 (2016)
50. Z. Liu, G. Han, Production of solid fuel biochar from waste biomass by low temperature pyrolysis. Fuel **158**, 159–165 (2015)
51. Q. Liu, Z. Xiong, S.S.A. Syed-Hassan, et al., Effect of the pre-reforming by Fe/bio-char catalyst on a two-stage catalytic steam reforming of bio-oil. Fuel **239**, 282–289 (2019)
52. C.R. Lohri, H.M. Rajabu, D.J. Sweeney, et al., Char fuel production in developing countries – A review of urban biowaste carbonization. Renew. Sust. Energ. Rev. **59**, 1514–1530 (2016)
53. J. Major, M. Rondon, D. Molina, et al., Maize yield and nutrition during 4 years after biochar application to a Colombian savanna oxisol. Plant Soil **333**(1–2), 117–128 (2010a)
54. J. Major, J. Lehmann, M. Rondon, et al., Fate of soil-applied black carbon: Downward migration, leaching and soil respiration. Glob. Chang. Biol. **16**(4), 1366–1379 (2010b)
55. K. Maliutina, A. Tahmasebi, J. Yu, et al., Comparative study on flash pyrolysis characteristics of microalgal and lignocellulosic biomass in entrained-flow reactor. Energy Convers. Manag. **151**, 426–438 (2017)
56. P. McKendry, Energy production from biomass (part 1): Overview of biomass. Bioresour. Technol. **83**(1), 37–46 (2002a)
57. P. McKendry, Energy production from biomass (part 2): Conversion technologies. Bioresour. Technol. **83**(1), 47–54 (2002b)
58. P. McKendry, Energy production from biomass (part 3): Gasification technologies. Bioresour. Technol. **83**(1), 55–63 (2002c)
59. E. Medina, C. Romero, M. Brenes, Residual olive paste as a source of phenolic compounds and triterpenic acids. Eur. J. Lipid Sci. Technol. **120**(4), 1700368 (2018)
60. Ministère de l'écologie du développement durable des transports et du logement (MEDDTL) and Ministère de l'économie, des finances et de l'industrie (MEFI)-République Française, Rapport sur l'industrie des énergies décarbonnées en 2010 (2011), https://www.ladocumentationfrancaise.fr/var/storage/rapports-publics/134000060.pdf. Accessed 25 May 2018
61. T.H. Mwampamba, M. Owen, M. Pigaht, Opportunities, challenges and way forward for the charcoal briquette industry in Sub-Saharan Africa. Energy Sustain. Dev. **17**(2), 158–170 (2013)
62. S. Nilsson, A. Gómez-Barea, D. Fuentes-Cano, et al., Gasification of biomass and waste in a staged fluidized bed gasifier: Modeling and comparison with one-stage units. Fuel **97**, 730–740 (2012)
63. T. Nur, A.W. Syahputra, Integrated biomass pyrolysis with organic Rankine cycle for power generation. IOP Conf. Ser. Mater. Sci. Eng. **308**(1), 012030 (2018)
64. O. Onay, O.M. Kockar, Slow, fast and flash pyrolysis of rapeseed. Renew. Energy **28**(15), 2417–2433 (2003)

65. P. Oswal, A. Rana, R. Cataluna Veses, et al., Waste derived biochar based bio nanocomposites: Recent progress in utilization and innovations, in *Modified Biopolymers. Challenges and Opportunities*, ed. by D. Pathania, G. Sharma, A. Kumar, (Nova Science Publishers Inc., New York, 2017)
66. J. Park, J. Meng, K.H. Lim, et al., Transformation of lignocellulosic biomass during torrefaction. J. Anal. Appl. Pyrolysis **100**, 199–206 (2013)
67. M. Puig-Gamero, J. Argudo-Santamaria, J.L. Valverde, et al., Three integrated process simulation using aspen plus®: Pine gasification, syngas cleaning and methanol synthesis. Energy Convers. Manag. **177**, 416–427 (2018)
68. N.A. Qambrani, M.M. Rahman, S. Won, et al., Biochar properties and eco-friendly applications for climate change mitigation, waste management, and wastewater treatment: A review. Renew. Sust. Energ. Rev. **79**, 255–273 (2017)
69. K. Qian, A. Kumar, H. Zhang, et al., Recent advances in utilization of biochar. Renew. Sust. Energ. Rev. **42**, 1055–1064 (2015)
70. S. Ren, H. Lei, L. Wang, et al., Hydrocarbon and hydrogen-rich syngas production by biomass catalytic pyrolysis and bio-oil upgrading over biochar catalysts. RSC Adv. **4**(21), 10731–10737 (2014)
71. H. Ritchie, M. Roser, CO_2 and other greenhouse gas emissions (2018), https://ourworldindata.org/co2-and-other-greenhouse-gas-emissions. Accessed 28 Sept 2018
72. M.P. Robbins, G. Evans, J. Valentine, et al., New opportunities for the exploitation of energy crops by thermochemical conversion in northern Europe and the UK. Prog. Energy Combust. Sci. **38**(2), 138–155 (2012)
73. P. Romero-Lankao, J.B. Smith, D.J. Davidson, et al., North America, in *Climate Change 2014: Impacts, Adaptation, and Vulnerability. Part B: Regional Aspects. Contribution of Working Group II to the Fifth Assessment Report of the Intergovernmental Panel on Climate Change*, ed. by V. R. Barros, C. B. Field, D. J. Dokken, et al., (Cambridge University Press, Cambridge, 2014)
74. P. Sarkar, S.G. Sahu, N. Chakraborty, et al., Studies on potential utilization of rice husk char in blend with lignite for cocombustion application. J. Therm. Anal. Calorim. **115**(2), 1573–1581 (2014)
75. S. Şensöz, Slow pyrolysis of wood barks from "Pinus brutia" Ten. and product compositions. Bioresour. Technol. **89**(3), 307–311 (2003)
76. S. Sohi, E. Lopez-Capel, E. Krull, et al., Biochar, climate change and soil: A review to guide future research. CSIRO Land Water Sci. Rep. **5**(09), 17–31 (2009)
77. S. Sohi, E. Krull, E. Lopez-Capel, et al., A review of biochar and its use and function in soil. Adv. Agron. **105**, 47–82 (2010)
78. C. Steiner, W.G. Teixeira, J. Lehmann, et al., Long term effects of manure, charcoal and mineral fertilization on crop production and fertility on a highly weathered Central Amazonian upland soil. Plant Soil **291**(1–2), 275–290 (2007)
79. Y.L. Tan, A.Z. Abdullah, B.H. Hameed, Catalytic fast pyrolysis of durian rind using silica-alumina catalyst: Effects of pyrolysis parameters. Bioresour. Technol. **264**, 198–205 (2018)
80. The World Bank, Solid waste management (2018), http://www.worldbank.org/en/topic/urbandevelopment/brief/solid-waste-management. Accessed 20 Sept 2018
81. H.N. Tran, H.P. Chao, S.J. You, Activated carbons from golden shower upon different chemical activation methods: Synthesis and characterizations. Adsorpt. Sci. Technol. **36**(1–2), 95–113 (2018)
82. M. Tripathi, J.N. Sahu, P. Ganesan, Effect of process parameters on production of biochar from biomass waste through pyrolysis: A review. Renew. Sust. Energ. Rev. **55**, 467–481 (2016)
83. W.P.M. van Swaaij, W. Palz, *Biomass Power for the World: Transformations to Effective Use* (Pan Stanford Publishing, Boca Raton, 2015)
84. L. Van Zwieten, S. Kimber, S. Morris, et al., Effects of biochar from slow pyrolysis of papermill waste on agronomic performance and soil fertility. Plant Soil **327**(1–2), 235–246 (2010)
85. Ville de La Tuque, *Affaire/Projet de développement/Bioénergie* (Ville de La Tuque, Québec, 2018). http://www.ville.latuque.qc.ca/fr/affaires/bioenergie-la-tuque/. Accessed 23 Nov 2018

86. Y. Wang, Q. Wu, D. Duan, et al., Co-pyrolysis of microwave-assisted acid pretreated bamboo sawdust and soapstock. Bioresour. Technol. **265**, 33–38 (2018a)
87. P. Wang, G. Wang, J. Zhang, et al., Co-combustion characteristics and kinetic study of anthracite coal and palm kernel shell char. Appl. Therm. Eng. **143**, 736–745 (2018b)
88. D. Woolf, Biochar as a soil amendment: A review of the environmental implications (2008), http://orgprints.org/13268/. Accessed 20 May 2018
89. N. Xiao, H. Luo, W. Wei, et al., Microwave-assisted gasification of rice straw pyrolytic biochar promoted by alkali and alkaline earth metals. J. Anal. Appl. Pyrolysis **112**, 173–179 (2015)
90. X. Xiong, I.K.M. Yu, L. Cao, et al., A review of biochar-based catalysts for chemical synthesis, biofuel production, and pollution control. Bioresour. Technol. **246**, 254–270 (2017)
91. X. Yang, H. Wang, P.J. Strong, et al., Thermal properties of biochars derived from waste biomass generated by agricultural and forestry sectors. Energies **10**(4), 469 (2017)
92. D. Yao, Q. Hu, D. Wang, et al., Hydrogen production from biomass gasification using biochar as a catalyst/support. Bioresour. Technol. **216**, 159–164 (2016)
93. Q. Yi, F. Qi, G. Cheng, et al., Thermogravimetric analysis of co-combustion of biomass and biochar. J. Therm. Anal. Calorim. **112**(3), 1475–1479 (2013)
94. B. Yousaf, G. Liu, Q. Abbas, et al., Systematic investigation on combustion characteristics and emission-reduction mechanism of potentially toxic elements in biomass-and biochar-coal co-combustion systems. Appl. Energy **208**, 142–157 (2017)
95. M. Zhai, L. Guo, Y. Wang, et al., Process simulation of staging pyrolysis and steam gasification for pine sawdust. Int. J. Hydrog. Energy **41**(47), 21926–21935 (2016)
96. A. Zheng, L. Jiang, Z. Zhao, et al., Catalytic fast pyrolysis of lignocellulosic biomass for aromatic production: Chemistry, catalyst and process. Wiley Interdiscip. Rev. Energy Environ. **6**(3), e234 (2016)

Chapter 14
An Investigation into the Potential of Maggot Oil as a Feedstock for Biodiesel Production

J. M. Shabani, O. O. Babajide, and O. O. Oyekola

14.1 Introduction

Biodiesel is considered as an alternative fuel that could mitigate the global challenges of climate change as a result of environmental pollution caused by the overutilization of petroleum-derived fuels [11, 20]. Currently, biodiesel production is envisaged to meet the rising demand of green/renewable fuel so much that the market trend is predicted to evolve in the nearest future [17, 25]. The biodiesel industry however is known to be burdened with feedstock cost which conventionally contributes to more than 80% of production costs [10, 11]. The use of refined vegetable oils as conventional feedstock for biodiesel production has resulted in controversial disputes with respect to food vs. fuel. Food versus fuel debate was borne out of a discussion highlighting the risk and disadvantages of diverting farmland or crops for biofuels production to the detriment of continuous food supply. This concern, in our view, is considered well justified especially with respect to the complexity and uncertainty of a large number of environmental impact assessment factors that significantly affect final biodiesel product price. In view to avoiding this scenario, the production of biodiesel in a sustainable manner using MagOil could present a panacea, far from creating food shortages. Reliable production and distribution of biodiesel presents good opportunities for sustainable energy production in Africa as it offers the prospect of real market competition and oil price

J. M. Shabani · O. O. Oyekola (✉)
Chemical Engineering Department, Cape Peninsula University of Technology, Cape Town, South Africa
e-mail: oyekolas@cput.ac.za

O. O. Babajide
Mechanical Engineering Department, Cape Peninsula University of Technology, Cape Town, South Africa

M. O. Daramola, A. O. Ayeni (eds.), *Valorization of Biomass to Value-Added Commodities*, Green Energy and Technology,
https://doi.org/10.1007/978-3-030-38032-8_14

moderation, a sustainable energy source that need not compete with land/food crops, solves the problems of the waste generated by local communities, creates jobs for the poor where previously were none, and offers minimal carbon footprint, a huge plus for mitigating the effects of climate change, as well as other beneficial social and environmental advantages. It is also pertinent to report that there are over one million known insect species which constitute over 60% of the global diversity [15]. Insects are historically known to have contributed significantly to human nutrition. Firstly, the main reason is they are human-dependent sources of protein and secondly because they contain a considerable amount of fat. This fact constitutes the major criteria for boosting their selection for biodiesel production. In light of the above and in view of current literature search, insects assessed for viable feasibility in biodiesel production include *Sarcophaga crassipalpis*, *Calliphora vicina*, caterpillars, *G. melloriella*, *Arophalus rusticus*, etc. [21]. In a study by Yang et al. [26], the use of lipid of *Boettcherisca peregrina* insect was reported to produce a high-quality biodiesel. There has been however very little work on insect-based oil for biodiesel production due to limitations of mass production, restricting their value for use mainly as animal feeds and for biodiesel mini-scale experimental purposes. Lipid of the black soldier fly investigated in this study stands exceptional in this regard, since it is available on large-scale production and is a by-product of an innovative research hub (AgriProtein) and a world leading waste-to-fly protein manufacturing company in Cape Town, South Africa. This study reports the potential of MagOil (a lipid by-product of a commercially produced insect protein feed) as a low-cost feedstock of value for biodiesel production.

14.2 Maggot Oil

14.2.1 Properties and Environmental Production Benefit

Maggot oil commercially denoted in this study as MagOil is a lipid derived from black soldier fly (BSF) during the larval stage of development [1]. The oil on the average constitutes a 35–45% by mass of the whole larva insect and appears as a solid fat at room temperature [12, 18, 23]. BSF is naturally found at waste disposal sites where it feeds on organic wastes and thereafter decomposes the waste [14]. It has been estimated that 100 g of waste consumed by BSF may produce 20% larva from the fed insects [8]. Food waste constitutes among the highest forms of domestic waste globally [22], and the disposal of this food waste creates a significant impact on the environment. The use of BSF for the purpose of decomposing biodegradable food waste components greatly influences current waste management practices directed toward a highly effective waste management strategy.

14.2.2 Maggot Oil Production and Sustainability Concerns

AgriProtein is the world's first commercial producer of BSF-derived protein meal, an innovative, natural, and cost-effective alternative to fishmeal. It has been estimated that an amount of nearly 1300 tonnes of MagOil is estimated to be produced yearly as a by-product from the industrial production of the protein meal in South Africa [1]. Other than serving waste management purposes, the use of BSF simultaneously is considered beneficial for MagOil production on a large scale as it guarantees sustainability of the oil as a feedstock for biodiesel production. MagOil therefore as compared to other biodiesel feedstock presents a promisingly more available and sustainable oil that can be used as an alternative to conventional feedstock. Previously, the abundant generation of MagOil has not been profitable because the oil has mainly been treated as an industrial waste. Other reasons for MagOil possessing a low to limited market value is due to the fact that the oil has only been reported for its sole application as an animal feed or flavoring in fish, poultry, and pet diet [1, 8]. Current interventions presented in this study aim to relay MagOil's potential as a sustainable biodiesel feedstock, hence making it more profitable in addition to offering a closed cycle waste beneficiation.

14.2.3 Composition of Maggot Oil

MagOil is comparable to other triglycerides containing fatty acid chains of myristic, palmitic, stearic, oleic, linolenic, and lauric acids [23, 24]. They are distributed majorly into medium chains of saturated fatty acids in which lauric acid is found to be the dominant fatty content [22, 23] (Table 14.1).

A crucial factor that greatly impacts the transesterification conversion of triglycerides to biodiesel is the fatty acid content of the biodiesel feedstock [11]. According to Surrendra et al. [22], oil substrates with high concentrations of saturated fatty acid (SFA) produce biodiesel with poor cold plugging point, while oils predominant with polyunsaturated fatty acid (PUFA) tend to result in low oxidative stable biodiesel. MagOil contains a balanced concentration between SFA and PUFA (neither high concentration of both, ~70% and ~13%) (Table 14.1), which makes it a suitable feedstock for biodiesel production. The solid state of MagOil at room

Table 14.1 Fatty acid composition of maggot oil

Fatty acid composition (wt.%)							
C12:0	C14:0	C16:0	C18:0	C18:1	C18:2	C18:3	Reference
35.6	5.77	15.6	2.76	18.2	14.3	0.99	Agriprotein [1]
18.89	9.91	20.96	6.5	22.54	12.67	–	Wang et al. [24]
44.9	8.3	13.5	2.1	12.0	9.9	0.1	Surendra et al. [22]
38.43	12.33	15.71	2.95	8.81	0.23	–	Ushakova et al. [23]

temperature is triggered by the high proportion of SFA [22]. The main properties of MagOil crucial to its suitability as a feedstock for biodiesel production have not yet been reported in open literature. The significance of density, viscosity, acid, and saponification values with respect to its considerable potential for biodiesel production is highlighted in this paper. Acid value of the oil (expressed in mg KOH to neutralize 1 g of oil) represents the concentration of acid present in the oil, and this indirectly is an indication of the overall concentration of free fatty acid present. Refined vegetable oils usually contain low free fatty acid (FFA) (0.5–1.5 wt.%), whereas animal fats and waste oils have been generally reported for high FFA(s) of above 1.5 to 15–20 wt.% [7, 11]. Saponification value (SV) of oil is associated with the length of fatty acid chains present in the oil, which in turn measures the average molecular weight of the oil. Oils with long-chain fatty acid are associated with low SV and vice versa. Density of a substrate oil is an indication of the degree of mixing or breaking up of the oil (in alcohol) during transesterification which indirectly may have an impact on the temperature required for conversion and amount of alcohol required and slightly on other influential factors of transesterification conversion of the substrate oil into biodiesel [7]. Density is also an indication of the fatty acid profile of the oil sample. The degree of saturation of waste oils such as MagOil containing high SFA is generally anticipated to have a high density, and with respect to the derived biodiesel, density influences the degree of fuel mixing, degree of atomization, injection timing, and injection spray pattern of the oil when utilized in diesel engine cylinders. On the other hand, oil viscosity is a measure of the resistance of oil flow, accounting for the oil flow characteristics and related fluid flow parameters. MagOil is anticipated to consist of high FFA which would likely affect both the yield and quality of biodiesel, and it is expected to present a high viscosity value due to its high number of carbon atoms and the molecular weight of SFA [7]. Consequently, with this information in hindsight, the characteristic properties of MagOil is investigated and reported.

14.2.4 Application of Maggot Oil in Biodiesel Production

In current literature, reviewed studies conducted using maggot oil for biodiesel production were found to be limited. Leong et al. [12] conducted a feasibility study of maggot oil conversion to biodiesel on a laboratory scale using sulfuric acid as a catalyst, and the oil proved to be a feasible feedstock by yielding up to 48.46% FAME. In a two-step lab-scale transesterification performed by Li et al. [14] over sulfuric acid and sodium hydroxide, respectively, the conversion of MagOil yielded a 93% FAME. A feasibility study of biodiesel production using lipids of *Hermetia illucens* larva fed with organic waste by Leong et al. [12] reported the types of waste fed to the larva and the quantity influenced the fatty acid composition. The highest FAME yield obtained in this study was 48.46% from feeding the larva with fruit waste. Overall, fatty acid methyl ester derived from larvae lipids has proven to be feasible for biodiesel production. In this study however, in the synthesis of biodiesel from maggot oil, a fly ash synthesized zeolite catalyst was utilized as a heterogeneous catalyst concomitantly with conventional homogeneous catalysts.

14.3 Experimental

14.3.1 Materials

Crude maggot oil (denoted as MagOil$^{(a)}$) and purified maggot oil (MagOil$^{(b)}$) were supplied by AgriProtein (Pty) Ltd., located in Cape Town, South Africa. Ethanol (96%), hydrochloric acid (32%), methanol (99.5%), and potassium iodide (99%) were supplied by Science World. Diethyl ether, sulfuric acid (98%), and potassium hydroxide pellets (99%) were supplied by Sigma Aldrich. Petroleum ether (40–60%) and iodine monochloride (98%) were obtained from Merck, while chloroform, toluene, hexane (96%), sodium hydroxide pellets (98%), and starch soluble were supplied by Kimix Chemicals. Isopropanol (99.9%), phenolphthalein indicator (1%), sodium sulfate anhydrous (98%), glacial acetic acid (98%), and sodium thiosulfate were supplied by Cleansafe Chemicals. Deionized water was obtained through a laboratory water distiller at the Chemical Engineering Lab at CPUT. Refined sunflower oil was purchased from a local store in Cape Town, South Africa.

14.3.2 Experimental Procedure

14.3.2.1 Fatty Acid Content Determination

Fatty acid content of maggot oil (MagOil) was determined according to standard three-stage method developed by Agrifood analysis station (CPUT, South Africa). The three broad stages involved included the fatty acid digestion, fatty acid extraction, and fatty acid methylation. In the first digestion stage, 0.2 g of MagOil was placed in a 100 ml test tube, and to it 2 ml of internal standard (undecanoic acid), 2 ml of ethanol, and 10 ml of 32% HCl were added. The obtained mixture was tightly sealed in the tube and transferred in a water bath where it was heated for fatty acid digestion, at 80 °C for 40 min. The heated mixture was then allowed to cool at room temperature. In the second stage of extraction, 25 ml of diethyl ether and 25 ml of petroleum ether were added separately to the cooled mixture and thoroughly shaken for 5 min after each addition. The resultant mixture was left to settle overnight. The separated layers, containing the digested fatty acids and the extraction solvents, were decanted into 250 ml beakers, respectively. The layer was then subjected to fume cupboard for 30 min to allow for solvent evaporation. 3 ml of diethyl ether and 3 ml of chloroform were then added to the remainder fatty acid sample to activate solubility and then transferred into a 15 ml test tube. It was dried with nitrogen at 40 °C. Lastly, 2 ml H_2SO_4 and 1 ml toluene were added to the fatty acid sample at 100 °C for 45 min. The resultant mixture containing the fatty acids in ester form was cooled to room temperature. To this mixture, 5 ml of deionized water and 1 ml of hexane was added and allowed to settle so as to isolate the unreacted sulfuric acid and methanol. The top layer containing the methylated fatty acids was then dried with anhydrous Na_2SO_2. The left over top being the digested and methylated fatty acid sample was then transferred into a vial and sent for GC analysis. The average molecular weight of the characterized MagOil was determined using the online Triglyceride Molecular Weight Calculator [3].

14.3.2.2 Procedure for Determination of Acid Value

Acid value of the oil was characterized through the titrimetric method of potassium hydroxide solution into the oil. 2 g of the oil (M) was placed into an Erlenmeyer flask, and to this 10 ml of isopropanol was added in order to extract and dissolve the FFA content in the oil. The mixture was swirled for a few seconds, followed by addition of three droplets of phenolphthalein indicator. 5 ml of 0.1 N (N) potassium hydroxide solution (KOH) prepared was charged into the burette and titrated into the oil-containing mixture (while swirling) until the end point was attained (indicated by pink persistent color for at most 30 s). The volume of the KOH solution consumed/titrated (*V*) was recorded and used to determine the acid value (AV), as per Eq. (14.1):

$$\text{Acid value}\left(\text{mg KOH}\cdot\text{g}^{-1}\text{ oil}\right)=\frac{V\times N\times 56.1}{M} \tag{14.1}$$

14.3.2.3 Saponification Value Determination

Saponification value (SV) of the oil was determined by titration of the oil in excess alcoholic solution of potassium hydroxide (KOH). A small amount of the oil was mixed with 0.5 M KOH solution by mass ratio of 1:10, and the resultant mixture was heated to reflux for over an hour. The mixture was afterward cooled to room temperature. Droplets of phenolphthalein indicator were added to the mixture, followed by titration with a required volume of 0.5 M hydrochloric acid solution (V_{HCl}) until the end point. The same procedure was undergone to obtain the amount of hydrochloric acid required to neutralize the blank (V_{B}). The experimental procedure was duplicated for oil sample so as to obtain the average SV using Eq. (14.2):

$$\text{SV}\left(\text{mg KOH}\cdot\text{g}^{-1}\text{ oil}\right)=\frac{\left(V_{\text{B}}-V_{\text{HCL}}\right)\times 0.5\times 56.1}{\text{g of oil sample}} \tag{14.2}$$

14.3.2.4 Density Measurement

The density of the given oil was measured using an Anton Paar portable density meter DMA.

MagOil as a fat solid state at room temperature was first preheated to 27 °C so as to convert it to liquid form suitable for measurement. An approximately 1 ml of the oil sample was then drawn into a 2.5 ml syringe and this was injected into the instrument. The measurement was immediately initiated, and the reading was allowed to stabilize within a few seconds and then recorded. The density of the oil was measured at 27 °C and thereafter duplicated at 40 °C.

14.3.2.5 Oil Viscosity

The viscosity of MagOil was determined using an Anton Paar Rheometer, of RheolabQc model, operating on a rotational principle of a spindle. 20 ml of the oil sample was used for the measurement at a temperature set from 40 to 60 °C. The instrument was coupled with RheoCompass software for data recording at various time and temperature points within the set range. The measurement was duplicated for both the crude and refined MagOil samples.

In addition to above properties, refractive index (n) and iodine value (IV) of the oil were also determined. The former was determined using a Misco Digital Refractometer (#PA203 model) and the latter according to Wij's method [6].

14.3.3 Characterization

14.3.3.1 Fatty Acid Characterization

The MagOil was compositionally analyzed in a pre-calibrated HP88 GC of 7890B GC system coupled with a flame ionization detector (GC-FID) and equipped with a polar capillary column that detects C_1–C_{12} as well as C_{12}–C_{23} hydrocarbons. The set injection volume of 1 μl of the sample was injected into the GC by a hydrogen carrier gas flowing at the rate of 40 ml/min at 240 °C injection temperature. The initial oven temperature was set at 100 °C for 5 min and then raised to 250 °C at the rate of 3 °C/min held for an additional period of 15 min. The various fatty acid constituents in the oil, as pre-calibrated by FAME standards, were detected at various retention times with respective weight proportions. The weight % (wt.%) of each FFA constituent was calculated using the respective peak area on the chromatograph as shown in Eq. (14.3):

$$\text{wt.\%FFA} = \frac{\text{Area of FFA peak on the chromatograph}}{\text{Sum of chromatogram peak areas} - \text{peak area of internal standard}} \times 100 \quad (14.3)$$

The measurements of all the abovementioned properties, including their procedures, were repeated on the obtained biodiesel products.

14.3.3.2 Catalyst Characterization

Heterogeneous catalyst (HS-zeolite) was synthesized from fly ash via a direct hydrothermal method [16]. Samples of the produced HS-zeolite were characterized mineralogically, morphologically, compositionally, and structurally using the X-ray diffraction (XRD), energy-dispersive spectroscopy (EDS), scanning electron microscopy (SEM), and Fourier transform infrared (FTIR).

14.3.4 Transesterification of Maggot Oil

14.3.4.1 Homogeneous Catalysis

The characterized MagOil was then used to further attest its potential as a feedstock for biodiesel production. Transesterification reaction tests were conducted in a 250 ml three-neck round bottom flask equipped with a reflux condenser and heated over a laboratory hotplate. An amount of 20 g of the oil was transferred and preheated into the 250 ml flask; and to this a mixture of 0.3 g KOH homogeneous catalyst and 15 ml methanol (1.5% catalyst weight equivalent to the mass of the oil and methanol/oil molar ratio of 15:1) was added. Homogeneous NaOH catalyst was similarly also used in a separate reaction. The transesterification reaction was carried out at 60 °C, under stirring by aid of a magnetic bar at 600 RPM, for a period of 1.5 h. After the reaction, the resultant product mixture was transferred into a separatory funnel in which it was allowed to settle for a minimum of 30 min. The observed top layer containing biodiesel (FAME) was drawn out of the funnel, washed excessively with hot deionized water at 60 °C so as to remove unreacted methanol, retained catalyst, and other impurities. The remainder biodiesel was then dried under heat and use of anhydrous sodium sulfate; afterward the dried and cooled sample was weighed prior to storage and characterization. The mass of the biodiesel obtained over the mass of the feedstock oil were used to determine the yield of biodiesel as presented in Eq. (14.4). The obtained biodiesel products were further characterized for qualitative properties:

$$\text{BD yield}(\%) = \frac{\text{Mass of biodiesel produced}}{\text{Mass of maggot oil feedstock}} \quad (14.4)$$

14.3.4.2 Heterogeneous Catalysis

The procedure highlighted in Sect. 14.3.5.1 under fixed conditions was repeated using the synthesized heterogeneous zeolite catalyst (HS-zeolite) to further assess the transesterification of MagOil. The aim of this was to enable a continuous process in a bid to eliminate the washing step peculiar to homogeneous catalysis. An additional reaction test using sunflower oil under the same procedure and conditions was conducted over the heterogeneous zeolite catalyst for comparative purposes.

14.3.4.3 Two-Stage Transesterification of Maggot Oil for Biodiesel Production

Two-stage transesterification of maggot oil was carried out under the setup and conditions highlighted in Sect. 14.3.5.1. In this case, 1% H_2SO_4 solution was used as catalyst in the first esterification stage for 45 min, followed by transesterification of the esterified oil over synthesized HS-zeolite for another 45 min. The produced mixture containing biodiesel was subsequently treated by applying similar procedure described in previous experiments.

14.4 Results and Discussion

The results of physicochemical properties of the characterized maggot oil, as well as the catalytic transesterification tests investigating its potential for biodiesel production, are discussed in Sects. 14.4.1 and 14.4.2.

14.4.1 Fatty Acid Composition of Maggot Oil

The maggot oil was found to be composed of the following distinct fatty acids; lauric, myristic, palmitic, palmitoleic, stearic, oleic, and arachidonic acid. This resultant chemical composition affirms the similarity of maggot oil to various triglyceride and conventional oils as feedstock for biodiesel production [13]. The relative proportion of the abovementioned fatty acids is presented in Table 14.2. Lauric acid in both crude and refined MagOil was found to be the major component in the oil sample constituting a weight content of between 37% and 42%, followed by palmitic (20.9–23.8 wt.%), oleic (14.6–17.8 wt.%), and linoleic acid (7.9–8.4 wt.%), which is similar to maggot oil composition as reported in literature (Table 14.2).

Table 14.2 Average fatty acid content of crude and refined MagOil

Fatty acid (wt.%)	Structure	MagOil[a]	MagOil[b]
Lauric	C12:0	41.5	37.14
Myristic	C14:0	7.44	7.1
Palmitic	C16:0	20.91	23.84
Palmitoleic	C16:1	3.28	3.16
Stearic	C18:0	3.21	2.52
Oleic	C18:1	14.57	17.81
Linoleic	C18:2	7.93	8.44
Linolenic	C18:3	–	–
Arachidic	C20:0	1.54	–

[a]Crude maggot oil
[b]Purified maggot oil

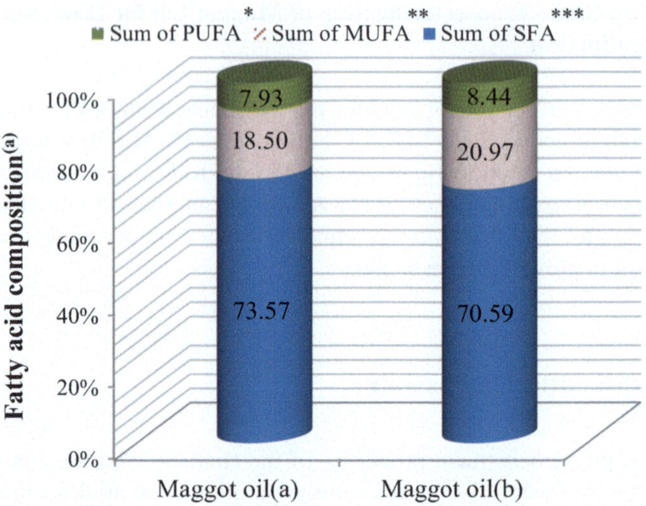

Fig. 14.1 Fatty acid profile of the characterized crude and purified maggot oil used in the current study. [(a)]Composition based on degree of fatty acid saturation. *Polyunsaturated fatty acid, **mono-unsaturated, ***saturated fatty acid

It is worthy of mention that the maggot oil sample characterized in this study was found to further contain distinct amounts of palmitoleic acid (C16:1), revealing high concentration of saturated fatty acid (SFA) with relatively low concentration of polyunsaturated fatty acid (PUFA). These prevailing solid-state phenomena indicate that the characterized MagOil sample is an insect derivative. More importantly, the equilibrium in concentrations of both the SFA and PUFA established in the characterized MagOil results (Fig. 14.1) guarantees its suitability as a starting feedstock for biodiesel production [19, 22]. Based on known fatty acid composition of the maggot oil, the average molecular weight of the characterized oil, using an online Triglyceride Molecular Weight Calculator [3] and as to equation reported by Da Cunha et al. [4], was determined as 812.55 g/mol.

14.4.2 Acid and Saponification Values of Maggot Oil

The acid value of both the crude and purified MagOil samples presented in Table 14.3 varied between 6.73 and 10.1 mg KOH/g of the oil, respectively. These values with respect to oleic acid are equivalent to 3.38 and 5.08 wt.% FFA, classifying the oil as a high free fatty acid containing triglyceride [11]. As a result, the characterized MagOil due to the high concentration of FFA may negatively impact the quality and yield of biodiesel if used in the crude state as a feedstock for biodiesel production. Thus, a pretreatment step was conducted via esterification,

Table 14.3 Acid and saponification value of maggot oil

	Acid value (mg KOH·g^{-1} oil)		
	Non-esterified	Esterified	Saponification value (mg KOH·g^{-1} oil)
MagOil[a]	10.1	–	–
MagOil[b]	7.2	1.4	176.43

[a]Crude maggot oil
[b]Purified maggot oil

Table 14.4 Kinematic viscosity and density of BSF maggot oil at various temperatures

	Kinematic viscosity (mm^2·s^{-1})					Density (g·cm^3)	
Temperature (°C)	20	25	30	35	40	27	40
MagOil[a]	102.51	100.15	91.46	82.43	57.27	0.913	0.875
MagOil[b]	78.36	75.97	68.03	60.35	43.16	–	0.883

[a]Crude maggot oil
[b]Purified maggot oil

tremendously reducing the acid value of the MagOil by 80% which resulted to a value of 1.4 mg kOH/g of the oil (equivalent to 0.7 wt.% FFA). This low acid value is comparable and falls within the recommended acid value range of common biodiesel feedstock [5, 7, 13]. Since it has been proven that the acid value of MagOil can be successfully minimized, the purified MagOil further promises a full potential as a suitable feedstock for biodiesel production.

Saponification value (SV) of the characterized maggot oil was found to be 176.43 mg KOH/g of oil (Table 14.3). This high SV is as a result of the presence of medium- to short-chain fatty acids. Other consequences of this composition are flow restriction, high viscosity of the oil, and high molecular weight (812.55 g/mol). The high SV of maggot is also congruent with the high acid value presented in Table 14.3. SV of conventional oils have been reported in the range of above 180–847.3 g mg KOH/g of oil [2].

14.4.3 Viscosity and Density of Maggot Oil

The kinematic viscosity values of MagOil at various temperatures are presented in Table 14.4. Maggot oils in both crude and refined forms at standard ASTM temperature of 40 °C are characterized by high viscosity of 57.27 and 43.16 mm^2/s, respectively. The viscosity of crude maggot oil far exceeds those of commonly used biodiesel feedstock [9, 13], whereas that of refined maggot oil presents a lower viscosity of 43.16 mm^2/s. The latter is comparable to different conventional oil feedstocks that fall between the viscosity range of 35–45 mm^2/s. The high viscosity of crude maggot oil indicates that this oil cannot be applied for direct use as a fuel in diesel engine. Being much higher at room temperature also implies that the oil will be associated with ineffective mixing during transesterification. In this scenario, thermal pretreatment at a high-temperature reaction medium coupled with a higher methanol/oil ratio may be required to achieve higher conversion and yield of biodiesel from the crude maggot oil. The resultant high viscosity of both crude and refined oil is also in direct correlation with high FFA content obtained. For this reason, an attempt to reduce both properties was conducted on the refined maggot oil by esterification (Table 14.3).

The density of maggot oil was 0.913 g/cm^3 at 27 °C and reduced to 0.875 and 0.883 g/cm^3 at an increased temperature of 40 °C. The density at 27 °C is within the range of conventional biodiesel feedstock density measurement values [9]. This indicates that an effective breakage and mixing potential of the oil with alcohol will minimize the viscosity and produce a high yield of biodiesel provided that the oil is preheated.

14.4.4 Transesterification of Maggot Oil via Homogeneous Catalysis

In this study, maggot oil was transesterified using potassium hydroxide (KOH) and sodium hydroxide (NaOH) as catalysts. These catalysts were prepared to be in the same phase as the oil sample. The oil marked as "Maggot$^{(b)}$" in Tables 14.2 and 14.3 was selected for all transesterification tests in this study, due to its relatively lower acid value and viscosity for better suitability of the reaction. Results showed a 70% biodiesel yield from using KOH and a 65.5% biodiesel yield from using NaOH (Fig. 14.2). This result highlights the potential of maggot oil as a sustainable feedstock for biodiesel production via homogeneous catalysis.

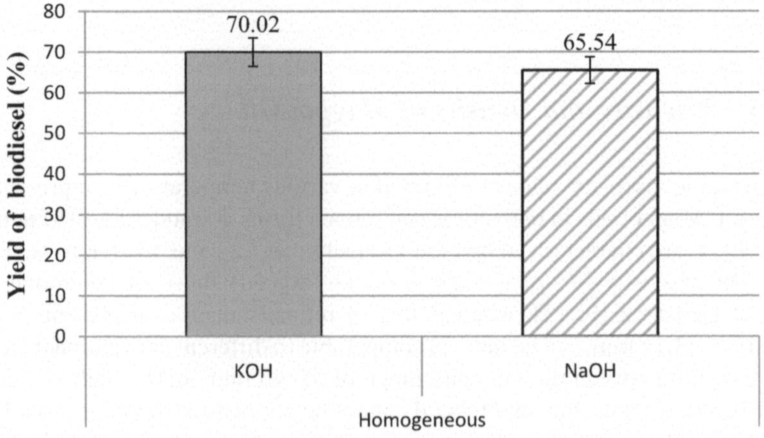

Fig. 14.2 Biodiesel yield from maggot oil using homogeneous catalysts

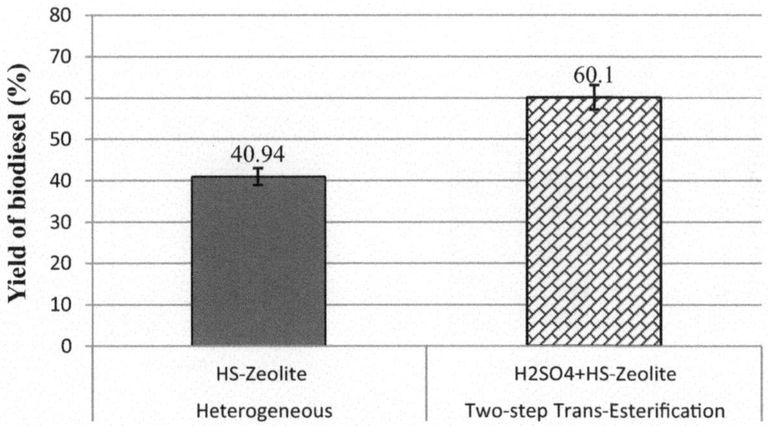

Fig. 14.3 Biodiesel yield from maggot oil via single and two-step catalyzed heterogeneous transesterification

14.4.5 Heterogeneous and Two-Stage Catalyzed Transesterification of Maggot Oil

The synthesized heterogeneous catalyst (HS-zeolite) in this study on the other hand resulted in a relatively low yield of biodiesel (40.94%) (Fig. 14.3). This can be attributed to the fact that MagOil contains a high FFA value beyond the recommended level of 2 wt.% (3.38–5.08 wt.%). Often basic heterogeneous catalysts are known to be sensitive to FFA [11]. Active sites of the catalyst are impaired during transesterification reactions consequently leading to lower biodiesel yields. A two-stage transesterification of maggot oil was therefore adopted in order to overcome this challenge. The yield of biodiesel was enhanced to 60.1% upon esterification of MagOil with 1% sulfuric acid, followed by the transesterification step using HS-zeolite as catalyst under fixed conditions. This resultant yield indicates the reduction of FFA in the oil to 0.7 wt.% upon the esterification step.

14.4.6 A Comparative Study of Maggot Oil and Sunflower Foil for Biodiesel Production

The conversion potential of maggot oil over the heterogeneous HS-zeolite catalyst was further compared with refined sunflower oil for biodiesel production. Sunflower oil gave a better potential for use as a biodiesel feedstock as compared to MagOil (Fig. 14.4). The lower yield obtained from maggot oil is attributable to higher FFA content. The overall low yield of biodiesel obtained by both feedstocks, particularly as reflected by sunflower oil, gives an indication of the poor catalytic performance

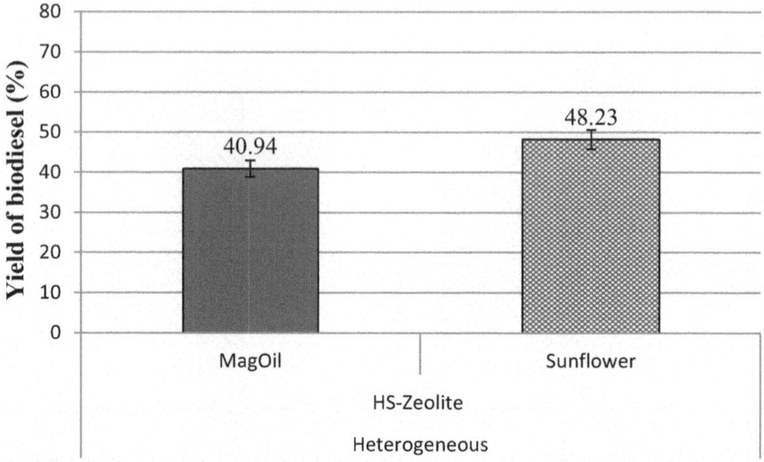

Fig. 14.4 Yield of biodiesel from maggot oil and sunflower oil

of the HS-zeolite synthesized as a heterogeneous catalyst in this research study. Sunflower oil is a refined vegetable oil with low and recommended FFA content of less than 1.5 wt.% [13], which at milder condition over NaOH catalyst has generally resulted in biodiesel yield of far greater than (over 90%) the one obtained from MagOil over the same catalyst (Fig. 14.4). Nonetheless, the 10% difference in biodiesel yield from both feedstocks serves as an indication of the relative potential of MagOil for biodiesel production (Fig. 14.4). In reference to the above finding, it is recommended to focus more on improving the catalytic performance of HS-zeolite other than the oil characteristic.

14.4.7 Physicochemical Properties of Maggot Oil-Biodiesel Produced

The specifications of biodiesel FAME produced in the current study are presented in Table 14.5. The values obtained were measured according to ENS and ASTM methods in compliance with the specification standards. A drastic decrease in the acid value to acceptable level of below 0.5 mg KOH/g was obtained for maggot oil-biodiesel using all catalysts tested. Decrease in saponification value to between 124.33 and 148.59 mg KOH/g of oil in the respective FAME produced depicts longer-chain fatty acid and lower molecular weight of the respective biodiesel FAME as reported (Table 14.5). Decrease in molecular weight is reflective of decreased viscosity from the higher viscous maggot oil to each biodiesel obtained; this is in congruence with decrease in the value of the refractive index (Table 14.5).

The acceptable ester content of 94.67% was observed in the biodiesel obtained using KOH as catalyst. Values of physicochemical properties obtained for biodiesel

Table 14.5 Physicochemical properties of biodiesel produced compared to standard specifications

		Obtained from current study using respective catalyst			Biodiesel standard	
Biodiesel properties	MagOil[a]	KOH	NaOH	HS[b]	ENS[c]	ASTM[d]
Acid value (mg KOH/g)	7.2	0.35	0.28	0.53	0.5 max	0.8 max
Saponification value (mg KOH/g)	176.43	146.64	124.33	148.59	–	–
Ester content (% m/m)	–	94.67	98.3	97.50	96.5	–
Iodine value (g of I_2/100 g)	–	82.63	69.02	50.17	–	130
Density at 40 °C (g/ml)	0.883	0.874	0.862	0.907	0.86–0.90	–
Kinematic viscosity at 40 °C (mm²/s)	43.16	4.39	3.8	5.7	3.5–5.0	1.9–6.0
Refractive index	1.4649	1.4457	1.4456	1.441	–	1.479

[a]Refers to properties of MagOil
[b]Biodiesel properties obtained over HS-zeolite via single-step transesterification reaction
[c]ENS14214
[d]ASTMD675

Table 14.6 Ester composition of maggot oil-biodiesel produced

		Composition (wt.%) from different catalysts		
Fatty acid esters	Structure	KOH	NaOH	HS
Methyl decanoate	C11:0	0.69	0.73	0
Methyl laurate	C13:0	30.64	28.28	26.47
Methyl myristate	C15:0	6.30	5.81	13.26
Methyl palmitate	C17:0	27.30	25.20	23.88
Methyl palmitoleate	C17:1			
Methyl stearate	C19:0	9.41	8.69	24.95
Methyl oleate	C19:1	13.94	12.87	11.45
Methyl linoleate	C19:2	9.98	9.22	–
Methyl linolenate	C19:3	–	–	–
Methyl arachidate	C21:0	1.73	1.60	–
Mw[a] of biodiesel obtained (g/mol)		260.52	259.96	261.83

[a]Molecular weight, determined as to equation reported by Da Cunha et al. [4]

samples, using NaOH and heterogeneous hydroxy sodalite (HS) catalyst slightly exceeded the standard limit (Table 14.5). Iodine values (IV) of between 50.17 and 82.63 I_2/100 g which fall within the acceptable standards depict a high degree of unsaturation of the respective biodiesel produced, and this in turn is an indication of high oxidation and reactivity fuel property of the obtained biodiesel samples. The fatty acid profiles of the produced biodiesel esters (FAME), using different catalysts is reported in Table 14.6. Methyl laurate (C12:0) was found to constitute the major proportion in all FAME produced, and this is as similar reflection as its parent lauric acid in the initial maggot oil feedstock reported (Table 14.2). The proportions of the resultant constituent esters in each biodiesel produced however are lower than those of the precursor fatty acids in the feedstock. This difference accounts for successful and effective conversion of parent fatty acids into their respective esters in the biodiesel products. Further on, the resultant and relatively low molecular weight of each FAME produced (Table 14.6) as compared to the initial feedstock oil (812.55 g/mol) also attest for successful oil conversion to biodiesel. While the potential of BSF maggot oil has been proven for biodiesel production, biodiesel produced using conventional homogeneous catalyst generally exhibits better quality (in reference to standard specifications) than that obtained through heterogeneous catalyzed transesterification.

14.5 Conclusion

MagOil proved to be a suitable and a potential feedstock for biodiesel production in view of the oil characteristics and in application. The oil presented a balance of dominant saturated fatty acid (SFA) and a low concentration of polyunsaturated fatty acid (PUFA) establishing its fitness for biodiesel production. The maggot oil presented physicochemical properties similar to most biodiesel feedstock: a saponification value of 176.43 mg KOH/g of oil, an average molecular mass of 812.55 g/mol, and a viscosity and density value of 43.16 mm^2/s and 0.883 g/cm^3 respectively, at 40 °C. The potential application of maggot via the homogeneously catalyzed transesterification reaction resulted in 65.5% and 72% biodiesel yields for NaOH and KOH, respectively. On the other hand, 40.94% and 60% biodiesel yields were obtained over HS-zeolite heterogeneous catalyst via a single followed and a two-stage transesterification reaction, respectively. The quality of biodiesel obtained from the above instances conformed to the EN and ASTM specification standards. Further, catalyst modification and testing studies are currently being carried out to improve the catalytic activity of the heterogeneous HS-zeolite catalyst synthesized in this study, with the bid to enhance the yield and quality of biodiesel obtained. Additional physicochemical and fuel property characterizations of the produced biodiesel are recommended for optimization studies and suitability for usage.

Acknowledgement The authors are grateful to Professor Leslie Petrik, grant holder, Environmental and Nano Sciences Research Group, University of Western Cape, South Africa, for research funding, mentorship, and collaboration; Agrifood Station at the Food Technology Department, Cape Peninsula University of Technology, South Africa, for GC analysis; iThemba LABS for catalyst characterization analysis; and AgriProtein company for industrial research collaboration.

References

1. AgriProtein, Our products (2017), http://agriprotein.com/our-products. Accessed 22 Oct 2018
2. S. Baroutian, M.K. Aroua, A.A. Raman, et al., RBD palm olein-based methyl/ethyl esters. J. Oil Palm Res. **21**, 659–666 (2009)
3. Biodiesel Education Program, Triglyceride Molecular weight calculator (2017), https://biodieseleducation.org/Production/MolecularweightCalculator.html. Accessed 26 Feb 2019

4. M.E. Da Cunha, L.C. Krause, M.S.A. Moraes, et al., Beef tallow biodiesel produced in a pilot scale. Fuel Process. Technol. **90**(4), 570–575 (2009)
5. M. Jahirul, W. Koh, R. Brown, et al., Biodiesel production from non-edible beauty leaf (Calophyllum inophyllum) oil. Energies **7**(8), 5317–5331 (2014)
6. A.A. Japir, S. Jumat, D. Darfizzi, et al., Physicochemical characteristics of high free fatty acid crude palm oil. EDP Sci. **24**, D506 (2017). https://doi.org/10.1051/ocl/2017033
7. P. Jayasinghe, K. Hawboldt, A review of bio-oils from waste biomass. Renew. Sust. Energ. Rev. **16**(1), 798–821 (2012)
8. J.S. John, Fly larvae to biofuel (2009), https://www.greentechmedia.com/articles/read/fly-larvae-to-biofuel-5923. Accessed 2 Feb 2019
9. A. Karmakar, S. Karmakar, S. Mukherjee, Properties of various plants and animals feedstocks for biodiesel production. Bioresour. Technol. **101**(19), 7201–7210 (2010)
10. Y.G. Keneni, J. Mario Marchetti, et al., Oil extraction from plant seeds for biodiesel production. AIMS Energy **5**(2), 316–340 (2017)
11. M.K. Lam, K.T. Lee, A.R. Mohamed, Homogeneous, heterogeneous and enzymatic catalysis for transesterification of high free fatty acid oil to biodiesel. Biotechnol. Adv. **28**(4), 500–518 (2010)
12. S.Y. Leong, S.R.M. Kutty, A. Malakahmad, et al., Feasibility study of biodiesel production using lipids of Hermetia illucens larva fed with organic waste. Waste Manag. **47**, 84–90 (2016)
13. D.Y. Leung, X. Wu, M.K.H. Leung, A review on biodiesel production using catalyzed transesterification. Appl. Energy **87**(4), 1083–1095 (2010)
14. Q. Li, L. Zheng, H. Cai, et al., From organic waste to biodiesel. Fuel **90**(4), 1545–1548 (2011)
15. F. Manzano-Agugliaro, M.J. Sanchez-Muros, F.G. Barroso, et al., Insects for biodiesel production. Renew. Sustain. Energy Rev. **16**(6), 3744–3753 (2012)
16. N.M. Musyoka, L.F. Petrik, G. Balfour, et al., Synthesis of hydroxy sodalite from coal fly ash using waste industrial brine solution. J. Environ. Sci. Health **46**(14), 1699–1707 (2011)
17. R.L. Naylor, M.M. Higgins, The rise in global biodiesel production. Glob Food Sec. **16**, 75–84 (2018). https://doi.org/10.1016/j.gfs.2017.10.004
18. D. Oonincx, A. Van Huis, J.J.A. Van Loon, Nutrient utilisation by black soldier flies fed with chicken, pig, or cow manure. J. Insects Food Feed **1**(2), 131–139 (2015)
19. W. Robin, Biodiesel from larva of flies: Maggot power (2017), https://www.eniday.com/en/technology_en/biodiesel-from-larva-of-flies. Accessed 22 Jan 2019
20. N. Saifuddin, A. Samiuddin, P. Kumaran, A review on processing technology for biodiesel production. Trends Appl. Sci. Res. **10**(1), 1–37 (2015)
21. D.S. Saunders, Larval diapause duration and fat metabolism in three geographical strains of the blow fly, Calliphora vicina. J. Insect Physiol. **46**, 509–517 (2000)
22. K.C. Surendra, O. Robert, K. Jeffery, et al., Bioconversion of organic wastes into biodiesel and animal feed via insect farming. Renew. Energy **98**, 197–202 (2016)
23. N.A. Ushakova, E.S. Brodskii, A.A. Kovalenko, et al., Characteristics of lipid fractions of larvae of the black soldier fly Hermetia illucens. Dokl. Biochem. Biophys. **468**, 209–212 (2016)
24. C. Wang, L. Qian, W. Wang, et al., Exploring the potential of lipids from black soldier fly. Renew. Energy **111**, 749–756 (2017)
25. World Bank, GEM commodities database, commodity (2017), http://data.worldbank.org. Accessed 15 Oct 2018
26. S. Yang, Q. Li, Q. Zeng, et al., Conversion of solid organic wastes into oil via Boettcherisca peregrine (Diptera: Sarcophagidae) larvae and optimization of parameters for biodiesel production. PLOS One **7**(9), 1–7 (2012)

Chapter 15
Biomass Conversion by Pyrolysis Technology

T. E. Odetoye and J. O. Titiloye

15.1 Introduction

Increasing global demand for energy and the need to save the environment have necessitated more research on alternative energy [18]. Biomass is now considered as an alternative and renewable source of fuel to meet the world's primary energy needs due to its renewability and environmental friendliness. Biomass has received more attention in recent times which resulted in the development of various methods for converting biomass to energy [41]. Energy derived from biomass, sometimes termed 'Bioenergy' is obtained by converting the stored chemical energy within biomass into heat and other useful energy sources. While there are different technologies of converting biomass into energy, pyrolysis is regarded as one of the promising methods. These technology methods have remained dominant [14] because they have a potential of producing long-chain hydrocarbons which can be transformed to synthetic diesel or aviation fuel [39].

15.2 Methods of Biomass Conversion

Biomass conversion methods are classified into three main categories which include biological, chemical and thermochemical processes (Fig. 15.1). Among the biological processes are anaerobic digestion and fermentation of biomass, with anaerobic

T. E. Odetoye (✉)
Department of Chemical Engineering, University of Ilorin, Ilorin, Nigeria
e-mail: odetoye.te@unilorin.edu.ng

J. O. Titiloye
Chemical and Environmental Engineering, College of Engineering, Swansea University, Swansea, UK
e-mail: j.o.titiloye@swansea.ac.uk

M. O. Daramola, A. O. Ayeni (eds.), *Valorization of Biomass to Value-Added Commodities*, Green Energy and Technology,
https://doi.org/10.1007/978-3-030-38032-8_15

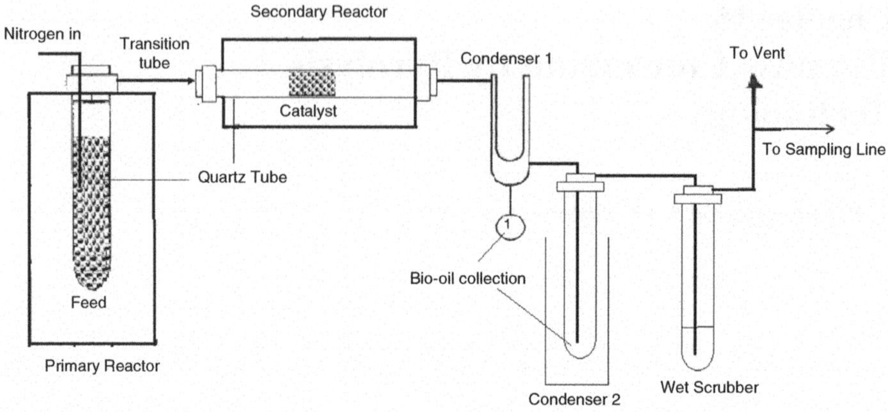

Fig. 15.1 Typical diagram showing slow pyrolysis set up [1]

digestion occurring by enzymatic action of bacteria and yeast to form fuels such as bioethanol and methane. The process is usually conducted under relatively low temperature conditions (<70 °C). The type of biomass that is used as a substrate for fermentation is usually rich in carbohydrates and has high moisture content [10]. Prior to fermentation, feedstock with high lignocellulosic content usually undergoes pre-treatment, such as delignification hydrolysis and depolymerisation. The major disadvantage of biological conversion process is the relatively slow reaction rates (running in hours and days) and low gas yield.

Chemical method includes transesterification route, such as conversion of vegetable oils into biodiesel. Vegetable oils that are triglycerides of fatty acids use transesterification process for conversion to ethyl or methyl esters in the presence of ethanol or methanol and hydroxide catalyst. Some of the disadvantages of the transesterification route include the need for methanol recycle and the generation of wastewater which pollute the environment.

Thermochemical methods for biomass conversion include liquefaction, combustion, gasification and pyrolysis [41]. This process relies on heat which may be accompanied by chemical reactions for converting biomass feedstock to liquid or gaseous fuels. Thermochemical methods are appropriate for biomass materials which have relatively lower moisture contents and high lignin contents. Such biomasses with high lignin contents are usually resistant to enzymatic actions.

Liquefaction is a process that generates a liquid from thermal decomposition of solid biomass. Liquefaction can be direct or hydrothermal process [15]. The liquid produced can further be upgraded for heat and chemical use. Liquefaction is suitable for processing wet biomass. Combustion involves generation of heat and power from thermal decomposition of biomass in the presence of excess oxygen. The heat produced cannot be stored hence it is usually utilised instantly for heat or power generation.

Gasification of biomass is done by converting solid biomass into gaseous bioenergy carrier by partial oxidation in oxygen or air under very high temperature condition. A blend of flue gases produced through this process contains CO, CH_4, CO_2, H_2 and N_2. These gaseous products obtained from gasification process are source of commercially important industrial chemicals. However, gasification is characterised with complex equipment and high maintenance cost. Pyrolysis that usually occurs in the absence of oxygen is regarded as a proficient process for biomass conversion [18]. It is widely accepted as a method which achieves efficient utilisation of energy content of biomass compared to other conversion technologies. Pyrolysis is preferable to gasification and hydrothermal liquefaction due to the advantage of direct liquid fuel production and absence of costly high-pressure system [3]. Pyrolysis thermally upgrades wastes into higher calorific value [17]. It offers the opportunity of chemicals and food by-products.

15.2.1 Pyrolysis as a Biomass Conversion Method

Pyrolysis is a process which entails the thermal decomposition of biomass in oxygen free environment, and most often is carried out in nitrogen environment to provide the needed inert atmosphere [16]. It is usually the initial stage of a thermochemical process, followed by combustion or gasification process. The main products resulting from biomass pyrolysis are solids, liquids and gases, all in varying proportion [18]. Pyrolysis process involves decomposing the biomass components by removing the moisture contents followed by condensation of the volatiles into liquid fuels [8]. During this process, there are several competing factors that come into play with an overall influence on the relative quantities of the yields based on the type of pyrolysis.

15.2.2 Types of Pyrolysis

Pyrolysis can be categorised mainly into slow, fast and intermediate pyrolysis. The heating rates, temperature, vapour residence time and the final yield of the liquid, gaseous and solid products obtainable are factors which distinguish pyrolysis types [40].

15.2.2.1 Slow Pyrolysis

Slow pyrolysis (Fig. 15.1) operates with a moderate temperature (300 °C) up to 400 °C and produces similar yield of liquid oils, solid char and gases. Slow pyrolysis is the usual pyrolysis practice where heating rate is kept at a slow pace of around 0.1–1 °C/second. The residence time of the vapour ranges between 5 and 30 minutes giving it sufficient time to form its products. Intermediate pyrolysis in contrast to slow pyrolysis operates at around 450 to 500 °C. The residence time of the vapour is usually few minutes and the reactions usually occur under controlled heating rates [40].

15.2.2.2 Intermediate Pyrolysis

This form of pyrolysis appears to be a comparatively new procedure which is operational within the conditions of the conservative, slow and fast pyrolysis techniques [46]. The residence time of the solid is varied and the biomass material form can also range from powder size to wood chips size. Haloclean® process [30] is an example of such technology. Schematic diagram of Haloclean pyrolysis pilot plant and Pyroformer patented by Hornung and Apfelbacher are shown in Figs. 15.2 and 15.3 [20, 21].

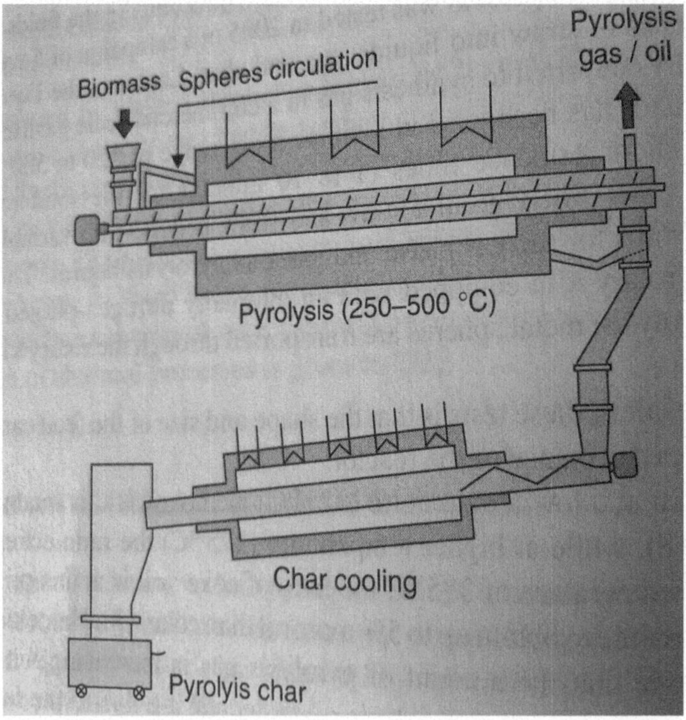

Fig. 15.2 Schematic setup of Haloclean pyrolysis plant [21]. (Used with permission from the publisher)

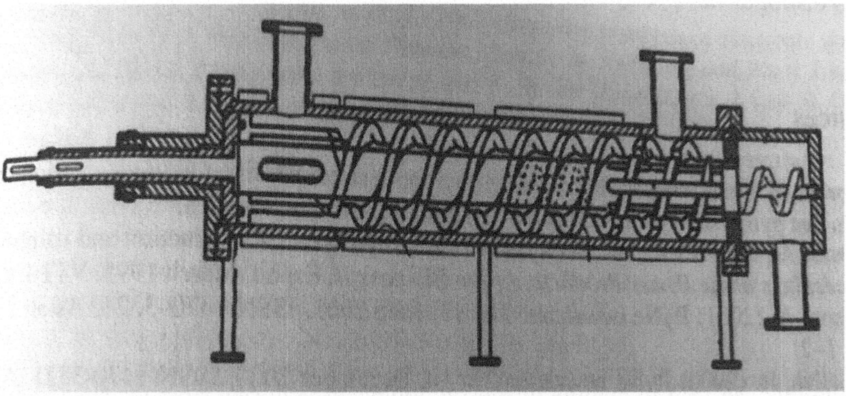

Fig. 15.3 Schematic diagram of the Pyroformer intermediate pyrolysis reactor [21]. (Used with permission from the publisher)

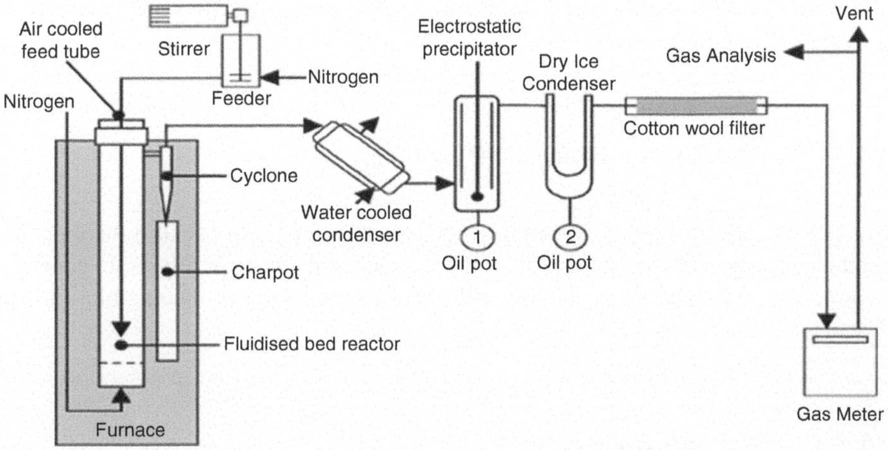

Fig. 15.4 Fast pyrolysis set-up [1]

15.2.2.3 Fast Pyrolysis

Fast pyrolysis (Fig. 15.4) gives a relatively higher yield of bio-oil of about 80 wt.% when compared to slow or intermediate process. It thus has a very fast heating rate with temperature of around 500 °C and the residence time of the vapour being often below 2 seconds followed by a fast cooling of the vapours to form liquids. The technology and varying reactor configurations for pyrolysis reaction are discussed below.

Fast pyrolysis of biomass which is usually achieved with high heating rate as well as short reaction times is known to maximise liquid yield. Bio-oil yield obtainable is up to 76 wt.% on dry basis in addition to solid char and gas production. The liquid is considered to be very valuable with potential for varying applications as it can be stored, transported easily, used in power stations for heat generation, upgraded to various hydrocarbon fuels and valuable source of precursors to different industrial chemicals.

In general, bio-oil product obtained from pyrolysis has dark brown colour, it is highly viscous with varying water contents that can be as high as 20%. It consists of mixtures of organic compound acids, aldehydes, sugars and furans, phenolic compounds and aromatic acids, all are derived from lignin decomposition fractions [31]. It also contains a high proportion of oxygen, hence making it very unstable with a tendency to polymerise when exposed to light or heat or left to stand for a period of time. This indicates the thermodynamic instability of the oil and this can lead to change of bio-oil composition during storage. The pyrolysis liquid is also known to contain organic acids resulting from hemi-cellulose contents of biomass, hence, slightly corrosive with fine particulate char. Although there is a relatively high proportion of water constituent in the product, the possibility of phase separation is usually very low and varies with type of biomass materials used in processing.

15.3 Feedstock for Biomass Pyrolysis

Consideration of the choice of biomass feedstock is important because the yield of product usually depends on the type of biomass utilised during pyrolysis. In general, organic materials have been suitably utilised as feedstock for producing bio-oil [5]. These are given below.

15.3.1 Agricultural Wastes

Examples of agricultural wastes include cassava rhizomes and stalk, corn cobs, cotton trash, cowpea husk, fruit/nut shells, rice husks, grasses and animal wastes [1].

15.3.2 Industrial Biomass Wastes

These include sawdust and bark left over from lumbering activities, sugar cane bagasse from sugar processing industries, distiller grain from beverage industry and paper sludge from paper mills [40].

15.3.3 Domestic Biomass Residues

Other feedstocks potentially available are domestic biomass wastes which can help reduce waste volumes of domestic refuse that includes papers, kitchen wastes and fruit peels [11].

15.3.4 Algae

Another upcoming biomass feedstock source is algae and various seaweeds which have been suggested as a potential alternative feedstock that do not compete with land use for food growing [35]. However, when algae ponds are required to grow significant amount of algae for fuel, land is utilised.

15.4 Factors Affecting Product Yield

The varieties of feedstocks ranging from solid biomass to microalgae biomass species have varying lignocellulosic composition on which the pyrolysis product compositions depend. Biomass obtained from animal wastes will certainly have different composition from that of wood, husks and straws. Biomass with relatively high ash matter will lead to high char content and low liquid product yield. Inorganic constituents in ash, such as Ca, K and Na, can act as a catalyst during the pyrolysis process towards favouring biochar production [31].

The feedstock particle sizes also have an effect on the pyrolysis yield [38]. Smaller particle sizes of feedstocks are more desirable especially for fast pyrolysis and for drying pre-treatment for reducing moisture content. Moisture content is another factor which determines the yield of liquid product. The moisture present in biomass contributes to water constituent in the liquid product, hence, moisture content in feedstock ought to be as small as possible to minimise such water content in bio-oil [45].

The structural composition of biomass requires a certain temperature for pyrolysis reaction to occur, hence, the degradation temperature is crucial to enable proper conversion of feedstock into required products. Koukios [26] confirmed this with a schematic outline of available thermal stability regimes of each of the lignocellulosic contents.

The time taken for the biomass feedstock to undergo complete conversion (i.e. residence time) in the reactor is another important variable for pyrolysis feedstock. As all feedstocks are available in different forms and shapes, it is vital for them to be processed accordingly to satisfy the limitations of the reactor feed system.

Table 15.1 Process conditions for selected biomass feedstock for pyrolysis

	Rice husk [19]	Brunei rice husk (IP) [2]	African Moringa [24]	Indian Jatropha [24]	Cassava rhizome [34]	*Parinari polyandra* (IP) [32]
Reactor type	Fluidised	Fixed	Fluidised	Fluidised	Fluidised	Fixed
Feed rate	90–150 g/h	100 g/h	90 g/h	90 g/h	140 g/h	100 g/h
Optimum temperature (°C)	400–450	450	500	500	477	450
Heating rate (°C/min)	–	25	>80	>80	>25	25
Feed particle size (μm)	–	1000	355–849	355–849	355–500	350–800
Holding time (s)	–	–	0.76	0.76	0.74	–
Purge/fluidising gas	Nitrogen	Nitrogen	Nitrogen	Nitrogen	Nitrogen	Nitrogen
Liquid yield (%)	50	42.17	55.43	52.29	64.9	36.5

IP Intermediate pyrolysis

Biomass particles are known to be non-homogeneous, hence, samples sometimes need to be milled and sieved for characterisation and subsequent reactions. Process conditions for selected biomass feedstock for pyrolysis are shown in Table 15.1.

15.5 Products of Biomass Pyrolysis and Their Applications

Three main products can be obtained from biomass pyrolysis process. They are pyrolysis oil, char and gases.

15.5.1 Pyrolysis Oil

Oil resulting from biomass pyrolysis (bio-oil) has a deep brownish colour. It is a liquid fuel that differs from vegetable oil or petroleum oil in its composition [5]. It contains valuable bio-chemicals that can serve as basis for bio-refineries. Bio-oil is CO_2 neutral; it produces about 50% of the nitrogen oxide emissions when compared with fossil fuels. It gives less SO_2 emissions during combustion. Bio-oil can undergo the usual storage and transportation procedures just like the petroleum products and is used in turbines and typical diesel engines for heat and power generation purposes [5].

Bio-oils are formed by fast degradation of cellulose, hemicellulose and lignin amidst a swift temperature rise. Bio-oil is associated with two phases: the aqueous phase includes the low molecular weight and an organic phase contains the aromatics of high molecular weight. Bio-oil consists of about 21–26% water, 6–11% organic acids, 6–11% non-polar hydrocarbons, 6–10% anhydrosugars (e.g. levoglucosan) and 24–31% pyrolytic lignin derivatives polymer. The combustion properties of the bio-oil are mainly determined by the physico-chemical properties.

Typical bio-oil has a density higher than that of water, 1.2 kg/litre. It has a heating value of up to 19 GJ/tonne, which is about half the heating value of diesel. The ash content of 0.02 wt.% is relatively less than 0.01% for diesel [7, 13]. However, bio-oil undesirably contains about 25% water and it is also acidic in nature, having a pH of 2–3 while diesel has a pH of 5 [7]. This acidic property of the oil plays a role during its storage and transportation.

Pyrolysis oil utilisation for transportation has a benefit over fossil fuels usage. When spillage occurs, bio-oil divides into a heavier inert organic layer that will sink and an aqueous layer that will be diluted as well as bio-degradable. Pyrolysis oil does not spread over water in thin layer like petroleum, resulting in environmental pollution [7].

For analysis of bio-oil, the main properties that are tested for according to American Society for Testing and Materials (ASTM) standards are gross heat of combustion, water content, pyrolysis solids content, kinematic viscosity, density, sulphur content, ash content, pH, flash point and pour point which are also considered for fossil fuels [25].

Bio-oil being also abundant in carbon can be refined similarly to crude oil [23]. It is relatively easier to transport and store, making it an advantage to function as a potential material that can be processed in the petroleum refineries. It can be used as substitutes for diesel in many static applications. Other chemicals that can be obtained from pyrolysis oil include food flavourings, resins, agro-chemicals, fertilisers and emissions control agents. Presently, research efforts are towards upgrading pyrolysis oil to transportation fuels as well as optimising its yield [9].

15.5.2 Biochar

In addition to the liquid produced, char or biochar is also produced during pyrolysis reaction [33]. It has a fine-grained, highly porous nature, being about 65–76% carbon by weight, 5–12% ash and less than 2% moisture [28]. It has heating value of around 28–30 GJ/tonne. Char consists mainly of carbon and hydrogen. It exhibits catalytic activity at the secondary cracking in the vapour phase. Char production is also found to increase the viscosity of the bio-oil during storage [7]. It can be burned for energy or used for plant nourishment [42]. Biochar has also been considered as a means of tackling the global warming problem by utilising it for carbon sequestration [29, 36].

15.5.3 Gases

Most of the gases produced during pyrolysis reactions are non-condensable. They consist mainly of carbon monoxide (CO), carbon dioxide (CO_2), hydrogen (H_2), methane (CH_4) and other hydrocarbons (short-chained alkanes and olefins).

These gases are valued as biosyngas which is quite similar to syngas (a mixture of H_2 and CO) after cleaning. Syngas is generally used as precursors for different chemicals and fuels [43]. Bio-syngas (predominantly a mixture of CO and H_2) is obtained from biomass as a by-product of pyrolysis or by gasification. The gas is then cleaned and converted to hydrocarbons of variable chain length using a gas-to-liquid technology involving Fischer–Tropsch process to produce diesel [12].

15.6 Biomass Pyrolysis Reactors

An extensive variety of reactor types and configurations have been developed with different approaches for fast pyrolysis technology [6]. However, there are challenges encountered for maximising the liquid product yield in pyrolysis. This is attributed to unsuitable pyrolysis reactor configurations and process conditions. Fluidised bed reactors have been widely researched [37] and are the most generally accepted type of reactor for fast pyrolysis. The fluidised bed reactors are based on the principle of fluidisation. This is when solid particles are suspended in a stream of upward flowing gas or liquid display fluid-like properties [22, 27]. It is this characteristic of fluidised beds that favours fast pyrolysis process as this ensures a high heat and mass transfer rates between the solid particles and gaseous fluidising medium.

Fluidised reactor (Fig. 15.5) also provides an efficient way of quick separation of char from the pyrolysis vapours and ensures that the char is not retained in the reactor bed. The char most often can act as vapour cracking catalyst thereby aiding secondary reactions of the volatiles. However, one major drawback of the fluidised bed reactors is the unavoidable escape of the char into the bio-oil thereby reducing the oil stability and encourages polymerisation of the product leading to more viscous oil. Table 15.2 shows typical characteristics of fast pyrolysis reactor types.

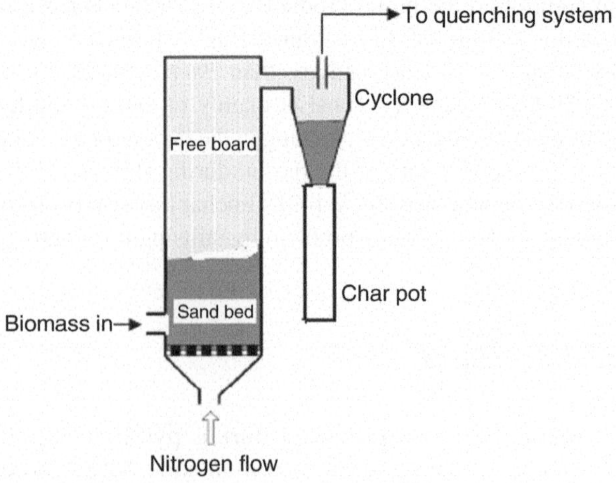

Fig. 15.5 Schematic diagram of a fluidised bed reactor

Table 15.2 Fast pyrolysis reactor types [7]

Reactor type	Characteristics
Fluidised bed	Heated recycle gas. High rates of heat transfer, efficient mixing, smaller particle size of less than 2 mm, partial gasification
Ablative	Reactor wall heating. Large particle size, around 5 mm, high char abrasion, sophisticated reactor
Circulating fluidised bed	In-bed gasification of char to heat sand. Relatively high rates of heat transfer, high char abrasion, solid recycle, complex design
Entrained flow	Minimal heat transfer rates, particle size of less than 2 mm, limited gas–solid mixing
Rotating cone	Wall and sand heating
Transported bed	Re-circulated hot gas heated by char combustion
Vacuum moving bed	Direct contact with hot surface

15.7 Evolving Trends in Biomass Pyrolysis

Recent pyrolysis research work includes pre-treatment of pyrolysis feedstock, gas phase fluidised-bed pyrolysis unit, the use of catalytic esterification and partial hydrogenation processes and blending of pyrolysis oil into diesel, building pilot plants for demonstration of commercial pyrolytic processes, upgrading pyrolysis oils for integration into refineries and microwave-activated pyrolysis process at low temperature [41]. There have been various efforts towards the use of catalysts to chart the pyrolysis pathways to favour bio-oil yields over gas or char. The investigations include the introduction of acid or base catalysts through impregnation of the biomass materials so as to increase the bio-oil yield. Other catalytic processing involves the adaptation of the catalytic cracking of petroleum using heterogeneous acid catalysts.

Different products upgrading process methods are applicable to bio-oil to meet up with the current standards and fuel quality. Among these upgrading methods are hydrotreatment, catalytic cracking, supercritical fluids, esterification and emulsification methods. Rabinovich et al. [35] used a combined approach for modelling fast pyrolysis of biomass particles under fluidised bed condition. Optimal temperatures (corresponding to bio-oil yield) for various particle sizes of wood biomass and various values of its moisture were determined. Wang and co-workers [44] reported that a major barrier in biomass fast pyrolysis is the shortage of raw biomass due to seasonality and regionality. However, utilisation and processing of bio-oil as an alternative resource of chemical materials and transportation fuels were recommended [44]. Upcoming and non-conventional biomass resources like Jatropha fruit shell (Fig. 15.6) and *Parinari polyandra* Benth [31] have been reported to be potential renewable energy sources for bio-oil production.

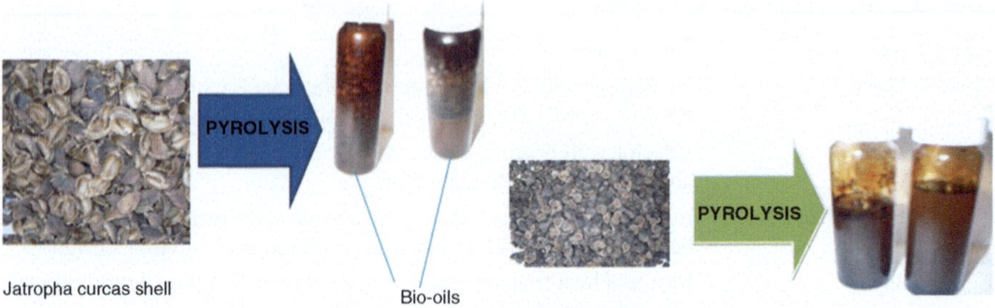

Fig. 15.6 Jatropha shell and Parinari fruit shell pyrolysis oils

15.8 Conclusion, Challenges and Future Outlook

For optimum product yields, pyrolysis process requires a dry feedstock and more often its products need to be upgraded to conform to current usage standard comparable to fossil fuel properties. Hydrothermal liquefaction method appears to be competing well with pyrolysis as it claims to be more economical and environmentally friendlier in its approach [4]. It can handle both dry and wet feedstocks, hence, saving costs, energy and time for drying.

However, pyrolysis technologies are easier to set up than gasification and liquefaction methods. Pyrolysis offers good biorefining opportunities but more economic methods of upgrading the liquid fuel needs to be adopted to make pyrolysis a competitive method for the future.

References

1. M.S. AbuBakar, Catalytic Intermediate Pyrolysis of Brunei Rice Husk for Bio-Oil Production. Dissertation, Aston University (2013)
2. M.S. AbuBarkar, J.O. Titiloye, Catalytic pyrolysis of rice husk for bio-oil production. J. Anal. Appl. Pryol. **103**, 362–368 (2013)
3. G. Aguilar, P. Muley, C. Henkel, D. Boldor, Effects of biomass particle size on yield and composition of pyrolysis bio-oil derived from Chinese tallow tree (Triadica Sebifera L.) and energy cane (Saccharum complex) in an inductively heated reactor. AIMS Energy **3**(4), 838–850 (2015)
4. H.A. Baloch, S. Nizamuddin, M.T.H. Siddiqui, S. Riaz, A.S. Jatoi, D.K. Dumbrea, N.M. Mubarak, M.P. Srinivasan, G.J. Griffin, Recent advances in production and upgrading of bio-oil from biomass: A critical overview. J. Environ. Chem. Eng. **6**, 5101–5118 (2018)
5. D. Bradley, Climate change solutions European market study for bio-oil (pyrolysis oil). IEA Bioenergy Task 40- Bio-trade report, 1–85 (2006)
6. A.V. Bridgwater, Review of fast pyrolysis of biomass and product upgrading. Biomass Bioenergy **38**, 68–94 (2012)
7. A.V. Bridgwater, G.V. Peacocke, Fast pyrolysis processes for biomass. Renew. Sust. Energ. Rev. **4**(1), 1–73 (2000)

8. G. Chang, Y. Huang, J. Xie, H. Yang, H. Liu, X. Yin, C. Wu, The lignin pyrolysis composition and pyrolysis products of palm kernel shell wheat straw, and pine sawdust. Energ. Convers. Manage. **124**, 587–597 (2016)
9. R. Conti, N. Jäger, J. Neumann, A. Apfelbacher, R. Daschner, A. Hornung, Thermocatalytic reforming of biomass waste streams. Energy Technol. **5**, 104–110 (2017)
10. J.A.V. Costa, H. Treichel, V. Kumar, A. Pandey, Advances in solid-state fermentation, in *Current Developments in Biotechnology and Bioengineering. Current Advances in Solid-State Fermentation*, ed. by A. Pandey, C. Larroche, C. R. Soccol, (Elsevier, 2018), pp. 1–17
11. S. Czernik, A.V. Bridgwater, Applications of biomass fast pyrolysis oil, in *Fast Pyrolysis of Biomass. A Handbook*, ed. by A. V. Bridgwater, vol. 3, (CPL Press, Newbury, 2005), pp. 105–120
12. P. Doss, A.V. Bridgwater, *Synthesis of Hydrocarbon Via Fischer-Tropsch Process, Proceeding of the Bioten Conference on Biomass, Biomass Energy* (CPL Press, Newbury, 2010), pp. 689–693
13. E4tech, RE-CORD and WUR "From the Sugar Platform to biofuels and biochemicals". Final report for the European Commission, contract No.ENER/C2/423-2012/SI2.673791 Accessed online on 20th January 2020 from https://ec.europa.eu/energy/sites/ener/files/documents/EC%20Sugar%20Platform%20final%20report.pdf. (2015)
14. R. Garcia, C. Pizarro, A.G. Lavin, J.L. Bueno, Characterization of Spanish biomass waste for energy use. Bioresour. Technol. **103**, 249–258 (2012)
15. A.R.K. Gollakota, N. Kishore, S. Gu, A review on hydrothermal liquefaction of biomass. Renew. Sust. Energ. Rev. **81, Part 1**, 1378–1392 (2018)
16. C.E. Greenhalf, D.J. Nowakowski, A.B. Harms, J.O. Titiloye, A.V. Bridgwater, Sequential pyrolysis of willow SRC at low and high heating rates – Implications for selective pyrolysis. Fuel **93**, 692–702 (2012)
17. S. Guran, Sustainable waste-to-energy technologies: Gasification and pyrolysis, in *Sustainable Food Waste-To-energy Systems*, (Academic Press, 2018), pp. 141–158
18. R.E. Guedes, A.S. Luna, A.R. Torres, Operating parameters for bio-oil production in biomass pyrolysis: A review. J. Anal. Appl. Pyrol. **129**, 134–149 (2018)
19. H.S. Heo, H.J. Park, J.I. Dong, S.H. Park, S. Kim, D.J. Suh, Y.W. Suh, S.S. Kim, Y.K. Park, Fast pyrolysis of rice husk under different reaction conditions. J. Ind. Eng. Chem. **16**(1), 27–31 (2010)
20. A. Hornung, A. Apfelbacher, Patent GB 2460156 (2009)
21. A. Hornung, *Transformation of Biomass; Theory to Practice* (Wiley, 2014), pp. 108–109
22. J.R. Howard, *Fluidized Bed Technology- Principles and Applications* (Taylors and Francis, 1989), pp. 2–14
23. G.W. Huber, S. Iborra, A. Corma, Synthesis of transportation fuels from biomass: Chemistry, catalysts and engineering. Chem. Revs. **106**(9), 4044–4098 (2006)
24. A. Kalgo, The Development and Optimisation of a Fast Pyrolysis Process for Bio-oil Production. Dissertation, Aston University (2010)
25. K.H. Khor, K.O. Lim, Z.A. Zainal, Characterization of bio-oil: A by-product from slow pyrolysis of oil palm empty fruit bunches. Am. J. Appl. Sci. **6**(9), 1647–1652 (2009)
26. E. Koukios, Progress in thermochemical, solid-state refining of biomass – From research to commercialization, in *Advances in Thermochemical Biomass Conversion*, ed. by A. Bridgwater, 2nd edn., (Springer, Netherlands, 1993), pp. 1678–1692
27. D. Kunii, O. Levenspiel, Fluidization engineering, in *Chemical Engineering Series*, ed. by H. Brenner, 2nd edn., (Butterworth-Heinemann, Massachusetts, 1991), pp. 1–7
28. J. Lehmann, A handful of carbon. Nature **444**(7141), 143–144 (2007)
29. S. Lehmann, J. Gaunt, M. Rondon, Bio-char sequestration in terrestrial ecosystems – A review. Mitig. Adapt. Strat. Gl. **11**, 403–427 (2006)
30. R Mai, H. Leibold, H. Seifert, Hot gas filtration and haloclean pyrolysis process operation experience and scale-up. In: 7th International Symposium on Gas Cleaning at High Temperatures (GCHT-7). Newcastle, Australia (2008)
31. T.E. Odetoye, M.S. AbuBarkar, J.O. Titiloye, Thermochemical characterization of Parinari polyandra Benth fruit shell for bio-oil production. Ind. Crop. Prod. **44**, 62–66 (2013)

32. T.E. Odetoye, K.R. Onifade, M.S. AbuBakar, J.O. Titiloye, Pyrolysis of Parinari polyandra Benth fruit shell for bio-oil production. Biofuel. Res. J. **3**, 85–90 (2014)
33. T.E. Odetoye, M.S. AbuBarkar, J.O. Titiloye, Thermochemical characterization of Nigerian *Jatropha curcas* fruit and seed residues for biofuel production. Energ. Ecol. Envion. **3**(6), 330–337 (2018)
34. A. Pattiya, Catalytic Pyrolysis of Agricultural Residues for Bio-Oil Production. Dissertation, Aston University (2007)
35. O.S. Rabinovich, V.A. Borodulya, L.M. Vinogradov, V. Korban, Fast pyrolysis of an ensemble of biomass particles in a fluidized bed. J. Eng. Phys. Thermophys. **83**(4), 742–752 (2010)
36. M. Rondon, J. Ramirez, J. Lehmann, Charcoal additions reduce net emissions of greenhouse gases to the atmosphere. In: Proceedings of the 3rd USDA Symposium on Greenhouse Gases and Carbon Sequestration. Baltimore, MD, 21–24 March 2005 (2015)
37. R. Suntivarakorn, W. Treedet, P. Singbua, N. Teeramaetawat, Fast pyrolysis from napier grass for pyrolysis oil production by using circulating fluidized bed reactor: Improvement of pyrolysis system and production cost. Energy Rep. **4**, 565–575 (2018)
38. D.V. Suriapparao, R. Vinu, Effects of biomass particle size on slow pyrolysis kinetics and fast pyrolysis product distribution. Waste Biomass Valori. **9**(3), 465–477 (2018)
39. J.O. Titiloye, M.S. AbuBarkar, T.E. Odetoye, Thermochemical characterisation of agricultural wastes from West Africa. Ind. Crop. Prod. **47**, 199–203 (2013)
40. J.O. Titiloye, T.E. Odetoye, Renewable energy from biomass and biofuels: Opportunities and challenges in Nigeria, in *Proceedings of 2nd International Conference and Workshop on Renewable Energy*, ed. by A. Willoughby, (Covenant University, Ota, 2013)
41. M.N. Uddin, K. Techato, J. Taweekun, M.M. Rahman, M.G. Rasul, T.M.I. Mahlia, S.M. Ashrafur, An overview of recent developments in biomass pyrolysis technologies. Energies **11**(11), 3115 (2018)
42. F.P. Vaccari, S. Baronti, G.L. LugatoE, S. Calstadi, F. Fornasier, F. Miglietta, Biochar as a strategy to sequester carbon and increase yield in durum wheat. Eur. J. Agron. **34**(4), 231–238 (2011)
43. L. Wang, C.L. Weller, D.D. Jones, M.A. Hanna, Contemporary issues in thermal gasification of biomass and its application to electricity and fuel production. Biomass Bioenergy **32**, 573–581 (2008)
44. D.H. Wang, R. Xiao, H.Y. Zhang, G.Y. He, Comparison of catalytic pyrolysis of biomass with MCM-41 and CaO catalysts by using TGA–FTIR analysis. J. Anal. Appl. Pryol. **89**(2), 171–177 (2010)
45. R.J.M. Westerhof, N.J.M. Kuipers, S.R.A. Kersten, W.P.M. van Swaaij, Controlling the water content of biomass fast pyrolysis oil. Ind. Eng. Chem. Res. **46**(26), 9238–9247 (2007)
46. Y. Yang, J.G. Brammer, M. Ouadi, J. Samanya, A. Hornung, H.M. Xu, Y. Li, Characterisation of waste derived intermediate pyrolysis oils for use as diesel engine fuels. Fuel **103**(0), 247–257 (2013)

Chapter 16
Pyro-gasification of Invasive Plants to Syngas

N. M. Okoro, K. G. Harding, and M. O. Daramola

16.1 Introduction

The thermochemical conversion of biomass into gaseous, liquid and solid fuels remains a broad research area which has garnered increasing interest in recent years despite the long-term existence of thermochemical conversion technologies dating back to the 1800s. This renewed interest has largely been influenced by the present global environmental and energy crises leading, to increased pressure for the adoption of renewable energy technologies. There is an urgent need to reduce the global carbon footprint, minimize the dependence on the depleting fossil fuels and foster energy independence. This can be actualized by creating localized access to a variety of available energy resources, in order to checkmate the unstable supply and rising price of crude oil and natural gas.

Among other renewable energy technologies, biomass thermochemical conversion remains one of the most attractive options due to the following advantages:

(i) The global availability, accessibility and sustainable supply of the biomass feedstock.
(ii) The maturity of the conversion technologies applied.
(iii) The energy-storage capabilities of biomass as feedstock.
(iv) The low net carbon emission resulting from the conversion of biomass to heat, power and transport fuels.

N. M. Okoro · K. G. Harding (✉)
School of Chemical and Metallurgical Engineering, Faculty of Engineering and the Built Environment, University of the Witwatersrand, Johannesburg, South Africa
e-mail: Kevin.Harding@wits.ac.za

M. O. Daramola
Department of Chemical Engineering, Faculty of Engineering, Built Environment and Information Technology, University of Pretoria, Hatfield, Pretoria, South Africa

M. O. Daramola, A. O. Ayeni (eds.), *Valorization of Biomass to Value-Added Commodities*, Green Energy and Technology,
https://doi.org/10.1007/978-3-030-38032-8_16

(v) The relatively cheaper capital involved in constructing bioenergy plants as compared to other renewable energy technologies.

Gasification, pyrolysis, liquefaction and combustion are the four key thermochemical technologies utilized for the conversion of low-grade carbonaceous materials such as biomass wastes into useful energy. However, while gasification, pyrolysis and liquefaction convert these carbonaceous materials into more useful, versatile fuels and chemical feedstock convenient for a wide range of applications, combustion is mostly confined to the production of heat energy from such materials. Moreover, combustion which involves the oxidation of the carbonaceous material to release its energy also produces very harmful and undesirable gaseous products with no useful energy value. This is contrary to the more environmentally benign pyrolysis, gasification and liquefaction processes which transform the energy stored in biomass into cleaner, more suitable liquid, solid and gaseous energy carriers for downstream applications.

Biomass such as invasive alien plants (IAPs) are a unique set of herbaceous and lingo-cellulosic shrubs and trees, renowned for their fast growth rate, ability to quickly invade natural habitats and disrupt the ecosystem, by dominating and displacing indigenous plants and animals alike. Their impacts are severe in regions like Australia, southern Africa, Asia and the Americas. The socio-economic impacts which also include wild fires, erosion, droughts and threats to agricultural crops and livestock, caused by the invasion of these species, have prompted the adoption of intensive control measures by government-funded environmental conservation agencies in the affected regions. The most common and effective of these control measures is the mechanical 'slash and burn' or 'slash and dump' which generates massive amounts of biomass wastes. Other common practices include chemical methods such as the use of herbicides and biological method which employs a 'pest attack' technique to curb their growth [1, 37]. In South Africa, it is estimated that over 165 million tonnes of woody biomass wastes from IAPs is obtainable within 44 million hectares of invaded land [8], which amounts to 36% of South Africa's total landscape. Opportunities, therefore, abound in the conversion of these abundant woody biomass wastes which are continuously being generated from the control of IAPs for biofuel production. Several studies on their impacts, control and feasibility for use as feedstock for energy production – particularly from a socio-economic standpoint – have been reported [5, 7, 9, 20, 23, 27]. However, there is limited knowledge of their properties for use as feedstock for thermochemical conversion processes. These properties include but are not limited to their heating values, energy density, proximate composition, ultimate composition (with more emphasis on nitrogen and sulphur compositions). In addition, are the potential char or ash yields of the feedstock when pyrolysed.

In addition to these is their potential to generate syngas with undesirable high amounts of tar during thermochemical conversion. Syngas with high tar composition is largely unsuitable for downstream application. Table 16.1 shows the allowable impurities and tar compositions of syngas acceptable for downstream applications [29]. Tar generation, caused by the release of volatiles during biomass thermal decomposition, compromises the purity of the syngas product and clogs

Table 16.1 Allowable amounts of tar and impurities in syngas for downstream applications [29]

	Application		
Syngas impurities	Internal combustion engines	Gas turbine	Fuel cell
Particles (mg/Nm^3)	<50	<30	–
Particulate matter (μm)	<100	<5	<1
Tar (mg/Nm^3)	<10	<5	<1
Alkali metals (mg/Nm)	–	<0.24	–

fuel transfer channels. Tar, as defined by the International Energy Agency Gasification Task meeting [11, 24], is a classification of all organic compounds capable of boiling at temperatures above that of benzene. They are the condensable compounds (condensing at <150 °C) contained in products of biomass pyrolysis and gasification [31]. Tars generated from biomass usually contain a variety of aromatic compounds, including benzene in varying proportions.

Other notable undesirable compounds which also affect the quality of syngas are nitrogen and sulphur compounds formed from the chemical reaction of nitrogen and sulphur elements contained of the biomass feedstock.

During a biomass gasification process, the pyrolysis phase is the most critical because it forms the primary products (tar, char and gas) which determine the quality of the final syngas product. The amount of tar and char formed in this early phase is largely dependent on the structural composition (cellulose, hemicellulose and lignin) of the biomass feedstock which when decomposed is responsible for the release of tar-forming volatiles and char residue. High-temperature tar cracking during gasification or tar cleaning of the syngas produced after gasification is common practice. However, *pyro-gasification* – a technique of extracting tar from the gasifier at the point of its formation prior to commencement of the char gasification phase – also has the potential to minimize tar content in the syngas prior to its release from the gasifier. This could aid in improving syngas quality and thus minimize the need for expensive downstream syngas cleaning processes.

Therefore, in the context of utilizing invasive alien plants for pyro-gasification, this chapter focuses on highlighting those properties of IAPs which either promote or limit their use as feedstock. In addition, pyro-gasification as a primary tar removal method will be discussed. However, for a better understanding of the pyro-gasification process, a recap of reactions involved in conventional gasification and pyrolysis of biomass is presented in Sect. 16.2.

16.2 Thermochemical Conversion Processes

For the purpose of this chapter, this section briefly describes the gasification and pyrolysis processes which are established technologies widely employed for conversion of biomass feedstock such as those from IAPs.

16.2.1 Conventional Biomass Gasification Process

Gasification is the thermochemical conversion of carbonaceous material to producer gas, which is a mixture of combustible gases comprising syngas (30–60% CO and 25–30% H_2), 0–5% CH_4, 5–15% CO_2 and other gaseous compounds [30]. Gasification of biomass for the production of syngas is a complex process comprising series of chemical reactions occurring at different stages of the process as shown in Fig. 16.1. Biomass exhibits high thermochemical reactivity and ignition stability, which make them more suitable than coals for conversion to gaseous fuels with minimal combustion propensity [34]. However, associated factors (such as biomass feedstock physical and chemical properties as well as the gasifier type) and process parameters (such as temperature, pressure, equivalence ratio and type of oxidant used) can affect syngas composition and amount of impurities present. The use of air as oxidant or gasifying agent can significantly reduce syngas lower heating values (LHV) due to the presence of nitrogen, while oxygen when used as an oxidant can increase syngas LHV to 15 MJ/Nm^3. Meanwhile, a hydrogen-rich syngas with LHV of up to 20 MJ/Nm^3 can be generated if steam is used as the preferred oxidant [3]. In addition to outlining the different stages of gasification, Fig. 16.1 summarizes their associated chemical reactions and resultant products.

16.2.1.1 Drying Phase

Drying occurs as the first phase of a biomass gasification process, and the heat required for this process is supplied either externally from a furnace or internally from the combustion of tars and gases occurring at a later stage. It is an endothermic

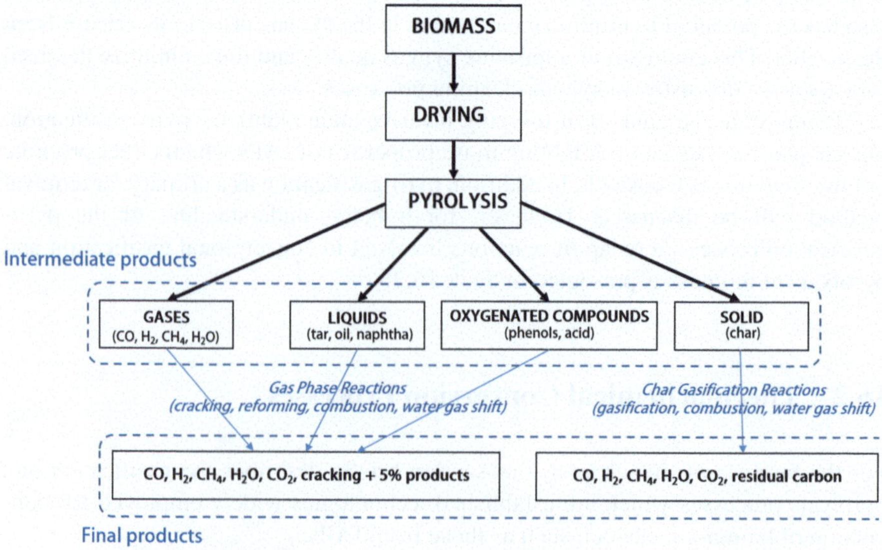

Fig. 16.1 Biomass gasification process and products. (Adapted from Sikarwar et al. [30])

reaction occurring at a maximum temperature input of up to 110 °C, whereby moisture is expelled from the biomass in form of steam as represented in Eq. 16.1:

$$\text{Wet Biomass} + \text{Heat} \rightarrow \text{Dry Biomass} + H_2O(g) \quad (16.1)$$

Depending on the biomass type, plant species and storage conditions, biomass can contain moisture of up to 31% of its total weight [34]. High amount of moisture in the biomass feedstock lowers gasification efficiency by decreasing the temperature in the gasification zone. For optimum efficiency of the gasification process, moisture contents below 15% have been reported suitable for biomass feedstock [34]. Drying is also commonly employed independently as a pre-treatment process for upgrading fresh solid biomass prior to use as feedstock for other thermochemical conversion.

16.2.1.2 Thermal Decomposition Phase

This phase of the gasification process is critical to the quality of the final syngas product as well as the kinetics of the whole process. Also known as the pyrolysis phase, the products formed, to a large extent, affect the syngas composition and quality. Similar to drying, it comprises an endothermic decomposition reaction, of which the moisture-free biomass feedstock is heated at pyrolysis temperatures between 200 °C and 600 °C in an oxygen-deficient atmosphere. As represented in Eq. 16.2, the products obtained during the pyrolysis of wood include mainly gaseous compounds, condensed tar in bio-oil (also known as pyrolysis oil) and a carbon-rich solid char residue:

$$\text{Dry biomass} + \text{Heat} \rightarrow H_2(g) + CO(g) + CO_2(g) + CH_4(g) + H_2O(g) + \text{TAR} + \text{CHAR} \quad (16.2)$$

During gasification, the amount of tar formed in the biomass decomposition phase is mostly a function of the amount of polysaccharides (cellulose and hemicellulose) composition of the feedstock. The expulsion of tar-forming volatiles from the decomposing biomass is to a great extent a result of the complete decomposition of its hemicellulose and cellulose structural components occurring mostly at pyrolysis temperatures. Therefore, biomass with greater proportions of these polysaccharides than lignin is expected to produce more tar or bio-oil than char in this phase.

16.2.1.3 Partial Oxidation Phase

This phase involves the partial oxidation or combustion of the char and other combustible pyrolysis products in limited oxygen concentrations to generate heat, CO and other gaseous products. However, the exothermic reaction may become a complete oxidation with higher heat energy released, if the oxygen concentration is high. As shown in Eq. 16.3, the use of oxygen as oxidizer in low concentration results to a partial oxidation of the carbon in the char residue to form CO. Meanwhile, at higher concentrations, the oxygen as represented in Eq. 16.4 may lead to a complete combustion of carbon, thereby forming CO_2 with higher amounts of heat energy released.

Partial oxidation of carbon at low oxygen concentration [19]:

$$C + \frac{1}{2}O_2 \leftrightarrow CO \qquad -268\,\text{MJ / kg mole} \tag{16.3}$$

Complete oxidation also may occur at a higher oxygen concentration [19]:

$$C + O_2 \leftrightarrow CO_2 \qquad -406\,\text{MJ / kg mole} \tag{16.4}$$

Depending on the gasifying agent used (steam, oxygen or/and air), the composition of the syngas produced may differ. Eq. 16.5 shows steam at higher concentrations supports hydrogen formation and therefore is widely preferred for the oxidation of carbon:

$$C(s) + 2H_2O(g) \leftrightarrow CO_2(g) + 2H_2(g) \tag{16.5}$$

16.2.1.4 Reduction Phase

This phase involves the formation of syngas through a series of endothermic and exothermic reactions (Eqs. 16.5–16.8) between the gasification agent, carbon and other gaseous compounds present in the gasifier. The reactions in this phase occur at higher gasifier temperatures of up to 1000 °C. Those involving steam as gasifying agent in this phase include (Eqs. 16.6 and 16.7) [19]:

Water gas reaction:

$$C(s) + H_2O(g) \leftrightarrow CO(g) + H_2(g) \qquad +118\,\text{MJ / kg mole} \tag{16.6}$$

Water gas shift reaction:

$$CO(g) + H_2O(g) \leftrightarrow CO_2(g) + H_2(g) \qquad -42\,\text{MJ / kg mole} \tag{16.7}$$

Equation 16.6 represents the water gas reaction of carbon in the char with superheated steam, producing a mixture of carbon monoxide and hydrogen also known as 'water gas' or syngas. The carbon monoxide produced is further oxidized to carbon dioxide with the production of more hydrogen in a water gas shift reaction.

More so, a methane formation reaction can occur between carbon monoxide and excess hydrogen produced to form methane and steam:

Methane formation:

$$CO(g) + 3H_2(g) \leftrightarrow CH_4(g) + H_2O(g) \qquad -88\,\text{MJ / kg mole} \tag{16.8}$$

Among all exothermic reactions, the complete combustion of carbon with oxygen (Eq. 16.4) gives off the most energy. As earlier stated, the heat released in this phase can be transferred for the biomass drying and decomposition phases.

16.2.1.5 Tar Cracking

When gasification processes are allowed to extend at elevated temperatures above 1000 °C, the tar formed during the decomposition phase is cracked to further release useful gaseous compounds as shown below:

$$TAR \rightarrow CO_2(g) + CO(g) + H_2(g) + CH_4(g) + C_nH_m(g) \qquad (16.9)$$

16.2.2 *Conventional Biomass Pyrolysis*

Biomass pyrolysis as a completely independent thermochemical process from gasification is ideally carried out in an absolute inert atmosphere made possible by the flow of an inert gas such as nitrogen, helium or argon through the reactor at low to medium temperatures. Similar to the biomass decomposition phase of the gasification process, the products of pyrolysis include gases, tars in bio-oil and char. Biomass pyrolysis incorporates both drying and volatilization as described below [17]:

Drying:

$$\text{Wet Biomass} + \text{Heat} \rightarrow \text{Dry Biomass} + H_2O(g) \qquad (16.10)$$

Valorization:

$$\text{Dry Biomass} \leftrightarrow H_2 + CO + CO_2 + CH_4 + H_2O(g) + TAR + CHAR \qquad (16.11)$$

The drying of the moist biomass feedstock occurs mostly around 105–110 °C, and this is gradually succeeded by the expulsion of volatiles as the gasifier temperature rises from 200 to 600 °C.

During the valorization stage, the heavy tars generated with the gaseous mix condense to form part of the liquid product known as bio-oil, which also contains oils and moisture previously trapped in the biomass. Meanwhile the gaseous products released during pyrolysis contain a mix of key gases, including H_2, CO, CO_2 and CH_4.

As previously mentioned, the yields of either of these products (solid, liquid or gas) depend on the structural composition of the feedstock as well as the process parameters such as the heating rate and the final pyrolysis temperature. With regard to these parameters mentioned, three pyrolysis methods exist, namely, fast pyrolysis, intermediate pyrolysis and slow pyrolysis. Table 16.2 compares the product yields of the three pyrolysis types with those obtained from gasification [32].

Therefore, increasing the heating rate during this process maximizes the production of bio-oils but at the expense of char yield. However, the char formed, though at a minimum yield, has been observed to be more reactive than that formed at lower heating rates [26].

Table 16.2 Three modes of pyrolysis and their products in comparison to gasification [32]

Process	Liquid (bio-oil)	Solid (bio-char) (%)	Gas (syngas) (%)
Fast pyrolysis: Moderate pyrolysis temperature (~500 °C), short hot vapour residence time (<2 s) (high heating rate)	75% tar (25% water)	12	13
Intermediate pyrolysis: Low to moderate pyrolysis temperature, moderate hot vapour residence time (moderate heating rate)	50% tar (50% water)	25	25
Slow pyrolysis: Low to moderate temperature, long residence time	30% tar (70% water)	35	35
Gasification: High temperature (>800 °C), long vapour residence time	5% tar (5% water)	10	85

One important feature of the pyrolysis technology is the capability of separating and extracting the condensed tar produced from the biomass. This therefore leaves behind a relatively tar-free, carbon-rich char residue. This volatile-free char generated can then be gasified separately for the production of syngas with much reduced tar content. Such is the basis on which pyro-gasification is being developed as discussed in Sect. 16.3.

16.3 Tar Reduction Methods and Pyro-gasification

16.3.1 Tar Reduction Methods

Techniques for tar reduction in gasification as reported in literature can be broadly categorized into primary methods (pre-gas production tar removal) and secondary methods (post-gas production tar removal) as summarized in Fig. 16.2 [31].

The secondary methods developed in literature basically involve the transfer of the tar-laden syngas produced from a gasifier to a separate cleaning reactor for tar separation or tar cracking processes. Wet gas cleaning as classified under secondary methods (Fig. 16.2) is based on tar condensation at low temperatures. This method involves the use of water and venture scrubbers to condense the tar by passing the impure syngas through it. Increasing the water flowrate of the scrubber has been reported to reduce the tar content in the syngas after cleaning [31]. However, in hot gas cleaning, the syngas produced in a primary gasifier is transferred to a separate secondary high-temperature gasifier for further cracking the tar contained in the presence of a catalyst at elevated temperatures. Though regarded as a very expensive procedure, the hot gas cleaning at elevated temperatures between 500 and 900 °C has been found most effective in tar reduction and improvement of syngas heating value, whereby a 98% tar to hydrogen conversion efficiency was attained [31].

On the other hand, the primary methods involve employing the reduction of tar as gasification progresses inside a reactor prior to the production of syngas and may

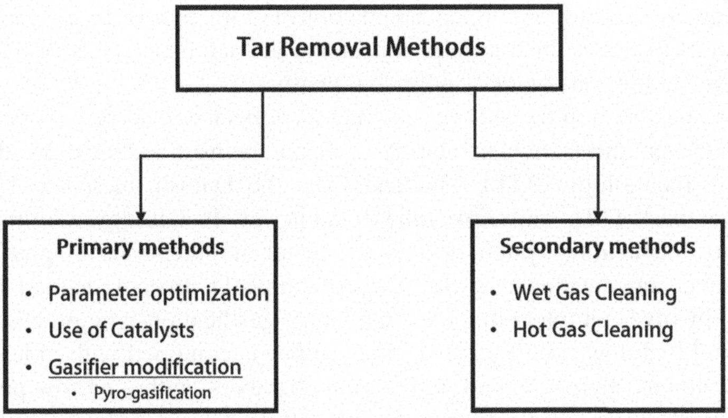

Fig. 16.2 Tar removal methods with focus on pyro-gasification

not require syngas transfer to a secondary purification reactor. This can be achieved by the following:

(a) Selection of ideal reactor operating parameters such as temperature, equivalence ratio, oxidizing agent/biomass ratio, steam/biomass ratio and feedstock residence time;
(b) Use of various catalysts such as activated carbon, calcined dolomite and Y-zeolites for tar cracking; and/or
(c) Gasifier modification; whereby conventional designs of gasifiers are altered to minimize tar mixing with syngas during gasification [31].

16.3.2 Pyro-gasification for Tar Reduction

The gasification of biomass is comprised of various superimposing phases and chemical reactions which make each phase difficult to control and optimize independently in a conventional single-stage reactor. This also makes it difficult to prevent the mixing of volatiles and char produced during the biomass decomposition phase. The contact between tar-forming volatiles and char negatively affects the gasification-phase process and contaminates the syngas produced from char gasification [30]. Ideally, the gasification of char in a volatile-free atmosphere would enhance syngas purity, char rate of reaction and efficiency as well as minimize tar formation.

Pyro-gasification can be classified under the gasifier modification option categorized under the secondary tar-removal methods. It can be defined as a hybrid or multi-stage thermochemical process primarily involving a combination of pyrolysis and gasification in one continuous process but controlled independently to restrict tar from being gasified alongside char. This thereby ensures only a carbon-rich char residue with negligible tar composition is left to be gasified as the conversion process progresses to the partial-oxidation and reduction phases. The aim of adopting this technique is to gain some independent control of the pyrolysis and gasification-phase kinetics to ensure the biomass conversion at each phase is carried out under the optimized conditions of each individual phase.

The pyro-gasification technology may not have been extensively reported in literature; however, pilot studies relating to it are being conducted by the Asian Institute of Technology (AIT), Thailand [31], the Danish Technical University (DTU), Denmark [13], and Karlsruhe Institute of Technology, Germany [30]. Several pyro-gasification system designs or reactor modifications are possible.

A system design developed by the Danish Technical University named the *Viking* (Fig. 16.3) involves an integrated two-stage pyro-gasification-type process consisting of a modified down-draft gasifier. The gasifier is equipped with independently controlled biomass pyrolysis and gasification chambers, with a narrow partial oxidation channel, all merged into a single overall unit as described on its process schematic diagram (Fig. 16.4). It was designed to first dry and pyrolyse the biomass feedstock in the cylindrical pyrolysis chamber of the system in an inert atmosphere. A rotating screw conveyor is installed in this cylindrical chamber to transfer the biomass feedstock at controlled speeds across the cylinder as it pyrolyses to form char and tar. The feedstock inlet end of the cylindrical pyrolysis chamber is heated to around 50 °C for drying the feedstock as it is fed into the pyrolysis chamber. Temperature increases as the feedstock moves across the pyrolysis cylinder towards

Fig. 16.3 The Viking gasifier [13]

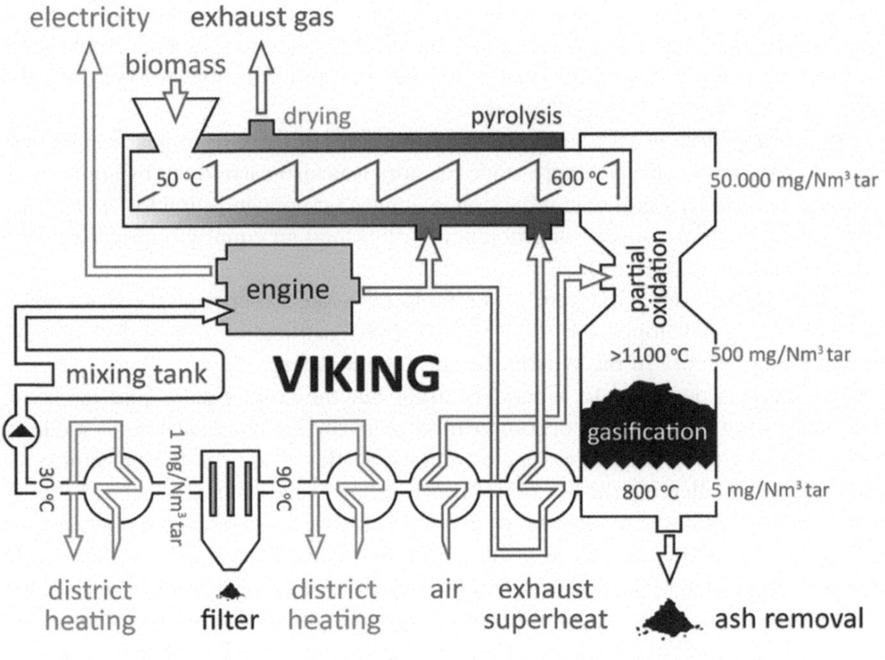

Fig. 16.4 Design diagram for the Viking demonstration and research plant [2]

the char and tar products outlet point at the other end of the pyrolysis cylinder, where the pyrolysis temperature gets to about 600 °C. The heating rate is controlled by the speed of the screw conveyor transferring the feedstock from the low temperature end to the high temperature end of the cylindrical pyrolysis chamber. Meanwhile, beyond this point of exiting the pyrolysis chamber, the mixture of char and tar is allowed to pass through a narrow partial oxidation corridor supplied by an air flow. This allows for an oxidation of the tar by air, resulting to a considerable reduction of up to 500 mg/Nm^3 in tar volume. This exothermic oxidation reaction also provides the heat required for the endothermic gasification reaction of the char when it is eventually deposited in the gasification chamber. Steam may be used as the oxidizer in the gasification chamber. With this system, the operating conditions of each phase can be controlled independently and optimized to minimize tar. The producer gas exiting the system has been observed to contain about 32% hydrogen and 16% carbon monoxide with negligible tar and methane composition. Currently, the system capacity is rated at about 200 kW$_e$ with a potential scale-up to about 500 kW$_e$ [2, 30].

McKendry [19] also reported on a previously proposed process similar to pyro-gasification, which entails a two-stage, two-reactor hybrid process for the improvement of syngas quality. Its process involves an initial pyrolysis of the biomass feedstock in the first reactor heated externally at temperatures of up to 600 °C, thereby forming char and releasing tar-laden gases which further react with steam fed into the reactor to crack the tars. Meanwhile, in the second-phase gasification reactor, only char transferred from the pyrolysis reactor is gasified to produce syngas which is suitable for use in spark ignition gas engines. Another recent study on pyro-gasification involving a simultaneous production of syngas and biochar from hazelnut pruning using a downdraft gasifier has been reported to improve syngas LHV from 3.84 MJ/Nm^3 to 4.45 MJ/Nm^3 [6].

An investigation into the strengths and limitations of these pyro-gasification systems over the conventional single-stage reactors was also carried out by simulation. The results reveal a significant reduction in syngas tar concentration to 10 mg/Nm^3, a high char thermal conversion efficiency of 98% and an equally high gasification efficiency of about 81% [30].

Figure 16.5 shows a diagram of a simple lab-scale experimental reactor setup currently being developed for a study on the pyro-gasification of woody IAP sawdust at the University of the Witwatersrand, Johannesburg.

The setup is designed to separately drain out the condensed liquid tar being released during the biomass decomposition or pyrolysis phase at low to medium temperatures from the reactor. One advantage of this design is the possibility of extracting and collecting the bio-oil for downstream processing, while the char residue is retained in the reactor, to be gasified. Moreover, the process parameters such as heating rate and temperature in the initial pyrolysis phase can be optimized to improve either char or tar/bio-oil yield (i.e. by slow or fast pyrolysis) prior to char gasification. It also allows for the use of an inert gas during pyrolysis phase to ensure the pyrolysis phase is carried out in an absolute inert atmosphere while heating rate and pyrolysis temperature are optimized.

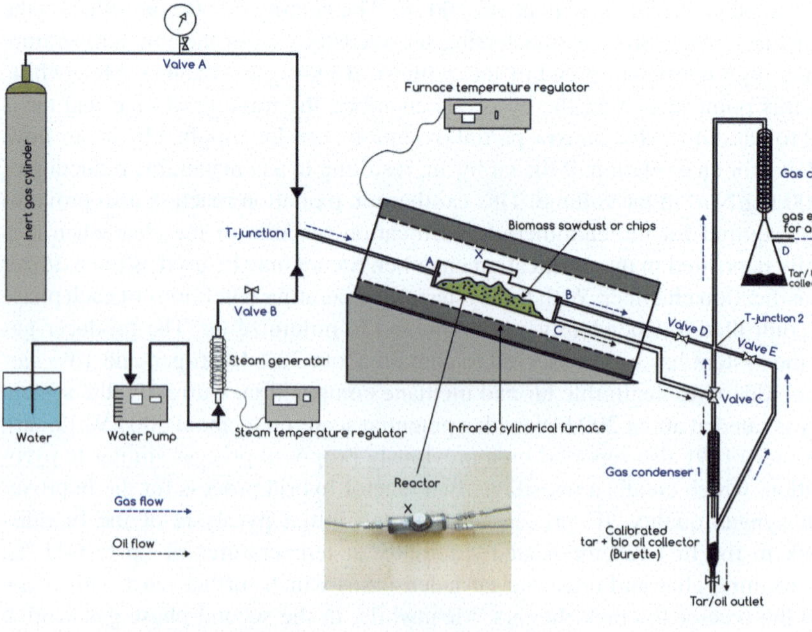

Fig. 16.5 Lab-scale pyro-gasification experimental setup [25]

The system (Fig. 16.5) consists of a cylindrical stainless steel reactor placed in an electrical infrared tube furnace and connected to a two-way input line at *T-junction 1*. The setup is operated by feeding small-sized wood chips or sawdust of known volume through the feeder *x* into the cylindrical reactor. The reactor is closed by fastening the feeder bolt and ensuring the reactor is air-tight. The reactor is then placed inside the tube furnace inclined at an angle of about 35° to the horizontal. This is to ensure a free flow of any condensed oil through the outlet pipe under gravity.

To activate the drying and pyrolysis phase, *Valve A* is opened to allow inert gas (nitrogen/argon/helium) flow into the reactor through the inlet point A to create an inert atmosphere while the pyrolysis temperature and heating rate are set. At the same time, *Valve D* is closed to ensure liquid and gaseous pyrolysis products are released from the reactor only through the outlet point *C*. Furnace temperature is increased from room temperature through to 105 °C (for drying) and then to about 600 °C (for pyrolysis). The heating rate is set at 20 °C/min or more to maximize tar/bio-oil yield via fast pyrolysis (Table 16.2). With only the solid char residue retained in the reactor, the bio-oil and pyrolysis gas released during this phase are drained out under gravity through the outlet point *C* towards the three-way *Valve C*. By setting the three-way *Valve C* in the direction of the *Gas condenser 1*, the tar contained in the pyrolysis gas is condensed to bio-oil and collected in the *calibrated tar collector (burette)* for volume measurement. Meanwhile, from this point, the lighter uncondensed gaseous products proceed towards the *Gas condenser 2* (via T-junction 2) for further condensation. After the final pyrolysis temperature is met at around 600 °C to 800 °C, the inert gas is withdrawn by closing *Valve A*.

To proceed to the gasification phase, super-heated steam generated by the steam generator at high temperatures of up to 550 °C is allowed into the reactor by opening *Valve B* and at the same time, increasing the furnace temperature to higher gasification temperatures of up to 1000 °C. The wet producer gas generated from passing steam through the char particles in the reactor at elevated temperatures exits the reactor through Outlets *B* and *C* and proceeds through *Valves C* and *D* towards the *Gas condenser 2*. While *Valve D* is left open to allow the flow of the producer gas towards the *Gas condenser 2* (via T-junction 2), *Valve E* is closed to prevent a back flow of the wet producer gas towards the bio-oil collector. Meanwhile, setting *Valve C* to allow the flow of producer gas only towards the *Gas condenser 2* (via T-junction 2) also prevents gas back-flow towards the *Gas condenser 1*. This ensures the condensed water from the wet gas produced by the steam gasification is prevented from diluting the bio-oil previously collected in the burette during the pyrolysis phase. Finally, the producer gas exiting the system at the outlet point *D* is collected for analysis. Studies on the above setup are presently on-going to identify areas for improvement and optimization.

In recent pyro-gasification tests conducted with this experimental setup on some IAP sawdust samples, an average of about 2–4 mL of condensed bio-oil was obtainable from sawdust volume of 10 g. The gas chromatography analysis of the producer gas collected shows syngas concentration of 50–60%, with H_2 concentration of 22–40% and CO concentration of 4–20%. Optimum syngas production was identified to occur at higher temperatures between 900 °C and 1000 °C. Operational issues were however identified during these tests and have been highlighted in Sect. 16.5.

One important advantage of pyro-gasification over conventional gasification is the potential of the former to prevent tar produced during the pyrolysis phase to be gasified alongside char. This thereby minimizes the amount of tar in the syngas produced. In addition, bio-oil can be retrieved during its operation and upgraded for use as feedstock for other downstream applications.

16.4 Challenges of Pyro-gasification of Invasive Alien Plants (IAPs)

Several species of invasive alien plants, which include woody and non-woody types, exist in nature. However, the most common are those with some commercial value and those identified as *category 1 & 2* invaders with the most severe impact on the environment [16]. In most instances where their impacts have exceeded their socio-economic benefits, their control or outright eradication via mechanical slash and burn/dump has been the most preferred solution [37]. Despite the recent studies [7, 20, 23] supporting the economic viability, and sustainability of using biomass wastes derived from IAP control as feedstock for biofuel conversion, there exists a gap in the knowledge of their suitability for thermochemical processes.

With the use of data gathered from lab experiments and literature, this section highlights some identified properties of selected invasive species, observed to either promote or limit their use as feedstock for pyro-gasification. Five woody IAPs – namely Eucalyptus, Radiate pine, Black wattle, Poplar and Jacaranda – have been selected for discussion. Table 16.3 summarizes their properties as obtained from lab experimental tests by the author. Meanwhile, the properties of 13 non-woody IAPs – namely Giant reed, Lantana, Pickerel weed, Castor-oil plant, Sweet prickly pear, Bugweed, Saltbush, Inkberry, Cassia, *Chromoleana*, Water hyacinth, Queen of the night and Sisal, as obtained from literature [21] – are also summarized in Table 16.4 for comparison. All species listed here are classified as either *Category 1 or 2* invaders which are referred to as extremely invasive and most threatening to the environment. They therefore represent those species which are subject to frequent control, from which an abundance of biomass wastes is continually being generated.

16.4.1 High Ash Tendency

Most IAPs are characterized by their high ash content which may be detrimental to the overall efficiency of their gasification process. Some examples of woody IAPs with higher ash contents exceeding the German and Austrian national maximum recommended values for wood fuels (Table 16.3) are the Radiate pine (4.76%),

Table 16.3 Properties of woody invasive alien plants in comparison to European standards for wood fuels

	Proximate analysis %				Ultimate analysis %					HHV (Exp.) (MJ/kg)	Bulk density (kg/m^3)	Energy density (GJ/m^3)	Fuel ratio	Char yield %
Biomass	MC**	VM	FC*	Ash	C	H	N	S	O*		ϱ_B	*ED*	*FR*	
Eucalyptus spp.	9.13	49.88	39.98	1.01	47.82	5.37	0.08	0.00	46.73	18.84 ± *0.04*	1010	19.03	0.80	40.99
Black wattle	8.60	54.73	33.94	2.73	43.41	5.62	0.32	0.00	50.66	18.09 ± *0.18*	870	15.74	0.62	36.67
Poplar	12.21	56.56	30.2	1.04	42.34	5.82	0.27	0.00	51.58	18.80 ± *1.10*	500	9.40	0.53	31.24
Radiate pine	6.69	43.49	45.06	4.76	40.37	4.91	0.19	0.00	54.54	18.85 ± *0.10*	840	15.83	1.03	49.82
Jacaranda	9.12	57.91	30.76	2.21	44.23	5.73	0.18	0.00	49.87	18.43 ± *0.10*	510	9.40	0.53	32.97
[a]*Austria ONORM M7135*				*≤0.5*			*<0.3*	*<0.04*		*>18.0*				
[b]*German DIN 51731/ DINplus*				*<1.5/<0.5*			*<0.3*	*<0.08/0.04*		*17.5–19.5/ >18*				

*MC*** air-dried biomass moisture content, *VM* volatile matter, *FC** fixed carbon by difference, *O** oxygen content by difference

[a]Austrian national standard for wood fuels (Austria ONORM M7135)

[b]German national standards for wood fuels (German DIN 51731/DINplus)

Black wattle (2.73%) and Jacaranda (2.21%). However, much higher amounts of ash ranging from 3.4% to 17.1% can be found in non-woody invasive plants (Table 16.4). Such high amounts of ash found in these species could be as a result of the high concentrations of mineral matter common in soils where they thrive. Factors such as high population density of invasive plants in the same area, with several plants competing for ground water and nutrients; low ground water levels and low mean annual rainfall, lead to decreased soil dilution capacity. With such low soil dilution capacity, IAPs are expected to absorb a high concentration of mineral matter from the soil.

A common occurrence in combustion and gasification of biomass which is also expected in pyro-gasification is the deposition of layers of ash residues containing reaction-inhibiting silicone particles on the char surface. These ash layers block char pores and inhibit the free transfer of hot oxidizing gases into its inner carbon particles, thereby minimizing the rate of char partial oxidation. Problems such as poor energy conversion efficiency, catalyst deactivation due to ash fouling, agglomeration due to low melting point of biomass ash, sintering, aggregation, deposition and eventual corrosion of the gasifier can arise as a result of high amounts of ash in the feedstock. Moreover, ash in high amounts reduces the calorific value of the feedstock and thus result to a reduced syngas calorific value [19, 34, 36].

16.4.2 Moisture Content

IAPs are renowned for their high ground water absorption rates which result to their high moisture contents (MC). The highest MC observed among the woody invasive plants is 12% for the Poplar, which is below the recommended value of 15% MC for feedstock conversion to syngas [34]. However, extremely high moisture contents within the range of 49.2–94.7% were observed in the non-woody species. This is in corroboration with literature whereby higher amounts of moisture have been found to be more in fibrous, fast growing non-woody shrubs than in large trees [2, 6, 13, 26]. Moisture content of invasive trees and shrubs is highest in their trunks and stems which contain fibres for transporting and storing water and nutrients. However, in most cases due to their equally high rates of transpiration, their leaves and branches may contain the least moisture and thus do not normally require as much drying as their stems and trunks.

High moisture composition in biomass feedstock of small particle sizes can result in agglomeration of its particles during loading on to the reactor. This can be more severe if the feedstock is milled to sawdust particles. The use of feedstock with moisture contents above 15% can result in a reduction in the gasifier temperature through evaporation, particularly in the oxidation and gasification phases, which lead to incomplete gasification of the products formed during the decomposition phase [34]. High moisture content also leads to the production of higher fractions of H_2 than CO (Eq. 16.5) which encourages increased methane formation. However, high amounts of H_2 and CH_4 with lesser amounts of CO in the product gas

Table 16.4 Properties of non-woody invasive alien plants (Adapted from [21])

Biomass	Proximate analysis (%)			Ultimate analysis					HHV (Exp.) (MJ/kg)	Bulk density (kg/m^3)	Energy density (GJ/m^3)
	MC	VM	Ash	C (%)	N (%)	S (ppm)	Si (ppm)	Cl (ppm)		ϱ_B	*ED*
Giant reed	49.2 ± 1.2	97.0 ± 0.7	3.4 ± 0.6	51.9	0.9	1566.6	91.8	2308.8	17.1 ± 0.2	60.48	1.034
Lantana	73.6 ± 2.2	83.4 ± 7.1	5.8 ± 0.2	57.0	2.6	2138.0	270.7	3108.0	16.9 ± 0.3	60.75	1.027
Pickerel weed	84.3 ± 0.1	91.2 ± 0.6	6.9 ± 0.3	46.2	2.2	1127.3	199.3	16747.6	15.9 ± 0.3	28.56	0.454
Castor-oil plant	84.3 ± 1.5	96.8 ± 1.1	5.6 ± 0.9	56.3	5.8	3609.2	53.5	6322.6	16.4 ± 0.5	63.75	1.045
Sweet prickly pear	92.4 ± 0.1	90.7 ± 0.2	8.3 ± 0.8	50.7	0.9	935.7	64.5	15682.1	16.0 ± 0.4	216.46	3.463
Bugweed	65.7 ± 3.3	95.8 ± 0.7	4.1 ± 1.0	43.3	3.0	1528.9	31.6	7690.1	16.9 ± 0.2	42.11	0.712
Saltbush	54.9 ± 1.9	86.2 ± 0.2	14.2 ± 0.3	44.3	1.5	1900.7	43.7	1642.8	16.1 ± 0.4	167.25	2.692
Inkberry	70.9 ± 0.6	93.4 ± 0.9	6.3 ± 0.3	56.6	2.0	2314.7	284.9	4031.5	19.3 ± 0.9	138.67	2.676
Cassia	70.0 ± 0.2	94.1 ± 0.5	6.2 ± 0	54.6	2.3	1693.8	221.9	4422.2	16.9 ± 0.1	80.39	1.359
Chromoleana	61.6 ± 2.0	94.4 ± 1.1	4.7 ± 0.5	58.6	1.4	1579.9	24.0	9004.3	17.2 ± 0.1	108.05	1.858
Water hyacinth	94.7 ± 0.1	85.5 ± 0.8	17.1 ± 1.2	36.7	3.2	2262.1	41.7	16925.3	13.9 ± 0.1	44.21	0.615
Queen of the night	87.5 ± 0.2	84.6 ± 0.4	16.6 ± 0.4	43.1	0.6	1990.7	117.9	603.8	13.3 ± 0.1	168.71	2.244
Sisal	83.3 ± 0.3	91.3 ± 0.6	9.2 ± 0.2	60.4	0.8	557.4	108.2	692.6	17.4 ± 0	110.01	1.914

MC moisture content, *VM* volatile matter content

may result in the production of syngas with a reduced heating value. Meanwhile, if utilized for combustion purposes, biomass feedstock with high moisture contents is usually difficult to ignite and emit high amounts of particulate matter as a result of incomplete combustion.

Therefore, since moisture supports microbial activity, biomass from non-woody IAPs may be more suited for biochemical conversion processes (fermentation and anaerobic digestion). Moreover, since their use as feedstock for thermochemical processes will require adequate and energy-intensive pre-treatment such as drying or torrefaction. Such feedstock also proves to be very expensive to handle due to the added weight caused by high moisture content.

16.4.3 Volatile Matter and Fixed Carbon Composition

Invasive alien plants are a mix of species with relatively high volatile matter contents and moderate amounts of fixed carbon. This limits their char yield but maximizes gas and bio-oil yield when pyrolysed. In general, non-woody species lack the required amounts of fixed carbon expected for considerable char yields suitable for pyro-gasification, which is an evidence of their low lignin composition. This, however, is contrary to woody IAPs which contain considerable amounts of fixed carbon and in turn yield adequate char residue necessary for gasification during the gasification phase of a pyro-gasification process (Table 16.3).

IAPs such as the eucalyptus, castor oil plant and inkberry naturally produce high amounts of strong-smelling ornamental oils while still alive and retain these oils in their woods even when cut down and used as feedstock. These oils are thereby released in large volumes during the pyrolysis phase of the gasification process. In addition, non-woody IAPs contain extremely high amounts of tar-forming volatiles which when condensed also add up to the volume of bio-oil produced in the pyrolysis phase. Such feedstock is suitable for bio-oil production via fast pyrolysis and also for gasification with high temperature tar cracking capabilities. Their use for pyro-gasification, however, may be limited due to an expected lesser char yield as opposed to the high amounts of tars generated in the pyrolysis phase.

16.4.4 Energy Density

Most IAPs – both woody and non-woody species, generally exhibit low bulk densities and low to medium higher heating values (HHVs) which translate to low energy densities. The energy densities of the non-woody IAPs were, however, observed to be very low (0.454–3.463 GJ/m^3), as opposed to those of the woody IAPs. The woody IAPs such as the *Eucalptus* (19.03 GJ/m^3), Black wattle (15.74 GJ/m^3) and radiate pine (15.83 GJ/m^3) exhibited high energy densities suitable for thermochemical conversion.

Low energy density of a feedstock results to poor conversion efficiency and decreased feedstock residence time which is detrimental to the economics of the process as a whole. However, the energy density of the biomass feedstock can be improved upon by torrefaction or conversion to char or bio-oil. For instance, the energy densities of straw (2 GJ/m^3) and woodchips (8 GJ/m^3) were observed to increase to 30 GJ/m^3 and 26 GJ/m^3 when converted to bio-oil and char-oil slurry [30].

16.4.5 Nitrogen and Sulphur Compounds' Formation

Syngas produced from gasification normally contains nitrogen and sulphur compounds which are formed primarily from the oxidation of nitrogen and sulphur present in the feedstock. However, high sulphur concentrations are more commonly found in coals and almost negligible in biomass.

The high nitrogen composition in most IAPs makes them less desirable for pyro-gasification and other thermochemical processes, as incombustible N_2, NH_3, HCN, NO_x (NO, NO_2) compounds are formed during conversion. Nitrogen present in the feedstock is oxidized to form NO and NO_2 which when released to the atmosphere dissolve in rain water to cause acid rain. Acid rain is known to be toxic to soil and aquatic organisms as it increases the soil and water acidity [28]. Furthermore, nitrogen can be partially converted to ammonia (NH_3) and hydrogen cyanide (HCN) which if contained in syngas can poison catalysts used for downstream applications [4]. Meanwhile, sulphur compounds SO_2 and SO_4, just as NO_x, can also result to acid rain formation, if found in the feedstock.

Nitrogen concentrations were particularly very high in the non-woody IAPs and lesser in the woody IAPs. The nitrogen composition of the non-woody IAPs was within the range of 0.6–5.8%, which was much higher than the maximum recommended values of 0.3% by the Austria ONORM M7135 and German DIN 51731/DINplus. Sulphur was observed to be negligible in the woody IAPs but found in considerable concentrations in the non-woody IAPs. The non-woody IAPs in general exhibited high concentrations of silicon, chlorine and sulphur elements which can cause chemical reactions that form slag and corrode reactor linings.

16.5 Possible Challenges of Scaling Up Pyro-gasification Systems

Due to a lack of extensive operation and study of existing pyro-gasification pilot plants reported, there is insufficient information to truly ascertain the challenges of scaling up the systems to commercial status. However, based on preliminary investigations reported so far [2, 13, 30, 33], this chapter outlines some possible bottlenecks which may hinder its development.

16.5.1 System Complexity

So far, existing pyro-gasification systems are very complex, consisting of several interconnected units and processes. A notable instance is the Viking, which requires the recycling of heat expelled for the partial oxidation channel, char gasifier and engine exhaust channelled for use in pyrolysis zone and for district heating. Inasmuch as this is economically very attractive, the complex piping connections and moving parts of the screw conveyer in the pyrolysis chamber will require periodic rigorous and expensive maintenance operations to maintain the overall system efficiency. The removal of ash deposits also proved to be problematic, as the Viking required regular removal of ash deposits from the grate which often resulted to a pressure drop across the system [13].

16.5.2 Oil Blockage in Pipes

Blockages in the pipes channelling bio-oil released during the pyrolysis phase of a pyro-gasification process are usually expected. This, however, depends on the viscosity of the bio-oil. For instance, after some tests were conducted with the pyro-gasification experimental setup (Fig. 16.5), it was observed that the bio-oil extraction pipe connected to the exit at point C of the reactor was often blocked by very viscous bio-oil residue. Such blockages were caused when the bio-oil flowing through the pipe flowed from the hot section of the pipe inserted in the cylindrical furnace to the cooler section of the pipe away from the heat of the furnace. This oil, coming in contact with the colder surface of the pipe, increasingly became more viscous prior to reaching the condenser 1, thereby reducing its flow and subsequently causing blockages. This required the passing of compressed air through the pipe to blow off the trapped bio-oil in the pipe after each run. However, a more permanent solution would be to increase the diameter of the bio-oil pipe and wrap it with heating coils to maintain high temperatures along the pipe. This is expected to preserve the high temperature of the oil and maintain a low viscosity as it flows towards the *Gas condenser 1*.

16.5.3 Feedstock Compatibility

One common problem with biomass thermochemical conversion is the wide variation in feedstock properties. Biomass feedstock properties vary widely, based on type, plant species, geographical, geological and climatic conditions of their source. This limits the use of pyro-gasification systems for the conversion of a wide variety of biomass feedstock and also hinders the possible blending of different feedstock for co-gasification. Moreover, pyro-gasification may be best suited for woody feedstock with moderate to high fixed carbon contents which translates to char yields sufficient for gasification in the gasification phase of the process. Thus, fibrous feedstock with high moisture and volatile matter composition may be difficult to utilize unless pre-treated adequately to meet standards. The pyrolysis chamber of the Viking reactor was, however, reported to handle feedstock with moisture contents as high as 45%. In the case of woody biomass feedstock, milling to a suitable particle size may be required. It was observed that blockages occurred in the feedstock conveyor of the Viking by larger wood particles stuck in the conveyor, which led to unplanned shut-downs. In addition, feedstock agglomeration may occur frequently during the transfer of biomass particles from the inlet point (i.e. drying point) through to the char and tar exit point in the pyrolysis chamber. This could result to a reduction in the overall system efficiency.

16.5.4 Downstream Syngas Application

Syngas produced from pyro-gasification systems may require additional cleaning prior to use as feedstock for downstream processes or applications like fuel cells, gas turbines, internal combustion engines with slim tar and particulates tolerance thresholds (Table 16.1). This may result to additional installation and operational costs asides that incurred in setting up the pyro-gasification plant. The ignition of the three-cylinder spark ignition engine connected to the Viking proved difficult and thus required the use of natural gas to start up before switching to the syngas supply. In addition, engine power was observed to have dropped, while a conversion efficiency of 28% from syngas to mechanical energy was attained [13].

16.6 Conclusion and Recommendations

The fast growth rates and invasiveness of IAPs have led to an increased interest in their potential use as feedstock for biofuel production. Inasmuch as harnessing biomass wastes generated from the continuous control of IAPs for thermochemical conversion processes create a commercially viable avenue for localized off-grid energy generation, their unique physical and chemical properties may be detrimental to such purposes. Properties such as heir relatively high ash, moisture, nitrogen and sulphur compositions are key properties of biomass known to reduce the overall process efficiency and quality of syngas produced. Fixed carbon and volatile matter content are also important properties which determine the suitability of a feedstock for thermochemical conversion purposes. These attributes were rightly exhibited by the woody IAPs due to their significant fixed carbon content but were found deficient in the non-woody species. Ash, nitrogen and sulphur were also observed to be much lower in the woody IAPs and significantly higher in the non-woody IAPs. The woody IAPs, due to their superior HHV and bulk densities, also exhibited higher energy densities necessary for the economics and efficiency of the process.

High tar generation during thermochemical conversion of IAPs is also expected, due to the high volatile matter content exhibited mostly by the non-woody IAPs. Tar concentration in syngas is known to be a common problem with biomass gasification and several studies have focused on developing methods to reduce its composition. Pyro-gasification – classified under the pre-gas production tar removal methods – was identified as a possible solution to tar removal. Reported examples of new pilot plants with significant reductions in syngas tar composition were highlighted. However, bottlenecks like the operational and installation costs of the complex system components and piping, feedstock compatibility and quality of the syngas for downstream application may hinder the scale-up of this technology to full commercialization.

In conclusion, the woody IAPs can be considered the most suitable types of IAPs for pyro-gasification and thermochemical conversion processes in general, due to their favourable properties. However, for non-woody IAPs, the high moisture content and volatile matter content, which is a reflection of high cellulose and hemicellulose contents in the biomass, make them most suitable feedstock for biochemical conversion than thermochemical processes. It is recommended that more research into the suitability of a wider variety of IAPs feedstock for thermochemical processes is required to promote their use for energy production. However, new and improved designs of pyro-gasification systems, as well as more research into existing pyro-gasification systems are encouraged to ascertain the viability of the technology in real-world situations. Moreover, limiting the system to the conversion of a particular type or species of biomass feedstock would be disadvantageous to its application and commercialization in different regions where biomass properties are dissimilar.

References

1. D. Aar, N. Cape, Alien Invasive Plant Management Plan – Longyuan Mulilo Maanhaarberg Wind Energy Facility (2014), https://sahris.sahra.org.za/sites/default/files/additionaldocs/12.%20Appendix%2011%20Alien%20Invasive%20Plant%20Management%20Plan_0.pdf. Accessed 6 June 2019
2. J. Ahrenfeldt, T.P. Thomsen, U. Henriksen, L.R. Clausen, Biomass gasification cogeneration – A review of state of the art technology and near future perspectives. Appl. Therm. Eng. **50**(2), 1407–1417 (2013). https://doi.org/10.1016/j.applthermaleng.2011.12.040
3. P. Basu, *Biomass Gasification and Pyrolysis Practical Design* (Elsevier, Kidlington, Oxford, 2010)
4. K.M. Broer, R.C. Brown, The role of char and tar in determining the gas-phase partitioning of nitrogen during biomass gasification. Appl. Energ. **158**, 474–483 (2015). https://doi.org/10.1016/j.apenergy.2015.08.100
5. J. Chamier, K. Schachtschneider, D.C. Maitre, et al., Impacts of invasive alien plants on water quality, with particular emphasis on South Africa. Water SA **38**(2), 345–356 (2012). https://doi.org/10.4314/wsa.v38i2.19
6. A. Colantoni, L. Longo, F. Gallucci, et al., Pyro-gasification of hazelnut pruning using a downdraft gasifier for concurrent production of syngas and biochar. Contemp. Eng. Sci. **9**(27), 1339–1348 (2016). https://doi.org/10.12988/ces.2016.68142

7. W. De Lange, Incorporating stakeholder preferences in the selection of technologies for using invasive alien plants as a bio-energy feedstock: Applying the analytical hierarchy process. J. Environ. Manag. **99**, 76–83 (2012)
8. Department of Environmental Affairs South Africa, Assessment of the potential to produce biochar and its application to South African soils as a mitigation measure. 84 (2015), https://www.environment.gov.za/sites/default/files/reports/biocharreport2015.pdf. Accessed 15 May 2019
9. R. Early, B.A. Bradley, J.S. Dukes, et al., Global threats from invasive alien species in the twenty-first century and national response capacities. Nat. Commun. **7** (2016). https://doi.org/10.1038/ncomms12485
10. J.L. Easterly, M. Burnham, Overview of biomass and waste fuel resources for power production. Biomass Bioenergy **10**(2–3), 79–92 (1996). https://doi.org/10.1016/0961-9534(95)00063-1
11. J. Fjellerup, J. Ahrenfeldt, U. Henriksen, et al., Formation, decomposition and cracking of biomass tars in gasification. Renew. Energ. **30** (2005). http://linkinghub.elsevier.com/retrieve/pii/S0960148104002927. Accessed 18 May 2019
12. T.L. Greaver, T.J. Sullivan, J.D. Herrick, et al., Ecological effects of nitrogen and sulfur air pollution in the US: What do we know? Front. Ecol. Environ. **10**(7), 365–372 (2012). https://doi.org/10.1890/110049
13. U. Henriksen, J. Ahrenfeldt, T.K. Jensen, et al., The design, construction and operation of a 75kW two-stage gasifier. Energy **31**(10–11), 1542–1553 (2006). https://doi.org/10.1016/j.energy.2005.05.031
14. K.A. Hunter, P.S. Liss, V. Surapipith, et al., Impacts of anthropogenic SOx, NOx and NH3 on acidification of coastal waters and shipping lanes. Geophys. Res. Lett. **38**, 2–7 (2011). https://doi.org/10.1029/2011GL047720
15. B. Jenkins, L. Baxter, T. Miles, et al., Combustion properties of biomass. Fuel Process. Technol. **54**(1–3), 17–46 (1998). https://doi.org/10.1016/S0378-3820(97)00059-3
16. P. Koch, Utilization of southern pines (Vol. 1). U. S. Department Of Agriculture Forest Service For Serv Agric Handbook 420:535–574. (1972), https://www.srs.fs.usda.gov/pubs/ah/ah420_vol1.pdf. Accessed 5 June 2019
17. W. Li, Q. Li, R. Chen, et al., Investigation of hydrogen production using wood pellets gasification with steam at high temperature over 800 C to 1435 C. Int. J. Hydrogen. Energ. **39**, 5580–5588 (2014). https://doi.org/10.1016/j.ijhydene.2014.01.102
18. C. Mabhaudhi, Impact of SOx and NOx Deposition on the Leaching Behaviour of Soils and Water Quality in the South African Highveld. Dissertation, University of KwaZulu-Natal (2014)
19. P. McKendry, Energy production from biomass (part 3): Gasification technologies. Bioresour. Technol. **83**(1), 55–63 (2002). https://doi.org/10.1016/S0960-8524(01)00120-1
20. M. Mandlakazi, Evaluation of the Potential of Non-woody Invasive Plant Biomass for Electricity Generation. Dissertation, Stellenbosch University (2016)
21. M. Mandlakazi, C. Ham, M. Meincken, Characteristics of selected non-woody invasive alien plants in South Africa and an evaluation of their potential for electricity generation. J. Energy South Afr. **28**(3), 92 (2017). https://doi.org/10.17159/2413-3051/2017/v28i3a1896
22. A. Monti, N. Di Virgilio, G. Venturi, Mineral composition and ash content of six major energy crops. Biomass Bioenergy **32**(3), 216–223 (2008). https://doi.org/10.1016/j.biombioe.2007.09.012
23. W. Mugido, J. Blignaut, M. Joubert, et al., Determining the feasibility of harvesting invasive alien plant species for energy. S. Afr. J. Sci. **110**(11), 1–6 (2014). https://doi.org/10.1590/sajs.2014/20130397
24. J.P.A. Neeft, H. Knoef, P. Onaji, *Behaviour of Tar in Biomass Gasification Systems: Tar Related Problems and their Solutions* (Utrecht, 1999)
25. N.M. Okoro, K.G. Harding, M. Daramola, Fuel Examination of Invasive Alien Plants and Tropical Hardwoods for Pyro-gasification Feedstock. Paper presented at the 2nd International Conference on Bioresource Technology for Bioenergy, Bioproducts & Environmental Sustainability, Melia, Sitges, 16–19 September, 2018 (2018)

26. T. Pattanotai, H. Watanabe, K. Okazaki, Gasification characteristic of large wood chars with anisotropic structure. Fuel **117**, 331–339 (2014). https://doi.org/10.1016/j.fuel.2013.09.030
27. C.F. Pratt, K.L. Constantine, S.T. Murphy, Economic impacts of invasive alien species on African smallholder livelihoods. Glob. Food Secur. **14**, 31–37 (2017). https://doi.org/10.1016/j.gfs.2017.01.011
28. E.M. Remedio, An analysis of sustainable fuelwood and charcoal production systems in the Philippines: A Case Study, in *Criteria and Indicators for Sustainable Woodfuels: Case Studies from Brazil, Guyana, Nepal, Philippines and Tanzania*, ed. by M. Trossero, (Food and Agricultural Organization of the United Nations, Rome, 2009)
29. S.K. Sansaniwal, K. Pal, M.A. Rosen, et al., Recent advances in the development of biomass gasification technology: A comprehensive review. Renew. Sust. Energ. Rev. **72**, 363–384 (2017). https://doi.org/10.1016/j.rser.2017.01.038
30. V.S. Sikarwar, M. Zhao, P. Clough, et al., An overview of advances in biomass gasification. Energy Environ. Sci. **9**(10), 2939–2977 (2016). https://doi.org/10.1039/c6ee00935b
31. R.N. Singh, S.P. Singh, J.B. Balwanshi, et al., Tar removal from producer gas: A review. Res. J. Eng. Sci. **3**(10), 16–22 (2014). www.isca.me. Accessed 18 March 2019
32. S.P. Sohi, E. Krull, E. Lopez-Capel, et al., A review of biochar and its use and function in soil. Adv. Agron. **105**(1), 47–82 (2010). https://doi.org/10.1016/S0065-2113(10)05002-9
33. J. Søren, U. Birk, Formation, Decomposition and Cracking of Biomass Tars in Gasification. Technical University of Denmark. (2005), https://orbit.dtu.dk/files/3220398/pyrotar2.pdf. Accessed 31 March 2019
34. J.G. Speight, Production of syngas, synfuel, bio-oils, and biogas from coal, biomass, and opportunity fuels, in *Fuel Flexible Energy Generation: Solid, Liquid and Gaseous Fuels*, (CD&W Inc., Laramie, 2015), pp. 145–174. https://doi.org/10.1016/B978-1-78242-378-2.00006-7
35. S.V. Vassilev, D. Baxter, C.G. Vassileva, An overview of the behaviour of biomass during combustion: Part II. Ash fusion and ash formation mechanisms of biomass types. Fuel **117**, 152–183 (2014). https://doi.org/10.1016/j.fuel.2013.09.024
36. J.F. Vélez, F. Chejne, C.F. Valdés, et al., Co-gasification of Colombian coal and biomass in fluidized bed: An experimental study. Fuel **88**(3), 424–430 (2009). https://doi.org/10.1016/j.fuel.2008.10.018
37. B.W. Van Wilgen, G.G. Forsyth, D.C. Le Maitre, et al., An assessment of the effectiveness of a large, national-scale invasive alien plant control strategy in South Africa. Biol. Conserv. **148**(1), 28–38 (2012). https://doi.org/10.1016/j.biocon.2011.12.035

Chapter 17
Valorisation of Human Excreta for Recovery of Energy and High-Value Products: A Mini-Review

T. O. Somorin

17.1 Introduction

The use of human excreta as fertiliser has been around for as long as we know [62]. Until the 1900s, it was socially acceptable in many Nordic countries to use human excreta from dry pit latrines for arable farming [29]. The use of night soils (biosolids collected from cesspools, pit latrines, septic tanks) on agricultural fields was commonly practised in countries like China, Vietnam, Japan and India [55], but the increasing use of chemical fertiliser, invention of the modern flush toilet and growing environmental and health concerns have caused such practices to face near-extinction [21]. The current urban sanitation systems are on the verge of facing a similar fate because the linear approach to managing waste raises environmental concerns, particularly in resource and nutrient recovery. For example, the conventional flush toilet requires at least 2 L of freshwater per flush [31] to convey waste from individual units to a centralised treatment plant and via a broad network of sewer infrastructures that combine storm drain and commercial wastewater with domestic sewers. These processes require a considerable amount of energy, capital investment and land space, which is particularly challenging for developing countries. Often, communities do not benefit from centralised water and energy services and settlements are densely populated, clustered or distant apart. Even when infrastructures are present, systems are largely dysfunctional, due to improper use and maintenance of sewer networks. In many communities, proper waste treatment and disposal methods are rarely followed, and toilet facilities are often shared, limited or unimproved [74]. There are reports of pit latrines and sewers leaking into groundwater sources [49], illicit dumping of sludges retrieved from septic tanks and pit

T. O. Somorin, PhD (✉)
University of Strathclyde, Glasgow, UK
e-mail: tosin.somorin@strath.ac.uk

M. O. Daramola, A. O. Ayeni (eds.), *Valorization of Biomass to Value-Added Commodities*, Green Energy and Technology,
https://doi.org/10.1007/978-3-030-38032-8_17

latrines into the environment [52] and practices of open defecation. The development of decentralised sanitary solutions is, therefore, a priority and opportunities to manage, recover and utilise human waste are being explored globally.

There are various propositions for the development of on-site sanitation systems, solutions that can safely collect and treat human waste in situ and by-products of value. Technological solutions are expected to operate without depending on an external supply of water, energy and infrastructures. Multiple tangible benefits, for example, clean water, nutrients, fuels, are demanded from the recycling of waste resources. The values of end-products are projected to exceed the cost needed to transform waste and systems must not compete with universal human needs, for example, nutrient recycling for food production or strain the natural environment. This mini-review summarises the effort to derive value from human waste. Valorisation of waste is examined in the context of nutrient recovery and for advanced fuels, heat and/or electricity. Focus is given to low-cost approaches and technological solutions that offer ecological benefits and opportunities to recover nutrients and/or value-added products. The barriers and opportunities for the adoption of on-site sanitation and appropriate technologies are discussed to inform future research programmes.

17.2 Human Excreta as a Resource

17.2.1 Human Urine

Urine contains large amounts of organic solutes and inorganic salts [28]. On the average, a healthy adult produces 0.6–2.6 L of urine per day [58], which contains about 300–2200 mg/L of potassium and 150–1800 mg/L of phosphorus [63]. Urea is the most predominant form of organic nitrogen (N), making up 50% of the organic solids and 75–90% of nitrogenous fractions [58]. The rest are mainly organic and inorganic salts, for example, phosphorus (P) in the form of superphosphate, potassium (K) as ionic salts and organic ammonium salts [40]. The dry solids contain about 14–18% N, 13% carbon (C), 3.7% P and 3.7% K [58], nutrients that are readily accessible to plant and microbes. The release of nutrients offers several advantages: it encourages the formation of microbiota and humus, improves water retention and adsorption in soils and expands soil structures [64]. Low concentration of metals in contrast to synthetic fertiliser is also considered a benefit as it minimises toxicity [63], but crop yield depends on the proportion of nutrients. Typically, nutrients become available when organics such as urea breaks down into constituent parts: ammonium (NH_4^+), ammonia (NH_3), bicarbonate (HCO_3^-) and carbonate (CO_3^{2-}). The proportion of NH_4^+ and NH_3 progressively changes due to increasing pH, leading to the release or loss of nutrients [26]. According to Rose et al. [59], body water balance, for example, perspiration, respiration and urination, affects the partitioning of nutrients in urine. This is influenced by factors such as fluid intake, diet particularly protein, salt and water consumption, diseases, physical exercises and environmental factors. Low concentrations of essential macro and micro-nutrients in urine can hinder growth and increase susceptibility to diseases. The inappropriate use and/or disposal can cause algal bloom, eutrophication, fish death and human intoxication [23, 34, 78]. There are reports on urea – N and P pollution in rivers, which are attributed to leaching of nutrients through soils into surface water and groundwater sources [15, 70]. For these reasons, recent focus is increasingly directed at nutrient extraction rather than direct use of urine.

17.2.2 Human Faeces

Faeces contain about 75 wt.% water and 25 wt.% dry solids [67]. An average healthy adult produces 150–250 g of faeces, but this can vary widely from 15 to 1505 g/cap/day [58], depending on age, dietary intake, geographical location, economic status, ethnicity, gender, health conditions etc. About 50 wt.% of the water in faeces is said to be found in bacterial cells and in complex biofilm matrix – mainly exopolysaccharides. Solids are primarily composed of undigested fat, protein and carbohydrate, but the proportion of bacterial biomass can be up to 50 wt.%, making up most of the protein [58]. The composition of undigested fat varies from 2% to 8% (wet basis) with fatty acids and phosphoglycerides as by-products of bacterial and body metabolism. Undigested carbohydrates make up about 25% of the solids and are dietary products, unlike protein that is formed from several sources, including dietary protein, nucleic acids, bacterial metabolism and intestinal cells. Figure 17.1 outlines the physical and chemical characteristics of human faeces and urine. These resources are useful for increasing soil fertility and crop yield [3, 33, 45, 60] but pose potential health risks due to large numbers of enteric pathogens. The threats to direct use of wastes can also include foul odour, flies' infestation, helminth risks, itchiness and foot rot [12]. There have been several cases of transmission of helminth eggs [9] and concern on accumulation and amplification of pollutants in the food chain due to the presence of organic pollutants, pharmaceutical residues and steroid hormones [16]. Like urine, the chemical composition can vary in individuals, leading to yield variations. As such recent efforts are exploring ways to safely sanitise faecal sludge in a way that it improves global sanitation goals, reduces environmental pollution and brings economic gains. Source separation of human waste, stabilisation of faecal sludge by composting and co-digestion with organic materials are being explored as opposed to direct use of wastes on farms [72].

17.3 Source Separation of Human Wastes

The concept of separating human excreta from other waste streams is common to all ecological sanitation systems. Toilets are designed to receive human waste as a mixed stream or split-collect the waste such that urine is collected separately from faecal matter and without the addition of rinse water [61]. Processes are said to follow a 'sanitise and recycle' model where waste streams undergo a form of biological treatment to limit pathogens and recovered bio-solids can be used as nutrients.

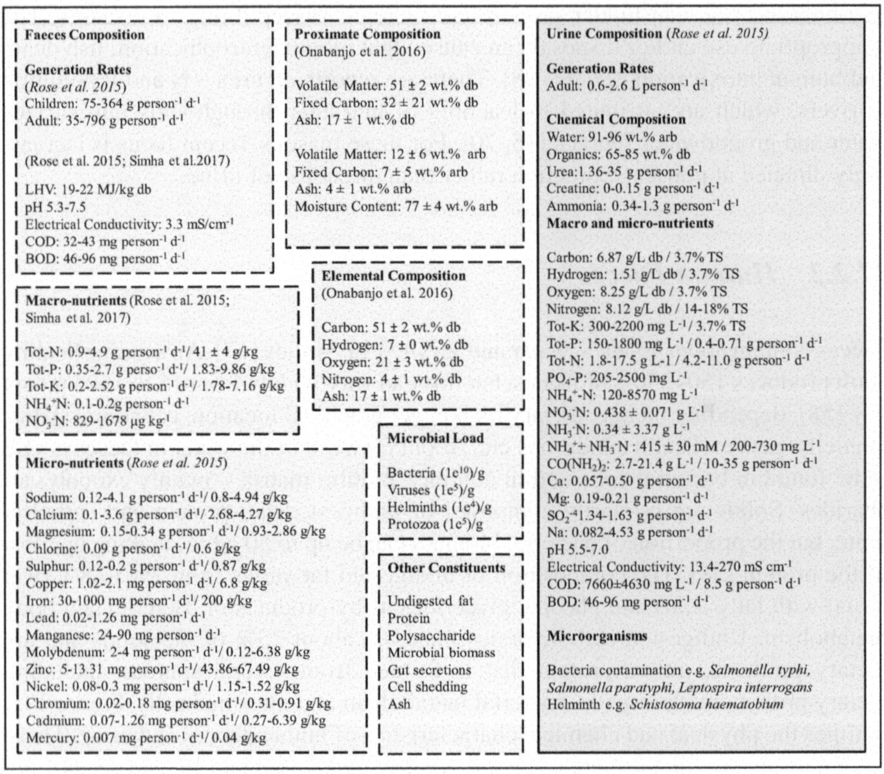

Fig. 17.1 Physical and chemical characteristics of human faeces and urine

The approach is increasingly becoming important because of the growing awareness that nutrient recovery from wastewater is costly and unsustainable. According to Spångberg et al. [79], human urine contributes more than 80% of the total N and more than 50% of the total P and K in wastewater, although the volume fraction of the urine treated is <1%. While efforts are increasingly tailored to recover nutrients from wastewater, these often cannot be achieved and if at all, <10% because of biological decomposition, chemical pollution and mass dilution of waste streams. Sections 17.3.1, 17.3.2, 17.3.3, and 17.3.4 describe some of the ecological sanitation systems (EcoSans) and the benefits and challenges for their use in source separation of human wastes.

17.3.1 Composting Toilets

Composting toilets are referred to as dry toilets because they do not require water, and as biological systems rely on microorganisms to decompose faecal matter to a useable organic form, known as compost. There are several variations of composting toilets: compact or large, with single or multiple composting tanks, electric or manual toilets and waterless or water-based, with source separation of urine or mixed reception of human wastes [2]. Typically, the system consists of (i) a user-interface 'toilet bowl' which can be plastic, ceramic or fibreglass, (ii) composting tank and (iii) accessories such as connecting vent pipe and drain that removes odour and excess leachates. These components can be disconnected from water and wastewater infrastructures. Like conventional flush toilets, the faecal waste is deposited in the toilet but then transferred into the composting tank where it is digested under aerobic conditions. In the process, carbon dioxide (CO_2), water and heat are generated. The heat generated raises the temperature of the matrix and improves moisture evaporation. It also maintains suitable growth conditions for thermophilic microorganisms. Factors such as moisture, temperature, aeration, pH and porosity affect the yield of products [77]. To encourage the growth and survival of microorganisms, a well-balanced aeration and moisture control system is required at relatively high temperatures of 50–60 °C [64].

Co-composting with organic materials such as wood/sawdust, agricultural wastes, food wastes or organic fractions of Municipal Solid Waste (MSW) is recommended. This is because faecal sludge is rich in nutrient but high in moisture content. However, organic materials such as wood dust and organic fractions of municipal solids wastes have a bulky matrix and, when mixed with faecal sludge, porosity and water retention improve, which maintains carbon-to-nitrogen ratio at optimum levels. The system benefits from deactivation of pathogens and helminth eggs (usually within 3 weeks of operation) and the resulting compost is humus-rich and malodour free. The process is, however, a disadvantage because it is not instant: it requires about 10–12 weeks, depending on the scale and operating conditions. It requires external energy supply to keep temperatures and aeration stable, except for manual operation. It requires technical skills, which is often lacking in rural settings. The two main challenges with the use of human compost for agriculture are toxicity and pathogenicity. While human faeces have low toxicity, the biochemical processes that occur during compost can cause the accumulation of toxic compounds along the food chain and toxicity to humans. Chemicals such as antibiotics and hormones can reduce the effectiveness of microorganisms to degrade organics in time. Microorganisms can have an antagonistic relationship with other organisms, leading to inhibition, nutrient depletion and death of indigenous organisms and growth of unwanted species. This could lead to the transmission of pathogens and disease. While temperature is a useful measure to determine the safety of compost, it does not accurately measure the entire compost or bioactivity unfolding in the compost [2], as such mixing to maintain uniform temperatures, and combinatory tests are suggested for safe quality. In open systems, the volatilisation of ammonia cannot be avoided, which poses risks to the environment.

Some composting toilets employ earthworms by a process known as vermicomposting. Unlike conventional methods, vermicomposting does not require high temperatures and the end-products are earthworm biomass and vermi-rich compost, which are useful for farming and for soil conditioning [30, 37]. The earthworms can decompose a wide range of organic materials, but moisture level needs to be maintained at 50–60% [64]. The process is more rapid, easily controllable, energy-efficient and cost-effective. It ensures complete removal of pathogens and helminth

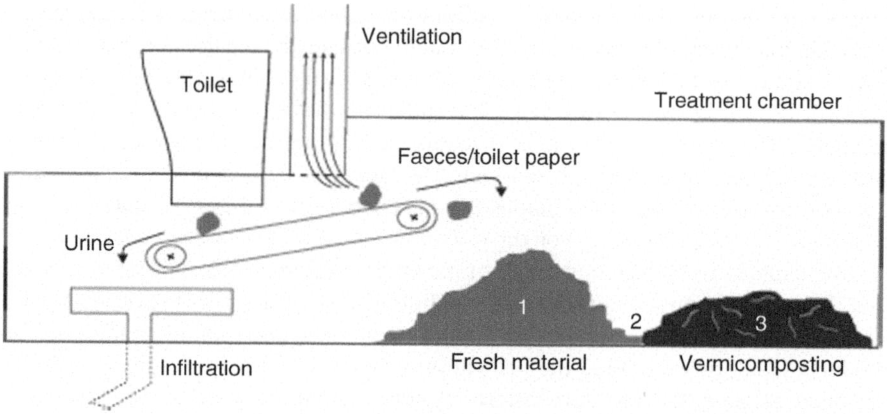

Fig. 17.2 Pictorial view of vermicomposting toilet [37]

eggs [5, 75]; however, technical skills are required for operation and maintenance. Furthermore, there is increasing interest in the use of black soldier fly, *Hermeticia illucens* L. for decomposing human waste [6]. The larvae feed on the sludge and in the process converts it to compost. The residue is highly rich in animal protein, about 32–64% of the dry solids. The larvae can serve as feed for poultry or fish. The process reduces the load of pathogenic microorganisms and helminth eggs, although further treatment is still required for compost of safe quality. Figure 17.2 shows a vermicomposting toilet with source separation of urine and faeces, followed by conversion of sludge to compost.

17.3.2 Urine Diverting Dehydration Toilets

These are also waterless toilets with a unique user-friendly interface that separate faeces and urine at the point of deposition [39, 44]. The setup consists of (i) a squatting pan or toilet seat that is designed to separate faeces and urine at source, (ii) vaults for dry faeces storage. These can be double dehydration units or single interchangeable containers, (iii) a ventilation pipe to remove odour and moisture from the vault, (iv) an anal cleaning area with a separate collection of wash water, (v) a piping system and container for urine collection and (vi) a storage container with dry cover material. To prevent odour and flies, the faecal solids are collected in ventilated vaults or containers. Dehydration occurs via moisture evaporation and ventilation and because of the addition of dry cover material, for example, ash, lime, following every deposition. The system relies on dehydration and segregation, as such vaults and receivers need to be properly designed. There is reduced pathogen due to dehydration, but this is subject to holding time and other operational factors. The solids need further treatment to inactivate pathogens and helminths eggs.

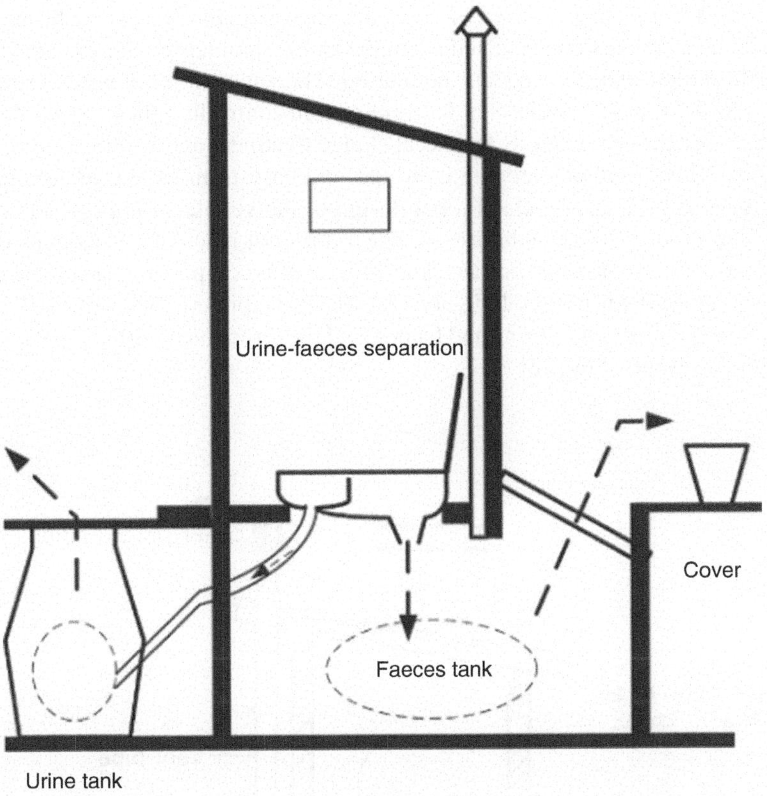

Fig. 17.3 Pictorial view of a urine diverting dehydration toilet [31]

The vault can be connected to a composting unit or to a shallow pit for mineralisation or periodically emptied for external treatments. Figure 17.3 shows a typical Urine Diverting Dehydration Toilet (UDDT) with source separation of waste and collection of the sludge material in a vault/tank.

17.3.3 Ventilated Improved Pit Latrine

These toilets consist of (i) an enclosed superstructure with gap for air circulation and facing the prevailing wind; (ii) a pit for storage of human waste and connected to a ventilation pipe; (iii) squatting pan, toilet seat or slab support structure that is centred over the covered pit; and (iv) a ventilation pipe that forces the flow of air from the outside via the pit and through the pipe to the atmosphere. The continuous movement of air removes odour from the pit to the atmosphere. Integrated flyscreen in the pipe discourages flies and the hole allows free passage of air. The system relies on the movement of air as such the superstructure needs to be properly designed. Foul odours are expected when the ambient air temperature is colder than the air in the pit and prevents air circulation. The superstructure needs to be kept dark to ensure the effectiveness of the vent pipe in controlling flies, except mosquitos. These are proven methods for faecal sludge treatment and widely used in many countries. These methods are simple to use and easy to maintain, often requiring emptying every 12–24 months. Proper design of the ventilated improved pit (VIP) latrine can ensure sludge stabilisation, but it does not inactivate pathogens or helminth eggs. Post-treatment such as composting, vermicomposting, anaerobic digestion (AD) or drying is required. Figure 17.4 shows the pictorial view of a VIP latrine with a mixed collection of urine and faeces, and the direction of airflow that removes odour and prevents flies [22].

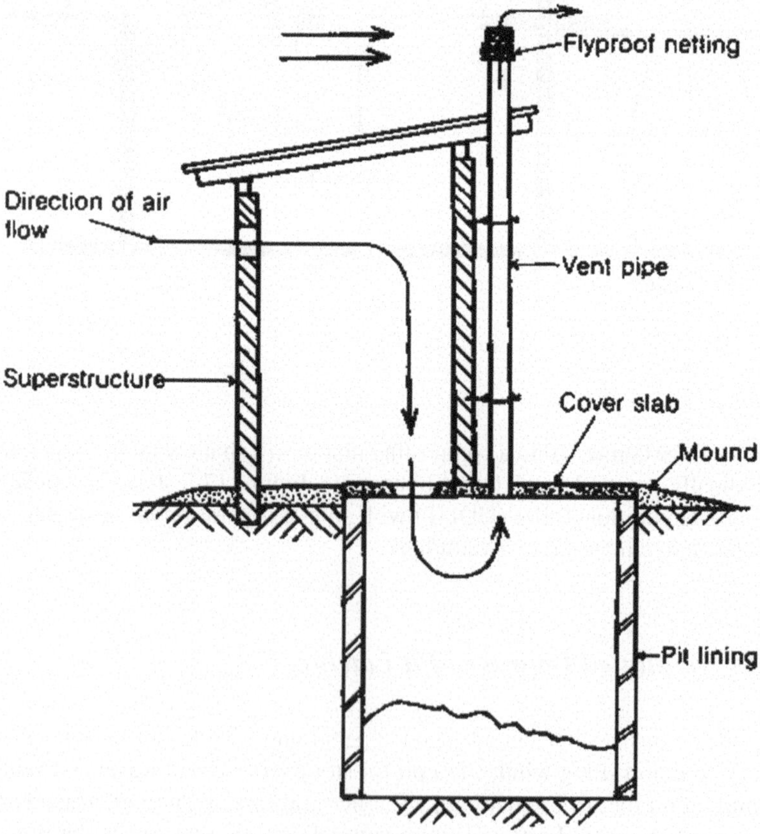

Fig. 17.4 Pictorial view of a ventilated improved pit latrine [22]

17.3.4 Biogas Toilets

These toilets can be integrated with dry toilets, for example, urine-diverting dehydrated toilet. The process is well-suited for co-digestion with animal manure, food waste, sewage sludge and organic waste from municipal solid wastes. It relies on the actions of anaerobic microorganisms to decompose the organic materials to gas (mainly CH_4 and CO_2) and digestate, a process known as anaerobic digestion (AD). AD is a naturally occurring process for human faeces, from the human gut to septic tanks, although the end-products of natural processes are directly released into the environment. The process requires the absence of oxygen (O_2) and optimum temperatures of 25–40 °C for the growth of mesophilic organisms (wet digestion conditions) or 50–60 °C for selective growth of thermophilic organisms (dry digestion conditions). The digestate from the wet mesophilic process is high in water content: nearly a slurry if not dewatered. The digestate and those from the thermophilic processes are rich in N, P, K, magnesium (Mg) and sulphur, and valuable to produce fertiliser. The gas is methane-rich but needs to be upgraded due to moisture and unwanted gases such as CO_2 and hydrogen sulphide (H_2S) that is present. The gas recovered is scrubbed with water and dried and then compressed to natural gas quality. The gas can then be used for cooking, lightning or electricity. While anaerobic digestion is largely suggested for treating human faeces, there is limited information on their application [13]. Few studies [53, 65] that have examined the anaerobic digestion of undiluted human faeces show that up to 0.16–0.5 m^3 biogas can be obtained per kilogram of undiluted faeces but the yield can be inhibited when faeces are co-digested with wood/straw or urine. This is because straw is difficult to decompose, and urine increases alkalinity. Hence, source separation of faeces and urine and co-digestion with organic materials such as food wastes are recommended. The use of advanced processes for concentrating the solids, for example, thickening, has improved gas yield [53]. The process does not inactivate all the pathogenic microorganisms and helminth eggs, so digestate requires further treatment. It is expensive to build and can fail if poorly maintained, as such technical skills are required for operation and maintenance. The system requires land space and on-site sludge management for produced effluent. All these limit their application in rural and peri-urban areas. Figure 17.5 shows the pictorial view of a Biogas-based latrine with mixed collection of waste.

These EcoSans systems focus on deriving nutrients for agriculture or for enriching the soil. But, like conventional systems, EcoSans are prone to abuse and could lead to severe environmental pollution if improperly installed, used or emptied. They do need to be designed and operated appropriately to be considered hygienic. There are instances of technology abandonment of UDDT, for example, in Burkina Faso, due to system failure and progressive lack of interest to maintain and repair facilities. Some of the situations were caused by inappropriate designs, poor performance or considered to be against sociocultural beliefs and norms [54]. There were instances of refusal, hesitation and unwillingness to use the by-products for agriculture.

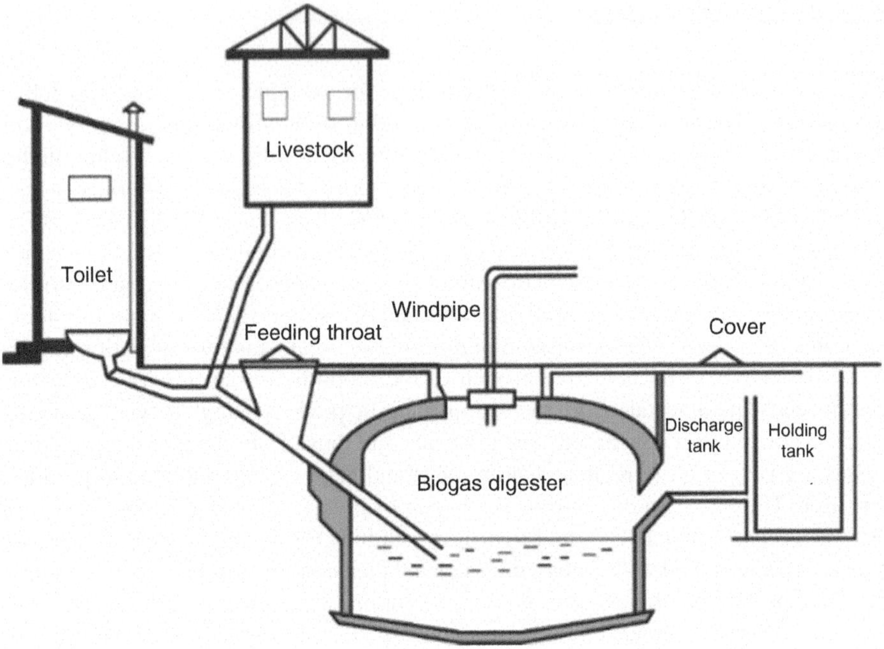

Fig. 17.5 Pictorial view of a biogas toilet [31]

There were cases of gas leakage, insufficient gas and odours resulting from biogas latrines [48], which stemmed out of lack of technical skills to operate and maintain these systems. Instances are cited on the frequent misuse of facilities, blockages with urine precipitates and faeces, overfilling of the vaults, improper segregation of urine and faeces, odour, flies, lack of spare parts for certain toilet models, for example, fans on chimney pipes. While most developed countries can benefit from the minimum regulatory requirement for toilet certification, communities that lack sanitation often must rely on user experiences to certify the use of a product. Thus, there is a need to tailor these sanitary solutions to user's needs and expectations, not only based on what is needed but aspired. This includes design considerations for system operation and maintenance, user interaction that promotes ease-of-use, understanding of how to operate and maintain the technology and accessibility to repair and parts. Pre-feasibility assessment is not only required; long-term project support is essential. A one-size-fit model would not be appropriate, even when communities are clustered on the basis of needs. Community listening programs and approaches that involve the user in the design of the technology have proven to be useful [10]. Other factors such as political, legal and sociocultural complexities play a role in adoption, thus the development and deployment of sanitation technologies are beyond solving a local need. Table 17.1 highlights the opportunities and barriers to adoption of EcoSans.

Table 17.1 Summary of opportunities and threats in source separation of human wastes

Advantages	Disadvantages
Approach to faecal sludge management (FSM) promotes nutrient recovery, extraction and recycling. Appropriate use of methods improves sludge stabilisation and reduces pathogen risks. Provides basic sanitation and reduced options for open defecation. Provides financial incentives by providing alternatives to mineral fertiliser and resources, hence reduced requirement for chemical fertilisers. Low-cost installation compared to conventional sanitary systems Minimal energy requirements and applicable in low-income countries	The technical complexity of biodigesters. Manual emptying poses risk to pit emptiers Continuous education needed to raise public awareness, ensure the correct use of facilities and foster a change in attitude towards human wastes. Sludge and effluents often need further treatment The fate of pharmaceutical chemicals in treated wastes and their impact on the ecosystem is not clear Poor public acceptance of faecal sludge products and limited understanding of market potentials Associated costs of emptying pits on a regular basis, e.g. every 2 years limit the proper use of facilities Complementary nutrients might be required due to low concentrations of certain micro-nutrients
Opportunities	Threats
Nutrient use for fertiliser to increase crop yield and reduce the cost for agricultural farming Use of recovered nutrients in soils to improve soil moisture, water retention and soil structure The reduced waste volume reduces storage capacities, transportation and processing requirements Opportunities for local businesses, e.g. faecal char and briquette production, fertiliser production Use of biogas for power generation: reduces deforestation from the exploitation of wood biomass Use of faecal char for fuels and biochar	Uncertainty on long-term effect of utilising waste. Bio-accumulation of micropollutants and bio-amplification in food chain Misuse of facilities leads to environmental pollution Transportation over long distances to recover sludge Unwillingness to pay for infrastructure Increased sludge production but lacking infrastructures, e.g. dewatering and energy system Capital costs for installing new sanitary systems

17.4 Solid–Liquid Separation

Solid–liquid separation limits moisture in recovered solids and facilitates the removal of solids from effluent [63, 64]. To ensure safe and hygienic treatment of human waste and to improve the potentials of supplementing soils with nutrients and organic matter, processes involving sludge thickening, coagulation and flocculation and dewatering are often recommended [43]. This is because moisture levels in recovered faecal solids are still often as much as 70 wt.% and for systems that receive mixed waste streams, moisture can be up to 97 wt.%, although source separation of waste streams limits the addition of water from other waste streams. For off-grid, low-income settings, the use of drying beds, geobags and Imhoff tanks [64] is often practised.

Industrial processes involving belt filters, centrifugation, vacuum filtration, filter presses and sludge conditioning by heat can also be applied, but these have limited application in low-income countries, particularly in rural communities where they are most needed. In this respect, low-cost sludge thickening approaches involving the addition of lime, sawdust or fly ash and the use of drying beds is a common practice. Studies by Cofie et al. [12] showed that drying beds effectively dewatered and removed helminth eggs from faecal sludges, but results varied, depending on the quality of the filtering medium, degree of stabilisation of faecal sludges, loading rates, bed height and on external conditions (e.g. rainfall and ambient temperature). Seck et al. [80] showed that drying rates improved in drying beds when faecal sludges were mixed during loading but covering the bed provided no significant additional benefits. Typically, drying beds are designed as receiving troughs with open bed of sand and gravel through which effluents percolate and moisture evaporates. They can be covered or uncovered, mixed or unmixed and planted or unplanted [69, 80]. Long residence times, high land space requirement and the need to treat the resulting effluent limit their application. With geobags and Imhoff tanks, land requirements are minimal, but pathogens are only slightly reduced, and solid and liquid waste streams require further treatment. A number of research activities are ongoing to effectively dewater and dry faecal sludges such that the by-products can be safely used as a fertiliser or further converted to fuel, heat and/or electricity. The removal of moisture reduces unpleasant odours, eliminates pathogens and improves longevity and quality of end-products. The reduction of waste volume limits storage, transportation and process requirements. The choice of treatment will depend on costs, space availability, location, quantity of wastes to be treated, requirements for the end-products etc. Preliminary screening and treatment processes might be required to prevent unwanted materials, for example, plastics, textiles. Thermal processes and solar dryers are being developed to provide these added benefits (Table 17.2).

17.5 Processing of Liquid Waste Streams for Nutrient Recovery

The effluent from sludge-dewatering processes (e.g. percolate from drying beds) and urine from source-separated waste streams can be processed to extract nutrients. Methods such as membrane filtration and precipitation are often mentioned in literature for selective removal of enteric pathogens and nutrients [17, 28]. Methods such as ozonation [8], adsorption/ion-exchange [32], ammonia stripping [38], reverse osmosis and forward osmosis [27] have been used for effluent treatment, but limited work has been done in source-separated human waste. Membrane filtration concentrates effluents across a semi-permeable membrane and, depending on size and configuration, selected nutrients can be recovered from the permeate. The method offers opportunities to accumulate, release and extract nutrients. The drawback of membrane filtration is that it is prone to fouling, which also reduces life span and increases energy use. The retentate accumulates useful nutrients and pollutants, which can pro-

Table 17.2 Examples of solid–liquid processes for faecal sludge treatment [64]

Sludge dewatering	Design criteria	Removal efficiency	Description	Advantages	Disadvantages
Unplanted drying bed	100–200 kg TS/m^2/year	>95% SS 70–90% COD 100% HE	A shallow filter filled with sand and gravel with an under-drain to collect leachate	Low-cost. Good dewatering efficiency. No energy requirement. It can be constructed locally. Suitable for peri-urban and rural communities	High land requirement. Long residence times. Promote odours and flies. Labour-intensive. Limited reduction of pathogens. Further treatment required for both solid and effluent waste streams. 0.05 m^2/capita/10-day for unplanted drying beds and 4000 m^2/MLD
Planted drying bed	<250 kg TS/m^2/year	96–99% SS 95–98% COD 70–80% TS	Constructed wetland with aquatic plants, bacteria, fungi and algae that can decompose and extract nutrients from waste	Cost-effective. Easy to operate and can handle high loading rates. Improved sludge treatment to unplanted beds	
Centrifugation	Depending on the amount of waste treated	85–99% SRE	Mechanical dewatering relies on centrifugal force of rotation, and the difference in density between the solid and liquid waste. Solids accumulate and liquids can be decanted	Compact with minimal land space requirement. High solids recovery efficiency and suitable for different sludge types and composition. Suited for urban areas	External energy required. Technical skills required. High power consumption. High operational and maintenance costs. High noise levels. Specialist knowledge for maintenance
Settling-thickening tank	SAR: 0.13 m^3/m^3 of raw faeces. Residence time: >4 hours	57% SS 24% COD 12% BOD 44% HE	An effluent-holding tank where FS enters at one end and supernatant exits at the other end. Settleable solids concentrate at the bottom of the tank. Scum floats on the surface. Lime/ammonia may be added to reduce pathogens and odours and to precipitate chemical compounds	Relatively robust. 0.006 m^2/capita land required Continuous processes requiring minimal operating and maintenance requirements. Suitable for peri-urban and rural communities	It requires lime/ammonia addition to improve efficiency and to reduce odour. Pathogen concentrations are only slightly reduced. Further treatment required for both solid and effluent waste streams

(continued)

Table 17.2 (continued)

Sludge dewatering	Design criteria	Removal efficiency	Description	Advantages	Disadvantages
Imhoff tank	–	50–70% SS 30–50% BOD	A two-story tank mechanism that utilises the force of gravity for the separation of solids. The process is based on sedimentation	Relatively low land requirement (600 m^2/MLD). Low-cost operation and maintenance. Ideal for urban areas including densely populated regions	Requires land space and structure with depth; hence not suitable for high water table areas; low reductions of pathogens; effluent and scum require further treatment
Geobags	–	–	Permeable textiles made into geotube containers Geo-bags used for dewatering the wet sludge	Economical viable, no excessive and constant labour, no frequent maintenance required. Suitable for peri-urban and rural communities	Drying required prior to composting to reduce pathogens and helminths

TS total solids, *SS* suspended solids, *BOD* biological oxygen demand, *COD* chemical oxygen demand, *HE* helminth egg, *SAR* solid accumulation rate, *SRE* solid recovery efficiency, *MLD* mega litres per day, *FS* faecal solids mote nutrient loss [43, 46]. Precipitation is well applied for wastewater treatment and relies on the addition of a chemical compound (coagulant) to separate nutrients from waste streams [26]. The most applied method of precipitation in urine involves the addition of Mg salts to urine waste streams in order to remove nutrients such as P. The mixture yields precipitate, which could be Mg ammonium phosphate (MAP) and Mg potassium phosphate (MPP) in the absence of ammonium. Other ionic salts can also be targeted including calcium phosphate, aluminium phosphate, sodium phosphate or aluminium phosphate [26]. Ultimately, precipitate can be used as fertiliser, which reduces the environmental concerns on pharmaceutical products and residues in urine. The slow tendency to release nutrients is considered an advantage for farming in areas close to water sources and for certain crop types, for example, sugar beets [56]. The struvite is colourless and odourless, as such effort to use it for fertiliser production is expected to improve low social acceptance for direct use of human urine. Other advantages include low capital requirements, ease of application and the purity of precipitate. The drawback of this approach involves the associated costs for the use of salts, recovery of nutrients and production of fertiliser, values that are similar for synthetic fertilisers. Also, the nutrient recovery process focuses on superphosphate and ammonium N, as such other organic nutrients can be disregarded, leading to nutrient loss.

17.6 Solids Processing for Energy and Value-Added Products

17.6.1 Thermal Processes

Human faeces can be subjected to heat, with or without the presence of oxygen, to produce energy carriers such as biochar and biogas [1, 4] and/or to generate heat and/or electricity [25, 50]. A brief description of appropriate technologies for processing recovered solids from human waste is provided in Sections 17.6.1, 17.6.2, 17.6.3, 17.6.4, and 17.6.5 with details on potential routes for recovery of value-added products and conditions for conversion of wastes via thermal and biological processes. Some of the methods have been applied in principle, and others are proposed and being developed; hence, advantages and disadvantages are discussed in the context of potential application in faecal waste treatment.

17.6.2 Pyrolysis

Pyrolysis is traditionally used in the production of charcoal and has been in existence before the widespread of coal [7], as such, an old and proven technology. It requires oxygen-depleted environment and moderate temperatures of 300–650 °C to thermally degrade carbon-based materials to char, oil and gaseous end-products. The process is increasingly becoming important because oil yield of up to 75 wt.%

can be obtained [81]. Moderate temperatures, rapid heating rate of up to 1000 °C/min and short vapour residence time of <2 seconds are conditions that favour the yield of the bio-oil. The rapid heating breaks down organic compounds while short residence time ensures that the recovered vapour does not further decompose, and secondary reactions and intermediate products are avoided. To favour the production of gas, high temperatures and long residence time are necessary; however, gaseous products are atypical. For char yield, low temperatures and long vapour residence time are recommended, a process that is defined as slow pyrolysis and of great interest in faecal sludge management [20]. Faecal char can be used in traditional furnaces and kiln, and as a substitute fuel for domestic heating and cooking. Since char is mainly composed of carbon with less oxygen and hydrogen, it is a useful form of carbon sequestration. This can improve aeration and soil quality and enhance water and nutrient retention for plant growth. The material is also considered useful for soil reclamation and remediation because of their absorptive properties. All these benefits will need to be proven, as there are debates on the use of faecal char for agriculture. The nutrient in char is said to be depreciated when compared to compost because the organic matter is lost during heating. There are studies that have shown that bio-char from other biomass can suppress crop yield [66], as such further research is required to confirm the benefits from faecal char. The yield and proportions of end-products depend on a number of factors, including temperature, heating rate, pressure, residence time and feedstock parameters such as size, composition, moisture content and type. To demonstrate the use of pyrolysis for waste conversion and the effect of operating conditions, Ward et al., Liu et al. [42] and Gold et al. [24] investigated the pyrolysis of faecal sludge for char production. Temperatures between 300 °C and 750 °C and heating rates up to 30 °C/min were considered. The faecal char had Higher Heating Value (HHV) of 25.6 MJ/kg at 300 °C, a value that is comparable to the HHV of sub-bituminous coal but decreased with increasing pyrolysis temperature. The char was further converted to faecal briquette by grinding and combining the char with materials such as molasses and lime. On a commercial scale (Fig. 17.6), three pyrolysis plants in Warangal, Wal and Narasapur, India, treat 15,000 L of septic tank per day and generate biochar and pasteurised liquid.

17.6.3 Hydrothermal Carbonisation

Hydrothermal carbonisation (HTC), which is another form of pyrolysis, can process wet organic materials under relatively mild temperatures of 180–250 °C and sub-critical pressure – conditions that enhance hydrochar formation and deactivate pathogens. It is one of the methods that are suited for human waste because it can handle highly moist feedstocks (<50 wt.%) with little or no pre-treatment requirement as opposed to conventional methods that rely on partially dried or dried materials. The end-products are hydrochar, aqueous liquid and gas. The aqueous

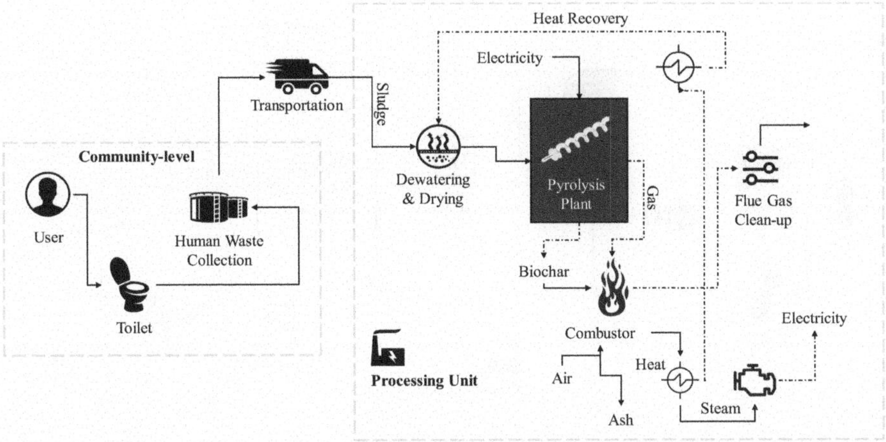

Fig. 17.6 Process flow for faecal sludge management via pyrolysis

liquid is a mixture of dissolved organic and inorganic compounds with nutrients that are valuable for soil conditioning but cannot be directly released into the environment without prior treatment. The hydrochar is enriched with carbonaceous compounds, highly porous and valuable for soil conditioning [20]. The drawback of this technology is that it requires high residence time (up to 12 hours) and relies on temperature gradients from convection and conduction for heat transfer and high pressures (up to 30 bar). The heating regime is non-selective, as such uncontrolled temperatures lead to uneven heating. The energy yield of the hydrochar is relatively low for a highly energy-intensive process. The aqueous product that is formed as part of the carbonisation cannot be released into the environment without treatment and these reduce process efficiency and environmental gains. To reduce heating time, Afolabi et al. [1] proposed the use of microwave-assisted HTC, considering temperatures lower than 200 °C and residence time of 0.5–2 hours, as it offers a more precise heating profile for sludge treatment. The conditions reduce residence time of the fuel and improve uniform heating. Fakkaew et al. [18] proposed a two-staged HTC, where the direct conversion of the biomass that is governed by devolatilisation, intracellular condensation, dehydration and decarboxylation, is separated from downstream aqueous conversion processes such as hydrolysis, dehydration, decarboxylation, fragmentation, polymerisation and aromatisation. Studies by Koottatep et al. [36] are ongoing to apply additives and catalysts to accelerate thermal chemical reactions and to improve yield. Other studies by Danso-Boateng et al. [14] and Fakkaew et al. [19] show that the technology shows great promise but, all the technical concepts are at laboratory scales: none is yet to be commercialised. Figure 17.7 shows the process flow of managing faecal sludge using hydrothermal carbonisation.

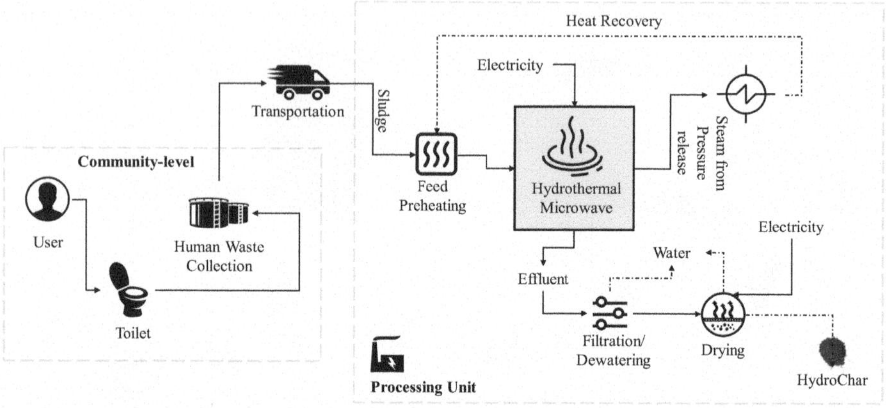

Fig. 17.7 Process flow for faecal sludge management via hydrothermal carbonisation

17.6.4 Gasification

Gasification operates at relatively high temperatures of more than to 1000 °C. Unlike pyrolysis that requires no oxidising agent, gasification requires a limited amount of oxidant (e.g. air, steam, nitrogen, CO_2, oxygen or a combination of these), to convert carbon-based material to char and gaseous end-products. Tar is produced as a black, viscous liquid but not desirable. The char that results from the process can be further valorised to improve tar cracking and convert unreacted carbon. This process has been studied using thermodynamic equilibrium model [51], and in a 10-kW gasifier to investigate the maximum amount of sludge permitted from reliable operation and their performance with other feedstocks. While the process is feasible thermodynamically, there were problems with 100% use of faecal sludge in some of the gasification experiments: issues relating to agglomeration and clinkering. Operational faults and technical failures were attributed as the main cause of the problem, not necessarily the use of faecal sludge. The use of raw faecal sludge was not a feasible option; hence, prior treatment and pelletisation were recommended to improve the heating value of the fuel. Other recommendations involve (a) source separation of faecal solids to reduce high ash content, especially in toilets where ash and lime addition are encouraged; (b) careful design of the feeding system to prevent crushing faecal pellets; (c) the use of additives to minimise clinker formation; and (d) co-pelletising faecal solids with other waste streams. Liu et al. [41] proposed an advanced gasification process that combines plasma gasification and microwave technology to convert human faeces to syngas, which is then converted to electricity via solid oxide fuel cells. To make the system energy sufficient, part of the electricity produced from the fuel cell was used to power the plasma gasifier while the heat recovered from the syngas and exhaust gas was used for drying the waste. The technology is still under development, hence not on a commercial scale. Figure 17.8 shows the process flow at community scales using plasma gasification.

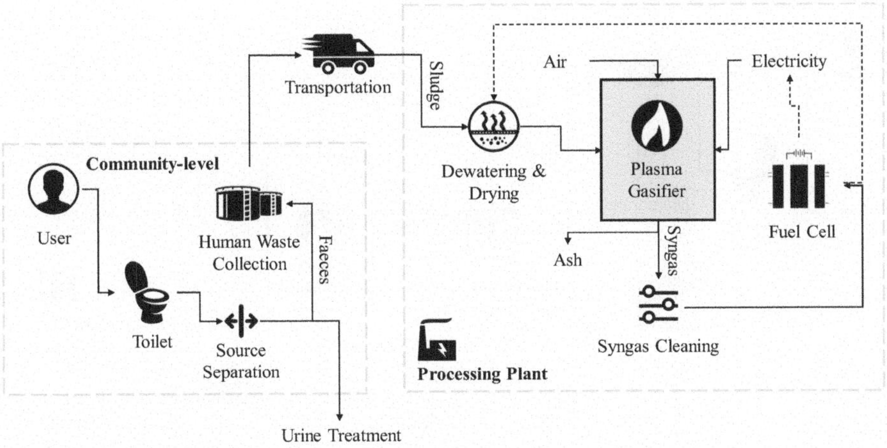

Fig. 17.8 Process flow for faecal sludge management via plasma gasification

For domestic sanitary applications, Jurado et al. [35] designed and commissioned a flexible updraft and downdraft gasifier for continuous conversion of human faeces. The system depended mainly on gasification for converting the fuel to ash. In the downdraft configuration, the faecal biomass enters from the fuel hopper via the rotary valve while the gasifying medium (air) enters the reactor slightly above the grate. In the updraft configuration, the faecal biomass and ash have a similar pathway; however, the primary air enters from below the grate while the product gas exits from the top of the unit. The reactor also maintained other flexible options such as gas re-circulation path, air bypass mode and variable air heater settings. It was however limited to a few hours of operation due to the challenges of fuel bridging/channelling in the fuel hopper, poor movement and exit of the flue gas, and ash, as well as complex operation, as such, improved design was proposed. Further development has been completed; however, this is yet to be in commercial operation.

17.6.5 Combustion

Combustion is a proven technology that holds great promise for faecal biomass valorisation. The process embodies heat and excess oxygen to thermal degrade materials and can be slow or rapid. The slow combustion, also known as smouldering, occurs at moderate temperatures of 250–700 °C without visible flame, but with progressive heat release. The fast process follows high temperatures (in excess of 1000 °C) with flame propagating outwards. Unlike pyrolysis that is largely an endothermic process and dominated by Boudouard reaction, combustion processes are exothermic and occur in the presence of sufficient oxidant and heat. The excess oxidant ensures complete conversion of the carbon-based material to CO_2 and H_2O, although other gaseous products such as CO, NO_x and SO_x are formed. Studies by

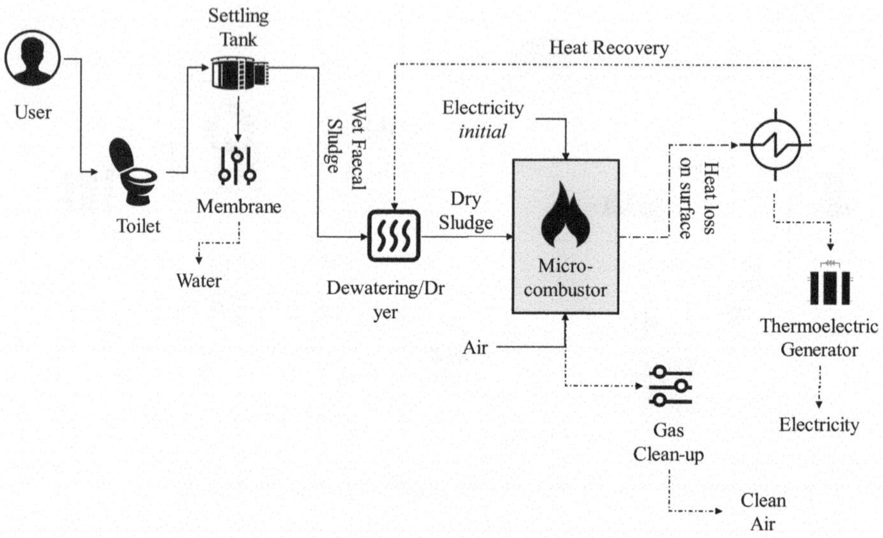

Fig. 17.9 Process flow for faeces combustion at household scale

Yerman et al. [76] have shown that moist faeces can be treated at low temperatures and at various operating conditions using smouldering technology and synthetic and dog faeces. Operational parameters such as sand pack height and sand-to-faeces mass ratio need to be optimised. Flaming combustion has been demonstrated at laboratory scales by Onabanjo et al. [51]. Their studies show that human faeces with moisture content below 60 wt.% can be treated, provided operating temperatures are higher than 600 °C. The direct use of the material will require prior treatment and the minimum acceptable blend for treating moist faeces can improve by blending with wood dust. Figure 17.9 shows the process flow of managing faecal sludge at the household level using combustion. On a large scale (Fig. 17.10), a large combustion-based facility known as 'Omniprocessor' was installed in Dakar, Senegal, to process faecal sludge and produce water and electricity. The unit produces about 10 ML of water per day with net electricity of 100 kW. A larger model is being developed to handle waste for 100,000 people with net electricity of 250 kW. The plant dries the faecal sludge, and the heat generated is used for producing electrical energy, part of which is used for drying and purifying water.

17.6.6 Biological Processes

There are several biological processes that can be applied for faecal sludge treatment. These processes can be integrated with on-site sanitation facilities as discussed in Sect. 17.3.2 or developed as a stand-alone unit. Typically, processes rely on the ability of biological organisms to break down complex organic materials. For instance,

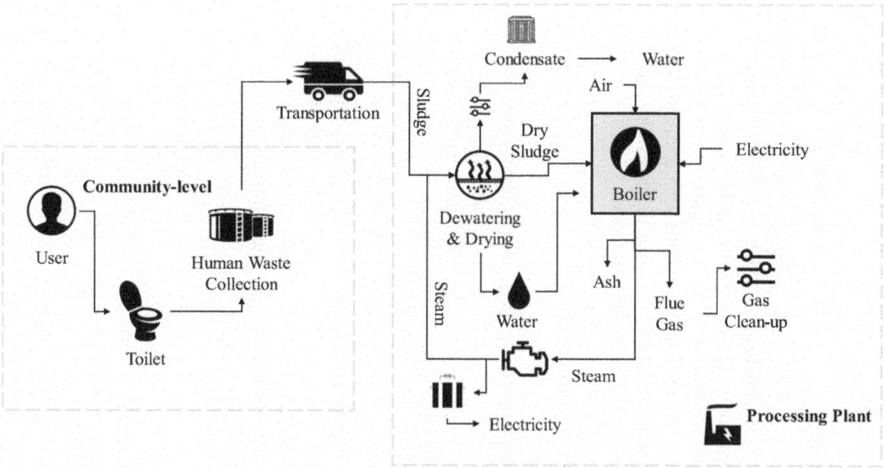

Fig. 17.10 Process flow for large-scale combustion facility

anaerobic digestion depends on methanogens and the absence of oxygen to break down organic matter. Biogas (mainly carbon dioxide and methane) can be recovered for subsequent conversion to heat and/or electricity and soluble nutrients for agricultural farming [61]. Other biological processes integrate composting [11, 77] and vermicomposting [75] to decompose organic materials. The decomposition allows the recovery of soluble nutrients from the digestate, provided decomposition is done in a controlled environment as found in a digester/biogas plant. In an open system, volatile nitrogen forms are lost to the atmosphere and nutrients can also be lost in leachate [26], if not recovered. The added benefits of biological processes include the reduction of odours, inactivation of certain pathogens at temperatures higher than 60 °C, decomposition of organic pollutants and further conversion of organic materials. This is particularly important for processes designed for nutrient recovery and for concentrated solid fractions. Factors such as retention time, feedstock composition, process temperature and feeding rate affect the yield and quality of end-products. The main disadvantages for biological processes are incomplete removal of heavy metals, micro-pollutants and other recalcitrant; hence, these can accumulate in the digestate and limit their use. The process routes for the valorisation of source-separated solids are summarised in Table 17.3.

17.7 Conclusion

Human excreta (urine and faeces) can be recovered and converted to value-added products, for example, struvite for fertiliser production and faecal char for heat applications. However, in many parts of the world, these rich organic nutrient sources are disposed of inappropriately in the environment, where they pose risk to

Table 17.3 Process route for valorisation of source-separated human wastes (faeces and urine)

Valorisation route	Description	Design criteria	Product recovery (efficiency)	Advantages	Disadvantages
Liquid processing					
Membrane Filtration Ultrafiltration (UF) Microfiltration (MF) Nanofiltration (NF) Reverse osmosis (RO) Forward osmosis (FO)	Concentrate effluents across a semi-permeable membrane. Can remove suspended solids microorganisms and organic molecules and ions	UF: 0.001–0.1 μm MF: 0.1–10 μm RO: 0.0001 μm	Permeate rich in nutrients. UF (90–99%) MF (85–95%) NF (75–90%) RO (60–90%)	A simple physical process which does not require the addition of chemicals Relatively low operating and maintenance costs compared to other recovery routes Nutrients can be used as fertiliser	Prone to fouling: reduced life span, efficiency and increased energy use Pollutants accumulation and partial nutrient loss in retentate
Chemical precipitation, e.g. struvite	Addition of salts, e.g. magnesium, calcium aluminium, sodium	–	Recovery of phosphate, e.g. Mg ammonium phosphate, Mg potassium phosphate	The precipitate can be used as fertiliser	Cost of salts can be expensive Partial nutrient loss in the effluent
Ammonia stripping via air/steam *Ozonation* *Photocatalytic oxidation* *Electrochemical oxidation*	Removes ammonia from liquid waste streams. Can be applied directly to digesters to reduce ammonia accumulation. Ammonia is oxidised into nitrate or nitrogen.	–	Ammonia	Stripped ammonia can be transformed into ammonium salts for chemical industries or fertiliser use	pH adjustment required Relatively high energy consumption for the value of the ammonia recovered

Solids processing					
Pyrolysis	Thermal degradation of carbon-based materials in the absence of oxygen and at moderate temperatures	Inert conditions. Temperatures: 300–850°C Pressure: 1 bar	Bio-oil Char Heat Syngas (VOCs, CH_4, CO, N_2, H_2)	Less oxygen – fewer emissions Useful products – gases, oils and char Bio-oil can be used as feedstock for chemicals Char can be used as soil conditioner Syngas can be upgraded to chemicals (e.g. alcohols, alkanes) or used for fuels and energy	Endothermic process, the energy required Undermines nutrient recycling and recovery Oil need for further upgrading to be used in internal combustion energy; hence, energy consumption can exceed the value of recovery. Product recovery and net energy efficiency depend on system configuration and feedstock composition. Pre-treatment (e.g. drying) required for moist fuel
Hydrothermal carbonisation	Processes organic materials under relatively mild temperatures and sub-critical pressure	Temperatures of 180–250°C. Pressure (up to 30 bar)	Hydrochar, aqueous liquid, gas	It can handle highly moist feedstocks (<50 wt.%) with little or no pre-treatment requirement. The aqueous liquid is a mixture of dissolved organic and inorganic compounds with nutrients that are valuable for soil conditioning. Hydrochar is enriched with carbonaceous compounds, highly porous and valuable for soil conditioning.	High residence time (up to 12 hours) Uncontrolled temperatures lead to uneven heating. The energy yield of the hydrochar is relatively low for a highly energy-intensive process. The aqueous product cannot be released into the environment without treatment.

(continued)

Table 17.3 (continued)

Valorisation route	Description	Design criteria	Product recovery (efficiency)	Advantages	Disadvantages
Combustion	Process embodies heat and excess oxygen to thermal degrade materials. Oxidising environment, excess oxidant (above stoichiometric). Smouldering at temperatures of 250–700°C. flame combustion in excess of 1000°C.	Temperatures: 500–1200 °C, depending on the application Pressure: 1 bar solid fuels with a high higher heating value above 14 MJ/kg	Heat Gas (N_2, O_2, CO_2 H_2O) Ash	Technology is more mature than alternatives. Commercially deployed at small and large scales. Can process fuel with low heating value.	Heat is the main end-product and product recovery depends on conversion efficiency and application. Ash disposal might be required if products contain pollutants. Gas requires further processing to avoid pollutant emissions. Pre-treatment (e.g. drying) required for moist fuel.
Gasification	Reducing, partial oxidising environment – sub-stoichiometric.	Temperatures: 500–1500°C, depending on the application Pressure: 1–45 bar	Heat Gas (VOCs, CH_4, CO, N_2, H_2) Ash	Syngas can have heating value as much as 23 MJ/Nm^3, depending on fuel composition Potential to produce hydrogen from renewable sources	Ash disposal might be required if products contain pollutants. Ash contains a relatively low level of carbon; thus, energy loss. Pre-treatment (e.g. drying) required for moist fuel.
Anaerobic digestion	Methanogens break down organic matter in the absence of oxygen	Relatively high temperatures of 50–60°C	CH_4, CO, Digestate	Biogas (mainly carbon dioxide and methane) can be recovered for subsequent conversion to heat and/or electricity and soluble nutrients for agricultural farming	Technical skills required. Capital costs for installation and maintenance of bio-digestors. Sensitive to feedstock. Digestate requires further processing.

Composting	Microorganisms (thermophilic and mesophilic) break down organic matter in the presence of oxygen	Factors such as moisture, temperature, aeration, pH and porosity. Relatively high temperatures of 50–60°C	Heat CO, H_2O Compost	Deactivation of pathogens and helminth eggs. It requires about 10–12 weeks, depending on the scale and operating conditions. Aeration requires external energy supply to keep temperatures and process stable. The low investment cost for open systems. Effective fertiliser Can be applied at a small scale.	Chemical pollutants can accumulate in compost and thus toxicity. Microorganisms also have an antagonistic relationship with other organisms, leading to inhibition, nutrient depletion and death of indigenous organisms and growth of unwanted species. Chemicals, such as antibiotics and hormones, can reduce the effectiveness of microorganisms to degrade organics in time. In open systems, the volatilisation of ammonia cannot be avoided, which poses risks to the environment.
Vermicomposting/fly larvae composting	The process employs biological organisms e.g. earthworms or flies to break down organic matter	Moisture level needs to be maintained at 50–60%. Moderately low temperatures of 35°C	Heat CO, H_2O Compost	Compost rich in nutrients for farming and soil conditioning. The process is more rapid, easily controllable, energy-efficient and cost-effective. The residue is highly rich in animal protein, about 32–64% of the dry solids.	Technical skills are required for operation and maintenance. A load of pathogenic microorganisms and helminths eggs are present and further treatment is required for compost of safe quality.

aquatic life, human health and the environment. This mini-review summarises the effort to derive value from human waste. Focus is given to low-cost technological solutions that offer ecological benefits and opportunities to recover nutrients and/or value-added products. Source separation of human wastes using EcoSans systems provides opportunities to recover and recycle nutrients; thus, reduced pathogen risks and the requirement for chemical fertilisers, but if improperly installed and inappropriately used, can lead to environmental pollution. Large amounts of sludge often result from these systems which require further treatment before it can be applied in practice. Biological processes can reduce odours and inactivate some pathogens, but micro-pollutants and other recalcitrant, for example, pharmaceutical residues, can accumulate in the digestate, limiting their use. The controlled environment is needed to avoid the loss of nutrients but value-added products such as biogas can be recovered for heat and/or electricity. Thermal methods, for example, pyrolysis, gasification, hydrothermal liquefaction, can also be applied, but nutrient recovery is limited; hence, focus could be given to energy, fuels and/or chemicals. Irrespective of technological solution and product recovered, processes must not put strain on already limited resources and values of end-products must exceed the cost needed to transform waste. This can only be achieved if all benefits (e.g. avoided environmental health impacts, land-use savings, cost savings, water-use savings, nutrient recycling) are accounted. Poor public acceptance and limited understanding of market potentials can limit the use of faecal sludge products. Continuous education is necessary to raise public awareness, ensure the correct use of sanitary facilities and foster a change in attitude towards human wastes. Appropriate low-cost dewatering and energy conversion systems are needed at domestic and community levels for sludge treatment and to increase the adoption of EcoSans systems. Other resource routes can be explored such as the use of faecal solids as construction materials, for example, bio-brick formation. Further research is required to understand the long-time effect of utilising faecal waste streams, the fate of pharmaceutical products and bio-accumulation and bio-amplification of micro-pollutants along the food chain.

References

1. O.O. Afolabi, M. Sohail, C.P. Thomas, Microwave hydrothermal carbonization of human biowastes. Waste Biomass Valoriz. **6**(2), 147–157 (2015)
2. C.K. Anand, D.S. Apul, Composting toilets as a sustainable alternative to urban sanitation–A review. Waste Manag. **34**(2), 329–343 (2014)
3. E. Andersson, Turning waste into value: Using human urine to enrich soils for sustainable food production in Uganda. J. Clean. Prod. **96**, 290–298 (2015)
4. D. Andriani, A. Wresta, A. Saepudin, et al., A review of recycling of human excreta to energy through biogas generation: Indonesia case. Energy Procedia **68**, 219–225 (2015)
5. O. Bajsa, J. Nair, K. Mathew, et al., Vermiculture as a tool for domestic wastewater management. Water Sci. Technol. **48**(11–12), 125–132 (2004)
6. I.J. Banks, W.T. Gibson, M.M. Cameron, Growth rates of black soldier fly larvae fed on fresh human faeces and their implication for improving sanitation. Tropical Med. Int. Health **19**(1), 14–22 (2014)

7. P. Basu, *Biomass Gasification and Pyrolysis: Practical Design and Theory* (Ac Press, Boston, 2010)
8. C. Bougrier, C. Albasi, J.P. Delgenès, et al., Effect of ultrasonic, thermal and ozone pre-treatments on waste activated sludge solubilisation and anaerobic biodegradability. Chem. Eng. Process. Process Intensif. **45**(8), 711–718 (2006)
9. E.J. Carlton, Y. Liu, B. Zhong, et al., Associations between schistosomiasis and the use of human waste as an agricultural fertilizer in China. PLoS Negl. Trop. Dis. **9**(1), e0003444 (2015)
10. R.C. Carter, S.F. Tyrrel, P. Howsam, The impact and sustainability of community water supply and sanitation programmes in developing countries. Water Environ. J. **13**(4), 292–296 (1999)
11. O. Cofie, D. Kone, S. Rothenberger, et al., Co-composting of faecal sludge and organic solid waste for agriculture: Process dynamics. Water Res. **43**(18), 4665–4675 (2009)
12. O.O. Cofie, G. Kranjac-Berisavljevic, P. Drechsel, The use of human waste for peri-urban agriculture in Northern Ghana. Renew. Agric. Food Syst. **20**(2), 73–80 (2005)
13. J. Colón, A.A. Forbis-Stokes, M.A. Deshusses, Anaerobic digestion of undiluted simulant human excreta for sanitation and energy recovery in less-developed countries. Energy Sustain. Dev. **29**, 57–64 (2015)
14. E. Danso-Boateng, R.G. Holdich, G. Shama, et al., Kinetics of faecal biomass hydrothermal carbonisation for hydrochar production. Appl. Energy **111**, 351–357 (2013)
15. J.J. Drewry, L.T. Newham, R.S. Greene, A review of nitrogen and phosphorus export to waterways: Context for catchment modelling. Mar. Freshw. Res. **57**(8), 757–774 (2006)
16. B.I. Escher, W. Pronk, M.J. Suter, Monitoring the removal efficiency of pharmaceuticals and hormones in different treatment processes of source-separated urine with bioassays. Environ. Sci. Technol. **40**(16), 5095–5101 (2006)
17. B. Etter, E. Tilley, R. Khadka, Low-cost struvite production using source-separated urine in Nepal. Water Res. **45**(2), 852–862 (2011)
18. K. Fakkaew, T. Koottatep, C. Polprasert, Effects of hydrolysis and carbonization reactions on hydrochar production. Bioresour. Technol. **192**, 328–334 (2015)
19. K. Fakkaew, T. Koottatep, C. Polprasert, Faecal sludge treatment and utilization by hydrothermal carbonization. J. Environ. Manag. **216**, 421–426 (2018)
20. K. Fakkaew, T. Koottatep, T. Pussayanavin, et al., Hydrochar production by hydrothermal carbonization of faecal sludge. J. Water Sanit. Hyg. De. **5**(3), 439–447 (2015)
21. D.T. Ferguson, Nightsoil and the 'Great Divergence': Human waste, the urban economy, and economic productivity, 1500–1900. J. Glob. Hist. **9**(3), 379–402 (2014)
22. R. Franceys, J. Pickford, R. Reed, et al., *A Guide to the Development of on-Site Sanitation* (World Health Organization, Geneva, 1992)
23. P.M. Glibert, J. Harrison, C. Heil, Escalating worldwide use of urea–a global change contributing to coastal eutrophication. Biogeochemistry **77**(3), 441–463 (2006)
24. M. Gold, M. Cunningham, M. Bleuler, et al., Operating parameters for three resource recovery options from slow-pyrolysis of faecal sludge. J. Water Sanit. Hyg. Dev. **8**(4), 707–717 (2018)
25. D.P. Hanak, A.J. Kolios, T. Onabanjo, et al., Conceptual energy and water recovery system for self-sustained nano membrane toilet. Energy Convers. Manag. **126**, 352–361 (2016)
26. R. Harder, R. Wielemaker, T.A. Larsen, Recycling nutrients contained in human excreta to agriculture: Pathways, processes, and products. Crit. Rev. Environ. Sci. Technol. **49**(8), 695–743 (2019)
27. H. Hasar, S.A. Unsal, U. Ipek, et al., Stripping/flocculation/membrane bioreactor/reverse osmosis treatment of municipal landfill leachate. J. Hazard. Mater. **1711-3**, 309–317 (2009)
28. H. Heinonen-Tanski, A. Sjöblom, H. Fabritius, Pure human urine is a good fertiliser for cucumbers. Bioresour. Technol. **98**(1), 214–217 (2007)
29. H. Heinonen-Tanski, C. van Wijk-Sijbesma, Human excreta for plant production. Bioresour. Technol. **96**(4), 403–411 (2005)
30. G.B. Hill, S.A. Baldwin, Vermicomposting toilets, an alternative to latrine style microbial composting toilets, prove far superior in mass reduction, pathogen destruction, compost quality, and operational cost. Waste Manag. **32**(10), 1811–1820 (2012)

31. M. Hu, B. Fan, H. Wang, et al., Constructing the ecological sanitation: A review on technology and methods. J. Clean. Prod. **125**, 1–21 (2016)
32. V.J. Inglezakis, S.G. Poulopoulos, *Adsorption, Ion Exchange and Catalysis*, vol 3 (Elsevier, Amsterdam, 2006), pp. 498–520
33. P.K. Jensen, P.D. Phuc, L.G. Knudsen, et al., Hygiene versus fertiliser: The use of human excreta in agriculture–a Vietnamese example. Int. J. Hyg. Environ. Health **211**(3–4), 432–439 (2008)
34. P.T. Johnson, A.R. Townsend, C.C. Cleveland, Linking environmental nutrient enrichment and disease emergence in humans and wildlife. Ecol. Appl. **20**(1), 16–29 (2010)
35. N. Jurado, T. Somorin, A.J. Kolios, et al., Design and commissioning of a multi-mode prototype for thermochemical conversion of human faeces. Energy Convers. Manage. **163**, 507–524 (2018)
36. T. Koottatep, K. Fakkaew, N. Tajai, et al., Sludge stabilization and energy recovery by hydrothermal carbonization process. Renew. Energy **99**, 978–985 (2016)
37. C.H. Lalander, G.B. Hill, B. Vinnerås, Hygienic quality of faeces treated in urine diverting vermicomposting toilets. Waste Manag. **33**(11), 2204–2210 (2013)
38. X. Lei, N. Sugiura, C. Feng, T. Maekawa, Pretreatment of anaerobic digestion effluent with ammonia stripping and biogas purification. J. Hazard. Mater. **145**(3), 391–397 (2007)
39. J. Lienert, T.A. Larsen, High acceptance of urine source separation in seven European countries: A review. Environ. Sci. Technol. **44**(2), 556–566 (2009)
40. B.B. Lind, Z. Ban, S. Bydén, Volume reduction and concentration of nutrients in human urine. Ecol. Eng. **16**(4), 561–566 (2001)
41. M. Liu, T. Woudstra, E.J. Promes, System development and self-sustainability analysis for upgrading human waste to power. Energy **68**, 377–384 (2014)
42. X. Liu, Z. Li, Y. Zhang, et al., Characterization of human manure-derived biochar and energy-balance analysis of slow pyrolysis process. Waste Manag. **34**(9), 1619–1162 (2018)
43. C.M. Mehta, W.O. Khunjar, V. Nguyen, Technologies to recover nutrients from waste streams: A critical review. Crit. Rev. Environ. Sci. Technol. **45**(4), 385–427 (2015)
44. N. Mkhize, M. Taylor, K.M. Udert, Urine diversion dry toilets in eThekwini Municipality, South Africa: Acceptance, use and maintenance through users' eyes. J. Water Sanit. Hyg. De. **7**(1), 111–120 (2017)
45. P.N. Mnkeni, L.M. Austin, Fertiliser value of human manure from pilot urine-diversion toilets. Water SA **35**(1), 133–138 (2009)
46. E. Monfet, G. Aubry, A.A. Ramirez, Nutrient removal and recovery from digestate: A review of the technology. Biofuels **9**(2), 247–262 (2018)
47. F.A. Monhol, M.F. Martins, Cocurrent combustion of human feces and polyethylene waste. Waste Biomass Valoriz. **6**(3), 425–432 (2015)
48. A. Nakagiri, C.B. Niwagaba, P.M. Nyenje, Are pit latrines in urban areas of Sub-Saharan Africa performing? A review of usage, filling, insects and odour nuisances. BMC Public Health **16**(1), 120 (2015)
49. M. Odagiri, A. Schriewer, M.E. Daniels, et al., Human fecal and pathogen exposure pathways in rural Indian villages and the effect of increased latrine coverage. Water Res. **100**, 232–244 (2016)
50. T. Onabanjo, A.J. Kolios, K. Patchigolla, An experimental investigation of the combustion performance of human faeces. Fuel **184**, 780–791 (2016)
51. T. Onabanjo, K. Patchigolla, S.T. Wagland, Energy recovery from human faeces via gasification: A thermodynamic equilibrium modelling approach. Energy Convers. Manag. **118**, 364–376 (2015)
52. K.D. Orner, J.R. Mihelcic, A review of sanitation technologies to achieve multiple sustainable development goals that promote resource recovery. Environ. Sci. Water Res. Technol. **4**(1), 16–32 (2018)
53. J.A. Park, J.M. Hur, B.S. Son, et al., Effective treatment of night soil using anaerobic sequencing batch reactor (ASBR). Korean J. Chem. Eng. **18**(4), 486–492 (2001)

54. N. Phaswana-Mafuya, N. Shukla, Factors that could motivate people to adopt safe hygienic practices in the Eastern Cape Province, South Africa. Afr. Health Sci. **5**(1), 21–28 (2005)
55. P.D. Phuc, F. Konradsen, P.T. Phuong, Practice of using human excreta as fertilizer and implications for health in Nghean Province, Vietnam. Southeast Asian J. Trop. Med. Public Health **37**(1), 222 (2006)
56. M.M. Rahman, M.A. Salleh, U. Rashid, et al., Production of slow release crystal fertilizer from wastewaters through struvite crystallization–A review. Arab. J. Chem. **7**(1), 139–155 (2014)
57. R. Reed, Why pit latrines fail: Some environmental factors. Waterlines **13**(2), 5–7 (1994)
58. F. Ronsse, S. Van Hecke, D. Dickinson, et al., Production and characterization of slow pyrolysis biochar: Influence of feedstock type and pyrolysis conditions. GCB Bioenergy **5**(2), 104–115 (2013)
59. C. Rose, A. Parker, B. Jefferson, et al., The characterization of feces and urine: A review of the literature to inform advanced treatment technology. Crit. Rev. Environ. Sci. Technol. **45**(17), 1827–1879 (2015)
60. O. Semalulu, M. Azuba, P. Makhosi, et al., Potential for reuse of human urine in peri-urban farming, in *Innovations as Key to the Green Revolution in Africa*, (Springer, Dordrecht, 2011), pp. 651–660
61. S. Semiyaga, M.A. Okure, C.B. Niwagaba, Decentralized options for faecal sludge management in urban slum areas of Sub-Saharan Africa: A review of technologies, practices and end-uses. Resour. Conserv. Recycl. **104**, 109–119 (2015)
62. L. Shiming, The utilization of human excreta in Chinese agriculture and the challenge faced (2002). http://www.ecosanres.org/pdf_files/Nanning_PDFs/Eng/Luo%20Shiming%2010_C11rev.pdf
63. P. Simha, M. Ganesapillai, Ecological Sanitation and nutrient recovery from human urine: How far have we come? A review. Sustain. Environ. Res. **27**(3), 107–116 (2017)
64. S. Singh, R.R. Mohan, S. Rathi, Technology options for faecal sludge management in developing countries: Benefits and revenue from reuse. Environ. Technol. Innov. **7**, 203–218 (2017)
65. J.R. Snell, Anaerobic digestion: III. Anaerobic digestion of undiluted human excreta. Sew. Work. J. **15**(4), 679–701 (1943)
66. S.P. Sohi, E. Krull, E. Lopez-Capel, R. Bol, A review of biochar and its use and function in soil, in *Advances in Agronomy*, (Academic Press, 2010), pp. 47–82
67. T.O. Somorin, A.J. Kolios, A. Parker, et al., Faecal-wood biomass co-combustion and ash composition analysis. Fuel **203**, 781–791 (2017)
68. E.H. Sonko, M. Mbéguéré, C. Diop, Environmental impact of recycling nutrients in human excreta to agriculture compared with enhanced wastewater treatment. Sci. Total Environ. **493**, 209–219 (2014)
69. E.H.M. Sonko, M. Mbéguéré, C. Diop, et al., Effect of hydraulic loading frequency on performance of planted drying beds for the treatment of faecal sludge. J. Water Sanit. Hyg. De. **4**(4), 633–641 (2014)
70. B. Sun, L. Zhang, L. Yang, Agricultural non-point source pollution in China: Causes and mitigation measures. Ambio **41**(4), 370–379 (2012)
71. K.M. Udert, M. Wächter, Complete nutrient recovery from source-separated urine by nitrification and distillation. Water Res. **46**(2), 453–464 (2012)
72. L. Van Zwieten, S. Kimber, S. Morris, et al., Effects of biochar from slow pyrolysis of papermill waste on agronomic performance and soil fertility. Plant Soil **327**(1–2), 235–246 (2010)
73. M. Winker, B. Vinnerås, A. Muskolus, et al., Fertiliser products from new sanitation systems: Their potential values and risks. Bioresour. Technol. **100**(18), 4090–4096 (2009)
74. World Health Organization, *Joint Water Supply and Sanitation Monitoring Programme, 2017. Progress on Sanitation and Drinking Water: Joint Monitoring Programme 2017 Update and SDG Baselines* (World Health Organization, Geneva, 2017)
75. K.D. Yadav, V. Tare, M.M. Ahammed, Vermicomposting of source-separated human faeces for nutrient recycling. Waste Manag. **30**(1), 50–56 (2010)
76. L. Yermán, R.M. Hadden, J. Carrascal, Smouldering combustion as a treatment technology for faeces: Exploring the parameter space. Fuel **147**, 108–116 (2015)

77. M.L. Zavala, N. Funamizu, T. Takakuwa, Biological activity in the composting reactor of the bio-toilet system. Bioresour. Technol. **96**(7), 805–812 (2005)
78. K. Finlay, A. Patoine, D.B. Donald, M.J. Bogard, P.R. Leavitt, Experimental evidence that pollution with urea can degrade water quality in phosphorus-rich lakes of the Northern Great Plains. Limnology and Oceanography, **55**(3), 1213–1230 (2010)
79. J. Spångberg, P. Tidåker, H. Jönsson, Environmental impact of recycling nutrients in human excreta to agriculture compared with enhanced wastewater treatment. Science of the Total Environment, 493, 209–219 (2014).
80. A. Seck, M. Gold, S. Niang, M. Mbéguéré, C. Diop, L. Strande, Faecal sludge drying beds: increasing drying rates for fuel resource recovery in Sub-Saharan Africa. Journal of Water, Sanitation and Hygiene for Development, **5**(1), 72–80 (2014)
81. A.V. Bridgwater, D. Meier, D. Radlein, An overview of fast pyrolysis of biomass. Organic Geochemistry, **30**(12), 1479–1493 (1999)

Chapter 18
Butanol as a Drop-In Fuel: A Perspective on Production Methods and Current Status

B. Ndaba, R. Adeleke, R. Makofane, M. O. Daramola, and M. Moshokoa

18.1 Introduction

The depletion of fossil fuel reserves and the adverse effects caused by burning of such fuels are a major global economic concern. In addition, supplies of fossil fuels are not renewable and may be limited in the near future, therefore, alternative solutions are required. One alternative is to develop sustainable methods based on renewable biomass, for instance, biomass-derived chemicals such as biofuels [95]. Biomass are organic materials such as grass, agricultural crops, wood and algae as well as their remains after processing, including animal waste. These organic materials are formed through biological photosynthesis from readily available atmospheric CO_2, sunlight and water. It has been established that approximately 170 billion metric tons of biomass are produced per year through photosynthesis with 75% accounting for carbohydrates [75]. Therefore, biomass is abundant, sustainable and inexpensive for the production of many biofuels [39].

Biofuels are comprised of fuels that are produced from sugars, and these include bioethanol, biobutanol and biogas as well as oil-derived fuels such as biodiesel [8].

B. Ndaba
Agricultural Research Council – Soil, Climate and Water (ARC-SCW), Pretoria, South Africa

R. Adeleke (✉)
Unit for Environmental Science and Management, North-West University, Potchefstroom Campus, Potchefstroom, South Africa
e-mail: Rasheed.Adeleke@nwu.ac.za

R. Makofane
Institute for the Development of Energy for African Sustainability, University of South Africa, Pretoria, South Africa

M. O. Daramola · M. Moshokoa
Department of Chemical Engineering, Faculty of Engineering, Built Environment and Information Technology, University of Pretoria, Hatfield, Pretoria, South Africa

M. O. Daramola, A. O. Ayeni (eds.), *Valorization of Biomass to Value-Added Commodities*, Green Energy and Technology,
https://doi.org/10.1007/978-3-030-38032-8_18

Any type of biomass rich in cellulose, hemicelluloses or high oil content has a potential to produce such fuels. A large number of technologies based on biological, thermal and chemical processes have been developed for biomass valorisation [69, 84, 107].

Currently, conversion of biomass to butanol is one of the most attractive and promising approaches (see Fig. 18.1; [39]). Butanol is a butyl alcohol that comprises of four carbon atoms. This fuel has been mainly utilised as an extractant, solvent, chemical intermediate and in cosmetics as well as pharmaceutical industries [23]. Currently, butanol interest has focused on its use as a fuel [8]. There are generally two pathways of butanol production. These are: butanol from biomass, whereby the product is termed biobutanol, and one from fossil fuels, which is termed petro-butanol; however, both these fuels have the same chemical properties.

Butanol is an alternative alcoholic fuel to ethanol because of its properties such as higher energy content, lower volatility and it being less hygroscopic [23, 48, 72]. In addition, it is less damaging to vehicle engines and is more appropriate as a drop-in fuel through existing petrol pipelines. Butanol also has a lower Reid vapour pressure compared to ethanol, thus making it less explosive [65]. When used as a blend with other fuels, *n*-butanol can be mixed with petrol in higher ratios compared to

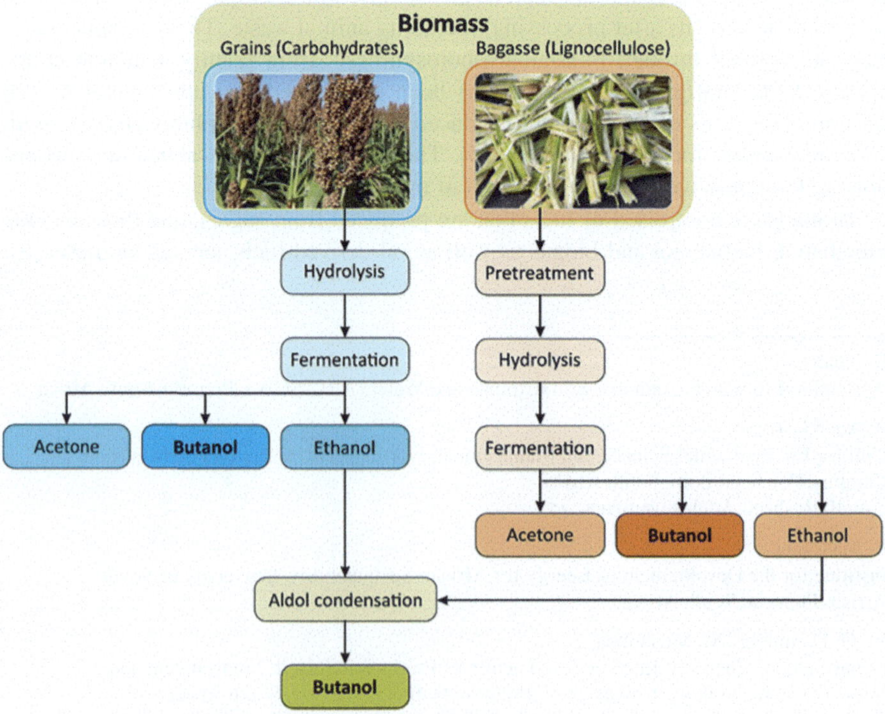

Fig. 18.1 Production pathways of butanol from biomass

ethanol without any engine modification as the air–fuel ratio and energy content are closer to that of petrol [12, 87].

Additionally, *n*-butanol's high octane ratio is an advantage to the internal combustion engines. For example, a lower octane rating fuel is more susceptible to engine knocking (extremely rapid and spontaneous combustion by compression), thereby damaging the engine [86]. The Environmental Energy Company (USA) concluded that *n*-butanol has a potential to be used as a replacement for petrol or can be blended with diesel without alterations to vehicle engines [10].

In a study conducted by Miller and co-workers, a vehicle engine running on butanol–diesel as a fuel was tested in 1981. From their findings, it was shown that up to 40 vol% of *n*-butanol can be added to diesel fuel with an engine running on pure diesel. The tests also showed that NO_x emission decreased significantly with addition of butanol in the diesel fuel, because higher enthalpy of vaporisation lowers in-cylinder temperature.

18.1.1 Economic Outlook of Butanol Production

Most butanol-producing companies today mainly produce butanol derived from a petrochemical reaction (petro-butanol). Examples of these companies include BASF SE (Germany), The Dow Chemical Company (USA), BASF-YPC Ltd. (China), OXO Corporation (USA), Sasol Ltd. (South Africa) and CNPC (China). These are considered as the industry scale main manufacturers of *n*-butanol globally. A few identified pilot scales that produce butanol from biomass are in China and Brazil (Brotas-SP); the latter produces butanol from sugarcane bagasse and other non-food feedstock [19].

18.1.2 Current Status of Butanol Production

The global butanol mark is expected to increase by 2020 as estimated by the Nexant and Chemical strategies [68]. High cost in acetone–butanol–ethanol (ABE) fermentation for butanol production is one of the major drawbacks in the commercialisation of the process. This high cost can be attributed to a number of factors such as low volumetric productivity, low yields, inhibitory effects of products during pretreatment and the toxicity of butanol to the microbes producing it [57]. Generally, butanol yields from lignocellulosic biomass range from 0.15 to 0.25 butanol/g sugar, while from microalgae range from 0.02 to 0.29 g butanol/g sugar [28, 79, 80, 103].

Due to such drawbacks, there are tensions between sources for the development of bioenergy and their sustainability. Feedstock for first-generation biofuels for example are food sources, which put pressure and also pose a threat to food security. More concerns thus arise towards environmental impacts, agricultural land competition and water resources. The issues around sustainability of the first-generation biofuels were discussed by Mohr and Raman [64] based on the policies and data affecting the UK, but which can be applicable to other parts of the world as well. They argued that the controversies surrounding the first-generation biofuels were necessary as they fulfilled an important sustainability assessment function for second-generation feedstock and other upcoming generational feedstock. They further pointed out that the three pillars of sustainability forming the socio-technical system, namely economic, environmental and social, are more intensely interconnected, and that an artificial separation causes problems.

According to Bailey [6], biofuels offer important opportunities mostly for the poor developing countries. Job creation increases in such instances when previously stagnant agricultural sectors get more investments for the production of biofuel feedstock. As cited by Bailey in the Oxfam briefing note, the European Commission published its Renewable Energy Roadmap in 2007 that proposed a mandatory target of a 10% increase in biofuel usage by 2020 from the then 1% contribution [11]. Apart from the benefits of job creation mentioned above, the European Union (EU) substantiated its proposal as an attempt to reduce the greenhouse gases emission. However, Bailey [6] argued that carbon savings differ considerably depending on numerous factors, and also upon the analysis of the life cycle it was found that feedstocks grown in tropical regions offer better carbon savings compared to those produced in Europe.

Many southern African countries have therefore been investing much effort in the agricultural production of such feedstock. These however pose a threat to the social and economic pillars of sustainability. Vulnerable people and their well-being in developing countries may suffer due to increased possibilities of land grabbing, exploitation and an increase in food pricing. Because biomass has a direct impact on the economy and the livelihood of people, more effort was put in research to reduce the food–fuel competition. As first-generation biofuel evolved to second generation, non-food lignocellulosic biomass were considered. The land competition between food and fuel was still present; however, second-generation feedstock promised a higher energy content compared to first-generation fuel crops, meaning that less biomass would be used [24].

As the bio-butanol production is developing, there is also a growing interest on the use of catalytic routes to produce *n*-butanol. This is because there are technical and economic challenges associated with the fermentation process; these include finding efficient microorganisms that can effectively convert fermentable sugars to bio-butanol as the main product. The chemical route usually involves utilisation of catalysts for effective conversion of ethanol to *n*-butanol for relatively higher yield. This process is mainly conducted using ethanol as a substrate; this can be biomass- or petroleum-derived ethanol. Catalytic processes for *n*-butanol production have been well documented [74, 98].

18.2 Biochemical Production Pathways of Butanol

The acetone–butanol–ethanol (ABE) fermentation process occurs in two important stages and can be used to produce *n*-butanol from renewable resources [1]. The first stage is acidogenesis, which is characterised by microbial growth and an accumulation of acetic and butyric acids. The second stage is called solventogenesis, where the acids re-assimilate into the acetone, butanol and ethanol solvents. Other parallel fermentation products are carbon dioxide and hydrogen. This process makes use of first-generation (grainy part of the plant) or second-generation (non-food biomass) feedstock to produce butanol [80].

18.2.1 Butanol Production Pathway from First-Generation Feedstock

First-generation biomass resources used as feedstock to biorefineries can be any organic matter that is renewable, available on a recurring basis, used for consumption by humans and also contains carbohydrates (starch; [60, 61]). Examples include maize, cassava, sorghum and potatoes [109]. The production pathway can be through a chemical or a biochemical processes [109].

First-generation biofuels are produced through well-understood technologies and processes, the first one being saccharification whereby carbohydrates are broken down to simple sugars, followed by fermentation to form acetone, butanol and ethanol. This process was first demonstrated by Louis Pasteur in the 1800s. He gave a lecture titled 'Lactate fermentation', which was later published as a historical account from personal knowledge. In this lecture, he demonstrated the role of microorganisms in metabolic processes [40, 58]. The ratios of the three main products of ABE fermentation are typically 3:6:1, with butanol making a 3% final con-

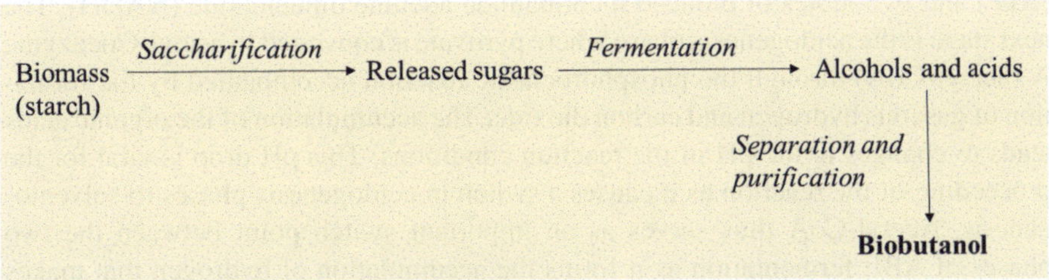

Fig. 18.2 Biofuel production flow diagram from biomass [56]

centration [58]. The main purpose of distillation is to separate the desired product from all other by-products formed. The schematic diagram for biobutanol production from first-generation biomass is demonstrated in Fig. 18.2. The type of biomass, microorganism species used for fermentation and separation process can be windows through which can be looked at for improvement in the output of biobutanol production [34, 36, 81].

18.2.1.1 Feedstock for First-Generation Biobutanol Production

The future and economics of any biofuel heavily depends on the type and suitability of the feedstock chosen to be used. A suitable feedstock is readily available, renewable and has lower economics involved for production process with high output levels [34]. First-generation feedstock was considered unethical because they were competing with food supply and agricultural land [71]. In the twentieth century, however, food industry wastes are used as feedstock for ABE fermentation.

18.2.1.2 ABE Fermentation

ABE fermentation is favoured due to its inexpensive and readily available raw materials. Like other fermentations, it is also distinguished into three different types namely batch, fed-batch and continuous mode of fermentation, which are differentiated by the time when substrates are added to bioreactors, the period of fermentation and the time when products are collected. Apart from the three major products formed through this process, two other intermediate metabolites are formed – acetic acid and butyric acid. The process thus undergoes two phases, namely acidogenesis and solventogenesis. The metabolic pathway is shown in Fig. 18.3. Glucose as the starting material is derived from the pre-treatment stage of feedstock whereby hydrolysis by enzymes of feedstock occurs [21, 110].

The resulting glucose is further metabolised through the Embden–Meyerhof–Parnas (EMP) pathway, forming two moles of pyruvate per glucose molecule. The other by-products formed during this stage are two moles of adenosine triphosphate (ATP) and two moles of reduced nicotinamide adenine dinucleotide (NADH). The next stage is the acidogenesis phase where pyruvate is converted to acetyl-Coenzyme A (acetyl-CoA) through the phosphoroclastic reaction accompanied by the formation of gaseous hydrogen and carbon dioxide. The accumulation of the organic acids leads to changes in the pH of the reaction conditions. This pH drop is vital for the proceeding of the reaction as it causes a switch in acidogenesis phases to solventogenesis. Acetyl-CoA thus serves as an important switch point between the two phases of ABE fermentation as it forms the accumulation of hydrogen that makes the reaction environment to be acidic. A side reaction from the major pathway forms ethanol by reduction of acetyl-CoA. It is believed that ethanol plays a role in regeneration of NAD^+ when acetone starts forming [21, 22].

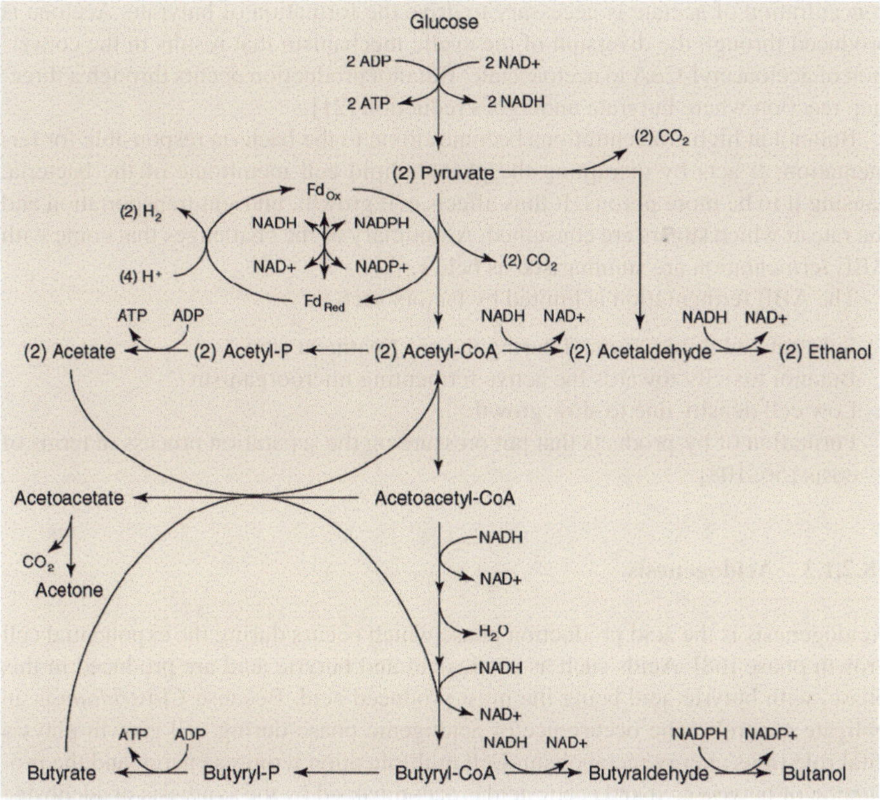

Fig. 18.3 Metabolic pathway of ABE fermentation by *C. acetobutylicum* [108]

The morphology of the microbes used also changes throughout the fermentation process, and thus can be used to determine the stage at which the fermentation process is at. The bacteria are short, rod-shaped and motile at the beginning of fermentation. They grow longer and rapidly divide at the early stage of exponential growth phase. At the late exponential phase, they become short and many club-shaped bacteria appear. The late phase or the ending of fermentation contains bacterial cells that are even shorter and some starts forming spores. The strains then enter an exponential growth phase where Acetyl-CoA further forms other intermediates (acetate, acetyl phosphate) catalysed by acetokinase and phosphotransacetylase enzymes respectively.

The acidic environment then triggers the enzymes responsible for solventogenesis to occur whereby the acid starts diminishing. Factors that control and trigger the enzymatic reactions in solventogenesis include the concentration of acetic and butyric acids, low pH conditions of the medium [20], the level of nutrients in the growth medium as well as the morphology of the bacteria [91]. The ability of microbes to switch to solventogenesis in acidic conditions is its adaptability response to avoid inhibitory effects. A cyclic reaction forms the key product, acetoacetyl-CoA, through two moles of acetyl-CoA to produce butyrate. A high concentration of acetate is necessary to drive the formation of butyrate. Acetone is produced through the diversion of the cyclic mechanism that results in the conversion of acetoacetyl-CoA to acetoacetate. Butanol production occurs through a three-step reaction where butyrate undergoes reduction [21].

Butanol at high concentrations becomes toxic to the bacteria responsible for fermentation. It acts by disrupting the phospholipid cell membrane of the bacteria, causing it to be more porous. It thus affects cell growth, nutrient transportation and the rate at which sugars are consumed. A summary of the challenges that come with ABE fermentation are summarised as below.

The ABE fermentation is limited by factors such as:

- Substrate inhibitors formed during the pre-treatment step
- Butanol toxicity towards the active fermenting microorganism
- Low cell density due to slow growth
- Formation of by-products that put pressure on the separation process in terms of costs [56, 108]

18.2.1.3 Acidogenesis

Acidogenesis is the acid production phase which occurs during the exponential cell growth phase [68]. Acids such as acetic acid and butyric acid are produced in this phase, with butyric acid being the most produced acid. Because *Clostridium* is an obligate anaerobe, the occurrence of acidogenic phase during cell growth plays a vital role in its energy metabolism. Cell multiplication requires energy and the production of butyric acid and acetic acid is accompanied by the synthesis of adenosine triphosphate (ATP; [89]). The production of acids and consequently the decrease in pH continues until a threshold that the *Clostridium* spp. can tolerate is exceeded. According to Maddox et al. [55], undissociated acid concentration above 57 mmol/L is known to be the threshold. At this acid concentration, the bacterial growth is inhibited resulting in lower acid production and consequently the end of the exponential growth phase [55, 89]. Because ABE fermentation is a biphasic process, the end of exponential phase triggers a change in metabolism, from acid production phase to solvent production phase. However, it is possible that excess acid production and consequently growth inhibition can happen without a significant change in metabolism and this is referred to as an 'acid crash' [55].

18.2.1.4 Solventogenesis

Solventogenesis is a solvent production phase in which the re-assimilation of acids into acetone–butanol–ethanol production is triggered by a certain concentration of undissociated acids, mainly undissociated butyric acid at a pH of 5.5 [55, 68]. This shift in mechanism is regulated at the genetic level and highly depends on the internal and external pH of the cell. Thus, this phase may only start when the pH of the media reaches a steady critical point and there is a shift from acid to solvent production

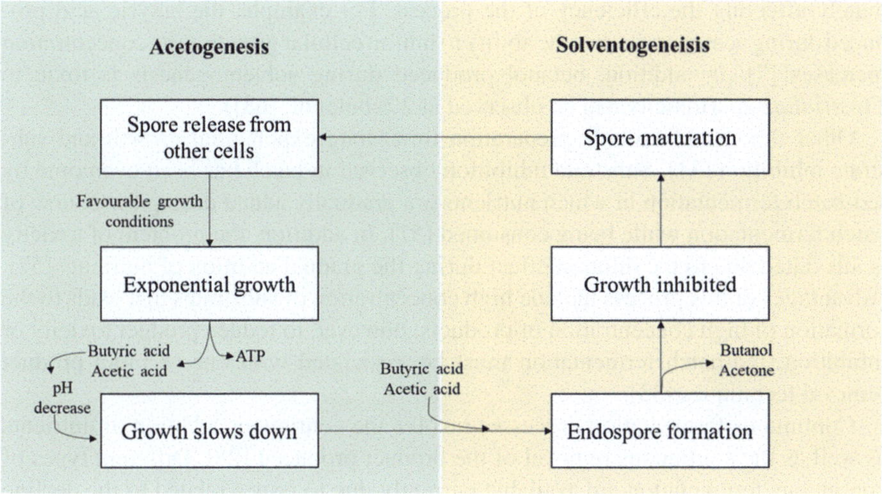

Fig. 18.4 *Clostridium* life cycle for the production of acids and solvents

[35]. These are the known to be the external factors that lead to solventogenesis [15]. Thus, the acid production slows down followed by their re-assimilation into butanol and acetone. Therefore, the pH of the fermentation media also plays an important role in the commencement, effectiveness and duration of solventogenesis. For example, at pH below 4.5, the duration of solventogenesis and effectiveness is reduced [68].

Solventogenesis is carried out by the enzyme aldehyde dehydrogenase (ALDH), butanol dehydrogenase (BDH) and alcohol dehydrogenase (ADH). ALDH catalyses the conversion of butyric acid and acetic acid to their respective aldehydes such as butyraldehyde, while ADH or BDH catalyses the final step, and the conversion of butyraldehyde to butanol by BDH depends upon the presence of NAD(P)H [7]. Solventogenesis is followed by sporulation – a process in which cells enter a dormant state and lose their solvent-producing ability [77]. Solvent production occurs during the onset of sporulation, and the overall ABE fermentation lasts for 36–72 hours [33, 89] (Fig. 18.4).

18.2.1.5 Fermentation Techniques

The ABE fermentation process is conducted under strictly anaerobic conditions using batch fermentation, fed-batch fermentation, continuous or semi-continuous fermentation, immobilised cell continuous fermentation, continuous flash fermentation, etc. [49]. Batch reactors are the most reported fermentation technique due to the simplicity of the process [48]. In batch, the process is conducted from start to finish in a single reactor without the addition of substrate or removal of the products until the end of the fermentation process. However, because the products produced remain in the reactor until the end of the process, it leads to growth inhibition, ultimately affecting the efficiency of the process. For example, the butyric acid produced during acetogenesis has the ability to inhibit cellular growth as its concentration increases [7]. In addition, butanol produced during solventogenesis is toxic to *Clostridium* sp. (toxicity can be observed at 2% butanol; [68]).

Other drawbacks include preparation time, long exponential growth and substrate inhibition [31]. Substrate inhibition observed in batch has been overcome by fed-batch fermentation in which nutrients are gradually added during the course of batch fermentation while being consumed [57]. In addition, the problem of toxicity is alleviated due to the dilution effect during the gradual addition of nutrients [57]. Advantages of this process include high concentration of substrates that leads to the formation of high concentration of products; however, to reduce product toxicity or inhibition, fed-batch fermentation must be associated with one or more product removal techniques [33].

Continuous fermentation processes involve the continuous addition of nutrients as well as the continuous removal of the product produced [28]. Different types of continuous fermentation are available currently due to issues related to the decline in solvent production because bacteria can lose their ability to produce solvent (the process is irreversible) within few weeks (2–3 weeks) after the commencement of continuous fermentation [7]. In comparison to batch and fed-batch, continuous fermentation has a number of advantages such as minimising exponential phase and high butanol yield and productivity [50]. Furthermore, advantages of continuous fermentation can be related to the continuous use of recycled or immobilised cells [31].

18.2.1.6 Microbes for Biobutanol Production

Microorganisms have been used for producing fermented food and drinks for human consumption for thousands of years, and now they are being adequately used also as powerhouses for biofuel production. The type of microorganism to facilitate the fermentation process is as important as selection of the type of biomass [56, 109]. Biobutanol fermentation occurs under anaerobic conditions by means of solventogenic bacteria. In 1914, Weizmann discovered *Clostridium acetobutylicum*. It has been labelled the Weizmann organism and has unique properties, like its ability to digest complex sugars and starchy feedstock during the ABE fermentation. For the purposes of ABE fermentation, the species of this genus are characterised according to their carbohydrate metabolism and biochemistry, energy metabolism, biomass they are likely to digest adequately (carbon sources), growth requirements and also genetic aspects [108]. The culture of *C. acetobutylicum*, for example, can digest substrates containing either simple or complex saccharides but does not have the ability to effectively ferment cellulose. The type of substrate and the concentration of the desired product through the ABE fermentation also play a role in directing the choice of genus of *Clostridium* to use [43, 56]. Apart from the Weizmann organism, other *Clostridium* genus organisms that were equally capable to perform the ABE fermentation were discovered through technologies such as deoxyribonucleic acid (DNA) fingerprinting and ribonucleic acid (RNA) sequencing. The discovered bacteria include *C. beijerinckii, C. saccaroperbutyl acetonicum* and *C. saccharoaceto butylicum.* For improved functionality of the microbes, their genes can be manipulated to rectify the impediments like inhibition and toxicity of butanol at high concentrations. This manipulation occurs through mutagenesis, which can occur either through chemical or physical processes. Examples of mutagens include ultraviolet (UV) radiation and exposure to certain chemicals like butanol [43].

China has been on the forefront of butanol synthesis at an industrial level and has invested in the improvement and effectiveness of the *Clostridia* through its different research groups. The Shanghai Cooperative Bio-butanol Group (SCBG; Shanghai Institute of Plant Physiology and Ecology, Chinese Academy of Sciences) engineered *C. acetobutylicum* EA2018 through chemical mutagenesis breeding of a soil-screened wild-type *C. acetobutylicum.* This strain improved the ratio of butanol in solvents by 70% compared to the original strain. Li Yin's group (Institute of Microbiology, Chinese Academy of Sciences) engineered *C. acetobutylicum* DSM1731 whose metabolism was improved to produce more solvents, that had higher butanol yield and that was more tolerant to solvents. The Qingdao Institute of Bioenergy and Bioprocess Technology within the same academy engineered *C. beijerinckii* ATCC55025 as the research strain, which is an asporogenic mutant of *C. beijerinckii* ATCC4259 [21].

18.2.2 Butanol Production Pathway from Second-Generation Feedstock

Lignocellulosic feedstock is used as second-generation feedstock for butanol production [7, 16, 68]. They are termed second generation because they are non-edible and do not compete and raise concerns related to the use of first-generation feedstocks for butanol production [68]. In addition, the use of lignocellulosic feedstocks is usually associated with high substrate availability, low cost and low generation of greenhouse emissions. According to Bharathiraja et al. [7], lignocellulosic biomass can be categorised into three groups: virgin biomass (e.g. trees and grasses), waste biomass (e.g. waste products from industrial activities) and energy crops (e.g. switchgrass).

The composition of the lignocellulose include hemicellulose, cellulose and lignin, where hemicellulose and cellulose are the two important biopolymers [4]. Important sugars required for butanol production are polymerised to cellulose and hemicellulose [5]. However, these two important biopolymers are protected from degradation by the lignin, which has a rigid structure with the ability to resist degradation while keeping the plant cell wall stable [7]. An intact lignocellulosic biomass cannot be used in the production of butanol using biochemical methods. This means that microbial communities such as *clostridia*, known to carry out butanol production, will not be able to access the fermentable sugars required. Therefore, the availability of these fermentable sugar polymerised in cellulose and hemicellulose is key in the process and modification of the lignocellulose [5]. When lignocellulose is modified, the hemicellulose and cellulose are disassociated from the lignin, and then sugars can be extracted, hydrolysed and subsequently fermented into butanol and other fermentation products [7, 33]. Lignocellulose modification is achieved by pre-treatment and hydrolysis.

18.2.2.1 Pre-treatment and Hydrolysis

Challenges caused by the lignocellulosic structure in the production of butanol can be circumvented by pre-treatment, which is a method used for destroying the lignin that protects hemicellulose and cellulose from degradation [32]. It is widely known that pre-treatment of the lignocellulose has a number of advantages, of interest; by breaking the protective barrier (lignin), the surface area and porosity of the cellulose and hemicellulose is increased, enabling high enzyme penetration into the cell wall for hydrolysis [62, 68]. Hydrolytic enzymes are used in hydrolysis – a stage in which extracted sugars such as carbohydrates (polysaccharides) are converted into simple monosaccharides. In addition, conversion of extracted sugars into simple monosaccharides can also be obtained through acid hydrolysis [7].

The operating condition and parameters during pre-treatment of lignocellulose are important and considered to be the ‘heart’ of hydrolysis, and even slight changes in this operation conditions may either lead to high sugar recovery or the production of high inhibitory compounds. Thus, pre-treatment and hydrolysis determine the efficiency and success of the fermentation process. Different pre-treatment and hydrolysis methods have been explored for sugar release from lignocellulose. These include physical (steam treatment), chemical (acid or alkaline treatment) and the use of enzymes (cellulases, hemicellulases and ligninases; [4]). In addition, for more pre-treatment efficiency, a combination of any of these three methods can be used. However, only the methods mainly used for lignocellulose pre-treatments for butanol production are elaborated below.

Acid Hydrolysis

The use of chemicals for modification of lignocellulosic biomass is referred to as acid pre-treatment or acid hydrolysis. This method is widely used for lignocellulosic biomass; however, due to the high generation of inhibitory compounds, it is currently becoming less attractive [46]. Chemical pre-treatment is divided into two categories: acid hydrolysis and alkaline hydrolysis [3]. Acid hydrolysis recovers the cellulose and converts it to sugars and it is divided into dilute and concentrated acids, with each having its own effects on lignocellulose modification [7]. Sulphuric acid (H_2SO_4) between 0.5% and 1% is used as a dilute acid at temperature between 140 and 190 °C. This dilute acid hydrolysis can disrupt and partially dissolve lignin to increase the availability of cellulose to enzyme degradation. In addition, this method is associated with high yields of xylose from hemicellulose and glucose from cellulose [52]. However, in the process of hydrolysing cellulose to sugar such as glucose, some of the sugars produced can be converted to compounds that may inhibit the microbes involved in the fermentation process [52]. Other disadvantages of dilute hydrolysis include corrosivity and neutralisation of the acid, which increases the cost of butanol production [7]. In concentrated acid hydrolysis, low temperatures are used with about 70% of the acid. In comparison to dilute acid, cellulose recovery and conversion to sugar is much higher. However, just like dilute acid hydrolysis, as a disadvantage, neutralisation of the solution is required with large amounts of alkaline used, which ultimately leads to high cost of the end product produced. A major disadvantage of acid hydrolysis is the production of high inhibitory compounds and the partially dissolved lignin that remains is the solution [88]. Examples of these inhibitory by-products include aliphatic carboxylic acids, phenolic compounds and furans, which are known to inhibit ABE fermentation [41].

Alkaline Hydrolysis

Sodium hydroxide (NaOH), aqueous ammonia and lime are examples of chemicals used during alkaline hydrolysis and they are known to improve the biodegradation of cellulose [7, 73]. Unlike the acid hydrolysis, which disrupts or partially dissolves the lignin, alkaline selectively removes the lignin and minor parts of hemicellulose [57]. In comparison to acid hydrolysis, alkaline hydrolysis results in slightly lower sugar recovery but the hydrolysate is less inhibitory to fermentation microorganisms due to reduced formation of inhibitory compounds [92], although inhibitory compounds such as acetic acid, hydroxy acids, dicarboxylic acids and phenolic compounds are produced [41]. According to Noomtim and Cheirsilp [73], application of 4% NaOH to palm empty fruit bunches removed over 50% of the lignin. Although efficient in lignin removal, the use of NaOH pre-treatment is associated with high cost. Lime is another type of alkaline used in the hydrolysis of lignocellulose. In comparison to NaOH, lime is less expensive, while the mechanism (removal of lignin) is similar to that of NaOH [54]. The benefits of over-liming include the ability to remove inhibitors affecting butanol production [54]. In addition, this process can also be used in combination with acid hydrolysis to remove inhibitory compounds produced.

Other Pre-treatment Methods

Steam pre-treatment method is one of the methods commonly used in lignocellulose treatment. It is also referred to as ‘autohydrolysis’ due to the changes that the method causes to the lignocellulosic biomass [46]. In this process, high pressure saturated steam at temperatures between 160 and 260 °C are used briefly (few seconds to minutes; [46]). The saturated steam expands the cell wall of the lignocellulose resulting in increased accessibility of the enzymes for cellulose hydrolysis. The process is referred

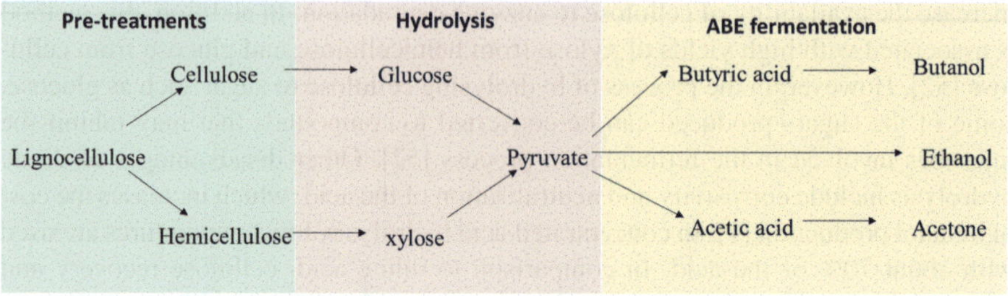

Fig. 18.5 Simplified second-generation substrate conversion into butanol

to as 'autohydrolysis' because, during this pre-treatment method, the acetic acid produced from the acetyl groups from hemicellulose carries out the hydrolysis of hemicellulose into the respective sugar monomers [66]. Resident time, temperature, biomass size and moisture content are some of the factors that can affect the efficiency of steam re-treatment. Addition of chemicals such as H_2SO_4 can be used to enhance the method. One of the drawbacks of this method is the possibility of the formation of inhibitory compounds at high temperatures and incomplete digestion of lignin–carbohydrate matrix as well as decreased sugar yield due to the need to wash the hydrolysate [3].

Enzymatic hydrolysis uses different enzymes such as cellulose, xylase, amylase, hemicellulases and ligninases [57]. The efficiency and selectiveness of enzymatic hydrolysis make this treatment superior in comparison to acid hydrolysis. According to Maiti et al. [57], enzymatic hydrolysis can obtain cellulose hydrolysis of close to 100%. However, enzymatic hydrolysis requires long running time and are expensive [26]. In addition, according to Bharathiraja et al. [7], this method has not yet been used independently for butanol production; however, it is usually coupled with other pre-treatment methods (Fig. 18.5).

18.2.3 Butanol from Third-Generation Feedstock

One of the major challenges in butanol production is the cost of butanol and for this reason, the search for feedstocks continues as a way to alleviate costs (Wang et al. 2017). Microalgae emerged as a promising source of biomass for butanol production and is referred to as third-generation feedstock. Like lignocellulosic biomass, microalgae can be converted to butanol via fermentation [47]. Microalgae are a wide variety of autotrophic organisms that grow through photosynthesis [37]. They are described as unicellular or multicellular organisms that can be either autotrophic or heterotrophic or a combination of both. Examples include *Arthrospira platensis*, *Nannochloropsis* sp., *Dunaliella tertiolecta*, *Dunaliella salina Teod*, *Galdieria partita Sentz* and *Cosmarium* sp. [25]. They are characterised by high growth rate in comparison to terrestrial plants used as lignocellulosic biomass. Unlike feedstock of the second generation, microalgae can be grown in seawater or wastewater, so they do not compete with agricultural crops for land. Microalgae consist of important compounds such as carbohydrates, lipids and proteins. Carbohydrates contain the lowest energy contents, however they are the most important source of biomass for biofuel production [38]. Carbohydrates in microalgae serve as energy reserves, mainly polyglucans (such as starch) in the chloroplasts. Another function includes forming part of the structural component of the cell wall (cellulose; Wang et al. 2017). The composition of carbohydrates, lipids and proteins differs in microalgae, so it is essential to select microalgae strains with high productivity in biomass and carbohydrate. In addition, the selected strains' carbohydrate content can be influenced by cultivation conditions such as nutrient starvation (Vitovaet al. 2015). Sulphur starvation, for example, has been described as the best strategy for microalgae starch accumulation [9, 101].

18.2.3.1 Pre-treatment of Microalgae

Microalgae contain polymers that are easily degraded due to their lack of hemicellulose and since carbohydrates mainly comprise of cellulose in cell walls and starch in lipids [59]. However, microalgae pre-treatment may be necessary to improve the conversion of microalgae into biofuels [59]. This, therefore, relates to reduced pre-treatment costs due to the use of mild pre-treatment methods. In addition, the use of mild pre-treatment methods reduces the formation of inhibitory compounds in comparison to the pre-treatment of lignocellulosic biomass [102]. In a study conducted by Wang et al. [102], the pre-treatment of microalgae using 3% H_2SO_4 produced trace (0.5 g/L) amounts of inhibitory compounds such as furfural and 5-hydroxymethyl furfural. This concentration of inhibitory compounds is too low to negatively affect butanol production, because *Clostridia* spp. can tolerate at least 1 g/L furfural and 2 g/L 5-hydroxymethyl furfural [102] (Wang et al. 2017).

Currently, thermal, mechanical and chemical methods and enzymes are used for microalgae pre-treatment, with thermal and mechanical pre-treatment being the most efficient [60, 61, 76]. Application of thermal and mechanical pre-treatments on microalgae leads to cell disruption; however, it does not lead to high sugar yield in comparison to mild chemical hydrolysis (such as dilute acid hydrolysis), which recovers high sugar yield. To increase the sugar yield, thermal or mechanical pre-treatment may be combined with enzymatic pre-treatment methods [100]. The advantage of using enzymes with thermal or mechanical pre-treatment methods is that inhibitory compounds are not generated in the process (Wang et al. 2017). For microalgae, the use of enzymatic pre-treatment methods is effective in combination with other methods. However, due to cost, for upscaling, the addition of enzymes might not be necessary [13]. Thermal and dilute acid hydrolyses of microalgae yield maltose, glucose and xylose, and *Clostridium* sp. are known to efficiently use all these sugars [13]. The production of butanol using microalgae also uses the ABE fermentation process and the same *Clostridium* sp. listed in section 'biochemical route' is used. Once the sugars are converted to the central metabolite, pyruvate (Fig. 18.1), the ABE fermentation as described occurs (Wang et al. 2017).

18.2.4 Challenges Encountered in ABE Fermentation

Although butanol, as an advanced liquid fuel, has several advantages compared to ethanol, the ABE fermentation process is associated with a number of challenges at upstream (pre-treatment and hydrolysis), fermentation stage (solvent production) and downstream (product recovery; [68]). Here, a summary of the challenges encountered is discussed and some of their possible solutions are given. These challenges result in the low yield of butanol, which is the biggest obstacle in the commercialisation of ABE fermentation [28]. In lignocellulosic biomass, the purpose of pre-treatment and hydrolysis is to unravel the lignocellulosic complex polymeric structure and the release of simple sugar [83]. However, during this process, a wide range of compounds with the ability to inhibit bacteria involved in the fermentation process is formed [57]. These inhibitory compounds can be produced from either cellulose, hemicellulose or lignin. The expected results from hydrolysis is that cellulose and hemicellulose are hydrolysed to glucose and xylose respectively as well as other monosaccharides [57]. However, some of the pre-treatment methods (such as acid hydrolysis) used are non-selective, leading to the conversion of glucose, xylose and other monosaccharide sugars into heterocyclic compounds (examples are listed under acid and alkaline hydrolyses; [57]). To list a few, these compounds have adverse effects on enzymes required during exponential phase, affect cell membrane's ability to serve as a selective barrier and decrease cell pH causing cell activity inhibition and ultimately cell lysis [57]. Thus, the removal of these inhibitory compounds is necessary. Detoxification methods such as evaporation, over-liming, neutralisation, microbial detoxification using genetically modified strains, etc. are available [57].

In view of the challenges, ABE fermentation by *Clostridial* species is a self-inhibition process, in which the produced product, that is, butanol, is toxic to the bacteria producing it [45]. This toxicity is one of the major limiting factors during ABE fermentation and yield of butanol [68]. For maximum butanol recovery, wild-type *Clostridium* strains that can tolerate 12–13 g/L butanol for conventional ABE fermentation has been reported [40]. Recent developments in genetic and metabolic engineering are also being applied to the butanol-producing microorganisms to overcome various limitations such as low butanol titre, yield and productivity. Construction of *Clostridium* mutant strains and other microorganisms is vital for improving butanol refinery on an industrial scale [28]. Another new technology to improve the microbial efficiency of butanol production is antisense RNA technology used in the production of mutants for enhanced ABE fermentation [97]. Butanol's toxicity is related to its hydrophobic nature [2, 7]. It causes the cell to lyse due to increased cell membrane fluidity [2]. Once butanol enters the cell membrane, it disrupts the phospholipid components thereby making the cell membrane more permeable to adenosine diphosphate and some ions [2, 7]. Thus, in some studies, butanol tolerance has been stimulated by externally supplementing saturated fatty acids to the growth medium, subsequently leading to the enrichment of indigenous saturated fatty acids of the bacterial cell membrane [40]. Currently, genetically modified microorganisms are being developed with the focus on increasing their tolerance to high butanol concentration. This is done to develop better microbial strains that can tolerate high concentrations of butanol and other solvent stress. This can potentially mitigate the future fuel crisis, as fossil fuel reserves decrease, and their adverse effects when used. This will aid in increasing the process efficiency and butanol yield during ABE fermentation [2]. Another development in the genetic/metabolic engineering of butanol-producing bacteria is the reduction or elimination of acetone and ethanol (minor products) production while increasing butanol production [2]. Other challenges of the ABE fermentation process include the bacteriophage infection, although few reports on this are available [68]. Examples of these bacteriophages include *Siphoviridae* and *Podoviridae*. Their effects include slow bacterial growth (and ultimately premature cell lysis) and the reduction in solvent production [40].

In third-generation feedstock such as microalgae, the most energy-rich compound is lipids followed by proteins and the least is carbohydrates [63]. High carbohydrate content in microalgae is very important for their use in butanol production. Low yields of butanol have been reported when microalgae are used as a feedstock due to lower initial sugar in the hydrolysate and this is the main limiting factor in butanol production (Wang et al. 2017). Once the carbohydrate content is high, as well as the use of efficient pre-treatment methods, high sugar recovery can be obtained. Therefore, the cultivation of microalgae under various conditions is known to enhance carbohydrate content. Additionally, the selection of microalgae strains based on high carbohydrate content and cultivation conditions that favour high carbohydrate production is very important.

Based on the current limitations of ABE, alternative routes to ABE for butanol production are still being developed. Catalytic conversions of bio-based/petrochemical-based ethanol into butanol are some of the methods being developed [18]. Even though these two types of ethyl alcohol are produced differently, their properties are the same after purification.

18.3 Catalytic Conversion of Biomass-Derived Ethanol to Butanol

The process of increasing carbon atoms of alcohol is termed Guerbet reaction. During this reaction, a primary or secondary alcohol reacts with itself or another alcohol to produce an alcohol with higher number of carbon atoms in the presence of a catalyst. Guerbet reaction involves different reactions, with dehydrogenation as an initial stage, followed by aldol condensation, dehydration and, finally, hydrogenation of the unsaturated aldehyde (Fig. 18.6). There are different types of heterogeneous catalysts that are currently used for Guerbet reaction. These include MgO [70], $MgO\text{-}Al_2O_3$ mixed oxide catalysts [82] and hydroxyapatite catalysts [96] that can be applied in both continuous and batch processes. Some of these studies involving the different types of catalysts emphasised on the importance of surface properties such as acidity and basicity because these affect the main reaction steps.

During Guerbet reaction there are several other by-products formed, such as ethyl acetate, 1,1-diethoxyethane, acetaldehyde and ethylene, probably formed during

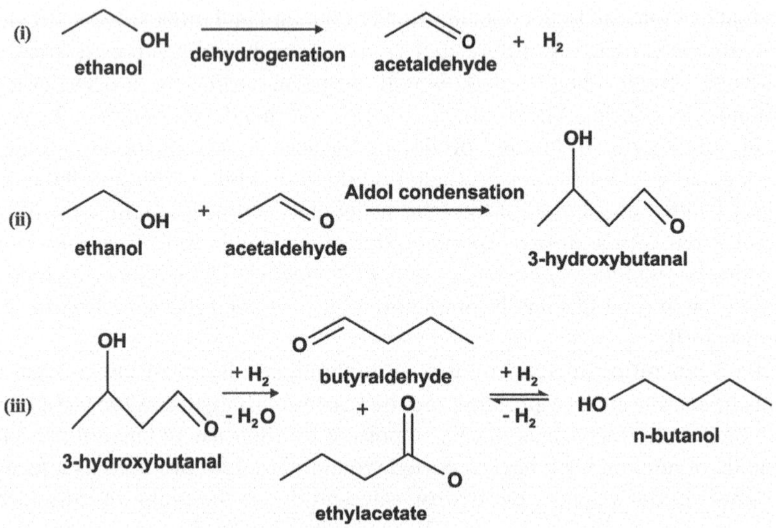

Fig. 18.6 The catalytic ethanol dehydrogenation to *n*-butanol [70]

condensation, dehydration or dehydrogenation reactions of ethanol or acetaldehyde, depending on the nature of the catalysts used [85]. The upgrading of bio-ethanol to *n*-butanol and other alcohols by homo- or heterogeneous catalysts is valuable because of the advantageous properties that butanol has than ethanol. Although this process is favoured over ABE fermentation, there are also challenges associated with the process. These mainly include the thermodynamic challenges in the dehydrogenation step of ethanol to form acetaldehyde and the formation of by-products due to the uncontrolled aldol reaction involving highly reactive acetaldehyde [14].

18.3.1 Dehydrogenation

This method makes use of ethanol as a starting material to form acetaldehyde in the presence of a catalyst. This is the initial step of ethanol conversion to acetaldehyde. Dehydrogenation of ethanol to produce acetaldehyde is an equilibrium reaction that is executed under controlled conditions, to allow further condensation of the acetaldehyde product to form by-products such as hydroxybutyraldehyde (aldol), ethyl acetate and subsequent hydrogenated products due to the presence of hydrogen by-product formed. The theory of the dehydrogenation of alcohols has been reported to occur through the formation of a complex between the alcohol and the catalyst [17], as illustrated in equation (i) of Fig. 18.6. There are a number of catalysts that have been studied for this purpose. Among them are catalytic materials such as compounds of iron, cobalt, nickel, manganese and zinc, in the order of their increasing dehydrogenation potential for producing acetaldehydes [93, 104].

18.3.2 Aldol Condensation

Aldol condensation is an important step towards formation of intermediates for formation of butanol. This step represents carbon–carbon formation in nature and/or in synthetic chemistry. During this step, it was reported that when acetaldehyde reacts with a base catalyst at room temperature, self-condensation takes place to form 3-hydroxybutanal (in Fig. 18.6, equation (ii)) and the product was named 'aldol' [30]. It was also observed that, in the presence of heat, aldol is further dehydrated to form unsaturated aldehydes such as butyraldehyde and ethyl acetate, which are presented in equation (iii) of Fig. 18.6.

18.3.3 Dehydration and Hydrogenation

Dehydration is a chemical reaction involving water removal from an organic molecule. Contrarily, hydrogenation is whereby a substance combines with water [30]. This is the step where water is removed from hydroxybutanal over solid acid sites to form aldehydes and later hydrated to form butanol and water as a final product in a closed system [105, 106]. The water contained in the solution is not required to run a motor vehicle engine; hence, butanol separation techniques have to be considered to recover butanol.

18.4 Butanol Separation Techniques

18.4.1 Recovery Through Distillation

Distillation is one of the traditional separation methods that have been used for years to separate homogenous solution components of different boiling points. Products with higher boiling points require more energy to separate them from other products whereas those with lower boiling points separate from the solution faster and hence do not consume a lot of energy. The vapours are collected and are allowed to condense at lower temperatures back to their liquid form [67]. The boiling point of butanol is higher than water and therefore becomes costly to separate it. Due to this, it is economically costly to use distillation to recover butanol due to its higher processing costs with lower yields [43, 78]. Patraşcu and his co-workers studied the optimisation of butanol recovery, where they made several improvements to the process. They first incorporated a decanter as the first unit of separation and therefore avoiding the pre-concentration step as suggested by van de Merwe et al. [99]. To avoid ethanol accumulation at the *n*-butanol/water recycle loop, they incorporated a column that specifically separates ethanol. They further improved the process intensification method by using a dividing-wall column.

18.4.2 Recovery by Gas Stripping

This method separates volatile compounds by lowering either pressure, heat or the inert gas used, although some industrial use have reported using a combination of the above factors that seemed to be effective. The column in which the solution is introduced allows for a separation of compounds at different eluding times. Gas stripping uses less energy compared to distillation and therefore can be of good use to industrial application [43]. However, it has low selectivity towards butanol separation as it removes a large amount of water with butanol, and therefore needs more effort to separate butanol from water. The advantages of gas stripping include operational simplicity, use of cheap equipment or chemicals and harmlessness to active bacteria [27]. The used inert gas can be regenerated as well for reuse. For in situ gas stripping process, the fermentation gases produced, CO_2 and H_2, can be used as stripping gas. The ABE fermentation solvents produced are captured by nitrogen or fermentation gases that are bubbled through the broth. The captured solvents are then separated from the gas by passing the mixture through a condenser for further processing, which also regenerates the gases used for separation [78].

Depending on the type of feedstock used, foaming may occur that can disrupt the fermentation process [29]. Thorough knowledge of the optimised gas stripping conditions as well as how they will affect the dynamics of ABE fermentation process are vital for the improvement of the integrated fermentation-gas stripping process [53]. Since the biggest problem with this method is the water that gets removed with butanol, a method that could form separate phases of the aqueous and organic solutions of the two can further reduce the costs of separation and increase its selectivity for butanol.

18.4.3 Recovery by Pervaporation

The separation of liquid mixtures through a membrane (porous/non-porous) is solely based on the differences in vaporisation properties of each individual liquid. The driving force for this separation method is differences in chemical potential between the two membrane sides, unlike distillation that is based on volatility. Pervaporation can be integrated with distillation to provide intensification and energy integration and thus provide a clean technology as well as a potential to save energy costs because it functions at low heat and pressures. An in situ bioreactor–pervaporation system can overcome the chemical inhibition and therefore favour productivity [94]. It selectively removes toxic components and impurities towards the active bacteria by diffusion through a membrane. The one side of the membrane is kept at low pressure to condense the permeated compounds, which is a disadvantage of this method due to high maintenance of such conditions. Selecting a suitable membrane structure is a crucial part of this process to allow effective separation of compounds. The affinity of the membrane plays a role in the selectivity of the separation towards the desired product. Molecules having higher affinity towards the membrane adsorb and diffuse through the membrane and those with lower affinity are retained. The pore size of the membrane also plays a role in the separation selectivity. Silicate rubber sheets and polydimethylsiloxane (PDMS) membranes are usually used for pervaporation due to their hydrophobic properties [94]. Organic liquid membranes of oleyl alcohol (OA) were more selective towards acetone and *n*-butanol recovery from the simulated fermentation broths [42, 94].

18.4.4 Recovery Using Liquid–Liquid Extraction

This has become a current interest in the separation of butanol from the broth using ionic liquids – acknowledging the difficulty of removing the main products of ABE fermentation from the fermentation broth. Ionic liquids are organic salts that exist in liquid state at room temperatures, have very low vapour pressure and are not soluble in water. While contemporary salts melt at 100 °C, the ionic salts used for this purpose melt at temperatures below 20 °C and are therefore called room temperature ionic liquids (RTIL). They exist as ions and their combination can be specifically designed to suit process requirements [43, 90]. The chosen extraction solvent needs to have a high affinity for the fermented products, especially butanol in this instance [29]. It is therefore these properties that render them suitable for the extraction of butanol from the fermentation broth since extraction occurs at liquid state. The ionic liquids can be used for separation by introducing them in the bioreactor or outside the bioreactor. The ionic liquid added to the bioreactor should be verified first that it is not toxic towards the active bacteria. The extracted compounds in the ionic liquid solution are further extracted by one of the other extracting methods mentioned and the ionic liquid gets regenerated [44]. Kubiczek and Kaminski studied the phase equilibrium of ionic solution containing five components with the ABE products. It was found that temperature and the amount of extractant determine the efficiency of extraction. When the solution was used at quantities comparable to the feedstock, the extraction efficiency was 50–65% at room temperature for butanol and acetone, but the separation efficiency of ethanol was found to be poor.

18.4.5 Recovery by Adsorption

During adsorption, the ABE broth passes through a filtering membrane for separation of butanol from water, other by-products and solids. The solids are then reused in the fermenter to form surplus butanol. In the adsorption columns, hydrophobic adsorbent such as silica are mostly used, and the substances that are adsorbed by the column are butanol, ethanol, acetone and some water [51]. After adsorption has taken place, butanol can then be desorbed by increasing temperature to approximately 200 °C. The process has been proven to have low energy costs compared to distillation. Qureshi et al. [81] reported that energy consumption decreased from 73.3 MJ/kg to 8.2 MJ/kg during distillation and adsorption, respectively.

18.5 Conclusion

Butanol has been shown to have better properties than ethanol in terms of engine performance as an advanced renewable fuel for replacing petrol or to form blends with fossil fuels. Concerning its production, butanol can be formed from biomass or bioethanol, and methods have been studied to produce this type of alcohol from different substrates. However, the main drawback for biobutanol production is that it is not as efficient as the production of ethanol. Therefore, the biochemical method is not economically feasible in terms of market demand of butanol as a fuel. The use of second-generation feedstock (lignocellulosic materials) might be suitable for an economical acceptable approach since the substrate does not compete with food for human consumption. Nonetheless, development of strains for efficient fermentation processes and high butanol yield are still required for market acceptance of butanol. Currently, chemical synthesis to form butanol is in use. Both methods, either microbial or catalytic, still require recovery processes for obtaining pure butanol; hence, researchers have also been looking at ways to improve the separation efficiency of butanol as a product.

References

1. M.H. Abd-Alla, A.W.E. El-Enany, Production of acetone-butanol-ethanol from spoilage date palm (*Phoenix dactylifera* L.) fruits by mixed culture of *Clostridium acetobutylicum* and *Bacillus subtilis*. Biomass Bioenergy **42**, 172–178 (2012). https://doi.org/10.1016/j.biombioe.2012.03.006
2. N. Abdehagh, F.H. Tezel, J. Thibault, Separation techniques in butanol production: challenges and developments. Biomass Bioenergy **60**, 222–246 (2014). https://doi.org/10.1016/J.BIOMBIOE.2013.10.003
3. V.B. Agbor, N. Cicek, R. Sparling, A. Berlin, D.B. Levin, Biomass pretreatment: fundamentals toward application. Biotechnol. Adv. **29**, 675–685 (2011). https://doi.org/10.1016/j.biotechadv.2011.05.005
4. B. Antizar-Ladislao, J.L. Turrion-Gomez, Second-generation biofuels and local bioenergy systems. Biofuels Bioprod. Biorefin. **2**, 455–469 (2008). https://doi.org/10.1002/bbb.97
5. S.C. Baek, Y.J. Kwon, Optimization of the pretreatment of rice straw hemicellulosic hydrolyzates for microbial production of xylitol. Biotechnol. Bioprocess Eng. **12**, 404–409 (2007)
6. R. Bailey, Biofuelling poverty: why the EU renewable fuel target may be disastrous for poor people. Oxfam Policy Pract. Clim. Change Resilience **3**(1), 89–100 (2007)
7. B. Bharathiraja, J. Jayamuthunagai, T. Sudharsanaa, A. Bharghavi, R. Praveenkumar, M. Chakravarthy, D. Yuvaraj, Biobutanol – an impending biofuel for future: a review on

upstream and downstream processing techniques. Renew. Sust. Energ. Rev. **68**, 788–807 (2017). https://doi.org/10.1016/j.rser.2016.10.017

8. K. Biernat, Introductory chapter: prospective biofuels, in *Biofuels-State of Development*, ed. by K. Biernat, (IntechOpen, 2018). https://doi.org/10.5772/intechopen.72955
9. I. Brányiková, B. Maršálková, J. Doucha, T. Brányik, K. Bišová, V. Zachleder, M. Vítová, Microalgae-novel highly efficient starch producers. Biotechnol. Bioeng. **108**, 766–776 (2011). https://doi.org/10.1002/bit.23016
10. K. Brekke, Butanol: an energy alternative? Ethanol Today. 36–39 (2007)
11. T. Brussels, Report on the roadmap for renewable energy in Europe. Committee on Industry, Research and energy, European parliament (2007)
12. J. Campos-Fernandez, J.M. Arnal, J. Gomez, M.P. Porado, A comparison of higher alcohols/diesel fuel blends in a diesel engine. Appl. Energy **95**, 267–275 (2012). https://doi.org/10.1016/j.apenergy.2012.02.051
13. Y.A. Castro, J.T. Ellis, C.D. Miller, R.C. Sims, Optimization of wastewater microalgae saccharification using dilute acid hydrolysis for acetone, butanol, and ethanol fermentation. Appl. Energy **140**, 14–19 (2015). https://doi.org/10.1016/j.apenergy.2014.11.045
14. S. Chakraborty, P.E. Piszel, C.E. Hayes, R.T. Baker, W.D. Jones, Highly selective formation of n-butanol from ethanol through the Guerbet process: a tandem catalytic approach. J. Am. Chem. Soc. **137**, 14264–14267 (2015). https://doi.org/10.1021/jacs.5b10257
15. C.-K. Chen, H.P. Blaschek, Effect of acetate on molecular and physiological aspects of *Clostridium beijerinckii* NCIMB 8052 Solvent production and strain degeneration. Appl. Microbiol. Biotechnol. **65**, 499–505 (1999). https://doi.org/10.1007/s002530051504
16. L. Cheng, P.Y. Che, B.Y. Chen, W.J. Lee, C.Y. Lin, J.S. Chang, Biobutanol production from agricultural waste by an acclimated mixed bacterial microflora. Appl. Energy **100**, 3–9 (2012). https://doi.org/10.1016/j.apenergy.2012.05.042
17. J.M. Church, H.K. Joshi, Acetaldehyde by dehydrogenation of ethyl alcohol. Ind. Eng. Chem. **43**, 1804–1811 (1951). https://doi.org/10.1021/ie50500a035
18. S. Cimino, L. Lisi, S. Romanucci, Catalysts for conversion of ethanol to butanol: effect of acid-base and redox properties. Catal. Today **304**, 58–63 (2018). https://doi.org/10.1016/j.cattod.2017.08.035
19. W.R. da Silva Trindade, R.G. dos Santos, Review on the characteristics of butanol, its production and use as fuel in internal combustion engines. Renew. Sust. Energ. Rev. **69**, 642–651 (2017). https://doi.org/10.1016/j.rser.2016.11.213
20. H.W. Doelle, *Bacterial Metabolism* (Academie Press, New York, 1975)
21. H. Dong, W. Tao, L. Yang, F. Gong, Y. Zhang, Y. Li, Butanol. J. Ind. Microbiol. Biotechnol. **36**, 1127–1138 (2009). https://doi.org/10.1007/s10295-009-0609-9
22. H. Dong, Y. Zhang, Z. Dai, Y. Li, Engineering *Clostridium* strain to accept unmethylated DNA. PLoS One **5**(2), 1–8 (2010). https://doi.org/10.1371/journal.pone.0009038
23. P. Durre, Biobutanol: an attractive biofuel. Biotechnol. J. **2**, 1525–1534 (2007). https://doi.org/10.1002/biot.200700168
24. K. Dutta, A. Daverey, J.G. Lin, Evolution retrospective for alternative fuels: first to fourth generation. Renew. Energy **69**, 114–122 (2014). https://doi.org/10.1016/j.renene.2014.02.044
25. E.N. Efremenko, A.B. Nikolskaya, I.V. Lyagin, et al., Production of biofuels from pretreated microalgae biomass by anaerobic fermentation with immobilized Clostridium acetobutylicum cells. Bioresour. Technol. **114**, 342–348 (2012). https://doi.org/10.1016/j.biortech.2012.03.049
26. T.S. El-Tayeb, A.A. Abdelhafez, S.H. Ali, E.M. Ramadan, Effect of acid hydrolysis and fungal biotreatment on agro-industrial wastes for obtainment of free sugars for bioethanol production. Braz. J. Microbiol. **43**(4), 1523–1535 (2012). https://doi.org/10.1590/S1517-83822012000400037
27. T.C. Ezeji, N. Qureshi, H.P. Blaschek, Production of acetone, butanol and ethanol by *Clostridium beijerinckii* BA101 and in situ recovery by gas stripping. World J. Microbiol. Biotechnol. **19**(6), 595–603 (2003). https://doi.org/10.1016/J.COPBIO.2007.04.002

28. T.C. Ezeji, N. Qureshi, H.P. Blaschek, Bioproduction of butanol from biomass: from genes to bioreactors. Curr. Opin. Biotechnol. **18**(3), 220–227 (2007). https://doi.org/10.1016/j.copbio.2007.04.002
29. A. Friedl, Downstream process options for the ABE fermentation. FEMS Microbiology Letters, **363**(9), 1–5 (2016). https://doi.org/10.1093/femsle/fnw073
30. J.M. Gallo, J. Bueno, U. Schuchardt, Catalytic transformations of ethanol for biorefineries. J. Braz. Chem. Soc. **25**, 2229–2243 (2014). https://doi.org/10.5935/0103-5053.20140272
31. L. Gholizadeh, L.G. Baroghi, Enhanced butanol production by free and immobilized *Clostridium* sp. cells using butyric acid as co-substrate. Thesis. University of Borås (2009)
32. L.D. Gottumukkala, B. Parameswaran, S.K. Valappil, K. Mathiyazhakan, A. Pandey, R.K. Sukumaran, Biobutanol production from rice straw by a non acetone producing *Clostridium sporogenes* BE01. Bioresour. Technol. **145**, 182–187 (2013). https://doi.org/10.1016/j.biortech.2013.01.046
33. L.D. Gottumukkala, K. Haigh, J. Görgens, Trends and advances in conversion of lignocellulosic biomass to biobutanol: microbes, bioprocesses and industrial viability. Renew. Sust. Energ. Rev. **76**, 963–973 (2017). https://doi.org/10.1016/j.rser.2017.03.030
34. E.M. Green, Fermentative production of butanol-the industrial perspective. Curr. Opin. Biotechnol. **22**(3), 337–343 (2011). https://doi.org/10.1016/j.copbio.2011.02.004
35. L.M. Harris, R.P. Desai, N.E. Welker, et al., Characterization of recombinant strains of the *Clostridium acetobutylicum* butyrate kinase inactivation mutant: need for new phenomenological models for solventogenesis and butanol inhibition? Biotechnol. Bioeng. **67**(1), 1–11 (2000). https://doi.org/10.1002/(sici)1097-0290(20000105)67:1<1::aid-bit1>3.3.co;2-7
36. M.G.N. Hartmanis, T. Klason, S. Gatenbeck, Uptake and activation of acetate and butyrate in *Clostridium acetobutylicum*. Appl. Microbiol. Biotechnol **20**, 66–71 (1984). https://doi.org/10.1007/bf00254648
37. R. Harun, M. Singh, G.M. Forde, M.K. Danquah, Bioprocess engineering of microalgae to produce a variety of consumer products. Braz. J. Microbiol. **14**, 1037–1047 (2010). https://doi.org/10.1016/J.RSER.2009.11.004
38. S.H. Ho, C.Y. Chen, J.S. Chang, Effect of light intensity and nitrogen starvation on CO_2 fixation and lipid/carbohydrate production of an indigenous microalga Scenedesmus obliquus CNW-N. Bioresour. Technol. **113**, 244–252 (2012). https://doi.org/10.1016/J.BIORTECH.2011.11.133
39. M.F. Ibrahim, S.W. Kim, S. Abd-Aziz, Advanced bioprocessing strategies for biobutanol production from biomass. Renew. Sust. Energ. Rev. **91**, 1192–1204 (2018). https://doi.org/10.1016/j.rser.2018.04.060
40. D.T. Jones, D.R. Woods, Acetone-butanol fermentation revisited. Microbiol. Rev. **50**(4), 484–524 (1986)
41. L.J. Jönsson, C. Martín, Pretreatment of lignocellulose: formation of inhibitory by-products and strategies for minimizing their effects. Bioresour. Technol. **199**, 103–112 (2016). https://doi.org/10.1016/j.biortech.2015.10.009
42. G. Jyoti, A. Keshav, J. Anandkumar, Review on pervaporation: theory, membrane performance, and application to intensification of esterification reaction. J. Eng. (2015). https://doi.org/10.1155/2015/927068
43. W. Kamiński, E. Tomczak, A. Górak, Biobutanol – production and purification methods. Ecol. Chem. Eng. **18**(1), 31–37 (2011)
44. A. Kubiczek , W. Kamiński, Ionic liquids for the extraction of n-butanol from aqueous solutions. Proceedings of ECOpole, **7**(1), 125–131 (2013). https://doi.org/10.2429/proc.2013.7(1)016
45. A. Kujawska, J. Kujawski, M. Bryjak, W. Kujawski, ABE fermentation products recovery methods—A review. Renewable and Sustainable Energy Reviews, **48**, 648–661 (2015). https://doi.org/10.1016/J.RSER.2015.04.028
46. A. K. Kumar, S. Sharma, Recent updates on different methods of pretreatment of lignocellulosic feedstocks: a review. Bioresources and Bioprocessing, **4**(1), (2017). https://doi.org/10.1186/s40643-017-0137-9.

47. A.M. Lakaniemi, O.H. Tuovinen, J.A. Puhakka, Anaerobic conversion of microalgal biomass to sustainable energy carriers–a review. Bioresour. Technol. **135**, 222–231 (2013). https://doi.org/10.1016/J.BIORTECH.2012.08.096
48. S.Y. Lee, J.H. Park, S.H. Jang, et al., Fermentative butanol production by Clostridia. Biotechnol. Bioeng. **101**, 209–228 (2008). https://doi.org/10.1002/bit.22003
49. S.Y. Li, R. Srivastava, S.L. Suib, Y. Li, R.S. Parnas, Performance of batch, fed-batch, and continuous A–B–E fermentation with pH-control. Bioresource technology, **102**(5), pp.4241–4250 (2011). https://doi.org/10.1016/j.biortech.2010.12.078
50. L. Li, H. Ai, S. Zhang, S. Li, Z. Liang, Z.-Q. Wu, J.-F. Wang, Enhanced butanol production by coculture of Clostridium beijerinckii and Clostridium tyrobutyricum. Bioresour. Technol. **143**, 397–404 (2013)
51. D. Liu, Y. Chen, F.Y. Ding, T. Zhao, J.L. Wu, T. Guo, H.F. Ren, B.B. Li, H.Q. Niu, Z. Cao, X.Q. Lin, Biobutanol production in a Clostridium acetobutylicum biofilm reactor integrated with simultaneous product recovery by adsorption. Biotechnology for biofuels, **7**,5 (2014). https://doi.org/10.1186/1754-6834-7-5
52. T.A. Lloyd, C.E. Wyman, Combined sugar yields for dilute sulfuric acid pretreatment of corn stover followed by enzymatic hydrolysis of the remaining solids. Bioresour. Technol. **96**, 1967–1977 (2005). https://doi.org/10.1016/j.biortech.2005.01.011
53. G. Lodi, L.A. Pellegrini, Recovery of butanol from ABE fermentation broth by gas stripping. Chem. Eng. Trans. **49**, 13–18 (2016)
54. H. Ma, L. Oxley, J. Gibson, W. Li, A survey of China's renewable energy economy. Renew. Sust. Energ. Rev. **14**, 438–445 (2010). https://doi.org/10.1016/J.RSER.2009.07.029
55. I.S. Maddox, E. Steiner, S. Hirsch, S. Wessner, The cause of "acid crash" and "acidogenic fermentations" during the batch Acetone-Butanol- Ethanol (ABE-) fermentation. J. Mol. Microbiol. Biotechnol. **2**, 95–100 (2000)
56. M.K. Mahapatra, A. Kumar, A short review on biobutanol, a second generation biofuel production from lignocellulosic biomass. J. Clean Energy Technol. **5**, 27–30 (2017). https://doi.org/10.18178/jocet.2017.5.1.338
57. S. Maiti, G. Gallastegui, S.J. Sarma, S.K. Brar, Y. Le Bihan, P. Drogui, M. Verma, A re-look at the biochemical strategies to enhance butanol production. Biomass Bioenergy **94**, 187–200 (2016). https://doi.org/10.1016/j.biombioe.2016.09.001
58. K. Manchester, Louis Pasteur, fermentation, and a rival. South Afri. J. Sci. **103**(9), 377–380 (2007)
59. G. Markou, I. Angelidaki, D. Georgakakis, Microalgal carbohydrates: an overview of the factors influencing carbohydrates production, and of main bioconversion technologies for production of biofuels. Appl. Microbiol. Biotechnol. **96**, 631–645 (2012). https://doi.org/10.1007/s00253-012-4398-0
60. A.D. Marques, F.P. Pinto, O.Q. Araújo, Assessment of methods to pretreat microalgal biomass for enhanced biogas production, in *Biomass and Green Chemistry*, (Springer, Cham, 2018a), pp. 394–404
61. A.D.L. Marques, F.P. Pinto, O.Q.F. Araújo, Assessment of methods to pretreat microalgal biomass for enhanced biogas production. J. Sust. Dev. Energy Water Environ. Syst. **6**, 394–404 (2018b). https://doi.org/10.13044/j.sdewes.d5.0193
62. H. Meehnian, A.K. Jana, M. Maiti, Effect of particle size, moisture content , and supplements on selective pretreatment of cotton stalks by Daedalea flavida and enzymatic saccharification. 3 Biotech **6**, 1–13 (2016). https://doi.org/10.1007/s13205-016-0548-x
63. T. Menegol, A.B. Diprat, E. Rodrigues, R. Rech, Effect of temperature and nitrogen concentration on biomass composition of Heterochlorella luteoviridis. Food Sci. Technol. **37**, 28–37 (2017). https://doi.org/10.1590/1678-457x.13417
64. A. Mohr, S. Raman, Lessons from fist generation biofuels and implications for the sustainability appraisal of second generation biofuels. Energy Policy **63**, 114–122 (2013). https://doi.org/10.1016/j.enpol.2013.08.033
65. A. Morone, R.A. Pandey, Lignocellulosic biobutanol production: gridlocks and potential remedies. Renew. Sust. Energ. Rev. **37**, 21–35 (2014). https://doi.org/10.1016/j.rser.2014.05.009

66. N. Mosier, C. Wyman, B. Dale, R. Elander, Y.Y. Lee, M. Holtzapple, M. Ladisch, Features of promising technologies for pretreatment of lignocellulosic biomass. Bioresource Technology, **96**(6), 673–686 (2005). https://doi.org/10.1016/J.BIORTECH.2004.06.025
67. S.N. Naik, V.V. Goud, P.K. Rout, A.K. Dalai, Production of first and second generation biofuels: a comprehensive review. Renewable and Sustainable energy reviews, **14**(2), 578–597 (2010). https://doi.org/10.1016/j.rser.2009.10.003
68. S. Nanda, D. Golemi-Kotra, J.C. McDermott, A.K. Dalai, I. Gökalp, J.A. Kozinski, Fermentative production of butanol: perspectives on synthetic biology. New Biotechnol. **37**, 210–221 (2017). https://doi.org/10.1016/j.nbt.2017.02.006
69. B. Ndaba, I. Chiyanzu, S. Marx, n-Butanol derived from biochemical and chemical routes: a review. Biotechnol. Rep. **8**, 1–9 (2015). https://doi.org/10.1016/j.btre.2015.08.001
70. A.S. Ndou, N. Plint, N.J. Coville, Dimerisation of ethanol to butanol over solid base catalysts. Appl. Catal. **251**, 337–345 (2003). https://doi.org/10.1016/S0926-860X(03)00363-6
71. J. Niemistö, P. Saavalainen, E. Pongrácz, et al., Biobutanol as a potential sustainable biofuel-assessment of lignocellulosic and waste-based feedstocks. J. Sust. Dev. Energy Water Environ. Syst. **1**(2), 58–77 (2013). https://doi.org/10.13044/j.sdewes.2013.01.0005
72. P.S. Nigam, A. Singh, Production of liquid biofuels from renewable resources. Prog. Energy Combust. Sci. **37**, 52–68 (2011)
73. P. Noomtim, B. Cheirsilp, Production of butanol from palm empty fruit bunches hydrolyzate by *Clostridium acetobutylicum*. Energy Procedia **9**, 140–146 (2011). https://doi.org/10.1016/j.egypro.2011.09.015
74. S. Ogo, A. Onda, K. Yanagisawa, Selective synthesis of 1-butanol from ethanol over strontium phosphate hydroxyapatite catalysts. Appl. Catal. **402**, 188–195 (2011). https://doi.org/10.1016/j.apcata.2011.06.006
75. C.M. Osmundsen, S. Dahl, E. Taarning, J. Rossmeisl, *Catalytic Conversion of Carbohydrates* (Department of Physics, Technical University of Denmark, Denmark, 2013)
76. F. Passos, I. Ferrer, Influence of hydrothermal pretreatment on microalgal biomass anaerobic digestion and bioenergy production. Water Res. **68**, 364–373 (2015). https://doi.org/10.1016/J.WATRES.2014.10.015
77. P. Patakova, M. Linhova, M. Rychtera, et al., Novel and neglected issues of acetone–butanol–ethanol (ABE) fermentation by clostridia: clostridium metabolic diversity, tools for process mapping and continuous fermentation systems. Biotechnol. Adv. **31**(1), 58–67 (2013). https://doi.org/10.1016/j.biotechadv.2012.01.010
78. I. Patraşcu, C.S. Bîldea, A.A. Kiss, Eco-efficient butanol separation in the ABE fermentation process. Seapartion and purification technology, **117**(13), 49–61 (2017). https://doi.org/10.1016/j.seppur.2016.12.008
79. N. Qureshi, T.C. Ezeji, Butanol, 'a superior biofuel' production from agricultural residues (renewable biomass): recent progress in technology. Biofuels Bioprod. Biorefin. **2**, 319–330 (2008). https://doi.org/10.1002/bbb.85
80. N. Qureshi, B.C. Saha, B. Dien, R.E. Hector, M.A. Cotta, Production of butanol (a biofuel) from agricultural residues: Part I-use of barley straw hydrolysate. Biomass Bioenergy **34**, 559–565 (2010). https://doi.org/10.1016/j.biombioe.2009.12.024
81. N. Qureshi, M.A. Cotta, B.C. Saha, Bioconversion of barley straw and corn stover to butanol (a biofuel) in integrated fermentation and simultaneous product recovery bioreactors. Food Bioprod. Process. **92**, 298–308 (2014). https://doi.org/10.1016/j.fbp.2013.11.005
82. K.K. Ramasamy, M. Gray, H. Job, D. Santosa, X.S. Li, A. Devaraj, A. Karkamkar, Y. Wang, Role of calcination temperature on the hydrotalcite derived MgO–Al_2O_3 in converting ethanol to butanol. Top. Catal. **59**, 46–54 (2016). https://doi.org/10.1007/s11244-015-0504-8
83. R. Ravindran, A.K. Jaiswal, A comprehensive review on pre-treatment strategy for lignocellulosic food industry waste: challenges and opportunities. Bioresour. Technol. **199** (2016). https://doi.org/10.1016/j.biortech.2015.07.106
84. T. Riittonen, E. Toukoniitty, D.K. Madnani, A.R. Leino, K. Kordas, M. Szabo, A. Sapi, K. Arve, J. Wärnå, J.P. Mikkola, One-pot liquid-phase catalytic conversion of ethanol to 1-butanol over

aluminium oxide-the effect of the active metal on the selectivity. Catalysts **2**, 68–84 (2012). https://doi.org/10.3390/catal2010068
85. T. Riittonen, K. Eränen, P. Mäki-Arvela, A. Shchukarev, A.R. Rautio, K. Kordas, N. Kumar, T. Salmi, J.P. Mikkola, Continuous liquid-phase valorization of bio-ethanol towards bio-butanol over metal modified alumina. Renew Energy **74**, 369–378 (2015). https://doi.org/10.1016/j.renene.2014.08.052
86. J. Rudloff, J.M. Zaccardi, S. Richard, J.M. Anderlohr, Analysis of pre-ignition in highly charged SI engines: emphasis on the auto-ignition mode. Proc. Combust. Inst. **34**, 2959–2967 (2013). https://doi.org/10.1016/j.proci.2012.05.005
87. S. Sarathy, S. Vranckx, K. Yasunanga, M. Mehl, P. Osswald, W.K. Metcalfe, C.K. Westbrook, W.J. Pitz, K. Kohse-Hoingaus, R.X. Fernandes, H.J. Curran, A comprehensive chemical kinetic combustion model for the four butanol isomers. Combust. Flame **159**, 2028–2055 (2012). https://doi.org/10.1016/j.combustflame.2011.12.017
88. T.J. Schwartz, M. Lawoko, Removal of acid-soluble lignin from biomass extracts using Amberlite XAD-4 resin. Bioresources **5**(4), 2337–2347 (2010)
89. R. Sillers, A. Chow, B. Tracy, E.T. Papoutsakis, Metabolic engineering of the non-sporulating, non-solventogenic Clostridium acetobutylicum strain M5 to produce butanol without acetone demonstrate the robustness of the acid-formation pathways and the importance of the electron balance. Metab. Eng. **10**(6), 321–332 (2008). https://doi.org/10.1016/j.ymben.2008.07.005
90. M. Solana, N. Qureshi, A. Bertucco, F. Eller, Recovery of butanol by counter-current carbon dioxide fractionation with its potential application to butanol fermentation. Materials **9**(7), 1–11 (2016)
91. M.J. Spivey, The acetone-butanol-ethanol fermentation. Process Biochem. **13**(11), 2–4 (1978)
92. S. Sun, X. Cao, R. Sun, The role of pretreatment in improving the enzymatic hydrolysis of lignocellulosic materials. Bioresour. Technol. **199**, 49–58 (2016). https://doi.org/10.1016/j.biortech.2015.08.061
93. S. Tayrabekova, P. Mäki-Arvela, M. Peurla, P. Paturi, K. Eränen, G.E. Ergazieva, A. Aho, D.Y. Murzin, K. Dossumov, Catalytic dehydrogenation of ethanol into acetaldehyde and iso-butanol using mono-and multicomponent copper catalysts. C. R. Chim. **21**, 194–209 (2018). https://doi.org/10.1016/j.crci.2017.05.005
94. A. Thongsukmak, K.K. Sirkar, Pervaporation membranes highly selective for solvents present in fermentation broths. J. Membr. Sci. **302**(1–2), 45–58 (2007). https://doi.org/10.1016/j.memsci.2007.06.013
95. K. Tomishige, Y. Nakagawa, M. Tamura, Production of diols from biomass, in *Production of Platform Chemicals from Sustainable Resources*, ed. by Z. Fang, 7th edn., (Springer, Singapore, 2017), pp. 343–373. https://doi.org/10.1007/978-981-10-4172-3
96. T. Tsuchida, S. Sakuma, T. Takeguchi, W. Ueda, Direct synthesis of n-butanol from ethanol over nonstoichiometric hydroxyapatite. Ind. Eng. Chem. Res. **45**, 8634–8642 (2006). https://doi.org/10.1021/ie0606082
97. S.B. Tummala, S.G. Junne, E.T. Papoutsakis, Antisense RNA downregulation of coenzyme A transferase combined with alcohol-aldehyde dehydrogenase overexpression leads to predominantly alcohologenic Clostridium acetobutylicum fermentations. Journal of Bacteriology, **185**, 3644–3653 (2003). https://doi.org/10.1128/JB.185.12.3644-3653.2003
98. M. Uyttebroek, W. Van Hecke, K. Vanbroekhoven, Sustainability metrics of 1-butanol. Catal. Today **239**, 7–10 (2015). https://doi.org/10.1016/j.cattod.2013.10.094
99. A.B. Van der Merwe, H. Cheng, J.F. Görgens, J.H. Knoetze Comparison of energy efficiency and economics of process designs for biobutanol production from sugarcane molasses. Fuel, **105**, 451–458 (2013). https://doi.org/10.1016/j.fuel.2012.06.058
100. H. van der Wal, B.L. Sperber, B. Houweling-Tan, R.R. Bakker, W. Brandenburg, A.M. López-Contreras, Production of acetone, butanol, and ethanol from biomass of the green seaweed Ulva lactuca. Bioresour. Technol. **128**, 431–437 (2013). https://doi.org/10.1016/j.biortech.2012.10.094

101. M. Vitova, K. Bisova, S. Kawano, et al., Accumulation of energy reserves in algae: from cell cycles to biotechnological applications. Biotechnol. Adv. **33**(6), 1204–1218 (2015). https://doi.org/10.1016/j.biotechadv.2015.04.012
102. Y. Wang, W. Guo, C.L. Cheng, S.H. Ho, J.S. Chang, N. Ren, Enhancing bio-butanol production from biomass of Chlorella vulgaris JSC-6 with sequential alkali pretreatment and acid hydrolysis. Bioresour. Technol. **200**, 557–564 (2016). https://doi.org/10.1016/j.biortech.2015.10.056
103. Y. Wang, S.H. Ho, H.W. Yen, D. Nagarajan, N.Q. Ren, S. Li, Z. Hu, D.J. Lee, A. Kondo, J.S. Chang, Current advances on fermentative biobutanol production using third generation feedstock. Biotechnology advances, **35**(8), pp.1049–1059 (2017). https://doi.org/10.1016/j.biotechadv.2017.06.001
104. R.D. Weinstein, A.R. Ferens, R.J. Orange, P. Lemaire, Oxidative dehydrogenation of ethanol to acetaldehyde and ethyl acetate by graphite nanofibers. Carbon **49**, 701–707 (2011). https://doi.org/10.1016/j.carbon.2010.10.027
105. R.M. West, E.L. Kunkes, D.A. Simonetti, J.A. Dumesic, Catalytic conversion of biomass-derived carbohydrates to fuels and chemicals by formation and upgrading of mono-functional hydrocarbon intermediates. Catal. Today **147**, 115–125 (2009a). https://doi.org/10.1016/j.cattod.2009.02.004
106. R.M. West, D.J. Braden, J.A. Dumesic, Dehydration of butanol to butene over solid acid catalysts in high water environments. J. Catal. **262**, 134–143 (2009b). https://doi.org/10.1016/j.jcat.2008.12.009
107. M.K. Yang, S. Keinänen, M. Vepsäläinen, J. Romar, H. Tynjälä, P.U. Lassi, A. Pappinen, The use of (green field) biomass pretreatment liquor for fermentative butanol production and the catalytic oxidation of biobutanol. Chem. Eng. Res. Des. **92**, 1531–1538 (2014). https://doi.org/10.1016/j.cherd.2014.01.015
108. L. Yerushalmi, Physiological aspects of the acetone-butanol fermentation. Dissertation, McGill University (1985)
109. S. Zafar, The concept of Biorefinery (2018), https://www.bioenergyconsult.com/biorefinery/. Accessed 20 Sept 2018
110. H.S. Zeng, C.R. He, A. Yen, et al., Assessment of acidified fibrous immobilization materials for improving acetone-butanol-ethanol (ABE) fermentation. Fermentation **3**(1), 1–9 (2016). https://doi.org/10.3390/fermentation3010003

Chapter 19
Biochar as an Adsorbent: A Short Overview

A. T. Akintola, E. T. Akinlabi, and S. O. Masebinu

19.1 Introduction

The rise in the level of industrialization and excessive usage of chemicals has increased the likelihood of environmental pollutions, through the release of toxic pollutants into the environment [43, 109, 126]. Similarly, rising human population has led to a consequential rise in various human activities that produce a huge amount of organic and inorganic wastes, adding more to environmental pollution, and thereby threatening the well-being of humans, together with the ecosystem [45, 124, 126]. In lieu of this, it is very important to either reduce or eliminate some of these toxic pollutants from the environment. Various techniques such as reduction, precipitation, oxidation, ion exchange, coagulation, biological treatment, membrane separation and adsorption processes have been suggested and utilized in the treatment of several environmental pollutants [1, 65, 125, 126, 136, 166]. Nevertheless, compared to adsorption process, the other treatment methods are characterized by drawbacks of huge capital and operating cost, and secondary sludge generation, leading to disposal and new pollutant problems [29, 74, 76, 126, 166]. Adsorption process, on the contrary, has been reported as the most appropriate, attractive and broadly utilized method in the reduction and elimination of these toxic pollutants, due to its high effectiveness, ease of operation, ease of process control, ease of regeneration, technological flexibility, affordable energy requirement, eco-friendliness and economic benefit [6, 44, 52, 57, 59, 63, 74, 123, 126, 129, 142]. In recent times, there is a gradual move from the use of existing adsorbents

A. T. Akintola (✉)
Department of Chemical Engineering Technology, University of Johannesburg, Johannesburg, South Africa

E. T. Akinlabi · S. O. Masebinu
Department of Mechanical Engineering Science, University of Johannesburg, Johannesburg, South Africa

M. O. Daramola, A. O. Ayeni (eds.), *Valorization of Biomass to Value-Added Commodities*, Green Energy and Technology,
https://doi.org/10.1007/978-3-030-38032-8_19

like activated carbon and silica gel, to utilizing biochar as an adsorbent in different adsorption processes. Biochar is a possible engineered adsorbent for various environmental applications [17]. The cost-effectiveness and wide range of its sustainable feedstock make it a possible alternative to different commercial adsorbents [143]. Similarly, the strength and simplicity of modifying some of its property have attested to it being the choicest adsorbent [25]. Biochar as an adsorbent has been reported effective in various applications, such as in the treatment of wastewater, lessening of greenhouse gases in soil, in soil amendment and more recently in anaerobic digestion process stabilization [94, 143]. The importance of using biochar in the place of other adsorbents was attested to with the recent rise in the number of literatures on the utilization of biochar as an adsorbent. This was validated using the Scopus database as a bibliometric tool. There was a document search using 'biochar as an adsorbent' as the 'article title, abstract, and keyword', with limits to papers published between the years 2001 and 2018. The document type was selected to be 'all (including journal articles, reviews, book chapters, conference papers)', and the access type was selected as 'all, including both open access and others'. A total of 664 documents [open access (71 documents) and others (593 documents)] came out on the basis of the search criterion. From the search, there was a drastic rise in the number of papers published. The number was from 1 in the year 2001 to 221 in the year 2018 (Fig. 19.1), with the most published work on 'biochar as an adsorbent' reportedly published in journal articles (Fig. 19.2). The goal of this overview is to discuss about adsorption concept, types of adsorption process, its various areas of application, different adsorbents and the possibility of utilizing biochar as a

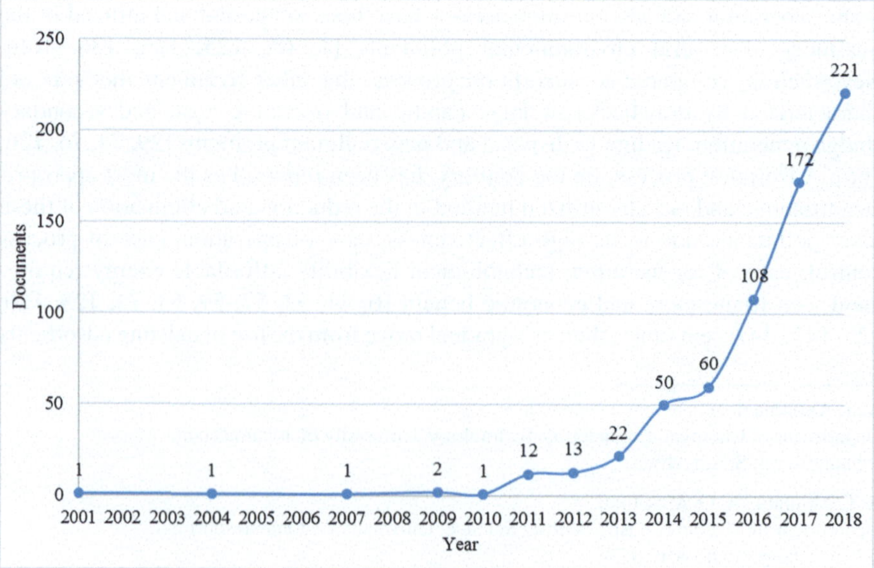

Fig. 19.1 Documents on 'biochar as an adsorbent' by publication numbers (2001–2018)

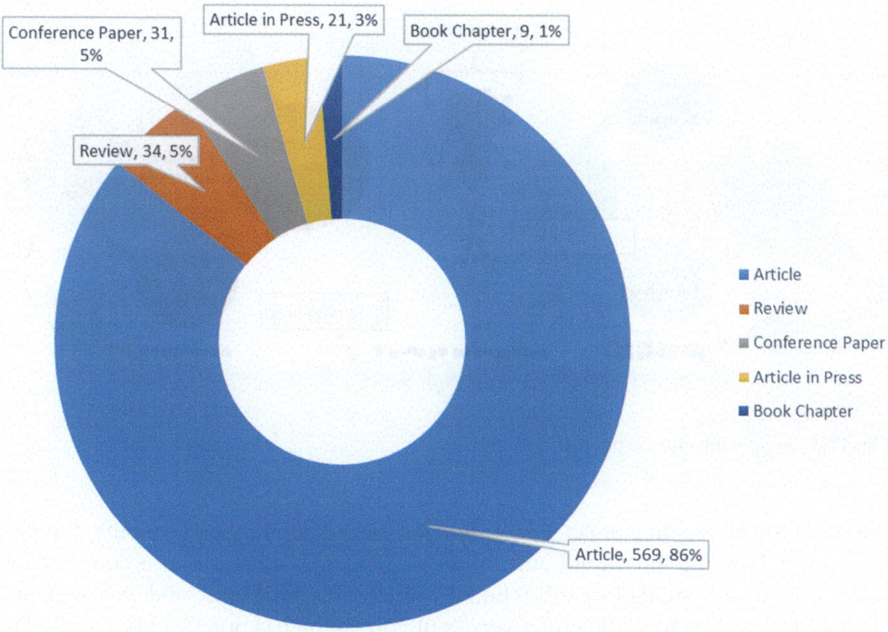

Fig. 19.2 Documents on 'biochar as an adsorbent' by publication type (2001–2018)

substituent to commercial adsorbents, biochar production processes and mechanisms involved in the adsorption of pollutants onto biochar.

19.2 Adsorption

According to studies [106, 121, 149], adsorption is a process that takes place when a solute (either gas or liquid) accrues on a surface, creating a film of molecules, ions or atoms on it. It is also the accumulation of molecules of a specific substance (liquid or gas) at the interfacial layer of another substance (mostly solid), such that there is always a concentration difference [24, 47, 99, 141, 164]. Furthermore, adsorption as a process is characterized by a build-up of molecular concentration of liquids or gaseous substance on a solid surface, with the initial unsaturated surface of the solid becoming saturated, thus presents it as a surface phenomenon [21, 100, 122, 141, 143, 149]. This process is overseen by either specific or vague interactions between atoms present on the surface of the solid substance and that of the molecules approaching it [149]. This is a major characteristic that differentiates adsorption from absorption [47, 149]. The particular surface where other substances are accrued to is known as an adsorbent [21, 36, 72, 100, 106, 122, 149, 153], the substance that becomes concentrated on this surface is referred to as an adsorbate [21, 24, 36, 72, 100, 106, 122, 149, 153]. See Fig. 19.3 for the simple adsorption process.

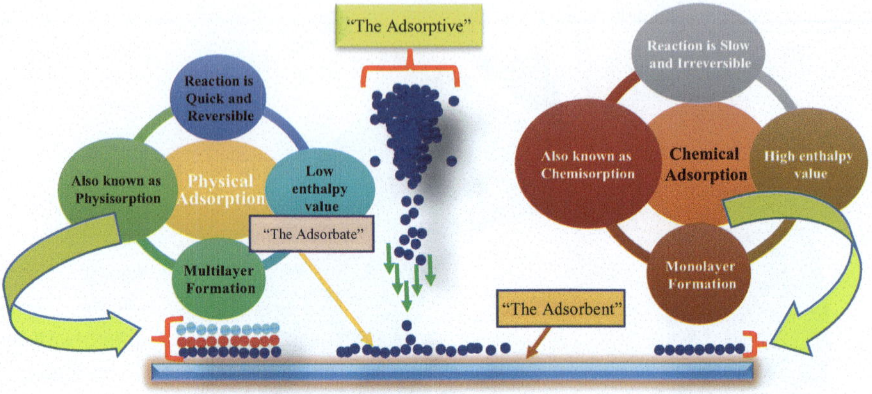

Fig. 19.3 Simple adsorption process

Based on the phases in contact, adsorption can be studied in these systems: liquid–gas, liquid–liquid, solid–liquid and solid–gas [24, 99], although on a commercial scale, it is mostly studied in solid–liquid [66, 68, 70, 155] and solid–gas systems [110, 111]. Furthermore, the effectiveness of adsorption has practically presented it useful in various domestic and industrial applications.

19.2.1 Types of Adsorption Process

Adsorption processes have been reported to depend on the attractive force between the surface of the adsorbate and the adsorbent [21, 100, 106, 122, 143, 149, 153]. Equally, this force depends on the intrinsic characteristics of the adsorbent and the adsorbate in contact. Owing to this, adsorption process could be either physical adsorption or chemical adsorption [21, 24, 72, 100, 106, 122, 149, 153]. For the former process, the pull between the adsorbate and the adsorbent is governed by weak van der Waal forces, while for chemical adsorption process, there is always a chemical reaction that results in chemical bonding [24, 72, 100, 122, 143, 149, 153]. Also, chemisorption could be through the following mechanism, including ion exchange, surface complexation and precipitation [143]. Other major differences between physical and chemical adsorption processes are presented in Table 19.1.

Similarly, it is crucial to understand that sometimes, the two processes can take place either concurrently or alternatively at favourable conditions [24, 99, 153], with physical adsorption always preceding the chemical process. The strength of each adsorption method is also a direct function of the heat generated during the process [149]. In addition, the two adsorption processes increase with rise in the adsorbent's surface area [141].

Table 19.1 Variations between physical and chemical adsorptions

Properties	Physical adsorption	Chemical adsorption	Reference(s)
Alternative name	Physisorption	Chemisorption	NPTEL [106]; Surface Chemistry [141]; Dąbrowski [24]; Webb [153]; Christmann [21]; Musin [100]; Vera Bolis [149]; Saad [122]
Layers formed	Multiple layers of the adsorbate	A single layer of the adsorbate	Dąbrowski [24]; Webb [153]; Saad [122]
Type of reaction	Quick and reversible	Slow and irreversible	Surface Chemistry [141]; Dąbrowski [24]; Webb [153]; Musin [100]; Saad [122]
Specificity	It is not specific with adsorbent type and surface. It could take place on nearly all solid surface	It is specific with adsorbent type and surface	Dąbrowski [24]; Webb [153]; Musin [100]; Vera Bolis [149]; Saad [122]
Occurrence	Takes place at a lower or near critical temperature of the adsorbate	Takes place at a high temperature, typically higher than the adsorbate's critical temperature	Dąbrowski [24]; Musin [100]
Activation energy	No energy is needed for activation	A particular amount of energy is necessary for activation	Surface Chemistry [141]
Enthalpy of adsorption ($kJmol^{-1}$)	It is characterized by a low enthalpy value, usually between 5 and 45	It is characterized by a high enthalpy value, usually between 80 and 400	Surface Chemistry [141]; Vera Bolis [149], Saad [122]

19.2.2 Applications of Adsorption Process

Adsorption's importance cannot be overestimated, due to its ease of operation, effectiveness, adaptability to various process applications, ease of sorbent handling, likelihood of regenerating used adsorbents and low cost [33, 143, 161, 168]. Previous researchers have identified some of the areas where adsorption is relevant [118]. The effectiveness of adsorption in the treatment of domestic wastes, and those generated from industrial processes has been mentioned, while adsorption method is reported to be effective in pollutant's eradication (heavy metals, nitrogen and phosphorus) from wastewater, owing to its ability to remove pollutants that are difficult for biological treatments [73, 98, 126, 127], and its potency being attested to in the purification of drinking water [98]. Similarly, researchers [100, 119] mentioned that adsorption can be utilized in the removal of organic pollutants like naphthalene and 1-naphthol, NO_x (nitrogen oxide(s)) and SO_2 (sulphur IV oxide) from gaseous substance, and also in the removal of water molecules from industrial gases such as O_2 (oxygen gas), N_2 (nitrogen gas) and hydrocarbons. Equally, Lv et al. [90] mentioned that adsorption can be efficiently utilized in removing and controlling spills from oil on the ocean surface. Adsorption process is effective not only in the removal of contaminants from gases, but also in the control of global warming. This is possible when adsorbents are correctly applied in reducing the amounts of CO_2 (carbon dioxide) and CH_4 (methane), together with N_2O (nitrogen oxide) present within the atmosphere. Interestingly, chlorofluorocarbon (CFC), one of the major causes of ozone layer depletion, can also be efficiently recovered using adsorption technology [98]. Furthermore, [106] noted that adsorption can be effectively used in heterogeneous catalysis and decolorization processes, while Jiang et al. [63] mentioned its effectiveness in reducing heavy metals and antibiotics pollution. Dissanayake Herath et al. [33] also observed and reported about the potential of utilizing adsorption process in applications that involve the removal of glyphosate, an organophosphate herbicide. Other applications of adsorption process are in surface corrosion, surface protection, use and formation of nanoparticles, making of a semiconductor device, structural support and adjustment of two-dimensional (2D) material properties, fixing complex three-dimensional (3D) problems like protein folding and structure of the polymer, medical analytical treatment and technologies involving prosthetics [118].

19.2.3 Adsorbents

This is the exposed solid surface where other molecules are accumulated during the adsorption process [122, 149]. Adsorbents can be identified by the size of their pore diameter; they include micropores (diameter of its pore is less than 2 nm), mesopores (between 2 and 50 nm) and macropores (more than 50 nm) [55, 72, 135, 149]. Similarly, they are usually in round pellets, cylindrical bars, mouldings and monolith form [122]. Some key properties of good adsorbents are high selectivity ability, high adsorption ability, good physical characteristic (large pore size, large pore volume, large surface area and small particle size), strength (excellent chemical and thermal strength), better kinetic features, huge abrasion resistance, regeneration potential and reduced cost of production [35, 95, 100, 122, 143, 144].

19.2.4 Commercial Adsorbents

Commercial adsorbents are classified as (1) compounds that consist of oxygen, with a hydrophilic and polar nature. Examples are silica gels and zeolites; (2) carbon-containing compounds, with a hydrophobic and non-polar nature. Examples are activated carbon and graphite; and (3) compounds made of polymeric substance, which are either polar or non-polar, with a porous-like matrix arrangement [38, 122]. Activated carbon is a commonly utilized commercial adsorbent. It is produced by charring its source material, followed by either chemical or thermal method activation. The activation step is mainly carried out to achieve a preferred chemical pore structure [11, 156]. The large amount of micropores, mesopores and large surface area [37, 122, 156], rich surface functional groups on activated carbon makes it efficient in a broad array of applications as an adsorbent [37, 73]. Activated carbon can be used in taking out of organic, non-polar substance, and of waste gas in wastewater treatment [122]. Molecular sieve is another commercial adsorbent. It is also known as zeolites [11]. It has a crystalline-like and very porous structure. Its pore system is always three-dimensional, with equal diameters [11, 26]. Some of its distinct properties are its thermal strength and high adsorption ability at low temperatures. Molecular sieves can be utilized in the taking out of organic and inorganic substances from liquids and gases, likewise in the removal of water from gaseous substance [57, 100]. Silica gel has a porous and amorphous-like structure of silica (SiO_2), with a huge network of interconnected tiny micropores. Silica gel can be produced through either the polymerization of silicic acid or the accumulation of colloidal silica particles with uniform sizes [156] and can be utilized as a desiccant for water vapor, organic solvents and non-polar liquids [57, 100]. Activated alumina is another adsorbent produced from hydrated alumina or gibbsite, either by removing the water of hydration using heat or chemical activation, within a well-controlled environment [11, 100, 156]. It is a hydrophilic substance, with a good pore arrangement [100]. Activated alumina can be used in catalysis process, oxygenates and mercaptan removal from the feed streams of hydrocarbons, and in removing the ions of fluorine from water [11].

19.2.5 The Need for a Low-Cost Adsorbent (Biochar)

Most of the materials used as adsorbents consist of natural materials, synthetic materials and wastes. Examples are naturally formed clays and waste (like nutshell) [83], industrial wastes (like fly ash and oven slag) [12, 83], oxides and hydroxides being synthesized from metal (like activated oxide of aluminium and coarse ferric hydroxide) [40, 83]. Nonetheless, there is a huge drawback of high production cost, environmental unfriendliness, inconsistent performance and regeneration issues associated with their use [77, 83]. Therefore, a low-cost, yet effective material is needed as an adsorbent. Numerous attempts have been made to produce such materials for pollutants elimination in various applications [73, 126]. Some of the materials whose cost are low, and that can be used for the uptaking of various pollutants, are algae, carbon-derived cloth, peat, agricultural derivatives, lignin, biomaterials, bagasse fly ash, slag from blast furnace, red mud, sand coated and treated with iron oxide, and biochar [97]. Nonetheless, biochar is projected as an efficient low-cost adsorbent and has been reported to possess similar properties to activated carbon [15, 143]. It is also being accepted as an ecologically friendly and efficient sorbent to lessen organic and inorganic pollutants, thus a tool to reduce the threat pollutants pose on the global environment [17]. Some properties that present biochar as a viable substitute for commercial adsorbents are abundant starting materials such as sawdust, rice straw, bamboo, rice husk, safflower seed, pinewood, olive husk, faecal carbon and sludge carbon, and other lignocellulose plant matter [3, 5, 20, 89]. Others are less energy production requirements, reduced production cost (its overall cost is reported to be less expensive, in contrast to activated carbon's), environmental friendliness (its production from waste biomass is an indirect method of waste management), production simplicity, high adsorption potential and desirable physicochemical surface characteristics [1, 33, 56, 61, 77, 83, 91, 102, 104, 112, 143, 152, 163, 171].

19.3 Biochar

Biochar has been referred to as a coarse material, which is rich in carbon [56, 102]. Its production is by biomass heating in a reduced or no-oxygen condition [20, 37, 62, 75, 102]. Biochar is surplus in carbon, and its production temperature range is typically between 300 and 900 °C [62, 78, 102, 129]. It is also referred to as 'biological charcoal' with varying size, depending on the size of the initial biomass. In relation to charcoal, the notable difference between biochar and charcoal is its production within a controlled condition, which is done majorly to obtain the conversion of most of its carbon content into various useful products. Unlike charcoal, biochar possesses minerals like calcium (Ca), potassium (K), phosphorus (P) and nitrogen (N) [62], and is free of the uncondensable and condensable gases associated with its production. This makes biochar distinct and more effective in numerous applications.

19.3.1 Biochar Production

Biochar production is via the thermochemical conversion of biomass. Thermochemical conversion process depends on the chemical reactions that occur within controlled temperatures and pressures, temperature most especially, and could be pyrolysis, gasification or combustion process [2, 9, 39, 84, 128, 145, 148, 167], however, within varied oxygen conditions. Pyrolysis process involves the breakdown of a solid material into smaller bits, usually solid (referred to as char), liquid (referred to as bio-oil or tar) and uncondensable gas (referred to as syngas), with the aid of heat [2, 9, 32, 170]. The formation of biochar, in addition to other products, makes pyrolysis unique [60]. Gasification involves the breakdown of a solid matter by oxidizing it partially with oxygen or via steam reformation method, with the main aim of gas production and trace fraction of solid and bio-oil. Combustion involves complete burning in air or oxygen-rich atmosphere. It is basically an exothermic reaction, with the heat generated being utilized in steam production or in generating thermal power [39, 128, 148]. The basic difference between pyrolysis, gasification and combustion thermochemical process, in relation to oxygen usage is shown below; see Fig. 19.4, where the outer environment contains a limitless amount of oxygen. Nevertheless, each of the process still requires a source of heat, more so, with pyrolysis being reported as the most efficient technique for biochar production [6].

For biochar production, the feedstocks are basically organic, and usually known as biomass [107, 163]. Nevertheless, the common type is lignocellulosic biomass, owing to their reduced cost, plenteousness, accessibility, adaptability and renewability [81, 113, 130, 131, 133]. It is being presented as a possible substitute to fuels from fossil source, more so, in the making of bio-fuels and various valued products [41, 81, 130, 167]. Lignocellulosic biomass is likewise a complex polymeric substance that is made of fibres, with three different main compositions. They are hemicellulose, cellulose, and lignin, which respond differently to heat treatment [50, 101, 113, 116, 131, 157, 167], and with a varied amount in different lignocellulosic biomass [50, 131, 132]. In general, the quantity of cellulose, hemicellulose, together with lignin in most lignocellulosic biomass ranges in 30%–60%, 20%–40% and 15%–25%, respectively [50, 131]. Good examples of lignocellulosic biomass are wood such as Leucaena wood, pine wood, spruce [8, 17, 54, 101, 103, 173]; grasses

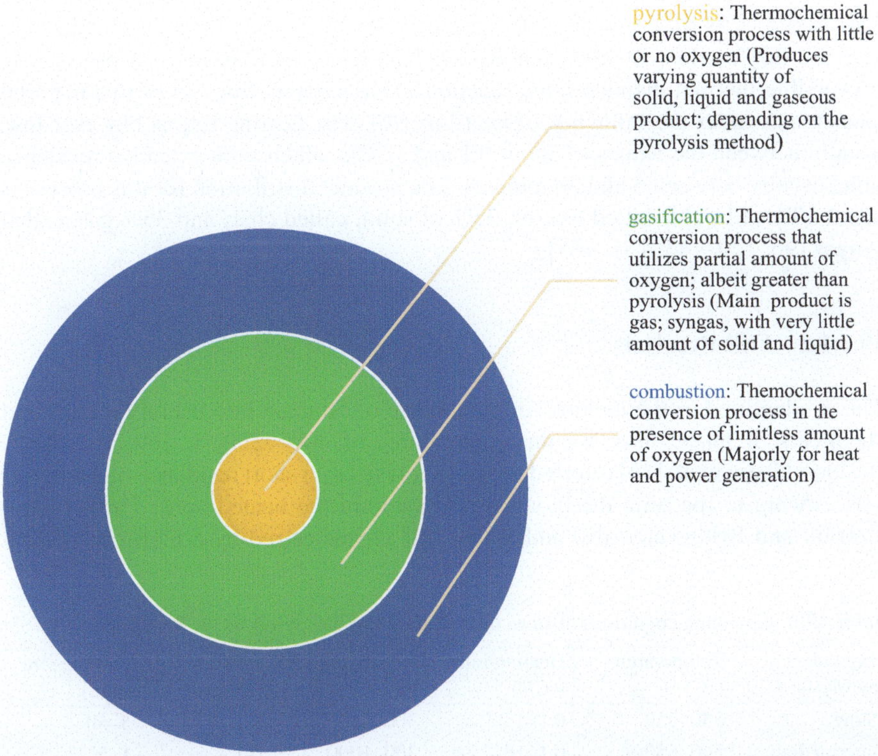

Fig. 19.4 Thermochemical conversion process as related to oxygen usage

[8, 17, 54]; wood waste such as bamboo leaves, sawdust and aspen [54, 81]; remains of crop [17, 22]; pine needle [17, 18]; agricultural wastes such as rice straw and husk, rice hull, wheat straw, fruit peel, corn stover, corncob, bagasse from sugarcane, peels of sugarcane, remains of coffee ground [17, 54, 81, 87, 101]; energy crops such as poplar and willow [101]; crab shell [25]; algal biomass [81]; and others such as refuse from paper mill and paper from municipal waste [101]. Based on the conditions and settings of production, pyrolysis comprises different types [34, 145]. Tripathi et al. [145] mentioned that pyrolysis could be specifically grouped into six, namely, slow, fast, intermediate, flash, vacuum, and hydro pyrolysis. However, fast and slow pyrolysis processes are the major types. During pyrolysis, the overall process conversion and the physiochemical characteristics of the products are mainly impacted by the nature of the feed, process conditions and reactor configurations [2, 3, 56, 62, 102]. In the next sub-section, a brief overview of slow, fast and intermediate pyrolysis processes will be discussed, with some of its operating conditions and distribution of product yield being presented in Table 19.2 and Fig. 19.5, respectively.

19.3.1.1 Slow Pyrolysis

This is known as the conventional method [2, 96, 145]. Most times, it is utilized to increase the fraction of solids (known as char) [2, 9, 16]. During this process, an irreversible thermal disintegration reaction of the organic fraction of this material takes place under the influence of heat [46, 96]. The heating rate is however low, usually between the values of about 0.1 and 1 °C/s, albeit with extended residence time, usually between 5 and 30 minutes. The product distribution for this process is about 30% of liquid, called bio-oil; 35% of solid, called char; and 35% gas, called syngas [96, 120, 145, 151, 165].

19.3.1.2 Fast Pyrolysis

This is identified as the method for the change of biomass into useful biofuels and chemicals. It takes place within temperatures of 400–1250 °C [69, 105, 145]. Besides, Gvero et al. [46] referred to fast pyrolysis as 'a short residence timed pyrolysis', owing to the time the biomass spent within the heated zone. Furthermore, Czernik and Bridgwater also added that it is characterized by a high rate of heat

Table 19.2 Operating conditions of slow, intermediate and fast pyrolysis [7, 30, 67, 80, 145, 165]

Type of pyrolysis	Temperature (°C)	Heating rate (°C/s)	Residence time (s)	Pressure (MPa)	Particle size (mm)
Slow	550–950	0.1–1.0	300–550	0.1	5–50
Intermediate	500–650	0.1–10	300–1000	0.1	1–5
Fast	400–1250	10–200	0.5–10	0.1	<1

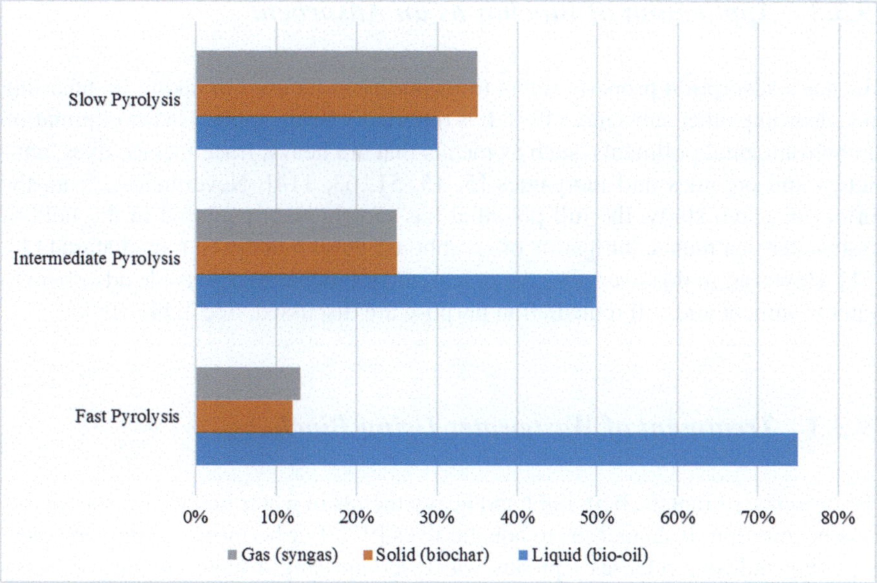

Fig. 19.5 Different pyrolysis process yield. (Data acquired from [31, 102, 107, 165])

transfer [23], typically around 300 °C/minute [42]. The residence time is generally within a few seconds or less, most times between 0.5 and 10 seconds. The main target is its liquid with the rationale behind using it being connected to its potential of generating a huge amount of liquids (bio-oil and tar), of about 60–75 wt.%, that could be utilized in an extensive array of applications, or even as an effective carrier of energy [9, 10, 23, 46]. The other co-products of fast pyrolysis are solid char, together with non-condensable gases, whose compositions are about 15–25 wt.% and 10–20 wt.%, respectively, with respect to the precursor material [46].

19.3.1.3 Intermediate Pyrolysis

The rationale behind this technique is at striking a balance between the amount of solid and liquid that is being produced. This is different from slow and fast pyrolysis process, whose main targets are the solid and liquid products, respectively. One of the good sides to this type of pyrolysis process is its ability to lessen the production of tars with high molecular mass and to generate chars which are dry and fit for agricultural purposes, or even in the production of energy, more so, with bio-oil of high quality. For this process the process conditions are typically temperature range of about 500–650°C, a rate of heating of 0.1–10°C/minute, and residence time of 300–1000 seconds. The products are usually distributed into 40–60 wt.% of liquid, 20–30 wt.% of uncondensable gases and 15–25 wt.% of solid [49, 92, 145].

19.3.2 Application of Biochar as an Adsorbent

Biochar's adsorption property refers to its ability to be used in taking in, hoarding and adsorbing other substances [97]. It is reportedly being applied in the elimination of environmental pollutants, such as metals that are heavy, trace metals, dyes, antibiotics and organics and inorganics [6, 15, 51, 63, 114]. Nevertheless, from the current research study, the full potential has mostly been explored in the field of wastewater treatments, mitigation of greenhouse gases and in soil remediation [143, 151]. However, in this overview, the potentials of biochar as a possible adsorbent in water treatment and soil remediation purpose are discussed. See Table 19.3.

19.3.3 Treatment of Wastewater Using Biochar

The presence of metals, both light and heavy metals in water bodies, irrespective of its concentration, is dangerous to aquatic lives [97]. Furthermore, mining and electroplating industry releases aqueous wastes comprising a huge amount of heavy metals such as uranium, mercury, cadmium and copper into the environment, which if not well treated may have a dangerous effect on water and aquatic lives. Several approaches have been utilized in treating pollutants present in wastewater; some are membrane separation, oxidation, photochemical degradation, electrochemical treatment, biological treatment and adsorption [4, 19, 79, 82, 138, 168, 172]; nevertheless, adsorption has grown into being the mainly utilized method for the taking out of pollutants in water [97]. The treatment of wastewater is reported to require a large amount of activated carbon [97], with an overall high cost, thereby giving rise to the need for an alternative adsorbent. To replace commercial adsorbent like activated carbon in water treatment, an adsorbent like biochar is required [97], and its utilization as a replacement wastewater treatment of is drawing rising attention [93, 151]. Biochar was effectively utilized in the removal of cation (e.g. ammonium ions) and anion (e.g. phosphate ions) out of aqueous mixtures [27, 170], and has also been efficient in the taking out of pollutants that are organic and inorganic, from water [31, 162]. Diethyl phthalate (DEP), a form of phthalic acid esters (PAEs), was successfully adsorbed utilizing biochar [64]. Dai et al. [25] also attested to using calcium-surplus biochar obtained from crab shell in removing dyes from wastewater, and Chen et al. [20] used modified biochar derived from *Enteromorpha prolifera* in the adsorption of chromium (Cr(VI)) from wastewater. Akech et al. [1] also employed biochar in adsorbing a hazardous chemical called tetrakis (hydroxymethyl) phosphonium chloride from water. Interestingly, biochar has also been utilized successfully in the adsorption of humic acids [77, 146], tannic acids and oestrogen [77], all from water, while Idrees et al. [58] used biochar derived from poultry manure and farmyard manure in the adsorption of manganese (Mn) from aqueous media. Similarly, bone-char was effective in the adsorption of As(V) within a pH of 2 and 5 in an aqueous solution [137]. Biochars from oak bark, pine bark, oak wood and pine wood were also applied in the removal of the toxic metals: arsenic,

Table 19.3 Pollutants adsorption by different biochars

SN	Biochar's feedstock	Pollutants being adsorbed	Application medium	References
1	Pine needle	Naphthalene, nitrobenzene, and m-dinitrobenzene	Water	Nartey and Zhao [18, 102]
2	Bamboo	Pentachlorophenol, sulfamethoxazole	Soil and water, respectively	Yao et al. [160]; Nartey and Zhao [102, 88]
3	Brazilian pepper wood	Sulfamethoxazole	Water	Yao et al. [160]; Nartey and Zhao [102]
4	Sugar bagasse	Sulfamethoxazole	Water	Yao et al. [160]; Nartey and Zhao [102]
5	Wheat straw	Hexachlorobenzene	Soil	De Wild et al. [28]; Nartey and Zhao [102]
6	Coconut coir	Chromium	Water	Shen et al. [134]; Patra et al. [112]
7	Corn straw	Copper and zinc	Water	Chen and Chen [17]; Patra et al. [112]
8	Alamo switch grass	Cadmium and copper	Water	Regmi et al. [117]; Patra et al. [112]
9	Pig manure	Cadmium, copper, lead and zinc	Water	Kołodyńska et al. [71]; Patra et al. [112]
10	Oak wood	Chromium VI	Water	Kołodyńska et al. [71]; Patra et al. [112]
11	Oak bark	Chromium VI	Water	Patra et al. [112]
12	Peanut straw	Copper	Water	Patra et al. [112]
13	Canola straw	Copper	Water	Patra et al. [112]
14	Rice straw	Aluminium	Water	Patra et al. [112]
15	Dairy manure	Lead, copper, zinc and cadmium	Water	Xu et al. [154]; Patra et al. [112]
16	Sugar beet tailings	Phosphate	Water	Yao et al. [159]; Patra et al. [112]
17	Sugarcane bagasse	Sulfamethoxazole	Water	Subhashini and Swamy [139]; Patra et al. [112]
18	Wood	Fluorinated herbicides	Water	Sun et al. [140]; Patra et al. [112]
19	Wood waste	Chromium (Cr (III)) and copper (cu (II))	Water	Zhang et al. [169]
20	Water caltrop shell	Chromium (Cr (III)) and copper (cu (II))	Water	Zhang et al. [169]
21	Cow manure	Tetracycline	Aqua-culture wastewater	Zhang et al. [168]
22	Crab shell	Dye	Wastewater	Dai et al. [25]
23	Rice husk	Lead, copper, zinc and cadmium	Water	Xu et al. [154]; Patra et al. [85, 112]

cadmium and lead (As3+, Cd2+, Pb2+) from water [97]. Similarly, biochars produced from wood bark, dairy manure, tailing of sugar beet, wood from pine and rice husk, at various process conditions, have been reportedly utilized in the removal of heavy metals such as arsenic (As), cadmium (Cd), chromium (Cr), mercury (Hg) and lead (Pb) from wastewater with positive results [115, 163]. Zhang et al. effectively utilized cow manure biochar in treating of aquaculture wastewater on farms to take out tetracycline [168]. In the adsorption of pollutants from water, biochar has been noted for the uptaking of 39% of pollutants that are organic, 46% of heavy metals and 13% of nitrogen, together with phosphorus, and 2% of others [31].

19.3.4 Biochar in Soil Remediation

The pollution of soils by a harmful substance (organic or inorganic) is unfavourable for crop cultivation and to human's health, and biochar has been used in lessening of pollutants present in the soil. Some of these pollutants are metals, including arsenic (As), chromium (Cr), copper (Cu), lead (Pb), and cadmium (Cd) [48, 53, 104, 169], which is a serious hazard to the health of humans and animals. However, previous research studies have reported success in using biochar as an efficient tool for remediating soils that is contaminated with pollutants [64, 150, 158]. Biochar in soils can function well as an adsorbent, assisting in averting nutrient loss and keeping soil water. Similarly, biochar-containing soils have been reported to possess a great affinity for organic pollutants, such as antibiotics and pesticides, which makes it effective in remediating pollutants available in soils [13, 14, 104, 147]. Similarly, biochar has been effectively used to adsorb polycyclic aromatic hydrocarbons (PAHs) [53], esters of phthalic acid [48, 53] and polychlorinated biphenyls (PCBs) [53], present in the soil.

19.3.5 Mechanism Governing Adsorption Using Biochar

Physisorption and chemisorption are the two major types of adsorption. While physisorption mechanism is mainly dependent on biochar's physical properties, chemisorption mechanism could be through chemically influenced precipitation, electrostatic interaction, ionic exchange and complexion with functional groups [14]. In overall, the physisorption and chemisorption mechanisms are dependent on pH value, amount of minerals, porosity, functional groups on the surface, chemical structure, active sites, particle size, stable molecular structure and specific surface area [6, 31, 102, 108], and a good understanding of how each mechanism affects the adsorption of the target adsorbate (pollutants) will assist in producing specific biochars for specific pollutants. There have been different observations on the principles and mechanisms governing the adsorption of pollutants onto biochar; some of them are mentioned here. Chen and Chen [17] observed that the adsorption mechanism of naphthalene and 1-naphthol in water was related to their physical properties (surface and structural). Furthermore Zheng et al. [170] noticed that the elimination of atrazine from a biochar-aqueous solution mixture rested on both the hydrophobic characteristic biochar, and on its particle size, with smaller particle size enhancing adsorption. Liu et al., likewise, observed that the up-taking of copper ions in an aqueous solution by biochar was governed by the mechanism of physisorption [86], while the adsorption of aromatic pollutants onto wood char's surface was reported to be a function of interactions of π-electrons [17, 173] and filling up of pores [17, 103]. Zhang et al. noted that the adsorption of Cr(III) and Cu(II) onto biochar gotten from wood and water caltrop shell was supported by the -COO- functional group on their surface [169]. Nie et al. also reported that the principle of tetracycline and Cu^{2+} ion adsorption and co-adsorption onto oxidized (using hydrogen peroxide H_2O_2) and non-oxidized bamboo biochar was generally based on the presence of -OH-functional group, π-electron presence, surface complexation, electrostatic attraction and presence of pores [104]. Dai et al. also, in their study, examined the mechanism of dye adsorption onto the surface of biochar derived from crab shell using zeta potential and fourier transform infrared (FTIR) spectra, and observed that electrostatic attraction, with hydrogen bonding and π–π interaction, was the driving force behind the adsorption process [25]. Furthermore, during tetracycline uptake from aquaculture wastewater using biochar made from cow manure, Zhang et al. noted that the mechanism of adsorption was most probably by pore filling, π–π EDA interactions, hydrogen bonding and electrostatic interactions between cow manure biochar and tetracycline. They added that via kinetics and isotherm study, cow manure biochar surface was heterogeneous, with the adsorption principle reported to be mainly via chemisorption [168]. From research studies on the mechanisms governing the adsorption potential of biochar, there hasn't been actually a universal mechanism that is generic to all pollutants; however, most of the factors aiding the mechanism are directly or indirectly related to biochar's physicochemical properties, hence, a need to understand the properties as it relates to each pollutant, and a meticulous approach in the production of biochar such that the chosen pyrolysis method and pyrolysis condition would optimize the required properties needed to promote specific biochar physicochemical properties for specific pollutant removal.

19.4 Conclusion

In this short overview, adsorption concept, types of adsorption and its applications were discussed. Furthermore, the recent shift from the use of commercial adsorbents to biochar was highlighted with much emphasis on the properties that make biochar desirable in adsorption processes. Areas where biochars have been successfully utilized as an adsorbent were mentioned. From this overview, it was noted that biochar's adsorption mechanism is majorly dependent on its physicochemical properties and can be improved if there is a proper understanding of the influence of physisorption and chemisorption mechanisms on target pollutants. Understanding these mechanisms will enhance the production of specially engineered biochars for specific pollutants adsorption.

References

1. S.R.O. Akech, O. Harrison, A. Saha, Removal of a potentially hazardous chemical, tetrakis (hydroxymethyl) phosphonium chloride from water using biochar as a medium of adsorption. Environ. Technol. Innov. **12**, 196–210 (2018). https://doi.org/10.1016/j.eti.2018.09.002
2. S. Al Arni, Comparison of slow and fast pyrolysis for converting biomass into fuel. Renew. Energy **124**, 197–201 (2018). https://doi.org/10.1016/j.renene.2017.04.060
3. D. Angin, Effect of pyrolysis temperature and heating rate on biochar obtained from pyrolysis of safflower seed press cake. Bioresour. Technol. **128**, 593–597 (2013). https://doi.org/10.1016/j.biortech.2012.10.150
4. E. Aydın, M. Şahin, E. Taşkan, et al., Chlortetracycline removal by using hydrogen based membrane biofilm reactor. J. Hazard. Mater. **320**, 88–95 (2016). https://doi.org/10.1016/j.jhazmat.2016.08.014
5. D. Ayhan, Effects of temperature and particle size on bio-char yield from pyrolysis of agricultural residues. J. Anal. Appl. Pyrolysis **72**, 243–248 (2004). https://doi.org/10.1016/J.JAAP.2004.07.003
6. A. B. Duwiejuah, S. J. Cobbina, N. Bakobie, Review of eco-friendly biochar used in the removal of trace metals on aqueous phases. Int. J. Environ. Bioremed. Biodegrad. **5**, 27–40 (2017). https://doi.org/10.12691/ijebb-5-2-1
7. M.-K. Bahng, C. Mukarakate, D.J. Robichaud, M.R. Nimlos, Current technologies for analysis of biomass thermochemical processing: a review. Anal. Chim. Acta **651**, 117–138 (2009). https://doi.org/10.1016/J.ACA.2009.08.016
8. L.C. Bornemann, R.S. Kookana, G. Welp, Differential sorption behaviour of aromatic hydrocarbons on charcoals prepared at different temperatures from grass and wood. Chemosphere **67**, 1033–1042 (2007). https://doi.org/10.1016/j.chemosphere.2006.10.052
9. A.V. Bridgwater, Renewable fuels and chemicals by thermal processing of biomass. Chem. Eng. J. **91**, 87–102 (2003). https://doi.org/10.1016/S1385-8947(02)00142-0
10. A.V. Bridgwater, Review of fast pyrolysis of biomass and product upgrading. Biomass Bioenergy **38**, 68–94 (2012). https://doi.org/10.1016/j.biombioe.2011.01.048
11. A. Buekens, N.N. Zyaykina, Adsorbents and adsorption processes for pollution control. J. Pollut. Control. Technol. **2**, 1–10 (2005)
12. M.Y. Can, E. Yildiz, Phosphate removal from water by fly ash: factorial experimental design. J. Hazard. Mater. **135**, 165–170 (2006). https://doi.org/10.1016/j.jhazmat.2005.11.036
13. K.B. Cantrell, P.G. Hunt, M. Uchimiya, et al., Impact of pyrolysis temperature and manure source on physicochemical characteristics of biochar. Bioresour. Technol. **107**, 419–428 (2012). https://doi.org/10.1016/j.biortech.2011.11.084
14. X. Cao, W. Harris, Properties of dairy-manure-derived biochar pertinent to its potential use in remediation. Bioresour. Technol. **101**, 5222–5228 (2010). https://doi.org/10.1016/J.BIORTECH.2010.02.052
15. X. Cao, L. Ma, B. Gao, W. Harris, Dairy-manure derived biochar effectively sorbs lead and atrazine. Environ. Sci. Technol. **43**, 3285–3291 (2009). https://doi.org/10.1021/es803092k
16. M. Carrier, T. Hugo, J. Gorgens, H. Knoetze, Comparison of slow and vacuum pyrolysis of sugar cane bagasse. J. Anal. Appl. Pyrolysis **90**, 18–26 (2011). https://doi.org/10.1016/j.jaap.2010.10.001
17. B. Chen, Z. Chen, Sorption of naphthalene and 1-naphthol by biochars of orange peels with different pyrolytic temperatures. Chemosphere **76**, 127–133 (2009). https://doi.org/10.1016/j.chemosphere.2009.02.004
18. B. Chen, D. Zhou, L. Zhu, Transitional adsorption and partition of nonpolar and polar aromatic contaminants by biochars of pine needles with different pyrolytic temperatures. Environ. Sci. Technol. **42**, 5137–5143 (2008). https://doi.org/10.1021/es8002684
19. J. Chen, Y.S. Liu, J.N. Zhang, et al., Removal of antibiotics from piggery wastewater by biological aerated filter system: treatment efficiency and biodegradation kinetics. Bioresour. Technol. **238**, 70–77 (2017). https://doi.org/10.1016/j.biortech.2017.04.023

20. Y. Chen, B. Wang, J. Xin, et al., Adsorption behavior and mechanism of Cr(VI) by modified biochar derived from Enteromorpha prolifera. Ecotoxicol. Environ. Saf. **164**, 440–447 (2018). https://doi.org/10.1016/j.ecoenv.2018.08.024
21. K. Christmann, *Practical Relevance of Adsorption Processes*, (2011) . http://www.fhi-berlin.mpg.de/acnew/department/pages/teaching/pages/teaching__wintersemester__2010_2011/klaus_christmann__adsorption__101105.pdf
22. Y. Chun, G. Sheng, G.T. Chiou, B. Xing, Compositions and sorptive properties of crop residue-derived chars. Environ. Sci. Technol. **38**, 4649–4655 (2004). https://doi.org/10.1021/es035034w
23. S. Czernik, A.V. Bridgwater, Overview of applications of biomass fast pyrolysis oil. Energy Fuels **18**, 590–598 (2004). https://doi.org/10.1021/ef034067u
24. A. Dąbrowski, Adsorption – from theory to practice. Adv. Colloid Interf. Sci. **93**, 135–224 (2001). https://doi.org/10.1016/S0001-8686(00)00082-8
25. L. Dai, W. Zhu, L. He, et al., Calcium-rich biochar from crab shell: an unexpected super adsorbent for dye removal. Bioresour. Technol. **267**, 510–516 (2018). https://doi.org/10.1016/j.biortech.2018.07.090
26. G. Davison, Sylobead, *Adsorbents for Process Applications* (Grace Davison, Columbia, 2010)
27. D. Day, R.J. Evans, J.W. Lee, D. Reicosky, Economical CO2, SOx, and NOx capture from fossil-fuel utilization with combined renewable hydrogen production and large-scale carbon sequestration. Energy **30**, 2558–2579 (2005). https://doi.org/10.1016/j.energy.2004.07.016
28. P.J. De Wild, W.J.J. Huijgen, H.J. Heeres, Pyrolysis of wheat straw-derived organosolv lignin. J. Anal. Appl. Pyrolysis **93**, 95–103 (2012). https://doi.org/10.1016/j.jaap.2011.10.002
29. A. Demirbas, Heavy metal adsorption onto agro-based waste materials: a review. J. Hazard. Mater. **157**, 220–229 (2008). https://doi.org/10.1016/j.jhazmat.2008.01.024
30. A. Demirbaş, G. Arin, An overview of biomass pyrolysis. Energy Sources **24**, 471–482 (2002). https://doi.org/10.1080/00908310252889979
31. Y. Deng, T. Zhang, Q. Wang, Biochar adsorption treatment for typical pollutants removal in livestock wastewater: a review, in *Engineering Applications of Biochar*, (Wu-Jang Huang, IntechOpen, London, 2017)
32. V. Dhyani, T. Bhaskar, A comprehensive review on the pyrolysis of lignocellulosic biomass. Renew. Energy **129**, 695–716 (2018). https://doi.org/10.1016/j.renene.2017.04.035
33. G.A. Dissanayake Herath, L.S. Poh, W.J. Ng, Statistical optimization of glyphosate adsorption by biochar and activated carbon with response surface methodology. Chemosphere **227**, 533–540 (2019). https://doi.org/10.1016/j.chemosphere.2019.04.078
34. S. Elkhalifa, T. Al-Ansari, H.R. Mackey, G. McKay, Food waste to biochars through pyrolysis: a review. Resour. Conserv. Recycl. **144**, 310–320 (2019). https://doi.org/10.1016/j.resconrec.2019.01.024
35. A.A. Fomkin, Nanoporous materials and their adsorption properties. Prot. Met. Phys. Chem. Surf. **45**, 121–136 (2009). https://doi.org/10.1134/s2070205109020014
36. A. Gabelman, Adsorption basics: Part 2. Chem. Eng. Prog. **113**, 38–45 (2017)
37. D. Gąsior, W.J. Tic, Biochar application in the mercury ions adsorption from aqueous solutions. Econ. Environ. Stud. **16**, 803–818 (2016)
38. S.M. Gawande, N.S. Belwalkar, A.A. Mane, Adsorption and its isotherm – theory. Int. J. Eng. Res. **6**, 312 (2017). https://doi.org/10.5958/2319-6890.2017.00026.5
39. S. Gent, M. Twedt, C. Gerometta, E. Almberg, Introduction to thermochemical conversion processes, in *Theoretical and Applied Aspects of Biomass Torrefaction: For Biofuels and Value-Added Products*, (Butterworth-Heinemann, Oxford, 2017), pp. 1–16
40. A. Genz, A. Kornmüller, M. Jekel, Advanced phosphorus removal from membrane filtrates by adsorption on activated aluminium oxide and granulated ferric hydroxide. Water Res. **38**, 3523–3530 (2004). https://doi.org/10.1016/j.watres.2004.06.006
41. D. Ghosh, D. Dasgupta, D. Agrawal, et al., Fuels and chemicals from lignocellulosic biomass: an integrated biorefinery approach. Energy Fuels **29**, 3149–3157 (2015). https://doi.org/10.1021/acs.energyfuels.5b00144

42. H.B. Goyal, D. Seal, R.C. Saxena, Bio-fuels from thermochemical conversion of renewable resources: a review. Renew. Sust. Energ. Rev. **12**, 504–517 (2008). https://doi.org/10.1016/j.rser.2006.07.014
43. V.K. Gupta, Suhas, Application of low-cost adsorbents for dye removal – a review. J. Environ. Manage. **90**, 2313–2342 (2009). https://doi.org/10.1016/j.jenvman.2008.11.017
44. V.K. Gupta, M. Gupta, S. Sharma, Process development for the removal of lead and chromium from aqueous solutions using red mud – an aluminium industry waste. Water Res. **35**, 1125–1134 (2001). https://doi.org/10.1016/S0043-1354(00)00389-4
45. V.K. Gupta, S. Agarwal, A.K. Bharti, H. Sadegh, Adsorption mechanism of functionalized multi-walled carbon nanotubes for advanced Cu (II) removal. J. Mol. Liq. **230**, 667–673 (2017). https://doi.org/10.1016/j.molliq.2017.01.083
46. P. Gvero, I. Mujanić, S. Prodanović, et al., Review of synthetic fuels and new materials production based on pyrolysis technologies, in *Advances in Applications of Industrial Biomaterials*, (Springer, Cham, 2017), pp. 65–85
47. Y. Hattori, K. Kaneko, T. Ohba, Adsorption properties, in *Comprehensive Inorganic Chemistry II (Second Edition): From Elements to Applications*, (Elsevier, Amsterdam, 2013), pp. 25–44
48. L. He, G. Gielen, N.S. Bolan, et al., Contamination and remediation of phthalic acid esters in agricultural soils in China: a review. Agron. Sustain. Dev. **35**, 519–534 (2015). https://doi.org/10.1007/s13593-014-0270-1
49. A. Hornung, A. Apfelbacher, S. Sagi, Intermediate pyrolysis: a sustainable biomass-to-energy concept-biothermal valorisation of biomass (BtVB) process. J. Sci. Ind. Res. (India) **70**, 664–667 (2011)
50. E. Hosseini Koupaie, S. Dahadha, A.A. Bazyar Lakeh, et al., Enzymatic pretreatment of lignocellulosic biomass for enhanced biomethane production-a review. J. Environ. Manag. **233**, 774–784 (2019). https://doi.org/10.1016/j.jenvman.2018.09.106
51. C.-H. Hsieh, W.-T. Tsai, S.-C. Liu, Preparation and fuel properties of biochars from the pyrolysis of exhausted coffee residue. J. Anal. Appl. Pyrolysis **93**, 63–67 (2012). https://doi.org/10.1016/j.jaap.2011.09.010
52. C. Huang, C. Huang, A.L. Morehart, The removal of Cu(II) from dilute aqueous solutions by Saccharomyces cerevisiae. Water Res. **24**, 433–439 (1990). https://doi.org/10.1016/0043-1354(90)90225-U
53. S. Huang, J. Bao, M. Shan, et al., Dynamic changes of polychlorinated biphenyls (PCBs) degradation and adsorption to biochar as affected by soil organic carbon content. Chemosphere **211**, 120–127 (2018a). https://doi.org/10.1016/j.chemosphere.2018.07.133
54. Y.F. Huang, P.T. Chiueh, W.H. Kuan, S.L. Lo, Product distribution and heating performance of lignocellulosic biomass pyrolysis using microwave heating. Energy Procedia **152**, 910–915 (2018b). https://doi.org/10.1016/j.egypro.2018.09.092
55. P. Hudec, Textúra tuhých materiálov. Charakterizá- cia adsorbentov a katalyzátorov fyzikálnou adsorpciou dusíka (Texture of solid materials. Characterisation of adsorbents and catalysts via physical nitrogen adsorption). Bratislava: STU. (in Slovak) (2012)
56. J. Hunt, M. Duponte, D. Sato, A. Kawabata, The basics of biochar: a natural soil amendment. Soil. Crop. Manag. **30**, 1–4 (2010)
57. E. Iakovleva, M. Sillanpää, The use of low-cost adsorbents for wastewater purification in mining industries. Environ. Sci. Pollut. Res. **20**, 7878–7899 (2013). https://doi.org/10.1007/s11356-013-1546-8
58. M. Idrees, S. Batool, H. Ullah, et al., Adsorption and thermodynamic mechanisms of manganese removal from aqueous media by biowaste-derived biochars. J. Mol. Liq. **266**, 373–380 (2018). https://doi.org/10.1016/j.molliq.2018.06.049
59. S. Iftekhar, D.L. Ramasamy, V. Srivastava, et al., Understanding the factors affecting the adsorption of Lanthanum using different adsorbents: a critical review. Chemosphere **204**, 413–430 (2018)

60. M. Inyang, B. Gao, P. Pullammanappallil, et al., Biochar from anaerobically digested sugarcane bagasse. Bioresour. Technol. **101**, 8868–8872 (2010). https://doi.org/10.1016/j.biortech.2010.06.088
61. W. Jan, Biochar application in the mercury ions adsorption from aqueous solutions. Econ. Environ. Stud. **16**, 803–818 (2016)
62. A. Jenkins, C. Jenkinson, Biochar basics, (2009). http://www.dpi.nsw.gov.au/research/topics/biochar. Accessed 28 May 2019
63. B. Jiang, Y. Lin, J.C. Mbog, Biochar derived from swine manure digestate and applied on the removals of heavy metals and antibiotics. Bioresour. Technol. **270**, 603–611 (2018). https://doi.org/10.1016/j.biortech.2018.08.022
64. F. Jing, S.P. Sohi, Y. Liu, J. Chen, Insight into mechanism of aged biochar for adsorption of PAEs: reciprocal effects of ageing and coexisting Cd2+. Environ. Pollut. **242**, 1098–1107 (2018). https://doi.org/10.1016/j.envpol.2018.07.124
65. P. Kaewsarn, Q. Yu, Cadmium(II) removal from aqueous solutions by pre-treated biomass of marine alga Padina sp. Environ. Pollut. **112**, 209–213 (2001). https://doi.org/10.1016/S0269-7491(00)00114-7
66. B.L. Karger, L.R. Snyder, C. Horvath, *An Introduction to Separation Science* (Wiley, New York, 1973)
67. K. Kebelmann, A. Hornung, U. Karsten, G. Griffiths, Intermediate pyrolysis and product identification by TGA and Py-GC/MS of green microalgae and their extracted protein and lipid components. Biomass Bioenergy **49**, 38–48 (2013). https://doi.org/10.1016/j.biombioe.2012.12.006
68. G.E. Keller, Industrial gas separation. ACS Symp. Ser **223**, 195 (1983)
69. A. Khosravanipour Mostafazadeh, O. Solomatnikova, P. Drogui, R.D. Tyagi, A review of recent research and developments in fast pyrolysis and bio-oil upgrading. Biomass Convers Bior. **8**, 739–773 (2018). https://doi.org/10.1007/s13399-018-0320-z
70. C.J. King, *Separation Processes*, 2nd edn. (Courier Corporation, New York, 2013)
71. D. Kołodyńska, R. Wnetrzak, J.J. Leahy, et al., Kinetic and adsorptive characterization of biochar in metal ions removal. Chem. Eng. J. **197**, 295–305 (2012). https://doi.org/10.1016/j.cej.2012.05.025
72. M. Kralık, Adsorption, chemisorption, and catalysis. Chem. Pap. **68**, 1625–1638 (2014)
73. G.Z. Kyzas, J. Fu, K.A. Matis, The change from past to future for adsorbent materials in treatment of dyeing wastewaters. Materials (Basel) **6**, 5131–5158 (2013). https://doi.org/10.3390/ma6115131
74. Y.F. Lam, L.Y. Lee, S.J. Chua, et al., Insights into the equilibrium, kinetic and thermodynamics of nickel removal by environmental friendly *Lansium domesticum* peel biosorbent. Ecotoxicol. Environ. Saf. **127**, 61–70 (2016). https://doi.org/10.1016/j.ecoenv.2016.01.003
75. J. Laufer, T. Tomlinson, Biochar field studies: an IBI research summary. Soil Sci. Plant. Nutr. **61**, 873–884 (2013)
76. X.J. Lee, L.Y. Lee, B.Y.Z. Hiew, et al., Evaluation of the effectiveness of low cost adsorbents from oil palm wastes for wastewater treatment. Chem. Eng. Trans. **56**, 937–942 (2017). https://doi.org/10.3303/CET1756157
77. D.-J. Lee, Y.-L. Cheng, R.-J. Wong, X.-D. Wang, Adsorption removal of natural organic matters in waters using biochar. Bioresour. Technol. **260**(201807), 413–416 (2018). https://doi.org/10.1016/j.biortech.2018.04.016
78. J. Lehmann, S. Joseph, *Biochar for Environmental Management: Science and Technology* (Earthscan, London, 2009)
79. N. Le-Minh, S.J. Khan, J.E. Drewes, R.M. Stuetz, Fate of antibiotics during municipal water recycling treatment processes. Water Res. **44**, 4295–4323 (2010). https://doi.org/10.1016/j.watres.2010.06.020
80. L. Li, J.S. Rowbotham, C.H. Greenwell, P.W. Dyer, An introduction to pyrolysis and catalytic pyrolysis: versatile techniques for biomass conversion, in *New and Future Developments in Catalysis: Catalytic Biomass Conversion*, (Elsevier, Amsterdam, 2013), pp. 173–208

81. X. Li, R. Xu, J. Yang, et al., Production of 5-hydroxymethylfurfural and levulinic acid from lignocellulosic biomass and catalytic upgradation. Ind. Crop. Prod. **130**, 184–197 (2019). https://doi.org/10.1016/j.indcrop.2018.12.082
82. S. Liang, H. Lin, X. Yan, Q. Huang, Electro-oxidation of tetracycline by a Magnéli phase Ti 4 O 7 porous anode: kinetics, products, and toxicity. Chem. Eng. J. **332**, 628–636 (2018). https://doi.org/10.1016/j.cej.2017.09.109
83. T. Liao, T. Li, X. Su, et al., La(OH) 3 -modified magnetic pineapple biochar as novel adsorbents for efficient phosphate removal. Bioresour. Technol. **263**, 207–213 (2018). https://doi.org/10.1016/j.biortech.2018.04.108
84. X. Liu, X.T. Bi, Removal of inorganic constituents from pine barks and switchgrass. Fuel Process. Technol. **92**, 1273–1279 (2011). https://doi.org/10.1016/j.fuproc.2011.01.016
85. Z. Liu, F.S. Zhang, Removal of lead from water using biochars prepared from hydrothermal liquefaction of biomass. J. Hazard. Mater. **167**, 933–939 (2009). https://doi.org/10.1016/j.jhazmat.2009.01.085
86. Z. Liu, F.S. Zhang, J. Wu, Characterization and application of chars produced from pinewood pyrolysis and hydrothermal treatment. Fuel **89**, 510–514 (2010). https://doi.org/10.1016/j.fuel.2009.08.042
87. Z. Liu, Y. Li, F. Yu, et al., Co-pyrolysis of oil sand bitumen with lignocellulosic biomass under hydrothermal conditions. Chem. Eng. Sci. **199**, 417–425 (2019). https://doi.org/10.1016/j.ces.2018.12.052
88. L. Lou, L. Luo, Q Yang, et al., Release of pentachlorophenol from black carbon-inclusive sediments under different environmental conditions. Chemosphere. **88**(5), 598–604 (2012).
89. A.C. Lua, T. Yang, J. Guo, Effects of pyrolysis conditions on the properties of activated carbons prepared from pistachio-nut shells. J. Anal. Appl. Pyrolysis **72**, 279–287 (2004). https://doi.org/10.1016/j.jaap.2004.08.001
90. N. Lv, X. Wang, S. Peng, et al., Study of the kinetics and equilibrium of the adsorption of oils onto hydrophobic jute fiber modified via the sol-gel method. Int. J. Environ. Res. Public Health **15**, 969 (2018). https://doi.org/10.3390/ijerph15050969
91. H. Madzaki, W.A.W.A.B. Karimghani, Nurzalikharebitanim, Azilbaharialias, Carbon dioxide adsorption on sawdust biochar. Procedia Eng **148**, 718–725 (2016). https://doi.org/10.1016/j.proeng.2016.06.591
92. A.S.N. Mahmood, J.G. Brammer, A. Hornung, et al., The intermediate pyrolysis and catalytic steam reforming of Brewers spent grain. J. Anal. Appl. Pyrolysis **103**, 328–342 (2013). https://doi.org/10.1016/j.jaap.2012.09.009
93. A.H. Malik, Z.M. Khan, Q. Mahmood, et al., Perspectives of low cost arsenic remediation of drinking water in Pakistan and other countries. J. Hazard. Mater. **168**, 1–12 (2009). https://doi.org/10.1016/j.jhazmat.2009.02.031
94. S.O. Masebinu, E.T. Akinlabi, E. Muzenda, A.O. Aboyade, A review of biochar properties and their roles in mitigating challenges with anaerobic digestion. Renew. Sust. Energ. Rev. **103**, 291–307 (2019). https://doi.org/10.1016/j.rser.2018.12.048
95. G. McKay, *Use of Adsorbents for the Removal of Pollutants from Wastewater* (CRC Press, Inc., Hong Kong, 1995)
96. D. Mohan, C.U. Pittman, P.H. Steele, Pyrolysis of wood/biomass for bio-oil: a critical review. Energy and Fuels **20**, 848–889 (2006)
97. D. Mohan, C.U. Pittman, M. Bricka, et al., Sorption of arsenic, cadmium, and lead by chars produced from fast pyrolysis of wood and bark during bio-oil production. J. Colloid Interface Sci. **310**, 57–73 (2007). https://doi.org/10.1016/j.jcis.2007.01.020
98. Motoyuki Suzuki, Application of adsorption technology for environmental control, in *Fundamentals of Adsorption*, ed. by M. D. LeVan, (Kluwer Academic Publishers, Boston, 1996). https://www.theseus.fi/bitstream/handle/10024/69655/Musin_Eduard.pdf?sequence=1&isAllowed=y
99. E. Musin, *Adsorption Modeling* (Mikkeli University of Applied Sciences, Helsinki, 2013a)
100. E. Musin, *Eduard Musin (A9838) T879SN Adsorption Modeling Environmental Engineering Adsorption Modelling*, (2013b)

101. S. Nanda, R. Azargohar, A.K. Dalai, J.A. Kozinski, An assessment on the sustainability of lignocellulosic biomass for biorefining. Renew. Sust. Energ. Rev. **50**, 925–941 (2015). https://doi.org/10.1016/j.rser.2015.05.058
102. O.D. Nartey, B. Zhao, Biochar preparation, characterization, and adsorptive capacity and its effect on bioavailability of contaminants: an overview. Adv. Mater. Sci. Eng. **2014**, 1–12 (2014). https://doi.org/10.1155/2014/715398
103. T.H. Nguyen, H.H. Cho, D.L. Poster, W.P. Ball, Evidence for a pore-filling mechanism in the adsorption of aromatic hydrocarbons to a natural wood char. Environ. Sci. Technol. **41**, 1212–1217 (2007). https://doi.org/10.1021/es0617845
104. T. Nie, P. Hao, Z. Zhao, et al., Effect of oxidation-induced aging on the adsorption and co-adsorption of tetracycline and Cu2+ onto biochar. Sci. Total Environ. **673**, 522–532 (2019). https://doi.org/10.1016/j.scitotenv.2019.04.089
105. M.L. Nieva, M.A. Volpe, E.L. Moyano, Catalytic and catalytic free process for cellulose conversion: fast pyrolysis and microwave induced pyrolysis studies. Cellulose **22**, 215–228 (2015). https://doi.org/10.1007/s10570-014-0484-z
106. NPTEL, Module 8: surface chemistry, (2018). https://nptel.ac.in/courses/122101001/downloads/lec-36.pdf. Accessed 14 Sep 2018
107. H.K. Nsamba, S.E. Hale, G. Cornelissen, R.T. Bachmann, Sustainable technologies for small-scale biochar production—a review. J. Sustain. Bioenergy Syst **05**, 10–31 (2015). https://doi.org/10.4236/jsbs.2015.51002
108. Y.B. Onundi, A.A. Mamun, M.F. Al Khatib, Y.M. Ahmed, Adsorption of copper, nickel and lead ions from synthetic semiconductor industrial wastewater by palm shell activated carbon. Int. J. Environ. Sci. Technol. **7**, 751–758 (2010). https://doi.org/10.1007/BF03326184
109. G. Ordorica-Garcia, A. Elkamel, P. Douglas, et al., Optimizing energy production with integrated CCS technology for CO2 emissions mitigation in the Canadian oil sands industry. Energy Procedia **1**, 3985–3992 (2009). https://doi.org/10.1016/j.egypro.2009.02.203
110. R.H. Ottawil, C.H. Rochester, ALSE, *Adsorption from Solution* (Academic Press, Ž, London, 1983)
111. Parfitt-Rochester, *Adsorption from Solution at the Solid/Liquid* (Academic Press, New York, 1983)
112. J.M. Patra, S.S. Panda, N.K. Dhal, Biochar as a low-cost adsorbent for heavy metal removal: a review. Int. J. Res. Biosci. **6**, 1–7 (2017)
113. H. Persson, W. Yang, Catalytic pyrolysis of demineralized lignocellulosic biomass. Fuel **252**, 200–209 (2019). https://doi.org/10.1016/j.fuel.2019.04.087
114. S. Pourkarimi, A. Hallajisani, A. Alizadehdakhel, A. Nouralishahi, Biofuel production through micro- and macroalgae pyrolysis – a review of pyrolysis methods and process parameters. J. Anal. Appl. Pyrolysis **1–19**, 104599 (2019). https://doi.org/10.1016/j.jaap.2019.04.015
115. K. Qian, A. Kumar, H. Zhang, et al., Recent advances in utilization of biochar. Renew. Sust. Energ. Rev. **42**, 1055–1064 (2015)
116. T. Qu, W. Guo, L. Shen, et al., Experimental study of biomass pyrolysis based on three major components: hemicellulose, cellulose, and lignin. Ind. Eng. Chem. Res. **50**, 10424–10433 (2011). https://doi.org/10.1021/ie1025453
117. P. Regmi, J.L. Garcia Moscoso, S. Kumar, et al., Removal of copper and cadmium from aqueous solution using switchgrass biochar produced via hydrothermal carbonization process. J. Environ. Manag. **109**, 61–69 (2012). https://doi.org/10.1016/j.jenvman.2012.04.047
118. J.R. Reimers, M. Li, D. Wan, et al., Surface adsorption, in *Non-Covalent Interactions in Quantum Chemistry and Physics*, 1st edn., (Elsevier Inc, Cambridge, 2017), pp. 387–416
119. T. C. G. Resources, *The Industrial Adsorbents Business: Commercial Strategy, Technical and R&D Assessment in Refining Chemicals/Syngas*, Natural Gas and Industrial Gases, (2013)
120. S.H. Russell, J.L. Turrion-Gomez, W. Meredith, et al., Increased charcoal yield and production of lighter oils from the slow pyrolysis of biomass. J. Anal. Appl. Pyrolysis **124**, 536–541 (2017). https://doi.org/10.1016/j.jaap.2017.01.028
121. D.M. Ruthven, *Principles of Adsorption and Adsorption Processes* (Wiley, New York, 1984)

122. B.E. Saad, Chapter 3 Principles of adsorption, (2016). http://shodhganga.inflibnet.ac.in/bitstream/10603/22665/8/08_chapter3.pdf. Accessed 14 Sep 2018
123. H. Sadegh, Development of graphene oxide from graphite: a review on synthesis, characterization and its application in wastewater treatment. Rev. Adv. Mater. Sci. **49**, 38–43 (2017)
124. H. Sadegh, K. Zare, B. Maazinejad, et al., Synthesis of MWCNT-COOH-Cysteamine composite and its application for dye removal. J. Mol. Liq. **215**, 221–228 (2016). https://doi.org/10.1016/j.molliq.2015.12.042
125. H. Sadegh, G.A.M. Ali, Z. Abbasi, M.N. Nadagouda, Adsorption of ammonium ions onto multi-walled carbon nanotubes. Stud. Univ. Babes-Bolyai Chem. **62**, 233–245 (2017a). https://doi.org/10.24193/subbchem.2017.2.18
126. H. Sadegh, M. Mazloumbilandi, M. Chahardouri, Low-cost materials with adsorption performance, in *Handbook of Ecomaterials*, (Springer International Publishing, Cham, 2017b), pp. 1–33
127. S. Sadri, B.B. Johnson, M. Ruyter-Hooley, M.J. Angove, The adsorption of nortriptyline on montmorillonite, kaolinite and gibbsite. Appl. Clay Sci. **165**, 64–70 (2018). https://doi.org/10.1016/j.clay.2018.08.005
128. P. Salatino, R. Solimene, Mixing and segregation in fluidized bed thermochemical conversion of biomass. Powder Technol. **316**, 29–40 (2017). https://doi.org/10.1016/j.powtec.2016.11.058
129. M.E. Saleh, A.A. El-Refaey, A.H. Mahmoud, Effectiveness of sunflower seed husk biochar for removing copper ions from wastewater: a comparative study. Soil Water Res. **11**, 53–63 (2016). https://doi.org/10.17221/274/2014-SWR
130. A. Sanchez, P. Hernández-Sánchez, R. Puente, Hydration of lignocellulosic biomass. Modelling and experimental validation. Ind. Crop. Prod. **131**, 70–77 (2019). https://doi.org/10.1016/j.indcrop.2019.01.029
131. C. Sawatdeenarunat, K.C. Surendra, D. Takara, et al., Anaerobic digestion of lignocellulosic biomass: challenges and opportunities. Bioresour. Technol. **178**, 178–186 (2015). https://doi.org/10.1016/j.biortech.2014.09.103
132. M. Schroyen, H. Vervaeren, H. Vandepitte, et al., Effect of enzymatic pretreatment of various lignocellulosic substrates on production of phenolic compounds and biomethane potential. Bioresour. Technol. **192**, 696–702 (2015). https://doi.org/10.1016/j.biortech.2015.06.051
133. N. Shabani, S. Akhtari, T. Sowlati, Value chain optimization of forest biomass for bioenergy production: a review. Renew. Sust. Energ. Rev. **23**, 299–311 (2013). https://doi.org/10.1016/j.rser.2013.03.005
134. Y.S. Shen, S.L. Wang, Y.M. Tzou, et al., Removal of hexavalent Cr by coconut coir and derived chars – the effect of surface functionality. Bioresour. Technol. **104**, 165–172 (2012). https://doi.org/10.1016/j.biortech.2011.10.096
135. K.S.W. Sing, D.H. Everett, R.A.W. Haul, et al., Reporting physisorption data for gas/solid systems with special reference to the determination of surface area and porosity. Pure Appl. Chem. **57**, 603–619 (1985). https://doi.org/10.1351/pac198557040603.
136. G. Singh, D. Rana, T. Matsuura, et al., Removal of disinfection byproducts from water by carbonized electrospun nanofibrous membranes. Sep. Purif. Technol. **74**, 202–212 (2010). https://doi.org/10.1016/j.seppur.2010.06.006
137. I.R. Sneddon, H. Garelick, E. Valsami-Jones, An investigation into arsenic(V) removal from aqueous solutions by hydroxylapatite and bone-char. Mineral. Mag. **69**, 769–780 (2005). https://doi.org/10.1180/0026461056950287
138. F.S. Souza, V.V. da Silva, C.K. Rosin, et al., Comparison of different advanced oxidation processes for the removal of amoxicillin in aqueous solution. Environ. Technol. (United Kingdom) **39**, 549–557 (2018). https://doi.org/10.1080/09593330.2017.1306116
139. V. Subhashini, A.V.V.S. Swamy, Phytoremediation of Pb and Ni contaminated soils using *Catharanthus roseus* (L.). Univers J Environ Res Technol **3**, 465–472 (2013)
140. K. Sun, M. Keiluweit, M. Kleber, et al., Sorption of fluorinated herbicides to plant biomass-derived biochars as a function of molecular structure. Bioresour. Technol. **102**, 9897–9903 (2011). https://doi.org/10.1016/j.biortech.2011.08.036

141. Surface Chemistry Adsorption (2019), https://www.toppr.com/guides/chemistry/surface-chemistry/adsorption/. Accessed 8 Jun 2019
142. Y. Suzuki, T. Kametani, T. Maruyama, Removal of heavy metals from aqueous solution by nonliving Ulva seaweed as biosorbent. Water Res. **39**, 1803–1808 (2005). https://doi.org/10.1016/j.watres.2005.02.020
143. R. Tareq, N. Akter, M.S. Azam, Biochars and biochar composites, in *Biochar from Biomass and Waste*, (Elsevier, Amsterdam, 2018), pp. 169–209
144. W.J. Thomas, B. Crittenden, *Adsorption Technology and Design* (Butterworth-Heinemann, Oxford, 1998)
145. M. Tripathi, J.N. Sahu, P. Ganesan, Effect of process parameters on production of biochar from biomass waste through pyrolysis: a review. Renew. Sust. Energ. Rev. **55**, 467–481 (2016). https://doi.org/10.1016/j.rser.2015.10.122
146. J.S. Tumuluru, C.T. Wright, J.R. Hess, K.L. Kenney, A review of biomass densification systems to develop uniform feedstock commodities for bioenergy application. Biofuels Bioprod. Biorefin. **5**, 683–707 (2011)
147. M. Uchimiya, I.M. Lima, K. Thomas Klasson, et al., Immobilization of heavy metal ions (Cu^{II}, Cd^{II}, Ni^{II}, and Pb^{II}) by broiler litter-derived biochars in water and soil. J. Agric. Food Chem. **58**, 5538–5544 (2010). https://doi.org/10.1021/jf9044217
148. G. van de Kaa, L. Kamp, J. Rezaei, Selection of biomass thermochemical conversion technology in the Netherlands: a best worst method approach. J. Clean. Prod. **166**, 32–39 (2017). https://doi.org/10.1016/j.jclepro.2017.07.052
149. Vera Bolis, Fundamentals in adsorption at the solid-gas interface. Concepts and thermodynamics, in *Calorimetry and Thermal Methods in Catalysis*, Springer Series in Materials Science, ed. by A. Auroux, (Springer, Berlin, Heidelberg, 2013)
150. F. Wang, H. Sun, X. Ren, et al., Effects of humic acid and heavy metals on the sorption of polar and apolar organic pollutants onto biochars. Environ. Pollut. **231**, 229–236 (2017). https://doi.org/10.1016/j.envpol.2017.08.023
151. H. Wang, X. Wang, Y. Cui, et al., Slow pyrolysis polygeneration of bamboo (*Phyllostachys pubescens*): product yield prediction and biochar formation mechanism. Bioresour. Technol. **263**, 444–449 (2018a). https://doi.org/10.1016/j.biortech.2018.05.040
152. T. Wang, J. Wu, Y. Zhang, et al., Increasing the chlorine active sites in the micropores of biochar for improved mercury adsorption. Fuel **229**, 60–67 (2018b). https://doi.org/10.1016/j.fuel.2018.05.028
153. P.A. Webb, *Introduction to Chemical Adsorption Analytical Techniques and Their Applications to Catalysis* (MIC Technical Publications, Norcross, 2003)
154. X. Xu, X. Cao, L. Zhao, Comparison of rice husk- and dairy manure-derived biochars for simultaneously removing heavy metals from aqueous solutions: role of mineral components in biochars. Chemosphere **92**, 955–961 (2013). https://doi.org/10.1016/j.chemosphere.2013.03.009
155. R.T. Yang, *Gas Separation by Adsorption Processes*, 1st edn. (Butterworth-Heinemann, Oxford, 1987)
156. R.T. Yang, *Adsorbents: Fundamentals and Applications* (Wiley, Hoboken, 2003)
157. H. Yang, R. Yan, H. Chen, et al., Characteristics of hemicellulose, cellulose and lignin pyrolysis. Fuel **86**, 1781–1788 (2007). https://doi.org/10.1016/j.fuel.2006.12.013
158. K. Yang, J. Yang, Y. Jiang, et al., Correlations and adsorption mechanisms of aromatic compounds on a high heat temperature treated bamboo biochar. Environ. Pollut. **210**, 57–64 (2016). https://doi.org/10.1016/j.envpol.2015.12.004
159. Y. Yao, B. Gao, M. Inyang, et al., Biochar derived from anaerobically digested sugar beet tailings: characterization and phosphate removal potential. Bioresour. Technol. **102**, 6273–6278 (2011). https://doi.org/10.1016/j.biortech.2011.03.006
160. Y. Yao, B. Gao, H. Chen, et al., Adsorption of sulfamethoxazole on biochar and its impact on reclaimed water irrigation. J. Hazard. Mater. **209–210**, 408–413 (2012). https://doi.org/10.1016/j.jhazmat.2012.01.046

161. S. Ye, G. Zeng, H. Wu, et al., Biological technologies for the remediation of co-contaminated soil. Crit. Rev. Biotechnol. **37**, 1062–1076 (2017). https://doi.org/10.1080/07388551.2017.1304357
162. Q. Yin, H. Ren, R. Wang, Z. Zhao, Evaluation of nitrate and phosphate adsorption on Al-modified biochar: influence of Al content. Sci. Total Environ. **631–632**, 895–903 (2018). https://doi.org/10.1016/j.scitotenv.2018.03.091
163. K.L. Yu, B.F. Lau, P.L. Show, et al., Recent developments on algal biochar production and characterization. Bioresour. Technol. **246**, 2–11 (2017). https://doi.org/10.1016/j.biortech.2017.08.009
164. A.C. Zettlemoyer, F.J. Micale, Adsorption from solution. Croat. Chem. Acta **42**, 247–263 (1970)
165. L. Zhang, C.C. Xu, P. Champagne, Overview of recent advances in thermo-chemical conversion of biomass. Energy Convers. Manag. **51**, 969–982 (2010). https://doi.org/10.1016/j.enconman.2009.11.038
166. S. Zhang, X. Yang, L. Liu, et al., Adsorption behavior of selective recognition functionalized biochar to Cd(II) in wastewater. Materials (Basel) **11**, 1–12 (2018). https://doi.org/10.3390/ma11020299
167. N. Zhang, L. Wang, K. Zhang, et al., Pretreatment of lignocellulosic biomass using bioleaching to reduce inorganic elements. Fuel **246**, 386–393 (2019a). https://doi.org/10.1016/j.fuel.2019.02.138
168. P. Zhang, Y. Li, Y. Cao, L. Han, Characteristics of tetracycline adsorption by cow manure biochar prepared at different pyrolysis temperatures. Bioresour. Technol. **285**, 121348 (2019b). https://doi.org/10.1016/j.biortech.2019.121348
169. Y.P. Zhang, V.S.K. Adi, H.L. Huang, et al., Adsorption of metal ions with biochars derived from biomass wastes in a fixed column: adsorption isotherm and process simulation. J. Ind. Eng. Chem. **76**, 240–244 (2019c). https://doi.org/10.1016/j.jiec.2019.03.046
170. W. Zheng, M. Guo, T. Chow, et al., Sorption properties of greenwaste biochar for two triazine pesticides. J. Hazard. Mater. **181**, 121–126 (2010). https://doi.org/10.1016/j.jhazmat.2010.04.103
171. Y. Zhou, B. Gao, A.R. Zimmerman, et al., Sorption of heavy metals on chitosan-modified biochars and its biological effects. Chem. Eng. J. **231**, 512–518 (2013). https://doi.org/10.1016/j.cej.2013.07.036
172. Y. Zhou, X. Liu, Y. Xiang, et al., Modification of biochar derived from sawdust and its application in removal of tetracycline and copper from aqueous solution: adsorption mechanism and modelling. Bioresour. Technol. **245**, 266–273 (2017). https://doi.org/10.1016/j.biortech.2017.08.178
173. D. Zhu, S. Kwon, J.J. Pignatello, Adsorption of single-ring organic compounds to wood charcoals prepared under different thermochemical conditions. Environ. Sci. Technol. **39**, 3990–3998 (2005). https://doi.org/10.1021/es050129e

Chapter 20
Development of Plastic Composite Using Waste Sawdust, Rice Husk and Bamboo in the Polystyrene-Based Resin (PBR) Matrix at Ambient Conditions

S. A. Abdulkareem, A. G. Adeniyi, M. K. Amosa, and S. A. Raji

20.1 Introduction

Plastic composites are relatively new engineering materials as compared to the long history of natural lumber or long-established wood composites like particleboard or fibreboard [32]. They come in various combinations of biomass fibres, thermoplastic resin and chemical additives based on the target modification and end users. The pre-treated biomass composition, usually in the form of flour/particles/fibres/sawdust, is usually mixed with thermoplastic materials under tailored combinations of heat and pressure to produce plastic composites. Additives are added for improved quality when required. Many researchers had worked on plastic composites. The commonest reported approach is flat-pressed method at various biomass-plastic ratios. The usual ranges of biomass fibre composition were put between 20% and 80% either as fillers or reinforcements [23, 27, 28]. Apart from the flat-pressed approach, the two principal techniques in the production of plastic composites, most especially at commercial quantities, are extrusion and injection moulding [24]. The extrusion practice gives continuous linear profiles by forcing a melted thermoplastic through a die, while the injection moulding process produces three-dimensional items with minimised stages of post-manufacturing.

S. A. Abdulkareem · A. G. Adeniyi (✉)
Chemical Engineering Department, Faculty of Engineering and Technology, University of Ilorin, Ilorin, Nigeria
e-mail: adeniyi.ag@unilorin.edu.ng

M. K. Amosa
Environmental Engineering and Management Research Group, Faculty of Environment and Labour Safety, Ton Duc Thang University, Ho Chi, Minh City, Vietnam

S. A. Raji
Civil Engineering Department, Faculty of Engineering and Technology, University of Ilorin, Ilorin, Nigeria

M. O. Daramola, A. O. Ayeni (eds.), *Valorization of Biomass to Value-Added Commodities*, Green Energy and Technology,
https://doi.org/10.1007/978-3-030-38032-8_20

In the literature, a number of investigations have been carried out on several types of biomass, such as kenaf [7], rice husk [5], coconut fibre [31], banana fibre [23], hemp [11], flax [37], sisal [17], bamboo [2], jute [26] and wood dust [4] to study the effect of these fibres on mechanical properties of composite materials. Singh et al. [35] studied the effect of surface treatment on mechanical properties of natural fibre-reinforced composites. In the study, different natural fibres like jute, banana and sisal were used as reinforcement and epoxy as matrix to make the natural fibre-reinforced polymer unidirectional composite with the help of hand lay-up and compression moulding. Mechanical properties, such as tensile strength, flexural strength and impact strength, for treated and non-treated natural fibre-reinforced polymer unidirectional composites were investigated and compared. It was found that tensile strength of composites improved due to incorporation of natural fibres to polymers. A considerable increase in tensile and flexural strength was observed with the use of surface treatment fillers. The study noted that jute-reinforced polymer composite showed the highest tensile strength. Sathiyamoorthy et al. [33] studied the mechanical behaviour of fibre-reinforced polymer composite at varying composition (25%, 30% and 35%) with silicon carbide at 4%, 8% and 12%, respectively. Tests, such as the compressive test, hardness test and the bending test, are carried out on the composite material. It was found that better mechanical properties were obtained for composites having 35% of fibre.

Arpitha and Yogesha [8] presented an overview of the mechanical properties, such as tensile, flexural, impact, fracture surface observations and corresponding modulus of elasticity of natural fibre-reinforced plastic composites. In the review, the authors noted that natural fibre may play an important role in developing biodegradable composites to resolve the current ecological and environmental problems. The overview showed that natural fibres possessed good mechanical properties which can be explored in the production of polymer composites with varying applications. Balaji and Jayabal [9] studied the modelling of mechanical behaviours of Zea fibre–polyester composites using artificial neural networks (ANN). The tensile, flexural and impact behaviours of Zea fibre-reinforced polyester composites were evaluated. Different proportion of fibre length, fibre content in weight percentage and moulding pressure were used as process variables in the fabrication of Zea–polyester composites. An artificial neural network algorithm was developed to predict the mechanical behaviour over the specified range of conditions, and the average absolute percentage error of less than 4% was observed for tensile, flexural and impact models. Better tensile, flexural and impact behaviour of 29.50 MPa, 39.55 MPa and 50.75 kJ/m^2, respectively, were obtained for the Zea–polyester composites.

Hemachandran et al. [17] studied the optimisation of tensile and impact behaviour of randomly oriented short sisal fibre-reinforced epoxy composites using response surface methodology. In the study, sisal-epoxy composites were fabricated with varying fibre length (of 10–75 mm) and fibre loading (of 10–50%) by weight as per response surface methodology and a user-defined design. Better value of tensile and impact behaviours was determined in non-woven randomly oriented sisal fibre-reinforced epoxy composites using response surface optimisation. Better values of tensile strength of 31.23 MPa and impact strength of 43.22 kJ/m^2 were obtained for the fibre length of 47.9 mm and fibre loading of 29.1% by weight. Abass [1] presented an experimental study which aimed at understanding the mechanical behaviour of orange peel (OP)-reinforced polyester composites. Composites having 2%, 4%, 6%, 8% and 10% weight fraction of orange peel were made using hand lay-up method. The fabricated composite samples were cut according to the ASTM standards for different experiments. Hardness test, impact test, compression test and tensile test were carried out on the samples. Maximum hardness, impact, compression and tensile were obtained for material prepared with the 10% reinforced OP polyester composite.

Of all the numbers of literatures documented on the production of plastic composites, none has documented the production of biomass-reinforced polystyrene composite using cold process contact moulding method. Hence, this chapter enumerates the study of mechanical properties of plastic composites developed from polystyrene and biomass wastes in an entire cold processing to form the basis of new materials and processes of interest for future applications. Studies on the development of biomass/polystyrene composites and their mechanical properties in non-thermal applications are limited. The studies discussed in this chapter demonstrate that waste polystyrene and biomass can be successfully used to fabricate polymer composites with physical properties. This capability combined with the ability of solvated polystyrene to fabricate thermoplastic composite without recourse to heat application could open up several new opportunities. However, in this chapter, the solvolysis of PS into polystyrene-based resin (PBR) was attempted following the report of Abdulkareem and Adeniyi [2]. It is a new approach of reducing large amount of PS waste into a small resinous volume of good adhesive and strong binding capacity. The synthesised PBR has been successfully used as binder in the production of particleboard composites [4] and metal-plastic composites [3] for conductive applications. In the present documentation, its potency in plastic composites development is explored using different biomass fillers from waste streams. These waste biomass particulates (sawdust, bamboo and rice husk) from the solid waste streams were used as fillers and PBR from recycled PS was employed as matrix to produce conductive plastic composites of varying particulates contents (10%, 20%, 30% and 40%) at room temperature. Mechanical properties of the emerged composites were investigated and compared.

20.2 Theoretical Review

20.2.1 *Composites*

When more than one materials of different properties and orientations are combined, they produce a composite [1]. Common examples are: a polymer resin versus lignocellulosic, sand/clay or metal particles combinations [13]. In the recent times, natural fibre-reinforced plastic composites have been the leading emerging engineering materials [8]. Their rates of recurrence and types of applications have grown progressively, penetrating, conquering and building new material markets prospect relentlessly. The existing plastic composites available in the engineering material markets and supply range from common recycled materials to smartly prepared components. As they range widely in types, they also range in production approaches. Consequently, they have established their worth as weight-saving materials in engineering applications; the current drawback is making the processing more optimal in terms of cost and environmental effectiveness. The efforts to produce economically and environmentally effective composites components have resulted in several innovative manufacturing techniques currently being used in the composites industry.

Likewise, the composite producing industries have started to recognise that the industrial applications of composites promise larger business opportunities in other sectors of economies than the aerospace sector previously targeted for composite application. This discovery has given rise to shift of composite applications from aircraft to other commercial uses in recent years. One of the recently developed approaches is the usage of PBR produced via solvolysis as the matrix for hosting different types of prepared biomass for plastic composite production; the biomass and the PBR are both gotten from the solid waste streams.

20.2.2 Waste

The sub-Saharan Africa and Africa in general has been termed as third world because of low industrial development, technology and low standard of living of the majority. Generation of solid waste in sub-Saharan Africa is approximately 62 million tonnes per year [18]. Per capita waste generation is generally low in this region, but spans a wide range, from 0.09 to 3.0 kg per person per day, with an average of 0.65 kg/capita/day [18]. This on one hand is a reason for low waste generation, but has led to little technological know-how on waste disposal and management. In sub-Sahara Africa, the most popular waste disposal method is incineration, therefore causing the average temperature of this already temperate region to further increase due to global warming caused by the resulting greenhouse gases. This work therefore is a recent contribution to the valuable waste disposal in this region.

20.2.2.1 Plastic Waste

The rapid increase in the generation of waste plastic in the world is attributable to economic growth, changing consumption and production patterns of plastic products. In the United States, for example, municipal waste generation rose from 88.1 million tonnes to 250.9 million tonnes in the past five decades [6], at 5% yearly increase [20]. This increase in generation has led to plastics waste becoming a key stream in solid waste, following food and paper wastes [16]. Even cities with low economic growth have started producing more plastic waste due to increased applicability of different types of plastics. This increase has challenged authorities like European Union (EU) to mandate that by 2020, all plastic waste must go to mechanical, thermal or chemical processing facilities, and no more waste will be allowed in landfills [20]. Plastics include polystyrene, polyethylene, polyvinyl-chloride, nylon-6,6, poly-ethylene terephthalate and polypropylene. The plastic in focus here is polystyrene because of its use in the manufacture of expanded Styrofoam, or more simply known as Styrofoam, a material used in packaging of food, delicate materials such as electronics, as insulators in buildings.

20.2.2.2 Styrofoam Waste

Styrofoam may pose little or no threat or health risk when in use, which is one major reason it has so much applications, but its effect on the environment when disposed could be devastating over a long period of time, as it is being produced and disposed at a very high rate [6].

20.2.3 Sawdust

Another great environmental issue is the disposal of sawdust which is the daily mounting waste in sawmills across the developing countries. In Nigeria, for example, a significant percentage of processed and unprocessed timber input in the Nigerian sawmills ends up as wood waste. Moreover, the volume of sawdust from these ends continues to increase due to a rise in lumber production to meet growing demand for wood products. Nigeria has not fully exploited the potential for recycling waste materials, especially sawdust. Sawdust heaps are considered waste and therefore are indiscriminately incinerated, making a significant contribution to the greenhouse gas emissions. In some areas, sawdust heaps when not incinerated are flushed into water ways, therefore posing a serious problem to aquatic life. Also due to its wide range of particle size (to about 0.1 μm), it can be inhaled when carried by wind, therefore causing respiratory complications. Sawdust disposal has been a menace especially in developing countries and needs a better means of management.

20.2.4 Rice Husk

Most countries especially the developing ones like Nigeria are very rich in agricultural fibre and eventual agricultural waste, this on the long run is being used as a fuel. The processing and production of rice lead to the generation of a large volume of waste/by-product stream in the form of husks. Rice husk is the by-product in rice milling operation with an approximately 20% of the total weight of the paddy grain being processed [30]. The components of this rice husk are therefore determined by the milling method employed. Despite the abundant nature of this waste products and its unique physical and chemical properties [29], it is however not being harnessed in Nigeria.

Research work to find ways and means to utilise rice husk for the production of valuable materials from biomass waste has been in progress for the past three decades or so [10, 15, 25]. However, because of its unique chemical composition, not many successful methods have been evolved. Also due to the high silica content of rice husk the conventional process of making particleboard was not successful.

20.2.5 Bamboo

Bamboo is a renewable raw material that is universally accepted for building construction. There are over 1000 species of bamboo located majorly in Asia, South America and parts of Africa [22]. Its waste from building construction can be reprocessed to powdered form for plastic composite development. It plays a fundamental role in industrial and domestic economics in many developing countries. Compared with some commercial wood species, bamboo exhibits equal or better physical and mechanical properties [19], which offer good potential for processing it into composites (bamboo-based panels) as a wood substitute. The bamboo-based industries have developed into a multi-million-dollar industry with their variety of products enjoying very high demand domestically as well as internationally [22].

20.3 Materials and Methods

20.3.1 Preparation of Biomass Particulates

Three different biomass particulates from different solid waste sources (Table 20.1) were obtained and dried in oven for 18 hours at 40 °C so as to remove free water present in it. The samples were sawdust [4], rice husk [5] and bamboo [2]. The dried sample was graded to obtain the powder of 150 μm in size.

20.3.2 Preparation of Polystyrene-Based Resin

The polystyrene-based resin (PBR) was produced according to Abdulkareem and Adeniyi [2] procedure with density of 855 kg/m^3. It is confirmed from the earlier claim that it hardens on exposure to air at room temperature within 48 hours when left uncovered.

Table 20.1 Description of biomass particulates

S/No	Biomass	Source of waste	Unit operations
1	Sawdust	Sawmill	Drying and sieving
2	Rice husk	Rice mill	Drying and sieving
3	Bamboo	Building scaffold	Grinding and sieving

20.3.3 Polystyrene-Based Resin Matrix

Polystyrene wastes were obtained in the university from the packages of the goods procured by the university. They were sorted, cleaned and dissolved in petroleum solvent via solvolysis, at room temperature. The solvent mass was weighed, at room temperature and the fragments of polystyrene was added under constant stirring conditions until the solvent became saturated with it [2, 4]. The temperature was maintained ambient until polystyrene was fully dissolved and the solvated polystyrene is formed. The density of the resultant resin was 855 kg/m^3 which re-solidify at room temperature within 48 hours when left uncovered. The physical properties description details are presented in Table 20.2.

20.3.4 Preparation of Plastic Composites

Each biomass particulate type of 150 μm in size was mixed with solvated PBR in a mixer by simple mechanical stirring in the percentage projected in Table 20.3 by weight. The mixing in each combination was done meticulously in 5 minutes to enable uniform composition at any point. The obtained mixture was further pressed using a single roller on the metallic plate to eliminate the possible voidage and to achieve a uniform thickness of 3 mm. Oil was rubbed on the metal surface to prevent the polymer from sticking and to achieve easy composites removal after curing. The plastic composite produced was then made to cure under ambient conditions for 7 days. Three types of plastic composites were produced in this sequence.

Table 20.2 Physical properties description details of the Styrofoam resins

Properties	Value
Volume of the polystyrene recycled (ml)	2007.37
Volume of the solvent spent (ml)	60
Volume of the polystyrene-based resin produced (ml)	73
Mass of the polystyrene recycled (g)	32.2
Mass of the solvent spent (g)	39.4
Mass of the polystyrene-based resin produced (g)	55.1
Mass of the escaped mass (gas) (g)	16.5
Density of the PBR produced (g/cm^3)	0.75

Table 20.3 Composition of the biomass plastic composites

Composites	Compositions
M1	100% PBR + 0% biomass particulates
M2	90% PBR + 10% biomass particulates
M3	80% PBR + 20% biomass particulates
M4	70% PBR + 30% biomass particulates
M5	60% PBR + 40% biomass particulates

20.3.5 Characterisation

20.3.5.1 Mechanical Testing

The tensile tests were conducted using a universal testing machine (UTM: M500–50) at room temperature, according to the ASTM D638–10 standard. The loading rate applied to measure the bond strength was controlled at 4 mm/min. It is fitted with load cell and extensometer to record the test load and elongation accurately. Tensile loads were applied till the failure of the sample and load-elongation curves were obtained for all the composite materials produced. Three specimens were used for all the tests and final results represent the average.

20.3.5.2 Water Absorption and Diffusion Coefficients

Water absorption of the samples of composites were determined according to the ASTM standard method [38]. The rectangular samples of 15.4 × 4.6 cm were soaked in water at room temperature (20–22 °C) for 10 days. Composite samples 20 × 20 × 4 mm^3 were cut off from composites developed, and their edges were sealed with PBR prior to water absorption testing and dried in an oven at 50 °C for 24 hours.

The water absorption experiments were carried out according to the following procedures. Firstly, the pre-dried composite samples were immersed fully into a water baths kept at room temperature. At regular intervals of the process the samples were removed from the water bath and wiped with tissue paper to remove surface water and immediately water uptake was measured gravimetrically by using an electronic balance (uncertainty *l* 0.001 g). Following, the sample composites were immersed in the water bath again to continue the sorption process until the equilibrium condition was reached. Each measure procedure of the absorbed water was done in less than 1 minute, so, water evaporation at the surface was insignificant. The results of absorbed moisture were presented as mass of absorbed water by dry composites mass. The moisture content was computed using Eq. (20.1) [36].

$$M(t) = \frac{W_t - W_0}{W_0} \times 100\% \tag{20.1}$$

where W_0 and W_t represent the dry weight of the composites samples ($t = 0$) and the wet weight at any specific time t, respectively. The diffusion coefficient, D_x, was calculated for each composite developed using Eq. (20.2) [34].

$$D_x = \pi \left[\frac{h}{4M_m} \right]^2 \left(\frac{M_2 - M_1}{\sqrt{t_2} - \sqrt{t_1}} \right)^2 \tag{20.2}$$

where '*Mm*' is the maximum percentage of moisture content, '*h*' is the sample thickness, 't_1' and 't_2' are the selected points in the initial linear portion of the plot of moisture absorption (*Mt*) versus immersion time *t* and 'M_1' and 'M_2' are the respective moisture content.

20.4 Results and Discussion

20.4.1 Mechanical Properties of Biomass-Polystyrene Composites

The mechanical analysis is the study of a material's behaviour when put under varying loads. It is considered in terms of many factors like force at peak, Young's modulus and elongation at break impact strength as a function of changing fibre content in the PBR matrix.

20.4.1.1 Effect of Biomass Content on Force at Peak of Biomass-Polystyrene Composite

The force at peak is the highest recorded direct force on the composite. Tables 20.4, 20.5 and 20.6 show the values of force at peak in relations to the sawdust, bamboo and rice husk content between 0% and 40%. The force at peaks increased as the percentage of each fibre type increases in the PBR matrix. The highest forces at peak were achieved at fibre content of 40% for each fibre type. It can then be inferred that a higher the fibre content enables the biomass fibre reinforced composite to withstand a larger force.

Table 20.4 Mechanical properties of sawdust fibre composite

Fibre content (wt.%)	Force at peak (*N*)	Young's modulus (*N*/mm^2)	Elongation at break (mm)
0	202.6	186.79	6.746
10	356.65	311.32	3.461
20	714.5	645.905	2.282
30	875	758.75	1.049
40	1070.7	942.923	0.836

Table 20.5 Mechanical properties of bamboo fibre composite

Fibre content (wt.%)	Force at peak (*N*)	Young's modulus (*N*/mm^2)	Elongation at break (mm)
0	316.056	310.4894	6.75
5	378.144	360.4656	2.86
10	458.016	400.1184	2.01
15	740.532	706.5993	1.76
20	1203.852	1144.5423	1.42

Table 20.6 Mechanical properties of bamboo fibre composite

Fibre content (wt.%)	Force at peak (*N*)	Young's modulus (*N*/mm^2)	Elongation at break (mm)
0	202.6	191.314	6.75
10	291.6	269.524	3.719
20	303	289.67	1.984
30	344.6	316.694	1.383
40	405.2	380.628	1.139

20.4.1.2 Effect of Fibre Content on Young's Modulus of Biomass-Polystyrene Composite

Young's modulus is the ratio between the applied stress and strain below a proportional limit. Also presented in Tables 20.4, 20.5 and 20.6 are the values of Young's modulus at the changing biomass fibre content between 0% and 40% fibre content. The similar trends were observed. By implication, a small percentage of the biomass fibre contained in the composite increases the hardness of the composite. Generally, Young's modulus of biomass fibres increase with increasing fibre content of the composites.

20.4.1.3 Effect of Fibre Content on Elongation at Break on Biomass-Polystyrene Composite

Elongation at break, also known as fracture strain, is the ratio between changed length and initial length after breakage of the test specimen. It is the capability of the bamboo fibre to resist changes of shape without crack formation. Tables 20.4, 20.5 and 20.6 show the elongation break against fibre content. At 0% fibre content, the elongation at break is observed to have the highest values in all the fibres considered while and at 40% fibre content, the elongation at break is at the smallest across the fibre types used for the plastic composite production. Generally, elongation at break of biomass fibre-based plastic composites decreases with increasing fibre content of the composites.

20.4.2 Water Absorption of Biomass-Polystyrene Composites

The influence of biomass content of the plastic composites on the water absorption of polystyrene/biomass composites is summarised in terms of diffusion coefficient in Table 20.7 and Fig. 20.1. The water absorption of all polystyrene/biomass composites type increased continuously as the biomass fibre increased. Normally, polystyrene does not have good water absorption. Biomasses were not considered as a hydrophilic material, but the significantly increased water absorption of both polystyrene/biomass composites was likely to be attributed to the many pores and gaps in the biomass structure. This makes the rate of water absorption to increase with increased

Table 20.7 Diffusion coefficients of water in biomass polystyrene composites

Biomass fibre content (%)	Diffusion coefficient (mm^2/s) bamboo	Diffusion coefficient (mm^2/s) wood	Diffusion coefficient (mm^2/s) rice husk
10	1.53×10^{-5}	2.09×10^{-5}	2.79×10^{-5}
20	3.64×10^{-5}	4.97×10^{-5}	6.62×10^{-5}
30	9.91×10^{-5}	1.23×10^{-4}	1.62×10^{-4}
40	9.75×10^{-5}	1.34×10^{-4}	1.77×10^{-4}

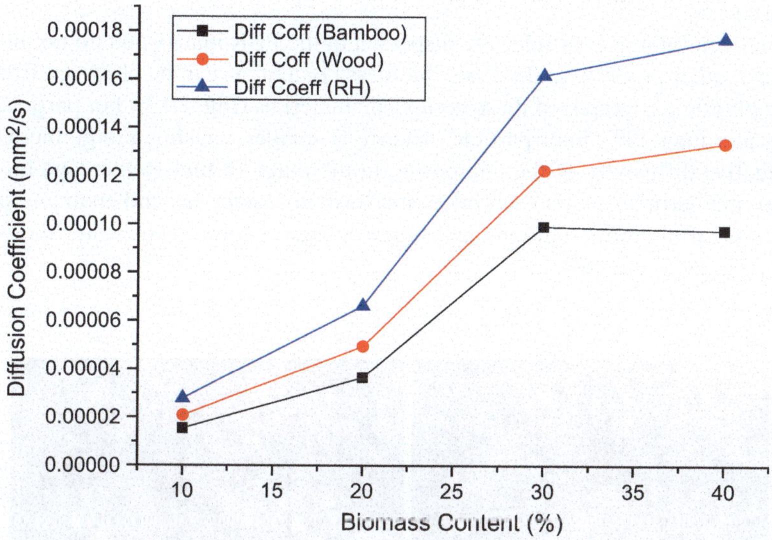

Fig. 20.1 Diffusion coefficients for water in biomass-polystyrene composites

amount of biomass; this is in agreement with Wang et al. [36]. However, the specific absorption capacity was equally confirmed, with the rice husk composites having the highest water absorption capacity. The influence of biomass content on the moisture diffusion rate in polystyrene/biomass composites is presented in Table 20.7.

The diffusion coefficient or diffusivity (D_x) of moisture absorption is the measurement of speed of diffusion. It characterises the ability of solvent molecules to move among the polymer segments [21]. From Eq. (20.2), values of D_x were evaluated for each composite developed and summarised in Table 20.7. The diffusion coefficient range for different biomass type was confirmed which is in conformity with the work of Wang et al. [36] with HDPE-rice husk composite. The rate diffusion was smallest with the composite of 10% biomass across board with higher plastic content which indicated smaller velocity of diffusion in interfacial gap between fibre and the PBR matrix [14]. However, the rate of diffusion increased as the percentage of biomass increased in the PBR matrix. This could be possible due to lower plastic content in the composite that could accelerate the diffusion in these cases. In other words, the diffusion coefficient increases with biomass fibre loading [12, 36].

20.4.3 Microstructure Analysis of Biomass-Polystyrene Composites

Micrographs reveal that there is a uniform distribution of biomass particulates at higher percentages of biomass in PBR matrixes. Furthermore, it can be seen from the optical micrographs that there is good dispersion of reinforcement in the matrix and the reinforcement particulates resulting in better load transfer from the matrix to reinforcement material as evident in the analysis of mechanical properties discussed.

Moreover, biomass particles are dispersed in the PBR matrix and do not interact with each other properly at the lower biomass content, as can be observed from the optical photomicrographs of these composite materials (Fig. 20.2). For particle content greater than 20%, inter-particle distance is smaller, causing a large increase in the effective diffusivity of the composite, in the range of high percentage biomass content, the particles touch each other and form agglomerates and chain, as shown in the optical photomicrographs and hence higher absorption of water molecules (Figs. 20.2, 20.3 and 20.4).

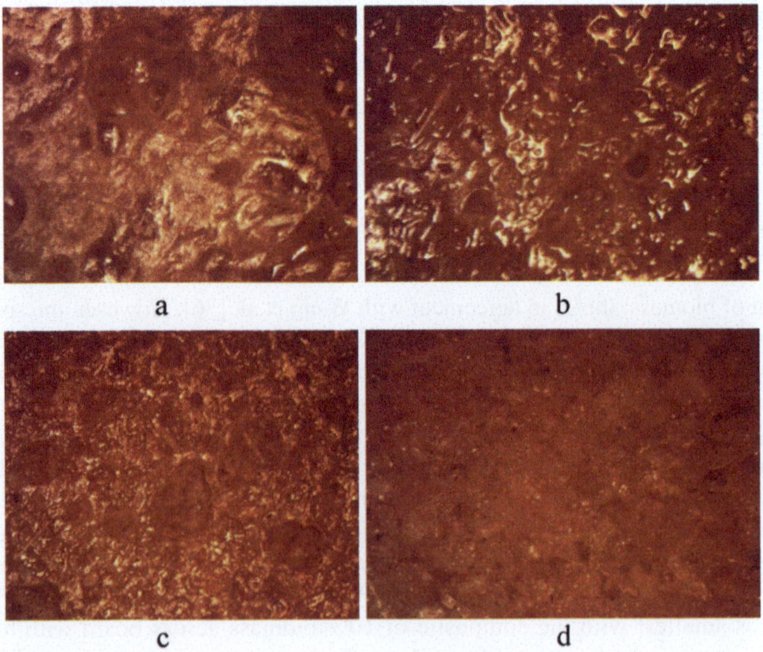

Fig. 20.2 Photomicrographs of bamboo fibre composite at different ratios of (**a**) 10%, (**b**) 20%, (**c**) 30% and (**d**) 40%

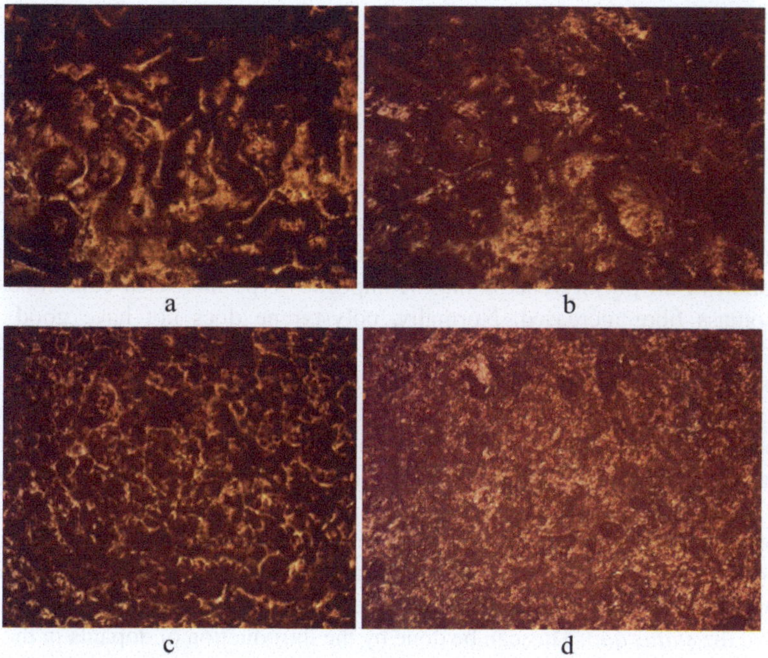

Fig. 20.3 Photomicrographs of polystyrene composites at 10% (**a**), 20% (**b**), 30% (**c**) and 40% (**d**). Plantain peel powder reinforcement

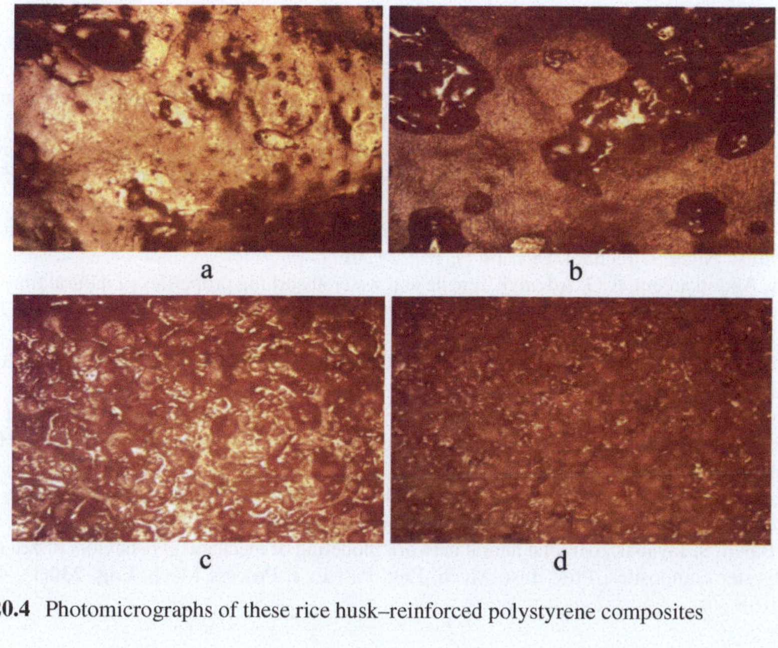

Fig. 20.4 Photomicrographs of these rice husk–reinforced polystyrene composites

20.5 Conclusions

From the mechanical analysis, it was observed that the force at peaks increased as the percentage of each fibre type increases in the PBR matrix (highest at 40% fibre content). It was then inferred that a higher the fibre content enables the biomass fibre-reinforced composite to withstand a larger force. Also Young's modulus of biomass fibres increased with increasing fibre content of the composites while elongation at break decreased with increasing fibre content of the composites. The water absorption of all polystyrene/biomass composites type increased continuously as the biomass fibre increased. Normally, polystyrene does not have good water absorption. Biomasses were not considered as a hydrophilic material, but the significantly increased water absorption of both polystyrene/biomass composites was likely to be attributed to the many pores and gaps in the biomass structure. Microstructural analysis revealed a uniform distribution of biomass particulates at higher percentages of biomass in PBR matrixes. Good dispersion of reinforcement in the matrix was also noticed resulting in better load transfer from the matrix to reinforcement material. PBR synthesised at room temperature was confirmed as a good matrix for biomass fillers like sawdust, rice husk and bamboo in the production of plastic composite. It is recommended that further studies be conducted to improve the *status quo*. This can be done by the introduction of dopants in the composites such as fly ash, kaolin, biochar.

References

1. R.U. Abass, Mechanical behavior of natural material (orange peel) reinforced polyester composite. Int. J. Eng. Sci. Res. Technol. **4**(1), 166–172 (2015)
2. S. Abdulkareem, A. Adeniyi, Production of particle boards using polystyreneand bamboo wastes. Niger. J. Technol. **36**(3), 788–793 (2017a)
3. S. Abdulkareem, M.K. Amosa, A. Adeniyi, Synthesis and structural analysis of Aluminium-filled polystyrene composites from recycled wastes. Environ. Res. Eng. Manag. **74**(2), 58–66 (2018)
4. S. Abdulkareem, S. Raji, A. Adeniyi, Development of particleboard from waste styrofoam and sawdust. Niger. J. Technol. Dev. **14**(1), 18–22 (2017)
5. S.A. Abdulkareem, A.G. Adeniyi, Tensile and water absorbing properties of natural fibre reinforced plastic composites from waste polystyrene and rice husk. Niger. J. Technol. Dev. **14**(1), 18–22 (2017b)
6. Agency, USEP, Municipal solid waste generation, recycling and disposal in the United States; Facts and figures for 2012 (2012)
7. H.M. Akil, M.F. Omar, A.A.M. Mazuki, S. Safiee, Z.A.M. Ishak, B.A. Abu, Kenaf fiber reinforced composites: a review. Mater. Des., 4107–4121 (2011). https://doi.org/10.1016/j.matdes.2011.04.008
8. G. Arpitha, B. Yogesha, An overview on mechanical property evaluation of natural fiber reinforced polymers. Mater. Today: Proc. **4**(2), 2755–2760 (2017)
9. N. Balaji, S. Jayabal, Artificial neural network modeling of mechanical behaviors of zea fiber–polyester composites. Proc. Inst. Mech. Eng. Part E: J. Process Mech. Eng. **230**(1), 45–55 (2016)

10. A. Bharadwaj, Y. Wang, S. Sridhar, V. Arunachalam, Pyrolysis of rice husk. Curr. Sci. **87**(7), 981–986 (2004)
11. R. Bhoopathi, M. Ramesh, R. Rajaprasanna, G. Sasikala, C. Deepa, Physical properties of glass-hemp-banana hybrid fiber reinforced polymer composites. Indian J. Sci. Technol. **10**(7), 1–7 (2017)
12. Y. Cao, S. Shibata, I. Fukumoto, Mechanical properties of biodegradable composites reinforced with bagasse fibre before and after alkali treatments. Compos. Part A: Appl. Sci. Manuf. **37**(3), 423–429 (2006)
13. D. Chandramohan, K. Marimuthu, A review on natural fibers. Int. J. Res. Rev. Appl. Sci. **8**(2), 194–206 (2011)
14. A. Espert, F. Vilaplana, S. Karlsson, Comparison of water absorption in natural cellulosic fibres from wood and one-year crops in polypropylene composites and its influence on their mechanical properties. Compos. Part A: Appl. Sci. Manuf. **35**(11), 1267–1276 (2004)
15. A. Ganesh, P.D. Grover, Combustion and gasification characteristics of rice husk. Fuel **71**(8), 889–894 (1992)
16. M.M. Guarav, K.N. Arunkumar, N.S. LinegeGaws, Conversion of LDPE plastic waste into liquid fuel by thermal degradation. Int. J. Mech. Prod. Eng. **2**(4), 104–107 (2014)
17. R. Hemachandran, M. Pugazhvadivu, S. Jayabal, Optimization of tensile and impact behaviours of randomly oriented short sisal fiber reinforced epoxy composites using response surface methodology. Int. J. ChemTech. Res. **9**(7), 660 (2016)
18. D. Hoornweg, P. Bhada-Tata, *What a Waste: A Global Review of Solid Waste Management*, vol 15 (World Bank, Washington, DC, 2012)
19. S. Jain, R. Kumar, U. Jindal, Mechanical behaviour of bamboo and bamboo composite. J. Mater. Sci. **27**(17), 4598–4604 (1992)
20. M. Johnson, S. Derrick, Pyrolysis. A method for Mixed Polymer Recycling. Micighan, Western Micighan University. pp 4–18 (2013)
21. S. Joseph, M. Sreekala, Z. Oommen, P. Koshy, S. Thomas, A comparison of the mechanical properties of phenol formaldehyde composites reinforced with banana fibres and glass fibres. Compos. Sci. Technol. **62**(14), 1857–1868 (2002)
22. H.A. Khalil, I. Bhat, M. Jawaid, A. Zaidon, D. Hermawan, Y. Hadi, Bamboo fibre reinforced biocomposites: a review. Mater. Des. **42**, 353–368 (2012)
23. U.K. Komal, V. Verma, T. Aswani, N. Verma, I. Singh, Effect of chemical treatment on mechanical behavior of banana fiber reinforced polymer composites. Mater. Today: Proc. **5**(9), 16983–16989 (2018)
24. F. La Mantia, M. Morreale, Green composites: a brief review. Compos. Part A: Appl. Sci. Manuf. **42**(6), 579–588 (2011)
25. K.S. Lin, H.P. Wang, C.-J. Lin, C.-I. Juch, A process development for gasification of rice husk. Fuel Process. Technol. **55**(3), 185–192 (1998)
26. F.S.d. Luz, L. Junior, E. Pereira, L.H.L. Louro, S.N. Monteiro, Ballistic test of multilayered armor with intermediate epoxy composite reinforced with jute fabric. Mater. Res. **18**, 170–177 (2015)
27. N. Lyn, *Effect of the chemical treatment on the inorganic content of Kenaf fibers and on the performance of Kenaf-polypropylene composites* (University of Waterloo, Waterloo, 2018)
28. R. Mangalaraja, T.M. Raj, D.R. Prabha, Statistical features of epoxy resin based hybrid composites reinforced with jute, banana and flax natural fibers. J. Chem. Technol. Metall. **54**(1), 35–47 (2019)
29. K. Mansaray, A. Ghaly, Physical and thermochemical properties of rice husk. Energy Sources **19**(9), 989–1004 (1997). https://doi.org/10.1080/00908319708908904
30. S. Nagrale, H. Hajare, P.R. Modak, Utilization of rice husk ash. Carbon **2**(6), 42 (2012)
31. N.A.S. Priya, P.V. Raju, P. Naveen, Experimental testing of polymer reinforced with coconut coir fiber composites. Int. J. Emerg. Technol. Adv. Eng. **4**(12), 453–460 (2014)
32. D.N. Saheb, J.P. Jog, Natural fiber polymer composites: a review. Adv. Pol. Technol.: J. Pol. Proc. Inst. **18**(4), 351–363 (1999)

33. V.P. Sathiyamoorthy, K.G. Aravind, P.K.C. Durai, Review on interface optimization and mechanical behaviour of natural fibre composite. Int. Res. J. Eng. Technol. **04**(10), 221–227 (2017)
34. C.H. Shen, G.S. Springer, Moisture absorption and desorption of composite materials. J. Compos. Mater. **10**, 1–20 (1976)
35. J.I.P. Singh, V. Dhawan, S. Singh, K. Jangid, Study of effect of surface treatment on mechanical properties of natural fiber reinforced composites. Mater. Today: Proc. **4**(2), 2793–2799 (2017)
36. W. Wang, M. Sain, P. Cooper, Study of moisture absorption in natural fiber plastic composites. Compos. Sci. Technol. **66**(3–4), 379–386 (2006)
37. L. Yan, N. Chouw, Behavior and analytical modeling of natural flax fibre-reinforced polymer tube confined plain concrete and coir fibre-reinforced concrete. J. Compos. Mater. **47**(17), 2133–2148 (2013)
38. American Society for Testing and Materials (ASTM). Evaluating properties of wood-based fiber and particle panel materials. ASTM D 1037–99. Vol. 04.10. ASTM, Philadelphia, PA. pp. 141–170 (1999)

Chapter 21
Development of an Integrated Process for the Production and Recovery of Some Selected Bioproducts From Lignocellulosic Materials

A. E. Taiwo, T. F. Madzimbamuto, and T. V. Ojumu

21.1 Introduction

Lignocellulosic biomass is a natural, sustainable and abundant raw material that could serve as an essential commodity for sustainable industrialization rather than causing environmental issues owing to its accumulation. A significant amount of such biomass are wastes generated as by-products from farming and agro-industries, animal husbandry, food processing, municipal, forestry residues, marine processing and bioprocessing industries [71, 139]. Lignocellulosics have gained continuous research attention, such that by applying various pre-treatment strategies to biomass, hydrolysis and/or fermentation have shown results in increasing the yield of sugars and/or other bioproducts, respectively [20, 135]. These products may serve as carbon sources for biofuels, industrial chemicals, pharmaceuticals and enzymes production [5, 109]. However, there is a tending rise on the usage of biomass as a source of energy and other manufacturing products, large percentage of which are still underutilized and are mostly disposed in a manner that may cause environmental or social problems [19, 70].

Cotton waste is one of the potential agricultural feedstocks commonly found in countries such as India, China, Brazil, Pakistan, Turkey and the United States, where cotton is being grown on a large scale [65]. The global 20-year average annual plantation of cotton was estimated to be around 33 million hectares and more than 100 countries in the world grow cotton [33]. As a result of this large-scale cotton production, million tonnes of cotton plant residue are generated from cotton harvesting and processing, which need a great deal of attention. This waste has been managed over the years using methods such as incineration, land application and

A. E. Taiwo · T. F. Madzimbamuto · T. V. Ojumu (✉)
Department of Chemical Engineering, Cape Peninsula University of Technology, Bellville, South Africa
e-mail: ojumut@cput.ac.za

M. O. Daramola, A. O. Ayeni (eds.), *Valorization of Biomass to Value-Added Commodities*, Green Energy and Technology,
https://doi.org/10.1007/978-3-030-38032-8_21

landfilling. Presently, the majority of the wastes from around the world, especially the wastes in farming operation (cotton wastes), are managed by land deposit [65]. This process is a threat to soil structure, which has the tendency of causing erosion over a long term. The prospects of the waste from cotton continue to attention of researchers, the importance and possibility of using this waste for several purposes became subject of debate over the last decade. The use of cotton waste as feed for livestock, fertilizer and feedstocks in production of paper, biogas and ethanol became prominent [116]. Despite different applications of cotton waste, today a vast amount of this waste is still disposed to the cultivating crop land. The chemical composition of cotton waste (Table 21.1) shows its vast potential as feedstock for production of several bioproducts. Buzała et al. [20] explained that the main driver in the application of lignocellulosic biomass for conversion to glucose in fermentation processes is the overall high percentage yield of cellulose and hemicellulose as compared to lignin in the biomass composition. These make cotton waste suitable for bioconversion of several bioproducts. While studies have shown the production of a wide array of bioproducts using lignocellulosic biomass [5, 7, 66], efficient recovery of these products remains a challenge, especially if a food or medicinal-grade product is envisioned. In a typical bioprocess operation, most products are usually in aqueous and dilute media [59]. This constitutes a problem in attaining a high recovery of bioproduct at a reduced cost of energy without compromising the standard of product quality. In addition, the complexity of the broth obtainable from fermentation and high purity required for the targeted products in commercial fermentation often involve a more complicated method of recovery, which may contribute to about 90% of the overall costs of production [32, 97]. Therefore, to address the problem caused by inhibitory activity of the targeted product in the broth, a wide range of product recovery techniques have been explored to enhance fermentation efficiency in the past decade [148]. Some of the reported separation techniques include salting-out extraction, distillation, ionic-liquid (IL) extraction, pervaporation, liquid–liquid extraction (LLE), perstraction, reverse osmosis, adsorption and gas-stripping [67, 91]. These techniques are time consuming and require a large

Table 21.1 Composition (ash, carbohydrates and lignin, dry weight) of cotton biomass

Cotton waste biomass	Cellulose % w/w	Hemicellulose % w/w	Lignin content % w/w	Ashes % w/w	Authors
Cotton stalk	50.21	24.89	24.9	2.96	Cengiz et al. [25]
Cotton seed hull	24.1	11.6	11.9	–	Li et al. [83]
Cotton seed cake	29.5	15.7	9.2	–	Viana et al. [137]
Cotton linters	83.9 ± 0.6	3.60 ± 0.60	5.9	1.5	Bezerra et al. [14]
Cotton gin	28.8 ± 0.18	25.5 ± 0.14	25.6 ± 0.22	10.4 ± 0.06	Chandrasekaran and Sivamani [26]

volume of solvent, toxicity of solvent remains a challenge and extraction is done at elevated temperature [68]. In addition, there is usually a probability of thermal degradation of the targeted compounds due to high temperature during most extraction processes which contaminate the final product [156] and, as a result, may compromise their suitability for food and pharmaceutical-grade products. This necessitates a need for a shift from the conventional solvent separation to green and non-toxic solvents for recovery of value-added products from fermentation broth [159]. Supercritical carbon dioxide (sc-CO_2) extraction is a green processing technique that has received a considerable interest because it is frequently operated at a lower temperature, for a short extraction period and not toxic as a solvent [87]. Effendi et al. [42] stated that carbon dioxide (CO_2) is undoubtedly the most increasingly used supercritical solvent, due to its significant benefits, such as low-cost price, ready availability in its pure form, safe operation and storage (non-combustible and non-volatile), eco-friendly, comparatively low pressure (7.38 MPa) and temperature (304.25 K) at critical conditions and an ease of solvent removal from the final product. The benefits and flexibility of supercritical systems in bioprocessing have been emphasized by Catchpole et al. [23], who reported that additional processing operation may be critical to fully integrate sc-CO_2 technique with bioprocess operations. This approach may allow the entire process to be adapted towards complete separation of the targeted value-added product. Thus, investigating an alternative method of extraction with high recovery yield and unique properties would be very desirable. This chapter reviews some fermentation processes for selected bioproducts with a view to presenting some experimental results from shake flask studies that may be used as a base case to develop an integrated bioprocess-supercritical carbon dioxide process for production of bioethanol, acetoin and vanillin as examples of bioproducts produced in relatively high, medium and low concentrations in fermentation broth, respectively.

21.2 Classification of Bioproducts

Vijayan et al. [138] have extensively categorized bio-based product recovery and purification into three sections: high value, low volume (HVLV), intermediate value, intermediate-volume (IVIV) and low value, high volume (LVHV). The ease of targeted product recovery and separation, and the economics of the whole process, is influenced by the concentration of the product in the fermentation broth. On this basis, bioethanol, acetoin and vanillin were chosen as examples of bioproducts produced in high, medium and low concentrations in fermentation broth, respectively, as shown in Fig. 21.1.

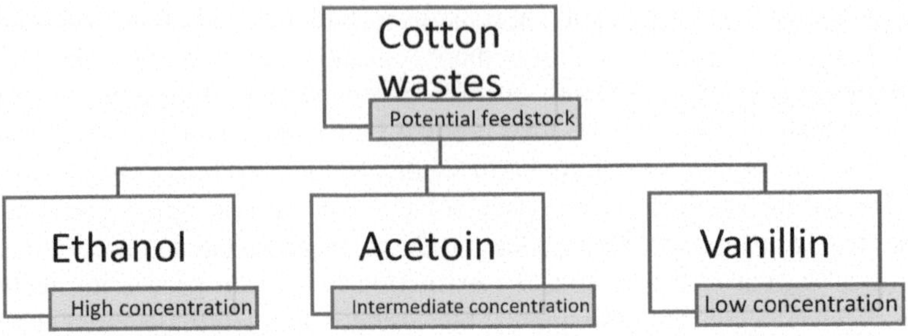

Fig. 21.1 Process integration of three-selected bioproduct from the fermentation broth

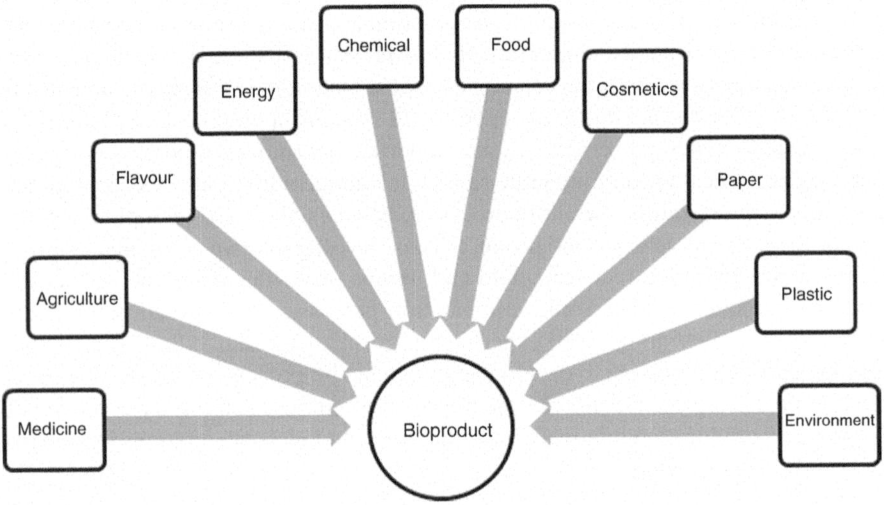

Fig. 21.2 Some sectors where fermentation products have found application

21.2.1 Application of Bioproducts

The development of bioproducts has touched almost every aspect of human life, from medicine to agriculture to the production of industrial products from different sectors which are being used to replace the synthetic products of petrochemical origin. Bioproducts are of great benefit to the nations of the world because these enhance the energy sector nationally, improve environmental friendliness, affect economic growth in the rural areas and advance the global bio-based market economy [120]. It has contributed to a more efficient use of natural resources, which is a necessary condition for creating a more sustainable economy. A schematic diagram is shown in Fig. 21.2, reflecting some of the sectors where bioproducts have found wide application.

21.3 Nature of Fermentation Broth

The fermentation broth is complex in nature as its composition is usually made up of different nutrients or supplements which microbes feed on to produce an array of products that may be of commercial importance [78]. The physiochemical or biochemical nature of the fermentation broth is as a result of different operations, interactions which are connected to the mass and heat transfers that arise within the fermenters [8, 93]. Rheological characteristics of most fermentation broths change significantly in the course of fermentation and the recovery process may have to deal with a viscous, highly non-Newtonian slurry as its feedstocks [36]. This attribute influences the flow behaviour of the fluids in fermenters and is a major factor affecting mass, heat and momentum transfer in a flow system [37, 95]. The nature and composition of fermentation broth are proportional to the efficiency of the fermentation process. Fermentation broths result from the differences in a wide range of microbes chosen and the adapted method of fermentation [140], which may be a complex solution of different types of bioactive compounds. The release of the broth from the aseptic environment of the fermenter or bioreactor automatically leads to a drastic change of conditions and expose the broth to a different mode of formation or contamination [41]. The targeted products from the broth could be intracellular or extracellular depending on the host cells or enzymes that secrete the desired products [103]. When the targeted product is extracellular, then there is a need to filter the biomass from the broth and disintegrate the product from the fluid, otherwise when intracellular product is considered, a cell disruption must be done and if the product is soluble in water, then the disruption should be carried out while the biomass is still in a slurry state [142]. The choice of a suitable method of recovery can be determined by the chemical stability and solubility as most of the biologically synthesized products have limited chemical stability. This will pose a constraint to temperatures, reactants, pH levels and other control parameters during fermentation.

21.4 Bioprocess-Supercritical Fluid Extraction Case Studies

21.4.1 Ethanol: High Volume Product (HVP) Concentration

The increase in oil prices and the foreseeing exhaustion of fossil fuels have stimulated the growth of exploration of biofuel as an alternative source of energy such as butanol and ethanol [3]. Production of bioethanol from starch, molasses and lignocellulose by fermentation technology is continuously increasing [57] with the view to improving production yield. According to Pieck et al. [104], research focused has been on proving the feasibility of separations as a way of overcoming the thermodynamic constraints on distillation operations of the water–ethanol azeotrope. Development of a separation technique that makes use of supercritical carbon dioxide may bring about efficient separation at reduced cost without denaturing the quality of the product. Supercritical carbon dioxide extraction (sc-CO_2) is becoming popular as an environment-friendly solvent used in the separation of biological products [35, 48]. The extraction of ethanol by sc-CO_2 has found wide applications in processes such as non-alcoholic beverages production [49] and ethanol recovery [58]. The equilibrium data for sc-CO_2 extraction of ethanol has been shown [18, 64, 84] with operating temperature and pressure of 333 K and 10.1 MPa, respectively. Güvenç et al. [58] investigate ethanol extraction in a continuous sc-CO_2 system using fermentation broth as the feed to be extracted. The authors studied the influence of CO_2 feed rate, period of extraction and pressure on the yield of extraction. Khosravi-Darani and Vasheghani-Farahani [73] report that sc-CO_2 for in situ extractive fermentation has been restrained by their inhibitory activity on the metabolism of a variety of microbes which is attributed to acidic pH that is inferred from the increased dissolution of CO_2 at high-pressure operating conditions. Furthermore, research on ternary system of CO_2–ethanol–water has been reported by many authors, resulting in the determination of the extraction and mass transfer rates in the separation system [13, 18, 85, 104]. Lang and Wai [80] report that the modelling of supercritical fluid extraction (SFE) processes will broaden the understanding of extraction mechanism and extraction operating conditions strategies. The research of Sovová [122] shows a review of different types of models applied to supercritical extraction in the last decades and the areas that demand attention. The technoeconomic analysis of ethanol has gained attention in areas such as feedstock supply system [2], inoculum propagation strategies [34], pre-treatment of biomass [10] and hydrolysis [115], with a few exceptions evaluating the technical and economic feasibility of the whole production. Pereira et al. [101] describe the technoeconomic appraisal of phorbol esters extraction from the seed cake of *Jatropha curcas* plant using supercritical dioxide. The author evaluates the production cost and concludes that supercritical technology is a promising technology when integrated with bioprocessing. It has been shown from a recent study that ethanol yield can be more than 80% even with low-cost medium and glucose that could be derived from lignocellulose as feedstock, which makes recovery of the ethanol broth technically feasible in supercritical CO_2 extraction [125]. Corn steep liquor, a cheap nitrogen source, was identified from the study as a potential nutritional supplement for commercial ethanol fermentation. Therefore, the fermentation broth could be used as a feasible feed in supercritical fluid extraction plant and the process integration of bioprocessing with high-pressure technology may be economical in the downstream recovery of this product.

21.4.2 Acetoin: Intermediate Volume Product (IVP) Concentration

Acetoin (3-hydroxy-2-butanone) is a natural occurring flavour commonly used as dietary supplements in wine, butter, strawberry and honey [146]. It is also a starting material for synthesizing specialty chemicals and widely used as an intermediate chemical which can be transformed to a broad range of chemicals [88]. Acetoin could be synthesized chemically; however, being a dietary supplement, the naturally produced product has a good commercial prospect. Microbial fermentation of acetoin is more preferred to chemical synthesis and enzymatic conversion, owing to abundant availability of its starting raw materials, moderate process conditions, environmental compatibility and its generally regarded safe measure as a food product [147]. The biotechnological production of acetoin investigation by different authors can be found in Table 21.2.

Glucose, which is obtainable from lignocellulose biomass, are potential substrates for acetoin production. It has been shown that carbon and nitrogen sources, which are a major determinant in acetoin bioproduction, can be produced using an inexpensive medium supplement [126]. More than 10 g/L yield of acetoin broth is obtained from shake flask study; this result shows it may be feasible to integrate bioprocessing and high-pressure technology to the production and recovery of acetoin.

Table 21.2 Biotechnological production of acetoin

Microorganism	Major substrate	Production scale	Maximum concentration	Reference
Lactococcus lactis subsp. lactis 3022	Glucose	100 mL	1.054 mol/L	Kaneko et al. [69]
Hanseniaspora guilliermondii	Glucose	–	367 mg/L	Teixeira et al. [129]
Bacillus subtilis CICC 10025	Molasses	5 L	35.4 g/L	Xiao et al. [146]
Bacillus licheniformis MEL09	Glucose	–	41.26 g/L	Liu et al. [88]
Serratia marcescens H32	Sucrose	3.7 L	60.5 g/L	Sun et al. [124]
Bacillus subtilis TH-49	Glucose	100 L	56.9 g/L	Xu et al. [147]
Bacillus amyloliquefaciens FMME044	Glucose	7 L	51.2 g/L	Zhang et al. [154]
Paenibacillus polymyxa CS107	Glucose	5 L	55.3 g/L	Zhang et al. [153]
B. subtilis SF4–3.	Glucose	5 L	48.9 g/L	Tian et al. [130]
Escherichia coli	Mesocarp fibre	Multistage bioreactor	29.9 g L – 1	Yusoff et al. [160]
Bacillus subtilis CICC 10025	Glucose	Shake flask	10.70 g/L	Taiwo et al. [126]

Source: Adapted and updated Sarma et al. [112]

Tian et al. [130] reported that acetoin and 2,3-butanediol are usually generated concurrently in its fermentation broth and that several efforts have been adopted to improve the yield and process recovery of acetoin through statistical optimization, two-stage speed control strategy in production. This necessitates a search for an alternative strategy, and the application of supercritical fluid extraction could prove to be efficient for acetoin recovery since it is preferred above the conventional extraction method. Supercritical fluid CO_2 extraction has so far not been reported to have been used for the recovery of acetoin. This could be due to lack of data on the phase properties of acetoin in a supercritical CO_2 system which was not made available in the literature until the work of Effendi et al. [42]. The author reported the solubility of acetoin in supercritical carbon (iv) oxide due to its low dipole nature and its behaviour as a fat-soluble solvent in supercritical condition. This was the first fundamental report on acetoin in the sc-CO_2 system. The research work of Lu et al. [89] shows the optimized processes for the downstreaming of different aromatic compounds in a synthetic broth of Zhenjiang aromatic vinegar of which acetoin was reported to be part of the recovered aromas. The technoeconomic feasibility of acetoin production is scarce in literature, as this could be due to much attention given to 2,3-butanediol production which is an analogue by-product in acetoin production. Koutinas et al. [79] conducted the bioprocess design and costing for 2,3-butanediol production. The authors found that minimum selling price of butanediol is higher than 1 $/kg, which is set as a selling target price, and the major reasons identified were complex nutrient supplement, raw material market price and the fermentation efficiency. Tian et al. [130] said that the commercial production of acetoin by microbial fermentation is yet to be promoted in the market. The authors were of the opinion that its industrialization will bring a vast economic change and social benefits to the consumers who desire a natural product. Thus, on the basis of the above, it can be deduced that the recovery of acetoin in a bioprocess-supercritical CO_2 system is worth investigating as it would find application in the chemical processing, food and fermentation industries. In addition, the economics of the process would reveal the feasibility of reducing the cost of production.

21.4.3 Vanillin: Low Volume Product (LVP) Concentration

An aromatic compound that imparts fruity fragrance in foods and beverages, making it appealing to the consumers [38]. Vanillin (4-hydroxy-3-methoxy benzaldehyde) is a commercially important flavour. While it can be produced from vanilla plants [107], its commercial production is met by synthetic chemical such as eugenol or guaiacol extract [21]. However, the high costs of vanilla flavour from plant origin along with the consumer preference for natural extract have encouraged the effort to explore bioprocessing route of production of vanillin by from other renewable sources. Vanillin and vanilla extracts have a projected annual total throughput of 16,000 metric tonnes, with an approximate value of $650 million [72]. Natural vanilla extract constitutes lower than 1% by volume of production, although it is more significant in monetary terms. Selling prices vary from about $1500 per kg for natural vanilla extract to $10–20 per kg for chemically produced vanillin [45]. It was reported that biological production of vanillin using a precursor that is chemically similar to vanillin, ferulic acid, can be economically feasible [28]. Ferulic acid is a phenolic compound found in the cell wall of lignocellulosic biomass, where it is covalently bounded with a variety of carbohydrates like glycoside conjugate, ester or amide [100]. Ferulic acid is a low-cost starting material for the production of bioactive compounds like vanillin when derived from lignocellulose wastes such as cotton, corncobs and rice bran. Therefore, environmental friendly routes and various agro-products have been investigated for vanillin production (Table 21.3). However, due to the complex nature of the fermentation broth and its host toxicity in the isolation of vanillin, there is a need for improvement in purification and recovery processes. Carroll et al. [22] reported that chemical volatility can be addressed using an organic bi-layer, gas stripping system but the technical and economic viability is not certain. Hence, Development of an alternative separation technique that will increase the microbial tolerance and productivity is necessary. Supercritical fluid treatment with CO_2 has been identified as a natural extraction process, and the products suitable for food industrial applications and thus carried the GRAS (Generally Regarded As Safe) status [40].

Liu et al. [86] describe the phase behaviour of solid vanillin in CO_2. This information is needed for the design of sc-CO_2 extraction process. Knez et al. [75] present vanillin and dense gases, fluorinated hydrocarbons and sulphur hexafluoride in

Table 21.3 Biotechnological production of vanillin

Microorganism	Major substrate	Production scale	Maximum concentration	Reference
Bacillus species	Eugenol and dimethyl sulfoxide	500 mL	0.32 mg/L	Sindhwani et al. [117]
Staphylococcus aureus	Ferulic acid and glucose	100 mL	45.7 mg/L	Sarangi et al. [111]
Bacillus aryabhattai BA03	Ferulic acid	250 mL	147.1 mg/L	Paz et al. [100]
Engineered *Saccharomyces cerevisiae*	Glucose	2 L	500 mg/L	Brochado et al. [17]
Aspergillus niger I-1472	Isoeugenol	250 ml	0.137 g/L	Tan et al. [127]
Escherichia coli strain JM109	Ferulic acid	100 mL	2.52 g/L	Barghini et al. [12]
Aspergillus niger CGMCC0774 and *Pycnoporus cinnabarinus* CGMCC1115	Ferulic acid (from waste residue of rice bran oil)	25 L	2.8 g/L	Zheng et al. [157]
Strain KK-02	Ferulic acid	50 L	15 g/L	Yiyong and Hong [149]
Streptomyces sp. strain V-1	Glucose, ferulic acid	500 mL	19.2 g/L	Hua et al. [63]

three-phase equilibria that was combined in a binary system. The author research data inform the behaviour of other compressed gases other than CO_2 in high-pressure technology and the solubility of vanillin. Nguyen et al. [96] investigated high-pressure extraction of oleoresin from vanilla beans at temperature of 307 K and pressures between 110 and 140 bar using CO_2. A further study was carried out on the supercritical CO_2 fractionation process of vanilla oleoresin to evaluate its extraction when temperature ranges from 310.5 to 348.3 K and at mild pressures of 140 and 180 bar, which increased vanilla oleoresin solubility in CO_2 at the highest working conditions [35]. The authors assessed the influence of extraction time and pre-treatment on the oleoresin yield and its composition. The work of the authors established the supercritical fluid operating conditions necessary for vanillin recovery.

Ferulic acid was biologically converted in a shake flask study using a low-cost microbial media to support and fulfil nutritional conditions for increased vanillin concentration in the fermentation broth; a yield of 386 mg/L was obtained from the study (data not yet published). Corn steep liquor was the cheap microbial media used in the shake flask study; it has been reported that its composition contains proteins, amino acids, minerals, vitamins and trace elements that support growth media in fermentation studies [131]. The percentage improved yield obtained in the shake flask study of vanillin compared with similar nutritional component in other studies and it would help in reducing the cost of production and recovery (Table 21.3). The optimum condition established from several authors shows the possibility of the fermentation broth in the SFE pilot plant for extraction of vanillin. In general, there is limited number of publications where supercritical fluid extraction system is incorporated with bioprocessing to isolate and purify vanillin. Technoeconomic feasibility report on natural vanillin is limited in most literature. This could be due to the high solubility of vanillin in water, which makes most separation techniques less efficient or feasible [21]. The market size of biovanillin was valued above \$11.5 million in 2015 and the industry was projected to make a profit of 13% compound annual growth rate (CAGR), with demand exceeding 500 tonnes by 2023 [53] as shown in Fig. 21.3.

21.5 Criteria for Targeted Product Recovery

The composition of bioproducts varies for different product separation, some of which are thermal stability, solubility, diffusivity, charge and isoelectric pH [60]. The factors that are considered when selecting a technology for the recovery of products from fermentation broths are partition coefficient, selectivity, biocompatibility, choice of solvent and waste generation management strategies [74]. The desired level of product purity, density, shape, recoverability and interfacial tension are also considered. Ghosh's [50] overview of biological separation reported that several substances which are usually impurities and mostly side products compete with the targeted bioproduct, and these make separation difficult because they often

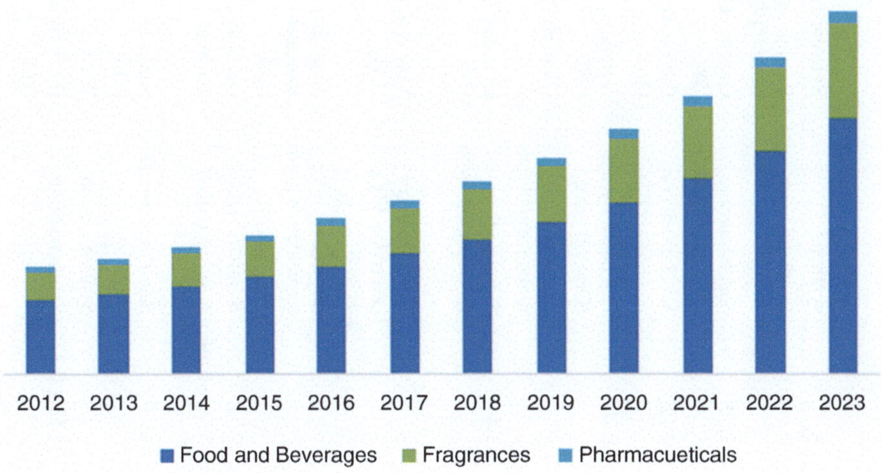

Fig. 21.3 US bio vanillin market size, by application, 2012–2023 (tonnes). (Source: Global Market Insight [53])

have the same physiochemical properties. Some of the main criteria for ease separation are hereby discussed:

(i) Partition coefficient (*k*): The presence of a third compound in a two-phase system causes the third compound to distribute itself between the two phases. The property that best describes the distribution of this third compound is known as the partition or distribution coefficient (Eq. 21.1) The regular practice used to estimate the partition coefficient is to quantify the equilibrium concentrations of the compound in two immiscible liquid phases that are contact [123]. The partition coefficient is important because it determines the choice of a solvent and the separation quality in the recovery of the targeted product [92]. The value (*k*) from Table 21.4 shows the separating capacity of different extraction techniques as reported by different authors. Gu [56] discusses that the distribution coefficient of the chosen solvent in extraction could be either smaller or greater than unity based on the penetrating ability of the solute to the solvent or aqueous phase of the system.

$$\text{Distribution coefficient} = \frac{\text{mass fraction in organic phase}}{\text{mass fraction in aqueous phase}} \tag{21.1}$$

(ii) Selectivity: The capacity of a solvent to separate a targeted product from water is best described as a separation constant or factor. It is defined as the ratio of the partition coefficient of the targeted product to that of the water (Eq. 21.2). In the process of extracting a targeted product from the fermentation broth, the separation factor greater than unity is an indication that the solvent ideally

Table 21.4 Some fermentation products and the separation techniques used in the recovery

S/N	Fermentation product	Separation technique	Extractant	(%) Recovery	*K*	Producing microbe/ synthetic	Reference
1	Ethanol	Membrane gas extraction	Propylene glycol	≤80	0.1–0.3	Synthetic broth	Bandini and Gostoli [9]
2	Ethanol	Supercritical fluid extraction	Carbon (iv) oxide	88.00	1.30[a]	*Saccharomyces cerevisiae (Y-567)*	Güvenç et al. [58]
3	Ethanol	Supercritical fluid extraction	Carbon (iv) oxide	78.10[a]	1.17[a]	*Saccharomyces cerevisiae (Y-567)*	Güvenç et al. [59]
4	Ethanol	Silicalite membrane	Zeolite	98.20	0.20	Dry baker's yeast (Oriental yeast)	Nomura et al. [98]
5	Ethanol	Extractive distillation	Ethylene glycol and calcium chloride	99.50	–	Simulated fermentation broth	Gil et al. [52]
6	Ethanol	Heat-integrated distillation (double effect distillation, vapor compression distillation)	Glycerol and Ethylene glycol.	74.00[a, b]	1.4	Simulated fermentation broth	Díaz and Tost [39]
7	Ethanol	Integrated-distillation membrane	Membrane-assisted vapour stripping	≥90.00	–	*Saccharomyces cerevisiae ATCC 4126*	Vane et al. [134]
8	Ethanol	Hybrid distillation and vapor permeation (DVP)	Carbon (iv) oxide	99.80	–	Simulated fermentation broth	Singh and Rangaiah [118]
9	Acetoin	Sugaring-out extraction	Ethyl acetate	61.2	0.61[a]	*B. subtilis DL01*	Dai et al. [30]
10	Acetoin	Salting-out extraction	Ethyl acetate and K_2HPO_4, ethanol	85.8 ± 0.37[a, b]	3.38 ± 0.19[b, a]	*B. subtilis DL01*	Dai et al. [31]
11	Acetoin	Salting-out extraction	Acetone/phosphate aqueous two-phase system.	96.4	22.3	*Serratia marcescens H32*	Sun et al. [124]
12	Acetoin	Adsorption	Hyper-cross-linked resin	≤98	–	Synthetic broth	Wu et al. [144]
13	Vanillin	Organophilic Pervaporation	Composite membrane	81.83[a, b]	1.18 ± 0.27	Synthetic broth	Brazinha et al. [16]

S/N	Fermentation product	Separation technique	Extractant	(%) Recovery	K	Producing microbe/ synthetic	Reference
14	Vanillin	Ion-exchange with neutralization	Polymeric resin Amberlite IR120H	–	–	Synthetic broth	Zabkova et al. [150]
15	Vanillin	Ultrafiltration membranes	Tubular ceramic	–	–	Kraft pulp liquor	Žabková et al. [151]
16	Vanillin	Adsorption–regeneration technique	Macroporous adsorption resins with cross-linked-polystyrene	≤95.6	–	Synthetic broth	Zhang et al. [152]
17	Vanillin	Solvent extraction	Octylamine	–	1.62 + 0.14	Synthetic broth	Tarabanko et al. [128]
18	Vanillin	Liquid membrane system	Tributyl phosphate	>60		–	Zidi et al. [158]
19	Vanillin	Adsorption	Non-polar Macroporous resin.	96.2		Synthetic broth (spent liquor)	Wang et al. [139]
20	Vanillin	Supercritical fluid extraction	Carbon (iv) oxide	84%	0.132	Synthetic broth (vanilla beans)	de la Vega et al. [35]

[a]Calculated data; K, Distribution coefficient
[b]Average value

extracts the product of interest rather than water [74]. It has been demonstrated that the extraction of the aroma from fermentation broth could be efficient by adjusting the selectivity of the extraction of PEA (2-phenyl ethyl alcohol) and ethanol in the extraction of aroma with supercritical CO_2 [46]. Groot et al. [55] were of the opinion that increased selectivity factor in separation could help in reducing the cost of recovery in extraction. The authors further justified by observation that high capacity of the solvent and selectivity of alcohol/water separation extraction would favour the extraction processes.

$$\text{Selectivity} = \frac{\text{distribution coefficient in organic phase}}{\text{distribution coefficient in aqueous phase}} \tag{21.2}$$

(iii) Biocompatibility: It is the term used in describing the present state of biomaterial within a physiological environment, without negatively and significantly affecting, any of the other body of environment and material [141]. The biocompatibility of 62 organic solvents has been tested with yeast cultures to determine experimentally the controlling parameters affecting the product recovery. The author reported that 15 of these solvents were proven to be biocompatible while 26 were completely toxic, the remaining showing different levels of inhibition [77]. The highlighted reasons for the discrepancy were a different type of microbial strain, medium formulation, an adaptation of microorganism to solvents, differences in solvents cell contact and the type of fermenter design. In the recent work of Zhang et al. [155], biocompatibility with the fermentative organism has been identified as criteria required for fermentative product recovery. Likewise, it has been revised that biocompatibility is highly strain-reliant and many solvents have been shown to be harmful to microbes like yeast and bacteria [76]. Therefore, biocompatibility is required in the choice of separation techniques in the recovery of specific biological product from the fermentation broth.

(iv) Nature of extractant or choice of solvent: The limitation of organic solvents in the recovery of the microbial product, when compared to the broad usage in chemical synthesis, could be attributed to the degradation of most of the recovered bioproduct [50]. Microbial products are usually exposed to denaturing and degradation that makes the compatibility of solvents selected a necessary factor to be considered in the recovery of the targeted product. Therefore, the nature of solvent is imperative because a solvent with low partition coefficient will not be effective in the recovery of a fermentation product and a solvent toxic with respect to the microbiological organism will inhibit cell growth and fermentation [113]. However, the choice of a suitable solvent is key to the development of downstream processing of bioproduct as it has implications on quality and yield of the targeted product [27]. Several solvents have been screened and modes of operation identified for recovery of various products from fermentation broths [74].

(v) Environmental acceptability: The compelling regulation, for example, the European Union (EU) environmental policy and legislation, has a goal and objectives set for the period of 2010–2050, to prioritize the reduction of hazardous solvents in industries. They stand against the common use of carcinogenic or toxic solvents that are detrimental to the health of end-users and also create a nuisance to the environment. Therefore, a solvent that will be used for recovery and separation of the product should be derived from either renewable resources or those that show better environmental, health and safety properties [29].

21.5.1 Separation Techniques for Targeted Fermentation Products

The basis for choosing a separation technique in bioproduct recovery is a function of the product composition (properties of the impurities and the producing microbes, cells or tissues) characteristics, desired purity, targeted yield and activity requirement. Most bioproduct recovery involves a series of unit operations for purification and concentration of the desired product. Solid materials suspended, such as cells, can be easily separated from fermentation broth through filtration or centrifugation technique, while the separation of solubilized impurities, especially organic acids and inorganic salts, requires much more complicated processes [60, 133]. Figure 21.4 depicts the different stages of separation techniques that may be required in a targeted product recovery. The primary recovery stage is mainly concerned with separation of cells or its debris to reduce the impurities, followed by the intermediate recovery stage that focuses on product concentration and the choice of the technique is a function of the targeted product characteristics. Sometimes, it may be required to put into consideration a protein refolding step if the targeted product is formed within the cells inclusion bodies (IBs) [60]. The final stage seems to be the most complex level of separation because it has been reported as difficult and expensive and requires different separation properties to attain the desired level of purity. The recovery of products from fermentation broths has been reported as the last difficulty to be overcome in closing the gap between laboratory scale and commercial production of most potential products in the broth [60, 61, 145]. Some of the recovery techniques that were used in the selected bioproduct earlier discussed have been shown in Table 21.4.

Ethanol is one of the most common alcohols produced by fermentation and its broth inhibition in yeast fermentation is usually around 5–10% per volume. This makes the feasibility of its recovery possible using supercritical CO_2 and its integrated approach feasible since a good volume of ethanol broth is obtained from lignocellulosic biomass feedstock. Essien and Pyle [44] emphasize that solvent extraction of ethanol is about 60% higher in investment costs than extracting using the distillation method. The authors further stressed that this shifts any advantage in energy and utility without considering the cost implication of the solvent. An hybrid separation process integrating, vapor stripping, compression and permeation

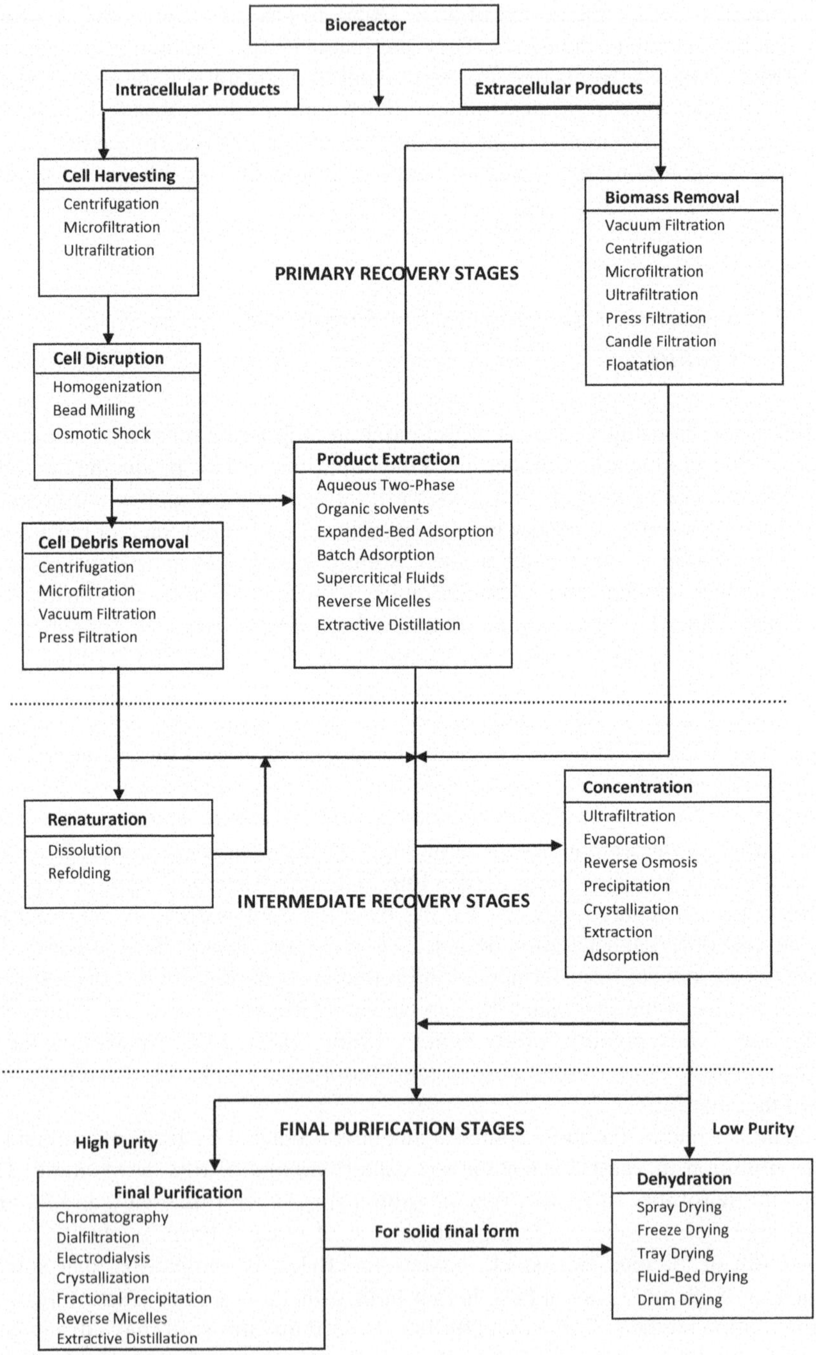

Fig. 21.4 Generalized block diagram of downstream processing. (Source: Harrison [60])

membrane was developed for the recovery of ethanol from fermentation broth [134]. The author's findings revealed that membrane-assisted vapor stripping could stand as a replacement to traditional distillation for separation of ethanol from fermentation broth. However, the microbial cells tolerance is greatly a concern, which is due to the toxicity of the cells to the extraction solvent. Also, most of the organic solvents used in extraction require a solute–solvent separation and invariably leave traces of the solvents in the products.

Acetoin separation forms an azeotropic mixture with water similar to alcohol, and its hydrophobic nature and miscibility make solvent extraction method a bit difficult. Several techniques have been used in acetoin recovery as shown in Table 21.4. Nevertheless, the question of toxic residues and environmental friendliness remains unanswered, although most of the methods devised by different authors yield a reasonable percentage yield, yet the quest by the consumer for natural acetoin void of any residue that will be detrimental to their health is still ongoing. Presently, the commercial acetoin available in the market is still produced using chemical pathway and solvents. Therefore, supercritical fluid extraction seems to be an end to the limitation imposed by solvent extraction, most importantly when a solvent like CO_2 is used in product recovery.

Vanillin production from carbon sources like vanilla beans, oxidized lignin solutions and acid has been reported (Table 21.3). Studies have been performed to understand the vanillin model solutions dissolving abilities, capabilities and other important vanillin-CO_2 system properties [121]. However, vanillin is usually of low concentrations in fermentation broth and the recovery its usually a challenge since a feasible concentration of broth is required to use supercritical fluid extraction technique. However, the concentration of the broth has been reported to be high as 19.2 g/L. Therefore, the feasibility of the recovery of vanillin from fermentation broth using supercritical CO_2 could be realizable if the chosen lignocellulosic biomass could offer a reasonable amount of lignin that could serve as a precursor for a broth concentration that could be as high as 5 g/L feed for the supercritical fluid extraction.

21.5.2 Integrated Bioprocess-Supercritical Extraction Techniques

The distillation method has been the dominant practice in most product recovery of bioproduct from the broth on a commercial scale [94]. In a current review paper by Singh and Rangaiah [119], the authors evaluated more than 50 published articles and chapters in a book for 8 years starting from the year 2008, categorizing separation technology used for recovery and dehydration of bioethanol. The authors reported that research trend shows keen interest of integrating distillation with other separation techniques but without a focus on process integration using supercritical carbon dioxide extraction technique. The authors further stressed that the future developments of separation should be directed at exploring efficient and eco-friendly solvents for bioproduct recovery and dehydration. Therefore, integrating supercritical carbon dioxide with other separating techniques will offer efficiency, economic feasibility and sustainability to the advancement of separation technology. The application of supercritical fluid extraction in bioprocessing may help in increasing the concentration of the targeted product. Khosravi-Darani and Vasheghani-Farahani [73] projected that targeted product recovery from a fermentation broth require alternative downstream processing methods rather than distillation commonly used by the chemical industry, this is needed for effective product concentration and without the use of a toxic solvent for recovery operation. The authors further stressed that the low concentrations of bioproduct make distillation not cost-effective due to the intense energy required to remove the bulk water from the broth. Also, reducing the number of separation techniques used in the recovery of product from the broth will save time and cost and improve the product recovery [54]. A study on biphasic hydrocarbon functionalization, catalysed by recombinant *Escherichia coli*, which faces the difficulty of stable emulsion formation during downstream processing using other separation technique has been addressed by the use of supercritical carbon dioxide [15]. The authors gave an account that using sc-CO_2 helps in reducing the number of steps required in downstream processing compared to other methods of separation since phase separation and product purification occur as part of the same process. In the study reported by Ütkür et al. [132] on the catalysis of quinaldine hydroxylation by *Pseudomonas putida*, product recovery (84%) was achieved with ease when supercritical carbon dioxide was coupled with the use of liquid–liquid extraction for phase separation and product recovery, respectively. The authors demonstrate that the concept of an integrated process with sc-CO_2 is an added advantage when compared with other techniques that were used solely in the recovery and dehydration.

21.6 Technoeconomic Feasibility of Product Recovery From the Broth: A Case of Integrated Bioprocess-Supercritical Extraction Technique

In bioprocessing, the effective cost evaluation of the whole production process is subject to the cost of downstream recovery. Hence, expensive cost of separating and purifying most biological processes limits their economic feasibilities [108]. Hence, there it becomes imperative to understand the cost implication of the chosen separating techniques used in the product recovery in other to ascertain its commercialization and to know if it will compete with existing petroleum-based product in the market. Perrut [102] reported that most industries believe that investment in supercritical fluid technology is very expensive when compared with other conventional low-pressure equipment, and the few who supported the technology said it could be cost-effective if it is restricted to high-added-value products. Although this has been proven to be untrue [6], yet the belief has limited the advancement in commercial separation technology, most importantly the change over from the conventional extraction methods which rely on distillation as compared to supercritical fluid extraction. Several authors have investigated the economic feasibility of supercritical fluid extraction of phenolic compounds from Jabuticaba skins and Brazilian plants [110, 136], polyphenol extracts from grape bagasse and seed extract [47, 106], defatting of annatto seeds [4] and cupuassu butter from defatted seeds [24]. However, most of these extracts are restricted to the locale of the country since biomass seasonal variation is country specific. The authors have shown a technoeconomical evaluation between the cost of manufacturing and the targeted yield of the product. Likewise, the experimental and industrial data of the extracts were economically evaluated using simulators like ASPEN Plus and, HYSYS(ASPEN TECHNOLOGY, INC., 20 Crosby Drive Bedford, Massachusetts 01730 USA), CHEMCAD (CHEMSTATIONS, INC., 3100 Wilcrest Drive Suite 300 Houston, TX 77042), PRO II(AVEVA GROUP PLC., High Cross Madingley Road Cambridge CB3 0HB, UK) and SuperPro Designer (INTELLIGEN, INC. 2326 Morse Avenue Scotch Plains, NJ 07076, United States) [24, 114] to determine the existing relationship between the process yield and recovery cost of bioproduct. The cost evaluation from the simulators showed that supercritical plant could be used to replace distillation if raw materials could be at lower cost [105]. However, the economic assessment of integrated bioprocess-supercritical carbon dioxide in the recovery of fermented product in the broth has been scarcely studied as compared to much literature on plant extracts. This could be due to the complexity of fermentation broth and the concentration of the targeted product, which could affect the potential application of recovery and purification of bioproducts using supercritical carbon dioxide [90]. To overcome this hurdle, various techniques have been adopted to improve the concentration of product in the broth which is paramount in the economics of bioproduct recovery, for example, the scale-up criterion and optimization studies [99], unique fermentation pathways [43], medium manipulation [125] and novel strain development [81, 82]. Therefore, if the downstream recovery of bioproducts from aqueous solution of high purity, productivity and low energy consumption is to be ascertained in supercritical carbon dioxide extraction plant, then capital cost and operating cost should be the driving key. The capital cost of an integrated bioprocess extraction is the summation of resources needed to purchase the equipment (capital investment) and the working capital (required amount for the start-up of the equipment). The operating cost is summative of all plant operational cost and amount required in recovering the capital investment. It can be simply referred to as the annual cost needed to produce the product of interest and pay back the cost used for investment [62]. An overview of the evaluation protocol needed for economic assessment is shown in Fig. 21.5.

There is a correlation between the selling price of a final product and its concentration in the starting for a broad range of products recovered from the fermentation broth [11]. The major objective of product recovery at the downstream operation is to ensure efficiency, productivity and sustainability of the targeted product at minimum recovery cost. However, most purification stages require a quite number of

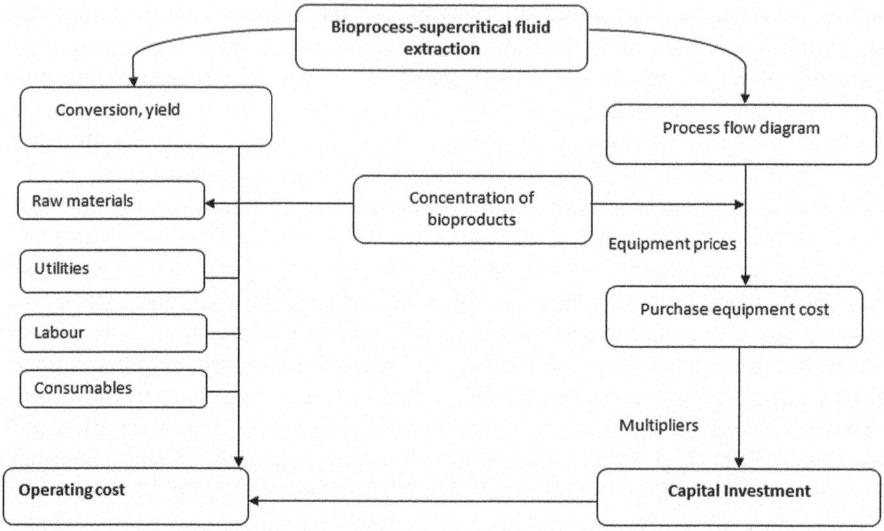

Fig. 21.5 An overview of the evaluation protocol needed for economic assessment of the bioprocess-supercritical extraction plant

steps to attain the desired level of purity which incurs additional operational cost on each purification step used in product recovery. Each step increase in purification process causes some loss of product cumulative yield and limit the total percentage of targeted product recovery [1]. Wrolstad et al. [143] made the same illustration clearer using the theoretical yield from multistep protein purifications. The authors assume a target protein yield of 75% at every four steps in a purification procedure and conclude that less than one-third of the desired protein still remain trapped at the completion of the whole process (Fig. 21.6). The illustration established that it is profitable to reduce the number of steps in purification and maintain a high yield at every possible step. Therefore, to obtain an optimum recovery that will be economically viable, it is advisable to reduce the purification steps needed to attain the desired product purity [11]. Thus, it can be argued that integrated bioprocess-supercritical carbon dioxide will help to improve the economic viability of bioproduct synthesis and reduce the cascade of equipment, which has been used previously to a single-step process. This argument was supported by Gifford et al. [51], who reported that supercritical fluid extraction uses half the energy of distillation to show its economic feasibility and it is a one-step process application.

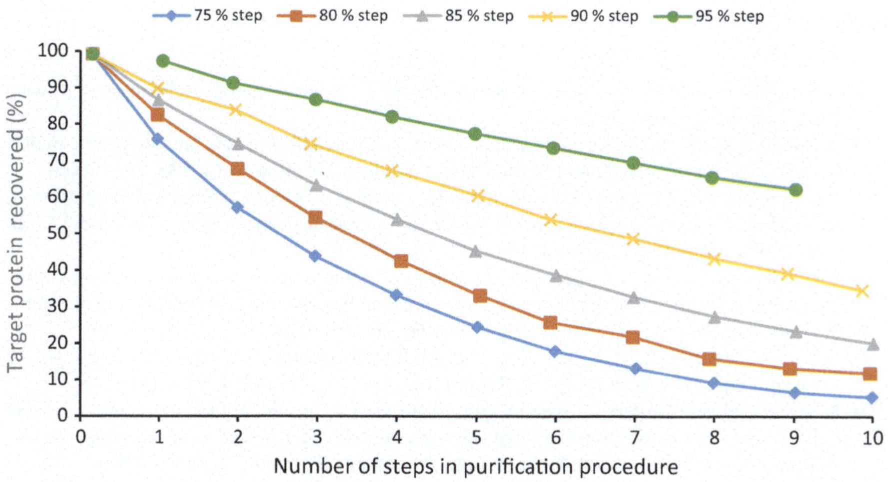

Fig. 21.6 Effect of the number of purification steps and percentage yield recovery of the overall process. (Source: Adapted from Wrolstad et al. [143])

21.7 Future Outlook

This review addresses the development of an integrated process for production and recovery of some selected bioproducts from substrates which may be derived from lignocellulosic materials. The chapter also attempts to explain developments with several strategies in the separation of bioproduct from the broth. Application of an integrated supercritical fluid extraction using carbon dioxide can be a useful alternative to conventional methods used in product recovery of fermentation broth. However, the cost of implementation and economic viability of supercritical fluid extraction when compared with the yield of bioproduct recovered have been the drawbacks to the integration of SFE in downstream processing. Although the commercial application of integrated supercritical carbon dioxide has been valued to be costly for non-food and non-drug bioproducts, the emerging trends in the industrial development of supercritical fluid requires critical analysis on a case-by-case basis with respect to desired product. The understanding of which is needed for attracting investment in separation of biological products. Therefore, integration of bioprocess-supercritical fluid extraction has potential for food, pharmaceutical, and chemical-producing industries in a sustainable manner. This calls for more appraisal of efficient processes for separation and recovery of bioactive compounds from dilute aqueous systems and for the isolation and concentration of targeted compounds. Supercritical fluids, most especially carbon dioxide, should be investigated, along with other processing alternatives, to establish the best fitted integrated process for different product recovery.

Acknowledgments The authors wish to acknowledge the financial support of Cape Peninsula University of Technology (CPUT) and National Research Foundation (NRF) to this project.

References

1. A.P.B. AB, *Protein Purification Handbook*, 18-1132-29 edn. (Amersham Pharmacia Biotech UK Limited, Uppsala, 1999)
2. A. Aden, T. Foust, Technoeconomic analysis of the dilute sulfuric acid and enzymatic hydrolysis process for the conversion of corn stover to ethanol. Cellulose **16**, 535–545 (2009)
3. V.I. Águeda, J.A. Delgado, M.A. Uguina, J.L. Sotelo, Á. García, Column dynamics of an adsorption–drying–desorption process for butanol recovery from aqueous solutions with silicalite pellets. Sep. Purif. Technol. **104**, 307–321 (2013)
4. C.L. Albuquerque, M.A.A. Meireles, Defatting of annatto seeds using supercritical carbon dioxide as a pretreatment for the production of bixin: Experimental, modeling and economic evaluation of the process. J. Supercrit. Fluids **66**, 86–95 (2012)
5. Z. Anwar, M. Gulfraz, M. Irshad, Agro-industrial lignocellulosic biomass a key to unlock the future bio-energy: A brief review. J. Radiat. Res. Appl. Sci. **7**, 163–173 (2014)
6. K.M. Ara, M. Jowkarderis, F. Raofie, Optimization of supercritical fluid extraction of essential oils and fatty acids from flixweed (*Descurainia sophia* L.) seed using response surface methodology and central composite design. J. Food Sci. Technol. **52**, 4450 (2015)
7. A. Arevalo-Gallegos, Z. Ahmad, M. Asgher, R. Parra-Saldivar, H.M. Iqbal, Lignocellulose: A sustainable material to produce value-added products with a zero waste approach—A review. Int. J. Biol. Macromol. **99**, 308–318 (2017)
8. S.H.M. Azhar, R. Abdulla, S.A. Jambo, H. Marbawi, J.A. Gansau, A.A.M. Faik, K.F. Rodrigues, Yeasts in sustainable bioethanol production: A review. Biochem. Biophys. Rep. **10**, 52–61 (2017)
9. S. Bandini, C. Gostoli, Ethanol removal from fermentation broth by gas membrane extraction. J. Membr. Sci. **70**, 119–127 (1992)
10. N.R. Baral, A. Shah, Techno-economic analysis of cellulose dissolving ionic liquid pretreatment of lignocellulosic biomass for fermentable sugars production. Biofuels Bioprod. Biorefin. **10**, 70–88 (2016)
11. J. Barford, *Fundamental and Applied Aspects of Animal Cell Cultivation* (NUS Press, Singapore, 1995)
12. P. Barghini, D. Di Gioia, F. Fava, M. Ruzzi, Vanillin production using metabolically engineered Escherichia coli under non-growing conditions. Microb. Cell Factories **6**(1), 13 (2007)
13. L. Bernad, A. Keller, D. Barth, M. Perrut, Separation of ethanol from aqueous solutions by supercritical carbon dioxide—Comparison between simulations and experiments. J. Supercrit. Fluids **6**, 9–14 (1993)
14. E.B. Bezerra, D.C. França, D.D.D.S. Morais, M.D.F. Rosa, J.P.S. Morais, E.M. Araújo, R.M.R. Wellen, Processing and properties of PCL/cotton linter compounds. Mater. Res. **20**, 317–325 (2017)
15. C. Brandenbusch, B. Bühler, P. Hoffmann, G. Sadowski, A. Schmid, Efficient phase separation and product recovery in organic-aqueous bioprocessing using supercritical carbon dioxide. Biotechnol. Bioeng. **107**, 642–651 (2010)
16. C. Brazinha, D.S. Barbosa, J.G. Crespo, Sustainable recovery of pure natural vanillin from fermentation media in a single pervaporation step. Green Chem. **13**, 2197–2203 (2011)
17. A.R. Brochado, C. Matos, B.L. Møller, J. Hansen, U.H. Mortensen, K.R. Patil, Improved vanillin production in baker's yeast through in silico design. Microb. Cell Factories **9**(1), 84 (2010)
18. M. Budich, G. Brunner, Supercritical fluid extraction of ethanol from aqueous solutions. J. Supercrit. Fluids **25**, 45–55 (2003)
19. K. Buzała, P. Przybysz, J. Rosicka-Kaczmarek, H. Kalinowska, Production of glucose-rich enzymatic hydrolysates from cellulosic pulps. Cellulose **22**, 663–674 (2015)
20. K.P. Buzała, H. Kalinowska, P. Przybysz, E. Małachowska, Conversion of various types of lignocellulosic biomass to fermentable sugars using kraft pulping and enzymatic hydrolysis. Wood Sci. Technol. **51**, 873–885 (2017)

21. M. Carlquist et al., Process engineering for bioflavour production with metabolically active yeasts–a mini-review. Yeast **32**, 123–143 (2015)
22. A.L. Carroll, S.H. Desai, S. Atsumi, Microbial production of scent and flavor compounds. Curr. Opin. Biotechnol. **37**, 8–15 (2016). https://doi.org/10.1016/j.copbio.2015.09.003
23. O. Catchpole et al., Integrated supercritical fluid extraction and bioprocessing. Am. J. Biochem. Biotechnol. **8**, 263–287 (2012)
24. R.N. Cavalcanti, C.L. Albuquerque, M.A.A. Meireles, Supercritical CO_2 extraction of cupuassu butter from defatted seed residue: experimental data, mathematical modeling and cost of manufacturing. Food Bioprod. Process. **97**, 48–62 (2016)
25. M. Cengiz, O.D. Dincturk, H.T. Sahin, Fractional extraction and structural characterization of opium poppy and cotton stalks hemicelluloses. Pharmacogn. Mag. **6**, 315 (2010)
26. A. Chandrasekaran, S. Sivamani, Statistical modeling and optimization of pretreatment for fermentable sugars production from cotton gin waste. Energ. Source. Part A **40**, 400–405 (2018)
27. A.R. Choudhury, P. Bhattacharjee, G.S. Prasad, Development of suitable solvent system for downstream processing of biopolymer pullulan using response surface methodology. PLoS One **8**, e77071 (2013)
28. A. Converti, B. Aliakbarian, J. Domínguez, G.B. Vázquez, P. Perego, Microbial production of biovanillin. Braz. J. Microbiol. **41**, 519–530 (2010)
29. M. Cvjetko Bubalo, S. Vidović, I. Radojčić Redovniković, S. Jokić, Green solvents for green technologies. J. Chem. Technol. Biotechnol. **90**, 1631–1639 (2015)
30. J.-Y. Dai, L.-H. Ma, Z.-F. Wang, W.-T. Guan, Z.-L. Xiu, Sugaring-out extraction of acetoin from fermentation broth by coupling with fermentation. Bioprocess Biosyst. Eng. **40**, 423–429 (2017a)
31. J. Dai, W. Guan, L. Ma, Z. Xiu, Salting-out extraction of acetoin from fermentation broth using ethyl acetate and K 2 HPO 4. Sep. Purif. Technol. **184**, 275–279 (2017b)
32. K.K. Darani, M.R. Mozafari, Supercritical fluids technology in bioprocess industries: A review journal of biochemical. Technology **2**, 144–152 (2010)
33. M. Dave, World cotton market (2018). https://cottonaustralia.com.au/cotton-library/fact-sheets/cotton-fact-file-the-world-cotton-market.
34. A.M. de Castro, J.A. López, L. dos Reis Castilho, D.M.G. Freire, Techno-economic analysis of a bioprocess for the production of multienzyme solutions from the cake of babassu industrial processing: Evaluation of five different inoculum propagation strategies. Biomass Convers. Biorefin. **4**, 237–247 (2014)
35. G.R. de la Vega, M.S. Cervantes, M.G. Alvarado, A. Romero-Martínez, P. Hegel, Fractionation of vanilla oleoresin by supercritical CO 2 technology. J. Supercrit. Fluids **108**, 79–88 (2016)
36. F.J. Dechow, *Separation and Purification Techniques in Biotechnology* (Noyes Publications, Park Ridge, 1989)
37. F.H. Deindoerfer, J.M. West, Rheological properties of fermentation broths. Adv. Appl. Microbiol. **2**, 265–273 (1960)
38. D. Di Gioia, L. Sciubba, L. Setti, F. Luziatelli, M. Ruzzi, D. Zanichelli, F. Fava, Production of biovanillin from wheat bran. Enzym. Microb. Technol. **41**, 498–505 (2007)
39. V.H.G. Díaz, G.O. Tost, Ethanol and isobutanol dehydration by heat-integrated distillation. Chem. Eng. Process. Process Intensif. **108**, 117–124 (2016)
40. B. Díaz-Reinoso, A. Moure, H. Domínguez, J.C. Parajó, Supercritical CO_2 extraction and purification of compounds with antioxidant activity. J. Agric. Food Chem. **54**, 2441–2469 (2006)
41. P.M. Doran, *Bioprocess Engineering Principles* (Academic Press, London, 1995)
42. C. Effendi, M. Shanty, Y.-H. Ju, A. Kurniawan, M.-J. Wang, N. Indraswati, S. Ismadji, Measurement and mathematical modeling of solubility of buttery-odor substance (acetoin) in supercritical CO 2 at several pressures and temperatures. Fluid Phase Equilib. **356**, 102–108 (2013)

43. M.-H. Eom, B. Kim, H. Jang, S.-H. Lee, W. Kim, Y.-A. Shin, J.H. Lee, Dynamic modeling of a fermentation process with ex situ butanol recovery (ESBR) for continuous biobutanol production. Energy Fuel **29**, 7254–7265 (2015)
44. D. Essien, D. Pyle, Fermentation ethanol recovery by solvent extraction Sep. Biotechnol **23**, 320–332 (1987)
45. Evolva, Vanillin: A sustainable production route. Evolva (2017). http://www.evolva.com/vanillin/
46. C.E. Fabre, J.S. Condoret, A. Marty, Extractive fermentation of aroma with supercritical CO2. Biotechnol. Bioeng. **64**, 392–400 (1999)
47. A.M. Farías-Campomanes, M.A. Rostagno, M.A.A. Meireles, Production of polyphenol extracts from grape bagasse using supercritical fluids: Yield, extract composition and economic evaluation. J. Supercrit. Fluids **77**, 70–78 (2013)
48. Y. Feng, D. Meier, Supercritical carbon dioxide extraction of value-added chemicals from slow pyrolysis liquids, in Wood 2015: Innovations in Wood Materials and Processes 37 (2015)
49. T. Fornari, E.J. Hernández, A. Ruiz-Rodriguez, F.J. Señorans, G. Reglero, Phase equilibria for the removal of ethanol from alcoholic beverages using supercritical carbon dioxide. J. Supercrit. Fluids **50**, 91–96 (2009)
50. R. Ghosh, *Principles of Bioseparations Engineering* (World Scientific Publishing Co Inc, Singapore, 2006)
51. M. Gifford, E. Biancani, W. Kearsley, W. Maluchnik, S. Farrell, M. Savelski, R. Hesketh, *Economic Feasibility Study on the Supercritical Fluid Extraction of Edible Oils* (Supercritical Fluid Technologies, Inc., USA, 2001)
52. I. Gil, A. Uyazán, J. Aguilar, G. Rodríguez, L. Caicedo, Separation of ethanol and water by extractive distillation with salt and solvent as entrainer: Process simulation. Braz. J. Chem. Eng. **25**, 207–215 (2008)
53. Global Market Insight, Bio vanillin market size by application, price trend, competitive market share and forecast, 2016–2023 (2016). https://www.gminsights.com/industry-analysis/bio-vanillin-market. Accessed Feb 2017
54. U. Gottschalk, *Process Scale Purification of Antibodies* (Wiley, Somerset, 2011)
55. W.J. Groot, H.S. Soedjak, P.B. Donck, R.G.J.M. van der Lans, K.C.A.M. Luyben, J.M.K. Timmer, Butanol recovery from fermentations by liquid-liquid extraction and membrane solvent extraction. Bioprocess Eng. **5**, 203–216 (1990). https://doi.org/10.1007/bf00376227
56. T. Gu, Liquid-liquid partitioning methods for bioseparations. Sep. Sci. Technol. **2**. Elsevier,, 329–364 (2000)
57. A. Gupta, J.P. Verma, Sustainable bio-ethanol production from agro-residues: A review. Renew. Sust. Energ. Rev. **41**, 550–567 (2015)
58. A. Güvenç, Ü. Mehmetoglu, A. Çalimli, Supercritical CO2 extraction of ethanol from fermentation broth in a semicontinuous system. J. Supercrit. Fluids **13**, 325–329 (1998). https://doi.org/10.1016/S0896-8446(98)00067-9
59. A. Güvenç, Ü. Mehmetoğlu, A. Çalimli, Supercritical CO_2 extraction of ethanol. Turk. J. Chem. **23**, 285–292 (1999)
60. R.G. Harrison, Bioseparation basics. Chem. Eng. Prog. **110**, 36–42 (2014)
61. R.G. Harrison, P. Todd, P.W. Todd, S.R. Rudge, D.P. Petrides, *Bioseparations Science and Engineering* (Topics in Chemical Engineering, New York, 2015)
62. E. Heinzle, A.P. Biwer, C.L. Cooney, *Development of Sustainable Bioprocesses: Modeling and Assessment* (Wiley, Hoboken, 2007)
63. D. Hua, C. Ma, L. Song, S. Lin, Z. Zhang, Z. Deng, P. Xu, Enhanced vanillin production from ferulic acid using adsorbent resin. Appl. Microbiol. Biotechnol. **74**, 783–790 (2007)
64. N. Ikawa, Y. Nagase, T. Tada, S. Furuta, R. Fukuzato, Separation process of ethanol from aqueous solutions using supercritical carbon dioxide. Fluid Phase Equilib. **83**, 167–174 (1993). https://doi.org/10.1016/0378-3812(93)87019-W

65. A. Isci, G. Demirer, Biogas production potential from cotton wastes. Renew. Energy **32**, 750–757 (2007)
66. F.H. Isikgor, C.R. Becer, Lignocellulosic biomass: A sustainable platform for the production of bio-based chemicals and polymers. Polym. Chem. **6**, 4497–4559 (2015)
67. P. Jiménez-Bonilla, Y. Wang, In situ biobutanol recovery from clostridial fermentations: A critical review. Crit. Rev. Biotechnol. **38**, 469–482 (2018)
68. G. Joana Gil-Chávez, J.A. Villa, J. Fernando Ayala-Zavala, J. Basilio Heredia, D. Sepulveda, E.M. Yahia, G.A. González-Aguilar, Technologies for extraction and production of bioactive compounds to be used as nutraceuticals and food ingredients: An overview. Compr. Rev. Food Sci. Food Saf. **12**, 5–23 (2013)
69. T. Kaneko, Y. Watanabe, H. Suzuki, Enhancement of diacetyl production by a diacetyl-resistant mutant of citrate-positive Lactococcus lactis ssp. lactis 3022 and by aerobic conditions of growth. J. Dairy Sci. **73**, 291–298 (1990)
70. S. Kaur, G. Dhillon, S. Brar, Increasing trend towards production of high value bioproducts from biomass feedstocks. Int. J. Waste Res. **3**, 2252–5211 (2013)
71. S. Kaur, G.S. Dhillon, S.J. Sarma, S.K. Brar, K. Misra, H.S. Oberoi, Waste biomass: A prospective renewable resource for development of bio-based economy/processes, in *Biotransformation of Waste Biomass into High Value Biochemicals*, (Springer, New York, 2014), pp. 3–28
72. C.R. Kennedy, *The Flavor Rundown: Natural vs. Artificial Flavors* (Harvard University: Science in the News, Boston, 2015), p. 21
73. K. Khosravi-Darani, E. Vasheghani-Farahani, Application of supercritical fluid extraction in biotechnology. Crit. Rev. Biotechnol. **25**, 231–242 (2005)
74. J.K. Kim, E.L. Iannotti, R. Bajpai, Extractive recovery of products from fermentation broths. Biotechnol. Bioprocess Eng. **4**, 1–11 (1999)
75. Ž. Knez, M. Škerget, A.P. Uzunalić, Phase equilibria of vanillins in compressed gases. J. Supercrit. Fluids **43**, 237–248 (2007)
76. B. Knutson, H. Strobel, S. Nokes, K. Dawson, J. Berberich, C. Jones, Effect of pressurized solvents on ethanol production by the thermophilic bacterium Clostridium thermocellum. J. Supercrit. Fluids **16**, 149–156 (1999)
77. F. Kollerup, A.J. Daugulis, Screening and identification of extractive fermentation solvents using a database. Can. J. Chem. Eng. **63**, 919–927 (1985)
78. J.M. Kordylas, Biotechnology for production of fruits, wines, and alcohol applications of biotechnology to traditional fermented foods: Report of an ad hoc panel of the board on science and technology for international development (National Academies Press (US), Washington (DC), 1992), pp. 170
79. A.A. Koutinas, B. Yepez, N. Kopsahelis, D.M. Freire, A.M. de Castro, S. Papanikolaou, I.K. Kookos, Techno-economic evaluation of a complete bioprocess for 2, 3-butanediol production from renewable resources. Bioresour. Technol. **204**, 55–64 (2016)
80. Q. Lang, C.M. Wai, Supercritical fluid extraction in herbal and natural product studies—A practical review. Talanta **53**, 771–782 (2001)
81. S.Y. Lee, H.U. Kim, Systems strategies for developing industrial microbial strains. Nat. Biotechnol. **33**, 1061–1072 (2015)
82. K.H. Lee, J.H. Park, T.Y. Kim, H.U. Kim, S.Y. Lee, Systems metabolic engineering of Escherichia coli for L-threonine production. Mol. Syst. Biol. **3**, 149 (2007)
83. X. Li, Y. Pang, R. Zhang, Compositional changes of cottonseed hull substrate during *P. ostreatus* growth and the effects on the feeding value of the spent substrate. Bioresour. Technol. **80**, 157–161 (2001)
84. J.S. Lim, Y.Y. Lee, H.S. Chun, Phase equilibria for carbon dioxide-ethanol-water system at elevated pressures. J. Supercrit. Fluids **7**, 219–230 (1994)
85. J.S. Lim, Y.-W. Lee, J.-D. Kim, Y.Y. Lee, H.-S. Chun, Mass-transfer and hydraulic characteristics in spray and packed extraction columns for supercritical carbon dioxide-ethanol-water system. J. Supercrit. Fluids **8**, 127–137 (1995)

86. J. Liu, Y. Kim, M.A. McHugh, Phase behavior of the vanillin–CO_2 system at high pressures. J. Supercrit. Fluids **39**, 201–205 (2006)
87. W. Liu, Y.-J. Fu, Y.-G. Zu, M.-H. Tong, N. Wu, X.-L. Liu, S. Zhang, Supercritical carbon dioxide extraction of seed oil from Opuntia dillenii Haw. and its antioxidant activity. Food Chem. **114**, 334–339 (2009)
88. Y. Liu, S. Zhang, Y.-C. Yong, Z. Ji, X. Ma, Z. Xu, S. Chen, Efficient production of acetoin by the newly isolated Bacillus licheniformis strain MEL09. Process Biochem. **46**, 390–394 (2011)
89. Z.M. Lu, W. Xu, N.H. Yu, T. Zhou, G.Q. Li, J.S. Shi, Z.H. Xu, Recovery of aroma compounds from Zhenjiang aromatic vinegar by supercritical fluid extraction. Int. J. Food Sci. Technol. **46**, 1508–1514 (2011)
90. S. Macedo et al., Recovery of wine-must aroma compounds by supercritical CO 2. Food Bioprocess Technol. **1**, 74–81 (2008)
91. S. Maiti, G. Gallastegui, S. Kaur Brar, Y. LeBihan, G. Buelna, P. Drogui, M. Verma, Quest for sustainable bio-production and recovery of butanol as a promising solution to fossil fuel. Int. J. Energy Res. **40**(4), 411–438 (2015)
92. J.-M. Menet, D. Thiebaut, *Countercurrent Chromatography* (CRC Press, Boca Raton, 1999)
93. G. Montague, A. Morris, A. Ward, Fermentation monitoring and control: A perspective. Biotechnol. Genet. Eng. Rev. **7**, 147–188 (1989)
94. D.M. Mousdale, *Biofuels: Biotechnology, Chemistry, and Sustainable Development* (CRC Press, Boca Raton, 2008)
95. J.M. Newton, J. Vlahopoulou, Y. Zhou, Investigating and modelling the effects of cell lysis on the rheological properties of fermentation broths. Biochem. Eng. J. **121**, 38–48 (2017)
96. K. Nguyen, P. Barton, J.S. Spencer, Supercritical carbon dioxide extraction of vanilla. J. Supercrit. Fluids **4**, 40–46 (1991)
97. J. Nielsen, Fermentation monitoring, design and optimization, in *Encyclopedia of Bioprocess Technology*, (Wiley, New York, 2008)
98. M. Nomura, T. Bin, S.-i. Nakao, Selective ethanol extraction from fermentation broth using a silicalite membrane. Sep. Purif. Technol. **27**, 59–66 (2002)
99. N.M. Nor, M.S. Mohamed, T.C. Loh, H.L. Foo, R.A. Rahim, J.S. Tan, R. Mohamad, Comparative analyses on medium optimization using one-factor-at-a-time, response surface methodology, and artificial neural network for lysine–methionine biosynthesis by Pediococcus pentosaceus RF-1. Biotechnol. Biotechnol. Equip. **31**, 1–13 (2017)
100. A. Paz, D. Outeiriño, R. Pinheiro de Souza Oliveira, J.M. Domínguez, Fed-batch production of vanillin by Bacillus aryabhattai BA03. New Biotechnol. **40**, 186–191 (2018). https://doi.org/10.1016/j.nbt.2017.07.012
101. C. Pereira, R. Rocha, F.L.P. Pessoa, M. Mendes, Phorbol esters extraction from Jatropha curcas seed cake using supercritical carbon dioxide, in III Iberoamerican Conference on Supercritical Fluids Cartagena de Indias (Colômbia) (2013)
102. M. Perrut, Supercritical fluid applications: Industrial developments and economic issues. Ind. Eng. Chem. Res. **39**, 4531–4535 (2000)
103. D. Petrides, Bioprocess design and economics. Biosep. Sci. Eng. **2**, 1–83 (2000)
104. C.A. Pieck, C. Crampon, F. Charton, E. Badens, Multi-scale experimental study and modeling of the supercritical fractionation process. J. Supercrit. Fluids **105**, 158–169 (2015)
105. I.M. Prado, C.L.C. Albuquerque, R.N. Cavalcanti, M.A.A. Meireles, Use of a Commercial Process Simulator to Estimate the Cost of Manufacturing (COM) of Carotenoids Obtained Via Supercritical Technology from Palm and. Paper Presented at the Ninety International Symposium on Supercritical Fluids, Arcachon, France (2009)
106. J.M. Prado et al., Supercritical fluid extraction of grape seed: Process scale-up, extract chemical composition and economic evaluation. J. Food Eng. **109**, 249–257 (2012)
107. H. Priefert, J. Rabenhorst, A. Steinbüchel, Biotechnological production of vanillin. Appl. Microbiol. Biotechnol. **56**, 296–314 (2001)

108. M.R. Pursell, M.A. Mendes-Tatsis, D.C. Stuckey, Effect of fermentation broth and biosurfactants on mass transfer during liquid–liquid extraction. Biotechnol. Bioeng. **85**, 155–165 (2004)
109. K. Rajendran, E. Drielak, V.S. Varma, S. Muthusamy, G. Kumar, Updates on the pretreatment of lignocellulosic feedstocks for bioenergy production–a review. Biomass Conver. Bioref. **4**, 1–13 (2018)
110. D.T. Santos, P.C. Veggi, M.A.A. Meireles, Optimization and economic evaluation of pressurized liquid extraction of phenolic compounds from jabuticaba skins. J. Food Eng. **108**, 444–452 (2012)
111. P.K. Sarangi, S. Nanda, H. Sahoo, Maximization of vanillin production by standardizing different cultural conditions for ferulic acid degradation. NY Sci. J. **3**, 77–79 (2010)
112. S.J. Sarma, G.S. Dhillon, K. Hegde, S.K. Brar, M. Verma, Utilization of agro-industrial waste for the production of aroma compounds and fragrances, in *Biotransformation of Waste Biomass into High Value Biochemicals*, (Springer, New York, 2014), pp. 99–115
113. M. Schaechter, *Encyclopedia of Microbiology* (Academic Press, Amsterdam, 2009)
114. W.D. Seider, J.D. Seader, D.R. Lewin, *Product & Process Design Principles: Synthesis, Analysis and Evaluation, (With CD)* (Wiley, New York, 2009)
115. A. Shah, M. Darr, A techno-economic analysis of the corn stover feedstock supply system for cellulosic biorefineries. Biofuels Bioprod. Biorefin. **10**, 542–559 (2016)
116. R. Sharma-Shivappa, Y. Chen, Conversion of cotton wastes to bioenergy and value-added products. Trans. ASABE **51**, 2239–2246 (2008)
117. G. Sindhwani, I. Uk, V. Aeri, Microbial transformation of eugenol to vanillin. J. Microbiol. Biotechnol. Res. **2**, 313–318 (2012)
118. A. Singh, G.P. Rangaiah, Process development and optimization of bioethanol recovery and dehydration by distillation and vapor permeation for multiple objectives. Differ. Evol. Chem. Eng. **6**, 289 (2017a)
119. A. Singh, G.P. Rangaiah, Review of technological advances in bioethanol recovery and dehydration. Ind. Eng. Chem. Res. **56**, 5147–5163 (2017b)
120. S.P. Singh, E. Ekanem, T. Wakefield Jr., S. Comer, Emerging importance of bio-based products and bio-energy in the US economy: Information dissemination and training of students. Int. Food Agribus. Manag. Rev. **5**, 14 (2003)
121. M. Škerget, L. Čretnik, Ž. Knez, M. Škrinjar, Influence of the aromatic ring substituents on phase equilibria of vanillins in binary systems with CO 2. Fluid Phase Equilib. **231**, 11–19 (2005)
122. H. Sovová, Modeling the supercritical fluid extraction of essential oils from plant materials. J. Chromatogr. A **1250**, 27–33 (2012). https://doi.org/10.1016/j.chroma.2012.05.014
123. W.H. Streng, *Characterization of Compounds in Solution: Theory and Practice* (Springer, Boston, 2012)
124. J. Sun et al., Enhanced acetoin production by Serratia marcescens H32 using statistical optimization and a two-stage agitation speed control strategy. Biotechnol. Bioprocess Eng. **17**, 598–605 (2012). https://doi.org/10.1007/s12257-011-0587-4
125. A. Taiwo, T. Madzimbamuto, T. Ojumu, Optimization of corn steep liquor dosage and other fermentation parameters for ethanol production by Saccharomyces cerevisiae type 1 and anchor instant yeast. Energies **11**, 1740 (2018a)
126. A.E. Taiwo, T.V. Ojumu, T.N. Madzimbamuto, Statistical optimization of acetoin production using corn steep liquor as a low-cost nitrogen source by Bacillus Subtilis CICC 10025, in *Renewable Resources and Biorefineries*, (IntechOpen, London, 2018b)
127. M. Tan, S. Liew, M.Y. Maskat, W.W. Aida, H. Osman, Optimization of vanillin production using isoeugenol as substrate by Aspergillus Niger I-1472. Int. Food Res. J. **22**, 1651 (2015)
128. V. Tarabanko, Y.V. Chelbina, V. Sokolenko, N. Tarabanko, A study of vanillin extraction by octylamine. Solvent Extr. Ion Exch. **25**, 99–107 (2007)

129. R.M. Teixeira, D. Cavalheiro, J. Ninow, A. Furigo Jr., Optimization of acetoin production by Hanseniaspora guilliermondii using experimental design. Braz. J. Chem. Eng. **19**, 181–186 (2002). https://doi.org/10.1590/S0104-66322002000200014
130. Y. Tian, Y. Fan, J. Liu, X. Zhao, W. Chen, Effect of nitrogen, carbon sources and agitation speed on acetoin production of Bacillus subtilis SF4-3. Electron. J. Biotechnol. **19**, 41–49 (2016). https://doi.org/10.1016/j.ejbt.2015.11.005
131. A. Tilay, M. Bule, U. Annapure, Production of biovanillin by one-step biotransformation using fungus Pycnoporous cinnabarinus. J. Agric. Food Chem. **58**, 4401–4405 (2010)
132. F.Ö. Ütkür, T. Thanh Tran, J. Collins, C. Brandenbusch, G. Sadowski, A. Schmid, B. Bühler, Integrated organic–aqueous biocatalysis and product recovery for quinaldine hydroxylation catalyzed by living recombinant Pseudomonas putida. J. Ind. Microbiol. Biotechnol. **39**, 1049–1059 (2012)
133. P. Van Hee, M. Hoeben, R. Van der Lans, L. Van der Wielen, Strategy for selection of methods for separation of bioparticles from particle mixtures. Biotechnol. Bioeng. **94**, 689–709 (2006)
134. L.M. Vane, F.R. Alvarez, L. Rosenblum, S. Govindaswamy, Efficient ethanol recovery from yeast fermentation broth with integrated distillation–membrane process. Ind. Eng. Chem. Res. **52**, 1033–1041 (2012)
135. S. Vani, R.K. Sukumaran, S. Savithri, Prediction of sugar yields during hydrolysis of lignocellulosic biomass using artificial neural network modeling. Bioresour. Technol. **188**, 128–135 (2015)
136. P.C. Veggi, R.N. Cavalcanti, M.A.A. Meireles, Production of phenolic-rich extracts from Brazilian plants using supercritical and subcritical fluid extraction: Experimental data and economic evaluation. J. Food Eng. **131**, 96–109 (2014)
137. P.T. Viana, F.A. Teixeira, A.J.V. Pires, G.G.P.D. Carvalho, Figueiredo MPD, J.H.A.D. Santana, Losses and nutritional value of elephant grass silage with inclusion levels of cottonseed meal. Acta Sci. Anim. Sci. **35**, 139–144 (2013)
138. S. Vijayan, D.P. Byskal, L.P. Buckley, Separation of oil from fried chips by a supercritical extraction process; an overview of bench-scale test experience and process economics. In *Supercritical Fluids Processing of Food and Biomaterials*, ed. by S. Rizvi (Chapman and Hall, Glasgow, 1994), pp. 74–92
139. Z. Wang, K. Chen, J. Li, Q. Wang, J. Guo, Separation of vanillin and syringaldehyde from oxygen delignification spent liquor by macroporous resin adsorption. Clean Soil. Air. Water **38**, 1074–1079 (2010)
140. P.N. Whittington, Fermentation broth clarification techniques. Appl. Biochem. Biotechnol. **23**, 91–121 (1990)
141. Williams DF (1981) Fundamental Aspects of Biocompatibility 1. CRC Press LLC, Boca Raton
142. K. Wilson, J. Walker, *Principles and Techniques of Practical Biochemistry* (Cambridge University Press, Cambridge, 2000)
143. R.E. Wrolstad et al., *Handbook of Food Analytical Chemistry, Volume 1: Water, Proteins, Enzymes, Lipids, and Carbohydrates* (Wiley, Hoboken, 2005)
144. J. Wu et al., Recovery of acetoin from the ethanol–acetoin–acetic acid ternary mixture based on adsorption methodology using a hyper-cross-linked resin. Ind. Eng. Chem. Res. **53**, 12411–12419 (2014)
145. W. Wu, K. Yenkie, C.T. Maravelias, A superstructure-based framework for bio-separation network synthesis. Comput. Chem. Eng. **96**, 1–17 (2017)
146. Z. Xiao, P. Liu, J.Y. Qin, P. Xu, Statistical optimization of medium components for enhanced acetoin production from molasses and soybean meal hydrolysate. Appl. Microbiol. Biotechnol. **74**, 61–68 (2007). https://doi.org/10.1007/s00253-006-0646-5
147. H. Xu, S. Jia, J. Liu, Production of acetoin by Bacillus subtilis TH-49, in *Consumer Electronics, Communications and Networks (CECNet), 2011 International Conference on, 16–18 April 2011*, (IEEE, Piscataway, 2011), pp. 1524–1527

148. S.-T. Yang, *Bioprocessing for Value-Added Products from Renewable Resources: New Technologies and Applications* (Elsevier, Burlington, 2011)
149. D. Yiyong, C. Hong, A new bioprocess to produce natural vanillin by microbial fermentation. Flavour Frag. Cosmet. **3**, 003 (2011)
150. M. Zabkova, E.B. da Silva, A. Rodrigues, Recovery of vanillin from Kraft lignin oxidation by ion-exchange with neutralization. Sep. Purif. Technol. **55**, 56–68 (2007)
151. M. Žabková, E.B. da Silva, A. Rodrigues, Recovery of vanillin from lignin/vanillin mixture by using tubular ceramic ultrafiltration membranes. J. Membr. Sci. **301**, 221–237 (2007)
152. Q.-F. Zhang, Z.-T. Jiang, H.-J. Gao, R. Li, Recovery of vanillin from aqueous solutions using macroporous adsorption resins. Eur. Food Res. Technol. **226**, 377–383 (2008)
153. L. Zhang, S. Chen, H. Xie, Y. Tian, K. Hu, Efficient acetoin production by optimization of medium components and oxygen supply control using a newly isolated Paenibacillus polymyxa CS107. J. Chem. Technol. Biotechnol. **87**, 1551–1557 (2012). https://doi.org/10.1002/jctb.3791
154. Y. Zhang, S. Li, L. Liu, J. Wu, Acetoin production enhanced by manipulating carbon flux in a newly isolated Bacillus amyloliquefaciens. Bioresour. Technol. **130**, 256–260 (2013)
155. S. Zhang, X. Huang, C. Qu, Y. Suo, Z. Liao, J. Wang, Extractive fermentation for enhanced isopropanol and n-butanol production with mixtures of water insoluble aliphatic acids and oleyl alcohol. Biochem. Eng. J. **117**, 112–120 (2017)
156. Q.-W. Zhang, L.-G. Lin, W.-C. Ye, Techniques for extraction and isolation of natural products: A comprehensive review. Chin. Med. **13**(20), 20 (2018)
157. L. Zheng, P. Zheng, Z. Sun, Y. Bai, J. Wang, X. Guo, Production of vanillin from waste residue of rice bran oil by Aspergillus Niger and Pycnoporus cinnabarinus. Bioresour. Technol. **98**, 1115–1119 (2007)
158. C. Zidi, R. Tayeb, N. Boukhili, M. Dhahbi, A supported liquid membrane system for efficient extraction of vanillin from aqueous solutions. Sep. Purif. Technol. **82**, 36–42 (2011)
159. V.G. Zuin, L.Z. Ramin, Green and sustainable separation of natural products from agro-industrial waste: Challenges, potentialities, and perspectives on emerging approaches. Top. Curr. Chem. **376**, 3 (2018)
160. M. Yusoff, H. Akita, M. Hassan, S. Fujimoto, M. Yoshida, N. Nakashima, & T. Hoshino, Production of acetoin from hydrothermally pretreated oil mesocarp fiber using metabolically engineered Escherichia coli in a bioreactor system. Bioresour. Technol. **245**, 1040–1048 (2017)

Chapter 22
Separation of Carboxylic Acids: Conventional and Intensified Processes and Effects of Process Engineering Parameters

V. M. Inyang and D. Lokhat

22.1 Introduction

The recent global research agenda highlighted industrial system development via process intensification in various sectors vis-à-vis chemical, water and food production as being crucial to achieving technological milestones in the envisaged 2050 [27]. The new trend in the utilization of renewable feedstock, wastes and by-products and studies in general green chemistry have made the production of chemicals through a bio-based route using cheap and available biomass materials to receive increasing attention. Many biorefinery processes employing especially the biochemical approach have challenges in the downstream separation and purification processes due to some factors such as product inhibition, low concentration feed and product yields [41]. Use of biomass materials that are cheap and abundantly available contributes to reducing the costs in producing desired products. Biotechnology is potentially presenting new, efficient and low-cost fermentation processes using biomass feedstock for chemical production. The economic impact of fermentation of bio-products is presently limited due to challenges in the recovery of products from aqueous solution. Therefore, the development of recovery techniques is essential to allow for chemicals obtained by fermentation to further penetrate into the industry [17].

Currently, acids derived from petrochemicals are produced at a cheaper rate compared to those derived through bio-based route. Therefore, research efforts must be expanded to discover more effective ways to decrease the cost of processing bio-based acids. More importantly, in the downstream processing, a significant amount

V. M. Inyang (✉) · D. Lokhat
Department of Chemical Engineering, University of KwaZulu-Natal, Howard College Campus, Durban, South Africa

M. O. Daramola, A. O. Ayeni (eds.), *Valorization of Biomass to Value-Added Commodities*, Green Energy and Technology,
https://doi.org/10.1007/978-3-030-38032-8_22

of the final processing costs are taken into account in the separation and purification of the required products [67, 92]. While several attempts have been made to improve bio-based separation processes to source different acids, no adequately successful technologies have been developed to industrial production level. In spite of the several separation processes proposed (chromatography, liquid–liquid extraction, ion exchange resins, precipitation, membrane separation) to overcome this bottleneck, limitations still remain in terms of lean production, waste generation, large energy input and material consumption [71, 92, 125].

A potential alternative process is using hybrid reactors in the direct conversion of acids in aqueous solutions to esters, where reaction and separation occur simultaneously. This intensified process has clear advantages over conventional processes by reducing the cost of production and providing esters as intermediates for other chemical syntheses. Although esterification is a well-known technology, focus on its operational performance still needs to be better understood to generate the necessary design and model tools, and consequently for its scale-up to commercial production. Thus, a major requirement for making the reactive separation process viable is the identification of suitable catalysts with high stability and activity, easy separation and recovery of products, and avoidance of equipment corrosion. Besides catalyst development, a better understanding of esterification reaction kinetics is necessary to enhance process chemistry analysis, reaction parameter optimization and equilibrium studies of the separation process. This information, in turn, will allow an assessment to be made on the potential industrial applicability of the overall design and development of a sustainable biorefinery approach to value-added production.

The development and implementation of these separation strategies are required for optimum and successful commercialization of biorefineries. Biorefineries are essentially the upstream, midstream and downstream processing of biomass to bio-based products including a collection of co-products and bioenergy (including chemicals and other materials; [39, 40]). Before any industrial commercialization can succeed, some technical challenges need to be overcome for the full utilization of bio-based products. Biomass feedstocks for production of chemicals and energy include starchy biomass (cassava, wheat/corn), sugarcane, lignocellulose biomass like agricultural residues (crop residues such as barley straw and sugarcane bagasse), municipal solid waste refined materials, such as fructose, glucose and sucrose, mixtures of these materials, post-fermentation liquor and the likes. In general, biomass-based industries have been moving towards the integrated biorefinery approach employing the three necessary sequential approaches/stages which include [25]:

1. Biomass separation into different components in the fractionation unit.
2. Conversion of fractions into useful products. For these secondary processes, three so-called platforms can be applied: chemical, biochemical and thermochemical processes (gasification and liquefaction).
3. Further downstream processing to value-added products. This can be achieved through three different platforms.

22.1.1 Biochemical Platform

The biochemical platform includes all biochemical conversion processes, including the production of fermentable sugars by saccharification of cellulose and starch, and the fermentation of sugars obtained from lignocellulosic biomass. One of the many biochemical conversions deals with sugars, where biomass is initially pre-treated and hydrolysed to mono-sugars that on further processing are fermented to biofuels (ethanol and butanol) or chemicals (e.g. succinic, butyric, lactic, malic, propionic, acid) depending on the biocatalysts that are employed.

22.1.2 Thermochemical Platform

Thermochemical platform biomass conversion includes biomass combustion for heat generation, bio-oil and bio-char, hydrothermal liquefaction to bio-oils as a key product, and biomass gasification to syngas. Syngas (CO and H_2) from biomass gasification can be converted further into a variety of diverse chemicals and fuels using appropriate catalysts and appropriate operating conditions.

22.1.3 Chemical Platform

Biorefinery comprises several other chemical conversion processes – for example, the manufacture of value-added chemicals such as butyric, succinic, malic and propionic acids, or formic, acetic and lactic acids [18]. A biorefinery can also utilize both thermochemical and biochemical or chemical and biochemical conversion approaches.

Therefore, the chemical industry based on biomass-derived materials will be developed on selective platform chemicals other than the petrochemical-based industry. There is no standard solution to define the 'optimum' biorefinery, but a highly efficient, economically viable and sustainable one can be realized by efficient downstream conversion processes, flexibility in equipment design, maximizing chemical energy utilization from feedstock and development of conversion capacities that are independent of specific fuels, materials and chemicals [16].

The implementation of the biorefinery concept brings about several advantages at various levels and for multiple stakeholders: From a national perspective, it may help to meet national energy needs, while reducing greenhouse gas (GHG) emissions [85]. From the industrial point of view, an integrated biorefinery reduces the risk of operation by increasing the number of products produced, aiming at various markets. Additionally, the return on investment is significantly improved when the processing plant operates continuously throughout the year.

In this chapter, the status and different recovery and purification processes for carboxylic acids, a platform chemical that can be obtained from biomass-derived materials, are summarized and presented. Conventional separation processes of carboxylic acid, which include membrane separation, precipitation, liquid–liquid extraction, distillation, chromatography and crystallization, have been reviewed and discussed. Each technology has its own limitation and no single method has proved to be simple, efficient and effective with regard to yields, purity, energy consumption and scale-up to commercial production. Therefore, improvements are still needed in the development of separation and purification processes, which deserves utmost attention for the promising biorefinery industry – hence the need for the intensified separation process and enhancement mechanisms with particular focus on process intensification of reactive extraction and reactive distillation processes as the most significant separation methods. Additionally, the reliability and potential of reactive separation processes which is promising and will enable higher efficiency and capacity are discussed. Downstream processing of carboxylic acids with different extractant types and diluent used, equilibrium and kinetic studies and models that are necessary for the overall process design of the reactive separation process and the effects of process parameters on the separation and purification of carboxylic acids are also discussed.

22.2 Carboxylic Acids/Platform Chemicals

Carboxylic acids such as butyric, lactic, malic, propionic, citric, lactic, succinic and itaconic acids and others are important chemical products and are often recovered from fermentation broths. Historically, raw materials for the production of carboxylic acids include animal fats, petroleum and vegetable oil sources in largely non-aqueous systems. In recent times, these carboxylic acids have been listed among the most desirable products which are manufactured from biomass feedstock through fermentation routes. In the later processing, it is produced from fermentation broths of dilute aqueous solutions [61]. A drawback of using fermentation of bio-based feedstocks for the production of chemicals is that an aqueous solution with inherently low product concentrations is achieved, and above 40% of the cost stems from downstream processing according to Straathof [105]. This has led to the advancement of existing processes as well as the development of new processes due to the current interest in bio-based economy [71], hence the need for an overview of recovery alternatives.

Most carboxylic acids have extensive applications in food, pharmaceuticals, detergents, surfactants and green solvent industries, and as raw materials for eco-friendly polymers [38]. Carboxylic acids are favourable intermediates in a bioprocessing multifaceted complex because the oxygen of the biomass is placed in a form that is useful for further reaction with many other products [60]. Typical carboxylic acids with early stage process development are shown in Table 22.1.

Table 22.1 Typical carboxylic acids of commercial interest for production by fermentation [71]

Molecular formula	Carboxylic acid	Status of biochemical production	Main application
$C_2H_4O_2$	Acetic	Industrial	Vinegar
$C_3H_4O_2$	Acrylic	Research	Polymers
$C_3H_6O_2$	Pyruvic	Research	Chemicals
$C_3H_6O_2$	Propionic	Design stage	Chemicals
$C_3H_6O_3$	D/L-Lactic	Industrial	Food, polymers
$C_3H_6O_3$	3-Hydroxy-propionic	Research	Polymers
$C_4H_4O_4$	Fumaric	Formerly industrial	Food, polymers
$C_4H_6O_4$	Succinic	Industrial	Polymers, chemicals
$C_4H_6O_5$	L-Malic	Research	Chemicals
$C_4H_8O_2$	Butyric	Design stage	Chemicals
$C_5H_6O_4$	Itaconic	Industrial	Polymers
$C_5H_8O_4$	Glutaric	Research	Polymers
$C_6H_4O_5$	2,5-Furan-dicarboxylic	Research	Polymers
$C_6H_8O_7$	Citric	Industrial	Food
$C_6H_{10}O_4$	Adipic	Design stage	Polymers
$C_6H_{10}O_7$	2-Keto-L-gulonic	Industrial	Vitamin C precursor
$C_6H_{12}O_7$	D-Gluconic	Industrial	Food

22.2.1 *Dissimilar Property Nature of Various Carboxylic Acids*

Physical properties of carboxylic acids are dissimilar and so recovery processes are different. But those carboxylic acids grouped on the basis of physical properties have a similar way of recovery. These dissimilar physical properties of various carboxylic acids are shown in Table 22.2. It is one of the bases for employing intensified separation approach, recovery and purification involving carboxylic acids. There is no particular recovery process for all carboxylic acid. Some carboxylic acids may be grouped based on their physical properties and they should be recoverable in a similar way, for example acetic, itaconic, propionic, succinic and butyric acids.

22.2.2 *Conventional Processes for Downstream Recovery of Carboxylic Acids*

Downstream processing accounts for a substantial part of the overall cost of production and requires a large amount of energy. To achieve optimum techno-economic feasibility and a functioning but sustainable biorefinery, a good knowledge of current and alternative separation techniques and generation of novel method approach are required. Most carboxylic acids are generated by solvent extraction from dilute aqueous solutions. They are also obtained as products from stable

Table 22.2 Dissimilar property nature of various carboxylic acids [71]

Acid name	pK_a values	Solubility in water (g/L)	Melting point (°C)	Boiling point (°C)
Acetic	4.75	Miscible	17	118
Butyric	4.81	Miscible	−8	163
Citric	3.14; 4.77; 6.39	~600	153	Decomposes
Fumaric	3.03; 4.44	6.3	Sublimes	200
Gluconic	3.60	Good	131	Decomposes
3-Hydroxypropionic	4.51	High	<25	Decomposes
Itaconic	3.85; 5.45	80–95	165 (decomposes)	
Lactic	3.86	High	53	Decomposes
Malic	3.40; 5.11	558	130	Decomposes >140
Propionic	4.87	Miscible	−21	141
Pyruvic	2.50	Miscible	12	Decomposes at 165
Succinic	4.16; 5.61	77	185–187	235

oxidation and by-products from organic wastes and aqueous streams. There are several potential environmental and industrial applications where carboxylic acids can be recovered from aqueous solutions. These include citric and lactic acid production by fermentation and carboxylic acids recovery from aqueous waste streams [61]. An important aspect of chemical industries and fermentation technology is efficient carboxylic acid separation from aqueous solution utilizing different solvents with attendant significant improvements. Extractions are carried out using different organic solvents and these are grouped into three types:

- Conventional solvents with oxygen and hydrocarbon content
- Oxygen-bearing extractants bonded with phosphorus
- Aliphatic amines with high molecular weight

Low distribution coefficients are obtained when the conventional solvents are employed for the extraction of carboxylic acid thus leading to inefficient extraction [48]. Organophosphate solvents such as trioctylphosphine oxide, tri-*n*-butyl phosphate and aliphatic amines for carboxylic acid extraction give higher distribution coefficient. A number of aliphatic amines have been employed in carboxylic acid extractions [44, 72].

To obtain an efficient and effective recovery process at an industrial level, the following requirements must be met [71]:

- Specified purity (99.5%) for dicarboxylic acids for polymerization use [84], since polymerization can be terminated using monocarboxylic acid
- High yields (90–100%) in the recovery process of the downstream processing
- Minimal energy utilization and waste production while recovering products and low chemical use
- Moderate cost of investment in the recovery equipment as a result of heat and mass transfer efficiency

Fermentation-derived chemicals such as carboxylic acids have difficulties related to the high recovery cost. Similar to those derived through chemical routes, final product purity and co-production prevention are key influencing factors. Also, an energy-efficient way of managing the dilute fermentation broth must be developed to enable scale-up to commercial production of carboxylic acid [42].

Various separation methods are used in the production of high-valued co-products from different feed streams in biorefineries. Huang and Ramaswamy [41] in their study articulated the fact that separation and purification processes are vital features of biorefinery operations; their optimum selection and design are catalysts in maximizing product yields and the improvement of overall process efficiencies. Modern biorefineries approach employed various methods of separation and purification to generate co-products of high value from input feed streams and finally end-products. The different conventional methods of separation are discussed in the following sections.

22.2.2.1 Membrane Separation

Because of its adaptability, selectivity, high purity and yields obtainable, membrane technology is another separation process used in the recovery of organic acids. With material technology and recovery process development, more attention has been given to membrane separation, particularly in the in-situ *product recovery* (ISPR) technology. Membranes are basically a thin natural or artificial impediment, which permits selective solvent or solute mass transport through the barrier, in order to attain enrichment objectives and physical separation. The major established membrane purification/filtration for carboxylic acid separation utilized for wastewater treatment include pervaporation, microfiltration, nanofiltration, ultrafiltration, electrodialysis and reverse osmosis [15, 75, 118]. For example, nanofiltration was employed in succinate recovery from replicated broth and a lower rejection to monovalent than the divalent ions were observed with the nanofiltration membrane [43]. González et al. [26] also employed nanofiltration to recover lactic acid from the clarified broth; the lactate ions to non-ionized lactic acid ratio have a significant effect on the permeate flux and rejection of the nanofiltration membrane. Lactate permeation decreased with pH and increased with pressure while an increase in pH and pressure also increased with the lactate rejection. Other methods for membrane separation can also be employed to recover organic acids from aqueous solution. The drawback for using membrane separation is the high consumption of energy with the attendant high cost of membranes although the method is highly efficient. In addition, the efficiency of membrane separation decreases with increase in the concentration of organic acids. Another drawback is membrane pollution that occurs during the process of ion exchange, which also leads to the formation of unwholesome by-products [9, 125].

22.2.2.2 Precipitation

This is a traditional method for organic acid recovery from the broth, which has been used for industrial separation of lactic and citric acids in the past century. It can efficiently recover organic acids from the bulk of fermentation broths making it quite competitive particularly in primary purification. Considering precipitation of calcium as an example, there are four steps used to separate organic acids: firstly, the filtration of fermentation liquid to remove impurities and obtain the mother liquor; second, the addition of $Ca(OH)_2$ or $CaCO_3$ to the liquor with agitation; third, the filtration of calcium salts of organic acid from the broth; and fourth, treatment of calcium salt with sulphuric acid and further purification to obtain the acid required [69, 125]. In the study to improve the conditions of citric acid recovery from fermentation mash, the optimum conditions were found to be 50 °C for 20 min with almost 100% yield [36, 81]. In another study, for isolation of lactic acid, it was observed that calcium lactate to sulphuric acid molar ratio played a vital role in the yield improvement. About 92% yield of lactic acid was obtained at optimized conditions [83]. As merit, precipitation has a preference for high selectivity and purity, no phase transition necessary as the main advantage, which hovers mainly on establishing the right product precipitant. The underlining factor is mainly the search for the right product precipitants. High consumption makes cost reduction difficult, hence the development of reusable precipitant is needed for competitiveness.

22.2.2.3 Chromatography

The method mainly focuses on the adsorption properties or ion exchange products from resin. Chromatography is an established method for purification of organic acids from the aqueous medium, specifically in the product refining. Insolubility in acids or organic solvents and their stability are some of the physical and chemical properties of resins. Resins have good organic acid selectivity and low energy consumption selectivity with no phase transition [89]. The most used resins in recovery processes are mainly ion exchange resin and macroporous adsorption resin [6]. Increased capacity, quicker recovery, low consumption regeneration and specificity of desired products are characteristics of resin sorbents [70]. In the study by Tong et al. [111], resins having weak anion exchange were employed in lactic acid purification. The result showed that yield improvement was optimum between pH of 5.0 and 6.0. The purity and yield of the recovery process were 92.2% and 82.6%, respectively. Poly (4-vinylpyridine) resin was also employed in lactic acid recovery and the purity and yield obtained were 88% and 95%, respectively [134]. XUS 40285 resin from 25 different sorbents showed outstanding performance with succinic acid stability either at neutral or acidic pH [32]. Co-products are not produced using this method and product yields are high. However, a significant quantity of waste liquor is usually produced while the elution process demands high salt consumption. Additionally, the ion exchange ability of the resin is usually weakened in the long run [1, 10].

22.2.2.4 Distillation

Distillation is a technique used in the separation of mixtures with differences in their component volatility or boiling points. It can be utilized both at the initial and final stages of separation and purification combined with crystallization; hence, it remains an important technology according to Errico and Rong [24]. Generally, at low organic acid concentrations, effective distillation is carried out but at high concentration or at the azeotropic point, it becomes inefficient [40].

Extractive Distillation

Extractive distillation using ionic liquid solvent was employed to separate organic acids, typically carboxylic acids [8]. Phosphonium ionic liquid was used in the extraction of butyric acid with up to 89% yield followed by regeneration using a short-path distillation carried out in two stages. The process offers a significant advantage in product recovery where a free acid is obtained instead of its salt. Distillation, which is a known separation method for volatile components, is often used in separation since the carbonyl group has a strong effect on the adsorb-electron and the boiling points of most organic acids are higher than water.

Molecular Distillation

Molecular distillation (vacuum distillation) is a distinct distillation process suitable for fractionation and chemical separation from pyrolysis of bio-oils and is performed below high-vacuum conditions [30, 31, 122]. Wang et al. [124] investigated crude biodiesel purification using molecular distillation and established a high-yield result of up to 98.32%. The molecular distillation condensation efficiency was enhanced using traditional vacuum distillation to eliminate a greater percentage of the water in the unrefined bio-oil followed by fractionation of molecular distillation. From the obtained results, the fractions that were distilled were rich in low molecular weight ketones and carboxylic acids; the residual fraction has little or no water with improvement in heating values of 21.29 MJ/kg and 22.34 MJ/kg for two different operating conditions (80 °C, 1600 Pa and 80 °C, 340 Pa), respectively. Vacuum distillation is generally employed in carrying out experiments and in industrial manufacturing in order to accrue high-level savings in general cost of production.

22.2.2.5 Liquid–Liquid Extraction

Liquid–liquid extraction is a standard operation in chemical engineering discipline for the separation of mixtures, based on their comparative solubility in two immiscible liquids, in chemical and biochemical industries [15, 67]. It can be employed in carboxylic acids separation [2, 80, 87, 96] from aqueous solutions and extracting compounds (toxic to microorganisms) from biomass hydrolysates [29]. Considerable research studies on liquid–liquid extraction have been conducted for the separation of carboxylic acids from aqueous solutions [40]. Chen et al. [13] developed an innovative recovery and purification process to obtain L-lactic acid with high quality without very low pressure. There was a significant improvement in the L-lactic acid yield and purity of about 61.73% and 91.6%, respectively. Advantages of liquid–liquid extraction include short cycle and a rapid mass transfer that occurs between phases. For effective and green separation, the required properties in solvent selection and formulation must be considered.

Solvent Selection Criteria

The criteria of effective solvent selection [40] are that the solvent:

- Be non-toxic to the microbes
- Have high stability, distribution coefficient and product selectivity
- Have aqueous phase low solubility
- Possess different broth density for easy phase separation
- Have low viscosity, low broth emulsification tendency, large interfacial tension and be of minimal costs

Ionic Liquid Extraction

Ionic liquids are salts consisting of ions that have developed rapidly in recent times [68, 106, 120]. Most of them are imidazolium [68, 120], quaternary phosphate [77, 87] or quaternary ammonium salt [82], which are non-flammable and non-volatile and liquids at different temperatures [20]. Ionic liquids are promising alternatives to conventional organic solvents because of the advantageous extractabilities of organic compounds [68, 106]. Oliveira et al. [87] employed phosphonium-based hydrophobic ionic liquids for L-lactic, succinic and L-malic acid separation from aqueous solutions. The results showed that the extraction ability of the ionic liquids was better than the traditional extractions using organic compounds. About 73% recoveries were achieved. Marták and Schlosser [77] demonstrated that industrially processed phosphonium-based ionic liquids have an improved performance than the conventional organic solvents used in lactic acid separation. Though many signs of progress were made using ionic liquids extraction, the only limitation has been that of the high cost of ionic liquids.

22.3 Process Intensification (PI)

Due to its sustainable and innovative potential for process improvement, studies on process intensification (PI) have received considerable attention in a chemical engineering discipline. Chemical processes with large equipment size lead to a large amount of energy consumption for process operation. The conventional processes often consist of large and cumbersome processes due to their old chemical engineering design [103] and can be replaced with the most efficient small unit operation [95, 103]. Thus, to achieve this, some unit operations can be integrated, combined and intensified for higher efficiency. For example, the reduction in the equipment size of methyl acetate production with 28 pieces of equipment to 3 pieces of equipment by combining reaction and separation can lead to decrease in energy consumption and cost of manufacturing by as much as 80%. Furthermore, since the intensified process has lower energy consumption than the conventional processes, there is a reduction in acidification pollution, global warming and depletion of fossil fuel and carbon emissions by a factor of 5 and 7, respectively [35]. Since the early introduction of process intensification in the 1970s, significant attempt has been made in the definition, application and development of process intensification by chemical engineers. Different definitions are occasionally mentioned in different studies. However, Van Gerven and Stankiewicz [117] opined that a universally accepted definition may be quite difficult to attain since it is a continuous growing field in chemical engineering [102, 132] and should answer the current needs of the global market and different stakeholders as well as those pertained to other fields.

Process intensification can be summarized as process development involving equipment (unit operation) size reduction that leads to enhancement in chemical reaction kinetics, energy efficiency, process safety, minimization of waste generation and overall capital cost reduction. Process intensification offers the most important drifts in recent process technology and chemical engineering. One fundamental component of process intensification is multifunctional reactors that combine unit operations that would traditionally be carried out separately in different equipment. Reaction and separation integration offer multifunctional reactors with significant refinement. This integration is mainly at the level of equipment; no additional functional interrelations are introduced between the operations and the fact that reactions are unaffected by the separation and vice versa. This combination is specifically aimed at improving energy management, reduced inventory and plant design. An example of process intensification is hybrid distillation at the plant level with different unit operations integration comprising of at least one unit operation, a conventional distillation column, so as to satisfy the separation job [5]. Membrane separation integration with conventional distillation can be considered to overcome certain barriers in thermodynamics such as azeotrope formation [73]. Another example of process intensification is the divided wall columns, [34, 74] and process intensification that enhance the reaction conversion through in situ product removal are membrane reactors [45, 115], reactive distillation columns [37, 100] and reactive extraction [131].

Chemical product manufacturing and processing requires intensified process development and improvement by reconsidering existing operation designs into more precise and efficient options. This process intensification encompasses combining distinct unit operations into one single unit such as reaction and separation – thus, resulting in a simpler, more economical, efficient and cleaner production process. There is significant mixing improvement that enhances reaction kinetics, heat and mass transfer, selectivity and yield of the overall process. This leads to equipment size reduction – the complexity of the process, facility footprints – thus minimizing risk and cost in chemical production facilities. Process performance optimization is the crux of PI and the focus is on reaction kinetics, mass and heat transfer and thermodynamics. The four fundamental principles for process intensification presented by Van Gerven and Stankiewicz [116] include:

- Effectively maximizing intramolecular and intermolecular outcomes (for instance, attaining kinetic regimes through a rigorous change of conditions to achieve higher selectivity and conversions)
- Uniform process occurrence for all molecules (for instance, plug flow reactor conversion with minimal and uniform heating)
- Optimizing the driving force while maximizing their individual surface areas (for instance, using microchannel designs to increase the surface area)
- Maximizing combined effects resulting from partial processes (for instance, product removals when formed thus altering the equilibrium of reaction)

Process intensification designs and strategies that successfully attain some/all of these optimal conditions at a molecular level will possibly be transformative. Reactors with adequate environment monitoring could considerably improve conversion, yield and also selectivity, which correspondingly would decrease energy, material requirements and carbon concentrations, reduce demands of purification, and minimize generation of wastes. In addition, PI technology know-how could facilitate product manufacture that could not be successfully achieved. PI approaches involve a combination of various steps of processing (e.g. reactive extraction and reactive distillation). Process intensification through reactive separation processes could be successfully applied to esterification with overwhelming results and outputs (Fig. 22.1).

22.3.1 Separation Through Process Intensification

Several unit operations have been applied by various researchers for carboxylic acids (e.g. lactic acid) recovery through fermentation routes such as membrane separation, precipitation, chromatography, distillation and extraction [33, 97, 134]. Process intensification can be used to improve separation processes that account for up to 60–80% of processing cost in chemical processes [93]. The following section is aimed at reviewing some of the intensified separation processes with major focus on reactive separation process intensification.

22.3.1.1 Reactive Extraction

Reactive extraction is the combination of reaction and separation in one single unit operation. This leads to the effectiveness of the operation either when the reaction significantly enhances separation through improved mass transfer rates or when the

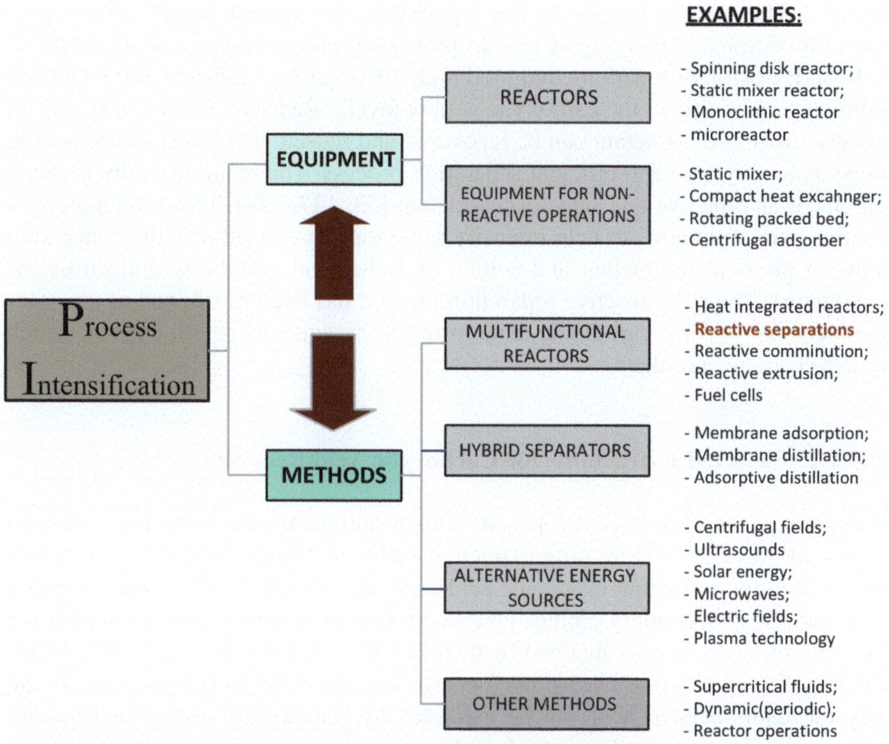

Fig. 22.1 Process intensification and successive components [104]

separation propels the reaction to higher conversions or both. This fusion of reaction and separation as a single unit operation is esteemed for its novelty and easiness. The low operating and investment cost also makes the operation desirable, which is accumulated on the commercial operation [131]. The processes associated with reactive extraction include the following; hydrolysis, ion complexation, dissociation and two-phase association and finally phase equilibrium.

Reactive extraction methods have gained much attention due to development of new improved processes and the decline of existing ones, and demand of low cost, high purity and environmentally safe products in response to excessive economic industrial pressure [53]. Reactive extraction also connects sources and sinks in chemical processes to improve reaction rates, selectivity and conversions. Since most of these chemical processes are driven by equilibrium, product removal as produced enhances reaction rates, increases conversion of feed, reduces the severity of reactions and provides milder operational conditions. Furthermore, reaction and separation processes facilitate increase in separation driving forces, equipment size reduction, utility cost reduction, safer equipment and removal of recyclable streams. The coupling also inhibits by-product reactions capable of exhibiting runaway behaviour, and separator design leads to inherent safety against serious process setbacks. The combination offers low equipment cost through recycle stream process elimination and merging of various process of pieces of equipment [130].

Reactive extraction with the required high distribution coefficient extractant is a promising technique in the carboxylic acid recovery. Reactive extraction is a clean process, since the extractant can be recovered and reused, and has spawned a wide scope as an effective and efficient separation process. The main difficulty is in the search for an effective and selective extractant [53, 127, 128]. The development of reactive extraction aims to help intensify separation and exemplify the connection between physical (extractant and solute) and chemical (solubility and diffusion) phenomena. Thus, the reactive separation process has been positioned as an effective recovery step for carboxylic acids from dilute aqueous solutions [12], which hitherto is quite herculean.

22.3.1.2 Selected Extractants for Carboxylic Acid Recovery

Several extractants have been proposed by many authors for the extraction of organic acids with the aim of increasing extraction yields and selectivity. Primary amines can be used for extraction of organic acids and are characterized by aqueous phase high solubility. Secondary amines give maximum distribution coefficients but the downstream regeneration obtained by distillation has the tendency to form amides. The most attractive extractive agents for carboxylic acids are the tertiary amines and only the undissociated acid can be extracted by primary, secondary and tertiary amines. Extraction of undissociated acid can be done using primary, secondary or tertiary amines while the dissociated and undissociated acid can be extracted using quaternary amines though regeneration by back extraction, which is difficult.

The extractant is the active constituent primarily responsible for carboxylate transfer to the diluent/solvent phase. Extractants having hydrophilic functional groups are designed with hydrophobic substituents, for example, long alkyl chains, to minimize their solubilities in water. Also, the alkyl substituent properties influence the chemistry of interaction, transport properties and extractant phase behaviour. Therefore, undesired mass transfer properties would be present if used pure. Hence, properties such as viscosity and interfacial tension can be improved by using a diluent with a positive impact on the mass transfer and phase separation. Examples of diluents include alcohols, halogenated hydrocarbons and alkanes.

Some extractants form complexes with acids that need to be solvated. If the diluent has no required solvation power as presented in Table 22.3, there will be the formation of complexes in a different third phase thus leading to difficulties in separation. A modifier, which generally improves complex solvation, is used in some cases. Modifiers are less cost-effective than diluents and have insufficient transport properties to be used exclusively with extractant [71]. The most commonly used modifiers are long chain alcohols. Modifiers enhance extraction and additionally influence the amine basicity thereby improving phase separation [59, 76]. Table 22.4 shows some extractants used in acid recovery together with their respective solvents/modifiers.

Table 22.3 Decreasing solvation power of extraction system diluents [107, 121]

Solvent type/class	Diluents
Alkanols	1-octanol, 1-decanol, 2-ethyl-1-hexanol
Halogenated proton donors	Dichloromethane, chloroform, 1,2-dichloroethane
Ketone and esters	Diisobutyl ketone, methyl isobutyl ketone. butyl acetate
Aliphatic hydrocarbons	Hexane, octane, dodecane
Alkyl aromatics	Xylene and toluene
Halogenated aromatics	Dichlorobenzene, chlorobenzene

22.3.1.3 Reactive Distillation

Reactive distillation, a coupling of reaction and distillation in a single unit operation offers multiple advantages over normal successive method of reaction and distillation in different units or other separation processes. This combination is an important concept in industrial applications of the multifunctional reactor. Some of the advantages offered by reactive distillation include enhanced selectivity, increase in conversion yields, heat control improvements, prevention of azeotropes and ease of separation, energy consumption reduction and separation of components with close boiling point resulting in a reduction in operating costs and capital investments. Integrating more functions into a single unit creates beneficial synergy for separations as the products are isolated and produced in situ (pulling the equilibrium conversion to completion). This in situ product recovery leads to complexation interactions between the vapour–liquid equilibrium (VLE), rates of mass transfer, chemical reaction kinetics and diffusion, posing a great design challenge and process synthesis of the system.

There are on-going research efforts on process modelling and simulation, process synthesis and hardware column design, due to reactive distillation being a new field. Reactive distillation in a continuous process was used for lactic acid recovery and consumes less energy with higher selectivity than conventional processes with discontinuity [64]. Process intensification using reactive extraction and reactive distillation approaches for recovery of carboxylic acids have been studied by various researchers and can be found in the literature. The techniques, solvents used, catalyst type, studies conducted and resultant outcomes/applications have been reviewed and summarized in Table 22.5, while Table 22.6 summarizes the various methods of recovery and their benefits and constraints.

22.3.1.4 In Situ Product Removal (ISPR)

ISPR is an express removal of organic acids as they are formed thus preventing subsequent holdups with medium components. It combines separation with fermentation, incorporating extraction, membrane and resin to achieve a continuous process. Furthermore, product removal can reduce toxicity to any microorganism.

Table 22.4 Primary recovery of carboxylic acids and selected extractants [71]

System	Extractant name	Functional group	Structure characteristics	Solvent (modifier)
Amine-based	N1923	Primary amine	Methyloctadecyl amine	1-Octanol Butyl acetate Hexane
	Primene® JM-T	Primary amine	Branched alkyl chains C16–C22	Kerosene
	Amberlite® LA-2	Secondary amine	Asymmetric alkyl chains C12–C15	Diethyl carbonate
				Methyl isobutyl ketone 1-Hexanol Kerosene (1-octanol)
	Tris(2-ethylhexyl) amine	Tertiary amine	2-Ethylhexyl alkyl chains	Kerosene
	Trihexylamine	Tertiary amine	Hexyl alkyl chains	1-Octanol
	Tri-*n*-octylamine (Alamine® 300)	Tertiary amine	Straight octyl chains	Kerosene Dodecane (1-decanol) *n*-Paraffins (isodecanol) Heptane (1-octanol, tripropylamine) Methyl isobutyl ketone Methyl isobutyl ketone (1-octanol) 1-Octanol
	Triisooctylamine	Tertiary amine	Isooctyl alkyl chains	Chloroform Heptane 1-Octanol
	Alamine® 336	Tertiary amine	Straight alkyl chains C8–C10	2-Octanol Kerosene 1-Octanol Decanol Cyclohexanone
Ionic liquids	Aliquat® 336	Quaternary ammonium salt	Linear alkyl chains C8–C10 and methyl substituent	2-Octanol Kerosene 1-Octanol Shellsol® A Dodecane (1-decanol) 1-Hexanol Hexane
	[Bmim][PF6]	Imidazolium salt	Butyl and methyl substituents	None, tributylphosphate

Table 22.4 (continued)

	[Bmim][BF4]	Imidazolium salt	Butyl and methyl substituents	None
	Trioctylamine-bis(2-ethylhexyl) phosphoric acid	Tertiary amine-organophosphate salt	Linear octyl chains mostly	Xylene
	Aliquat® 336 – bis(2-ethylhexyl) phosphoric acid	Quaternary ammonium-organophosphate salt	Linear alkyl chains C8–C10–C16 phosphate	Isopar® K
	Cyphos® IL104	Phosphonium-alkylphosphinate salt	Linear alkyl chains C10–C16 phosphinate	Dodecane
Neutral/ solvating	Tributylphosphate	Phosphate ester	Linear butyl chains	Dodecane Hexane Sunflower oil
	Tri-*n*-octylphosphine oxide	Organophosphorus oxide	Linear octyl chains	1-Octanol Hexane

The ISPR investigation focuses widely on inorganic acid separation from fermentation broths [15, 67, 90]. In situ lactic acid removal from fermentation broth was carried out by Ataei and Vasheghani-Farahani [3] using ion exchange resin and the extractive fermentation productivity was about 4.3 times greater than the conventional processes at a higher temperature. Several other studies are listed in Table 22.7. In situ product recovery makes continuous fermentation possible by instant acid removal from the broth. The pH is controlled and product inhibition minimized, leading to high feedstock utilization, product improvement, load and ideal cost reduction of downstream processing. An ideal in situ product recovery needs minimal chemical addition and energy consumption. ISPR and continuous fermentation are more cost-efficient and cost-effective than the conventional batch production and therefore require more research attention and possible adaptation [15].

22.3.2 *Equilibrium Studies Relating to Carboxylic Acid Separation*

The modelling of phase equilibria or general thermodynamic properties of systems of components such as alkanols and acids with potential for association via hydrogen bonding or normal dipole–dipole interaction remains a daunting challenge. Systems of this nature exhibit complex non-ideal aggregation behaviours that are complicated. Several descriptions for liquid–liquid equilibrium (LLE) and vapour–liquid equilibrium (VLE) of the alkanols-containing composition of mixtures have been presented using multi-scale associated concepts employing a generalized solvatochromic approach that uses linear solvation energy relationship (LSER). Other

Table 22.5 Some carboxylic acid recovery processes using reactive separation as process intensification approach developed by researchers

Carboxylic acid	Process intensification technique	Solvent employed	Catalyst	Comments/ remarks	References
Lactic acid	Reactive distillation – integration of reaction and distillation in one single piece of equipment	*n*-Butanol methanol	Amberlite catalyst, ion exchange resin	The study focused on the effect of feed concentration, mole ratio and catalyst loading/ weight on water removal from the aqueous lactic acid solution	Rao et al. [94], Komesu et al. [62] and Kumar et al. [63]
Succinic acid	Reactive extraction	Tri-*n*-octylamine (TOA) in 1-Decanol, 1-Octanol		The study focused on equilibria and temperature effect on succinic acid extraction, extraction efficiency and initial acid concentration	Eda et al. [22], Eda et al. [23], Umpuch et al. [112] and Orjuela et al. [88]
Acetic acid	Reactive distillation	Methanol	Ion exchange resin	Effect of changing various designs and operating parameters was studied, valid kinetic expression was developed and catalyst loading insensitivity of reaction rate was addressed	Singh et al. [101] and Grzenia et al. [28]
Itaconic acid	Reactive extraction	Amine-diluent combination		Equilibrium studies were conducted to evaluate extraction performance; the effect of medium components on extraction performance were also performed	Kaur and Elst [51]

(continued)

Table 22.5 (continued)

Carboxylic acid	Process intensification technique	Solvent employed	Catalyst	Comments/ remarks	References
Citric acid	Reactive extraction	Tri-*n*-butyl phosphate (TBP), Tri-*n*-octylamine (TOA) and Aliquat® 336 dissolved in three different diluents: butyl acetate, decanol and benzene		Isothermal batch experiments were performed for the equilibrium and kinetic studies; kinetics of extraction of citric acid was performed Extraction parameters were estimated by a differential evolution optimization technique	Thakre et al. [109] and Thakre et al. [108]
Formic acid		Trioctylamine TOA dissolved in various alcohols: isoamyl alcohol, hexan-1-ol, octan-1-ol, nonan-1-ol, decan-1-ol as diluents Alamine® 336 Sunflower oil		The difference between the physical extraction and reactive extraction were studied; equilibrium studies were also conducted	Uslu [113] and Martı [78]
Propionic acid	Reactive extraction	Tri-*n*-butyl phosphate (TBP) in petroleum ether, Aliquat® 336 in *n*-dodecane and 1-decanol		Equilibrium studies on distribution coefficient, equilibrium complexation constant, loading ratio and extraction efficiency were conducted	Keshav et al. [54], Keshav et al. [56, 57] and Kumar and Babu [65]
Levulinic acid		Tri-*n*-octylamine TOA in 1-octanol		The study focused on kinetics and extraction model development comprising of equilibrium complexation constant	Kumar et al. [66]

Table 22.6 Comparison of benefits and constraints of conventional and intensified process recovery of carboxylic acids

S/N	Process recovery	Benefits	Constraints	References
1	Solvent extraction	No further processing step is required	The quantified requirement of solvent is heeded at a higher cost; solvent toxicity; compound extraction; increased energy requirement and by-product formation	Seibert [98]
2	Vacuum distillation	Easy, simplified and well-established technology	Extensive energy requirement; the formation of by-products	Carlson [11]
3	Steam stripping	High purity is achieved	The energy requirement is extensive	Rackemann and Doherty [92] and Carlson [11]
4	Membrane separation	Continuous separation, mono-step separation of by-products creates room for enhanced productivity while minimizing undesirable by-product formation	Costly; membrane fouling	Den Boestert et al. [19]
5	Adsorption	Simple Ease of the auxiliary phase removal	Low adsorbent capacity limits use at industrial level and the susceptibility of adsorbents fouling leading to the limitation in lifetime operation of materials	Rackemann and Doherty [92]
6	In situ product removal (ISPR)	Improves the final acid concentration Base addition for controlled fermentation on the addition of base is prevented Reduction in waste generation Reduces end-product inhibition	A large quantity of acid is needed	López-Garzón and Straathof [71] and Ayoub [4]
7	Reactive extraction	Simple, efficient, high product yield, low energy consumption, economic and clean process	Search for efficient and selective extractant	Datta et al. [17], Hong et al. [38], Wasewar et al. [128, 129] and Wasewar et al. [130]

(continued)

Table 22.6 (continued)

S/N	Process recovery	Benefits	Constraints	References
8	Reactive distillation	High purification levels, low energy consumption, reduction in the number of equipment	Corrosion problems when using a homogeneous catalyst; search for efficient catalyst; applied to reversible chemical reactions	Rao et al. [94], Kumar et al. [63] and Kumar et al. [64]
9	Precipitation	Easy operation and applicable in chemical and industrial plants	The purity of product is low; consumption of a large amount of sulphuric acid; landfill disposal due to gypsum generation	Wasewar [125]

theoretical methods/approaches like activity coefficient model of universal functional group activity coefficient (UNIFAC), non-random two-liquid (NRTL), NRTL-Hayden O'Connell (NRTL-HOC), Rellich Kwong Soave-equation of state (RKS-EOS), cubic equation of state respectively have been applied widely to these systems. Senol [99] presented another group contribution method that has been used in the chemical industries extensively and successfully. It has been of interest in the development of several chemical processes. But in all these applications and adaptation, UNIFAC aboriginal model has been the most fascinating of the comprehensive nature of these parametric matrices that are significantly and easily obtained.

The prediction of LLE data by the UNIFAC method is the most effective even as UNIQUAC – universal quasichemical – and NRTL models have been employed in correlating experimental data. Uslu et al. [114] reported in their study that the model neither fitted the data qualitatively nor quantitatively despite the different techniques of iterations used with different values of correlation. So far, UNIFAC is the only model that has been reported to indicate the suitability of this model for these thermodynamic systems. The UNIFAC approach applies the activity coefficients, Y_i, for prediction. In liquid–liquid equilibria, the actions of component i in the mutual phases are equivalent and the mole fractions X_i^E, X_i^R of conjugate phases may be computed using Eq. (22.1):

$$\gamma_i^E X_i^E = \gamma_i^R X_i^R \tag{22.1}$$

where in γ_i^E and γ_i^R are the equivalent activity coefficients of component, i, in both raffinate and extract phases. The interface parameters amongst each of the main groups are employed in calculating component i activity coefficients. UNIFAC no doubt predicts temperately accurate the extraction equilibria of ternary systems and schemes. The procedure of LLE is dependent on the solubility of solutes in different organic solvents and for a specified system, the solute distribution between different solvents is determined at a partition ratio. This ratio invariably will not change if the system remains in equilibrium. The overall LLE system properties can be evaluated using the equilibrium partition ratio, distribution coefficient, separation factor and extraction factor.

Table 22.7 Literature data on intensified process separation of carboxylic acid from biomaterials with process performance at optimized conditions

Carboxylic acid produced	Separation type employed in the study	Process performance at optimized conditions	References
Lactic acid	In situ separation using reactive extraction; sunflower oil and Alamine® 336 in oleyl alcohol solvent was employed	25.59 g/dm^3 maximum yield was obtained using 15% (v/v) of the solvent at 37 °C with immobilized cells	Tik et al. [110]
Lactic acid	In situ extraction using Alamine® 336 in oleyl alcohol with kerosene as a diluent (20:40:40 wt%)	Lactate maximum yield of 67% that of theoretical yield (pH 5.0 at 43 °C)	Chen and Lee [14]
Butyric acid	Integrated extraction and pertraction using 20% w/w Hostarex A327 in oleyl alcohol	Up to 0.30 g/g sugar with 0.21 g/L/h productivity with a pH of 5.2 at 37 °C	Zigová et al. [135]
Propionic acid	Extractive separation using Adogen 283 (ditridecylamine) in oleyl alcohol	Propionate yield of 0.66 g/g substrate with 75 g/L product concentration and 90% purity (pH of 5.3 at 30 °C)	Jin and Yang [46]
Hexanoic and butyric acids	Pertraction with oleyl alcohol using 10% v/v trioctylamine	Increased productivity and conversion rates of the studied carboxylic acid by threefold when compared to the batch conventional process	Nelson et al. [86]
Acetic, propionic, valeric and butyric acid	In situ separation using conventional electrodialysis	Removal of about 99% volatile fatty acid from the broth containing up to 1200 mg/L each of the acid within 1 h	Jones et al. [47]
Lactic acid	Integrated process operation of electrodialysis with bipolar membranes (EDBM)	Up to 69.5% lactate recovery under an initial concentration of no less than 1 mol/L	Wang et al. [123]
Lactic acid	In situ separation with ion exchange resin using Amberlite IRA-96 combined with Amberlite IR-120	98.9% recovery of lactate with 99.17% purity; maximum loading of 210.46 mg/g	Bishai et al. [7]
Lactic, butyric and acetic acid	In situ separation using Amberlite IRA-67	Up to 74% acid removal from the Amberlite IRA-67 at pH of 3.3	Yousuf et al. [133]
Lactic acid	In situ lactic acid separation using ion exchange resin – Amberlite IRA-400, CL^-	A maximum yield of 0.85 g lactate/g substrate with the productivity of 0.984 g/L/h; the concentration of 37.4 g/L at 37 °C and pH of 6.1	Ataei and Vasheghani-Farahani [3]

22.4 Effects of Process Variables

This section explores the various effects of process parameters in process intensification of biomass-derived materials. Some of the process parameters of interest are physical equilibrium, chemical equilibrium, extraction kinetics, the effect of temperature, substrate, pH, back extraction equilibrium and kinetics, back extraction or regeneration, water co-extraction and toxicity. These are discussed and analysed.

22.4.1 *Physical Extraction*

Physical extraction (dimerization and ionization) involves the separation of solutes into substituted hydrocarbons (SHC) and non-reacting hydrocarbons (NHC) that are free of complexities. In physical extraction, factors responsible and considered according to some authors [58, 128] are:

(a) Aqueous phase ionization of acids
(b) Organic phase acid dimerization
(c) Partial acid dissociation between phases

Conventional solvents used for physical extraction can be polar or non-polar diluents, protic–aprotic diluents, inorganic–organic diluents such as polar–non-polar, protic–aprotic, inorganic–organic, natural solvents and so on. Distribution coefficient is presented with respect to dimerization coefficients according to the following equation [52]:

$$K_D^{\text{diluent}} = p + 2P^2 D[HA\}_{aq} \tag{22.2}$$

where

$$P = {[HA]_{org}}\Big/{[HA]_{aq}} \qquad D = {[HA]_{2,org}}\Big/{[HA\}^2_{org}}$$

22.4.2 *Chemical Extraction*

In chemical extraction (diffusion and solubility), the process involves the contacting of a second phase extractant that will reversibly react with the solute. The option of the complexing agent is dependent on its strength, specificity and whether it is yet reactively reversible with the solute. These complexing agents are usually viscous or solid in nature, and so they are dissoluble in low viscous and low molecular weight diluents. Equilibrium is attained and improved by the diluents through complex solvation and density control; also, interfacial tension and viscosity of the mixed solvent play crucial roles. To explain and describe the mechanism of chemical extraction using Eq. (22.3), the equilibrium constant according to Wasewar et al. [128] is presented in Equation as:

$$\text{Solute} + n.\text{Extractant} \leftrightarrow \text{Complex} \tag{22.3}$$

$$K_c = \frac{[\text{Complex}]}{[\text{Solute}][\text{Extractant}]^n} \tag{22.4}$$

22.4.3 Extraction Kinetics

The design of an efficient extraction process necessitates the knowledge and application of extraction reaction kinetics, the importance of which is predicated in that extractant depends on extractant type, diluents and the nature of the process that are explainable by the kinetic study [56, 127]. Keshav et al. [52] studied the reaction in a solution of two molecules and presented the three important steps involved:

(a) Reactant molecules' diffusion to one another
(b) Product diffusion away from each other
(c) Actual product diffusion from each other

22.4.4 Temperature Effect

Study on the effect of temperature is important in considering back extraction/regeneration step and operating temperatures. Commercial/industrial scale fermenters for carboxylic acid production are usually conditioned to operate within the temperature regime of 305–313 K. It is, therefore, possible for an extractant to be able to efficiently operate within this operating regime. As extraction is generally an exothermic process with the liberation of heat, a reduction in extraction (to about 50%) is anticipated as the temperature is intensified. However, the works of several authors have presented the fact that general decrease/step down in extraction is primarily a function of the extractant and the choice of diluents involved [55, 57].

22.4.5 pH Effect

Product inhibition is the main drawback in the fermentation of carboxylic acids. Simultaneous fermentation with end-product inhibition decreases the medium pH due to end production of acid – hence the need to add a neutralizing agent for optimal maintenance of the fermentation process to cancel out the presence of acid. Alternatively, product removal as they are formed (in situ) is a second important alternative that also increases the bioreactor performance and productivity. In reactive extraction, maintenance of high pH is achieved by in situ acid removal [126, 128]. Increase in pH decreases the extraction power in the case of tertiary amines with high molecular weight, for example Alamine® 336 and phosphorus-based extractant; another example is tri-*n*-octyl phosphoric acid in the case of phosphorus-based extractant while optimum pH is obtained with quaternary amine [55, 57, 130].

22.4.6 Effect of a Mixed System

Increase in acid extraction yield can be achieved by mixed extractant using synergetic extraction principle whereby the presence of one extractant can affect the increased performance of the other [59]. Several research studies have been carried out with the aim of improving the extraction of organic acids, especially in dilute media. These comprise using mixed extractant system with a single diluent or mixed diluent system in a single extractant system. To increase the stripping efficiency of carboxylic acids, mixed polar and non-polar diluents have been employed in the reactive extraction of these acids. A modifier can also be added as the active diluent that has the advantage of preventing the formation of a third phase that occurs generally when an extractant is used with an inert diluent. The extraction power of the extractants depends largely on the type of modifier used. Generally, polar diluents favour an increase in extraction since they help stabilize the formation of ion pair formed by solvation [59].

22.4.7 Effect of Substrates

In reactive extraction, it is necessary to explore the effect of substrate concentrations since 100% feed conversion is not achieved in the fermentation broth that usually contains around 40 g/L substrate source. Increase in viscosity can lead to a corresponding decrease in K_D value of the dilute aqueous phase that could change the interface surface tension leading to low complexation and hence low efficiency in extraction. However, when unconverted substrate conversion is left in the bioreactor (>40 g/L), lower K_D value would likely be expected [55]. It was observed that certain substrates, for example lactose, at any given concentration showed an insignificant decrease in K_D value. Thus, basically, acid extraction is unaffected by the source of substrates [130].

22.4.8 Water (Polar Component) Co-extraction

The type of acid and its concentration affects the mutual solubility between a particular solvent and an aqueous solution at a stable temperature. There is a substantial change in volume with weak organic acids caused by mutual solubility and the extent of the volume change is related to the water co-extraction and the acid. The increase in organic phase volume for Aliquat® 336 was about 2–5% with a corresponding decrease in the dilute aqueous phase while for tertiary amines and tributyl phosphate, no significant change in the volume was observed [57, 130]. For high concentration extraction with (>25%) amine in a diluent, an observation of a third phase was also made between the surface of the aqueous and organic phases. The change in volume is related to water co-extraction and depends on temperature, concentration and diluent type, which thus may affect the overall process economics. For example, during regeneration, pure acid is recovered from a dilute aqueous solution generated from the extract. The solubility of the acid is decreased by the stripping method. Generally, extraction using amine extractants gives higher selectivity of the acid over water as compared with conventional solvents. There is a minimum amount of water in the extract phase compared to that in an aqueous back extraction; therefore, there is little or no effect on the viability of the process [57, 130].

22.4.9 Regeneration of Acid and Back Extraction

The realization of the process of reactive extraction depends on complete acid recovery from the loaded organic phase. Next is the regeneration phase, which involves reaction reversal to recover the acid into a product phase and the recycling of free acid extractant. Various regeneration methods can be employed to back-extract the acid from the loaded organic phase using NaOH, HCl or trimethylamine (TMA) or by diluent and temperature swing [53, 130, 131]. TMA was found to be better amongst all, yielding approximately complete acid regeneration. Kinetics of acid regeneration was also analysed and a fast reaction was observed, which indicates that TMA was effective [130].

22.4.10 Toxicity

Toxicity of both the extractants and organic solvents to microbes is a problematic issue in extractive fermentation. The organic solvent can lead to physical, biochemical and microbial effects on the catalytic activity of the microorganisms. As microbes are key agents and are always present during fermentation, the degree of toxicity of the extractants and organic solvents to microbes depends on the different combination used. Toxic effects can be considerably reduced by avoidance of direct contact of the organism with the extractants [119, 129].

22.5 Kinetic Studies on Reactive Extraction of Some Carboxylic Acids

The overall design and modelling of a reactive extraction unit require both equilibrium and kinetic studies. A number of kinetic and equilibrium studies for carboxylic acid extractions are available in the literature with only limited kinetic studies. Jun et al. [50], Jun et al. [49] and Eda et al. [22] conducted kinetic and equilibrium studies for succinic acid extraction from aqueous solution with 1-octanol solutions of tri-*n*-octyl-amine (TOA) using a stirred cell with a microporous hydrophobic membrane. A Lewis-type stirred cell was used in equilibrium and kinetic studies on reactive extraction of pyruvic acid with trioctylamine in 1-octanol [79]. Kinetic measurements of citric acid extraction from aqueous solutions with trioctylamine in mixtures of isodecanol/*n*-paraffins were conducted in a cylindrical stirring vessel with a highly agitated system [91]. These investigative studies aimed at describing and analysing the kinetic mechanism of reactive extraction using basic reaction kinetic model, the reaction mechanism of acid–amine complexation and extraction theory. The intrinsic kinetic parameter estimation, for example reaction rate constants, and the order of reaction were obtained from experimental analysis. From the investigations, the reaction involving the carboxylic acid and the type of extractant depends not only on the extract or organic phase and aqueous phase compositions, but also on the system hydrodynamic parameters (phase volume ratio, temperature, agitation speed and interfacial area). In the study conducted by Jun et al. [49], it was observed in the aqueous phase that the rates of reaction were affected by pH and contamination present. At a pH greater than the acid p*Ka*, more dissociation occurred thus leading to reduced extraction efficiency. Therefore, an effective acid separation with the fermentation pH kept at a value less than the acid p*Ka* was recommended.

22.5.1 *Kinetic Model*

Doraiswamy [21] recommended an all-inclusive study on extraction theory in a stirred cell accompanied by chemical reaction to establish its effect on the specific reaction rate. A reactive system together with the aid of film and renewal theories with physicochemical and hydrodynamic parameters have been classified into four reaction regimes (very slow, slow, fast and instantaneous) subject to their relative diffusion and reaction rates. The value of physical mass transfer coefficient (kL) is essential in verifying the regime of reaction. This is achieved by performing physical extraction of acid with pure diluent from the aqueous phase. For a batch process, a differential mass balance yields Eq. (22.5):

$$V_{aq}\frac{dC_{org}}{dt} = K_l A_c\left(C^*_{org} - C_{org}\right) \tag{22.5}$$

where A_c is the interfacial area (m^2); V_{aq} is the aqueous phase volume (m^3); C^*_{org} is the equilibrium acid concentration in the organic phase.

The time-dependent acid concentration in the organic phase is obtained by integrating (22.6) as:

$$\ln\left(\frac{C^*_{org}}{C^*_{org} - C_{org}}\right) = \frac{K_l A_c}{V_{org}}t \tag{22.6}$$

A plot of $\ln\left(C^*_{org} / \left(C^*_{org} - C_{org}\right)\right)$ versus time (t) yields a straight line and the slope is used to estimate the coefficient of physical mass transfer (K_l).

A reversible reaction occurs between acid and extractant and can be avoided by measuring the initial reaction rate governed by the forward reaction. Thus, the initial rate of reaction, $R_{HC,0}$ mol. $m^{-2}S^{-1}$, is calculated using the following equation:

$$R_{HC,0} = \frac{V_{org}}{A_c}\left(\frac{dC_{HC,org}}{dt}\right)_{t=0} \tag{22.7}$$

$\left(\frac{dC_{HC,org}}{dt}\right)_{t=0}$ represents the initial slope of curve that represents the concentration in the organic phase versus time (t). The values of $R_{HC,0}$ are determined with different experimental conditions and used to determine the possible effect of the important process variables and to make a suitable deduction on the suitable reactive extraction kinetics. Consequently, to control the reaction regime, the effects of agitation speed (N) and the ratio of the volume of the phases $\left(\frac{V_{org}}{V_{aq}}\right)$ on the initial extraction rate must be examined. Following the suggestion by Doraiswamy [21], the reactive extraction of acid with an extractant in different diluent is governed by Eq. (22.8):

$$R_{HC,0} = K_{\alpha'\beta'}\left[\overline{HC}\right]^{\alpha'}\left[\overline{S}\right]^{\beta'} \tag{22.8}$$

where α' and β' are the orders of the reaction with respect to acid and extractant, respectively, and $K_{\alpha'\beta'}$ is the reaction rate constant.

For a (α', β') reaction occurring in the organic phase with a rate law shown in Eq. (22.9), and with a high excess of extractant, Hatta number (Ha) is given as a general expression:

$$Ha = \frac{\sqrt{\left(2/(\alpha'+1)\right)K_{\alpha'\beta'}\left[\overline{HC}\right]^{\alpha'-1}\left[\overline{S}\right]^{\beta'}D_{HC}}}{K_L} \tag{22.9}$$

D_{HC} represents the diluent acid diffusion coefficient. D_{HC} value is estimated using Eqs. (22.10) and (22.11).

$$D_{HC} = 7.4 \times 10^{-12} \frac{T\sqrt{M\psi}}{\eta\left(\forall^{acid}\right)^{0.6}} \tag{22.10}$$

$$D_{HC} = 10^{-11} \frac{T\sqrt{M}}{\eta\left(\forall^{diluent}\forall^{acid}\right)^{1/3}} \tag{22.11}$$

where ψ connotes diluent association factor; $\forall$denotes the component molar volume; T is equal to temperature (in °K); M and η represent molecular weight ($kg{\cdot}kmol^{-1}$) and viscosity ($kg{\cdot}m^{-1}{\cdot}s^{-1}$) of the diluent, respectively.

22.6 Industrial Applications of Intensified Separation Processes

Process intensification (PI) has a wide range of application owing to the fact that it can reduce the cost of inventory, and improve energy utilization and heat management. This application ranges from fine chemical and pharmaceutical industries to biofuels, petrochemicals and bulk chemicals, offshore processing and carbon capture. Improvement in yield, selectivity and processing time are of utmost importance for pharmaceuticals and chemical due to low cost of energy, which constitutes an insignificant percentage compared to production costs. Since petrochemicals and bulk chemicals are produced in large amounts, environmental impact and energy consumption reduction are significant incentives for technology innovation. Several applications of process intensification technologies are adopted in bulk chemicals, agro-allied production and petrochemicals that have been reviewed in this work particularly focusing on reactive distillations that have been implemented on large and commercial scales. Fine chemicals, often labelled as resource raw materials for speciality chemicals, such as renewables and allied chemicals, are manufactured in limited quantities, mostly batch-based processes. Despite the fact that the trend is drastically changing to continuous processes, the technologies of process intensification in these diverse areas have been successfully implemented. Potential benefits, however, are significant with a compelling overall reduction in general costs that are constantly decreasing with time. The application of intensified distillation systems should be favourable with the continual development of the biorefineries concept away from single to multiple product systems. Additionally, bio-based raw material integration into an existing typical plant will eventually promote hybrid processing or intensified systems. However, the implementation challenges of distillation systems to bio-based processes persist with respect to the operating conditions, such as systems with high viscosity, and solid systems handling (e.g. enzymes, cells), and require new intensified process development.

22.7 Conclusion and Outlook

Recovery of carboxylic acids from aqueous solutions, specifically of bio-based materials and processes, is a herculean task for bulk product implementation from renewable and green resources. It is the recurrent developmental progress for all industries to attain and be able to operate at low costs and higher efficiencies. There are economic considerations for these. Despite the diverse opportunities and options available, most processes of separation in the chemical industries have diverted attention to the innovative solution based on process intensification (PI). Reactive separation no doubt provides the most appropriate and strategic approach for separation of multicomponent mixtures, especially at dilute concentrations. In distillation-based separation processes, the use of alternative energy sources or the coupling of reaction and separation into one unit of reactive separation is the most fundamental method of PI application. Particularly, improvements in reactive separation processes have several advantages in green technology. These include reduction in energy requirements, improvement of the reaction rate, and productivity and selectivity enhancement that eventually leads to high effectiveness and efficiency of the separation processes. Several challenges still need to be overcome for these promising technologies to realize their full potential. Other separation technologies combined with reactive distillation with more integrated hybrid configurations offer great potential for future considerations of complementing the potential of the respective units. To realize the smooth and efficient development of separation processes, other factors such as new and effective solvents and catalysts, membrane materials, enhanced rotary machines, dependable control systems and inexpensive fabrication of equipment need utmost consideration.

To improve yields, productivity and subsequent scale-up to commercial scale, an intensified separation system needs to be adopted for carboxylic acid synthesis. An optimized reactive separation process, with a selection of appropriate extracting solvent in the separation process, which can be recycled efficiently, is promising for industrial production of carboxylic acids. Future research should be focused on energy input minimization, wastewater generation and improvement in carboxylic acid yields that will pave the way for optimal recovery. Kinetic and equilibrium studies are necessary to substantiate the chemical reaction pathways. Equilibrium data on chemical compounds and their mixtures play a significant role in separation processes in chemical industries as they are needed for efficient and effective design and operation of chemical processing plants. This information is limited in the literature. Process integration, recovery and purification of the product should be conducted to improve yield while minimizing the energy consumption. Finally, a thorough techno-economic analysis for sustainable process will be pivotal to successful process commercialization.

References

1. I.H. Aljundi, J.M. Belovich, O. Talu, Adsorption of lactic acid from fermentation broth and aqueous solutions on Zeolite molecular sieves. Chem. Eng. Sci. **60**(18), 5004–5009 (2005)
2. Y.S. Aşçi, I. Inci, A novel approach for itaconic acid extraction: Mixture of trioctylamine and tridodecylamine in different diluents. J. Ind. Eng. Chem. **18**(5), 1705–1709 (2012)
3. S.A. Ataei, E. Vasheghani-Farahani, In situ separation of lactic acid from fermentation broth using ion exchange resins. J. Ind. Microbiol. Biotechnol. **35**(11), 1229–1233 (2008)
4. P.M. Ayoub, Process for the reactive extractive extraction of levulinic acid. U.S. Patent 7,378,549 (2008)
5. D.K. Babi, P. Lutze, J.M. Woodley, R. Gani, A process synthesis-intensification framework for the development of sustainable membrane-based operations. Chem. Eng. Process. Process Intensif. **86**, 173–195 (2014)
6. M. Bishai, S. De, B. Adhikari, R. Banerjee, A platform technology of recovery of lactic acid from a fermentation broth of novel substrate Zizyphus oenophlia. 3 Biotech, **5**(4), 455–463 (2014). https://doi.org/10.1007/s13205-014-0240-y
7. M. Bishai, S. De, B. Adhikari, R. Banerjee, A platform technology of recovery of lactic acid from a fermentation broth of novel substrate Zizyphus oenophlia. 3 Biotech **5**(4), 455–463 (2015)
8. M. Blahušiak, Š. Schlosser, J. Cvengroš, Simulation of a new regeneration process of solvents with ionic liquid by short-path distillation. Sep. Purif. Technol. **97**, 186–194 (2012)
9. W. Boonkong, P. Sangvanich, A. Petsom, N. Thongchul, Comparison of an ion exchanger and an in-house electrodialysis unit for recovery of L-Lactic acid from fungal fermentation broth. Chem. Eng. Technol. **32**(10), 1542–1549 (2009)
10. X. Cao, H.S. Yun, Y.-M. Koo, Recovery of l-(+)-lactic acid by anion exchange resin Amberlite IRA-400. Biochem. Eng. J. **11**(2–3), 189–196 (2002). https://doi.org/10.1016/S1369-703X(02)00024-4
11. L.J. Carlson, Process for the manufacture of levulinic acid. U.S. Patent 3,065,263 (1962)
12. D. Cascaval, A.-I. Galaction, New extraction techniques on bioseparations: 1. Reactive extraction. Hemijska Industrija **58**(9), 375–386 (2004)
13. L. Chen, A. Zeng, H. Dong, Q. Li, C. Niu, A novel process for recovery and refining of L-lactic acid from fermentation broth. Bioresour. Technol. **112**, 280–284 (2012)
14. R. Chen, Y. Lee, Membrane-mediated extractive fermentation for lactic acid production from cellulosic biomass. Appl. Biochem. Biotechnol. **63**(1), 435 (1997)
15. K.-K. Cheng, X.-B. Zhao, J. Zeng, R.-C. Wu, Y.-Z. Xu, D.-H. Liu, J.-A. Zhang, Downstream processing of biotechnological produced succinic acid. Appl. Microbiol. Biotechnol. **95**(4), 841–850 (2012)
16. A. Corma, S. Iborra, A. Velty, Chemical routes for the transformation of biomass into chemicals. Chem. Rev. **107**(6), 2411–2502 (2007)
17. D. Datta, S. Kumar, H. Uslu, Status of the reactive extraction as a method of separation. J. Chem. **2015**, 1–16 (2015)
18. R.O. de Souza, L.S. Miranda, R. Luque, Bio (chemo) technological strategies for biomass conversion into bioethanol and key carboxylic acids. Green Chem. **16**(5), 2386–2405 (2014)
19. J.L.W.C. Den Boestert, J.P. Haan, A. Nijmeijer, Process for permeation enhanced reactive extraction of levulinic acid. U.S. Patent 7,501,062 (2009)
20. U. Domańska, E.V. Lukoshko, M. Królikowski, Measurements of activity coefficients at infinite dilution for organic solutes and water in the ionic liquid 1-butyl-1-methylpyrrolidinium tris (pentafluoroethyl) trifluorophosphate ([BMPYR][FAP]). Chem. Eng. J. **183**, 261–270 (2012)
21. L.K. Doraiswamy, Heterogeneous reactions: Analysis, examples, and reactor design. Gas-Solid Solid-Solid React **1**, 43 (1984)

22. S. Eda, T.P. Kumar, B. Satyavathi, P. Sudhakar, R. Parthasarathy, Recovery of succinic acid by reactive extraction using tri-n-Octylamine in 1-Decanol: Equilibrium optimization using response surface method and kinetic studies. IJCST **1**, 14 (2016)
23. S. Eda, R. Parthasarathy, T. Prathap Kumar, Reactive extraction of succinic acid from aqueous solutions using Tri-N-Octylamine (TOA) in 1-decanol: Equilibria and effect of temperature. (2015)
24. M. Errico, B.-G. Rong, Synthesis of new separation processes for bioethanol production by extractive distillation. Sep. Purif. Technol. **96**, 58–67 (2012). https://doi.org/10.1016/j.seppur.2012.05.022
25. B. Girisuta, Levulinic acid from lignocellulosic biomass. PhD thesis, Chemical engineering, University Library Groningen, 2007
26. M.I. González, S. Alvarez, F.A. Riera, R. Álvarez, Lactic acid recovery from whey ultrafiltrate fermentation broths and artificial solutions by nanofiltration. Desalination **228**(1), 84–96 (2008). https://doi.org/10.1016/j.desal.2007.08.009
27. A. Górak, A. Stankiewicz, Towards the sustainable world of 2050: European research agenda for process intensification. Chem. Ing. Tech. **84**(8), 1260–1260 (2012). https://doi.org/10.1002/cite.201250050
28. D.L. Grzenia, D.J. Schell, S.R. Wickramasinghe, Membrane extraction for removal of acetic acid from biomass hydrolysates. J. Membr. Sci. **322**(1), 189–195 (2008)
29. D.L. Grzenia, D.J. Schell, S.R. Wickramasinghe, Membrane extraction for detoxification of biomass hydrolysates. Bioresour. Technol. **111**, 248–254 (2012)
30. Z.-G. Guo, S.-R. Wang, Y.-Y. Zhu, Z.-Y. Luo, K.-F. Cen, Separation of acid compounds for refining biomass pyrolysis oil. J. Fuel Chem. Technol. **37**(1), 49–52 (2009)
31. Z. Guo, S. Wang, Y. Gu, G. Xu, X. Li, Z. Luo, Separation characteristics of biomass pyrolysis oil in molecular distillation. Sep. Purif. Technol. **76**(1), 52–57 (2010)
32. B.H. Davison, N.P. Nghiem, G.L. Richardson, Succinic acid adsorption from fermentation broth and regeneration. Appl. Biochem. Biotechnol **113–116**, 653–669 (2004). https://doi.org/10.1385/ABAB:114:1-3:653
33. V. Hábová, K. Melzoch, M. Rychtera, B. Sekavová, Electrodialysis as a useful technique for lactic acid separation from a model solution and a fermentation broth. Desalination **162**, 361–372 (2004)
34. I.J. Halvorsen, S. Skogestad, Energy efficient distillation. J. Nat. Gas Sci. Eng. **3**(4), 571–580 (2011)
35. G.J. Harmsen, G. Korevaar, S.M. Lemkowitz, Process intensification contributions to sustainable development. Chem. Ind. **98**, 495–522 (2004)
36. L.G. Heding, J.K. Gupta, Improvement of conditions for precipitation of citric acid from fermentation mash. Biotechnol. Bioeng. **17**(9), 1363–1364 (1975). https://doi.org/10.1002/bit.260170910
37. J. Holtbruegge, S. Heile, P. Lutze, A. Górak, Synthesis of dimethyl carbonate and propylene glycol in a pilot-scale reactive distillation column: Experimental investigation, modeling and process analysis. Chem. Eng. J. **234**, 448–463 (2013)
38. Y. Hong, W. Hong, D. Han, Application of reactive extraction to recovery of carboxylic acids. Biotechnol. Bioprocess Eng. **6**(6), 386–394 (2001). https://doi.org/10.1007/BF02932319
39. H.-J. Huang, S. Ramaswamy, C. Bergeron, D.J. Carrier, S. Ramaswamy, Separation and purification of phytochemicals as co-products in biorefineries, in *Biorefinery co-products*, (John Wiley & Sons, Ltd, Chichester, 2012), pp. 37–53. https://doi.org/10.1002/9780470976692.ch3
40. H.-J. Huang, S. Ramaswamy, U.W. Tschirner, B.V. Ramarao, A review of separation technologies in current and future biorefineries. Sep. Purif. Technol. **62**(1), 1–21 (2008). https://doi.org/10.1016/j.seppur.2007.12.011
41. H.J. Huang, S. Ramaswamy, Overview of biomass conversion processes and separation and purification technologies in biorefineries. Sep. Purif. Technol. Biorefineries, ch 1, 1–36 (2013). https://doi.org/10.1002/9781118493441

42. Y.S. Huh, Y.-S. Jun, Y.K. Hong, H. Song, S.Y. Lee, W.H. Hong, Effective purification of succinic acid from fermentation broth produced by Mannheimia succiniciproducens. Process Biochem. **41**(6), 1461–1465 (2006)
43. S. Hyeon Kang, Y. Keun Chang, Removal of organic acid salts from simulated fermentation broth containing succinate by nanofiltration. J. Membr. Sci. **246**(1), 49–57 (2005). https://doi.org/10.1016/j.memsci.2004.08.014
44. M. Ingale, V. Mahajani, Recovery of carboxylic acids, C2·C6, from an aqueous waste stream using tributylphosphate (TBP): Effect of presence of inorganic acids and their sodium salts. Sep. Technol. **6**(1), 1–7 (1996)
45. T. Inoue, T. Nagase, Y. Hasegawa, Y. Kiyozumi, K. Sato, M. Nishioka, S. Hamakawa, F. Mizukami, Stoichiometric ester condensation reaction processes by pervaporative water removal via acid-tolerant zeolite membranes. Ind. Eng. Chem. Res. **46**(11), 3743–3750 (2007)
46. Z. Jin, S.T. Yang, Extractive fermentation for enhanced propionic acid production from lactose by Propionibacterium acidipropionici. Biotechnol. Prog. **14**(3), 457–465 (1998)
47. R.J. Jones, J. Massanet-Nicolau, A. Guwy, G.C. Premier, R.M. Dinsdale, M. Reilly, Removal and recovery of inhibitory volatile fatty acids from mixed acid fermentations by conventional electrodialysis. Bioresour. Technol. **189**, 279–284 (2015)
48. R.S. Juang, R.T. Wu, Effect of a water-insoluble organic acid on amine extraction of acetic acid from aqueous solutions. Equilibrium studies. J. Chem. Technol. Biotechnol. **66**(2), 160–168 (1996)
49. Y.-S. Jun, E.Z. Lee, Y.S. Huh, Y.K. Hong, W.H. Hong, S.Y. Lee, Kinetic study for the extraction of succinic acid with TOA in fermentation broth; effects of pH, salt and contaminated acid. Biochem. Eng. J. **36**(1), 8–13 (2007). https://doi.org/10.1016/j.bej.2006.06.011
50. Y.S. Jun, Y.S. Huh, W.H. Hong, Y.K. Hong, Kinetics of the extraction of succinic acid with tri-n-octylamine in 1-Octanol solution. Biotechnol. Prog. **21**(6), 1673–1679 (2005)
51. G. Kaur, K. Elst, Development of reactive extraction systems for itaconic acid: A step towards in situ product recovery for itaconic acid fermentation. RSC Adv. **4**(85), 45029–45039 (2014)
52. A. Keshav, S. Chand, K.L. Wasewar, Equilibrium studies for extraction of propionic acid using tri-n-butyl phosphate in different solvents. J. Chem. Eng. Data **53**(7), 1424–1430 (2008a)
53. A. Keshav, S. Chand, K.L. Wasewar, Recovery of propionic acid from aqueous phase by reactive extraction using quarternary amine (Aliquat 336) in various diluents. Chem. Eng. J. **152**(1), 95–102 (2009a). https://doi.org/10.1016/j.cej.2009.03.037
54. A. Keshav, K. Wasewar, S. Chand, Reactive extraction of propionic acid using tri-n-butyl phosphate in petroleum ether: Equilibrium study. Chem. Biochem. Eng. Q. **22**(4), 433–437 (2008b)
55. A. Keshav, K.L. Wasewar, S. Chand, Extraction of propionic acid from model solutions: Effect of pH, salts, substrate, and temperature. AICHE J. **55**(7), 1705–1711 (2009b)
56. A. Keshav, K.L. Wasewar, S. Chand, Reactive extraction of propionic acid using Aliquat 336 in MIBK: Linear solvation energy relationship (LSER) modeling and kinetics study. J. Sci. Ind. Res. **68**(8), 708 (2009c)
57. A. Keshav, K.L. Wasewar, S. Chand, Recovery of propionic acid by reactive extraction-1. Equilibrium, effect of pH and temperature, water coextraction. Desalin. Water Treat. **3**(1–3), 91–98 (2009d)
58. A. Keshav, K.L. Wasewar, S. Chand, Recovery of propionic acid from an aqueous stream by reactive extraction: Effect of diluents. Desalination **244**(1–3), 12–23 (2009e). https://doi.org/10.1016/j.desal.2008.04.032
59. A. Keshav, K.L. Wasewar, S. Chand, H. Uslu, Effect of binary extractants and modifier–diluents systems on equilbria of propionic acid extraction. Fluid Phase Equilib. **275**(1), 21–26 (2009f). https://doi.org/10.1016/j.fluid.2008.09.012
60. C.J. King, Amine-based systems for carboxylic acid recovery. Chemtech **22**(5), 285–291 (1992)
61. C.J. King, L.J. Poole, Carboxylic acid sorption regeneration process. U.S. Patent 5,412,126 (1995)

62. A. Komesu, P. Martins Martinez, B. Lunelli, R. Filho, M. Maciel, Lactic acid purification by reactive distillation system using design of experiments. Chem. Eng. Process. **95**, 26–30 (2015). https://doi.org/10.1016/j.cep.2015.05.005
63. R. Kumar, S. Mahajani, H. Nanavati, S. B Noronha, Recovery of lactic acid by batch reactive distillation. Chem. Eng. Process. Process Intensif. **95**, 26–30 (2006a). https://doi.org/10.1002/jctb.1444
64. R. Kumar, H. Nanavati, S.B. Noronha, S.M. Mahajani, A continuous process for the recovery of lactic acid by reactive distillation. J. Chem. Technol. Biotechnol. **81**(11), 1767–1777 (2006b)
65. S. Kumar, B. Babu, Reactive extraction of propionic acid with aliquat 336 dissolved in 1-decanol and n-dodecane. J. Future Eng. Technol. **3**, 21–27 (2008)
66. T.P. Kumar, B. Vishwanadham, K.P. Rani, M. Mallikarjun, V.B. Rao, Reactive extraction of levulinic acid from aqueous solutions with tri-n-octylamine (TOA) in 1-octanol: Equilibria, kinetics, and model development. Chem. Eng. Commun. **198**(4), 572–589 (2010)
67. T. Kurzrock, D. Weuster-Botz, Recovery of succinic acid from fermentation broth. Biotechnol. Lett. **32**(3), 331–339 (2010)
68. H. Lateef, A. Gooding, S. Grimes, Use of 1-hexyl-3-methylimidazolium bromide ionic liquid in the recovery of lactic acid from wine. J. Chem. Technol. Biotechnol. **87**(8), 1066–1073 (2012). https://doi.org/10.1002/jctb.3843
69. S.C. Lee, H.C. Kim, Batch and continuous separation of acetic acid from succinic acid in a feed solution with high concentrations of carboxylic acids by emulsion liquid membranes. J. Membr. Sci. **367**(1), 190–196 (2011)
70. Q. Li, D. Wang, Y. Wu, W. Li, Y. Zhang, J. Xing, Z. Su, One step recovery of succinic acid from fermentation broths by crystallization. Sep. Purif. Technol. **72**(3), 294–300 (2010). https://doi.org/10.1016/j.seppur.2010.02.021
71. C.S. López-Garzón, A.J.J. Straathof, Recovery of carboxylic acids produced by fermentation. Biotechnol. Adv. **32**(5), 873–904 (2014). https://doi.org/10.1016/j.biotechadv.2014.04.002
72. S. Luque, J.R. Alvarez, C. Pazos, J. Coca, Recovery of valeric acid from aqueous solutions by solvent extraction. Solvent Extr. Ion Exch. **13**(5), 923–940 (1995)
73. P. Lutze, A. Gorak, Reactive and membrane-assisted distillation: Recent developments and perspective. Chem. Eng. Res. Des. **91**(10), 1978–1997 (2013)
74. G. Madenoor Ramapriya, M. Tawarmalani, R. Agrawal, Thermal coupling links to liquid-only transfer streams: A path for new dividing wall columns. AICHE J. **60**(8), 2949–2961 (2014)
75. F. Mao, G. Zhang, J. Tong, T. Xu, Y. Wu, Anion exchange membranes used in diffusion dialysis for acid recovery from erosive and organic solutions. Sep. Purif. Technol. **122**, 376–383 (2014)
76. M. Marinova, J. Albet, J. Molinier, G. Kyuchoukov, Specific influence of the modifier (1-decanol) on the extraction of tartaric acid by different extractants. Ind. Eng. Chem. Res. **44**(17), 6534–6538 (2005)
77. J. Marták, Š. Schlosser, Extraction of lactic acid by phosphonium ionic liquids. Sep. Purif. Technol. **57**(3), 483–494 (2007)
78. M.E. Martı, Recovery of formic acid by reactive extraction using an environmentally friendly solvent. Selcuk University Journal of Engineering, Science & Technology 5.1 26–37 (2017)
79. M.E. Marti, T. Gurkan, L.K. Doraiswamy, Equilibrium and kinetic studies on reactive extraction of pyruvic acid with trioctylamine in 1-octanol. Ind. Eng. Chem. Res. **50**(23), 13518–13525 (2011). https://doi.org/10.1021/ie200625q
80. G. Maurer, Modeling the liquid–liquid equilibrium for the recovery of carboxylic acids from aqueous solutions. Fluid Phase Equilib. **241**(1–2), 86–95 (2006). https://doi.org/10.1016/j.fluid.2005.11.005
81. B. Max, J.M. Salgado, N. Rodríguez, S. Cortés, A. Converti, J.M. Domínguez, Biotechnological production of citric acid. Braz. J. Microbiol. **41**(4), 862–875 (2010)
82. J.-P. Mikkola, P. Virtanen, R. Sjöholm, Aliquat 336®—A versatile and affordable cation source for an entirely new family of hydrophobic ionic liquids. Green Chem. **8**(3), 250–255 (2006)

83. D.-J. Min, K.H. Choi, Y.K. Chang, J.-H. Kim, Effect of operating parameters on precipitation for recovery of lactic acid from calcium lactate fermentation broth. Korean J. Chem. Eng. **28**(10), 1969 (2011)
84. M.T. Musser. Cyclohexanol and cyclohexanone. Ullmann's encyclopedia of industrial chemistry. (2000). https://doi.org/10.1002/14356007.a08_217.pub2
85. S. Naik, V.V. Goud, P.K. Rout, A.K. Dalai, Production of first and second generation biofuels: A comprehensive review. Renew. Sustain. Energy Rev. **14**(2), 578–597 (2010)
86. R.S. Nelson, D.J. Peterson, E.M. Karp, G.T. Beckham, D. Salvachúa, Mixed carboxylic acid production by Megasphaera elsdenii from glucose and lignocellulosic hydrolysate. Fermentation **3**(1), 10 (2017)
87. F.S. Oliveira, J.M. Araújo, R. Ferreira, L.P.N. Rebelo, I.M. Marrucho, Extraction of l-lactic, l-malic, and succinic acids using phosphonium-based ionic liquids. Sep. Purif. Technol. **85**, 137–146 (2012)
88. A. Orjuela, A.J. Yanez, L. Peereboom, C.T. Lira, D.J. Miller, A novel process for recovery of fermentation-derived succinic acid. Sep. Purif. Technol. **83**(0), 31–37 (2011). https://doi.org/10.1016/j.seppur.2011.08.010
89. C. Park, H.-G. Nam, K.B. Lee, S. Mun, Optimal design and experimental validation of a simulated moving bed chromatography for continuous recovery of formic acid in a model mixture of three organic acids from Actinobacillus bacteria fermentation. J. Chromatogr. A **1365**, 106–114 (2014)
90. E.C. Peterson, A.J. Daugulis, Demonstration of in situ product recovery of butyric acid via CO2-facilitated pH swings and medium development in two-phase partitioning bioreactors. Biotechnol. Bioeng. **111**(3), 537–544 (2014)
91. F.A. Poposka, K. Nikolovski, R. Tomovska, Kinetics, mechanism and mathematical modelling of extraction of citric acid with isodecanol/n-paraffins solutions of trioctylamine. Chem. Eng. Sci. **53**(18), 3227–3237 (1998)
92. D.W. Rackemann, W.O. Doherty, The conversion of lignocellulosics to levulinic acid. Biofuels Bioprod. Biorefin. **5**(2), 198–214 (2011)
93. A.J. Ragauskas, C.K. Williams, B.H. Davison, G. Britovsek, J. Cairney, C.A. Eckert, W.J. Frederick, J.P. Hallett, D.J. Leak, C.L. Liotta, The path forward for biofuels and biomaterials. Science **311**(5760), 484–489 (2006)
94. V.B. Rao, P.S. Kumar, C. Sailu, S.R.M. Rao, Recovery of lactic acid by reactive distillation. J. Appl. Sci. **14**(12), 1289 (2014)
95. D. Reay, C. Ramshaw, A. Harvey, *Process Intensification: Engineering for Efficiency, Sustainability and Flexibility* (Butterworth-Heinemann, Amsterdam, 2013)
96. G.G. Redhi, I. Bahadur, N.M. Xhakaza, Liquid–liquid equilibria measurements of ternary systems (acetonitrile+ a carboxylic acid+ dodecane) at 303.15 K. Fluid Phase Equilib. **388**, 1–5 (2015). https://doi.org/10.1016/j.fluid.2014.12.013
97. Y.-P. Ren, J.-J. Wang, X.-F. Li, X.-H. Wang, Reactive extraction of short-chain fatty acids from synthetic acidic fermentation broth of organic solid wastes and their stripping. J. Chem. Eng. Data **57**(1), 46–51 (2011)
98. F. Seibert, A method of recovering levulinic acid. WO 2010/030617 A1. (2010)
99. A. Senol, Optimum extraction equilibria of the systems (water+ carboxylic acid+ 1-hexanol/Alamine): Thermodynamic modeling. Fluid Phase Equilib. **360**, 77–87 (2013)
100. M. Shah, A.A. Kiss, E. Zondervan, A.B. de Haan, Evaluation of configuration alternatives for multi-product polyester synthesis by reactive distillation. Comput. Chem. Eng. **52**, 193–203 (2013)
101. A. Singh, A. Tiwari, S.M. Mahajani, R.D. Gudi, Recovery of acetic acid from aqueous solutions by reactive distillation. Ind. Eng. Chem. Res. **45**(6), 2017–2025 (2006). https://doi.org/10.1021/ie0505514
102. A. Stankiewicz, Reactive separations for process intensification: An industrial perspective. Chem. Eng. Process. Process Intensif. **42**(3), 137–144 (2003)
103. A. Stankiewicz, J.A. Moulijn, Process intensification. Ind. Eng. Chem. Res. **41**(8), 1920–1924 (2002). https://doi.org/10.1021/ie011025p
104. A.I. Stankiewicz, J.A. Moulijn, Process intensification: Transforming chemical engineering. Chem. Eng. Prog. **96**(1), 22–34 (2000)

105. A. Straathof, The proportion of downstream costs in fermentative production processes. Compr. Biotechnol. **2**, 811–814 (2011)
106. X. Sun, H. Luo, S. Dai, Ionic liquids-based extraction: A promising strategy for the advanced nuclear fuel cycle. Chem. Rev. **112**(4), 2100–2128 (2011)
107. J.A. Tamada, C.J. King, Extraction of carboxylic acids with amine extractants. 3. Effect of temperature, water coextraction, and process considerations. Ind. Eng. Chem. Res. **29**(7), 1333–1338 (1990)
108. N. Thakre, D. Datta, A. Prajapati, P. Chaudhari, D. Pal, Reactive extraction of citric acid using different extractants: Equilibrium, kinetics and modeling. Chem. Biochem. Eng. Q. **31**(4), 437–446 (2018)
109. N. Thakre, A.K. Prajapati, S.P. Mahapatra, A. Kumar, A. Khapre, D. Pal, Modeling and optimization of reactive extraction of citric acid. J. Chem. Eng. Data **61**(7), 2614–2623 (2016)
110. N. Tik, E. Bayraktar, Ü. Mehmetoglu, In situ reactive extraction of lactic acid from fermentation media. J. Chem. Technol. Biotechnol. **76**(7), 764–768 (2001)
111. W.-Y. Tong, X.-Y. Fu, S.-M. Lee, J. Yu, J.-W. Liu, D.-Z. Wei, Y.-M. Koo, Purification of l(+)-lactic acid from fermentation broth with paper sludge as a cellulosic feedstock using weak anion exchanger Amberlite IRA-92. Biochem. Eng. J. **18**(2), 89–96 (2004). https://doi.org/10.1016/S1369-703X(03)00170-0
112. C. Umpuch, S. Sakeaw, S. Kanchanatawee, K. Jantama, Removal of contaminated organic acids from simulated succinic acid fermentation broth by reactive extraction process: Single- and mixed-solute solution. Sep. Sci. Technol. **51**(4), 629–640 (2016)
113. H. Uslu, Reactive extraction of formic acid by using tri Octyl amine (TOA). Sep. Sci. Technol. **44**(8), 1784–1798 (2009). https://doi.org/10.1080/01496390902775893
114. H. Uslu, A. Gök, Ş.İ. Kırbaşlar, Phase equilibria of (water+levunilic acid+alcohol) ternary systems. Fluid Phase Equilib. **273**(1–2), 21–26 (2008). https://doi.org/10.1016/j.fluid.2008.08.004
115. D. Van Baelen, B. Van der Bruggen, K. Van den Dungen, J. Degrève, C. Vandecasteele, Pervaporation of water–alcohol mixtures and acetic acid–water mixtures. Chem. Eng. Sci. **60**(6), 1583–1590 (2005)
116. T. Van Gerven, A. Stankiewicz, Structure, energy, synergy, time—The fundamentals of process intensification. Ind. Eng. Chem. Res. **48**(5), 2465–2474 (2009a). https://doi.org/10.1021/ie801501y
117. T. Van Gerven, A. Stankiewicz, Structure, energy, synergy, time·The fundamentals of process intensification. Ind. Eng. Chem. Res. **48**(5), 2465–2474 (2009b)
118. L.M. Vane, A review of pervaporation for product recovery from biomass fermentation processes. J. Chem. Technol. Biotechnol. **80**(6), 603–629 (2005)
119. M.D. Waghmare, K.L. Wasewar, S.S. Sonawane, D.Z. Shende, Natural nontoxic solvents for recovery of picolinic acid by reactive extraction. Ind. Eng. Chem. Res. **50**(23), 13526–13537 (2011)
120. J. Wang, Y. Pei, Y. Zhao, Z. Hu, Recovery of amino acids by imidazolium based ionic liquids from aqueous media. Green Chem. **7**(4), 196–202 (2005)
121. K. Wang, Z. Chang, Y. Ma, C. Lei, S. Jin, Y. Wu, I. Mahmood, C. Hua, H. Liu, Equilibrium study on reactive extraction of propionic acid with N1923 in different diluents. Fluid Phase Equilib. **278**(1), 103–108 (2009a)
122. S. Wang, Y. Gu, Q. Liu, Y. Yao, Z. Guo, Z. Luo, K. Cen, Separation of bio-oil by molecular distillation. Fuel Process. Technol. **90**(5), 738–745 (2009b)
123. X. Wang, Y. Wang, X. Zhang, H. Feng, T. Xu, In-situ combination of fermentation and electrodialysis with bipolar membranes for the production of lactic acid: Continuous operation. Bioresour. Technol. **147**, 442–448 (2013)
124. Y. Wang, J. Nie, M. Zhao, S. Ma, L. Kuang, X. Han, S. Tang, Production of biodiesel from waste cooking oil via a two-step catalyzed process and molecular distillation. Energy Fuels **24**(3), 2104–2108 (2010)
125. K.L. Wasewar, Separation of lactic acid: Recent advances. Chem. Biochem. Eng. Q. **19**(2), 159–172 (2005)

126. K.L. Wasewar, A.B.M. Heesink, G.F. Versteeg, V.G. Pangarkar, Equilibria and kinetics for reactive extraction of lactic acid using Alamine 336 in decanol. J. Chem. Technol. Biotechnol. **77**(9), 1068–1075 (2002a)
127. K.L. Wasewar, A.B.M. Heesink, G.F. Versteeg, V.G. Pangarkar, Reactive extraction of lactic acid using alamine 336 in MIBK: Equilibria and kinetics. J. Biotechnol. **97**(1), 59–68 (2002b). https://doi.org/10.1016/S0168-1656(02)00057-3
128. K.L. Wasewar, D. Shende, A. Keshav, Reactive extraction of itaconic acid using quaternary amine Aliquat 336 in ethyl acetate, toluene, hexane, and kerosene. Ind. Eng. Chem. Res. **50**(2), 1003–1011 (2010)
129. K.L. Wasewar, D. Shende, A. Keshav, Reactive extraction of itaconic acid using tri-n-butyl phosphate and aliquat 336 in sunflower oil as a non-toxic diluent. J. Chem. Technol. Biotechol. **86**(2), 319–323 (2011)
130. K.L. Wasewar, A.A. Yawalkar, J.A. Moulijn, V.G. Pangarkar, Fermentation of glucose to lactic acid coupled with reactive extraction: A review. Ind. Eng. Chem. Res. **43**(19), 5969–5982 (2004)
131. K.L. Wasewar, C.K. Yoo, Intensifying the recovery of carboxylic acids by reactive extraction. In: Proceedings of 3rd International Conference on Chemistry and Chemical Engineering, June, 2012. pp. 29–30 (2012)
132. Ö. Yildirim, A.A. Kiss, E.Y. Kenig, Dividing wall columns in chemical process industry: A review on current activities. Sep. Purif. Technol. **80**(3), 403–417 (2011)
133. A. Yousuf, F. Bonk, J.-R. Bastidas-Oyanedel, J.E. Schmidt, Recovery of carboxylic acids produced during dark fermentation of food waste by adsorption on Amberlite IRA-67 and activated carbon. Bioresour. Technol. **217**, 137–140 (2016)
134. J. Zhou, W. Bi, K.H. Row, Purification of lactic acid from fermentation broth by spherical anion exchange polymer. J. Appl. Polym. Sci. **120**(5), 2673–2677 (2011)
135. J. Zigová, E. Šturdík, D. Vandák, Š. Schlosser, Butyric acid production by Clostridium butyricum with integrated extraction and pertraction. Process Biochem. **34**(8), 835–843 (1999)

Chapter 23
Advances in Engineering Strategies for Enhanced Production of Lipid in *Rhodosporidium* sp. from Lignocellulosics and Other Carbon Sources

R. Saini, K. Hegde, S. K. Brar, and C. R. Soccol

23.1 Introduction

In the current scenario, the primary source of energy is fossil fuels, and its total contribution is 80% in which 58% goes to transportation section [10]. Use of fossil fuel results in the release of a high amount of harmful gases, which leads to various negative effects on the environment such as biodiversity loss, receding of glaciers, the rise in sea level, and climate change. The depletion of fossil fuel and its increasing demand are now also affecting the global economy [3]. As an alternative to fossil fuel, biofuel has come up with promising applications. Scientists are tirelessly working for the enhanced biofuel production from sustainable resources [57]. In response to the increasing need for biofuel production, new processes, technologies, and renewable biological sources are being rapidly developed. Biofuels produced from biomass are known as solid, liquid, and gaseous fuels. The classification of biofuel in three different generations is based on the complex and chemical nature

R. Saini
Department of Civil Engineering, Lassonde School of Engineering, York University, North York, ON, Toronto, Canada

K. Hegde
INRS-ETE, University of Quebec, Quebec City, QC, Canada

S. K. Brar (✉)
Department of Civil Engineering, Lassonde School of Engineering, York University, North York, ON, Toronto, Canada

INRS-ETE, University of Quebec, Quebec City, QC, Canada
e-mail: satinder.brar@ete.inrs.ca; Satinder.Brar@lassonde.yorku.ca

C. R. Soccol
Department of Bioprocess Engineering and Biotechnology, Federal University of Paraná, Curitiba, Brazil

M. O. Daramola, A. O. Ayeni (eds.), *Valorization of Biomass to Value-Added Commodities*, Green Energy and Technology,
https://doi.org/10.1007/978-3-030-38032-8_23

of the biomass used, such as the first generation of biofuels, which comprises the biodiesel, that was produced from the crops. The second generation of biofuels, i.e., biohydrogen, bioethanol, and microbial lipid, has been produced from agricultural by-products such as lignocellulosic biomass. The third generation of biofuel such as biogas has attracted the use of seaweeds and marine resources [12, 17].

With the advancement of genetic and metabolic engineering, biofuel production has been greatly improved through microbial photosynthesis or microbial bioconversion of organic biomass. Conventionally, to develop superior microbial strains, the industries pass them through random mutagenesis and different screening processes [8, 23]. However, these methods were slow, uncontrolled, and unpredictable. Without the knowledge of the biochemistry of a microorganism, it may be difficult to develop a microorganism of desired characters through mutagenesis and random selection [37]. Best methods for mutagenesis are genetic and metabolic engineering [27, 37]. Engineering is being employed for modification of the microbial cells to express novel surface proteins or other internal substance to enhance the microbial biocatalytic activities [21, 31]. For instance, expression of specific enzymes on the microbial cells may improve the biofuel production through the enhanced conversion of starch into the sugar. Cells displaying the relevant surface enzymes along with advanced metabolic pathways could greatly improve the feedstock conversion into the biofuel. Metabolic engineering can bless the microorganism to tolerate high toxic concentrations, improved product yield, and the ability to utilize wide varieties of carbon substrate [64]. Desirable biochemical pathways can be enhanced by regulation and production of specific enzymes or by removing the inhibitory and side pathways. New metabolic pathways from one microorganism can also be engineered into another microorganism [31]. Overall, metabolic and genetic engineering allows us to do precise modifications in the microorganism without any accumulation of undesirable mutations as well as it leads to the consumption of substrates that were previously impossible to consume by the microorganism [31].

Over the past decade, many lipid-producing candidates has been exploited such as *Cryptococcus curvatus, Yarrowia lipolytica, Lipomyces starkeyi, Trichosporon fermentans*, and many more [36]. But, with the recognition of lipid accumulation ability of *Rhodosporidium* sp., it has now received increasing interest. It has a number of advantages over other lipid-producing candidates. For instance, wild-type *Y. lipolytica* can only grow on glycerol and glucose, but *Rhodosporidium toruloides* can consume cellobiose, sucrose, maltose, and glycerol [60] and has a high tolerance to a number of inhibitors [58]. Implementation of genetic and metabolic engineering to further enhance the *Rhodosporidium* sp. properties in terms of lipid accumulation, inhibitor tolerance, lipid conversion into advanced biofuels, and carbon consumption rate is under progress. Despite the number of published research papers claiming strains improvements, implementation of these engineered strains is still limited to the lab-scale biofuel production which could not meet the increasing demand of biofuels among society.

The aim of this chapter is thus to interpret and summarize the current trends of engineering strategies employed in *Rhodosporidium* for lipid production. Potential of *Rhodosporidium* sp. for the production of lipid using various carbon sources has also been briefly discussed.

23.2 Biochemistry of Lipid Accumulation

Oleaginous yeast is known to accumulate more than 20% lipid of their total cell weight such as *R. toruloides*, *Rhodotorula glutinis*, *Cryptococcus curvatus*, *Trichosporon fermentans*, and *Yarrowia lipolytica* [2]. In general, during the presence of an excess of nutrients in the media, these yeast divert the flux of carbon from energy production to lipid (TAGs) synthesis [2, 44]. Nitrogen starvation leads to inhibition of isocitrate dehydrogenase (ICDH), hence leading to citrate accumulation in mitochondria which then takes part in fatty acid synthesis (FAS) as shown in Fig. 23.1. Excess of citrate cleaved into oxaloacetate and acetyl-CoA by the ATP-citrate lyase (ACL) [50]. Acetyl-CoA, palmitoyl-CoA and stearoyl-CoA, then directed toward the FAS complex which further relocates to the endoplasmic reticulum, where it undergoes NADPH-dependent desaturation, and is then utilized for TAGs production. TAGs synthesis then yields diacylglycerol (DAG), phosphatidic acid (PA), and lysophosphatidic acid (LPA) through the Kennedy pathway and finally is stored in lipid droplets. NADPH for lipid synthesis is generally provided through oxidative pentose phosphate pathway or the malic enzyme (ME) [44]. Recently, it is shown that the oxidative pentose phosphate pathway is the primary NADPH source for the biosynthesis of fatty acid in *Y. lipolytica* [56] and presumably in *R. toruloides* [66].

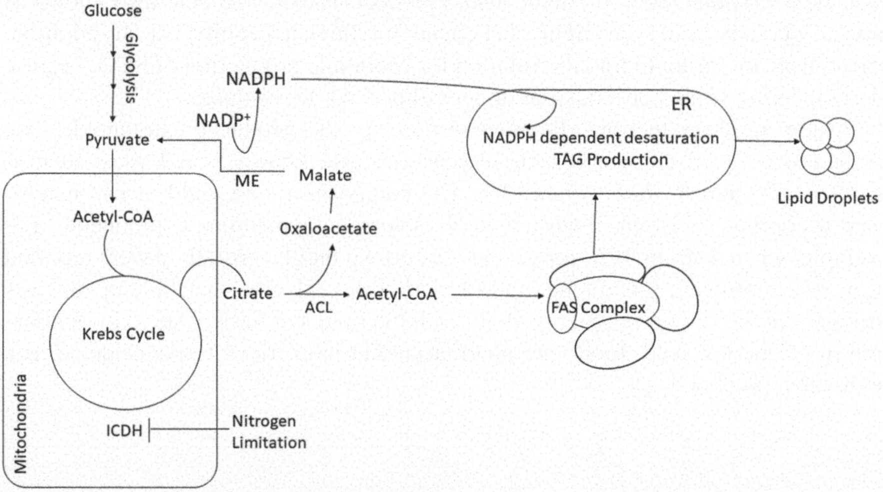

Fig. 23.1 An overview of lipid production pathway

23.3 Microbial Lipid

Microbial lipids can be used to produce a variety of compounds such as biodiesel, surfactants, waxes, lubricants, and creams [4, 5, 46, 61]. It is also known as single cell oil (SCO). In general, major lipid compositions produced by strains are stearic acid, linoleic acid, oleic acid, and palmitic acid. Microbial lipid has shown the similarities with the vegetable oil-derived fatty acids and thus can replace them for biodiesel production. The percentage of these fatty acids may vary because it depends on the type of microorganism, plant, and growth conditions, but types of fatty acid present are similar [65].

In general, the fatty acid profile of an organism varies due to the carbon/nitrogen (C/N) ratio, environmental factors, and media composition. For instance, Patel et al. [39] cultivated *R. toruloides* HIMPA1 on different types of carbon sources and reported that monounsaturated fatty acid contents were high in fructose and glucose-supplemented media, while in the presence of sucrose, leads to an increase in polyunsaturated fatty acid. In addition, the nitrogen source also plays an important role in the production of fatty acids. It was demonstrated that lipid produced in the rapeseed meal hydrolysate-containing media has lower palmitic acid and higher linoleic acid in comparison to those which grew in yeast extract [52]. Changing the environmental condition can also lead to a change in the fatty acid composition of lipid. For instance, Polburee et al. [41] reported a higher production of linolenic and linoleic acid when the two-stage cultivation system was employed with a shift in temperature from 30 to 25 °C. The fatty acid composition of lipids can also vary due to the change in the substrate as shown in Table 23.1. Besides, as a feedstock for biodiesel production, SCO has also many other applications. For instance, polyunsaturated fatty acids, i.e., linolenic acid and arachidonic acid, have a medical significance that is why they could be utilized as cocoa butter substitutes. Furthermore, fatty acids, such as arachidonic acid, linolenic acid, and docosahexaenoic acid also known as omega-3/6 fatty acid, are highly beneficial for human health [14]. In addition, researchers are trying to find the solution for economic production of SCO, i.e., use of engineering strains or industrial or agricultural waste residues.

Besides lipid productions, *Rhodosporidium* sp. also produced carotenoids. Due to the red color of carotenoids, these yeast are also known as red yeast such as *R. toruloides* and *R. diobovatum* [34]. The composition and yield of carotenoids vary according to strain, environmental factors, and medium components. For example, when a strain of *R. toruloides* was grown on glycerol, they were reported to produce torulene, β-carotene, and torularhodin [25]. The metabolism suggests that carotenoids are co-producing along with the lipid synthesis. Due to its application in cosmetics, feed, food, and pharmaceutical industries, it is considered as a high-value product.

Table 23.1 Fatty acid compositions of lipids produced in *Rhodosporidium* when grown on different substrates

Rhodosporidium strain	Substrate	Fatty acid composition (%)				References
		Palmitic acid	Oleic acid	Linoleic acid	Linolenic acid	
R. toruloides Y4	Cornstalk hydrolysate	30.0	51.0	6.3	ND	[62]
R. toruloides Y4	Glycerol	18	8.1	66.6	4.4	[52]
R. toruloides Y2	Glucose	0.9	49.9	13.6	ND	[68]
R. toruloides CCT	Molasses	24.7	50	12.5	2.2	[53]
R. toruloides DSM4444	Crude glycerol	27	37.8	15.8	2.2	[48]
R. toruloides AS 2.1389	Xylose	35.3	51.4	1.8	ND	[28]
R. toruloides AS 2.1389	Bagasse hydrolysate	29.31	49.36	9.62	2.26	[67]
R. toruloides AS 2.1389	Glucose	26.1	46.4	9.2	3.8	[59]
R. toruloides 2F5	Inulin	22.14	52.19	10.96	ND	[55]
R. kratochvilovae HIMPA1	Glucose	5.9	37.3	ND	0.2	[38]
R. kratochvilovae HIMPA1	Hemp seed extract	11.15	45.55	8.47	0.15	[38]
R. fluviale DMKU-RK253	Crude glycerol	37.5	33	13.4	3.9	[41]
R. diobovatum 08-225	Glycerol	28	33	20	3	[35]

ND not detected

23.4 Utilization of Carbon Sources for Lipid Accumulation

The cost and type of substrate for lipid accumulation in microorganism have a huge impact on its economic production. Different types of carbon sources, for example, glucose, biomass-derived sugars, glycerol, and organic acid, have been employed for the production of lipids. Glucose is generally an ideal carbon source for most microorganisms. For instance, *R. toruloides* was reported to accumulate 71.9 g/L of lipid and achieved a biomass of 106.5 g/L, and glucose to lipid conversion yield was 0.23 g/g, after 134 h of fermentation [26]. Exploiting the pure carbon substrates for biofuel production is economically unacceptable.

In addition, utilization of lignocellulosic biomass as feedstock for lipid production has drawn a lot of attention. Naturally, lignocellulosic biomass emanates mixture of carbons (e.g., pentoses and hexoses) at low cost, which can be exploited for economical biofuel production. In general, lignocellulosic biomass is composed of cellulose, hemicellulose, and lipid. It is very important to pretreat the biomass for the removal of lipid and to enhance the cellulose and hemicellulose accessibility by the enzymes. Their action leads to breakdown of polymeric sugars into pentose and

Table 23.2 Lipid accumulation by *Rhodosporidium* sp. using different types of carbon sources

Microorganism	Lignocellulosic biomass	Biomass (g/L)	Lipid (g/L)	References
Rhodosporidium toruloides DSM 4444	Glucose	37.2	24	[49]
	Glycerol and glucose	17.4	11	[6]
Rhodococcus opacus PD630	Kraft hardwood pulp	24.5	11.0	[22]
R. diobovatum	Crude glycerol	14.1	7.1	[35]
R. kratochvilovae	*Cassia fistula* L. fruit pulp	2.58	1.37	[40]
R. kratochvilovae HIMPA1	Xylose	6.9	3.84	[39]
	Glycerol	14.15	8	
R. toruloides 21167	Cassava starch	20.1	13.0	[13]
R. toruloides	Sugarcane bagasse	23.5	12.34	[67]
R. toruloides AS2. 1389	Acetic acid	4.35	2.10	[15]
	Glycerol	26.5	10.0	[58]
	Distillery wastewater	8.12	3.54	[30]
R. toruloides CCT 0783	Molasses	34.8	16	[53]
Rhodosporidium glutinis ATCC 204091	Wheat straw	13.8	3.45	[63]

hexose, which then could be utilized for microbial growth [45] (Table 23.2). But the pretreatment process also results in the production of various inhibitory compounds such as acetic acid, furfural, 5-hydroxymethylfurfural, formic acid, hydroxybenzaldehyde, and levulinic acid [36]. These compounds are reported to negatively interfere with microbial growth as well as fatty acid production [16, 42]. For instance, furfural inhibits the enzymes involved in fatty acid production such as glyceraldehyde 3-phosphate dehydrogenase, which leads to a reduction in lipid production. Chen et al. [7] demonstrated that 0.05% of furfural leads to cell growth arrest and inhibition of lipid production in *R. toruloides*. Hence, it is very important to develop the novel method of lignocellulosic biomass pretreatment with high cellulase availability but with less inhibitor production.

23.5 Engineering Strategies

Rhodosporidium has been studied and exploited for the production of lipid from wide varieties of the carbon substrate. The implementation of metabolic engineering allows the researchers to efficiently utilize the sustainable and complex carbon sources for lipid production. This is highly desirable for making the lipid production economically feasible. The general idea behind engineering is described in Fig. 23.2.

The *Rhodosporidium* sp. adaptability has been intensively studied, predominantly for the purpose of exploitation of various biomass-derived carbon sources. Culture derived from feedstock such as lignocellulosic biomass consists of a mixture of carbon sources (e.g., C5 and C6), along with the inhibitors in varying concentration.

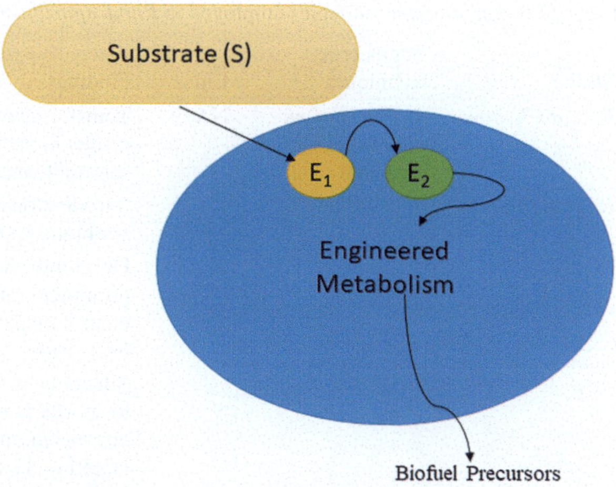

Fig. 23.2 Mechanism of engineered metabolism. Intracellular enzyme converts substrate into a product, which then enters the metabolic pathway of cell and eventually generates biofuel or its precursor

The effort was made based on mutagenesis and environment adaptation of sarin in the presence of stress condition. For instance, Qi et al. [43] mutated *R. toruloides* through temperature plasma mutagenesis, which resulted in increased tolerance to the inhibitors such as furfurals and aromatic compounds. RNA-sequencing assisted annotation, proteomic analysis, transcriptomic analysis, and genetic analysis allowed us to understand the molecular mechanism of lipid metabolism in *Rhodosporidium* [20, 47, 69]. There are similarities in pathways such as nitrogen metabolism-induced lipogenesis in *Y. lipolytica* and *Rhodosporidium* sp., but different metabolisms were also noticed. For instance, the oxidative pentose phosphate pathway was proved to be the sole NADPH supplier in *Y. lipolytica* [56], while *R. toruloides* have an alternative pathway for NADPH production, i.e., NADH-dependent isocitrate dehydrogenase [69]. There is another example in which inhibition of the same gene resulted in the opposite effect on both the abovementioned strains. Inhibition of *pex10*, a gene involved in fatty acid catabolism, resulted in increased accumulation of lipid in *Y. lipolytica*, while decreased lipid accumulation was seen in *R. toruloides* [65]. However, it still needs deep exploration of the metabolic pathway that's why inhibition of a similar gene resulted in a different effect.

Through the knowledge of proteomic and transcriptomic analysis, key genes involved in metabolic pathways can be identified and mutated for further strain engineering. Genetic tools have been progressively developed, including constitutive promoter and transformation for a set of gene expression, to establish the genetically tractable system. *Rhodosporidium* sp. transformation was rarely reported and was considered recalcitrant. PEG-mediated transformation was reported in *R. toruloides* [51] but was limited because of its unstable chromosomal integration, low efficiency, and auxotrophic selection [32]. The successful transformation was reported to be done with *Agrobacterium tumefaciens*-mediated transformation (ATMT). Abbott et al. [1] transformed *R. kratochvilovae* with the *URA5* gene [28, 29] and also integrated multiple genes in *R. toruloides* and obtained mutants resistant to the multiple antibiotics. Wang and co-workers evaluated the

Table 23.3 Engineering strategies employed in *Rhodosporidium* and their findings

Strain	Engineering techniques	Lipid	Findings	References
R. kratochvilovae	*Agrobacterium tumefaciens*-mediated transformation	NA	Transformation of *URA5* gene results in the introduction of the selectable marker	[1]
R. toruloides			Introduction of multiple antibiotic resistant genes	[28, 29]
			Development of 5 constitutive promoter system for increased control on desired gene expression	[54]
			Silencing of *KU70*, which leads to shorter homology sequence and increased the chance of targeting the desired gene	[18]
R. diobovatum	CRISPER- Cas9		Deletion of two Acetyl-CoA synthase leads to the inability of consuming carbon source.	[33]
R. toruloides IFO0880	*Agrobacterium tumefaciens*-mediated transformation	16.4 g/L	Overexpression of acetyl-CoA and diacylglycerol acyltransferase (DGA1) leads to an increase in lipid productivity	[66]
R. toruloides IFO0880		24.7 g/L	Overexpression of malic enzyme and stearoyl-CoA desaturase has significantly improved the lipid productivity	[65]
R. toruloides	Arabinose-inducible cre-lox recombination	39.5 g/L	Expression of diacylglycerol O-acyltransferase 1 and stearoyl-CoA desaturase-1 results in growth on non-detoxified wheat straw hydrolysate and high lipid productivity	[9]
R. toruloides	*Agrobacterium tumefaciens*-mediated transformation	50 g/L	Increased expression of 3-ketoacyl-CoA synthases gene resulted in the production of erucic acid and nervonic acid	[11]

NA data not provided in reference

functional evaluation of five constitutive promoters from *R. toruloides* [54]. Genetic tools also allowed to integrate novel gene into the microorganism or removal of the native gene which resulted in the deactivation of an enzyme related to a particular gene, hence causing a change in the metabolic pathway. For instance, Lee et al. [24] integrated the pleiotropic drug resistance and membrane transport encoding gene into *R. toruloides* from *Saccharomyces cerevisiae.* This resulted in the in situ exportation of carotenoids when the strain was cultured in the two-phase system. In another report, deletion of the gene in *R. toruloides* was demonstrated that resulted in the generation of *KU70* null mutant [18]. Removal of certain enzyme or overexpression of native enzyme-encoding genes, which involves in the production of side products or in the conversion of carbon flux toward the desired product, respectively, will result in shifting of the whole pathway toward the lipid production. For instance, Zhang et al. [66] overexpressed the diacylglycerol acyltransferase (DGA1) and acetyl-CoA carboxylase (ACC1) genes in *R. toruloides*, which resulted in increased production of the TAG. Hence, they were able to increase the lipid yield due to more carbon directed toward lipid synthesis. In addition, other metabolic enzymes were also explored. Zhang and co-worker found that overexpression of stearoyl-CoA desaturase and malic enzyme could significantly increase lipid titter (Table 23.3) [65].

In summary, strain engineering strategies include optimization of pathways, increased carbon source utilization, knockout or knock in of certain genes. Although a lot of work has already been done on a number of strains for increasing their lipid accumulation, or inhibitor tolerance, there is still need to explore the metabolic complexity of an organism, i.e., from the consumption of sugars till the production of bioproducts. The deeper understanding of lipid production pathways in *Rhodosporidium* will not only allow us to identify the key steps but also will help to develop the super mutant strains with the ability to utilize wide carbon sources, high lipid titer, and high tolerance to inhibitors.

23.6 Future of *Rhodosporidium* sp.

Rhodosporidium species are considered as promising candidates for lipid production, carotenoids, and fatty acid-derived products. Microbial lipid can be used to convert into biodiesel, but, currently, it's not viable because of high substrate cost and cost of other steps which are involved in fermentation [19]. According to techno-economic-based analysis, lipid production is majorly dependent on feedstock, process design, and fermenter [19]. In other words, to make the lipid production economical, more efforts need to be done to utilize renewable feedstock and for high lipid yield. Researchers have exploited the use of various waste-derived and renewable products to demonstrate the great potential of *Rhodosporidium* as a consumer of a wide range of resources. However, some of the raw materials require a certain unit operation to make carbon and nitrogen easily accessible for an organism. Study of metabolic behaviors of the organism can help us to understand the substrate utilization and can provide a path of strengthening the engineering strategies. The increasing knowledge of lipid metabolism and the use of genetic and metabolic tools in *Rhodosporidium* will lead to the development of desired strain or super mutant strain. Over the past few decades, passionate research done on *Y. lipolytica* made it a high producer of lipid- and fatty acid-derived chemicals. *Rhodosporidium* sp. might be the future of lipid production. Wild-type strains have already shown great potential, and with the addition of engineering, it will be the best. The capabilities of *Rhodosporidium* should be further exploited in the future.

23.7 Conclusion

With the use of genetic and metabolic engineering, researchers are capable of developing or mutant the strains as per societies' requirements. *Rhodosporidium* has come up with promising advantages over other available lipid-producing strains. The deeper understanding of metabolic pathways in *Rhodosporidium* will allow us to understand the mechanism of lipid production as well as the carbon channelization toward the desired product. However, still, there are many things to reveal about this strain which will come out when more and more work will be done. The cheapest source of the substrate is lignocellulosic biomass. It consists of both C5 and C6 carbons along with the inhibitors. *Rhodosporidium* has shown the great potential to grow on lignocellulosic biomass, and with the addition of engineering, it can do great. Less is known about *Rhodosporidium*; *the* researcher has tried to build the selection markers in this system to specify the targets. In the past, research has identified various bottlenecks to improve lipid titer, production rate, and yield. Now, this is the time to not only solve the previous problems but also to make lipid production more economical and feasible. In summary, *Rhodosporidium* needs further exploitation to enhance its capabilities of lignocellulosic biomass consumption, high productivity, titer, and tolerance to inhibitors.

Acknowledgment The authors are sincerely thankful to the Natural Sciences and Engineering Research (Discovery Grant 23451) (Strategic Project Grants) and for financial support. The views or opinions expressed in this article are exclusively those of the authors.

References

1. E.P. Abbott, G. Ianiri, R. Castoria, et al., Overcoming recalcitrant transformation and gene manipulation in Pucciniomycotina yeasts. Appl. Microbiol. Biotechnol. **97**, 283–295 (2013)
2. J.L. Adrio, Oleaginous yeasts: Promising platforms for the production of oleochemicals and biofuels. Biotechnol. Bioeng. **114**, 1915–1920 (2017)
3. A.K. Agarwal, Biofuels (alcohols and biodiesel) applications as fuels for internal combustion engines. Prog. Energy Combust. Sci. **33**, 233–271 (2007)
4. J.M. Ageitos, J.A. Vallejo, P. Veiga-Crespo, et al., Oily yeasts as oleaginous cell factories. Appl. Microbiol. Biotechnol. **90**, 1219–1227 (2011)
5. J. Blazeck, A. Hill, L. Liu, et al., Harnessing Yarrowia lipolytica lipogenesis to create a platform for lipid and biofuel production. Nat. Commun. **5**, 3131 (2014)
6. R.R. Bommareddy, W. Sabra, A.-P. Zeng, Glucose-mediated regulation of glycerol uptake in Rhodosporidium toruloides: Insights through transcriptomic analysis on dual substrate fermentation. Eng. Life Sci. **17**, 282–291 (2017)
7. X. Chen, Z. Li, X. Zhang, et al., Screening of oleaginous yeast strains tolerant to lignocellulose degradation compounds. Biotechnol. Appl. Biochem. **159**, 591 (2009)
8. P.M. Derkx, T. Janzen, K.I. Sørensen, et al., The art of strain improvement of industrial lactic acid bacteria without the use of recombinant DNA technology. Microb. Cell Factories **13**, S5 (2014)
9. T. Díaz, S. Fillet, S. Campoy, et al., Combining evolutionary and metabolic engineering in Rhodosporidium toruloides for lipid production with non-detoxified wheat straw hydrolysates. Appl. Microbiol. Biotechnol. **102**, 3287–3300 (2018)

10. J.C. Escobar, E.S. Lora, O.J. Venturini, et al., Biofuels: Environment, technology and food security. Renew. Sust. Energy Rev. **13**, 1275–1287 (2009)
11. S. Fillet, C. Ronchel, C. Callejo, et al., Engineering Rhodosporidium toruloides for the production of very long-chain monounsaturated fatty acid-rich oils. Appl. Microbiol. Biotechnol. **101**, 7271–7280 (2017)
12. N. Gaurav, S. Sivasankari, G.S. Kiran, et al., Utilization of bioresources for sustainable biofuels: A review. Renew. Sust. Energy Rev. **73**, 205–214 (2017)
13. Q. Gen, Q. Wang, Z.-M. Chi, Direct conversion of cassava starch into single cell oil by co-cultures of the oleaginous yeast Rhodosporidium toruloides and immobilized amylases-producing yeast Saccharomycopsis fibuligera. Renewable Energy **62**, 522–526 (2014)
14. C. Huang, X.-F. Chen, L. Xiong, et al., Single cell oil production from low-cost substrates: The possibility and potential of its industrialization. Biotechnol. Adv. **31**, 129–139 (2013)
15. X.-F. Huang, J.-N. Liu, L.-J. Lu, et al., Culture strategies for lipid production using acetic acid as sole carbon source by Rhodosporidium toruloides. Bioresour. Technol. **206**, 141–149 (2016)
16. L.J. Jönsson, C. Martín, Pretreatment of lignocellulose: Formation of inhibitory by-products and strategies for minimizing their effects. Bioresour. Technol. **199**, 103–112 (2016)
17. Q. Kang, L. Appels, T. Tan, et al., Bioethanol from lignocellulosic biomass: Current findings determine research priorities. Sci. World J. **2014**, 13 (2014)
18. C.M. Koh, Y. Liu, Moehninsi, M. Du, L. Ji, Molecular characterization of KU70 and KU80 homologues and exploitation of a KU70-deficient mutant for improving gene deletion frequency in Rhodosporidium toruloides. BMC Microbiol. **14**, 50 (2014)
19. A.A. Koutinas, A. Chatzifragkou, N. Kopsahelis, et al., Design and techno-economic evaluation of microbial oil production as a renewable resource for biodiesel and oleochemical production. Fuel **116**, 566–577 (2014)
20. S. Kumar, H. Kushwaha, A.K. Bachhawat, et al., Genome sequence of the oleaginous red yeast Rhodosporidium toruloides MTCC 457. Eukaryot. Cell **11**, 1083–1084 (2012)
21. K. Kuroda, M. Ueda, Arming technology in yeast—novel strategy for whole-cell biocatalyst and protein engineering. Biomol. Ther. **3**, 632 (2013)
22. K. Kurosawa, S.J. Wewetzer, A.J. Sinskey, Engineering xylose metabolism in triacylglycerol-producing Rhodococcus opacusfor lignocellulosic fuel production. Biotechnol. Biofuels **6**, 134 (2013)
23. J. Lee, Y.-S. Jang, S.J. Choi, et al., Metabolic engineering of *Clostridium acetobutylicum* ATCC 824 for isopropanol-butanol-ethanol fermentation. Appl. Environ. Microbiol. **78**, 1416–1423 (2012)
24. J.J.L. Lee, L. Chen, B. Cao, et al., Engineering Rhodosporidium toruloides with a membrane transporter facilitates production and separation of carotenoids and lipids in a bi-phasic culture. Appl. Microbiol. Biotechnol. **100**, 869–877 (2016)
25. J.J.L. Lee, L. Chen, J. Shi, et al., Metabolomic profiling of Rhodosporidium toruloides grown on glycerol for carotenoid production during different growth phases. J. Agric. Food Chem. **62**, 10203–10209 (2014)
26. Y. Li, Z. Zhao, F. Bai, High-density cultivation of oleaginous yeast Rhodosporidium toruloides Y4 in fed-batch culture. Enzym. Microb. Technol. **41**, 312–317 (2007)
27. J.C. Liao, L. Mi, S. Pontrelli, et al., Fuelling the future: Microbial engineering for the production of sustainable biofuels. Nat. Rev. Microbiol. **14**, 288 (2016)
28. J. Lin, S. Li, M. Sun, et al., Microbial lipid production by oleaginous yeast in d-xylose solution using a two-stage culture mode. RSC Adv. **4**, 34944–34949 (2014a)
29. X. Lin, Y. Wang, S. Zhang, et al., Functional integration of multiple genes into the genome of the oleaginous yeast Rhodosporidium toruloides. FEMS Yeast Res. **14**, 547–555 (2014b)
30. J. Ling, S. Nip, H. Shim, Enhancement of lipid productivity of Rhodosporidium toruloides in distillery wastewater by increasing cell density. Bioresour. Technol. **146**, 301–309 (2013)
31. D. Liu, T. Evans, F. Zhang, Applications and advances of metabolite biosensors for metabolic engineering. Metab. Eng. **31**, 35–43 (2015)

32. Y. Liu, C.M.J. Koh, L. Sun, et al., Characterization of glyceraldehyde-3-phosphate dehydrogenase gene RtGPD1 and development of genetic transformation method by dominant selection in oleaginous yeast Rhodosporidium toruloides. Appl. Microbiol. Biotechnol. **97**, 719–729 (2013)
33. Y. Liu, M. Zhang, T. Wang, et al., Two acetyl-CoA synthetase isoenzymes are encoded by distinct genes in marine yeast Rhodosporidium diobovatum. Biotechnol. Lett. **38**, 417–423 (2016)
34. I. Mannazzu, S. Landolfo, T.L. Da Silva, et al., Red yeasts and carotenoid production: Outlining a future for non-conventional yeasts of biotechnological interest. World J. Microbiol. Biotechnol. **31**, 1665–1673 (2015)
35. G. Munch, R. Sestric, R. Sparling, et al., Lipid production in the under-characterized oleaginous yeasts, Rhodosporidium babjevae and Rhodosporidium diobovatum, from biodiesel-derived waste glycerol. Bioresour. Technol. **185**, 49–55 (2015)
36. C.S. Osorio-González, K. Hegde, S.K. Brar, et al., Challenges in lipid production from lignocellulosic biomass using Rhodosporidium sp.; A look at the role of lignocellulosic inhibitors. Biofuels Bioprod. Biorefin. **13**(3), 740–759 (2018)
37. J.H. Park, S.Y. Lee, Towards systems metabolic engineering of microorganisms for amino acid production. Curr. Opin. Biotechnol. **19**, 454–460 (2008)
38. A. Patel, M. Pravez, F. Deeba, et al., Boosting accumulation of neutral lipids in Rhodosporidium kratochvilovae HIMPA1 grown on hemp (Cannabis sativa Linn) seed aqueous extract as feedstock for biodiesel production. Bioresour. Technol. **165**, 214–222 (2014)
39. A. Patel, V. Pruthi, R.P. Singh, et al., Synergistic effect of fermentable and non-fermentable carbon sources enhances TAG accumulation in oleaginous yeast Rhodosporidium kratochvilovae HIMPA1. Bioresour. Technol. **188**, 136–144 (2015a)
40. A. Patel, D.K. Sindhu, N. Arora, et al., Biodiesel production from non-edible lignocellulosic biomass of Cassia fistula L. fruit pulp using oleaginous yeast Rhodosporidium kratochvilovae HIMPA1. Bioresour. Technol. **197**, 91–98 (2015b)
41. P. Polburee, W. Yongmanitchai, K. Honda, et al., Lipid production from biodiesel-derived crude glycerol by Rhodosporidium fluviale DMKU-RK253 using temperature shift with high cell density. Biochem. Eng. J. **112**, 208–218 (2016)
42. R. Poontawee, W. Yongmanitchai, S. Limtong, Efficient oleaginous yeasts for lipid production from lignocellulosic sugars and effects of lignocellulose degradation compounds on growth and lipid production. Process Biochem. **53**, 44–60 (2017)
43. F. Qi, K. Yuki, W. Zitian, et al., Novel mutant strains of Rhodosporidium toruloides by plasma mutagenesis approach and their tolerance for inhibitors in lignocellulosic hydrolyzate. J. Chem. Technol. Biotechnol. **89**, 735–742 (2014)
44. C. Ratledge, Fatty acid biosynthesis in microorganisms being used for Single Cell Oil production. Biochimie **86**, 807–815 (2004)
45. Z. Ruan, M. Zanotti, Y. Zhong, et al., Co-hydrolysis of lignocellulosic biomass for microbial lipid accumulation. Biotechnol. Bioeng. **110**, 1039–1049 (2013)
46. C.D. Rutter, S. Zhang, C.V. Rao, Engineering Yarrowia lipolytica for production of medium-chain fatty acids. Appl. Microbiol. Biotechnol. **99**, 7359–7368 (2015)
47. J. Shi, H. Feng, J. Lee, et al., Comparative proteomics profile of lipid-cumulating oleaginous yeast: An iTRAQ-coupled 2-D LC-MS/MS analysis. PLoS One **8**, e85532 (2013)
48. L. Signori, D. Ami, R. Posteri, et al., Assessing an effective feeding strategy to optimize crude glycerol utilization as sustainable carbon source for lipid accumulation in oleaginous yeasts. Microb. Cell Factories **15**, 75 (2016)
49. S.S. Tchakouteu, K. Nikolaos, C. Afroditi, et al., Rhodosporidium toruloides cultivated in NaCl-enriched glucose-based media: Adaptation dynamics and lipid production. Eng. Life Sci. **17**, 237–248 (2017)
50. O. Tehlivets, K. Scheuringer, S.D. Kohlwein, Fatty acid synthesis and elongation in yeast. Biochim. Biophys. Acta. **1771**, 255–270 (2007)
51. M. Tully, H.J. Gilbert, Transformation of Rhodosporidium toruloides. Gene **36**, 235–240 (1985)

52. E. Uçkun Kiran, A. Trzcinski, C. Webb, Microbial oil produced from biodiesel by-products could enhance overall production. Bioresour. Technol. **129**, 650–654 (2013)
53. J.P.F. Vieira, J.L. Ienczak, C.E.V. Rossell, et al., Microbial lipid production: Screening with yeasts grown on Brazilian molasses. Biotechnol. Lett. **36**, 2433–2442 (2014)
54. Y. Wang, L. Xinping, Z. Sufang, et al., Cloning and evaluation of different constitutive promoters in the oleaginous yeast Rhodosporidium toruloides. Yeast **33**, 99–106 (2016)
55. Z.-P. Wang, W.-J. Fu, H.-M. Xu, et al., Direct conversion of inulin into cell lipid by an inulinase-producing yeast Rhodosporidium toruloides 2F5. Bioresour. Technol. **161**, 131–136 (2014)
56. T.M. Wasylenko, W.S. Ahn, G. Stephanopoulos, The oxidative pentose phosphate pathway is the primary source of NADPH for lipid overproduction from glucose in Yarrowia lipolytica. Metab. Eng. **30**, 27–39 (2015)
57. Y. Weldemichael, G. Assefa, Assessing the energy production and GHG (greenhouse gas) emissions mitigation potential of biomass resources for Alberta. J. Clean. Prod. **112**, 4257–4264 (2016)
58. J. Xu, D. Liu, Exploitation of genus Rhodosporidium for microbial lipid production. World J. Microbiol. Biotechnol. **33**, 54 (2017)
59. J. Xu, X. Zhao, W. Wang, et al., Microbial conversion of biodiesel byproduct glycerol to triacylglycerols by oleaginous yeast Rhodosporidium toruloides and the individual effect of some impurities on lipid production. Biochem. Eng. J. **65**, 30–36 (2012)
60. S.-J. Xue, Z. Chi, Y. Zhang, et al., Fatty acids from oleaginous yeasts and yeast-like fungi and their potential applications. Crit. Rev. Biotechnol. **38**, 1049–1060 (2018)
61. Z. Xue, P.L. Sharpe, S.-P. Hong, et al., Production of omega-3 eicosapentaenoic acid by metabolic engineering of Yarrowia lipolytica. Nat. Biotechnol. **31**, 734 (2013)
62. X. Yang, G. Jin, Z. Gong, et al., Recycling microbial lipid production wastes to cultivate oleaginous yeasts. Bioresour. Technol. **175**, 91–96 (2015)
63. X. Yu, Y. Zheng, K.M. Dorgan, et al., Oil production by oleaginous yeasts using the hydrolysate from pretreatment of wheat straw with dilute sulfuric acid. Bioresour. Technol. **102**, 6134–6140 (2011)
64. F. Zhang, S. Rodriguez, J.D. Keasling, Metabolic engineering of microbial pathways for advanced biofuels production. Curr. Opin. Biotechnol. **22**, 775–783 (2011)
65. S. Zhang, M. Ito, J.M. Skerker, et al., Metabolic engineering of the oleaginous yeast *Rhodosporidium toruloides* IFO0880 for lipid overproduction during high-density fermentation. Appl. Microbiol. Biotechnol. **100**, 9393–9405 (2016a)
66. S. Zhang, J.M. Skerker, C.D. Rutter, et al., Engineering Rhodosporidium toruloides for increased lipid production. Biotechnol. Bioeng. **113**, 1056–1066 (2016b)
67. X. Zhao, F. Peng, W. Du, et al., Effects of some inhibitors on the growth and lipid accumulation of oleaginous yeast Rhodosporidium toruloides and preparation of biodiesel by enzymatic transesterification of the lipid. Bioprocess Biosyst. Eng. **35**, 993–1004 (2012)
68. W. Zhou, W. Wang, Y. Li, et al., Lipid production by Rhodosporidium toruloides Y2 in bioethanol wastewater and evaluation of biomass energetic yield. Bioresour. Technol. **127**, 435–440 (2013)
69. Z. Zhu, S. Zhang, H. Liu, et al., A multi-omic map of the lipid-producing yeast Rhodosporidium toruloides. Nat. Commun. **3**, 1112 (2012)

Chapter 24
Biotechnological Strategies for Enhanced Production of Biofuels from Lignocellulosic Biomass

K. K. Brar, B. S. Chadha, S. K. Brar, and P. Singh

24.1 Introduction

Lignocellulosic biomass in the form of easily accessible agro-residues (sugarcane bagasse, rice straw, and corn-residue), energy crops, and forest biomass is the low-cost renewable feedstock [1]. Lignocellulosic biomass offers immense opportunities for its bioconversion to second-generation ethanol (2G) and other value-added products. Till 2012, various 2G ethanol demonstration plants were established, that is, Inbicon in Denmark, Iogen in Canada and the USA, SEKAB Biofuels & Chemicals AB in Sweden, and Chemtex's plant in Italy [2]. However, in last 3 years commercial-scale 2G ethanol plants have been commissioned in the USA (DSMPOET in South Dakota and DuPont in Iowa), but BETA Renewables (Italy) and DuPont (USA) have great challenges to make this technology mature and competitive [3, 4]. In India, Glycol India and Praj industries have been set up as small demonstration 2G ethanol plants and owing to government policy to enhance the 2G ethanol, at least 9 biorefineries have been planned to be set up by 2022.

K. K. Brar (✉)
Department of Microbiology, Guru Nanak Dev University, Amritsar, India

Center-Eau Terre Environnement, INRS-ETE, Université du Québec, Québec, QC, Canada

B. S. Chadha
Department of Microbiology, Guru Nanak Dev University, Amritsar, India

S. K. Brar
Center-Eau Terre Environnement, INRS-ETE, Université du Québec, Québec, QC, Canada

P. Singh
Ludhiana Beverages Pvt Ltd, Jaspalon, Doraha, Ludhiana, India

M. O. Daramola, A. O. Ayeni (eds.), *Valorization of Biomass to Value-Added Commodities*, Green Energy and Technology,
https://doi.org/10.1007/978-3-030-38032-8_24

Lignocellulosic biomass is a complex polymer of cellulose, hemicellulose, and lignin. The conversion of lignocellulosic biomass into biofuels depends on pretreatment, enzymatic hydrolysis, and fermentation technologies. Different pretreatment options are used to break down the intact and relatively recalcitrant structure of lignocellulosic biomass (physical, chemical, thermochemical, and biochemical methods), which accounts for >20% of the total cost involved in bioconversion of lignocellulosics into ethanol [4].

The deconstruction of cellulose and xylan into monomeric forms is primarily mediated through multicomponent cellulolytic/hemicellulolytic enzyme system that works in a synergistic manner [5]. The cellulase enzymes include endoglucanases, cellobiohydrolases/exoglucanase, and β-glucosidase/cellobiase [6]. Hemicellulolytic enzymes are ubiquitous and diverse in nature. The most important of these enzymes are β-1, 4-xylanase and β-xylosidase [7]. These cellulases/hemicellulases are collectively called as glycosyl hydrolases (GH) and grouped into 133 families in the continually updated CAZymes [8]. Recently, lytic polysaccharide monooxygenase (LPMOs), which are classified as Auxillary Activity 9 (AA9) proteins [9, 10], have also been included in CAZymes. The AA9 proteins are known to be active on crystalline surface of cellulose and carry out the puncturing of cellulose sheets through oxidation mechanism, thus making available sites for initiation of cellobiohydrolase action [11, 12]. Enzymatic hydrolysis of pretreated biomass is mostly carried out using commercial fungal enzymes derived from *Trichoderma ressei.* This organism is used as a platform for the production of more than 80 proteins by major manufacturing companies (Novozyme and Danisco-Genencor) [13]. Novozymes have cellulase preparations known as Cellic CTec2, improved version Cellic CTec3 in addition to Cellic HTec3 whih are capable of achieving higher levels of hydrolysis. However, the cost of enzymes is still a major bottleneck hindering the process of lignocellulosic biomass bioconversion [14, 15].

Other hindering factors are fermentation inhibitors (hydroxymethylfurfurals (HMF), furfurals, formic acid, and acetic acid), sugar, and ethanol concentration in the media, which hinder the ethanol yield. Researchers have applied different strategies (such as strain adaptation, mutagenesis, genome shuffling, and metabolic engineering) for improvement of strains to overcome all the hurdles. The main objective of this study is to evaluate different advanced research strategies carried out by researchers to enhance sugar as well as ethanol yield.

24.2 Bioprocessing of Lignocellulosic Biomass

Bioprocessing of lignocellulosic biomass into value-added products generally involves the following sequential steps: (a) pretreatment, (b) enzymatic hydrolysis/saccharification, and (c) fermentation [16] (Fig. 24.1).

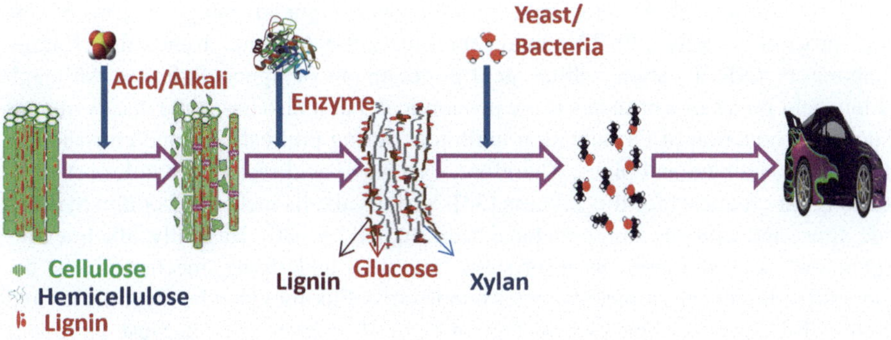

Fig. 24.1 Lignocellulosic process converting the biomass into biofuels and co-products. (Adapted from Xiao et al. [16])

24.2.1 Pretreatment Technologies

The aim of pretreatment is to distort and disengage the lignin and modify the structure of cellulose and hemicellulose to make them easily accessible to enzymes. An ideal pretreatment process should maximize the yield of monomeric sugars and also minimize energy consumption. Unfortunately, the existing technologies cannot satisfy all the criteria and is an area of continued research interventions.

First of all, lignocellulosic biomass is subjected to size reduction (10–30 mm) by mechanical methods such as chopping, grinding, or milling through which the surface area of biomass is increased. This method is followed by the energy-intensive step of reducing the feedstock from millimeter size to micrometers, which is unacceptable from the engineering viewpoint [17, 18]. Radiations such as microwave that can penetrate and heat the feedstock instantly have also been studied [19, 20]. Steam explosion (SE) of lignocellulosic biomass was developed by Iogen Corporation biorefinery in Canada and later adopted by Beta Renewable Company in Italy. This process involves steam explosion (SE) of biomass at temperature 160–260 °C corresponding to pressure of 0.69–4.83 MPa with a holding time of a few minutes to partially disrupt the structure of lignocellulosic biomass [19]. This disruption is catalyzed by acetic acid released from acetylated hemicelluloses and other organic acids such as levulinic and formic acids making the process auto-hydrolytic in nature [21]. Liquid hot water (LHW) is another hydrothermal pretreatment that can enhance sugar extraction [22, 23]. Dilute acid pretreatments have been intensively studied over the years with various feedstocks and reactors at different scales [24–27]. Dilute acid pretreatment followed by steam explosion is preferred by Abengoa Bioenergy, Spanish global biotechnology company for biofuel production. Sulfuric (H_2SO_4) and phosphoric (H_3PO_4) acids are widely used because they are relatively cheap and efficient in hydrolysis of lignocellulose. Hydrochloric (HCl) acid is more volatile, easier to recover, and attacks biomass better than H_2SO_4 [28], whereas nitric acid (HNO_3) possesses good cellulose to sugar conversion rates [29]. Use of low acid

(<1% w/v sulfuric/phosphoric) concentration has a major advantage in terms of cost and process severity [30–32]. In recent times, the National Renewable Energy Laboratory (NREL) started dilute acid pretreatment of lignocellulosics. Although dilute acid pretreatment seems more economical, but it also has some disadvantages such as production of fermentation inhibitors during pretreatment and corrosion of metallic containers and pipes. Therefore, expensive acid resistant stainless steel or coatings are required for this process [33]. Various alkalis including sodium hydroxide, lime, and aqueous ammonia have been studied [34–36]. Basically, alkaline pretreatment is a delignification process, and the underlying mechanism is the saponification of intermolecular ester bonds cross-linking xylan hemicelluloses and lignin [17]. However, this process cannot be used on large scale because large volumes of alkaline black liquor are produced, which not only affects microbial growth and fermentation but also raises an environmental concern. DuPont pretreats corn stover or switchgrass with dilute ammonia at low temperature for bioethanol conversion. An alternative strategy, that is, ammonia recycling percolation (ARP) process can overcome these disadvantages; however, the cost of ammonia recovery is high [37, 38]. Ammonia fiber explosion (AFEX) is a hybrid of the SE and ARP processes, in which biomass is pretreated with liquid anhydrous ammonia at mild temperatures (60–100 °C) and high pressure [39]. The use of ozone as a lysing agent reduces the lignin content and enhances digestibility of lignocellulosic material without generation of toxic residues [40]. This is an eco-friendly method; however, the cost factor is high. Organosolv process developed by Lignol in collaboration with the University of British Columbia for biorefining of softwood to ethanol and lignin processing is another approach that results in high grade lignin [41]. Solvent pretreatment is a fractionating process, in which organic or aqueous organic solvents such as methanol, ethanol, ethylene glycol, triethylene glycol, tetrahydrofurfuryl alcohol, glycerol, and acetone have been explored to extract lignin as well as hydrolyze hemicelluloses to render cellulose for enzymatic hydrolysis [42, 43]. The advantage of organic solvents over other chemical pretreatments is that they are relatively pure, and low-molecular-weight lignin can be recovered as a by-product. However, the high cost of organic solvents and the intensive energy consumption associated with solvent recovery make this strategy economically uncompetitive. Biological pretreatment is energy saving and environmental friendly over physical and chemical pretreatments. Biological pretreatment employs microorganisms that degrade lignocellulosic biomass at mild conditions without special requirements for equipment [44]. Particularly, basidiomycetes including white-rot fungi (e.g., *Phanerochaete chrysosporium*) and brown-rot fungi (e.g., *Fomitopsis palustris*) are the predominant species in lignocellulose degradation for the purpose of biofuel production. During pretreatment, these organisms secrete abundant ligninolytic enzymes including lignin peroxidase, manganese peroxidase, laccases, and other enzymes that degrade lignin [45–47]. However, biological pretreatment has apparent disadvantages as it requires longer residence time (several hours to few days) [48]. Most microorganisms consume some part of sugars by hydrolytic enzymes for their growth, which are released from hemicellulose and even cellulose [49]; this negatively affects the sugar yield of the process. Another disadvantage is to separate fungus from delignified biomass after biological pretreatment, which is cumbersome. Recently, Rajak and Banerjee [50] used hyperactive laccase produced from *Pleurotus* sp., for delignification of wasteland weed, *Saccharum spontaneum*. The maximum delignification obtained was 84.67% during 6.21 h of incubation.

24.2.2 Saccharification of Lignocellulosic Biomass

Enzymatic hydrolysis of lignocellulosic biomass is one of the key steps in 2G ethanol platform. These lignocellulosic biomasses are hydrolyzed into monomeric fermentable sugars for subsequent conversion to ethanol by ethanolgens, that is, yeast/bacteria [51–53]. Enzymatic saccharification is carried out separately at 45–50 °C followed by fermentation, which is termed as separate hydrolysis and fermentation (SHF) [180]. Since high temperature cannot be tolerated by microorganisms, ethanol fermentation is usually carried out at 30 °C [27, 54]. When cellulose hydrolysis and fermentation are carried out simultaneously, the phenomenon is known as simultaneous saccharification and fermentation (SSF) [55]. The SSF process is usually carried at higher temperature using thermotolerant yeast *Kluyeromyces marxianus* at 45 °C for achieving high rate of saccharification and fermentation. Similarly, researchers have also developed recombinant *E. coli* strain [56] and *Saccharomyces cerevisiae* [57] capable of utilizing both hexose and pentose sugars efficiently during either SSF or SHF.

The process of simultaneous saccharification and fermentation with the help of genetically engineered microbes that ferment xylose and glucose in the same medium is known as simultaneous saccharification and co-fermentation (SSCF) [58]. The major advantage of this technology is that SSF and SSCF can be performed in the same tank, which makes the entire process cheap, feasible, and cost-effective. Simultaneous saccharification, filtration, and fermentation (SSFF) process is an integrated process, wherein a membrane filtration chamber is placed between the saccharification and fermentation chambers. Most of the engineered yeast strains have a lower affinity to xylose than to glucose. Thus, xylose utilization begins after depletion of glucose from the medium [59]. SSFF overcomes the limitations of both SHF and SSF; it also allows the separate use of both hydrolytic enzymes and the fermenting microorganism at their optimum conditions. The effort to develop consolidated bioprocessing (CBP), also known as direct microbial conversion (DMC) that combines enzyme production, enzymatic saccharification, and fermentation in a single step [60] by microbial strains belonging to genus *Clostridium* [61–63], has been developed. The CBP concept has been adopted by Mascoma Corporation from SunOpta Inc. (http://www.mascoma.com/assets/Mascoma-SBI-September-1-2010.pdf). Many factors such as enzyme concentration (mg/g of substrate or FPU/g of substrate), enzyme adsorption, end-product inhibition and lignocellulosic biomass loading influence the rate of saccharification [20, 64]. Biomass loading rate is an important factor in enzymatic hydrolysis affecting the energy consumption and economic feasibility of lignocellulosic ethanol production. In the conventional process, biomass loading is low, which results in low ethanol concentration and high capital cost. From an economic point of view, when the concentration of ethanol after fermentation is lower than 7% (v/v), extensive energy must be consumed in subsequent distillation processes. However, it is difficult to achieve higher ethanol concentration than 7% (v/v) from lignocellulosic biomass through ordinary fermentation processes [65]. Enzymatic hydrolysis operated beyond 15% solids loading (w/w) is generally called high-solids enzymatic hydrolysis [52, 53]. Theoretically, ~14% (v/v) fermentable sugars must be available in the hydrolysate to achieve ethanol levels of approximately 7% (v/v). However, to achieve 14% sugars in hydrolysate generally 22–25% or even higher biomass loading rate is desired to achieve efficiency of enzymatic hydrolysis between 80 and 90% [53, 66]. The process of enzymatic hydrolysis of high biomass content in solid state has few or no free water, and the process is named solid-state enzymatic hydrolysis (SSEH). SSEH is a potential approach from the industrial point of view because SSEH can produce higher levels of fermentable sugars and ethanol concentrations, and thus it could reduce energy input, distillation cost, water consumption, and operating costs [53, 66]. Conditions of enzymatic hydrolysis for conversions of lignocellulosic biomass using different substrate loadings are given in Table 24.1. However, there are several technical challenges hindering the large-scale implementation of SSEH as high substrate concentration can cause water constraint and high viscosity, resulting in the poor efficiency of mass transfer. The enzymatic conversion is found inversely proportional to the solids/biomass concentration leading to excessive energy consumption. Another disadvantage of using high-solid loadings is product inhibition of, especially, the cellulolytic enzyme system [75, 76, 84, 85].

Several reactors have been reported to increase the efficiency of high biomass loading rate. Jorgensen et al. [86] developed a reactor system for enzymatic hydrolysis at 40% solids loading rate with an enzyme loading of 7 FPU/g dry matter for 96 h, which resulted in glucose conversion of 86%. Dasari et al. [87] employed 8 L scraped surface bioreactor equipped with horizontally rotating shaft and blades to provide mixing and prevent particle settling. In this system, enzymatic hydrolysis tests were performed for solids loading between 10% and 25%, and the efficiency factor is the highest for 20% solids loading. National Renewable Energy Laboratory (Golden, CO, USA) developed a semicontinuous conversion process, which achieved a capacity of 4000 L ethanol and operate at solids loading of >20% w/w [52]. Jilin province, China, established an operation integrating steam explosion and SSCF at high-solids loading >20% and enzyme loading was reduced by 25.0% using the synergistic enzyme systems.

The equipment included steam explosion reactors, simultaneous saccharification and fermentation reactors, and one ethanol distillation tower achieving more than 4% (w/v) ethanol concentration in the process. In addition, lignin plastic composite material and compressed natural gas were co-produced. As a result, this operation system reduced the cost of ethanol production and facilitated the industrial application of this process [88].

Fed-batch feeding scheme is considered as an alternative method for achieving high-solids loadings during enzymatic hydrolysis [89–92] as O_2 diffusion and mixing

Table 24.1 Conditions of enzymatic hydrolysis for conversion of lignocellulosic biomass using different substrate loadings

Pretreated lignocellulosic biomass	Solid loading for saccharification	Enzyme loading and hydrolysis conditions	Source of enzymes and hydrolysis conditions	Conversion efficiency (%)	References
Sugarcane bagasse (1% H_2SO_4)	2.5 (w/v)	Accellerase 1500–25 FPU/g solids, Novozyme 188–50 beta-glucanase units/g solids 50 °C, pH 5.0, 200 rpm, 72 h	Accellerase 1500 (Genecor, USA), Novozyme 188 (Novozyme, Denmark)	~40%	Rezende et al. [67]
Sugarcane bagasse (1.5 M NaOH, 60 °C, 3 h)	5% (w/v)	Crude enzyme extract – 25 FPU/g solids, 41 °C, pH 4.8, 130 rpm, 72 h	*Aspergillus niger*	Cellulose – 28% Hemicellulose – 83%	Shaibani et al. [68]
Sugarcane bagasse (NaOH)	9 + 8 + 7 + 6% (w/v)	9.6 FPU/g solids 50 °C, pH 5.0, 120 rpm, 144 h	Accellerase® 1500 (Genecor, China)	55%	Zhang et al. [31, 32]
Rice straw (1% H_2SO_4 + steam explosion)	2% (w/v)	Cellulase – 20FPU/g solids	Cellulase (Genencor, Spezyme CP)	Cellulose ~58.7% Hemicellulose~ 6.8%	Chen et al. [69]
Rice straw (2% H_2SO_4 + steam explosion)				Cellulose ~68.0% Hemicellulose~ 4.8%	
Rice straw (3% H_2SO_4 + steam explosion)				Cellulose ~72.5% Hemicellulose~ 4.7%	
Rice straw (0.5% NaOH)	25% (w/v)	9 FPU/ g solids 50 °C, pH 4.8, 120 rpm, 32 h	Enzymes from mutant *At9* (*Aspergillus terreus*)	59.29%	Dixit et al. [70]
Rice straw (0.5 M KOH, room temperature, 4 h + 0.1 N H_2SO_4, room temperature, 1 h)	10% (w/v)	Crude enzyme (loaded @ 2.0) 40 °C, pH 5.0, 2.5 h	Cellulase produced from *A. niger* BK01	Sugars yield – 23.78%	Aggarwal et al. [71]

(continued)

Table 24.1 (continued)

Pretreated lignocellulosic biomass	Solid loading for saccharification	Enzyme loading and hydrolysis conditions	Source of enzymes and hydrolysis conditions	Conversion efficiency (%)	References
Corncob (2% H_2SO_4)	17% (w/v)	Cellulase – 30 FPU/g solids + Cellobiase – 20 CBU/g solids 50 °C, pH 4.8, 120 rpm for 12 h	Cellulase GC220 (Genencor) and cellobiase (Novozyme 188)	91.39%	Xie et al. [72]
Corncob (2% NaOH)				86.46%	
Corncob (15% aqueous ammonia)				86.47%	
Corncob (2% dilute sulfuric acid – 15% aqueous ammonia)				85.40%	
Corn stover (aqueous ammonia)	20% w/w	Cellulase – 15 FPU /g solids, β-glucosidase – 30 CBU /g solids, xylanase – 7.5 mg /g solids 50 °C, pH 4.8, 200 rpm, 168 h	Novozyme 188 (Sigma–Aldrich), Accellerase 1500, and Multifect xylanase (Genencor)	~63%	Qin et al. [73]
Corn stover (ammonia fiber expansion)	10 + 10 (w/w)	15 FPU/g solids 50 °C, pH 5.0, 200 rpm, 72 h	Commercial enzymes CTec3 and HTec3 (Novozymes, Denmark)	~92%	Bals et al. [74]
Sweet sorghum bagasse (liquid hot water)	10 + 5 + 5% (w/w)	30 FPU/g solids 50 °C, pH 4.8, 100 rpm, 120 h	Cellulase produced from *Penicillium* sp. + Xylanase (Imperial Jade Biotechnology Co. Ltd., China)	60%	Wang et al. [75, 76]
	15 + 7.5 + 7.5% (w/w)			54%	
Wheat straw (dilute H_2SO_4 impregnation + steam explosion)	2% (w/w)	Celluclast 1.5 L – 15 FPU/g solids + Novozyme 188–18 IU/g solids 40 °C, pH 5.0, 96 h	Cellulase mixture of Celluclast 1.5 L and β-glucosidase (Novozymes A/S, Denmark) supplemented with the β-glucosidase preparation Novozym 188	Glucose (102%) and xylose (96%)	Linde et al. [77]

Wheat straw (fine grinding + wet oxidation)	2% (w/w)	Enzymes mixture Cellulase – 30 FPU/ g solids, β-glucosidase – 37 CBU/g solids 50 °C, pH 5.0, 750 rpm, 24 h	Celluclast 1.5 L derived from *T. reesei*, Novozyme 188 derived from *A. niger*	ND	Pedersen and Meyer [78]
Barley straw (steam)	10% (w/w)	Cellulase –7.5 FPU/g solids + β-glucosidase – 13 CBU/g solids 50 °C, pH 5.0, 75 rpm, 72 h	Celluclast 1.5 L® and Novozyme 188 (NS 188) (Novozymes A/S, Denmark)	73%	Rosgaard et al. [79]
	15% (w/w)			81%	
	5 + 5 + 5% (w/w)			65%	
	10 + 5% (w/w)			68%	
Barley straw (steam explosion)	15% (w/v)	Cellulase – 7 FPU/g solids + β-glucosidase – 8.4 IU/g Solids + xylanase-72 U /g solids 50 °C, pH 4.8, 150 rpm, 120 h	Celluclast 1.5 L FG, Novozym 188, and a mixture that included NS50013 (cellulase), NS50010 (β-glucosidase), and NS50030 (endo-xylanase) (Novozymes)	59%	Garcia-Aparicio et al. [80]
Rye straw (soda pulp)	17.5% (w/w)	Cellulase – 13 FPU/g cellulose + β-glucosidase – 35 CBU/g cellulose 45 °C, pH 5.0, 120 rpm, 48 h	Celluclast 1.5 L (Sigma–Aldrich) and Novozyme 188 (Novozymes Denmark)	40%	Ingram et al. [81]
Rye straw (liquid hot water)	17% (w/w)			65%	
Poplar (organosolv)	20% (w/w)	Cellulase – 20 FPU/g solids, β-glucosidase – 80 CBU/g solids 50 °C, pH 4.8, 200 rpm, 96 h	Celluclast 1.5 L (cellulase) and Novozym 188 (β-glucosidase) (Novozymes North America)	~ 83%	Zhang et al. [82]
Olive tree pruning biomass (liquid hot water)	20% (w/v)	Celluclast – 15 FPU/g solids β-glucosidase – 15 IU/g solids 50 °C, pH 4.8, 150 rpm, 72 h	Celluclast 1.5 (Novozymes A/S Denmark), Fungal b-glucosidase (Novozym 188, Novozymes A/S)	64%	Cara et al. [83]

occur more easily in less viscous saccharification medium/broth. Zhang et al. [31, 32] studied fed-batch approach for the conversion of NaOH-pretreated sugarcane bagasse and wheat straw. Pretreated biomass was fed into the reactor at 9%, 8%, 7%, and 6% solids for 48 h with sequential addition of cellulosic biomass and cellulases to achieve a final solids loading of 30% (w/v). Glucose conversion from wheat straw reached a maximum 60% after the first feeding, but decreased with each successive feeding. The glucose conversion of bagasse increased continuously over the course of hydrolysis reaction except the last feeding time (6% solids at 48 h). The final glucose conversion of the sugarcane bagasse was 55%. Yang et al. [91, 92] obtained high cellulose conversion rate (70.6%) with a high steam-exploded corn stover solid loading (30%) and enzyme loading (@ 20 FPU/g cellulose) in fed-batch approach. Wang et al. [75, 76] used fed-batch feeding scheme in which two additional feedings of sweet sorghum bagasse pretreated with liquid hot water were done at 24 and 48 h. The system containing 30% solids achieved high final sugar concentration ~115 g/L in fed-batch system. The conversion decreased with increasing solids loadings; however, the conversion of the 55% and 60% at 15% and 20% solid loadings, respectively, were achieved.

High cost of cellulase enzyme is one of the major barriers for commercialization of lignocellulosic bioethanol production [15, 93, 94]. Currently, the cellulase cost evaluation methods are controversial because some researchers claimed that the cellulase enzyme cost was only varied from $0.1 to $0.4/gal of ethanol; hence, current technology was already economically sound [95–97]. Others pointed out that the cost of cellulase enzyme was as high as up $0.69 to $1.47/gal of ethanol [15, 98]. High cellulase cost gave a negative evaluation on the feasibility of commercial cellulosic ethanol production. Therefore, on-site or near-site production of cellulase enzyme is proposed as a promising alternative for the significant reduction of enzyme cost below $0.3/gal of ethanol because of its simplified purification as well as the potential cheap carbon source utilization for their production [99]. According to Takimura et al. [100], upto 70% reduction of cellulase cost was achieved using the on-site produced cellulase enzyme as compared to purchased enzyme. Hence, on-site enzyme production should be explored in the industrial scale to yield an economically sound enzyme supply for the future cellulosic ethanol production.

24.2.3 Structural Modifications of Lignocellulosic Biomass

Advanced study of structural changes in the lignocellulosic biomass that occurred during pretreatment and enzymatic hydrolysis are depicted by scanning electron microscopy (SEM), Fourier transform infrared study (FTIR), and X-ray diffraction (XRD). Scanning electron microscopy represents ultra-structure of native lignocellulosic biomass such as sugarcane bagasse, corncob, rice straw, and wheat straw, which has rough, compact, ordered, and rigid fibril structures having thick-walled fiber cells interlinked with pith [101, 102]. Cruz et al. [103] observed that 1% H_2SO_4 pretreated sugarcane bagasse results in surface with elongated fibers, increase in piths, and the presence of small holes on surfaces. The dilute acid pretreatment increased the enzyme-accessibility to the cellulose, leading to an enhanced degradability of pretreated substrate. Zhang et al. [101] found that 88% formic acid in pretreated corncob displayed roughened and compact particles due to hydrogen bonding and removal of hemicellulose and lignin. After corncob pretreatment, crystallinity index (CrI) increased to 44.8% from 24.3% of native corncob, and FTIR analysis showed disappearance of 1732 cm^{-1} band which represents the presence of uronic ester groups of hemicellulose or hemicellulose–lignin complexes [104]. Kim et al. [105] studied rice straw morphology after combined sulfuric acid–aqueous ammonia pretreatment and showed comparatively large amounts of microfibers and destructive perforations of the lumen. Some particles were found attached to the external surface of fragments. Crystallinity index (CrI) increased from 35.42% to 60.23%. A similar trend was also observed in case of rice straw pretreated with ammonium hydroxide and sodium carbonate [106, 107].

Rocha et al. [108] found that sugarcane bagasse, after alkaline delignification, showed rough cell wall surface and unpacked fibers with an open structure. Singh et al. [109] observed that the silicon waxy structure is ruptured, and lignin-hemicellulosic complex of NaOH-microwave pretreated rice straw is broken down drastically. Previous studies have also shown that the surface of the samples treated with microwave-assisted organic acid became loose and irregular. Microwave-assisted $FeCl_3$ damaged the cell wall morphology and altered the fibrillar structure of rice straw [110]. Earlier studies show that oxalic acid fiber expansion (OAFEX) of lignocellulosic biomass disrupts the structure of fibers and pith by removing the hemicellulose fraction of the cell wall and other deposits. OAFEX leaves the overall structure disorganized and increases the surface area for enzymatic action [111, 112]. Momayez et al. [113] examined untreated rice straw had a packed structure covered by a layer of silica. Due to the treatment with 4% lactic acid at 140 °C, the silica layer was opened up, resulting in the increased surface area for the enzyme penetration.

Chandel et al. [102] observed that sequential acid–base pretreatment of sugarcane bagasse separated fibers from pith and caused loosening of the fibrous network due to the removal of hemicellulose during dilute acid pretreatment. After the sodium hydroxide pretreatment of cellulignin, lignin was removed substantially and the surface of the substrate appeared very smooth with the parallel sheaths and pores were seen. Enzymatic hydrolysis of acid–base pretreated substrate has resulted in complete destruction of the cellulose. The CrI of acid (58.82%) and alkali-pretreated sugarcane bagasse (71.87%) was comparatively higher than natural sugarcane bagasse (48.8%) showing a hike of CrI due to the sequential increment in cellulose content in these samples.

Brar et al. [114] analyzed that native corncob shows irregular morphology due to the presence of amorphous hemicelluloses on the surface. After acid pretreatment, corncob showed ordered and rigid fibril structure due to the exposure of crystalline cellulose as a result of hemicellulose removal (Fig. 24.2a, b). The pretreatment resulted in 16-fold increase in the surface area as well as in pore volume by 1.69. This was also confirmed by FTIR and XRD analysis (Fig. 24.2e, f). The maximum

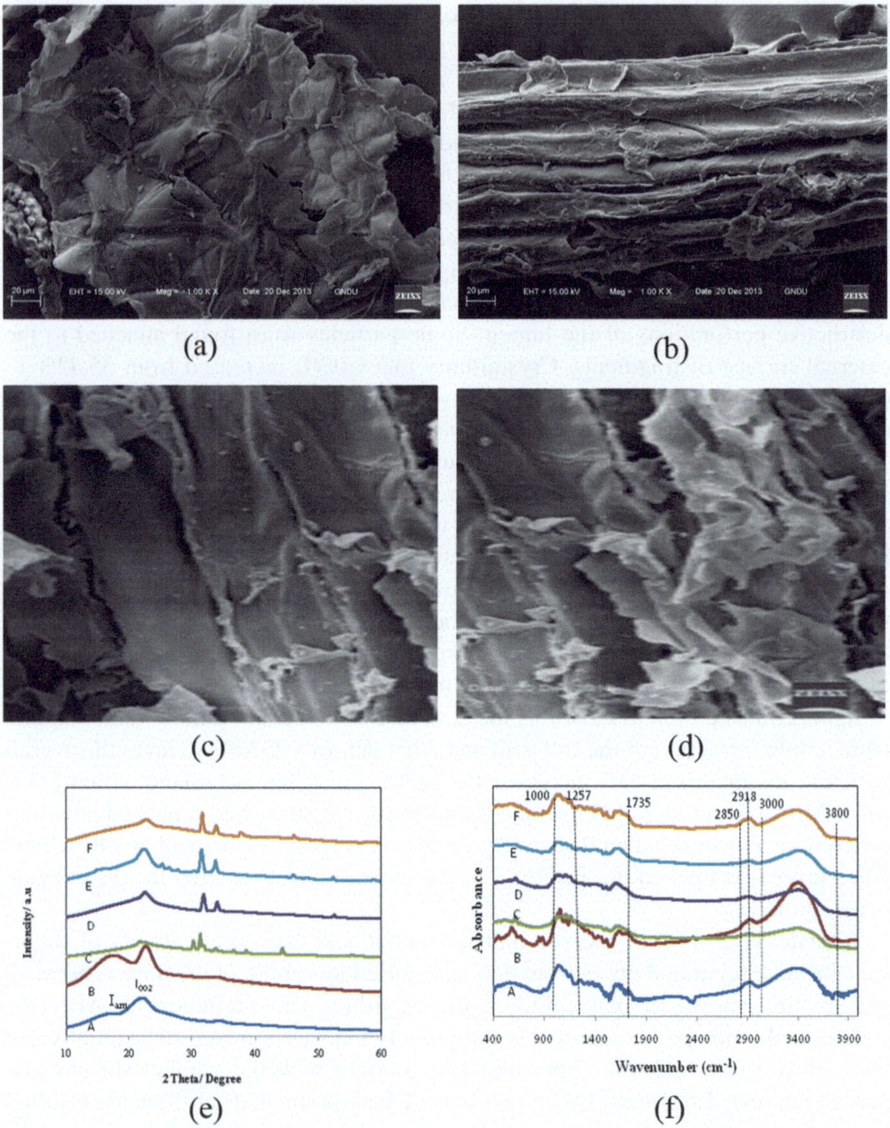

Fig. 24.2 Scanning electron microscopy (SEM) of (**a**) natural; (**b**) dilute sulfuric acid pretreated corncob and enzymatically hydrolyzed substrates; (**c**) Cellic CTec2; (**d**) *M. cinnamomea* (CM-10 T); (**e**) X-ray diffraction (XRD) analysis after acid pretreatment and saccharification of corncob with different cellulase enzymes; (**f**) Fourier transform infrared spectroscopy (FT-IR) analysis after acid pretreatment and saccharification of corncob with different cellulase enzymes. (Adapted from Brar et al. [114])

crystallinity index of pretreated corncob was obtained due to hemicellulose removal. Subsequent saccharification with cellulase preparations, Cellic CTec2 and *M. cinnamomea*, resulted in decreased crystallinity index value from 40.7% of pretreated corncob to 16% and 23%, respectively (Fig. 24.2c–e). In the case of *Scytalidium thermophilium* and *Aspergillus* sp., crystallinity index decreased to 36% and 28%, respectively. The FT-IR results indicated that the linkage between lignin and hemicellulose was cleaved by dilute acid treatment (Fig. 24.2f). After enzymatic hydrolysis, the peaks between 1000 and 1200 cm^{-1} disappeared; this indicated the hydrolysis of cellulose. A major effect was observed as a result of hydrolysis of acid pretreated corncob with Cellic CTec2 and *M. cinnamomea* cellulases when compared to other enzymes. The strains, *S. thermophilium* and *Aspergillus* sp., show very little changes in peaks at 1000 and 1200 cm^{-1} and at 3000 to 3800 cm^{-1} as compared to *M. cinnamomea*.

Chundawat et al. [115] observed decrease in ester carbonyl peak at 1720 cm^{-1} and the aldehyde peak at 1640 cm^{-1} due to the removal of hemicellulose and hemicellulose–lignin complexes in ammonia fiber explosion (AFEX) treated corn stover. There was also a change in the relative intensity of the peaks at 1670 and 1610 cm^{-1}, which are characteristic of amide linkages. Guilherme et al. [116] studied the effect of four pretreatments on sugarcane bagasse: combined acid and alkaline, combined hydrothermal and alkaline, alkaline, and hydrogen peroxide pretreatments. They observed more disorganized structure results from the application of the pretreatments compared to the native sugarcane bagasse. Cui et al. [117] evaluated that after combined nitric acid–sodium hydroxide pretreatment CrI increased to 58.3%. Wanitwattanarumlug et al. [118] studied microwave-assisted KOH (0.75–3%) effect on corncob and analyzed that structure was damaged: looked soft and several pores appeared on its surface. Pretreatment has resulted in increased surface area of corncob from 3.926 to 5.719 m^2/g and crystallinity index increased from 24.5% to 57.28% due to lignin and hemicellulose removal. Li et al. [119] treated corncob hydrothermally with solid acid catalyst (SO_4^{2-}/TiO_2-ZrO_2/La^{3+}) and found that fibrillar structure appeared on the corncob surface and vascular fibers exposed after hydrothermal pretreatment (HTP) at 160 °C for 1 h, which causes solubilization of hemicellulose and partial removal of lignin from cell wall. Brunauer–Emmett–Teller (BET) surface area of pretreated corncob (28.2739 m^2/g) was higher than that of the raw material (0.2702 m^2/g), which was attributed to the separation of individual fibers or precipitation of lignin and adsorption of catalyst. Higher values of lateral order index (LOI) and total crystallinity index (TCI) are indicative of biomass with higher crystallinity and more ordered structure of cellulose. However, both LOI and TCI decreased significantly after pretreatment from 0.959 to 0.3718 (LOI 1426/2917 cm^{-1}) and 0.4057 to 0.1937 (TCI 1373/2917 cm^{-1}).

Rajak and Banerjee [50] pretreated the wasteland weed, *Saccharum spontaneum* with hyperactive laccase produced from *Pleurotus* sp., for delignification. The raw substrate before pretreatment was in the form of a rigid and highly ordered surface structure. The surface structure was distorted in the enzyme-mediated delignification because of an enzymatic action on lignin, which further enhanced the surface area of cellulose, making it amenable for celluloytic enzymes. The cellulose crystallinity value of the raw substrate was observed to be 76.71%, which was increased to 85.26% in the delignified substrate and is consistent with the results of crystallinity increase in enzymatically treated *Bambusa bambos* (33%) than the raw sample (28.44%) observed by Kuila et al. [120]. Infrared spectra of the delignified sample were similar to that of raw spectra, which signifies that the delignification condition does not promote severe changes in the chemical structures of cellulose and hemicellulose (Table 24.2).

Table 24.2 Changes in ultrastructure of different lignocellulosics after bioprocessing

Feedstock	Pretreatment	Ultrastructral changes	References
Cotton	Formic acid (78.22%), water (17.78%), and hydrochloric acid (4%)	The CrI decreased from 91.7% (before pretreatment) to 75.8% (after pretreatment)	Sun et al. [104]
Rice straw	Sulfuric acid (1.21%) and aqueous ammonia (1.54–20.93%)	CrI increased from 35.42% (before pretreatment) to 60.23% (after pretreatment)	Kim et al. [105]
Rice straw	Ammonium hydroxide (27%)	CrI increased from 40.5% (before pretreatment) to 52.3% (after pretreatment)	Phitsuwan et al. [106]
Rice straw	Sodium carbonate (0.5 M)	CrI increased from 41.8% (before pretreatment) to 39.8% (after pretreatment)	Molaverdia et al. [107]
Sugarcane bagasse	Combined sulfuric acid (1%) and sodium hydroxide (1%)	CrI increased from 48.8% ((before pretreatment) to 58.82% (acid pretreatment) and 71.87% alkali pretreatment))	Chandel et al. [102]
Sugarcane bagasse	Combined sulfuric acid (2%) and sodium hydroxide (4%) Combined hydrothermal (20 bar) and sodium hydroxide (4%) Sodium hydroxide (4%) Hydrogen peroxide (8.15%)	CrI increased from 49.32% (before pretreatment) to 63.54% (after pretreatment) CrI increased from 49.32% (before pretreatment) to 69.96% (after pretreatment) CrI increased from 49.32% (before pretreatment) to 55.51% (after pretreatment) CrI increased from 49.32% (before pretreatment) to 60.59% (after pretreatment)	Guilherme et al. [116]
Sugarcane bagasse	Combined nitric acid (3%) and sodium hydroxide (1.5%) Enzymatic hydrolysis with cellulase from *Aspergillus niger*	CrI increased to 58.3% CrI decreased from 93.35enzyme hydrolysis as evidenced by a high saccharification rate of 77.86% within 30 h	Cui et al. [117]
Corncob	Formic acid (88%)	CrI increased from 24.3% (before pretreatment) to 44.8% (after pretreatment)	Zhang et al. [101]

(continued)

Table 24.2 (continued)

Feedstock	Pretreatment	Ultrastructral changes	References
Corncob	Microwave-assisted KOH (0.75–3%)	Surface area increased from 3.926 m^2/g to 5.719 m^2/g CrI increased from 24.5% (before pretreatment) to 57.28% (after pretreatment)	Wanitwattanarumlug et al. [118]
Corncob	Hydrothermally with solid acid catalyst (SO_4^{2-}/TiO_2-ZrO_2/La^{3+})	Surface area increased from 0.2702 m^2/g to 28.2739 m^2/g after pretreatment LOI and TCI decreased from 0.959 to 0.3718 (LOI 1426/2917 cm^{-1}) and 0.4057 to 0.1937 (TCI 1373/2917 cm^{-1}) after pretreatment	Li et al. [119]
Corncob	Sulfuric acid (1%) Saccharification/ enzymatic hydrolysis	16-fold increase in pore volume CrI increased from 30% (before pretreatment) to 40.7% (after pretreatment) In case of Cellic CTec2 *and Malbranchea cinnamomea* CrI decreased from 40.7% (after pretreatment) to 16% and 23%, respectively (after enzymatic hydrolysis) In case of *Scytalidium thermophilium* and *Aspergillus* sp., CrI decreased to 36% and 28%, respectively	Brar et al. [114]
Bambusa bambos	Laccase and cellulase produced from *Pleurotus* sp. and *Trichoderma reesei* Rut C30	CrI increased from 28.44% to 33%	Kuila et al. [120]
Saccharum spontaneum	Laccase produced from *Pleurotus* sp.	CrI increased from 76.71% to 85.26%	Rajak and Banerjee [50]

24.3 Fermentation of Hydrolysate into Bioethanol

Various hurdles affect the efficient ethanol production from lignocellulosic biomass such as the presence of various pretreatment-derived inhibitors, sugar concentration (glucose and mannose) as well as ethanol concentration that adversely affect growth and fermentation of naturally occurring microbes [121]. Baker's yeast *Saccharomyces cerevisiae* has many positive attributes, which makes it suitable for industrial ethanol production such as a high rate of ethanol production from glucose. However, this yeast species also has the limitation of not being able to ferment pentoses such as xylose or arabinose due to lack of the D-xylose metabolic pathway [122]. Even though some genera of yeasts such as *Schizosaccharomyces, Pichia, Candida*, and *Pachysolen* are capable of fermenting pentoses to ethanol having xylose reductase (XR), xylulokinase, and subsequent enzymes needed for a full xylose metabolic pathway [123]. However, they have poor ethanol yields and low ethanol tolerance as compared to glucose-fermenting yeasts, such as *S. cerevisiae*. Another reason is the presence of hexoses (mainly glucose and mannose) that compete with or inhibit xylose utilization. In addition, the presence of oxygen during xylose fermentation contributes to the concurrent production and utilization of ethanol, while considerable amount of xylose remains in the medium. Moreover, xylose-fermenting yeasts require semi-aerobic condition to ferment ethanol, while hexose-fermenting yeasts perform fermentation under anaerobic condition. Another problem is the occurrence of co-factor imbalance in xylose reductase (XR) that has a higher affinity for NADPH and xylitol dehydrogenase (XDH) that is only active with NAD. Some bacteria can metabolize xylose into ethanol, through HMP pathway, where xylose isomerase (XI) converts xylose into xylulose. In addition to *E. coli*, thermophilic anaerobic bacterial strains *Thermoanaerobacter ethanolicus, Thermoanaerobacterium saccharolyticum,* and *Clostridium thermotherum* can ferment xylose into ethanol. But, it is difficult to maintain anaerobic conditions in large-scale fermentation, which restricts the use of thermophilic anaerobes. *Kluveromyces marxianus* is a thermotolerant yeast that is capable of co-fermenting both hexose and pentose sugars and can survive the temperature of 42–45 °C [124].

Researchers have applied different strategies such as strain adaptation, mutagenesis, genome shuffling, and metabolic engineering for the improvement of strains to overcome all the abovementioned hurdles (Fig. 24.3).

Two metabolic pathways, either oxidoreductase or isomerase based, have been implanted into *S. cerevisiae* to enable xylose fermentation. Zhou et al. [125] constructed an efficient xylose-fermenting recombinant *S. cerevisiae* by increasing dosages of the xylose isomerase gene, enhanced xylulose kinase (XK) expression, and upregulated nonoxidative PPP. The high ethanol yield from xylose was achieved through overexpression of a mutant XI and *Scheffersomyces stipitis* TAL1 along with deletion of GRE3 and PHO13 [126]. Another robust ethanol-producing *S. cerevisiae* strain was engineered with a two-step oxidoreductase pathway consisting of NAD(P)H-linked xylose reductase (XR), xylitol dehydrogenase (XDH), and xylulokinase (XK) [127]. The hybrid strain has been developed by fusing protoplast of *S. cerevisiae* and xylose-fermenting yeasts such as *C. Shehatae, P. tannophilus,* and *P. stipitis* [128]. Li and Elledge [129] developed mutants *Z. mobilis* ZM4 *ΔadhA* and *Z. mobilis* ZM4 *ΔadhB* by knockout of *adhA* and *adhB* genes, using the RecET homologous recombination system for introducing heterologous genes responsible for metabolizing alternative carbon source and increasing ethanol yield. Chung et al. [130] developed a strain *C. bescii* by deletion of lactate dehydrogenase and heterologous expression of a *Clostridium thermocellum* bifunctional acetaldehyde/alcohol dehydrogenase. This strain gave 70% ethanol yield by direct conversion of switchgrass to ethanol representing a new paradigm for consolidated bioprocessing.

BC International Corporation (www.bcintlcorp.com) is commercially using genetically engineered *E. coli* that produces ethanol from biomass sugars.

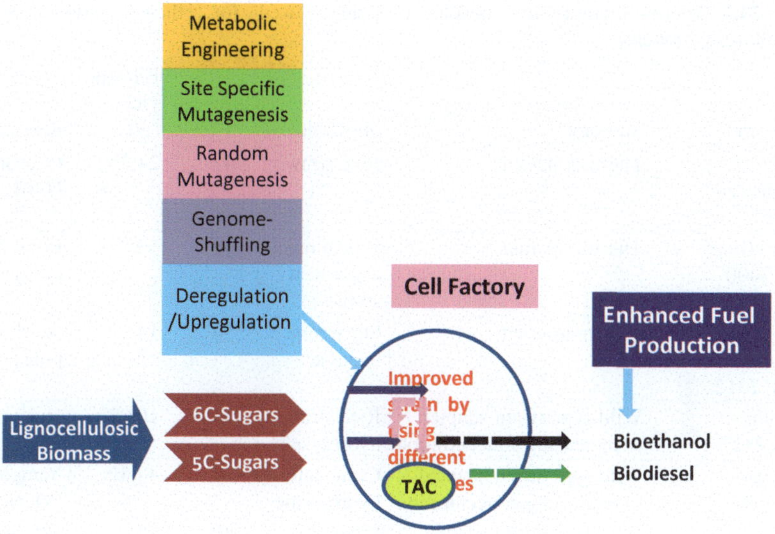

Fig. 24.3 Different molecular strategies to improve biofuel production

Modification of the phosphoenolpyruvate/sugar phosphotransferase system (PTS), showing improvement in sugar co-utilization in *E. coli* strains and an increment in the ethanol yields, was evaluated by Balderas-Hernández [131]. Jilani et al. [132] improved sugar utilization pattern of *E. coli* strain via adaptive evolution on glucose and xylose for 150 days, and one of the evolved strains fermented 113 (g/l) xylose and produced 51.1 (g/l) ethanol in 72 h with ethanol productivity of 1.4 (g/l/h) as compared to unevolved SSY13 strain. Yeast/bacterial strains adapted to tolerate fermentation inhibitors by repeated sub-culturing in a mixture of fermentation inhibitors of lignocellulosic hydrolysate have also been reported [133]. Yeast strain *S. stipitis* CBS 6054 was developed by sub-culturing on corn stalk hydrolysate and gave ethanol yield 0.47 (g/g of sugar) [91, 92].

Random mutagenesis is another approach used to obtain mutants of xylose-fermenting yeasts with improved tolerance to inhibitors in lignocellulosic hydrolysates. Bajwa et al. [134] employed successive rounds of UV mutagenesis followed by screening on a gradient plate of hardwood spent sulfite liquor (HWSSL) to successfully obtain mutants (PS301 and PS302) of *S. stipitis* NRRL Y-7124 wild type. These mutants were more tolerant of HMF and produced ethanol 6.4 (g/l) as compared to 3.3 (g/l) in wild strain. Harner et al. [181] obtained mutants of *P. tannophilus* NRRL Y-2460 with enhanced tolerance to HWSSL and acetic acid. Watanabe et al. [135] developed mutant (PET41) of *S. stipitis* NBRC1687 produced 44 (g/l) ethanol than WT, that is, 40 (g/l) (Table 24.3).

Native xylose-fermenting yeasts typically suffer from glucose repression and/or glucose inactivation. Some researchers have developed derepressed mutants of *P. lannophilus, S. stipitis, S. Shehatae,* and *Candida shehatae* by using glucose ana-

Table 24.3 Thermotolerant and developed strains used for ethanol production from lignocellulosic biomass

Organism	Strategy	Substrate	T (°C)	Ethanol yield (g/L)	References
Pichia kudriavzevii HOP-1	Thermotolerant	Rice straw	45	24.25	Oberoi et al. [136]
S. cerevisiae TMB 3400	Thermotolerant	Wheat meal and wheat straw	40	53.3	Erdei et al. [137]
Blastobotrys adeninivorans RCKP 2012	Thermotolerant	Sugarcane bagasse	50	14.05	Antil et al. [138]
S. stipitis BCRC21777	Wild-type strain adapted to increasing concentration	Rice straw hydrolysate	30	10.27	Huang et al. [139]
S. stipitis CBS 6054	Wild-type strain adapted to increasing concentration	Corn stalk hydrolysate	30	43.42	Yang et al. [91, 92]
S. stipitis NBRC1687	Wild-type strain adapted to increasing concentration	Ethanol (50–70 g/L)	30	44	Watanabe et al. [135]
S. cerevisiae GLBRC Y35	Incorporated XR and XDH genes from *S. stipitis* and overexpressed endogenous xylulokinase gene	Corn Stover	30	46	Jin et al. [140]

log 2-deoxyglucose (2-DG) and mutagenesis, which cannot be metabolized by yeasts. These mutants were able to co-utilize glucose and xylose compared to the WT [141].

Two genome-shuffled *Scheffersomyces stipitis* strains, GS301 and GS302, were developed by UV mutagenesis of the WT followed by genome shuffling. These strains completely utilized glucose and xylose in hydrolysate and produced 0.39–1.4% (w/v) ethanol [142]. Zhang and Geng [143] subjected *S. stipitis* CBS 6054 and *S. cerevisiae* ATCC 24860 to recursive recombination of the whole genomes followed by genome shuffling. The evolved strain (ScF2) produced 47 (g/l) of ethanol from 250 (g/l) xylose as compared to *S. stipitis* parental strain that produced 20 (g/l) ethanol. *Candida glabrata* is another promising candidate for developing the SSF process, as this yeast is more tolerant to both high temperature and high acid concentration along with superior ethanol-producing ability [144] (Table 24.3).

Fermentation of bioethanol can be carried out in batch, fed-batch, and continuous mode. Productivity of fed-batch fermentation can be increased by maintaining the substrate at low concentration, which allows the conversion of sufficient amount of fermentable sugars to ethanol. Fed-batch process has higher productivity, higher dissolved oxygen in medium, shorter fermentation time, and lower toxic effect of the medium components compared to other types of fermentation. The advantages of continuous system over batch and fed-batch system are higher productivity, smaller bioreactor volumes, and less investment and operational costs have been observed [145]. Li et al. [146] studied batch SSF of reed with *S. cerevisiae* ATCC24858 and achieved highest ethanol yield 55.5 (g/l) with a productivity of 0.57 (g/l/h). Olofsson et al. [147] observed batch SSF fermentation of wood chips with *S. cerevisiae* TMB3400 and found 32.9 (g/l) ethanol yield with 0.34 (g/l/h) productivity. Fed-batch SHCF of wheat meal and wheat straw with *S. cerevisiae* TMB 3400 gave ethanol yield 53.3 (g/l) with a productivity of 0.44 (g/l/h) [137]. Fermentation by fed-batch SSF of corn stover with Baker's yeast achieved ethanol yield 25.7 (g/l) with a productivity of 0.36 (g/l/h) [148]. Fermentation of Miscanthus with *S. cerevisiae* CHY1011 by continuous SSF feeding system provided sufficient time for liquefaction of the substrate by cellulases to achieve highest ethanol concentration of 69.2 (g/l) with a productivity of 1.24 (g/l/h) [149].

In a strategy, fermentation of mixture of glucose and xylose by a sequential application of *S. cerevisiae* and *P. stipitis* was employed, which included heat inactivation of *S. cerevisiae* cells before addition of *P. stipitis*. Similar strategy was followed for sequential fermentation of mixed sugars by *S. cerevisiae* and *P. tannophilus* in case of alkali-pretreated rice straw and ethanol production obtained 24.5% (w/v) [150]. Co-fermentation of rice hull hydrolysate (RHH) by *Saccharomyces cerevisiae* and *Spathaspora arborariae* in bioreactor under oxygen limitation produced ethanol and xylitol to final concentrations of 14.5 g/L and 3 g/L, respectively, have also been reported [151].

24.4 Utilization of Molasses Along with Lignocellulosic Biomass for Ethanol Production

Ethanol is produced in India by the fermentation of molasses [152]. The yield of sugar on average is approximately 105 kg per ton of sugarcane. About 40 kg of molasses (20 kg sugars) is produced per ton of cane from which about 10 L of ethanol can be obtained. If sugarcane is directly and fully used in ethanol manufacture, the yield of ethanol is 70 L per ton. About 30% of cane goes for making gur (jaggery) and khandsari (unrefined sugar). If there is no additional increase in khandsari demand, sugar and molasses production would increase. The present distiller capacity is 2.0 billion and looks to be sufficient for 5% blend until 2016–2017 [153]. The main constituents of molasses are chiefly sucrose (30–35%), glucose and fructose (10–25%), non-sugar compounds (2–3%), water, and mineral content. The total fermentable sugars in molasses ranged from 45% to 55% depend on cane variety, soil type, and nature of applied nutrients. Integrating the bioethanol production of lignocellulosic biomass with molasses as substrate would be expected to increase the ethanol yield. The concept is under trial in India and Brazil. Yu et al. [154] evaluated simultaneous saccharification and fermentation (SSF), separate hydrolysis and fermentation (SHF), and pre-saccharification simultaneous saccharification and fermentation (P-SSF) of the mixture of sugarcane bagasse whole pretreated slurry (WPS) and molasses (1:1) employing fed-batch approach and obtained 41.49 g/l of ethanol. Brar et al. [155] evaluated BOLT-ON technology for co-fermentation of sugars derived from molasses and enzyme hydrolysate of corncob obtained after saccharification with Cellic CTec2 benchmark as well as enzyme cocktails known as AT9DRCellicCTec2 and AS10DRCellicCTec2 that were prepared using Cellic CTec2 spiked with extracts of *Aspergillus terreus* (AS9DR) and *Achaetomium strumarium* (AS10DR), respectively. The enhanced ethanol yield was 48.9 g/l, which was made possible by a combination of approaches that were followed for utilizing agnostic lignocellulosics.

24.5 Bioconversion of Lignocellulosic Biomass into Biodiesel

The lower cost of lignocellulosic biomass as a raw material has attracted research interest in the production of microbial oils. It was found that different oleaginous microorganisms, mostly yeast or fungi, displayed the capabilities of utilizing the hydrolysate of lignocellulosic biomass such as wheat straw, sugarcane bagasse, rice straw, rice hulls, and corn stover for lipid accumulation. Oleaginous yeasts and filamentous fungi are known to accumulate lipids more than 25% of their biomass (dry weight) [156, 157]. *Candida* sp., *Cryptococcus* sp., *Lipomyces* sp., *Rhodosporidium* sp., *Rhodotorula* sp., *Trichosporon* sp., *Torulopsis* sp., *Yarrowia* sp., *Thamnidium* sp., *Mortierrela* sp., *Mucor* sp., *Aspergillus* sp., *Zygosaccharomyces* sp., *Zygorhynchus* sp., and *Pichia* sp. are more specifically involved [158] (Table 24.4).

Oleaginous yeasts and molds mainly accumulate triacylglycerides (TAGs) accounting for ~90% of their stored lipids. Oleaginous yeasts can produce lipids rich in polyunsaturated fatty acids (PUFAs) γ-linolenic acid (D6, 9, 12C, 18:3), dihomo-g-linolenic acid, arachidonic acid, docosahexaenoic acid, and eicosapentanoic acid. The main fatty acids accumulated by oleaginous yeasts are myristic (C14:0), palmitic (C16:0), palmitoleic (C16:1), stearic (C18:0), oleic (C18:1), and linoleic (C18:2) acids [172].

Researchers have also employed metabolic engineering approach to enhance the lipid accumulation in oleaginous strains. A common pathway involves (1) overexpression of key-enzymes acetyl-CoA carboxylase (ACC), diacylglycerol acyl transferase (DGAT), acetyl-CoA synthetase (ACS), glycerol-*sn*-3-phosphate acyl-transferase (GPAT), glycerol-3-phosphate dehydrogenase (GPDH), ATP citrate lyase (ACL; malic enzyme), glycerol-3-phosphate dehydrogenase (G3PDH); (2) downregulating the catabolism of fatty acids by inhibiting β-oxidation or lipase hydrolysis [173] (Table 24.5).

The overexpression of ACC1 in *Y. lipolytica* was shown to increase the lipid content by twofold as compared to the wild strain [174]. The deletion of the putative acyl-CoA synthetase YAL1 gene in *Y. lipolytica* by the copper-resistant CRF1 gene via homologous recombination resulted in increased saturated to unsaturated fatty acids ratio by sixfold, and the total lipid production of the mutant strain increased by 1.47-fold relative to the wild strain [175]. Tai and Stephanopoulos [174] combined the two genes ACC1 and DGA1 in a tandem gene construct for the co-expression,

Table 24.4 Microbial conversion of lignocellulosic biomass into oils

Microorganisms	Biomass (g/L)	SCO (g/L)	Substrate	Culture mode	References
Aspergillus oryzae A-4	6.7	1.70	Wheat straw, bran, orange peel, apple peel, sugar cane bagasse	Static	Hui et al. [159]
Aspergillus niger LFMB	5.4	3.1	Crude glycerol	Batch	Andre et al. [160]
Candida curvatta NRRL Y-1511	9.3	0.5	Raw glycerol	Batch	Chatzifragkou et al. [161]
Cryptococcus curvatus ATCC 20509	15.5	9.9	Sorghum hydrolysate, bagasse	Batch	Liang et al. [162]
Cunninghamella echinulata ATHUM 4411	12.1	3.8	Molasses	Batch	Chatzifragkou et al. [163]
Mortierella isabellina ATHUM 2935	5.6	3.6	Rice hull hydrolysate	Batch	Economou et al. [164]
Cryptococcus curvatus	17.2	3.35	Wheat straw	Batch	Yu et al. [165]
Yarrowia lipolytica	11.4	5.85	Sugarcane bagasse	Batch	Tsigie et al. [166]
Trichosporon fermentans	28.6	4.01	Rice straw	Batch	Huang et al. [139]
Trichosporon cutaneum	15.44	2.35	Corn stover	Batch	Huang et al. [167]
Trichosporon dermatis	24.4	4.01	Corncobs	Batch	Huang et al. [168]
Mortierella isabellina	18.2	4.57	Corn fiber	Batch	Xing et al. [169]
Rhodotorula glutinis	18.1	3.41	*Populus euramevicana* cv leaves	Batch	Tao et al. [170]
Aspergillus oryzae	NA	36.6 mg/g dry substrate	Wheat straw	Solid-state fermentation	Lin et al. [171]

which increased lipid content up to 41.4% in *Y. lipolytica*. Beopoulos et al. [176] performed an additional deletion of the POX1 to POX6 gene (genes encoded for six acyl-coenzyme A oxidase (*AOX*)) in the *Y. lipolytica* Dgut2 mutant strain, which led to a fourfold increase in lipid content principally for free fatty acid. Heterologous expression of acetyl-CoA carboxylase (ACC1) from the oleaginous fungus *Mucor rouxii* in the non-oleaginous yeast *Hansenula polymorpha* was able to achieve 40% increase in total fatty acid content [177]. Cellulolytic thermophile *Clostridium thermocellum* was engineered to express fatty acyl–acyl carrier protein reductase (ACR) and aldehyde-deformylating oxygenase (ADO) from *Synechococcus elongatus* PCC

Table 24.5 Heterologous overexpression and gene knockout for the improvement of the biosynthesis of lipids involving microorganisms

Enzymes/genes	Source species	Receiver species	Lipid content
ACS (FAA3) DGAT (DGA1)	*S. cerevisiae* Dsnf2	*S. cerevisiae* Dsnf 2	Total lipid content: 29.8% DW
ACC1 (ACC)	*Y. lipolytica*	*Y. lipolytica*	2 × lipid content
ACC1 (ACC)	*Mucor rouxii*	*Hansenula polymorpha*	1.4 × total FA content
Exo-inulase (INU1)	*Kluyveromyces marxianus* CBS 6556	*Y. lipolytica* ACA-DC SO104	Oil content: 50.6% DW
ΔME	*A. nidulans*	*A. nidulans*	−50% lipid content
MalA (ME)	*Mortierella alpina, Mucor circinelloides*	*Mucor circinelloides*	2.5 × lipid content
ACL	*A. oryzae*	*A. oryzae*	1.7 × fatty acid content
accA-D (ACC), tesA (thioesterase I) synthesis	*E. coli*	*E. coli*	6 × fatty acid
FAT	*Ricinus communis*	*E. coli* ML103	>2.0 g/L fatty acid content

7942. This strain produced both decanol and dodecanol in the organic overlay phase of the culture [178]. Engineering the overexpression of ALR homolog YbbO gene with the cyanobacterial acyl-ACP reductase (AAR) and modulation of fatty acid and phospholipid biosynthesis pathways in *E. coli* resulted in higher yields of long-chain (C14–C18) fatty alcohols (LCFA) under fed-batch conditions (2.0 g/L) [179].

24.6 Conclusion

Biofuels have been extensively investigated as alternatives to conventional plant-based fuels. Recent research targeted modification in central carbon metabolism, such as overexpression or deletion of specific genes; applying environmental stress can effectively enhance biofuel production. It is still problematic and challenging for biofuel to be commercially competitive over fossil fuel. To develop new strains with commercial potential, it is necessary to combine multiple genetic engineering approaches to optimize biofuel production.

Acknowledgment The financial support in the form of fellowship to KKB by UGC (UPE Grant-Environment Management) and by Department of Biotechnology and ICAR, Government of India to BSC is duly acknowledged.

References

1. S.S. Adav, A. Ravindran, S.K. Sze, Quantitative proteomic analysis of lignocellulolytic enzymes by *Phanerochaete chrysosporium* on different lignocellulosic biomass. J. Proteome **75**, 1493–1504 (2012)
2. J. Larsen, H.M. Stergaard, L. Thirup, Inbicon makes lignocellulosic ethanol a commercial reality. Biomass Bioenergy **46**, 36–45 (2012)
3. E. Warner, A. Schwab, *2016 Survey of Non-starch Alcohol and Renewable Hydrocarbon Biofuels Producers* (National Renewable Energy Laboratory, Golden, 2017), pp. 6–10
4. M. Valdivia, J.L. Galan, J. Laffarga, et al., Biofuels 2020: biorefineries based on lignocellulosic materials. Microb. Biotechnol. **9**, 585–594 (2016)
5. K. Igarashi, T. Uchihashi, A. Koivula, et al., Traffic jams reduce hydrolytic efficiency of cellulases on cellulose surface. Science **333**, 1279–1282 (2011)
6. C.M. Payne, B.C. Knott, H.B. Mayes, et al., Fungal cellulases. Chem. Rev. **115**, 1308–1448 (2015)
7. F.L. Motta, C.C.P. Andrade, M.H.A. Santana, A review of xylanase production by the fermentation of xylan: classification, characterization and applications, in *Sustainable Degradation of Lignocellulosic Biomass – Techniques, Applications and Commercialization*, ed. by A. Chandel, (IntechOpen, London, 2013), pp. 251–275
8. S.L. Marinai, Exploring methods for functional studies of CBM33-type lytic polysaccharide monooxygenases, Master's thesis, Norwegian University of Life Sciences, Ås, 2013
9. A. Levasseur, E. Drula, V. Lombard, et al., Expansion of the enzymatic repertoire of the CAZy database to integrate auxiliary redox enzymes. Biotechnol. Biofuels **6**, 1–14 (2013)
10. A. Villares, C. Moreau, C. Bennati-Granier, et al., Lytic polysaccharide monooxygenases disrupt the cellulose fibers structure. Sci. Rep. **7**, 40262 (2017)
11. P.V. Harris, D. Welner, K.C. McFarland, et al., Stimulation of lignocellulosic biomass hydrolysis by proteins of glycoside hydrolase family 61: structure and function of a large, enigmatic family. Biochemist **49**, 3305–3316 (2010)
12. S. Jung, Y. Song, H.M. Kim, et al., Enhanced lignocellulosic biomass hydrolysis by oxidative lytic polysaccharide monooxygenases (LPMOs) GH61 from *Gloeophyllum trabeum*. Enzym. Microb. Technol. **77**, 38–45 (2015)
13. R.N. Maeda, V.I. Serpa, V.A.L. Rocha, et al., Enzymatic hydrolysis of pretreated sugar cane bagasse using *Penicillium funiculosum* and *Trichoderma harzianum* cellulases. Process Biochem. **46**, 1196–1201 (2011)
14. X. Fang, S. Yano, H. Inoue, et al., Strain improvement of *Acremonium cellulolyticus* for cellulase production by mutation. J. Biosci. Bioeng. **107**, 256–261 (2009)
15. D. Klein-Marcuschamer, P. Oleskowicz-Popiel, B.A. Simmons, et al., The challenge of enzyme cost in the production of lignocellulosic biofuels. Biotechnol. Bioeng. **109**, 1083–1087 (2012)
16. L.P. Xiao, Z.J. Shi, F. Xu, et al., Hydrothermal treatment and enzymatic hydrolysis of *Tamarix ramosissima*: evaluation of the process as a conversion method in a biorefinery concept. Bioresour. Technol. **135**, 3–81 (2012)
17. Y. Sun, J.Y. Cheng, Hydrolysis of lignocellulosic materials for ethanol production: a review. Bioresour. Technol. **83**, 1–11 (2002)
18. A.T.W.M. Hendriks, G. Zeeman, Pretreatments to enhance the digestibility of lignocellulosic biomass. Bioresour. Technol. **100**, 10–18 (2009)
19. P. Kumar, D.M. Barrett, M.J. Delwiche, et al., Methods for pretreatment of lignocellulosic biomass for efficient hydrolysis and biofuel production. Ind. Eng. Chem. Res. **48**, 3713–3729 (2009)
20. P. Binod, K.U. Janu, R. Sindhu, et al., Hydrolysis of lignocellulosic biomass for bioethanol production, in *Biofuels*, (Academic Press, Burlington, 2011), pp. 229–250
21. L.P. Ramos, The chemistry involved in the steam treatment of lignocellulosic materials. Quim Nova **26**, 863–871 (2003)

22. C. Liu, C.E. Wyman, Partial flow of compressed-hot water through corn stover to enhance hemicellulose sugar recovery and enzymatic digestibility of cellulose. Bioresour. Technol. **96**, 1978–1885 (2005)
23. M. Pedersen, K.S. Johansen, A.S. Meyer, Low temperature lignocellulose pretreatment: effects and interactions of pretreatment pH are critical for maximizing enzymatic monosaccharide yields from wheat straw. Biotechnol. Biofuels **4**, 1–10 (2011)
24. D.J. Schell, J. Farmer, M. Newman, J.D. McMillan, Dilute–sulfuric acid pretreatment of corn stover in pilot-scale reactor. Appl. Biochem. Biotechnol. **105**, 69–85 (2003)
25. Y. Zhu, Y.Y. Lee, R.T. Elander, Dilute-acid pretreatment of corn stover using a highsolids percolation reactor. Appl. Biochem. Biotechnol. **117**, 103–114 (2004)
26. T.A. Lloyd, C.E. Wyman, Combined sugar yields for dilute sulfuric acid pretreatment of corn stover followed by enzymatic hydrolysis of the remaining solids. Bioresour. Technol. **96**, 1967–1977 (2005)
27. B.C. Saha, L.B. Iten, M.A. Cotta, et al., Dilute acid pretreatment, enzymatic saccharification and fermentation of rice hulls to ethanol. Biotechnol. Prog. **21**, 816–822 (2005)
28. A. Demirbas, Products from lignocellulosic materials *via* degradation processes. Energy Sources **30**, 27–37 (2008)
29. M. Tutt, T. Kikas, J. Olt, Influence of different pretreatment methods on bioethanol production from wheat straw. Agric. Res. Biosys. Eng. **1**, 269–276 (2012)
30. M. Galbe, G. Zacchi, A review of the production of ethanol from softwood. Appl. Microbiol. Biotechnol. **59**, 618–628 (2002)
31. Y. Zhang, Y.Y. Liu, J.L. Xu, et al., High solid and low enzyme loading based saccharification of agricultural biomass. BioResources **7**, 345–353 (2012)
32. Z. Zhang, I.M. O'Hara, W.O.S. Doherty, Pretreatment of sugarcane bagasse by acid-catalyzed process in aqueous ionic liquid solutions. Bioresour. Technol. **120**, 149–156 (2012)
33. E.W. Jennings, D.J. Schell, Conditioning of dilute-acid pretreated corn stover hydrolysate liquors by treatment with lime or ammonium hydroxide to improve conversion of sugars to ethanol. Bioresour. Technol. **102**, 1240–1245 (2011)
34. T.H. Kim, J.S. Kim, C. Sunwoo, et al., Pretreatment of corn stover by aqueous ammonia. Bioresour. Technol. **90**, 39–47 (2003)
35. S. Kim, M.T. Holtzapple, Lime pretreatment and enzymatic hydrolysis of corn stover. Bioresour. Technol. **96**, 1994–2006 (2005)
36. R. Gupta, Y.Y. Lee, Pretreatment of corn stover and hybrid poplar by sodium hydroxide and hydrogen peroxide. Biotechnol. Prog. **26**, 1180–1186 (2010)
37. J.S. Kim, H. Kim, J.S. Lee, et al., Pretreatment characteristics of waste oak wood by ammonia percolation. Appl. Biochem. Biotechnol. **148**, 15–22 (2008)
38. P. Alvira, E. Tomas-Pejo, M. Ballesteros, et al., Pretreatment technologies for an efficient bioethanol production process based on enzymatic hydrolysis: a review. Bioresour. Technol. **101**, 4851–4861 (2010)
39. F. Teymouri, L. Laureano-Perez, H. Alizadeh, et al., Optimization of the ammonia fiber explosion (AFEX) treatment parameters for enzymatic hydrolysis of corn stover. Bioresour. Technol. **96**, 2014–2018 (2005)
40. N. Eqra, Y. Ajabshirchi, M. Sarshar, Effect of ozonolysis pretreatment on enzymatic digestibility of sugarcane bagasse. Agric. Eng. Int. CIGR J. **16**, 151–156 (2014)
41. X. Pan, C. Arato, N. Gilkes, et al., Biorefining of softwoods using ethanol organosolv pulping: Preliminary evaluation of process streams for manufacture of fuel grade ethanol and co-products. Biotechnol. Bioeng. **90**, 473–481 (2005)
42. X. Zhao, K. Cheng, D. Liu, Organosolv pretreatment of lignocellulosic biomass for enzymatic hydrolysis. Appl. Microbiol. Biotechnol. **82**, 815–827 (2009)
43. N. Park, H.Y. Kim, B.W. Koo, et al., Organosolv pretreatment with various catalysts for enhancing enzymatic hydrolysis of pitch pine (*Pinus rigida*). Bioresour. Technol. **101**, 7046–7053 (2010)

44. J.-W. Lee, B.-W. Koo, J.-W. Choi, et al., Evaluation of waste mushroom logs as a potential biomass resource for the production of bioethanol. Bioresour. Technol. **99**, 2736–2741 (2007)
45. J.J. Yoon, C.J. Cha, Y.S. Kim, et al., The brown-rot basidiomycete *Fomitopsis palustris* has the endo-glucanases capable of degrading microcrystalline cellulose. J. Microbiol. Biotechnol. **17**, 800–805 (2007)
46. L.G. Ljungdahl, The cellulase/hemicellulase system of the anaerobic fungus *Orpinomyces* PC-2 and aspects of its applied use. Ann. N. Y. Acad. Sci. **1125**, 308–321 (2008)
47. M. Dashtban, H. Schraft, T.A. Syed, et al., Fungal biodegradation and enzymatic modification of lignin. J. Int. Biochem. Mol. Biol. **1**, 36–50 (2010)
48. W.V. Balan, L. da Costa Sousa, S.P.S. Chundawat, et al., Mushroom spent straw: a potential substrate for an ethanol-based biorefinary. J. Ind. Microbiol. Biotechnol. **35**, 293–301 (2008)
49. K.E. Hammel, Fungal degradation of lignin, in *Driven by Nature: Plant Litter Quality and Decomposition*, (CAB International, Wallingford, 1997), pp. 33–45
50. R.C. Rajak, R. Banerjee, Enzymatic delignification: an attempt for lignin degradation from lignocellulosic feedstock. RSC Adv. **5**, 75281–75291 (2015)
51. H. Jorgensen, J.B. Kristensen, C. Felby, Enzymatic conversion of lignocellulose into fermentable sugars: challenges and opportunities. Biofuels Bioprod. Biorefin. **1**, 119–134 (2007)
52. A. Modenbach, Sodium hydroxide pretreatment of corn stover and subsequent enzymatic hydrolysis: an investigation of yields, kinetic modeling and glucose recovery, Thesis and Dissertations-Biosystems and Agricultural Engineering, University of Kentucky, UKnowledge, 2013
53. R. Koppram, E. Tomas-Pejo, C. Xiros, et al., Lignocellulosic ethanol production at high-gravity: challenges and perspectives. Trends Biotechnol. **32**, 46–53 (2014)
54. L. Olsson, H.R. Soerensen, B.P. Dam, et al., Separate and simultaneous enzymatic hydrolysis and fermentation of wheat hemicellulose with recombinant xylose utilizing *Saccharomyces cerevisiae*. Appl. Biochem. Biotechnol. **129–132**, 117–129 (2006)
55. J.C. Lopez-Linares, I. Romero, C. Cara, et al., Experimental study on ethanol production from hydrothermal pretreated rapeseed straw by simultaneous saccharification and fermentation. J. Chem. Technol. Biotechnol. **89**, 104–110 (2014)
56. B.S. Dien, M.A. Cotta, T.W. Jeffries, Bacteria engineered for fuel ethanol production: current status. Appl. Microbiol. Biotechnol. **63**, 258–266 (2003)
57. D.L. Goncalves, A. Matsushika, B. Belisa, et al., Xylose and xylose/glucose co-fermentation by recombinant *Saccharomyces cerevisiae* strains expressing individual hexose transporters. Enzym. Microb. Technol. **63**, 13–20 (2014)
58. L.C. Teixeira, J.C. Linden, H.A. Schroeder, Simultaneous saccharification and co-fermentation of per-acetic acid-pretreated biomass. Appl. Biochem. Biotechnol. **84–86**, 111–127 (2000)
59. M.M. Ishola, M.J. Taherzadeh, Simultaneous saccharification, filtration and fermentation (SSFF): A novel method for bioethanol production from lignocellulosic biomass. Bioresour. Technol. **133**, 68–73 (2013)
60. J.D. Keasling, Synthetic biology and the development of tools for metabolic engineering. Metabo. Eng. **14**, 189–195 (2012)
61. M.J. Jin, V. Balan, C. Gunawan, et al., Consolidated bioprocessing (CBP) performance of Clostridium phytofermentanson AFEX-treated corn stover for ethanol production. Biotechnol. Bioeng. **108**, 1290–1297 (2011)
62. H. Akinosho, K. Yee, D. Close, et al., The emergence of *Clostridium thermocellum* as a high utility candidate for consolidated bioprocessing applications. Front. Chem. **2**, 1–18 (2014)
63. V. Mbaneme, M. Chinn, Consolidated bioprocessing for biofuel production: recent advances. Energy Emiss. Control Technol. **3**, 23–44 (2015)
64. Z.H. Liu, L. Qin, F. Pang, et al., Effects of biomass particle size on steam explosion pretreatment performance for improving the enzyme digestibility of corn stover. Ind. Crop. Prod. **44**, 176–184 (2013)

65. P.W. Madson, Ethanol distillation: the fundamentals, in *The Alcohol Textbook*, ed. by W. M. Ingledew, D. R. Kelsall, G. D. Austin, C. Kluspies, (Nottingham University Publishers, Nottingham, 2009), pp. 292–297
66. H.-Z. Chen, Z.-H. Liu, Enzymatic hydrolysis of lignocellulosic biomass from low to high solids loading. Eng. Life Sci. **17**, 489–499 (2016)
67. C.A. Rezende, M.A. de Lima, P. Maziero, et al., Chemical and morphological characterization of sugarcane bagasse submitted to a delignification process for enhanced enzymatic digestibility. Biotechnol. Biofuels **4**, 1–18 (2011)
68. N. Shaibani, S. Ghazvini, M.R. Andalibi, et al., Ethanol production from sugarcane bagasse by means of enzymes produced by solid state fermentation method. Int. J. Chem. Mol. Nucl. Mater. Met. Eng. **5**, 966–969 (2011)
69. W.H. Chen, B.L. Pen, C.T. Yu, et al., Pretreatment efficiency and structural characterization of rice straw by an integrated process of dilute-acid and steam explosion for bioethanol production. Bioresour. Technol. **102**, 2916–2924 (2011)
70. G. Dixit, A.R. Shah, D. Madamwar, et al., High solid saccharification using mild alkali-pretreated rice straw by hyper-cellulolytic fungal strain. Bioresour. Bioprocess. **2**, 1–8 (2015)
71. N.K. Aggarwal, V. Goyal, A. Saini, et al., Enzymatic saccharification of pretreated rice straw by cellulases from *Aspergillus Niger* BK01. 3 Biotech **7**, 1–10 (2017)
72. N. Xie, N. Jiang, M. Zhang, et al., Effect of different pretreatment methods of corncob on bioethanol production and enzyme recovery. Cellul. Chem. Technol. **48**, 313–319 (2014)
73. L. Qin, Z. Liu, M. Jin, et al., High temperature aqueous ammonia pretreatment and post-washing enhance the high solids enzymatic hydrolysis of corn stover. Bioresour. Technol. **146**, 504–511 (2013)
74. B.D. Bals, C. Gunawan, J. Moore, et al., Enzymatic hydrolysis of pelletized AFEX™-treated corn stover at high solid loadings. Biotechnol. Bioeng. **111**, 264–271 (2014)
75. W. Wang, X. Zhuang, Z. Yuan, et al., High consistency enzymatic saccharification of sweet sorghum bagasse pretreated with liquid hot water. Bioresour. Technol. **108**, 252–257 (2012)
76. L. Wang, R. Templer, R.J. Murphy, High-solids loading enzymatic hydrolysis of waste papers for biofuel production. Appl. Energy **99**, 23–31 (2012)
77. M. Linde, E. Jakobsson, M. Galbe, et al., Steam pretreatment of dilute H_2SO_4-impregnated wheat straw and SSF with low yeast and enzyme loadings for bioethanol production. Biomass Bioenergy **32**, 326–332 (2008)
78. M. Pedersen, A.S. Meyer, Influence of substrate particle size and wet oxidation on physical surface structures and enzymatic hydrolysis of wheat straw. Biotechnol. Prog. **25**, 399–408 (2009)
79. L. Rosgaard, S. Pedersen, A.S. Meyer, Comparison of different pretreatment strategies for enzymatic hydrolysis of wheat and barley straw. Appl. Biochem. Bioethanol. **143**, 284–296 (2007)
80. M.P. Garcia-Aparicio, J.M. Oliva, P. Manzanares, et al., Second-generation ethanol production from steam exploded barley straw by *Kluyveromyces marxianus* CECT 10875. Fuel **90**, 1624–1630 (2011)
81. T. Ingram, K. Wormeyer, J.C.I. Lima, et al., Comparison of different pretreatment methods for lignocellulosic materials. Part I: Conversion of rye straw to valuable products. Bioresour. Technol. **102**, 5221–5228 (2011)
82. X. Zhang, W.J. Qin, M.G. Paice, et al., High consistency enzymatic hydrolysis of hardwood substrates. Bioresour. Technol. **100**, 5890–5897 (2009)
83. C. Cara, I. Romero, J.M. Oliva, et al., Liquid hot water pretreatment of olive tree pruning residues. Appl. Biochem. Biotechnol. **137**, 379–394 (2007)
84. C. Felby, L.G. Thygesen, J.B. Kristensen, et al., Cellulose-water interactions during enzymatic hydrolysis as studied by Time Domain NMR. Cellulose **15**, 703–710 (2008)
85. P. Andric, A.S. Meyer, P.A. Jensen, et al., Reactor design for minimizing product inhibition during enzymatic lignocellulose hydrolysis II. Quantification of inhibition and suitability of membrane reactors. Biotechnol. Adv. **28**, 407–425 (2010)

86. H. Jorgensen, J. Vibe-Pedersen, J. Larsen, et al., Liquefaction of lignocellulose at high-solids concentrations. Biotechnol. Bioeng. **96**, 862–870 (2006)
87. R.K. Dasari, K. Dunaway, R.E. Berson, A scraped surface bioreactor for enzymatic saccharification of pretreated corn stover slurries. Energy Fuels **23**, 492–497 (2009)
88. H.Z. Chen, Z.H. Liu, Steam explosion and its combinatorial pretreatment refining technology of plant biomass to bio-based products. Biotechnol. J. **10**, 866–885 (2015)
89. D.B. Hodge, M.N. Karim, D.J. Schell, et al., Model-based fed-batch for high-solids enzymatic cellulose hydrolysis. Appl. Biochem. Biotechnol. **152**, 88–107 (2009)
90. R.P. Chandra, K. Au-Yeung, C. Chanis, et al., The influence of pretreatment and enzyme loading on the effectiveness of batch and fed-batch hydrolysis of corn stover. Biotechnol. Prog. **27**, 77–85 (2011)
91. J. Yang, X.P. Zhang, Q.A. Yong, et al., Three-stage enzymatic hydrolysis of steam-exploded corn stover at high substrate concentration. Bioresour. Technol. **102**, 4905–4908 (2011)
92. X.S. Yang, S.J. Zhang, Z. Zuo, et al., Ethanol production from the enzymatic hydrolysis of non-detoxified steam exploded corn stalk. Bioresour. Technol. **102**, 7840–7844 (2011)
93. K. Brijwani, H.S. Oberoi, P.V. Vadlani, Production of a cellulolytic enzyme system in mixed-culture solid-state fermentation of soybean hulls supplemented with wheat bran. Process Biochem. **45**, 120–128 (2010)
94. S. Chovau, D. Degrauwe, B.V. Bruggen, Critical analysis of techno-economic estimates for the production cost of lignocellulosic bio-ethanol. Renew. Sust. Energ. Rev. **26**, 307–321 (2013)
95. A. Aden, M. Ruth, K. Ibsen, et al., *Lignocellulosic Biomass to Ethanol Process Design and Economics Utilizing Co-current Dilute Acid Prehydrolysis and Enzymatic Hydrolysis for Corn Stover* (National Renewable Energy Laboratory, Golden, 2002)
96. A. Wingren, M. Galbe, C. Roslander, et al., Effect of reduction in yeast and enzyme concentrations in a simultaneous-saccharification-and-fermentation-based bioethanol process: technical and economic evaluation. Appl. Biochem. Biotechnol. **122**, 485–499 (2005)
97. P. Sassner, M. Galbe, G. Zacchi, Techno-economic evaluation of bioethanol production from three different lignocellulosic materials. Biomass Bioenergy **32**, 422–430 (2008)
98. F.K. Kazi, J.A. Fortman, R.P. Anex, et al., Techno-economic comparison of process technologies for biochemical ethanol production from corn stover. Fuel **89**, 20–28 (2010)
99. S.T. Merino, J. Cherry, Progress and challenges in enzyme development for biomass utilization. Adv. Biochem. Eng. Biotechnol. **108**, 95–120 (2007)
100. O. Takimura, T. Yanagida, S. Fujimoto, et al., Estimation of bioethanol production cost from rice straw by on-site enzyme production. J. Jpn. Pet. Inst. **56**, 150–155 (2013)
101. M. Zhang, W. Qi, R. Liu, et al., Fractionating lignocellulose by formic acid: characterization of major components. Biomass Bioenergy **34**, 525–532 (2010)
102. A.K. Chandel, F.A.F. Antunes, V. Anjos, et al., Multi-scale structural and chemical analysis of sugarcane bagasse in the process of sequential acidebase pretreatment and ethanol production by *Scheffersomyces shehatae* and *Saccharomyces cerevisiae*. Biotechnol. Biofuels **7**, 1–17 (2014)
103. G. Cruz, P.A.S. Monteiro, C.E.M. Braz, et al., Thermal and morphological evaluation of chemically pretreated sugarcane bagasse. Int. J. Chem. Mol. Eng. **7**, 435–440 (2013)
104. Y. Sun, L. Lin, C.S. Pang, et al., Hydrolysis of cotton fiber cellulose in formic acid. Energy Fuel **21**, 2386–2389 (2007)
105. S.B. Kim, S.J. Lee, J.H. Lee, et al., Pretreatment of rice straw with combined process using dilute sulfuric acid and aqueous ammonia. Biotechnol. Biofuels **6**, 2–11 (2013)
106. P. Phitsuwan, C. Permsriburasuk, S. Baramee, et al., Structural analysis of alkaline pretreated rice straw for ethanol production. Int. J. Polym. Sci. **2017**, 1–9 (2017)
107. M. Molaverdia, K. Karimia, S. Mirmohamadsadeghia, et al., High titer ethanol production from rice straw via solid-state simultaneous saccharification and fermentation by Mucor indicus at low enzyme loading. Energy Convers. Manag. **182**, 520–529 (2019)

108. G.J.M. Rocha, C. Martín, V.F.N. da Silva, et al., Mass balance of pilot-scale pretreatment of sugarcane bagasse by steam explosion followed by alkaline delignification. Bioresour. Technol. **111**, 447–452 (2012)
109. R. Singh, S. Tiwari, M. Srivastava, et al., Microwave assisted alkali pretreatment of rice straw for enhancing enzymatic digestibility. J. Energy **2014**, 1–7 (2014)
110. G. Gong, D. Liu, Y. Huang, Microwave-assisted organic acid pretreatment for enzymatic hydrolysis of rice straw. Biosyst. Eng. **107**, 67–73 (2010)
111. D.C. Li, A.N. Li, A.C. Papageorgiou, Cellulases from thermophilic fungi: insights and biotechnological potential. Enzym. Res. **2011**, 308730–308739 (2011)
112. A.K. Chandel, F.F.A. Antunes, V. Anjos, et al., Ultra-structural mapping of sugarcane bagasse after oxalic acid fiber expansion (OAFEX) and ethanol production by *Candida shehatae* and *Saccharomyces cerevisiae*. Biotechnol. Biofuels **6**, 1–15 (2013)
113. F. Momayez, K. Karimi, S. Karimia, et al., Efficient hydrolysis and ethanol production fromrice straw by pretreatment with organic acids and effluent of biogas plant. RSC Adv. **7**, 50537–50545 (2017)
114. K.K. Brar, S. Kaur, B.S. Chadha, A novel staggered hybrid SSF approaches for efficient conversion of cellulose/hemicellulosic fractions of Corncob into Ethanol. Renew. Energy **98**, 16–22 (2016)
115. S.P.S. Chundawat, B. Venkatesh, B.E. Dale, Effect of particle size based separation of milled corn stover on AFEX pretreatment and enzymatic digestibility. Biotechnol. Bioeng. **96**, 219–231 (2007)
116. A.A. Guilherme, P.V.F. Dantas, E.S. Santos, et al., Evaluation of composition, characterization and enzymatic hydrolysis of pretreated sugarcane bagasse. Braz. J. Chem. Eng. **32**, 23–33 (2015)
117. L. Cui, X. Wei, J. Li, et al., Structure and saccharification of sugarcane bagasse pretreated with acid coupled alkaline. Adv. Eng. Res. **134**, 104–107 (2017)
118. B. Wanitwattanarumlug, A. Luengnaruemitchai, S. Wongkasemjit, Characterization of corn cobs from microwave and potassium hydroxide pretreatment. Int. J. Mater. Metal. Eng. **6**, 354–358 (2012)
119. H. Li, A. Deng, J. Ren, C. Liu, et al., Catalytic hydrothermal pretreatment of corncob into xylose and furfural via solid acid catalyst. Bioresour. Technol. **158**, 313–320 (2014)
120. A. Kuila, M. Mukhopadhyay, D.K. Tuli, Accessibility of enzymatically delignified *Bambusa bambos* for efficient hydrolysis at minimum cellulase loading: an optimization study. Enzym. Res. **2011**, 1–8 (2011)
121. T.L. Richardson, N.K. Harner, P.K. Bajwa et al., Approaches to deal with toxic inhibitors during fermentation of lignocellulosic substrates, in *Sustainable Production of Fuels, Chemicals, and Fibers from Forest Biomass. ACS Symposium Series* (American Chemical Society, Washington, DC, 2011), pp. 171–202
122. B. Hahn-Hagerdal, K. Karhumaa, C. Fonseca, et al., Towards industrial pentose-fermenting yeast strains. Appl. Microbiol. Biotechnol. **74**, 937–953 (2007)
123. S.I. Mussatto, E.M.S. Machado, L.M. Carneiro, et al., Sugar metabolism and ethanol production by different yeast strains from coffee industry wastes hydrolysates. Appl. Energy **92**, 763–768 (2012)
124. S. Yanase, T. Hasunuma, R. Yamada, et al., Direct ethanol production from cellulosic materials at high temperature using the thermotolerant yeast *Kluyveromyces marxianus* displaying cellulolytic enzymes. Appl. Microbiol. Biotechnol. **88**, 381–388 (2010)
125. H. Zhou, J.S. Cheng, B.L. Wang, et al., Xylose isomerase overexpression along with engineering of the pentose phosphate pathway and evolutionary engineering enable rapid xylose utilization and ethanol production by *Saccharomyces cerevisiae*. Metab. Eng. **14**, 611–622 (2012)
126. S.-M. Lee, T. Jellison, H.S. Alper, Systematic and evolutionary engineering of a xylose isomerase-based pathway in *Saccharomyces cerevisiae* for efficient conversion yields. Biotechnol. Biofuels **7**, 1–8 (2014)

127. S. Kwak, Y. Jin, Production of fuels and chemicals from xylose by engineered *Saccharomyces cerevisiae*: a review and perspective. Microb. Cell Factories **16**, 1–15 (2017)
128. R. Kumari, K. Pramanik, Bioethanol production from *Ipomea Carnea* biomass using a potential hybrid yeast strain. Appl. Biochem. Biotechnol. **171**, 1771–1785 (2013)
129. M.Z. Li, S.J. Elledge, Harnessing homologous recombination in vitro to generate recombinant DNA via SLIC. Nat. Methods **4**, 251–256 (2007)
130. D. Chung, M. Cha, A.M. Guss, et al., Direct conversion of plant biomass to ethanol by engineered *Caldicellulosiruptor bescii*. Proc. Natl. Acad. Sci. U. S. A. **111**, 8931–8936 (2014)
131. V.E. Balderas-Hernández, V. Hernández-Montalvo, F. Bolívar, et al., Adaptive evolution of *Escherichia coli* inactivated in the phosphotransferase system operon improves co-utilization of xylose and glucose under anaerobic conditions. Appl. Biochem. Biotechnol. **163**, 485–496 (2011)
132. S.B. Jilani, S.S.K. Venigalla, A.J. Mattam, et al., Improvement in ethanol productivity of engineered *E. coli* strain SSY13 in defined medium *via* adaptive evolution. J. Ind. Microbiol. Biotechnol. **44**, 1375–1384 (2017)
133. N.K. Harner, X. Wen, P.K. Bajwa, et al., Genetic improvement of native xylose-fermenting yeasts for ethanol production. J. Ind. Microbiol. Biotechnol. **42**, 1–20 (2015)
134. P.K. Bajwa, T. Shireen, F. D'Aoust, et al., Mutants of the pentose-fermenting yeast *Pichia stipitis* with improved tolerance to inhibitors in hardwood spent sulfite liquor. Biotechnol. Bioeng. **104**, 892–900 (2009)
135. T. Watanabe, I. Watanabe, M. Yamamoto, et al., A UV-induced mutant of *Pichia stipitis* with increased ethanol production from xylose and selection of a spontaneous mutant with increased ethanol tolerance. Bioresour. Technol. **102**, 1844–1848 (2011)
136. H.S. Oberoi, N. Babbar, S.K. Sandhu, et al., Ethanol production from alkali-treated rice straw *via* simultaneous saccharification and fermentation using newly isolated thermotolerant *Pichia kudriavzevii* HOP-1. J. Ind. Microbiol. Biotechnol. **39**, 557–566 (2012)
137. B. Erdei, B. Franko, M. Galbe, et al., Separate hydrolysis and co-fermentation for improved xylose utilization in integrated ethanol production from wheat meal and wheat straw. Biotechnol. Biofuels **5**(1), 12 (2012)
138. P. Antil, R. Gupta, R. Kuhad, Simultaneous saccharification and fermentation of pretreated sugarcane bagasse to ethanol using a new thermotolerant yeast. Ann. Microbiol. **65**, 423–429 (2015)
139. C. Huang, M.H. Zong, H. Wu, et al., Microbial oil production from rice straw hydrolysate by *Trichosporon fermentans*. Bioresour. Technol. **100**, 4535–4538 (2009)
140. M.J. Jin, C. Sarks, C. Gunawan, et al., Phenotypic selection of a wild *Saccharomyces cerevisiae* strain for simultaneous saccharification and co-fermentation of AFEX pretreated corn stover. Biotechnol. Biofuels **6**, 1–14 (2013)
141. Y. Li, J.Y. Park, R. Shiroma, et al., Improved ethanol and reduced xylitol production from glucose and xylose mixtures by the mutant strain of *Candida shehatae* ATCC 22984. Appl. Biochem. Biotechnol. **166**, 1781–1790 (2012)
142. P.K. Bajwa, C. Phaenark, N. Grant, et al., Ethanol production from selected lignocellulosic hydrolysates by genome shuffled strains of *Scheffersomyces stipitis*. Bioresour. Technol. **102**, 9965–9969 (2011)
143. W. Zhang, A.L. Geng, Improved ethanol production by a xylose-fermenting recombinant yeast strain constructed through a modified genome shuffling method. Biotechnol. Biofuels **5**, 46–57 (2012)
144. I. Watanabe, T. Nakamura, J. Shima, Characterization of a spontaneous flocculation mutant derived from *Candida glabrata:* a useful strain for bioethanol production. J. Biosci. Bioeng. **107**, 379–382 (2009)
145. A. Jain, S.P. Chaurasia, Bioethanol production in membrane bioreactor (MBR) system: a review. Int. J. Environ. Res. Public Health **4**, 387–394 (2014)

146. H. Li, N.J. Kim, M. Jiang, et al., Simultaneous saccharification and fermentation of lignocellulosic residues pretreated with phosphoric acid-acetone for bioethanol production. Bioresour. Technol. **100**, 3245–3251 (2009)
147. K. Olofsson, M. Wiman, G. Liden, Controlled feeding of cellulases improves conversion of xylose in simultaneous saccharification and co-fermentation for bioethanol production. Biotechnolgy **145**, 168–175 (2010)
148. K. Ohgren, A. Rudolf, M. Galbe, Fuel ethanol production from steam pretreated corn stover using SSF at higher dry matter content. Biomass Bioenergy **30**, 863–869 (2006)
149. K.E. Kang, D.-P. Chung, Y. Kim, et al., High-titer ethanol production from simultaneous saccharification and fermentation using a continuous feeding system. Fuel **145**, 18–24 (2015)
150. M.K. Balasubramaniam, Potential utilization of rice straw for ethanol production by sequential fermentation of cellulose and xylose using *Saccharomyces cerevisiae* and *Pachysolen tannophilus*. Int. J. Sci. Eng. Technol. Res. **2**, 1531–1535 (2013)
151. L.R. Hickert, P.B. de Souza-Cruz, C.A. Rosa, et al., Simultaneous saccharification and co-fermentation of un-detoxified rice hull hydrolysate by *Saccharomyces cerevisiae* ICV D254 and *Spathaspora arborariae* NRRL Y-48658 for the production of ethanol and xylitol. Bioresour. Technol. **143**, 112–116 (2013)
152. R.K. Sukumaran, V.J. Surender, R. Sindhu, et al., Lignocellulosic ethanol in India: prospects, challenges and feedstock availability. Bioresour. Technol. **101**, 4826–4833 (2010)
153. J.B. Gonsalves, An assessment of the biofuels industry in India, in *United Nations Conference on Trade and Development (UNCTAD/DITC/TED/2006/6)* (UNCTAD, Geneva, 2006), pp. 1–42
154. N. Yu, L. Tan, Z.Y. Sun, et al., Production of bio-ethanol by integrating microwave-assisted dilute sulfuric acid pretreated sugarcane bagasse slurry with molasses. Biotechnol. Appl. Biochem. **185**, 191–206 (2017)
155. K.K. Brar, D. Agrawal, B.S. Chadha, et al., Evaluating novel fungal secretomes for efficient saccharification and fermentation of composite sugars derived from hydrolysate and molasses into ethanol. Bioresour. Technol. **273**, 114–121 (2019)
156. R. Subramaniam, S. Dufreche, M. Zappi, et al., Microbial lipids from renewable resources: production and characterization. J. Ind. Microbiol. Biotechnol. **37**, 1271–1287 (2010)
157. J. Ageitos, J. Vallejo, P. Veiga-Crespo, et al., Oily yeasts as oleaginous cell factories. Appl. Microbiol. Biotechnol. **90**, 1219–1227 (2011)
158. S. Papanikolaou, G. Aggelis, Lipids of oleaginous yeasts. Part II: technology and potential applications. Eur. J. Lipid Sci. Technol. **113**, 1052–1073 (2011)
159. L. Hui, C. Wan, D. Hai-tao, et al., Direct microbial conversion of wheat straw into lipid by a cellulolytic fungus of *Aspergillus oryzae* A-4 in solid-state fermentation. Bioresour. Technol. **101**, 7556–7562 (2010)
160. A. Andre, P. Diamantopoulou, A. Philippoussis, et al., Biotechnological conversions of biodiesel derived waste glycerol into added-value compounds by higher fungi: production of biomass, single cell oil and oxalic acid. Ind. Crop. Prod. **31**, 407–416 (2010)
161. A. Chatzifragkou, A. Makri, A. Belka, et al., Biotechnological conversions of biodiesel derived waste glycerol by yeast and fungal species. Energy **36**, 1097–1108 (2011)
162. Y.N. Liang, T.Y. Tang, T. Siddaramu, et al., Lipid production from sweet sorghum bagasse through yeast fermentation. Renew. Energy **40**, 130–136 (2012)
163. A. Chatzifragkou, S. Fakas, M. Galiotou-Panayotou, et al., Commercial sugars as substrates for lipid accumulation in *Cunninghamella echinulata* and *Mortierella isabellina* fungi. Eur. J. Lipid Sci. Technol. **112**, 1048–1057 (2010)
164. C.N. Economou, A. Makri, G. Aggelis, et al., Semi-solid state fermentation of sweet sorghum for the biotechnological production of single cell oil. Bioresour. Technol. **101**, 1385–1388 (2010)
165. X.C. Yu, Y.B. Zheng, K.M. Dorgan, et al., Oil production by oleaginous yeasts using the hydrolysate from pretreatment of wheat straw with dilute sulfuric acid. Bioresour. Technol. **102**, 6134–6140 (2011)

166. Y.A. Tsigie, C.Y. Wang, C.T. Truong, et al., Lipid production from *Yarrowia lipolytica* Po1g grown in sugarcane bagasse hydrolysate. Bioresour. Technol. **102**, 9216–9222 (2011)
167. X. Huang, Y. Wang, W. Liu, et al., Biological removal of inhibitors leads to the improved lipid production in the lipid fermentation of corn stover hydrolysate by *Trichosporon cutaneum*. Bioresour. Technol. **102**, 9705–9709 (2011)
168. C. Huang, X.F. Chen, L. Xiong, et al., Oil production by the yeast *Trichosporon dermatis* cultured in enzymatic hydrolysates of corncobs. Bioresour. Technol. **110**, 711–714 (2012)
169. D. Xing, A. Pan, D. Xue, et al., Biomass carbohydrates assimilation and lipid accumulation by *Mortierella isabellina*. Chin. J. Biotechnol. **26**, 189–193 (2010)
170. J. Tao, C.C. Dai, Q.Y. Yang, et al., Production of biodiesel with acid hydrolysate of *Populus euramevicana* cv leaves by *Rhodotorula glutinis*. Int. J. Green Energy **7**, 387–396 (2010)
171. H. Lin, W. Cheng, H.T. Ding, et al., Direct microbial conversion of wheat straw into lipid by a cellulolytic fungus of *Aspergillus oryzae* A-4 in solid-state fermentation. Bioresour. Technol. **101**, 7556–7562 (2010)
172. H. Uemura, Synthesis and production of unsaturated and polyunsaturated fatty acids in yeast: current state and perspectives. Appl. Microbiol. Biotechnol. **95**, 1–12 (2012)
173. M.H. Liang, J.G. Jiang, Advancing oleaginous microorganisms to produce lipid *via* metabolic engineering technology. Prog. Lipid Res. **52**, 395–408 (2013)
174. M. Tai, G. Stephanopoulos, Engineering the push and pull of lipid biosynthesis in oleaginous yeast *Yarrowia lipolytica* for biofuel production. Metab. Eng. **15**, 1–9 (2013)
175. G.S. Wang, J.W. Jae-Won Lee, et al., Dilute acid pretreatment of corncob for efficient sugar production. Biotechnol. Appl. Biochem. **163**, 658–668 (2011)
176. A. Beopoulos, Z. Mrozova, F. Thevenieau, et al., Control of lipid accumulation in the yeast *Yarrowia lipolytica*. Appl. Environ. Microbiol. **74**, 7779–7789 (2008)
177. R. Ruenwai, S. Cheevadhanarak, K. Laoteng, Overexpression of acetyl-CoA carboxylase gene of *Mucor rouxii* enhanced fatty acid content in *Hansenula polymorpha*. Mol. Biotechnol. **42**, 327–332 (2009)
178. S. Ichikawa, S. Karita, Bacterial production and secretion of water-insoluble fuel compounds from cellulose without the supplementation of cellulases. FEMS Microbiol. Lett. **362**, 1–24 (2015)
179. Z. Fatma, K. Jawed, A.J. Mattam, et al., Identification of long chain specific aldehyde reductase and its use in enhanced fatty alcohol production in *E. coli*. Metab. Eng. **37**, 35–45 (2016)
180. U.S. Department of Energy (DOE) Annual Report. Department of Energy (2003)
181. N.K. Harner, P.K. Bajwa, M.B. Habash, et al., Mutants of the pentose-fermenting yeast Pachysolen tannophilus tolerant to hardwood spent sulfite liquor and acetic acid. Anton. van Leeuw. **105**, 29–43 (2014)

Chapter 25
Application of Lifecycle Concepts in the Conversion of Biomass to Value-Added Commodities

I. S. Dunmade

25.1 Lifecycle Concepts

The term "lifecycle concepts" refers to various tools that are used to design or evaluate various stages in the lifecycle of a system and/or the entire lifecycle [7]. There are two categories of lifecycle tools, namely, lifecycle design tools (DFXs) and lifecycle assessment (LCA) tools (Fig. 25.1). Among the lifecycle design tools are design paradigms where X stands for materials, modularity, assembly, manufacturing, disassembly, use and re-use, remanufacturing, upgrading, multipurpose, multi-lifecycle, packaging, recycling, and so on. Similarly, there are four lifecycle assessment tools, namely, environmental lifecycle assessment, lifecycle costing, social lifecycle assessment, and lifecycle sustainability assessment. Lifecycle assessment, used to evaluate the (potential) environmental implications of taken specific step at any stage in the lifecycle of a system, is the most popular of the lifecycle tools. It is a robust tool that enables decision-makers to identify the hotspots along the value chain of a system, a product, or a process. This chapter is devoted to environmental lifecycle assessment.

25.2 Importance and Possible Areas of LCA Application Along the Biomass Value Chain

Biomass has been recognized as a viable alternative resource to fossil fuels for the production of fuel, power, electricity, bioproducts, industrially important chemicals, and many other value-added products. However, a number of pretreatment and

I. S. Dunmade (✉)
Faculty of Science & Technology, Mount Royal University, Calgary, AB, Canada
e-mail: idunmade@mtroyal.ca

M. O. Daramola, A. O. Ayeni (eds.), *Valorization of Biomass to Value-Added Commodities*, Green Energy and Technology,
https://doi.org/10.1007/978-3-030-38032-8_25

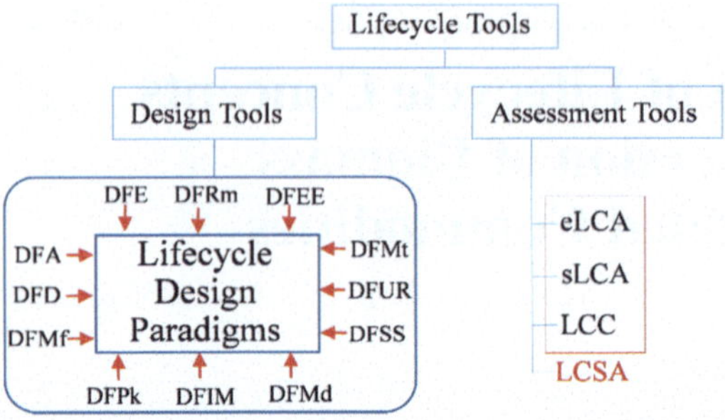

DFE Design for Environment DFRm Design for Remanufacturing DFMf Design for Manufacturing
DFD Design for Disassembly DFMt Design for Materials DFEE Design for Energy Efficiency
DFA Design for Assembly DFMd Design for Modularity DFSS Design for Simplicity & Safety
DFPk Design for Packaging DFUR Design for Use and Reuse DFIM Design for Installation & Maintenance
LCC Life Cycle Costing sLCA Social Life Cycle Assessment
eLCA Environmental Life Cycle Assessment LCSA Life Cycle Sustainability Assessment

Fig. 25.1 Two categories of lifecycle tools and their constituent elements. DFE Design for Environment, DFRm Design for Remanufacturing, DFMf Design for Manufacturing, DFD Design for Disassembly, DFMt Design for Materials, DFEE Design for Energy Efficiency, DFA Design for Assembly, DFMd Design for Modularity, DFSS Design for Simplicity and Safety, DFPk Design for Packaging, DFUR Design for Use and Re-use, DFIM Design for Installation and Maintenance, LCC Lifecycle Costing, sLCA Social Lifecycle Assessment, eLCA Environmental Lifecycle Assessment, LCSA Lifecycle Sustainability Assessment

conversion technologies are needed to transform biomass feedstock to various value-added products. Combustion, gasification, carbonization, co-firing, liquefaction, pyrolysis, and torrefaction are among the conversion technologies that are commonly used to convert biomass to value-added products [1–3, 12, 15, 16, 20–23, 26, 27].

The lifecycle of biomass conversion to value-added products starts from cultivation or gathering of the biomass feedstock through various complicated process steps to the utilization of the end products. These processes do require the use of resources such as land, water, and various chemicals either as catalysts or as reactants. To facilitate the forward reaction or enhance yields, a number of these processes usually require application of energy at some stages in the conversion process. Furthermore, co-products that may or may not have economic value are often produced in addition to the targeted value-added products. Consequently, the co-products with very little or no economic value have to be managed as wastes. Moreover, technology infrastructures for the conversion of biomass to value-added products also have to be designed, developed, and managed. The production and use of various types of chemicals as well as the production and use of energy from various sources in the conversion process do have some environmental, economic, and social sustainability impacts. The development and management of conversion technology infrastructure and the associated wastes also have associated ecological footprints and socioeconomic impacts. The extent of these impacts varies according to the volume/quantity of the resources used and how the wastes resulting from the conversion processes are managed. Being able to quantify the potential damage that any of the steps in the conversion process could cause is necessary, because it will make it possible to compare alternative options. This will facilitate the choice of the "best" option. Lifecycle assessment (LCA) is a robust/comprehensive evaluation tool for assessing the environmental burden of a product, process, or activity [11]. It enables the tool users to quantitatively evaluate potential environmental impacts of resources such as materials and energy that are deployed in the manufacturing of a product. In addition, it provides the platform to identify improvement opportunities so that decision-makers can choose the best course of action(s) to minimize the ecological footprint of their product, process, or activity [5, 6, 12, 13, 19].

Lifecycle analysis can therefore be used to evaluate every aspect of the biomass conversion process to value-added products. Specific possible areas of LCA applications include evaluation of the potential environmental impacts of energy crop cultivation, biomass harvesting and handling, biomass pretreatment processes, conversion processes, and various post-production processes. In other words, LCA can be used to determine hotspots in the lifecycle of value-added products right from the cultivation through various processes of biomass conversion to the end-of-life management of both the conversion technologies/facilities and of the value-added products. It can be used to identify improvement opportunities at various stages in the lifecycle of biomass conversion technologies. LCA can also be used to develop long-term policies regarding various aspects of biomass-value-added products' lifecycle. In addition, LCA can be used to evaluate the potential resource use effects associated with adopted or planned to be adopted policies and alternative waste management techniques. Moreover, LCA can be used to provide information to the stakeholders (i.e., farmers, distributors/transporters, marketers, processors, investors, the public, and other interest groups) to enable them make informed choices with regard to various aspects of biomass lifecycle management.

25.2.1 Current Areas of LCA Application in the Biomass Value Chain

Current levels of LCA utilization in biomass conversion to value-added products are limited to few case studies along the value chain. The focus of majority of the LCA studies on biomass to date is centered on the evaluation of environmental benefits of utilizing biomass as an alternative energy source to fossil fuel. Other studies assessed greenhouse gas (GHG) emissions from biomass production, biomass preprocessing methods, biomass conversion technologies, and/or utilization of various types of biomass feedstock for bioenergy and biofuel production. Lifecycle assessment covering the entire value chain is still lacking. There are still many other aspects of LCA studies that are yet to be conducted on the conversion of biomass to value-added commodities. Majority of the LCA studies on biomass involved case studies on one or two types of biomass feedstock pretreatment or conversion to biofuel. Most of them didn't cover the entire lifecycle [4, 8, 10, 18, 24, 25].

25.2.2 *Future Trends in LCA Applications Along the Biomass Conversion to Value-Added Commodities Lifecycle*

Future trends in LCA application could see massive use of lifecycle engineering concepts involving utilization of lifecycle design paradigms not only in the development of biomass conversion technologies but also in the entire value chain/lifecycle. The future would see simultaneous/concurrent use of lifecycle design concepts with other members of lifecycle tools such as environmental LCA, social LCA, lifecycle costing, and lifecycle sustainability assessment. Lifecycle assessment could possibly become a regulatory requirement for the approval of a biomass facility location and size. The current challenges in the biomass lifecycle assessment, especially in the developing countries, are twofold, namely, shortage of lifecycle assessment practitioners and non-availability of adequate geographical location-relevant data for lifecycle assessment. Accurate lifecycle analysis of potential consequences of decisions at any stage in the biomass value chain in the future would require availability of well-trained professionals for the job. There is therefore a need for increased and consistent offering of lifecycle assessment education in engineering colleges and universities at both undergraduate and postgraduate levels. The future of lifecycle concept application in the biomass industry will also require intensified effort in the development of more location-relevant biomass database for lifecycle assessment.

25.3 Potential Challenges in LCA Application and How to Overcome the Challenges

Although lifecycle assessment is a robust tool for evaluating potential environmental impacts of products and processes, it has some limitations and challenges. One of its challenges is that the assessment exercise can be time-consuming and resource-intensive. Commonly used softwares that help in reducing the time required for LCA analysis are SimaPro, GaBi, Umberto, EDIP, EIOLCA, IDEMAT, LCA-it, SPOLD, GEMIS, OpenLCA, and TEAM [9, 17]. LCA implementation challenges could also be partly due to the difficulty in obtaining requisite data for the exercise. At times, available data may be inaccurate and this may skew/influence the results. Data availability challenges are being ameliorated by the current trend in the development of locally relevant databases. Ecoinvent and US Life Cycle Inventory database are among the popular LCA databases. These software and databases facilitate reduction of the analysis time, make data available, and lead to better results than if fresh data have to be collected or when relying on foreign databases that may not be locally adequate.

25.4 How Lifecycle Analysis of Biomass Conversion to Value-Added Products Can Be Conducted

There are various approaches to lifecycle assessment (LCA) process, but the globally accepted approach to doing LCA follows the ISO LCA standard. The latest versions of ISO LCA standards are ISO 14040 and 14044 [13, 14]. The LCA process consists of four standardized steps, regardless of the chosen level of rigor. The four steps consist of goal and scope definition, lifecycle inventory, lifecycle impact analysis, and lifecycle interpretation. The four steps are illustrated in Fig. 25.2.

25.4.1 Goal and Scope Definition

25.4.1.1 Goal Definition

This is the stage of the assessment process where the purpose of the study is articulated. It is a statement of intent for the study, how the study will be done, and what will and will not be included. ISO 14044 stipulates that the analyst should state three things under goal definition, namely, (i) the intended application of the LCA, (ii) reasons for carrying out the study, and (iii) the intended audience that would use the results of the study.

There could be several reasons for implementing a LCA study. *Intended application* of LCA results can be classified into six categories, namely, product/process development, product/process improvement, product/process comparison, identification of resource use effects, developing long-term policy, and/or for public information. Since LCA is a decision-making aid, various categories of stakeholders may be the intended audience/user of the LCA results. These categories of stakeholders could include lawmakers at various levels of government, governmental agency managers, as well as corporate organizations and nongovernmental organizations' managers/policy makers. It could also include other categories of decision-makers, procurement officers, project managers, engineers, architects, manufacturing facility foreman, or production employees. Identification of the intended audience is crucial to the understanding of the LCA report and usefulness of the LCA results. The knowledge of the intended audience would enable the LCA practitioner to

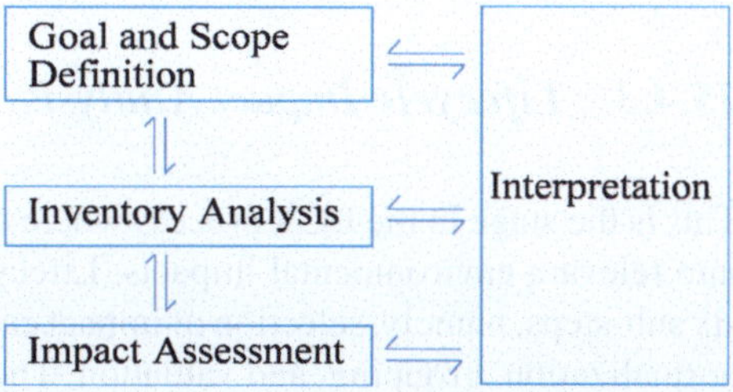

Fig. 25.2 An illustration of the four steps of LCA process

determine the appropriate language to use and the level of details to include in the report. It will also enable him/her to determine where in the report to devote a lot of his/her effort in order to facilitate an understanding of the results and enhance its usefulness.

25.4.1.2 Scope Definition

This part of the goal and scope definition involves articulation of the function of the system, the functional unit, the flow chart of the system being studied, and the system boundaries. *Functional unit* refers to the metric for measuring the performance of product, process, or system in both quantitative and qualitative terms. *Functional unit* facilitates appropriate comparison of products and enables stakeholders to determine the extent of improvement achieved in a product that has been redesigned. Scope definition also involves specifying the types of impacts that would be evaluated and the methodology of impact assessment. Moreover, scope definition includes the specification of the procedure that would be used to allocate environmental impacts. ISO standards specify that allocation should be avoided if possible, but where a system yields more than one output, the analyst should either expand the system boundaries or allocate the impacts to the outputs. Furthermore, data requirement, assumptions and limitations, and initial data quality requirements are specified in this section. In addition, the type of critical review and format of report required for the study are also specified under scope definition.

25.4.2 Lifecycle Inventory

This is the stage in the LCA process where the analyst compiles and quantifies the inputs and outputs at various points in the process with the aim of meeting the stated goal and scope. This stage usually involves planning where and how those data needed could be obtained and implementing the data collection. It also involves preparing the data collected in a form that could be utilized in the subsequent stages of the LCA. The kind of data collected could be either primary data or secondary data. Primary data are measured data and those data collected directly from the factory. Secondary data refers to data collected from reports and websites.

25.4.3 Lifecycle Impact Analysis

This is the stage in the LCA process where data collected and processed is modelled into relevant environmental impacts. Lifecycle impact analysis (LCIA) consists of six sub-steps, namely, selection of impact categories, classification, characterization, normalization, grouping, and valuation. The first three, according to ISO standards, are mandatory, while the other three are non-mandatory.

25.4.3.1 Selection of Relevant Impact Categories

LCIA starts with first selecting impact categories that meet the goal and scope of the study. Among the common impact categories that are often evaluated are global warming potential (aka climate change impact), ozone depletion potential, photochemical oxygen creation potential, acidification potential, eutrophication potential, human toxicological impact, eco-toxicological impact, abiotic resource depletion potential, biotic depletion impact, land use, and water footprint. However, a number of factors are often considered when selecting impact categories to be analyzed. The factors considered include practicality, independence from other impact categories, completeness with regard to the goals and scope of the study, and the relation to the characterization step.

25.4.3.2 Classification and Characterization

Selection of impact categories is followed by the classification step. *Classification* involves mapping relevant lifecycle inventory data into individual impact categories selected before they are quantitatively modelled in *characterization* step to obtain the severity of the process impact in terms of the individual impact category selected.

The severity of contribution of each step in a bioconversion process to specific impact category (such as climate change, ozone depletion, and so on) is determined by first multiplying the conversion factor of each emission for a specific impact category with the quantity of the emission involved. Results obtained in this first sub-step are summed for each impact category at each stage in the process to obtain the impact of each lifecycle stage with regard to that impact. Results obtained at each lifecycle stage are further summed to obtain the process lifecycle impact.

25.4.3.3 Normalization

This is a procedure that expresses potential impacts in ways that can be easily compared. It is used to evaluate the extent to which a stressor or the sum of environmental releases at a lifecycle stage contributes to an impact category. Normalization is done by dividing the indicator results with a selected reference value. Among the commonly selected reference values are:

(i) The total emissions or resource use for a given area. The area may be local, regional, or global in scale.
(ii) The total emissions or resource use for a given area on a per capital basis.
(iii) The highest value among alternative processes or products under consideration.

One of the main factors that are often considered in determining the most appropriate reference value to choose is the goal and scope of the LCA.

Moreover, normalized data can only be compared within an impact category because each impact category value is based on the characterization factors that were calculated using different scientific methods. It is just as we can only compare one currency with another currency but not currency with humidity, heights, weight, or light intensity.

25.4.3.4 Grouping and Weighting

Grouping involves categorization of the evaluated impacts into local, regional, and global impacts. For example, photochemical smog and aquatic toxicity are usually seen as local impacts while climate change and ozone depletion are global impacts. However, some impacts could be local, regional, and/or global. Examples of such impacts are resource depletion and human health.

Weighting, which is also referred to as *valuation*, involves the attachment of relative importance weight/value to each of the impact results. This is the most controversial step in lifecycle impact assessment because of its subjective nature. There is tendency for change in value judgments with location and time of the year. Consequently, it is seldom implemented. Valuation is done to express the relative preference by stakeholders. Various methods are used to determine the preference weights. Among such methods are authoritative panels, authorized goals or standards, and multi-criteria decision-making.

25.4.4 Lifecycle Interpretation

This is a systematic technique used to identify, quantify, check, and evaluate information from LCI and LCIA results and communicate them effectively. In other words, it is the stage at which the whole study is put into perspective and the analytic results of the LCI and LCIA stages are examined to determine their implications and significance. This step is taken with the consideration of the goal and scope of the study. There are three steps in the lifecycle interpretation process. The three steps are identification of environmentally significant issues, evaluation of results and process used, and conclusions that could be drawn from the results.

Environmentally significant issues are those data elements that contribute most to the LCI and LCIA results for each biomass conversion process. It is necessary to identify significant issues because of the extensive amount of data that were collected and the need to focus on those data elements that contribute significantly to the outcome of the results in view of the available limited time and resources.

Contribution analysis, dominance analysis, and anomaly assessment are three of the commonly used approaches to identify significant issues.

Contribution analysis method involves the comparison of the quantitative contribution of various stages of the process lifecycle, emissions, or resources consumed to the total result. It also includes an examination of the relevance of such contribution. Thus, contribution analysis is used to determine which process, emissions, resource consumption, and/or process lifecycle stages play significant roles in the LCA results.

Dominance analysis method involves quantitative or qualitative ranking of the LCI and LCIA results and identifying the significance of their contributions, while *anomaly assessment method*, based on previous experience, examines LCI and LCIA results and determines any unusual or surprising deviations from the expected or normal results. The significance of the deviations is then assessed for relevance.

The purpose of evaluation step in lifecycle interpretation is to establish confidence in the result of the LCA study. This step is undertaken in accordance with the goal and scope of the study, and it takes into account the final use of the study. This second step in lifecycle interpretation consists of three elements, namely:

(i) Conduction of qualitative check of all selected data and processes with the aim of determining the possible consequences of leaving out the information.
(ii) Evaluating implications of changes in data due to data methodological uncertainties. A systematic qualitative or quantitative analysis is used for the evaluation.
(iii) Discussion of the variations in the goal and scope of the study.

25.5 Levels of LCA Rigor

Lifecycle assessment (LCA) can be implemented at various levels of rigor (Fig. 25.3), namely, *conceptual LCA*, *streamlined LCA*, and *comprehensive LCA*.

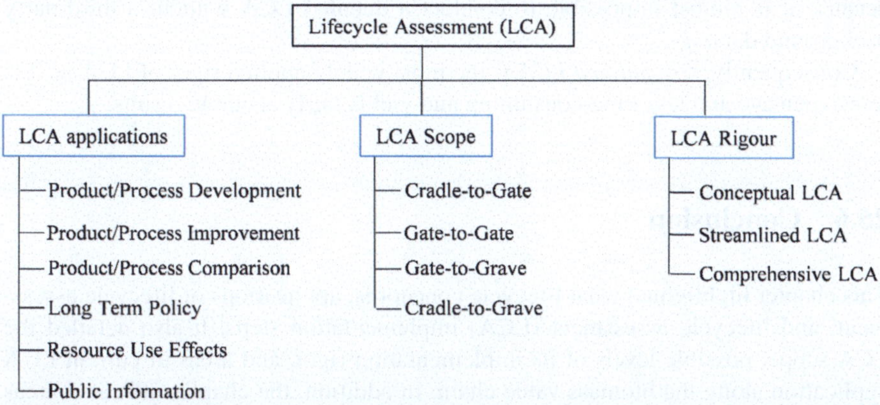

Fig. 25.3 Areas of LCA application, scope, and rigor

25.5.1 Conceptual LCA

Conceptual LCA, which is also known as *lifecycle thinking*, is often used to get a rough idea of potential environmental implication of a course of action or design. It is commonly used for screening various options before detailed design is carried out. Qualitative data and scoring models are often employed for this kind of LCA.

25.5.2 Streamlined LCA

This is a more rigorous LCA than the conceptual LCA. It is the most common LCA rigor done in practice. This kind of LCA is also referred to as *simplified LCA* or *abridged LCA*. It involves the collection and use of both primary and secondary data. *Primary data*, which is also called foreground data, refers to data that are specific to the project, product, process, or site. They are usually collected through the use of questionnaire administered to the stakeholders involved in the project. *Secondary data*, which is also known as background data, is obtained from databases or reports. The results obtained from this kind of analysis are usually comprehensive and are fairly reliable.

25.5.3 Comprehensive LCA

This is a *detailed* LCA involving basically the use of only primary data collected from the facility under study. It also includes measurements taken directly from the process. Essentially, it avoids the use of data from databases or from reports. As a result, it is the approach that yields the highest accurate results. However, it is not only the most time-consuming and costly but also not realistic. It is not practicable because it is almost impossible to conduct a detailed LCA without a third-party background data.

Consequently, *streamlined LCA* is the most widely applied rigor of LCA as it is less expensive and less time-consuming and yields fairly accurate results.

25.6 Conclusion

This chapter highlighted what lifecycle concept is, applications of lifecycle assessment, and lifecycle assessment (LCA) implementation steps. It also detailed the LCA scope, possible levels of its implementation rigor, and areas of current LCA application along the biomass value chain. In addition, the chapter indicated areas of future LCA use in the biomass chain. It can be concluded that there are lots of potentials for increased use of lifecycle concepts to improve the sustainability of biomass conversion to value-added commodities. There is also a need for lifecycle design and lifecycle assessment education to facilitate the use of these valuable tools.

References

1. S.A. Ashter, *Technology and Applications of Polymers Derived from Biomass* (2018). https://doi.org/10.1016/B978-0-323-51115-5.00005-0
2. M. Balat, M. Balat, E. Kirtay, H. Balat, Main routes for the thermo-conversion of biomass into fuels and chemicals. Part 1 & 2 Energy Convers. Manag. **50**(12), 3147–3168 (2009)
3. P. Basu, *Biomass Gasification, Pyrolysis and Torrefaction: Practical Design and Theory*, 3rd edn. (Academic Press, Cambridge, 2018)
4. P. Brassard, J.H. Palacios, S. Godbout, D. Bussieres, R. Lagace, J.-P. Larouche, F. Pelletier, Comparison of the gaseous and particulate matter emissions from the combustion of agricultural and forest biomasses. Bioresour. Technol. **155**, 300–306 (2014)
5. S.S. Da Silva, A.K. Chandel, S.R. Wickramasinghe, J.M.G. Dominguez, Fermentative production of value-added products from lignocellulosic biomass. J. Biomed. Biotechnol. (2012). https://doi.org/10.1155/2012/826162
6. I.S. Dunmade, Lifecycle assessment of a stapling machine. Int. J. Eng. Technol. **4**(1), 12–19 (2014). ISSN 2227-524X
7. I.S. Dunmade, Sustainable engineering: a vital approach to innovative product development and community capacity building. The 5th Inaugural Lecture of Covenant University, Ota, Ogun State presented on 19 Feb 2016 (2016)
8. I.R. Emery, N.S. Mosier, The impact of dry matter loss during herbaceous biomass storage on net greenhouse gas emissions from biofuels production. Biomass Bioenergy **39**, 237–246 (2012)
9. ESU, Software for calculating lifecycle assessment carbon footprint and other indicators (2018). Accessed online at http://esu-services.ch/software/
10. P.A. Fokaides, F. Christoforou, *Life Cycle Sustainability Aassessment of Biofuels* (2016). https://doi.org/10.1016/B978-0-08-100455-5.00003-5. Accessed 20 Feb 2019
11. J.B. Guinee, *Handbook on Lifecycle Assessment: Operational Guide to the ISO Standards* (Kluwer Academic Publishers, Boston, 2002)
12. S. Irmak, Biomass as raw material for production of high-value products (2017). https://www.intechopen.com/books/biomass-volume-estimation-and-valorization-for-energy/biomass-as-raw-material-for-production-of-high-value-products. Accessed 20 Feb 2019
13. ISO (International Standard Organization), ISO 14040: environmental management – lifecycle assessment – principles and framework (2006a). http://imsiran.ir/?wpfb_dl=15. Accessed 20 Feb 2019
14. ISO (International Standard Organization), ISO 14044: environmental management – lifecycle assessment – requirements and guidelines (2006b). http://wap.sciencenet.cn/home.php?mod=attachment&id=4637. Accessed 20 Feb 2019
15. S.B. Jones, J.E. Holladay, C. Valkenburg, D.J. Stevens, C.W. Walton, C. Kinchin, D.C. Elliott, S. Czernik, *Production of Gasoline and Diesel From Biomass via Fast Pyrolysis, Hydrotreating and Hydrocracking: A Design Case* (Pacific Northwest National Laboratory, Richland, 2009)
16. M. Khanna, B. Dhungana, J. Clifton-Brown, Costs of producing Miscanthus and switchgrass for bioenergy in Illinois. Biomass Bioenergy **32**(6), 482–493 (2008)
17. H. Levin, Life cycle assessment software, tools and databases (2018). http://www.building-ecology.com/sustainability/life-cycle-assessment/life-cycle-assessment-software. Accessed 20 Feb 2019

18. W. Liu, J. Wang, T.L. Richard, D.S. Hartley, S. Spatari, T.A. Volk, Economic and life cycle assessments of biomass utilization for bioenergy products. Biofuels Bioprod. Biorefin. **11**(4), 633–647 (2017)
19. R.K. Mishra, K. Mohanty, Pyrolysis kinetics and thermal behavior of waste sawdust biomass using thermogravimetric analysis. Bioresour. Technol. **251**, 63–74 (2018)
20. M. Patel, X. Zhang, A. Kumar, Techno-economic and life cycle assessment on lignocellulosic biomass thermochemical conversion technologies: a review. Renew. Sust. Energ. Rev. **53)**, 1486–1499 (2016). https://doi.org/10.1016/j.rser.2015.09.070
21. A. Pirraglia, R. Gonzalez, D. Saloni, J. Denig, Technical and economic assessment for the production of torrefied ligno-cellulosic biomass pellets in the US. Energy Convers. Manag. **66**, 153–164 (2013)
22. J. Sadhukhan, K.S. Ng, E.M. Hernandez, *Biorefineries and Chemical Processes: Design, Integration and Sustainability Analysis* (John Wiley & Sons, Inc, Hoboken, 2014)
23. B. Sharma, R.G. Ingalls, C.L. Jones, A. Khanchi, Biomass supply chain design and analysis: basis, overview, modeling, challenges, and future. Renew. Sustain. Energy Rev. **24**, 608–627 (2013)
24. S. Sokhansanj, S. Mani, A. Turhollow, A. Kumar, D. Bransby, L. Lynd, M. Laser, Large-scale production, harvest and logistics of switchgrass (Panicum virgatum L.) – current technology and envisioning a mature technology. Biofuels Bioprod. Biorefin. **3**(2), 124–141 (2009)
25. J.Z. Wu, J. Wang, Q. Cheng, D. DeVallance, Assessment of coal and biomass to liquid fuels in central Appalachia, USA. Int. J. Energy Res. **36**(7), 856–870 (2011)
26. C.E. Wyman, *Aqueous Pretreatment of Plant Biomass for Biological and Chemical Conversion to Fuels and Chemicals* (Wiley, Hoboken, 2013)
27. L. Zhang, C. Xu, P. Champagne, Overview of recent advances in thermo-chemical conversion of biomass. Energy Convers. Manag. **51**, 969–982 (2010)

Chapter 26
Sustainable Production of Value-Added Commodities from Biomass Feedstocks

I. S. Dunmade

26.1 Introduction

Biomass is a renewable resource that is widely available in almost all countries of the world in one form or another. It is not only abundant, but it could also be used for wide varieties of value-added commodities (Fig. 26.1). Thus, it is a suitable alternative to crude petroleum for varieties of products that we depend upon to maintain our standards of living. Biomass as a feedstock is derived from various sources. These sources include residuals of agricultural/forestry sector, wastes from households, food manufacturing wastes, bio-based market residuals, and biosolids from wastewater treatment facilities (Fig. 26.2). It is gratifying that these wastes have found extremely important uses to oil our economy, because it is these wastes that are the main causes of environmental pollution. They generally pose threats to human health and ecosystems well-being. Utilizing them by converting them to value-added commodities is a virile approach to diverting them from landfills, thereby protecting our surface water and groundwater from contamination. However, the various processes involved in the cultivation, gathering/harvesting, pretreatment, conversion, transportation/distribution, and the entire lifecycle management of biomass and its products have significant economic, social, and environmental consequences. While the intent of biomass conversion to value-added commodities is to meet human needs and to improve the living standard of the population, there are many possible unintended outcomes of biomass value chain. They include reduction in crop yields, loss of biodiversity, desertification, drought, groundwater and surface water contamination, high food prices, human health issues, economic hardships, and several ecosystem problems. These unintended outcomes may result from the choice of feedstocks, transportation involved, energy and water use,

I. S. Dunmade (✉)
Faculty of Science & Technology, Mount Royal University, Calgary, Canada
e-mail: idunmade@mtroyal.ca

M. O. Daramola, A. O. Ayeni (eds.), *Valorization of Biomass to Value-Added Commodities*, Green Energy and Technology,
https://doi.org/10.1007/978-3-030-38032-8_26

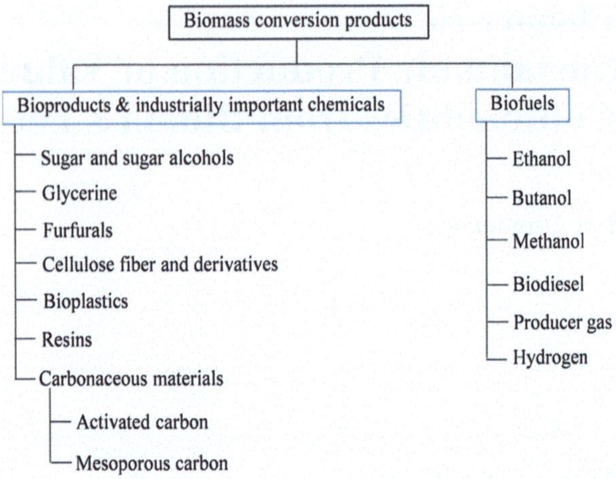

Fig. 26.1 Biomass conversion products. (Preparation was based on Irmak [21])

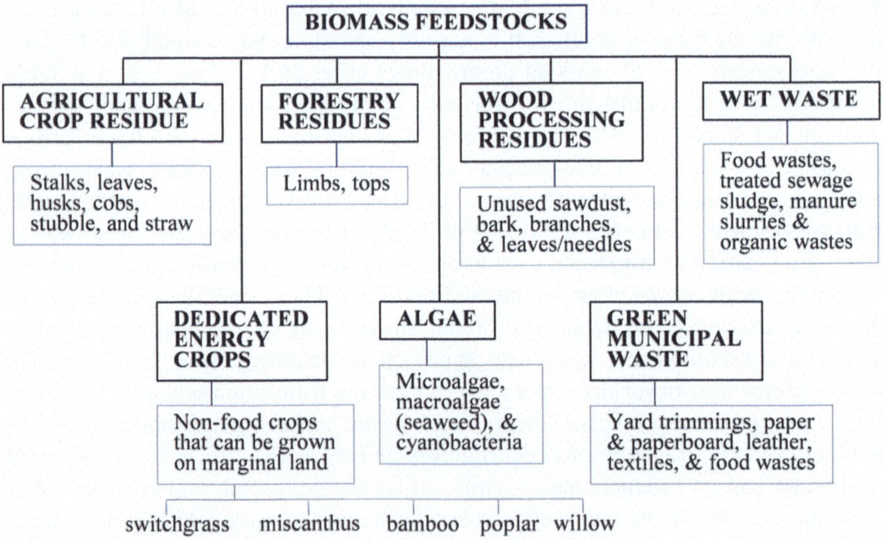

Fig. 26.2 Categories of biomass feedstock according to origin. (The diagram was prepared based on Ahorsu et al. [3] and Irmak [21])

choices of pretreatment and conversion technologies, as well as by-product and waste mismanagement or the consequent resource depletions. These potential problems necessitate sustainability consideration of various aspects of biomass production through its conversion to value-added commodities to the end-of-life management of the entire system [1, 9, 12, 18, 20, 21, 22].

According to Veleva et al. [32], "The concept of sustainable production emerged in 1992 at the United Nations Conference on Environment and Development

(UNCED) and is closely linked to the concept of sustainable development." The conference identified unsustainable pattern of production and consumption as one of the main causes of continuous global environmental deterioration. There was then a call to businesses, governments, and other stakeholders to take appropriate measures that would lead to more sustainable consumption and production patterns [28]. The importance of sustainable consumption and production was captured by Steinbach et al. [34] in the statement that "achieving Sustainable Consumption and Production (SCP) patterns has been recognized as an integral part of the 2030 Agenda for Sustainable Development. It is identified as a stand-alone Sustainable Development Goal (SDG 12) and as a central component of many of the 17 goals and 169 targets agreed in the agenda." But one may ask: What is sustainable production? What does it involve? And how does it apply to biomass conversion to value-added commodities? Sustainable production, according to LCSP [24], refers to the creation of goods and services using non-polluting processes and systems that are not only economically viable but also conserve energy and natural resources, safe and healthful for stakeholders, and socially and creatively rewarding for all workers. Sustainable production and consumption is also explained as a system that promotes social and economic development within the carrying capacity of ecosystems. It is a system that attempts to de-link economic growth from environmental degradation through improving efficiency and sustainable use of resources and production processes. It is also seen as a system of continuous economic and social progress that within the limits of the earth's ecosystems meets the needs and aspirations of everyone for a better quality of life at the present time and into the distant future. The motivation behind sustainable production concept is the long-term economic, social, and environmental benefits of production systems that satisfy the criteria. Thus, sustainable production goes beyond environmental issues. Virile and credible sustainable production indicators will include production measures and measures of relationship between production and environmental and socioeconomic systems [7, 32].

26.2 Sustainable Consumption and Production Concepts

According to Sabapathy [26], "there is a growing trend toward the adoption of sustainable production and consumption concept." It is a "continuous economic and social progress that respects the limits of the Earth's ecosystems, and meets the needs and aspirations of everyone for a better quality of life, now and for future generations to come." Sustainable consumption and production is gaining international interests because it is considered an essential requirement to achieve sustainable development. Consequently it becomes necessary for various aspects of our industrial activities to apply the concept in our match to achieve sustainable development in that sector [14, 29, 34].

The idea of sustainable consumption and production, according to Sabapathy [26], found its early expression in concepts like eco-efficiency, cleaner production, and factor-4 production. These concepts were promoted by businesses, governments, NGOs, and researchers. Researches in sustainable production and consumption started with a focus on energy use, food, buildings, and transportation because they constitute 70–80% of our economic activities which have dominant impact on our lives. In general, sustainable production and consumption seeks to optimize the utilization of raw materials and energy resources in our economic activities. Subsequent efforts have seen development of approaches to achieving sustainable consumption and production. There has also been widening of the sustainable production scope to include the evaluation of water resources use with the aim of ensuring judicious use of the scarce resource both quantitatively and qualitatively. Moreover, developments in sustainable consumption and production have also seen incorporation of its approaches into many governments' policies and frameworks.

In other words, there has been significant progress toward the adoption of sustainable consumption and production principles both in our lives and by various economic sectors. While significant progress has been made in the application of sustainable production principles in the manufacturing sector, some other economic sectors such as agri-industrial sector are yet to receive the same level of attention [19, 23]. There is therefore a need for continuous efforts in expanding the concept of sustainable production and consumption scope to the yet to be impacted sectors of the economy. Wang et al. [35] particularly highlighted the need for the development of sustainable agri-food production indicators to facilitate the measurement of sustainable production progression in the agri-industrial sector. Biomass value chain sub-sector should not be an exception in this drive.

26.3 How Sustainability Can Be Achieved Along the Biomass Value Chain

The main goals of sustainable consumption and production essentially have to do with achieving socioeconomic growth/development with minimum possible resource consumption and elimination of wastes. The ultimate goal is an assurance of high standard of living while protecting human health and preserving ecosystems well-being now and in the future. There are three main resources that require proper management throughout the biomass supply chain in order to achieve sustainable production goals (Fig. 26.3). They are materials, energy, and water resources. Energy is required in one form or another at almost all stages of the biomass value chain. Some bioconversion technologies such as thermochemical and hydrothermal processes require heat energy to facilitate the breakdown of lignin and to foster forward reaction with some chemicals. At some stages, electric energy is needed to power the machinery that reduces biomass size, for turning the biomass slush, to run the conveyor that carries the biomass from one point to another, and/or for other things. To achieve sustainability energy-wise, efforts need to be targeted at utilizing renewable sources of energy, reducing the amount of energy consumed, and

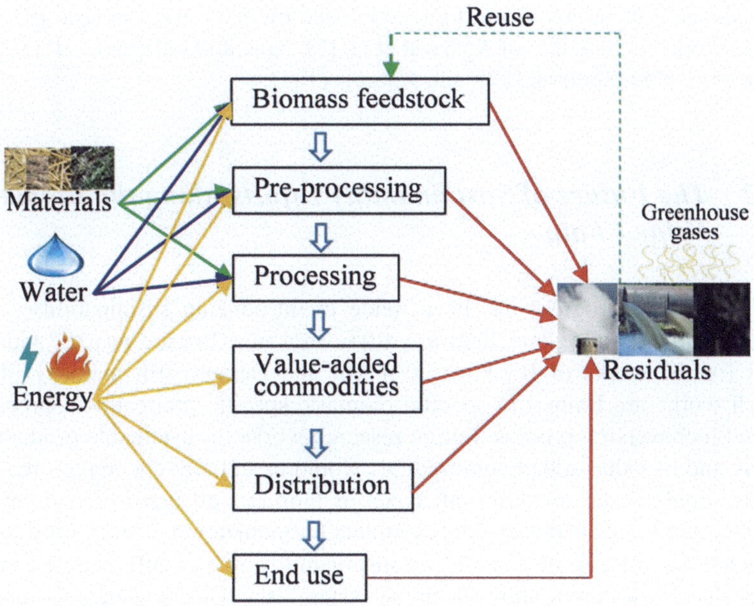

Fig. 26.3 An illustration of bioconversion value chain

eliminating wastage by reusing residual heat for space heating either for homes, for offices, or for greenhouse utilization. Materials, either in the form of feedstock or as catalysts, are needed at some stages of the biomass value chain; achieving sustainability in that aspect requires taking steps to avoid wastes and to minimize energy and water use in the cultivation, harvesting, gathering, pre-processing, and processing of the biomass materials. Some bioconversion technologies require enormous amount of water. Achieving sustainability in water use would involve minimizing water use, recycling used water, and/or utilizing processes that require little or no water at all. One of the simple, low-cost, and effective methods of achieving sustainability is proper maintenance of the bioconversion facility to avoid pipe leaks that could result in material, water, and/or heat loss [11, 17, 29, 31, 34, 36].

26.3.1 Current Efforts at Achieving Sustainability in the Biomass Value Chain

A number of scholars have reported research efforts being made at achieving sustainable production of value-added commodities from biomass conversion. Examples include a review of the significance and challenges of biomass as a suitable feedstock for bioenergy and biochemical production by Ahorsu et al. [3], a study on sustainable production of bioethanol using lipid-extracted biomass from *Scenedesmus dimorphus*, and a study on sustainable production of value-added chemicals and materials from lignocellulosic biomass by Demesa [9]. Other reported works include that of Agarwal et al. [2], Arevalo-Gallegos et al. [5], David [8], Ma et al. [25], Sheldon [27], and Xafenias [33].

26.3.2 The Future of Sustainability Efforts Along the Biomass Value Chain

While significant progress has been made in introducing sustainability concept along the biomass value chain, there are still several areas that are yet to be addressed. Except for the works of Bizikova et al. [6] and Demesa [9], majority of other research works are limited to specific biomass, specific pretreatment process, or specific bioconversion process. Future research works in sustainable production of biomass and of value-added commodities would essentially encompass the whole biomass supply chain involving all types of biomass, all types of bioconversion processes, and associated logistics. Complete harmonization of the various aspects of biomass value chain in terms of its sustainability will greatly increase the economic value of the commodities from the system. It will also enhance sociocultural well-being of the stakeholders and consumers and improve human health and ecosystem welfare.

26.4 Drivers, Techniques, and Enablers of Sustainable Consumption and Production

The realization of sustainable consumption and production goals in biomass value chain like any other industrial sector would necessarily involve the interaction of three categories of factors (illustrated in Fig. 26.4) [4, 13, 15, 16, 26, 30].

26.4.1 Drivers of Sustainable Biomass and Value-Added Commodities Production

Adoption of sustainable production principles in biomass conversion to value-added commodities would require taking a critical look at how biomass feedstocks and the value-added products of biomass conversion are produced and consumed. While attempting to reduce our ecological footprint to reduce damage to the environment is a good step, more pragmatic and effective changes would be needed. These changes would require rigorous evaluation of our values and systems that drive the demand and supply of bio-based commodities. It would also require the cooperation of all and sundry to shift from the current consumerism and economic benefit-driven

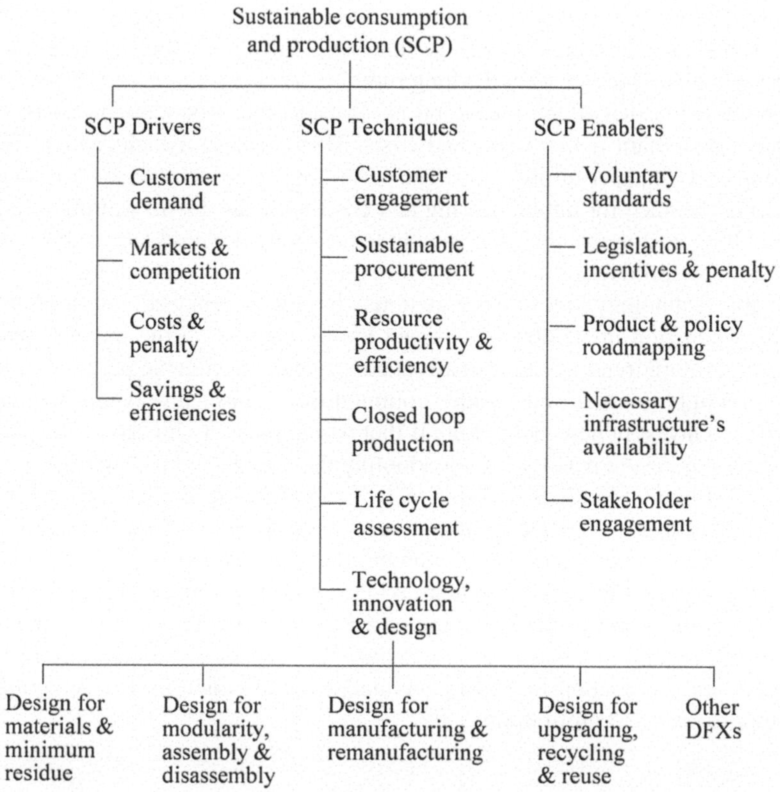

Fig. 26.4 An illustration of drivers, techniques, and enablers of sustainable production. (Source: Prepared from Dunmade [10] and Sabapathy [26])

approach to a sustainable biomass and allied production. There are a number of drivers that would encourage producers and processors in the biomass value chain to adopt sustainable production and consumption practices. These drivers include customer demand, available markets and fierceness of competition, costs of adopting sustainable production practices and penalties associated with not adopting sustainability concept, and savings accruable to utilizing sustainable production methods as well as their efficiencies.

The quantity and quality of any commodity produced and supplied by manufacturers and producers are largely determined by the appetite and tastes of consumers. Effective/dramatic changes to a product are driven by what consumers need, when and why they need it, and how they need it. In the same vein, the nature of a commodity's market, how large it is, existence of competition in the market, how fierce the competition is, availability of alternative markets, as well as consumers' willingness to pay higher prices for sustainably produced commodities would influence the business decision on whether to risk an investment on an untried innovative sustainable production method or not. A market where consumers can afford and are willing to pay the price for sustainability would encourage investors to invest on and utilize sustainable production system, knowing fully well that their investment would yield a good dividend in the long run.

Another factor that affects the adoption of sustainable production methods in the biomass value chain is the associated costs. This cost factor includes capital and operating costs, and permitting costs. The permitting costs include the costs incurred in meeting the need for public hearing and costs associated with waiting to get the permit. Other cost-associated issues include how long it would take to have return on sustainable production technology investment as well as the longevity of the sustainable technology lifecycle before it may have to be replaced. Furthermore, the kind of production method(s) being used by competitors in the industry and the anticipated future trend would also affect adoption of sustainable production in the conversion of biomass to value-added commodities. A business organization would be forced to invest in new technology if that would make it outperform its peers or if it would pay severe penalty for not adopting the use of such technology.

Since economic factor is a major factor in business, a biomass production or conversion facility would gladly invest in sustainable production technology that would lead to more efficient use of human, material, energy, and water resources. This is premised on the condition that taking such steps would result in significant cost savings and thereby translate to increased profitability. The efficiency in the use of material, energy, and water resources may be in terms of reduction in resource consumption or reduction in waste generated, both of which may have significant cost implications and profitability.

26.4.2 Sustainable Production Techniques for Biomass Conversion to Value-Added Commodities

Sustainable production practices come in various forms. The notable sustainable production practices that could be adopted in the biomass conversion to value-added products include customer engagement, sustainable procurement, and resource productivity and efficiency. Others include closed-loop production, lifecycle assessment, and technology, innovation, and design.

Customer engagement is paramount sustainable production and consumption because satisfying consumer needs would affect continuity in the demand for the value-added commodities produced from the biomass feedstock. Customers need to be carried along right from the onset of impending changes in the products and how they will be made. It is particularly important to know from consumers how the changes would affect them and to factor in their concerns in the final decision on the course of action. Sustainable procurement is another side of the same coin as customer engagement in that a bioprocess facility would need biomass feedstock in one form or the other. The facility could demand from its suppliers how it wants the feedstock to be produced. It may also want to know the level of sustainability practices adoption by its suppliers in view of the fact that that would in turn affect the sustainability profile of its own products. Some years ago a sustainability-minded manufacturing company requested me to help them develop a multi-criteria-based sustainability evaluation system for screening their suppliers. In the biomass value chain, the frequency and intensity of such demand on suppliers is expected to increase in the future as it is not only environmentally important, it is also socioeconomically necessary to enable a bioconversion facility to demonstrate good corporate citizenship, maintain public goodwill, and be reckoned as a sustainability leader. Among the parameters used for assessing sustainability of procurement is purchasing resources from local producers, thereby eliminating long distance transportation with its attendant greenhouse gases emission. Others include imbibing just in time procurement, thereby eliminating long storage that could have serious implication on quality, energy use, and other related issues.

Resource productivity and efficiency involve utilizing the same amount of feedstock to produce more volumes of value-added commodities. It may also come in the form of eliminating the use of certain catalyst or using smaller quantity than the usual. Furthermore, it may involve using smaller volume of water or less amount of energy for the production of the same quantity of value-added commodities. Generally it involves minimizing waste and attempting to ensure that every molecule of material and every erg of energy deployed translate to saleable product. Improving resource productivity and/or resource use efficiency often requires closing the production loop. Closed-loop production could be done within a production facility in the form of in-process recycling or in collaboration with other members of the biomass conversion value chain. A material, water, or energy resource that is unsuitable for facility A may be utilizable by facility B. A group of biomass production and processing facilities can enter into symbiotic relationships involving the use of each other's wastes and utilizing each other's residual energy/heat or gray water, either on a contract or voluntary basis. Doing so would reduce waste and increase profitability.

Lifecycle assessment is a set of robust tools usable for evaluating each stage in the biomass production and value-added commodities lifecycle to identify hotspots along the value chain and to assess improvement opportunities in the system. Lifecycle assessment can be used either as a stand-alone tool or in conjunction with design tools for systems optimization. For further details please refer to the chapter on lifecycle assessment of biomass in this book series [11].

There are several design paradigms that could result in technology innovations in the area of sustainable production of bioconversion facilities. Some of these design approaches are shown in Fig. 26.1. They include design for materials and minimum residues that favor the choice of materials that are not only cheaply and abundantly available in the locality but also require less water and energy resources to produce and process. The concept also emphasizes the selection of durable and non-toxic materials for a given task. In addition, the concept encourages the use of process that would lead to generation of minimum waste both at the production of the facility and during the operation lifetime of the facility. Design for modularity, assembly, and disassembly emphasizes putting the components that work together to perform the same function as a module and locating them at specific part of the machinery such that if/when there is a need for repair or replacement of such module, the operator/technician would require minimum time to repair or remove and replace the affected module. The design for assembly and disassembly aspect of the paradigm seeks to incorporate easy and fast ways to assemble and disassemble system's component parts. There are many other paradigms that seek to make things easier to implement in order to minimize waste, encourage reuse, and reduce repair and maintenance times, thereby facilitating optimal use of resources.

26.4.3 Enablers of Sustainable Production of Value-Added Commodities from Biomass Feedstock

The third leg in the tripod is the enablers of sustainable feedstock and value-added commodities productions. Standardization either at individual facility level or at the industrial group level is one of the factors that would facilitate cheaper facility maintenance costs and lower operations cost and improve the durability of biomass processing facility, thereby making the system more sustainable. Standardization also improves product quality as it makes it easier to measure conformity to industry standard.

Legislation, incentives, and penalty are enablers of sustainable production. They are necessary instruments to achieve new levels of improvement in sustainable production and consumption. Incentives encourage proactive firms to research and try sustainable production practices, while penalties would force unwilling organizations to adopt/conform to required standards.

Sometimes there is willingness to adopt innovative sustainable production practices, but such innovation may run aground if there is no enabling infrastructure for the functioning of sustainable production technology. The innovation may be short-lived if quality or size of the requisite infrastructure is inadequate for the proper functioning of the system. Utilizing sustainable technology under inadequate scenario may result in dismal failure of the system and could jeopardize future adoption of new innovative sustainable technologies [10].

26.5 Conclusion

Sustainable production of biomass and value-added commodities is possible. Efforts are being made by scholars, policy makers, and biomass supply chain actors to incorporate sustainability into biomass production, biomass conversion, and value-added commodities produced from various categories of biomass feedstocks. There are still a lot of opportunities for improvement. Collaborative efforts are needed to ensure sustainable production and consumption of various types of biomass and its end products. There is a need for critical examination of how and where productivity can be enhanced while waste is eliminated or reduced at each stage in the lifecycle of biomass production and conversion process. This examination can be done by looking at materials, energy, and water consumption at each stage of the biomass supply chain and implementing improvement opportunities.

References

1. S. Abinandan, S.R. Subashchandrabose, N. Cole, R. Dharmarajan, K. Venkateswarlu, M. Megharaj, Sustainable production of biomass and biodiesel by acclimation of nonacidophilic microalgae to acidic conditions. Bioresour. Technol. **271**, 316–324 (2019)
2. B. Agarwal, V. Ahluwalia, A. Pandey, R.S. Sangwan, S. Elumalai, Sustainable production of chemicals and energy fuel precursors from lignocellulosic fractions, in *Biofuels, Green Energy and Technology*, ed. by A. K. Agarwal et al., (Springer, Singapore, 2017). ISBN 978-981-10-3790-0
3. R. Ahorsu, F. Medina, M. Constanti, Significance and challenges of biomass as a suitable feedstock for bioenergy and biochemical production: a review. Energies **11**, 3366 (2018). https://doi.org/10.3390/en11123366
4. V. Allee, Reconfiguring the value network. J. Bus. Strateg. **21**(4), 36–39 (2000). https://doi.org/10.1108/eb040103
5. A. Arevalo-Gallegos, Z. Ahmad, M. Asgher, R. Parra-Saldivar, H.M.N. Iqbal, Lignocellulose: a sustainable material to produce value-added products with a zero waste approach – a review. Int. J. Biol. Macromol. **99**, 308–318 (2017)
6. L. Bizikova, L. Pinter, G. Huppe, H. Schandl (2015) Sustainable consumption and production indicators for the future SDGs. UNEP Discussion Paper – March 2015, https://sustainabledevelopment.un.org/content/documents/2301SCP%20indicators.pdf. Accessed 13 March 2019
7. Z. Chen, D. Wang, M. Sun, H.H. Ngo, W. Guo, G. Wu, W. Jia, L. Shi, Q. Wu, F. Guo, H.-Y. Hu, Sustainability evaluation and implication of a large scale membrane bioreactor plant. Bioresour. Technol. **269**, 246–254 (2018)
8. A. David, Thermophilic anaerobic digestion: enhanced and sustainable methane production from co-digestion of food and lignocellulosic wastes. Energies **11**(8), 2058 (2018). https://doi.org/10.3390/en11082058
9. A.G. Demesa, Towards sustainable production of value-added chemicals and materials from lignocellulosic biomass: carboxylic acids and cellulose nanocrystals. A doctoral dissertation submitted to Lappeenranta University of Technology, Finland (Lappeenrantaensis 819) (2018), http://lutpub.lut.fi/bitstream/handle/10024/158668/Abayneh%20Demesa%20A4.pdf?sequence=1&isAllowed=y. Accessed 13 March 2019
10. I.S. Dunmade, Indicators of sustainability: assessing the suitability of a foreign technology for a developing economy. Technol. Soc. **24**, 461–471 (2002)
11. I.S. Dunmade, The use of lifecycle management principles in Biosystems engineering: a pragmatic approach to solving Agri-industrial sustainability problems. J. Agric. Sci. Technol. A **2**(3), 357–362 (2012)
12. I.S. Dunmade, A study on the management of residuals in Alberta's agricultural and agri-processing industry. Int. J. Env. Protect. Policy **1**(3), 38–43 (2013a). https://doi.org/10.11648/j.ijepp.20130103.11
13. I.S. Dunmade, Sustainability issues in innovative waste reduction technology adoption and assimilation. Int. J. Env. Protect. Policy **1**(4), 59–67 (2013b)
14. I.S. Dunmade, Hybridizing renewable energy systems in Nigeria: a contextual framework for their sustainability assessment. Eur. J. Eng. Technol. **4**(5), 33–40 (2016a)

15. I.S. Dunmade, Sustainable engineering: a vital approach to innovative product development and community capacity building. The 5th Inaugural Lecture of Covenant University, Ota, Ogun State presented on 19 February 2016 (2016b)
16. I.S. Dunmade, O.S.I. Fayomi, Lifecycle engineering of infrastructure: an essential approach to engineering for a sustainable Africa. IOP Conf Series Material Science and Engineering 391 012010 (2018)
17. I.S. Dunmade, S. Onawumi, O. Ajayi, An approach to lifecycle design and sustainable management of Nigerian water infrastructure. Sustainab. Energ. Build. Res. Adv. **5**(1), 11–20 (2016)
18. A. Foukis, O.A. Gkini, P.Y. Stergiou, V.A. Sakkas, A. Dima, K. Boura, A. Koutinas, E.M. Papamichael, Sustainable production of a new generation biofuel by lipase-catalyzed esterification of fatty acids from liquid industrial waste biomass. Bioresour. Technol. **238**, 122–128 (2017)
19. K. Govindan, M. Hasanagic, A systematic review on drivers, barriers, and practices towards circular economy: a supply chain perspective. Int. J. Prod. Res. **56**(1–2), 278–311 (2018). https://doi.org/10.1080/00207543.2017.1402141
20. J.R. Hansen, R. Madsen, R. Fehrmann, C.H. Christensen, Fuel and chemicals from renewable alcohols: Part 1+2 (2008), http://orbit.dtu.dk/ws/files/4949309/Jeppe%20RH%20Appendix.pdf. Accessed on 13 March 2019
21. S. Irmak Biomass as raw material for production of high-value products (2017), https://www.intechopen.com/books/biomass-volume-estimation-and-valorization-for-energy/biomass-asraw-material-for-production-of-high-value-products. Accessed on 15 March 2019
22. F.H. Isikgor, C.R. Becer, Lignocellulosic biomass: a sustainable platform for production of bio-based chemicals and polymers (2018), https://arxiv.org/ftp/arxiv/papers/1602/1602.01684.pdf. Accessed 13 March 2019
23. U.U. Lorenz, S.S. Veenhoff, Integrated scenarios of sustainable food production and consumption in Germany. Sustainab. Sci. Pract. Policy **9**(2), 92–104 (2013)
24. LCSP, What is sustainable production? (2018), https://www.uml.edu/Research/Lowell-Center/About/Sustainable-Production-Defined.aspx. Accessed 15 March 2019
25. R. Ma, Y. Xu, Zhang X (2015) catalytic oxidation of biorefinery lignin to value-added chemicals to support sustainable biofuel production. ChemSusChem **8**(1), 24–51 (2015). https://doi.org/10.1002/cssc.201402503
26. A. Sabapathy A Business primer for sustainable consumption and production (2007), https://www.cisl.cam.ac.uk/resources/publication-pdfs/sustainable-consumption.pdf. Accessed 15 March 2019
27. R.A. Sheldon, Utilization of biomass for sustainable fuels and chemicals: molecules, methods and metrics. Catalyst Today **167**(1), 3–13 (2011)
28. United Nations, United Nations conference on environment and development, Rio de Janeiro, Brazil, in *In: Agenda 21: Programme of Action for Sustainable Development*, (United Nations, New York, 1992)
29. UN, Sustainable consumption and production (2018), https://sustainabledevelopment.un.org/topics/sustainableconsumptionandproduction. Accessed on 27 March 2019
30. UNEP, Sustainable Consumption and Production Global edition. *A Handbook for Policymakers* (2015), https://sustainabledevelopment.un.org/content/documents/1951Sustainable%20Consumption.pdf. Accessed 13 March 2019
31. USDE, Strategic plan for a thriving and sustainable bioeconomy (2016), https://www.energy.gov/sites/prod/files/2016/12/f34/beto_strategic_plan_december_2016_0.pdf. Accessed on 23 March 2019
32. V. Veleva, M. Hart, T. Greiner, C. Crumbley, Indicators of sustainable production. J. Clean. Prod. **9**, 447–452 (2001)
33. N. Xafenias, Sustainable production of value-added fuels and commodity products via microbial electrosynthesis (2018), https://www.chalmers.se/en/projects/Pages/Sustainable-productionof-value-added-fuels-and-commodity_1.aspx. Accessed on 23 March 2019

34. N. Steinbach, V. Palm, S. Constantino, Monitoring the Shift to Sustainable Consumption and Production Patterns in the context of the SDGs (2016), https://sustainabledevelopment.un.org/content/documents/2298SCP%20monitoring.pdf. Accessed on 13 March 2019
35. Wang C, .et al (2019) A literature review of sustainable consumption and production: a comparative analysis in developed and developing economies. J. Clean. Prod., 206: 741–754
36. J.Y. Zhu, X.J. Pan, Woody biomass pretreatment for cellulosic ethanol production: technology and energy consumption evaluation. Bioresour. Technol. **101**, 4992–5002 (2010)

Index

M. O. Daramola, A. O. Ayeni (eds.), *Valorization of Biomass to Value-Added Commodities*, Green Energy and Technology,
https://doi.org/10.1007/978-3-030-38032-8

B

D

R

T

U

V

The manufacturer's authorised representative in the EU is Springer Nature Customer Service Centre GmbH, Europaplatz 3, 69115 Heidelberg, Germany. If you have any concerns regarding our products, please contact ProductSafety@springernature.com

Printed and bound by CPI Group (UK) Ltd, Croydon, CR0 4YY
30/09/2026
02225647-0018